Recent Advances in Urban Ventilation Assessment and Flow Modelling

Recent Advances in Urban Ventilation Assessment and Flow Modelling

Special Issue Editors

Riccardo Buccolieri
Jian Hang

MDPI • Basel • Beijing • Wuhan • Barcelona • Belgrade

MDPI

Special Issue Editors

Riccardo Buccolieri
University of Salento
Italy

Jian Hang
Sun Yat-sen University
China

Editorial Office
MDPI
St. Alban-Anlage 66
4052 Basel, Switzerland

This is a reprint of articles from the Special Issue published online in the open access journal *Atmosphere* (ISSN 2073-4433) from 2017 to 2019 (available at: https://www.mdpi.com/journal/atmosphere/special_issues/urban_ventilation)

For citation purposes, cite each article independently as indicated on the article page online and as indicated below:

LastName, A.A.; LastName, B.B.; LastName, C.C. Article Title. *Journal Name* **Year**, *Article Number*, Page Range.

ISBN 978-3-03897-806-0 (Pbk)
ISBN 978-3-03897-807-7 (PDF)

Contents

About the Special Issue Editors

Riccardo Buccolieri has been an Assistant Professor of Atmospheric Physics at the University of Salento, Italy since November 2016. His research, both experimental and modelling, deals with the study of flow and pollutant dispersion in the urban environment as well as the effects of wind direction, morphology, and passive methods on city ventilation. He is a reviewer and Editorial Board Member of several international scientific journals and is an author of scientific articles and book chapters. He has taken part in several national and international projects, and collaborates with several international institutes, such as the Sun Yat-sen University (China), the Nanjing University (China), the University of Gavle (Sweden), and the CIEMAT (Spain), among others. He received his degree cum laude in Environmental Science from University of Salento (Lecce, Italy) in April 2005 and a Ph.D. in Environmental Geophysics from University of Messina, Italy in March 2009.

Jian Hang is a professor in the School of Atmospheric Sciences, Sun Yat-sen University. He got his Ph.D. degree in 2009 from the University of Hong Kong (HKU) and finished his postdoctoral research in 2012 at HKU. In the past decade, Prof. Hang significantly contributed to urban-built micro-climate research, including indoor ventilation and particle dispersion, city ventilation by various ventilation indices, numerical simulation of urban pollutant dispersion and exposure analysis, urban flow simulation by porous turbulence models, scale-model outdoor field measurement of urban turbulence and thermal environment. He has published 42 SCI papers, among which he is the first or corresponding author of 32 SCI papers, and 26 of them are in first regional SCI journals of Thomson Reuters, including Environmental Pollution (impact factor IF:5.1), Atmospheric Environment (IF:3.6), Science of the Total Environment (IF:4.9), Building and Environment (IF:4.1), and Boundary-Layer Meteorology (IF:2.6).

atmosphere

MDPI

Editorial

Recent Advances in Urban Ventilation Assessment and Flow Modelling

Riccardo Buccolieri [1],* and Jian Hang [2]

[1]   Dipartimento di Scienze e Tecnologie Biologiche ed Ambientali, University of Salento, S.P. 6
       Lecce-Monteroni, 73100 Lecce, Italy
[2]   School of Atmospheric Sciences, Sun Yat-sen University, Guangzhou 510275, China;
       hangj3@mail.sysu.edu.cn
*   Correspondence: riccardo.buccolieri@unisalento.it

Received: 12 March 2019; Accepted: 14 March 2019; Published: 16 March 2019

check for
updates

1. Introduction

The *Atmosphere* Special Issue "Recent Advances in Urban Ventilation Assessment and Flow Modelling" collects twenty-one original papers and one review paper published in 2017, 2018 and 2019 dealing with several aspects of ventilation in urban areas (https://www.mdpi.com/journal/atmosphere/special_issues/urban_ventilation).

The ventilation of cities is fundamental to the removal of heat and airborne pollutants. An increasing number of people are exposed to high temperatures and air pollution levels due to the ongoing intense urbanization. The study of ventilation becomes of critical importance as it addresses the capacity with which a built urban structure is capable of replacing the polluted air with ambient fresh air. Here, ventilation is recognized as a transport process that improves local microclimate and air quality and closely relates to the term "breathability" [1,2]. The efficiency at which street canyon ventilation occurs depends on the complex interaction between the atmospheric boundary layer flow and the local urban morphology [3].

This Special Issue includes contributions on recent experimental and modelling works, techniques and developments mainly tailored to the assessment of urban ventilation on flow and pollutant dispersion in cities. In the next section, the individual contributions to this issue are summarized and categorized into four broad topics: (1) outdoor ventilation efficiency and application/development of ventilation indices, (2) relationship between indoor and outdoor ventilation, (3) effects of urban morphology and obstacles to ventilation and (4) ventilation modelling in realistic urban districts. Please note that most of the papers cover more than one topic, but for simplicity each paper has been categorized in one of the above topics.

2. Summary of This Special Issue

2.1. Outdoor Ventilation Efficiency and Application/Development of Ventilation Indices

Ho et al. [4] employed reduced-scale physical models and wind tunnel experiments to study the dynamics over hypothetical areas (surface mounted ribs in crossflows) to enrich the understanding of the street-level ventilation mechanism. They found that the drag coefficient Cd is able to characterize the transport processes such that the street-level turbulent component of the air change rate (ACH) is proportional to the square root of the drag coefficient. This parameterization was then extended to formulate a new indicator, the vertical fluctuating velocity scale in the roughness sublayer (RSL) for breathability assessment in urban areas.

Wang et al. [5] found that the temporal decay profile of tracer gas concentration in urban canopy layer (UCL) models accords with the exponential decay law, and it is effective to introduce the concentration decay method into computational fluid dynamics (CFD) simulations to predict the net air change rate per hour (*ACH*) flushing UCL space and never returning. Street-scale, medium-dense (frontal area index λ_f = planar area index λ_p = 0.25), urban-like geometries were studied under neutral atmospheric conditions. They found that larger urban size attains smaller *ACH*. For square overall urban form, parallel wind attains greater *ACH* than non-parallel wind, but it experiences smaller *ACH* than the rectangular urban form under most wind directions. Open space increases *ACH* more effectively under oblique wind compared to parallel wind. *ACH* calculated by the concentration decay approach has been proven effective to evaluate the effects of urban morphologies on the overall UCL ventilation capacity induced by mean flows and turbulent diffusion, which seems to be a better ventilation index than *ACH* calculated by the volumetric flow rates integrating the normal mean velocity or fluctuation velocity across UCL boundaries.

The paper by You et al. [6] assessed regional spatial ventilation performance for optimizing residential building arrangements. The ventilation index net escape velocity (*NEV*) was employed to quantify the influence of design variation on ventilation efficiency within four typical spaces, which refer to space form, size and position changing. Spatial mean velocity magnitude (*VM*) and visitation frequency (*VF*) were also employed. Using CFD simulations, several multi-residential building arrangements, referring to building length, lateral spacing and layout variations, were investigated under the effect of surrounding buildings. Results show that *NEV* is useful to comprehensively reflect the pollutant removal ability of wind velocity and flow recirculation in different regional spaces. *VM* and *VF* are also useful indices that can reflect the air flow rates and recirculation phenomena of the calculated domains, respectively. Some design strategies are also provided.

Nguyen Van and De Troyer [7] have elaborated, via a computationally intensive analysis, a surrogate model to calculate wind pressure coefficient (*Cp*) values of several urban patterns. This multiple linear regression model, with different functions for different orientations, is based on other surrogate models developed based on field measurements, wind tunnel measurements and CFD simulations. The developed meta-model of *Cp* allows for the inclusion of the effect of the schematic urban environment. The regression functions of this meta-model show a high correlation with the values of the existing surrogate model. This meta-model is able to predict the *Cp* values on different surfaces very quickly and is thus very significant when a huge number of simulations is required. The developed surrogate model is fast, and can easily be integrated in a dynamic energy simulation tool like EnergyPlus for optimization of natural ventilation in urban areas.

Peng et al. [8] investigated a quantitative correlation between urban-like geometries of different building site coverage (*BSC*) and six ventilation indices. The floor area ratio (*FAR*) was kept constant at 5.0, while *BSC* gradually increased from 11% to 77% (building heights H decreased from 135 m to 21 m, respectively), resulting in a total of 101 asymmetrical idealized configurations. Results show that, among the indices investigated, the purging flow rate (*PFR*), the residence time (*TP*), and the mean age of air (τ_p) better correlate with urban morphological parameters. Further, when *FAR* is intermediate, *BSC*s (and building heights) may experience airflow channels by controlling the layout, which is an effective way to increase the overall ventilation and thus the ability of pollutant dispersion. These results show that by using *BSC* and *FAR* as morphological parameters, the ventilation performance of void spaces can be effectively improved in high-density cities by reducing the average height of buildings and designing the architectural layout appropriately in urban design.

2.2. Relationship between Indoor and Outdoor Ventilation

The review by Bo et al. [9] focuses on the studies dealing with the assessment of indoor–outdoor particulate matter air pollution. The aim was to review recent developments in site-specific approaches to evaluate emissions, concentrations and exchange parameters of ambient and non-ambient pollutants in indoor and outdoor contexts. A variety of research developed both in work and life environments and for engineering and epidemiological purposes were considered. From the different adopted methodological approaches and technologies, concentration indicators and ventilation exchange factors were analysed. Data and plots reassume the most common points between the different approaches and the strength or weakness of the researcher's choices. Thus, this review represents a step towards a comprehensive understanding of indoor and outdoor exposures and may stimulate the development of innovative tools for further epidemiological and multidisciplinary research.

Hong et al. [10] present results obtained from CFD simulations to determine the effects of four typical building-tree grouping patterns on the outdoor wind environment and the indoor/outdoor relationships for $PM_{2.5}$ as a result of partly wind-induced natural ventilation in Beijing; the relationship between the resulting wind pressure differences and indoor $PM_{2.5}$ concentrations; and which building-tree grouping pattern could provide the best ventilation potential or provide the lowest indoor $PM_{2.5}$ concentrations. Results clearly indicate that airflow and indoor/outdoor $PM_{2.5}$ dispersion strongly depend on the relationships of building layouts, tree arrangements, and orientation towards the prevailing wind.

Suszanowicz [11] presents the results of research on heat loss from various types of residential buildings through ventilation systems. A model of heat loss from the discharge of exhaust air outside through air ducts has since been developed. Experiments were conducted on three experimental systems of building ventilation: gravitational, mechanical, and supply-exhaust ventilation systems with heat recovery. Results show that mechanical intake–exhaust ventilation systems with heat recovery should be used in residential buildings or in communal residential buildings because they may meet the requirements of the air quality in residential spaces and at the same time minimize the loss of heat from the ventilated rooms, making it possible to reduce the heat demand for heating buildings by 9–12%.

2.3. Effects of Urban Morphology and Obstacles to Ventilation

Chew et al. [12] studied pedestrian-level wind enhancement in urban street canyons with wind catchers. Water channel experiments with idealized models of street canyons revealed that a wind catcher could enhance pedestrian-level wind speed by up to 2.5 times in two-dimensional canyons. CFD simulations are performed to extend the study to three-dimensional canyons and show that a wind catcher with closed sidewalls enhances pedestrian-level wind speed by up to 4 times. The findings encourage better designs of wind catchers, which have applications such as heat and pollutant dispersion in street canyons.

Dong et al. [13] analyzed the impact of seasonal changes, including solar radiation and anthropogenic heating effects, on an idealized urban canyon ventilation, as well as pollutant dispersion characteristics by CFD simulations. Results show a more evenly distributed surface temperature with relatively weak diurnal fluctuation in winter. The summer afternoon case shows a multi-vortex flow structure due to strong buoyance disturbance generated by heated adjacent building walls, with low near-ground velocity and poor pollutant removal performance. The proposed research method offers an effective solution for urban ventilation and wind path designs.

Kellnerová et al. [14] present an advanced statistical technique for qualitative and quantitative validation of large eddy simulations (LES) of turbulent flow within and above a two-dimensional street canyon. Time-resolved data from three-dimensional (3D) LES were compared with those obtained from time-resolved 2D particle image velocimetry (PIV) measurements. The standard validation approach based solely on time-mean statistics was extended by a novel approach based on analyses of the intermittent flow dynamics. While the standard Hit rate validation metric indicates not so good agreement between compared mean values of both the stream-wise and vertical velocity within the canyon canopy, the Fourier, quadrant and proper orthogonal decomposition (POD) analyses demonstrate very good LES prediction of highly energetic and dominant transient features in the flow. These findings indicate that although the mean values predicted by the LES do not meet the criteria of all the standard validation metrics, the dominant coherent structures are simulated well.

Kristóf and Papp [15], by assuming an analogy between heat and mass transport processes, utilized a graphics processing unit based software to model urban dispersion. The software allows for the modification of the geometry as well as the visualization of the transient flow and concentration fields during the simulation. By placing passive turbulence generators near the inlet, a numerical wind tunnel was created, capable of producing the characteristic velocity and turbulence intensity profiles of the urban boundary layer. The model results show a satisfactory agreement with wind tunnel experiments examining a single street canyon. The effect of low boundary walls placed in the middle of the road and adjacent to the walkways, as well as the impact made by the roof slope angle, were also investigated. The approach can be used in the early phase of urban design, by screening the concepts to be tested with high accuracy models.

The impacts of wind catchers were also investigated by Liu et al. [16] in short road tunnels. CFD simulations were performed for thirty-five cases with long and short wind catchers. The intake fraction (*IF*) was applied to assess the in-tunnel ventilation conditions and pollutant exposure. Results show that long-catcher designs experience poor ventilation due to extremely strong velocity reduction effects in the upstream region and large recirculation zones behind the catcher entrance. A downstream vortex could be found in short-catcher cases to help transport the pollutant from one pedestrian side to the other and from lower to upper levels. Consequently, pollutants accumulate at the left-top (or right-top) of the tunnel. Among all cases, the closer wind catchers were positioned at the tunnel entrance, so they can provide better ventilation for the inner-tunnel environment. Design of double short-catchers in parallel arrangement is thus recommended for providing natural ventilation in short road tunnels, with the smallest *IF* being only 61% of the base case.

2.4. Ventilation Modelling in Realistic Urban Districts

Gronemeier et al. [17] investigated the ventilation of Kowloon, Hong Kong, under unstable conditions by means of large eddy simulations (LES). The purpose was to show the differences in city ventilation occurring between a neutral and an unstable atmospheric stratification under weak-wind conditions. The often-used definition of velocity ratio was changed to better compare ventilation under different stratification. By using their LES model PALM, it was shown that the ventilation is altered significantly by different stratification. A correlation between the plan-area index and the ventilation, also found by other studies, was confirmed for the neutral case as well as the unstable case. However, in the unstable case, the correlation is larger. In weak-wind unstable conditions, the plan-area index has a high impact on the ventilation. In contrast to other studies, no correlation is found between ventilation and average building height. This might be due to the idealized cases which are often used in other studies. It is finally suggested that in air ventilation assessments not only neutral but also unstable atmospheric conditions should be investigated to allow city planners to better react to poor-ventilated areas within a city.

Liu et al. [18] performed a CFD study to explore the ventilation effectiveness on the microclimate and pollutant removal in the urban street canyon based on the rebuilt Southern New Town region in Nanjing, China, under parallel and perpendicular wind directions. A novel pressure coefficient was defined. Results reveal that there is little comfort difference under two ventilation patterns in the street canyon. Air stagnation occurs easily in dense building clusters, especially under the perpendicular wind. In addition, large pressure coefficients ($Cp > 1$) appear at the windward region, contributing to promising ventilation. The investigation of the air age shows that young air is distributed where the corresponding ventilation is favourable and the wind speed is large. Results can be useful in further city renovation for the street canyon construction and municipal planning.

Santiago et al. [19] investigated the role of trees on NOx pollutant dispersion in a real neighborhood in Pamplona (Spain). Aerodynamic and deposition effects were jointly studied by means of CFD modelling. Scenarios changing the tree-foliage and introducing new vegetation in a tree-free street with traffic are simulated. Results suggest the predominance of aerodynamic effects, which induce an increase of concentration, versus deposition. The distribution of pollutant is modified by the inclusion of new trees not only in that street but also in nearby locations. Therefore, the use of trees in streets with traffic cannot be considered as a general air pollution mitigation strategy. Some general recommendations can be provided but the suitable location of trees should be analyzed for each particular case.

Tan and Deng [20] investigated the natural ventilation potential of residential buildings based on a typical single-story house in the three most populous climate zones in Australia using the simulation software TRNSYS (Transient System Simulation Tool). Simulations were performed for all seasons in three representative cities, i.e., Darwin (hot humid summer and warm winter zone), Sydney (mild temperate zone) and Melbourne (cool temperate zone). Results reveal that natural ventilation potential is related to the local climate. The greatest natural ventilation potential is observed in Darwin, while the least natural ventilation potential is found in the Melbourne case. Moreover, summer and transition seasons (spring and autumn) are found to be the optimal periods to sustain indoor thermal comfort by utilising natural ventilation in Sydney and Melbourne. By contrast, natural ventilation is found throughout the whole year in Darwin. In addition, indoor thermal comfort can be maintained only by utilising natural ventilation for all cases during the whole year, except for the non-natural ventilation periods in summer in Darwin and winter in Melbourne. These findings could improve the understanding of natural ventilation potential in different climates, and are beneficial for the climate-conscious design of residential buildings in Australia.

Yuan et al. [21] employed the building energy optimization software BEopt to evaluate the influence of the inter-building effect (IBE) with a highly-reflective (HR) building envelope on building energy use in different scenarios. Results indicate that the building with surrounding buildings is more effective than that without surrounding buildings in terms of reducing annualized energy use, annualized utility bills and annualized energy related costs for five representative cities of Japan. Specifically, it is necessary to consider the influence of IBE, including mutual shading and mutual reflection within a network of buildings, when evaluating the energy consumption of buildings. In addition, the IBE with HR building envelope can better contribute to annualized energy savings and annualized utility bill savings.

Kurppa et al. [22] examined the impact of orientation and shape of perimeter blocks on the dispersion and ventilation of traffic-related emissions in a planned city boulevard. High-resolution simulations were conducted using the LES model PALM together with a Lagrangian particle model over a highly retailed representation of an 8 km^2 urban domain including street trees and forested areas. To include the role of meteorology, two contrasting inflow conditions with neutral and stable atmospheric stratification and different wind directions were applied. Pollutant ventilation is shown to improve and mean concentration at pedestrian level to decrease up to 9% by introducing variability in building height and limiting the length of street canyons along the boulevard with high traffic rates. However, the impact of smaller scale variability in building shape was shown to be negligible. This is the first high-resolution, neighbourhood-scale LES study applying sophisticated measures to provide realistic estimations for the removal of non-reactive gaseous air pollutants from the pedestrian level. The results are directly applicable by local urban planners but support urban planning in other cities as well.

Kwak et al. [23] investigated spatial and temporal variations of air pollutant concentrations in a street canyon in a central business district of Seoul, Republic of Korea, on multiple days based on complementary approaches using a mobile laboratory and a CFD model. In the emission and dispersion processes of on-road air pollutants, the high emitting vehicles (HEV) portion and the street-canyon ventilation are the determining factors of spatial and temporal variations in their on-road concentrations. Among the seven traffic compositions, RV/SUV appears to be the most responsible for poor air quality in the street canyon. Air quality at a signalized intersection is aggravated by up to 25% over that between signalized intersections due to the emission increase that is partially compensated by efficient lateral ventilation. Consequently, controlling the number of HEVs and the in-canyon ventilation near signalized intersections can effectively manage on-road air quality in street canyons.

Ming et al. [24] investigated flow and pollutant dispersion in a 3D geometrical model based on a street canyon section with a typical traffic tidal phenomenon in the 2nd Ring Road of Wuhan, China.

The number and driving speeds of vehicles on the road during different periods of one day were measured; then, the emission rate of pollutants in two parallel roads was calculated with the CFD simulation method. Simulation results indicate that in the 3D asymmetrical shallow street canyon, when pollution sources in the street are non-uniformly distributed and the wind flow is perpendicular to the street axis, a stronger source intensity near the leeward side than near the windward side will cause expansion of the pollution space even though the total source intensity remains equal. However, with increasing wind speeds and when the wind direction is not perpendicular to the street, the concentration of pollutants in the whole street shows a decreasing trend, i.e., the effect of the non-uniform distribution of the pollution source on pollutant dispersion characteristics is weakened. Additionally, the source intensity significantly impacts the level of pollutant concentration in the street canyon. With decreasing source intensity, the level of pollutant concentration at pedestrian breathing level decreased proportionally, suggesting that for the control of urban street vehicle pollution, a reduction of pollution source intensity is the most direct and effective way.

Nguyen et al. [25] present the results of a source apportionment module implemented in the SIRANE urban air-quality model tested on a real case study (the urban agglomeration of Lyon, France, for the year 2008) focusing on the NO_2 emissions and concentrations. This module uses the tagged species approach and includes two methods, named SA-NO and SA-NO_X, in order to evaluate the sources' contributions to the NO_2 concentrations in air. Results of a data assimilation method (SALS), which uses the source apportionment estimates to improve the accuracy of the SIRANE model results, are also presented. Results of the source apportionment with the SA-NO and SA-NO_X models are similar. Both models show that traffic is the main cause of NO_2 air pollution in the studied area. Results of the SALS data assimilation method highlight its ability in improving the predictions of an urban atmospheric model. Results suggest the usefulness of the source apportionment method for the assessment of emission reduction strategies at an urban scale.

3. Conclusions

This Special Issue assembles twenty-two contributions discussing a wide range of aspects on the ventilation of urban areas by international experts. Results and approaches presented and proposed will be of great interest to experimentalists and modelers and may constitute the starting point for the improvement of numerical simulations of flow and pollutant dispersion in the urban environment, for the development of simulation tools and the implementation of mitigation strategies. In this regard, we hope that the research presented here will stimulate new ideas and indicate future research directions.

Author Contributions: R.B. and J. H. contributed equally to the concept, creation and editing of this editorial.

Acknowledgments: We would like to thank all the contributors to this special issue, as well as the Editorial team of *Atmosphere*.

Conflicts of Interest: The authors declare no conflict of interest.

References

1.   Buccolieri, R.; Sandberg, M.; Di Sabatino, S. City breathability and its link to pollutant concentration distribution within urban-like geometries. *Atmos. Environ.* **2010**, *44*, 1894–1903. [CrossRef]
2.   Buccolieri, R.; Salizzoni, P.; Soulhac, L.; Garbero, V.; Di Sabatino, S. The breathability of compact cities. *Urban Clim.* **2015**, *13*, 73–93. [CrossRef]
3.   Di Sabatino, S.; Buccolieri, R.; Kumar, P. Spatial distribution of air pollutants in cities. In *Clinical Handbook of Air Pollution-Related Diseases*; Capello, F., Gaddi, A., Eds.; Springer: Cham, Switzerland, 2018.
4.   Ho, Y.-K.; Liu, C.-H. Street-level ventilation in hypothetical urban areas. *Atmosphere* **2017**, *8*, 124. [CrossRef]
5.   Wang, Q.; Sandberg, M.; Lin, Y.; Yin, S.; Hang, J. Impacts of urban layouts and open space on urban ventilation evaluated by concentration decay method. *Atmosphere* **2017**, *8*, 169. [CrossRef]

6. You, W.; Gao, Z.; Chen, Z.; Ding, W. Improving residential wind environments by understanding the relationship between building arrangements and outdoor regional ventilation. *Atmosphere* **2017**, *8*, 102. [CrossRef]

7. Nguyen Van, T.; De Troyer, F. New surrogate model for wind pressure coefficients in a schematic urban environment with a regular pattern. *Atmosphere* **2018**, *9*, 113. [CrossRef]

8. Peng, Y.; Gao, Z.; Buccolieri, R.; Ding, W. An investigation of the quantitative correlation between urban morphology parameters and outdoor ventilation efficiency indices. *Atmosphere* **2019**, *10*, 33. [CrossRef]

9. Bo, M.; Salizzoni, P.; Clerico, M.; Buccolieri, R. Assessment of Indoor-Outdoor Particulate Matter Air Pollution: A Review. *Atmosphere* **2017**, *8*, 136. [CrossRef]

10. Hong, B.; Qin, H.; Lin, B. Prediction of wind environment and indoor/outdoor relationships for $PM_{2.5}$ in different building–tree grouping patterns. *Atmosphere* **2018**, *9*, 39. [CrossRef]

11. Suszanowicz, D. Optimisation of heat loss through ventilation for residential buildings. *Atmosphere* **2018**, *9*, 95. [CrossRef]

12. Chew, L.W.; Nazarian, N.; Norford, L. Pedestrian-level urban wind flow enhancement with wind catchers. *Atmosphere* **2017**, *8*, 159. [CrossRef]

13. Dong, J.; Tan, A.; Xiao, Y.; Tu, J. Seasonal changing effect on airflow and pollutant dispersion characteristics in urban street canyons. *Atmosphere* **2017**, *8*, 43. [CrossRef]

14. Kellnerová, R.; Fuka, V.; Uruba, V.; Jurčáková, K.; Nosek, Š.; Chaloupecká, H.; Jaňour, Z. On street-canyon flow dynamics: Advanced validation of LES by time-resolved PIV. *Atmosphere* **2018**, *9*, 161. [CrossRef]

15. Kristóf, G.; Papp, B. Application of GPU-based Large Eddy Simulation in urban dispersion studies. *Atmosphere* **2018**, *9*, 442. [CrossRef]

16. Liu, S.; Luo, A.; Zhang, K.; Hang, J. Natural ventilation of a small-scale road tunnel by wind catchers: A CFD simulation study. *Atmosphere* **2018**, *9*, 411. [CrossRef]

17. Gronemeier, T.; Raasch, S.; Ng, E. Effects of unstable stratification on ventilation in Hong Kong. *Atmosphere* **2017**, *8*, 168. [CrossRef]

18. Liu, F.; Qian, H.; Zheng, X.; Zhang, L.; Liang, W. Numerical study on the urban ventilation in regulating microclimate and pollutant dispersion in urban street canyon: A case study of Nanjing New Region, China. *Atmosphere* **2017**, *8*, 164.

19. Santiago, J.-L.; Rivas, E.; Sanchez, B.; Buccolieri, R.; Martin, F. The impact of planting trees on NOx concentrations: The case of the Plaza de la Cruz neighborhood in Pamplona (Spain). *Atmosphere* **2017**, *8*, 131. [CrossRef]

20. Tan, A.; Deng, X. Assessment of natural ventilation potential for residential buildings across different climate zones in Australia. *Atmosphere* **2017**, *8*, 177. [CrossRef]

21. Yuan, J.; Farnham, C.; Emura, K. Inter-building effect and its relation with highly reflective envelopes on building energy use: Case study for cities of Japan. *Atmosphere* **2017**, *8*, 211. [CrossRef]

22. Kurppa, M.; Hellsten, A.; Auvinen, M.; Raasch, S.; Vesala, T.; Järvi, L. Ventilation and air Quality in city blocks using large-eddy simulation—urban planning perspective. *Atmosphere* **2018**, *9*, 65. [CrossRef]

23. Kwak, K.-H.; Woo, S.H.; Kim, K.H.; Lee, S.-B.; Bae, G.-N.; Ma, Y.-I.; Sunwoo, Y.; Baik, J.-J. On-road air quality associated with traffic composition and street-canyon ventilation: Mobile monitoring and CFD modeling. *Atmosphere* **2018**, *9*, 92. [CrossRef]

24. Ming, T.; Fang, W.; Peng, C.; Cai, C.; De Richter, R.; Ahmadi, M.H.; Wen, Y. Impacts of traffic tidal flow on pollutant dispersion in a non-uniform urban street canyon. *Atmosphere* **2018**, *9*, 82. [CrossRef]

25. Nguyen, C.V.; Soulhac, L.; Salizzoni, P. Source apportionment and data assimilation in urban air quality modelling for NO2: The Lyon case study. *Atmosphere* **2018**, *9*, 8. [CrossRef]

atmosphere

MDPI

Article

Street-Level Ventilation in Hypothetical Urban Areas

Yat-Kiu Ho [†] and Chun-Ho Liu *,[†]

Department of Mechanical Engineering, The University of Hong Kong, Hong Kong, China;
mea09ykh@connect.hku.hk
* Correspondence: liuchunho@graduate.hku.hk; Tel.: +852-3917-7901
† These authors contributed equally to this work.

Received: 11 May 2017; Accepted: 10 July 2017; Published: 16 July 2017

Abstract: Street-level ventilation is often weakened by the surrounding high-rise buildings. A thorough understanding of the flows and turbulence over urban areas assists in improving urban air quality as well as effectuating environmental management. In this paper, reduced-scale physical modeling in a wind tunnel is employed to examine the dynamics in hypothetical urban areas in the form of identical surface-mounted ribs in crossflows (two-dimensional scenarios) to enrich our fundamental understanding of the street-level ventilation mechanism. We critically compare the flow behaviors over rough surfaces with different aerodynamic resistance. It is found that the friction velocity u_τ is appropriate for scaling the dynamics in the near-wall region but not the outer layer. The different freestream wind speeds (U_∞) over rough surfaces suggest that the drag coefficient C_d (= $2u_\tau^2/U_\infty^2$) is able to characterize the turbulent transport processes over hypothetical urban areas. Linear regression shows that street-level ventilation, which is dominated by the turbulent component of the air change rate (ACH), is proportional to the square root of drag coefficient $\mathrm{ACH}'' \propto C_d^{1/2}$. This conceptual framework is then extended to formulate a new indicator, the vertical fluctuating velocity scale in the roughness sublayer (RSL) $\hat{w}''_{\mathrm{RSL}}$, for breathability assessment over urban areas with diversified building height. Quadrant analyses and frequency spectra demonstrate that the turbulence is more inhomogeneous and the scales of vertical turbulence intensity $\left\langle \overline{w''w''} \right\rangle^{1/2}$ are larger over rougher surfaces, resulting in more efficient street-level ventilation.

Keywords: air change rate (ACH); flow and turbulence profiles; hypothetical urban areas; street-level ventilation; ventilation assessment; wind-tunnel dataset

1. Introduction

Cities are growing [1], with over 50% of the global population currently residing in these areas [2]. Megacities might allow for more efficient energy consumption at the expense of diversified air-pollutant sources [3]. Knowledge accumulated to rectify these problems is therefore crucial to society. Concurrently, atmospheric flows, which cover a variety of length and time scales [4], are key factors governing the transport processes over urban areas with dynamics that strongly affect street-level air quality and pollutant removal. Engineering flows over rough surfaces are commonly used as the analytical platforms to enrich our fundamental understanding of urban atmospheric boundary layer (ABL) problems [5,6]. Typical applications include wind engineering for the built environment [7,8], particulate matter (PM) in street canyons [9], city breathability [10,11], and pedestrian wind comfort/safety [12] as well as guideline formulation [13]. Unlike their smooth-surface counterparts, the aerodynamic resistance induced by rough surfaces on turbulent boundary layers (TBLs) is less sensitive to the Reynolds number Re (= Uh/ν; where U is the characteristic velocity scale of flows, h the characteristic length scale of roughness elements and ν the kinematic viscosity). Instead, it is largely influenced by the roughness geometries of surfaces that are commonly measured by blockage ratio h/δ (where δ is the TBL thickness), friction velocity

u_τ ($= [\tau_w/\rho]^{1/2}$; where τ_w is the surface shear stress and ρ the fluid density), zero-plane displacement d_0 and roughness length z_0 [14,15]. They are therefore critical (roughness) similarity parameters which should be monitored carefully in urban ABL modeling [16,17]. However, the same roughness parameters for different rough surfaces do not necessarily imply the same flow properties [18,19]. Besides, their effect on street-level ventilation is less studied. This study is therefore conceived, using reduced-scale physical modeling, to examine how surface roughness quantitatively affects the dynamics and the subsequent influence on street-level ventilation, facilitating innovation of urban planning guidelines from the pollutant removal [20] as well as urban heat island [21] perspective.

Apart from field measurements and mathematical modeling, wind tunnel experiments are common laboratory solutions to the transport processes over various land features [22–24]. Unlike their open-terrain counterparts, urban surfaces absorb momentum from the mean flows, converting it to turbulence kinetic energy (TKE) [25]. Surface roughness enhances wake flows and turbulence intensities [26], we therefore hypothesize that aerodynamic resistance, which is measured by drag coefficient C_d ($= 2u_\tau^2/U_\infty^2$; where U_∞ is the prevailing wind speed) in this paper, could serve as an indicator of street-level ventilation in urban areas. In the urban climate community, street-level ventilation is commonly assessed by the transfer coefficient w_t/U_∞, where w_t is the passive-scalar transfer velocity [27,28]. The elevated transfer coefficient over urban areas is explained by the length-scale adjustment over rough surfaces [29]. The (time scale of) mass exchange between street canyons and the overlying urban-ABL flows is also measured by the cavity wash-out time [30]. Flushing, which refers to the the instantaneous, large-scale turbulence structures prevailing across the street canyons, plays a key role in aged air removal [31]. It is believed that the unsteady fluid exchanges between street canyons and the overlaying flows are mainly driven by the shear over the buildings [32]. Analogous to smooth-wall flows, ejection (Q2) [33] and sweep (Q4) [34] contribute most to the turbulent transport processes in near-surface region, reflecting the stronger momentum exchange over rough surfaces.

At high Reynolds number, the logarithmic law of the wall (log-law) is a fundamental part of mean-flow description [35] that applies to the inertial sublayer (ISL) over both smooth and rough surfaces [36]. Turbulence structures, such as fluctuating velocities and cross-correlations, are of the same type [37] over rough and smooth surfaces. While roughness effects are confined to the inner layer [38], recent studies revealed the roughness sublayer (RSL) in-between the ISL and roughness elements [39]. RSL scaling, because of the local dependence on individual roughness elements, is different from that in the ISL [40,41]. This feature, which is different from that over smooth surfaces, is attributed to the organized eddy structures in the near-surface region over rough surfaces [42]. An RSL velocity profile therefore departs from the conventional log-law relationship, eventually affecting the transport processes. An example is the surface fluxes of atmospheric constituents over urban areas [43]. In view of the diversified indicators for street-level ventilation and the recent findings in the RSL over urban canopy, a series of wind tunnel experiments are performed in attempt to refine the current indicator for street-level ventilation especially for flows over inhomogeneous buildings.

In this paper, we focus on the (rough) surface layer of the urban ABL in neighborhood scales [44] in attempt to examine how building morphology (e.g., regimes of skimming flow and isolated roughness) modifies the dynamics together with the implication to street-level ventilation. The functionality of drag coefficient C_d, which is commonly used to measure aerodynamic resistance, is explored to estimate street-level ventilation. In addition to flow statistics, this paper looks into the intermittency in an attempt to demystify the correlation between building roughness and street-level ventilation performance. This section introduces the background, reviews the literatures and defines the problem statement. The methodology and solution approach are detailed in Section 2. The results, including flow properties and turbulence profiles, are interpreted in Section 3 before we propose a new ventilation indicator and look into the ventilation mechanism. Finally, the conclusions are drawn in Section 4.

2. Methodology

Wind tunnel experiments are conducted to address the research problem in this paper. An infinitely large, idealized urban surface is simulated by gluing an array of identical rib-type roughness elements (two-dimensional, 2D, scenarios). In contrast to the majority of wind tunnel studies in building science, in which arrays of three-dimensional (3D) roughness elements in the form of cuboids are mostly used, 2D roughness elements are adopted in the current study for comparison with our previous large-eddy simulation (LES). Moreover, the aerodynamic resistance over ribs is larger than that over cuboids that widens the range of drag coefficient being tested in the ventilation estimate. Our hypothetical urban models, though simplified, enable the fundamental understanding of rough-surface dynamics, fostering the theoretical framework for street-level ventilation.

2.1. Wind Tunnel Infrastructure

The experiments are carried out in the open-circuit, isothermal wind tunnel in the Department of Mechanical Engineering, University of Hong Kong (Figure 1). Its test section is made of acrylic whose size is 6000 mm (length) × 560 mm (width) × 560 mm (height). The flows are driven by a three-phase, electricity-powered blower and the wind speed is controlled (by a damper) in the range of $0.5 \text{ m s}^{-1} \leq U_\infty \leq 20 \text{ m s}^{-1}$. A honeycomb is installed before the test section to straighten the flows as well as to reduce the turbulence intensity. Square aluminum tubes of size h (= 19 mm) are placed in crossflows at h apart in the first 2000 mm to initiate the TBL. The remaining 4000 mm downstream is reserved for the reduced-scale urban models.

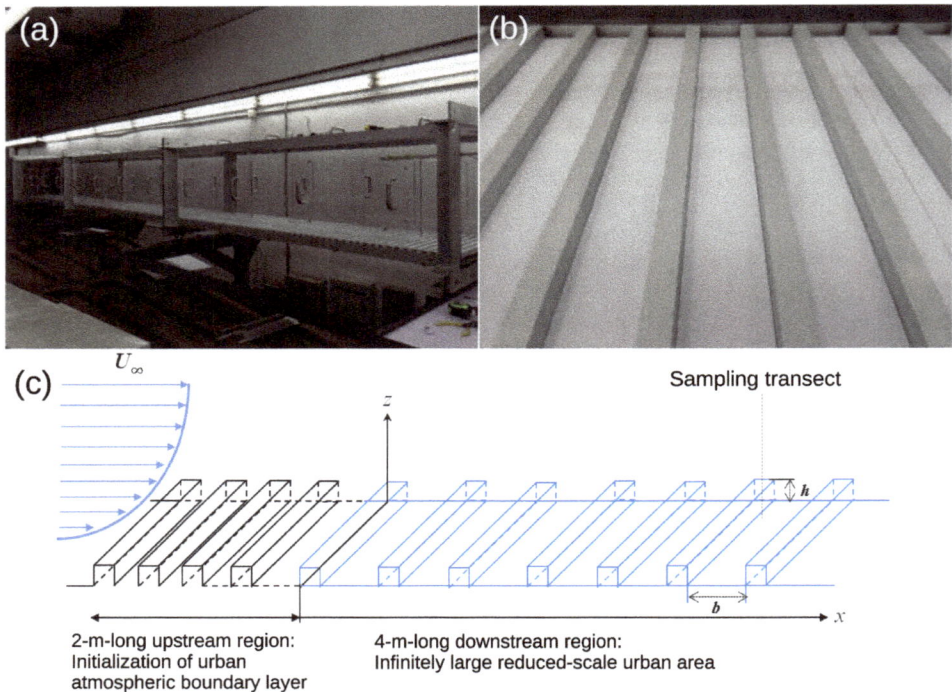

Figure 1. Apparatus used in the experiments. (**a**) Wind tunnel infrastructure in the Department of Mechanical Engineering, University of Hong Kong; (**b**) rough surfaces in the form of ribs in cross flows; and (**c**) schematic of the flow configuration.

2.2. Idealized Urban Surfaces

Hypothetical urban areas are fabricated by aligning square aluminum tubes of size h (= 19 mm) in crossflows. Eight types of rough surfaces in the form of ribs are adopted (Table 1). The wind-tunnel test floor is fully covered by baseboards (each 500 mm long × 560 mm wide × 5 mm thick) on which the aluminum tubes are glued. The streamwise extent of the array of aluminum ribs is 4000 mm that is sufficient for fully developed TBL flows. The aluminum tubes are 560 mm ($\approx 29h$) long, spanning the full width of the test section. They are placed at b (38 mm $\leq b \leq$ 228 mm) apart to adjust the aspect ratio (AR = h/b) in order to model the aerodynamic resistance of urban areas. The range of ARs covers the typical urban flow regimes (skimming flows, wake interference and isolated roughness) [45] that are analogous to the engineering rough-wall flow regimes (d-type and k-type) [46]. The test-section height is H (= 560 mm) so the roughness-element-height-to-test-section-height (blockage) ratio $h{:}H$ is bounded by 1:28. This small blockage ratio ensures sufficient headroom in the wind-tunnel, enabling proper TBL development across the test section [47].

2.3. Hot-Wire Anemometry

The wind tunnel is equipped with a computer-controlled traversing system whose positioning accuracy is 1 mm. The system interface is the National Instruments (NI) motion control unit for probing flow samples. The flows are measured by hot-wire anemometry (HWA). A thin platinum-plated tungsten wire (5×10^{-3}-mm diameter), whose resistance is sensitivity to heat, is soldered to a sampling probe that is connected to a bridge circuit. It is cooled down by forced convection in flows, leading to a change in electrical resistance (measurable at high frequency). The bridge circuit then feeds back to adjust the voltage supply in order to maintain the hot wire at constant temperature. The HWA sampling frequency is up to 10^3 Hz, facilitating the measurements of turbulence intensity and momentum flux.

In-house developed HWA and data acquisition systems are employed in this study. The hot wires are partly etched by copper-electroplating in which the effective sensing length is 2 mm. An X-wire design is used, which includes an angle between a pair of hot-wires of 100°. It measures two velocity components (streamwise u and vertical w) simultaneously and the large included angle ($\geq 90°$) reduces the potential error arising from the elevated turbulence level in the near-wall region. The hot-wire probe is connected to a constant-temperature anemometer which is basically the bridge circuit mentioned previously. The X-wire pair therefore consists of the parts of the resistance components in the circuit whose changes measure the flow velocities. A NI compactDAQ unit (NI cDAQ-9188), which has a processor for data conversion and temporary storage, is used to digitalize the analog signal. It is connected to a desktop computer via a Local Area Network (LAN) cable to avoid data loss. The data acquisition is then managed by the LabVIEW software to minimize the delay in real-time data transfer.

The universal conversion scheme for the 2-mm Institute of Sound and Vibration Research (ISVR) probe [48] is used to convert from voltage output to velocity reading. Each hot-wire probe has its own characteristic response so its thermal behavior is unique. A scaling coefficient is therefore required even it is operated at room temperature. Before measurement, the voltage output at calm-wind conditions is recorded which is then used to obtain a proper scaling for each hot-wire probe. All the hot-wire probes are also calibrated against uniform flows in the range of 1 m s$^{-1} \leq U_\infty \leq$ 11.5 m s^{-1} prior for quality assurance.

The wind speed adopted in the current wind tunnel experiments is in the range of 8 m s$^{-1} \leq U_\infty \leq$ 9.1 m s^{-1}. The Reynolds number based on freestream wind speed and TBL thickness Re$_\delta$ (= $U_\infty \delta / \nu$) is thus at least two orders of magnitude larger than the critical one (approximately 1200, irrespective of the state of the walls) [49] and hence the effect of molecular viscosity is negligible. The friction velocity measured is in the range of 0.362 m s$^{-1} \leq u_\tau \leq$ 0.671 m s^{-1}; the roughness Reynolds number Re$_\tau$ (= $u_\tau h / \nu$) is at least two orders of magnitude larger than unity [50] so the flows are fully developed. Therefore, our wind tunnel experiments are appropriate to model the TBL flows over rough surfaces. The parameters of the wind tunnel experiments are tabulated in Table 1.

Table 1. Configuration of the idealized urban surfaces and the flows in the wind tunnel experiments.

		Types of Idealized Urban Surface							
		A	B	C	D	E	F	G	H
Rib [mm]	Size h	19	19	19	19	19	19	19	19
	Separation b	38	57	76	95	114	152	190	228
Size of a repeating unit l ($= h + b$) [mm]		57	76	95	114	133	171	209	247
Aspect ratio AR ($= h : b$)		1:2	1:3	1:4	1:5	1:6	1:8	1:10	1:12
Boundary layer thickness	δ [mm]	244	248	283	284	294	294	304	304
	δ/h	12.84	13.05	14.89	14.95	15.47	15.47	16.00	16.00
Sampling location	x_{sample} [mm]	3705	3686	3609	3648	3590	3562	3571	3619
	x_{sample}/h	195	194	190	192	189	187	188	190
	x_{sample}/δ	15.18	14.86	12.75	12.85	12.21	12.12	11.75	11.90
Number of profiles in a repeating unit		7	7	7	7	7	9	9	9
Velocity [m s^{-1}]	Free-stream U_∞	8.0	8.4	8.5	8.5	8.5	8.4	9.1	9.0
	Mean U_{mean}	6.6	6.9	7.3	7.3	7.4	7.4	8.2	8.2
Friction velocity	u_τ [m s^{-1}]	0.453	0.516	0.556	0.592	0.598	0.598	0.645	0.671
	u_τ/U_∞	0.057	0.062	0.066	0.069	0.070	0.071	0.071	0.074
	u_τ/U_{mean}	0.069	0.075	0.077	0.081	0.081	0.081	0.079	0.082
Drag coefficient f ($= 2u_\tau^2/U_{\text{mean}}^2$) [$\times 10^{-3}$]		9.550	1.112	1.176	1.316	1.297	1.319	1.252	1.336
Reynolds number	Re_δ ($= U_\infty\delta/\nu$)	195,168	207,495	239,729	242,190	250,614	247,520	276,800	278,400
	Re_∞ ($= U_\infty h/\nu$)	15,200	15,900	16,100	16,200	16,200	16,000	17,300	17,400
	Re_{mean} ($= U_{\text{mean}}h/\nu$)	12,500	13,100	13,800	13,900	14,100	14,000	15,500	15,600
	Re_τ ($= u_\tau h/\nu$)	864	983	1060	1127	1138	1138	1229	1277

3. Results and Discussion

We look into the wind-tunnel measured velocity profiles to examine the flows and TBL characteristics before analyzing the dynamics and ventilation estimate in detail. Between seven and nine vertical transects are used to sample the flow data over a unit of the street canyon. The vertical spatial resolution is in the range of 1 mm $\leq \Delta z \leq$ 10 mm stretching in the wall-normal direction. The first point is 5 mm over the roughness elements and the sampling height is up to 300 mm, covering the entire TBL. Afterward, we derive the new indicator measuring street-level ventilation. Finally, we look into the intermittency using quadrant analysis and frequency spectrum to elucidate ventilation mechanism.

3.1. Thickness of TBL and ISL

In this study, the TBL thickness δ is defined at the height where the spatio-temporal average of momentum flux is asymptotically approaching zero $\langle \overline{u''w''} \rangle \big|_{z=\delta} \approx 0$ [51]. It in turn implies that the two velocity components u and w are no longer correlated. Here, angle brackets $\langle \phi \rangle$ and the overbar $\overline{\phi}$ represent spatial and temporal average of statistical flow properties, respectively. Double primes denote the deviation from the spatio-temporal average $\phi'' = \phi - \langle \overline{\phi} \rangle$. Under this circumstance, the dynamics are dominated by advection rather than the crosswind turbulent transport. This phenomenon can only be observed over the TBL where the intermittency resumes the prevailing flows. The TBL thickness observed in this paper is in the range of 200 mm $\leq \delta \leq$ 304 mm (Table 1).

The thickness of the inertial sublayer (ISL, where the logarithmic law of the wall applies) is determined by monitoring the momentum flux such that its variation in the wall-normal direction is less than a certain level. The vertical variation is measured by [52]:

$$\zeta = \frac{\langle \overline{u''w''} \rangle}{M_{\langle \overline{u''w''} \rangle}} . \tag{1}$$

Here $M_{\langle \overline{u''w''} \rangle}$ is the mean of the spatio-temporal average of momentum flux that is calculated by five-point moving average in the wall-normal direction. In this study, the range of ISL is defined at z where $0.95 \leq \zeta \leq 1.05$ after a series of sensitivity tests. The resolution of the current wind-tunnel measurements is too coarse to resolve the variation of ζ down to 0.01. The roughness parameters, such as displacement height d_0 and roughness length z_0, are determined subsequently.

3.2. Friction Velocity

Friction velocity u_τ is one of the key parameters in TBL flows over rough surfaces. It is the characteristic velocity to scale the near-surface turbulence in this paper. Among various methods, surface-level momentum flux is most commonly used to estimate the friction velocity [53]. For rough-surface flows, the turbulent momentum flux is much larger than its viscous counterpart [50] so it is reasonable to assume that $u_\tau = \langle -\overline{u''w''} \rangle^{1/2} \big|_{z=h}$ as the estimate to friction velocity.

3.3. Wind Speed Profiles

Figure 2 depicts the vertical profiles of spatio-temporally averaged wind speed $\langle \overline{u} \rangle$ over rib-type rough surfaces of different ARs. Although the range tested is narrow (from 1 : 12 to 1 : 1), the current wind-tunnel measured friction factor f generally increases with decreasing AR (widening the separation b between roughness elements; Table 1). The near-surface velocity gradient $\partial \langle \overline{u} \rangle / \partial z|_{z/\delta=0}$ over different rough surfaces shows a similar behavior, i.e., velocity increases more sharply with increasing aerodynamic resistance, signifying the elevated drag in the near-surface region (Figure 2). The length scales of flows are highly correlated with the velocity gradient at different elevation [54].

The transport processes are therefore enhanced because of the more uniform wind speed over rougher surfaces, extending to the outer layer close to $z = 0.5\delta$.

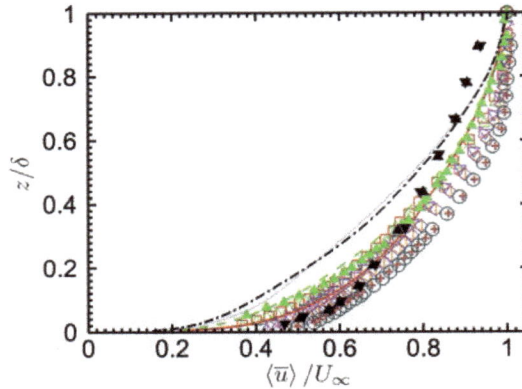

Figure 2. Dimensionless profiles of mean wind speed $\langle \overline{u} \rangle / U_\infty$ in wall-normal direction z/δ over ribs of aspect ratio AR = 1:2 ($\square$), 1:3 ($\triangle$), 1:4 (∇), 1:5 ($\triangleright$), 1:6 ($\triangleleft$), 1:8 ($\diamond$), 1:10 ($\circ$) and 1:12 (+) measured in the current wind tunnel experiments. Also shown are the LES results [55] for AR = 1:2 (——), 1:3 (— — — — — —), 1:5 (— · — · — · — · —) and 1:10 ($\cdots\cdots$) together with the measurements available in literature for AR = 1:3 ($\blacktriangle$) [56] and AR = 1:1 ($\blacktriangle$ and $\blacktriangledown$) [57].

The current wind-tunnel measured wind-speed profiles $\langle \overline{u} \rangle$ are compared well to those calculated by LES [55] of ARs = 1 : 2 and 1 : 3 (Figure 2). For the other two cases with ARs = 1 : 5 and 1 : 10, the LES-calculated near-surface velocity gradient is notably less than that of wind-tunnel data implying that the modeling turbulent transport processes are weaker than those measured in the wind tunnel experiments. Their surface velocity is so small that is slowed down by the sharp flow impingement on the windward wall of roughness elements. Apart from the smaller Reynolds number in the LES (by about two times), the difference could be attributed to the unavoidable turbulence production in the wind tunnel upstream, which, however, does not exist in the idealized horizontally homogeneous LES calculation. On top of roughness-generated turbulence, background turbulence enhances momentum transport so the the wind-tunnel mean wind speed is more uniform than its LES counterpart. The wind-speed profiles show a more favorable agreement in *d*-type flows than that in *k*-type. The LES spatial resolution could be an issue because a high spatial resolution is needed to resolve the flow impingement on the windward faces of roughness elements where recirculating flows dominate the dynamics.

Experimental results from other research groups are also adopted to verify the current wind-tunnel measurements (Figure 2). The experimental results agree reasonably well with each other in the near-surface region such that the data available in literature [56,57] fall within the range of current wind-tunnel measurements of *k*-type flows. A notable discrepancy is observed for z over 0.5δ (in the outer layer) that is attributed to the dissimilar modeling configuration in different wind-tunnel settings. The TBL thickness δ in [57] is around $8h$, more shallow than that of the current wind-tunnel measurements ($12.8\,\mathrm{h} \leq \delta \leq 16\,\mathrm{h}$) by over 30%, leading to a thinner near-surface region.

3.4. Turbulence Profiles

Vertical profiles of streamwise $\left\langle \overline{u''u''} \right\rangle^{1/2}$ and vertical $\left\langle \overline{w''w''} \right\rangle^{1/2}$ fluctuating velocities are illustrated in Figure 3a,b, respectively. Similar to most TBL studies of open-channel flows, the fluctuating velocities $\left\langle \overline{u_i''u_i''} \right\rangle^{1/2}$ are scaled by the friction velocity u_τ. Rough-surface turbulence is more isotropic compared with its smooth-surface counterpart [58] so the difference between the two components, streamwise and vertical, is less. Streamwise fluctuating velocity $\left\langle \overline{u''u''} \right\rangle^{1/2}$ decreases with increasing wall-normal distance (almost linearly). The profiles over different rough surfaces collapse well in the inner layer ($z \leq 0.5\delta$), demonstrating the dominance of roughness-generated turbulence and the appropriateness of the velocity scale. Aloft the near-surface flows in $z > 0.5\delta$, the dimensionless streamwise fluctuating velocity over different rough surfaces shows a mild dissimilarity that is caused by the background turbulence in the wind tunnel and the uncertainty of velocity scales. Rib-type surfaces are installed upstream to initialize TBL development in which the turbulence levels are relatively higher in the flows over smoother surfaces (larger ARs). Hence, the dimensionless streamwise fluctuating velocity $\left\langle \overline{u''u''} \right\rangle^{1/2} / u_\tau$ decreases with increasing AR in $0.5\delta \leq z \leq 0.8\delta$. Besides, the friction velocity u_τ, which decreases with smoother surfaces, is not the most appropriate characteristic scale to normalize the velocity in the outer region, leading to the discrepancy.

The current streamwise fluctuating velocity $\left\langle \overline{u''u''} \right\rangle^{1/2}$ profiles compare more favorably with other wind-tunnel measurements available in literature [56,57] than do the LES. The different behavior observed is attributed to the dissimilar flow configurations in the physical and mathematical models. The LESs are calculated over idealized geometry in which the prevailing flows are driven by a uniform background pressure gradient. The friction velocity u_τ is calculated by the pressure gradient in the LES. The dynamics are therefore generally in line with those observed in theoretical open-channel flows. On the other hand, the flows in wind tunnels are driven by upstream speed that fall into the category of TBL flows. The flows are, though gradually, unavoidably developing over the urban models.

The current wind-tunnel measured profiles of vertical fluctuating velocity profiles $\left\langle \overline{w''w''} \right\rangle^{1/2}$ show a behavior different from that of their streamwise counterparts. The surface-level vertical fluctuating velocities over different rough surfaces are comparable with each other ($0.9u_\tau \leq \left\langle \overline{w''w''} \right\rangle^{1/2} \leq 1.1u_\tau$), supporting the scaling using friction velocity. Differences are generally grouped into two categories according to the nature of drag force. In the skimming flow or wake interference regimes (AR in the range of $1:8$ to $1:2$), the mean flows seldom descend from the TBL core down to the street canyon so the broad peak of vertical fluctuating velocity is elevated analogous to that over smooth surfaces [59]. On the other hand, because of the flow entrainment from the TBL core down to the street canyons, the drag mechanism in the flows in isolated roughness regime (AR in the range of 1:12 to 1:10), is dominated by flow impingement on the windward walls of roughness elements. The vertical fluctuating velocity is thus peaked at the surface level instead. It decreases thereafter with increasing wall-normal distance that ends up with a lower turbulence level compared with that in *d*-type flows. The smaller dimensionless vertical fluctuating velocity in the outer layer is also partly attributed to the larger friction velocity in *k*-type flows. The agreement in the vertical fluctuating velocity among the two experimental measurements and the LES results is good, though the *k*-type flows in wind tunnel show a lower dimensionless vertical fluctuating velocity.

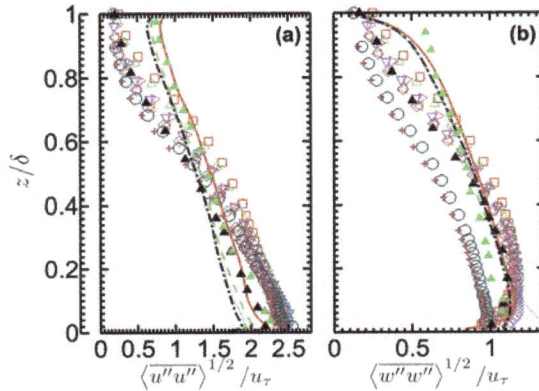

Figure 3. Dimensionless profiles of (**a**) streamwise $\left\langle \overline{u''u''} \right\rangle^{1/2} / u_\tau$ and (**b**) vertical $\left\langle \overline{w''w''} \right\rangle^{1/2} / u_\tau$ fluctuating velocities in wall-normal direction z/δ over ribs of aspect ratio AR = 1:2 (□), 1:3 (△), 1:4 (▽), 1:5 (▷), 1:6 (◁), 1:8 (◇), 1:10 (○) and 1:12 (+) measured in the current wind tunnel experiments. Also shown are the LES results [55] for AR = 1:2 (———), 1:3 (— — — — —), 1:5 (— · — · — · — · —) and 1:10 (· · · · · ·) together with the measurements available in literature for AR = 1:3 (▲) [56] and AR = 1:1 (▲) [57].

Figure 4 compares the dimensionless profiles of spatio-temporally averaged momentum flux $\left\langle \overline{u''w''} \right\rangle / u_\tau^2$ over different rough surfaces obtained in the current wind-tunnel measurements. Negative momentum flux signifies that the streamwise momentum is transported downward by the vertical fluctuating velocity. Hence, prevailing flows play key roles in near-surface turbulence generation and the associated transport processes. Apart from the mild surface-level reduction, the magnitude of momentum flux obtained from the two approaches consistently decreases with increasing wall-normal distance. Slight dissimilarity is observed. Because of the idealized configuration, the theoretical solution to the momentum flux in forced open-channel flows is a linear function of wall-normal distance. In the current wind tunnel experiments, on the other hand, a constant-flux region (in the inner layer) up to $z/\delta \approx 0.2$ is clearly observed, assembling the general behavior in the atmospheric surface layer (ASL) where the logarithmic law of the wall applies [60]. Moreover, the decreasing rate of momentum flux slows down so the wind-tunnel measured profiles are not as linear as the theoretical ones in the outer layer $z > 0.5\delta$. The over-predicted turbulence is due to the background turbulence level in the wind tunnels. It is noteworthy that 3D roughness elements produce turbulence scales of the order of the roughness height h while the motions generated by 2D roughness elements may be much larger due to the width of the roughness elements [61]. Rib-type roughness elements deviate from the similarity in the outer layer in the cases with 3D roughness and smooth walls [62]. While the authors are aware of the dissimilarity representing urban areas, the use of rib-type roughness elements in the current wind-tunnel experiments facilitates the comparison with the LES [55] in which a configuration of ribs in crossflows was adopted to reduce the computation load by ensemble averaging in the homogeneous spanwise direction.

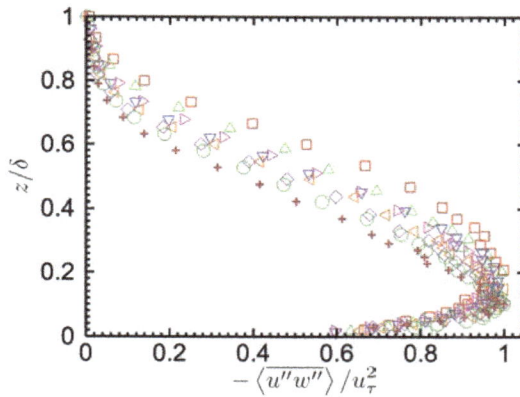

Figure 4. Dimensionless profiles of momentum flux $\left\langle \overline{u''w''} \right\rangle / u_\tau^2$ in wall-normal direction z/δ over ribs of aspect ratio AR = 1:2 ($\square$), 1:3 ($\triangle$), 1:4 ($\triangledown$), 1:5 ($\triangleright$), 1:6 ($\triangleleft$), 1:8 ($\diamond$), 1:10 ($\circ$) and 1:12 ($+$) measured in the current wind tunnel experiments.

3.5. Street-Level Ventilation Estimate

Roughness sublayer (RSL) dynamics over rough surfaces are crucial to surface-level turbulence generation. In the RSL, individual roughness elements have their own effects on the flows aloft, complicating the turbulence structure such that the conventional understanding of ISL no longer fully describes the dynamics and the ventilation mechanism. One of the examples is the TKE redistribution from vertical to spanwise components, resulting in decreasing anisotropy and turbulence scales [63,64]. While the RSL dynamics were examined in details in our previous study [65], such as the velocity profiles and length scale in RSL, this paper focuses on the practical significance of RSL particularly related to the street-level ventilation estimate.

For idealized urban street canyons, the air change rate [66],

$$\text{ACH} = \int_\Gamma w|_{z=h} \ \mathrm{d}x \, , \tag{2}$$

was proposed, where Γ is the width of a street canyon, in order to measure street-level ventilation by comparing the aged air removal (or the fresh air entrainment) [67]. It was found that the street-level ventilation is largely governed by its turbulent component ACH″ (over 70%). Recently, analytical solutions and mathematical modeling consistently showed that ACH″ exhibits a linear correlation with the square root of the drag coefficient [68]

$$\text{ACH}'' \propto C_d^{1/2} \, . \tag{3}$$

It is hence proposed that the performance of street-level ventilation can be estimated once the drag coefficient of a specific urban area is available. However, ACH″ originally proposed by [67] is calculated along the roof level of buildings of uniform height only. This definition of ventilation indicator is hardly implemented over urban areas practically such as building height variability. Moreover, street-level ventilation is governed by RSL dynamics but not those along building-roof level. Under this circumstance, a new indicator, which is able to handle urban areas with diversified building height and elevated RSL turbulence intensity, is proposed in this paper. The technical details are reported below.

In view of the importance of RSL dynamics, we switch the focus from building-roof level to the RSL. A new indicator, namely the RSL vertical velocity scale:

$$\widehat{w}''_{\text{RSL}} = \frac{\int_{\Omega_{\text{rsl}}} w''_+ \, d\Omega}{\int_{\Omega_{\text{rsl}}} d\Omega} \tag{4}$$

is therefore proposed as a ventilation indicator over urban areas. Here, Ω_{rsl} is the RSL domain and the subscript $+$ denotes the upward flows only. The effect of RSL turbulence on street-level ventilation is thus included. Figure 5 expresses the RSL vertical fluctuating velocity scale $\widehat{w}''_{\text{RSL}}$ as a function of the square root of drag coefficient $C_d^{1/2}$. Similar to the indicator defined in [67], a linear correlation is revealed in which the correlation coefficient is up to 0.92. This linear correlation suggests that the drag coefficient of urban areas could be used to parameterize the street-level ventilation performance.

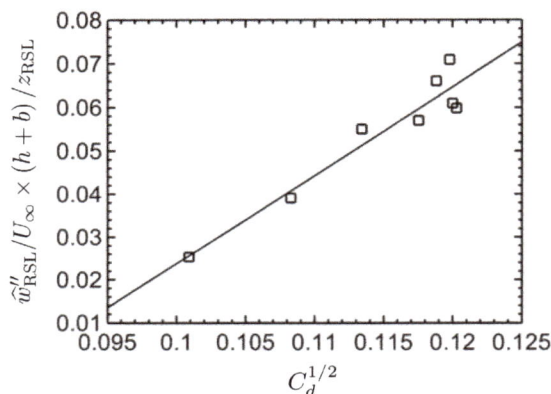

Figure 5. Vertical fluctuating velocity scale in the roughness sublayer (RSL) $\widehat{w}''_{\text{RSL}}$ plotted against the square root of drag coefficient $C_d^{1/2}$. Also shown is the linear regression $y = 2.0477x - 0.181$ whose correlation coefficient is $R^2 = 0.9193$.

3.6. Quadrant Analysis

In view of the important intermittency in the ventilation mechanism reflected by $\widehat{w}''_{\text{RSL}}$, Figure 6 compares the contributions from different quadrants to the turbulent momentum flux $u''w''$. Our definition of joint probability density function (JPDF) $P(u'', w'')$ and the covariance integrand $u''w'' P(u'', w'')$ of the fluctuating velocities u'' and w'' are based on those suggested by [69],

$$\left\langle \overline{u''w''} \right\rangle = \int_{-\infty}^{+\infty} u''w'' P(u'', w'') \, du'' dw'' . \tag{5}$$

The JPDF is calculated according to the ratio of the occurrences of individual quadrants to the total number of data samples. The quadrants are defined in Table 2. The JPDF measures the frequency of occurrence while the covariance integrand measures the strength of events. Figure 6 shows a wider range of turbulent intensity relative to the friction velocity u''_i/u_τ for flows over street canyons of AR = 1:2 (d-type flows) compared with AR = 1:12 (k-type flows). Flows over both street canyons of AR = 1:2 and 1:12 exhibit more frequent events of sweep Q4 and ejection Q2, in line with the laboratory measurements over a single street canyon available in the literature [70]. For d-type flows over street canyons of AR = 1:2, $P(u'', w'')$ illustrates a similar shape over building roofs and street canyons such that stronger sweeps Q4 and ejection Q2 are more frequent than outward interaction

Q1 and inward interaction Q3. On the other hand, as shown by the JPDF, extreme events occur more frequently, which is likely attributed to the accelerating (decelerating) flow entrainment (removal), demonstrating their influence on street-level ventilation. The covariance integrand over the building roofs and street canyons are similar for flows over AR = 1:2, implying a more homogeneous transport process. In contrast, extreme events, as signified by the elevated covariance integrand (by five times) are observed for flows over street canyons of AR = 1:12, resulting in the more efficient street-level ventilation in wider streets. A wider street thus favors extreme events (stronger updraft and downdraft) for ventilation enhancement.

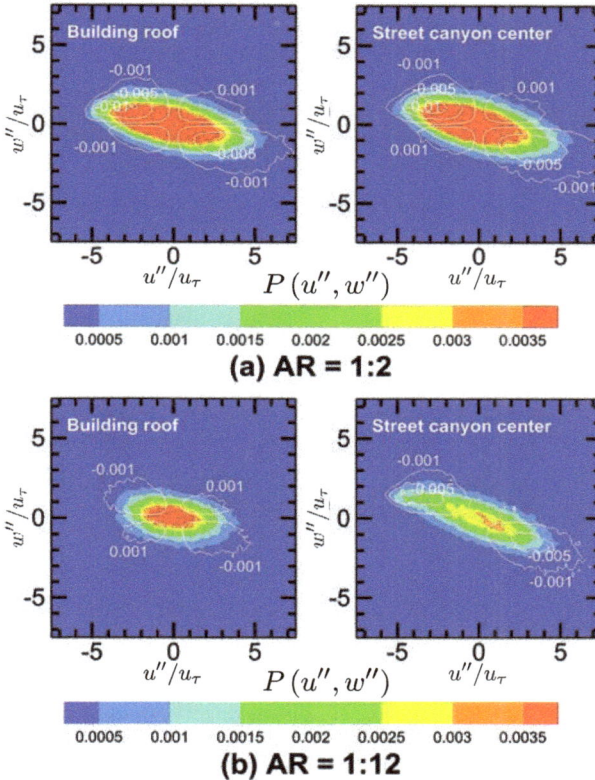

Figure 6. Shaded contours of joint probability density function (JPDF) $P(u'', w'')$ and contours of covariance integrand $u''w''P(u'', w'')$ at roof level $z/h = 0.26$ over street canyons of hypothetical urban areas of AR = (**a**) 1:2 and (**b**) 1:12.

Table 2. Quadrants for vertical turbulent momentum flux $u''w''$.

Quadrants	Events	u''	w''
1	Outward interaction	$+$	$+$
2	Ejection	$-$	$+$
3	Inward interaction	$-$	$-$
4	Sweep	$+$	$-$

3.7. Frequency Spectrum

Fast Fourier transform (FFT) is used to convert the time traces of velocity from time to frequency domain. Figure 7 shows that the roof-level spectra over the arrays of hypothetical urban areas exhibit a conventional inertial subrange with a −5/3 slope. The energy spectra cover almost five orders of magnitude of spatial/temporal scales. The low-frequency fraction ($fh/u_* \leq 10$) of streamwise turbulence scales is obviously stronger than its vertical counterpart by an order of magnitude. Contributions from high-frequency fractions are about the same, demonstrating the isotropic nature of small-scale turbulence. This finding also concurs the importance of extreme events in a street-level ventilation mechanism which is discussed in Section 3.6 previously. No notable difference in streamwise turbulence spectra over buildings and street canyons is observed. However, a mild difference is shown in the vertical turbulence spectra. The low-frequency fraction over building roofs is slightly higher than that over street canyons and is attributed to the higher level of velocity shear near the solid boundary, enhancing mechanical turbulence generation. Moreover, it is interesting that the low-frequency fraction of turbulence-intensity spectra over street canyons of AR = 1:12 is higher than that of 1:2. Hence, the turbulence scales governing street-level ventilation in wider street canyons are stronger.

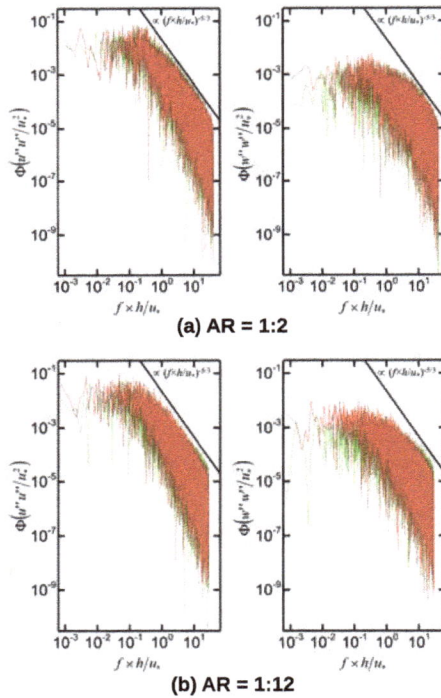

Figure 7. Frequency spectra of dimensionless streamwise $\Phi \left(u''u''/u_\tau^2 \right)$ and vertical $\Phi \left(w''w''/u_\tau^2 \right)$ turbulence intensities at roof level $z/h = 0.26$ over building roof (red) and street canyon center (green) over street canyons of hypothetical urban areas of AR = (**a**) 1:2 and (**b**) 1:12.

4. Conclusions

In view of the importance of air quality, a series of wind tunnel experiments are sought to elucidate the mechanism of street-level ventilation in urban areas. Vertical profiles of mean wind speed $\langle \overline{u} \rangle$ illustrate that the near-surface velocity gradient $\partial \langle \overline{u} \rangle / \partial z|_{z=0}$ is large, leading to the drag and enhanced transport processes. Examination of the profiles of fluctuating velocities $\left\langle \overline{u_i'' u_i''} \right\rangle^{1/2}$ and momentum flux $\left\langle \overline{u'' w''} \right\rangle$ suggests that friction velocity u_τ is an appropriate quantity to scale the inner-layer turbulence quantities but not those in the outer layer. In the spatio-temporally averaged turbulence statistics, the streamwise fluctuating velocity $\left\langle \overline{u'' u''} \right\rangle^{1/2}$ is peaked at the roof level while the vertical fluctuating velocity $\left\langle \overline{w'' w''} \right\rangle^{1/2}$ has relatively broad maxima. These observations are consistent with previous findings in literature [71]. A new indicator, the vertical fluctuating velocity scale in roughness sublayer $\widehat{w}''_{\text{RSL}}$, is proposed to measure the street-level ventilation performance over urban areas of different aerodynamic resistance and building height variability. Major benefits include its applicability to areas with inhomogeneous buildings and the (linear) correlation with (the square root of) drag coefficient that facilities sensitivity tests in design stage. Nevertheless, additional tests are required to unveil the limitations and drawbacks in a practical perspective. Because street-level ventilation is mainly driven by turbulence, quadrant analysis and frequency spectrums are carried out. Similar to the flows over smooth surfaces, vertical turbulent transport is dominated by sweep and ejection. The inertial subrange is exhibited in the wind tunnel measurements as well.

Acknowledgments: The first author wishes to thank the Hong Kong Research Grants Council (RGC) for financially supporting his study through the Hong Kong PhD Fellowship (HKPF) scheme. This project is partly supported by the General Research Fund (GRF) of RGC HKU 17210115.

Author Contributions: Yat-Kiu Ho and Chun-Ho Liu conceived and designed the experiments; Yat-Kiu Ho performed the experiments; Yat-Kiu Ho and Chun-Ho Liu analyzed the data and wrote the paper.

Conflicts of Interest: The authors declare no conflict of interest.

References

1. Deweerdt, S. The urban downshift. *Nature* **2016**, *531*, S52–S53.
2. Wigginton, N.S.; Fahrenkamp-Uppenbrink, J.; Wible, B.; Malakoff, D. Cities are the future. *Science* **2016**, *352*, 904–905.
3. Parrish, D.D.; Zhu, T. Clean air for megacities. *Science* **2009**, *326*, 674–675.
4. Fernando, H.J.S. Fluid dynamics of urban atmospheres in complex terrain. *Annu. Rev. Fluid Mech.* **2010**, *42*, 365–389.
5. Claus, J.; Krogstad, P.A.; Castro, I.P. Some measurements of surface drag in urban-type boundary layers at various wind angles. *Bound. Layer Meteorol.* **2012**, *145*, 407–422.
6. Ricci, A.; Burlando, M.; Freda, A.; Repetto, M.P. Wind tunnel measurements of the urban boundary layer development over a historical district in Italy. *Build. Environ.* **2017**, *111*, 192–206.
7. Tse, K.T.; Hitchcock, P.A.; Kwok, K.C.S.; Thepmongkorn, S.; Chan, C.M. Economic perspectives of aerodynamic treatments of square tall buildings. *J. Wind Eng. Ind. Aerodyn.* **2009**, *97*, 455–467.
8. Aly, A.M. Atmospheric boundary-layer simulation for the built environment: Past, present and future. *Build. Environ.* **2014**, *75*, 206–221.
9. Stabile, L.; Arpino, F.; Buonanno, G.; Russi, A; Frattolillo, A. A simplified benchmark of ultrafine particle dispersion in idealized urban street canyons: A wind tunnel study. *Build. Environ.* **2015**, *93*, 186–198.
10. Panagiotou, I.; Neophytou, M.K.A.; Hamlyn, D.; Britter, R.E. City breathability as quantified by the exchange velocity and its spatial variation in real inhomogeneous urban geometries: An example from central London urban area. *Sci. Total Environ.* **2013**, *442*, 466–477.
11. Hang, J.; Wang, Q.; Chen, X.; Sandberg, M.; Zhu, W.; Buccolieri, R.; Di Sabatino, S. City breathability in medium density urban-like geometries evaluated through the pollutant transport rate and the net escape velocity. *Build. Environ.* **2015**, *94*, 166–182.
12. Blocken, B.; Stathopoulos, T.; van Beeck, J.P.A.J. Pedestrian-level wind conditions around buildings: Review of wind-tunnel and CFD techniques and their accuracy for wind comfort assessment. *Build. Environ.* **2016**, *100*, 50–81.

13. Kubota, T.; Miura, M.; Tominaga, Y.; Mochida, A. Wind tunnel tests on the relationship between building density and pedestrian-level wind velocity: Development of guidelines for realizing acceptable wind environment in residential neighborhoods. *Build. Environ.* **2008**, *43*, 1699–1708.

14. Jiménez, J. Turbulent flows over rough walls. *Annu. Rev. Fluid Mech.* **2004**, *36*, 173–196.

15. Kamruzzaman, M.; Djenidi, L.; Antonia, R.A.; Talluru, K.M. Drag of a turbulent boundary layer with transverse 2D circular rods on the wall. *Exp. Fluids* **2015**, *56*, 121–129.

16. Hagishima, A.; Tanimoto, J.; Nagayama, K.; Meno, S. Aerodynamic parameters of regular arrays of rectangular blocks with various geometries. *Bound. Layer Meteorol.* **2009**, *132*, 315–337.

17. Karimpour, A.; Kaye, N.B.; Baratian-Ghorghi, Z. Modeling the neutrally stable atmospheric boundary layer for laboratory scale studies of the built environment. *Build. Environ.* **2012**, *49*, 203–211.

18. Krogstad, P.-Å.; Antonia, R.A. Surface roughness effects in turbulent boundary layers. *Exp. Fluids* **1999**, *27*, 450–460.

19. Antonia, R.A.; Krogstad, P.-Å. Turbulence structure in boundary layers over different types of surface roughness. *Fluid Dyn. Res.* **2001**, *28*, 139–157.

20. Mirzaei, P.A.; Haghighat, F. Pollution removal effectiveness of the pedestrian ventilation system. *J. Wind Eng. Ind. Aerodyn.* **2011**, *99*, 46–58.

21. Mirzaei, P.A.; Haghighat, F. A procedure to quantify the impact of mitigation techniques on the urban ventilation. *Build. Environ.* **2012**, *47*, 410–420.

22. Kanda, I.; Yamao, Y. Velocity adjustment and passive scalar diffusion in and above an urban canopy in response to various approach flows. *Bound. Layer Meteorol.* **2011**, *141*, 415–441.

23. Carpentieri, M.; Hayden, P.; Robins, A.G. Wind tunnel measurements of pollutant turbulent fluxes in urban intersections. *Atmos. Enviorn.* **2012**, *46*, 669–674.

24. Chung, J.; Hagishima, A.; Ikegaya, N.; Tanimoto, J. Wind-tunnel study of scalar transfer phenomena for surfaces of block arrays and smooth walls with dry patches. *Bound. Layer Meteorol.* **2015**, *157*, 219–236.

25. Raupach, M.R.; Hughes, D.E.; Cleugh, H.A. Momentum absorption in rough-wall boundary layers with sparse roughness elements in random and clustered distributions. *Bound. Layer Meteorol.* **2006**, *120*, 201–218.

26. Tachie, M.F.; Bergstrom, D.J.; Balachandar, R. Roughness effects in low-Re_θ open-channel turbulent boundary layers. *Exp. Fluids* **2003**, *35*, 338–346.

27. Pascheke, F.; Barlow, J.F.; Robins, A.G. Wind-tunnel modelling of dispersion from a scalar area source in urban-like roughness. *Bound. Layer Meteorol.* **2008**, *126*, 103–124.

28. Ikegaya, N.; Hagishima, A.; Tanimoto, J.; Tanaka, Y.; Narita, K.-I.; Zaki, S.A. Geometric dependence of the scalar transfer efficiency over rough surfaces. *Bound. Layer Meteorol.* **2012**, *143*, 357–377.

29. Barlow, J.F.; Harman, I.N.; Belcher, S.E. Scalar fluxes from urban street canyons. Part I: Laboratory simulation. *Bound. Layer Meteorol.* **2004**, *113*, 369-385.

30. Salizzoni, P.; Soulhac, L.; Mejean, P. Street canyon ventilation and atmospheric turbulence. *Atmos. Environ.* **2009**, *43*, 5056–5067.

31. Takimoto, H.; Sato, A.; Barlow, J.F.; Ryo M.; Inagaki, A.; Onomura, S.; Kanda, M. Particle image velocimetry measurements of turbulent flow within outdoor and indoor urban scale models and flushing motions in urban canopy layers. *Bound. Layer Meteorol.* **2011**, *140*, 295–314.

32. Perret, L.; Savory, E. Large-scale structures over a single street canyon immersed in an urban-type boundary layer. *Bound. Layer Meteorol.* **2013**, *148*, 111–131.

33. Djenidi, L.; Antonia, R.A.; Anselmet, F. LDA measurements in a turbulent boundary layer over a *d*-type rough wall. *Exp. Fluids* **1994**, *16*, 323–329.

34. Schultz, M.P.; Flack, K.A. Outer layer similarity in fully rough turbulent boundary layers. *Exp. Fluids* **2005**, *38*, 328–340.

35. Smits, A.J.; McKeon, B.J.; Marusic, I. High-Reynolds number wall turbulence. *Annu. Rev. Fluid Mech.* **2011**, *43*, 353–375.

36. Squire, D.T.; Morrill-Winter, C.; Hutchins, N.; Schultz, M.P.; Klewicki, J.C.; Marusic, I. Comparison of turbulent boundary layers over smooth and rough surfaces up to high Reynolds numbers. *J. Fluid Mech.* **2016**, *795*, 210–240.

37. Volino, R.J.; Schultz, M.P.; Flack, K.A. Turbulence structure in rough- and smooth-wall boundary layers. *J. Fluid Mech.* **2007**, *592*, 263–293.

38. Flack, K.A.; Schultz, M.P. Roughness effects on wall-bounded turbulent flows. *Phys. Fluids* **2014**, *101305*, doi:10.1063/1.4896280.
39. De Ridder, K. Bulk transfer relations for the roughness sublayer. *Bound. Layer Meteorol.* **2010**, *134*, 257–267.
40. Kastner-Kelin, P.; Rotach, M.W. Mean flow and turbulence characteristics in an urban roughness sublayer. *Bound. Layer Meteorol.* **2004**, *111*, 55–84.
41. Smeets, C.J.P.P.; van den Broeke, M.R. The parameterisation of scalar transfer over rough ice. *Bound. Layer Meteorol.* **2008**, *128*, 339–355.
42. Castro, I.P.; Cheng, H.; Reynolds, R. Turbulence over urban-type roughness: Deductions from wind-tunnel measurements. *Bound. Layer Meteorol.* **2006**, *118*, 109–131.
43. Mihailovic, D.T.; Rao, S.T.; Hogrefe, C.; Clark, R.D. An approach for the aggregation of aerodynamic surface parameters in calculating the turbulent fluxes over heterogeneous surfaces in atmospheric models. *Environ. Fluid Mech.* **2002**, *2*, 339–355.
44. Britter, R.E.; Hanna, S.R. Flow and dispersion in urban areas. *Annu. Rev. Fluid Mech.* **2003**, *35*, 469–496.
45. Oke, T.R. Street design and urban canopy layer climate. *Energy Bldg.* **1988**, *11*, 103–113.
46. Perry, A.E.; Schofield, W.H.; Joubert, P.N. Rough wall turbulent boundary layers. *J. Fluid Mech.* **1969**, *37*, 383–413.
47. Baetke, F.; Werner, H.; Wengle, H. Numerical simulation of turbulent flow over surface-mounted obstacles with sharp edges and corners. *J. Fluid Mech.* **1990**, *35*, 129–147.
48. Bruun, H.H. Interpretation of a hot wire signal using a universal calibration law. *J. Sci. Instrum.* **1971**, *4*, 225–231.
49. Aydin, E.M. Leutheusser, H.J. Plane-Couette flow between smooth and rough walls. *Exp. Fluids* **1991**, *11*, 302–312.
50. Snyder, W.H.; Castro, I.P. The critical Reynolds number for rough-wall boundary layers. *J. Wind Eng. Ind. Aerodyn.* **2002**, *90*, 41–54.
51. Garratt, J.R. *The Atmospheric Boundary Layer*; Cambridge University Press: Cambridge, UK, 1994; p. 316.
52. Cheng, H.; Castro, I.P. Near wall flow over urban-like roughness. *Bound. Layer Meteorol.* **2002**, *104*, 229–259.
53. Ho, Y.-K. Wind-Tunnel Study of Turbulent Boundary Layer over Idealised Urban Roughness with Application to Urban Ventilation Problem. Ph.D. Thesis, The University of Hong Kong, Hong Kong, China, 2017; 237p.
54. Takimoto, H.; Inagaki, A.; Kanda, M.; Sato, A.; Michioka, T. Length-scale similarity of turbulent organized structures over surfaces with different roughness types. *Bound. Layer Meteorol.* **2013**, *147*, 217–236.
55. Wu, Z.; Liu, C.-H. Time scale analysis of chemically reactive pollutants over urban roughness in the atmospheric boundary layer. *Int. J. Environ. Pollut.* **2017**, accepted.
56. Burattini, P.; Leonardi, S.; Orlandi, P.; Antonia, R.A. Comparison between experiment and direct numerical simulations in a channel flow with roughness on one wall. *J. Fluid Mech.* **2008**, *600*, 403–426.
57. Rafailidis, S. Influence on building area density and roof shape on the wind characteristics above a town. *Bound. Layer Meteorol.* **1997**, *85*, 255–271.
58. Keirsbulck, L.; Labraga, L.; Mazouz, A.; Tournier, C. Influence of surface roughness on anisotropy in a turbulent boundary layer flow. *Exp. Fluids* **2002**, *33*, 497–499.
59. Nourmohammadi, K.; Hopke, P.K.; Stukel, J.J. Turbulent air flow over rough surfaces II. Turbulent flow parameters. *J. Fluids Eng.* **1985**, *107*, 55–60.
60. Panofsky, H.A. The atmospheric boundary layer below 150 meters. *Annu. Rev. Fluid Mech.* **1974**, *107*, 147–177.
61. Volino, R.J.; Schultz, M.P.; Flack, K.A. Turbulence structure in a boundary layer with two-dimensional roughness. *J. Fluid Mech.* **2009**, *635*, 75–101.
62. Volino, R.J.; Schultz, M.P.; Flack, K.A. Turbulence structure in boundary layers over periodic two- and three-dimensional roughness. *J. Fluid Mech.* **2011**, *676*, 172–190.
63. Smalley, R.J.; Leonardi, S.; Antonia, R.A.; Djenidi, L.; Orlandi, P. Reynolds stress anisotropy of turbulent rough wall layers. *Exp. Fluids* **2002**, *33*, 31–37.
64. Reynolds, R.T.; Castro, I.P. Measurements in an urban-type boundary layer. *Exp. Fluids* **2008**, *45*, 141–156.
65. Ho, Y.-K.; Liu, C.-H. A wind tunnel study of flows over idealised urban surfaces with roughness sublayer corrections. *Theor. Appl. Climatol.* **2016**, 1–16, doi:10.1007/s00704-016-1877-8.
66. Liu, C.-H.; Leung, D.Y.C.; Barth, M.C. On the prediction of air and pollutant exchange rates in street canyons of different aspect ratios using large-eddy simulation. *Atmos. Environ.* **2005**, *39*, 1567–1574.

67. Ho, Y.-K.; Liu, C.-H.; Wong, M.S. Preliminary study of the parameterisation of street-level ventilation in idealised two-dimensional simulations. *Build. Environ.* **2015**, *89*, 345–355.

68. Liu, C.H.; Ng, C.T.; Wong, C.C.C. A theory of ventilation estimate over hypothetical urban areas. *J. Hazard. Mater.* **2015**, *296*, 9–16.

69. Wallace, J.M. Quadrant analysis in turbulence research: History and evolution. *Annu. Rev. Fluid Mech.* **2016**, *48*, 131–158.

70. Immer, M.; Allegrini, J.; Carmeliet, J. Time-resolved and time-averaged stereo-PIV measurements of a unit-ratio cavity. *Exp. Fluids* **2016**, *57*, 101–118.

71. Hong, J.; Katz, J.; Schultz, M.P. Near-wall turbulence statistics and flow structures over three-dimensional roughness in a turbulent channel flow. *J. Fluid Mech.* **2011**, *667*, 1–37.

atmosphere

MDPI

Article

Impacts of Urban Layouts and Open Space on Urban Ventilation Evaluated by Concentration Decay Method

Qun Wang [1], Mats Sandberg [2], Yuanyuan Lin [1], Shi Yin [3] and Jian Hang [1,*]

[1] School of Atmospheric Sciences, Sun Yat-Sen University, Guangzhou 510275, China; wangqun_sysu@163.com (Q.W.); linyy8@mail2.sysu.edu.cn (Y.L.)
[2] Laboratory of Ventilation and Air Quality, University of Gävle, SE-80176 Gävle, Sweden; Mats.Sandberg@hig.se
[3] Department of Mechanical Engineering, The University of Hong Kong, PokFuLam Road, Hong Kong, China; shiyin103@gmail.com
* Correspondence: hangj3@mail.sysu.edu.cn; Tel.: +86-20-8411-0375

Received: 20 August 2017; Accepted: 6 September 2017; Published: 11 September 2017

Abstract: Previous researchers calculated air change rate per hour (ACH) in the urban canopy layers (UCL) by integrating the normal component of air mean velocity (convection) and fluctuation velocity (turbulent diffusions) across UCL boundaries. However they are usually greater than the actual ACH induced by flow rates flushing UCL and never returning again. As a novelty, this paper aims to verify the exponential concentration decay history occurring in UCL models and applies the concentration decay method to assess the actual UCL ACH and predict the urban age of air at various points. Computational fluid dynamic (CFD) simulations with the standard k-ε models are successfully validated by wind tunnel data. The typical street-scale UCL models are studied under neutral atmospheric conditions. Larger urban size attains smaller ACH. For square overall urban form ($Lx = Ly = 390$ m), the parallel wind ($\theta = 0°$) attains greater ACH than non-parallel wind ($\theta = 15°$, $30°$, $45°$), but it experiences smaller ACH than the rectangular urban form ($Lx = 570$ m, $Ly = 270$ m) under most wind directions ($\theta = 30°$ to $90°$). Open space increases ACH more effectively under oblique wind ($\theta = 15°$, $30°$, $45°$) than parallel wind. Although further investigations are still required, this paper provides an effective approach to quantify the actual ACH in urban-like geometries.

Keywords: small open space; air change rate per hour (ACH); concentration decay method; urban age of air; computational fluid dynamic (CFD) simulation

1. Introduction

The increase in number of vehicles in cities and the ongoing urbanization worldwide are causing more concerns about urban air pollution. The urban canopy layer (UCL) is defined as the outdoor air volume below the rooftops of buildings [1]. Raising UCL ventilation capacity by the surrounding atmosphere with relatively clean air has been regarded as one effective approach to diluting environmental pollutants [2–8] and improving the urban thermal environment in the hot summer [9,10] as well as reducing human exposure to outdoor pollutants [11–13].

UCL ventilation is strongly correlated to urban morphologies. For two-dimensional street canyons [14–19], street aspect ratio (building height/street width, H/W) is the first key parameter to affect the flow regimes and UCL ventilation. Four flow regimes have been classified depending on different aspect ratios (H/W) [14–16], i.e., the isolated roughness flow regime (IRF, in which the aspect ratio is less than 0.1 to 0.125), the wake interference flow regime (WIF, with an aspect ratio of 0.1 to 0.67), the skimming flow regime (SF, with an aspect ratio of 0.67 to 1.67), and the multi-vortex regime in two-dimensional deep street canyons with two or more vortices as $H/W > 1.67$. For three-dimensional UCL geometries, the plan area index λ_p and frontal area index λ_f [4–6,9,10] are regarded as the key urban parameters to influence the flow and pollutant dispersion in urban areas. In addition, the other factors including overall urban form and ambient wind directions [2,3,6–9], building height variations [5,20–25], thermal buoyancy force for weak-wind atmospheric conditions [10,18,26–30], etc. also play significant roles in the flow and pollutant dispersion in UCL models.

In recent years, various ventilation indices have been applied for UCL ventilation assessment, such as pollutant retention time and purging flow rate [2,7,23], age of air and ventilation efficiency [3–7,23], in-canopy velocity and exchange velocity [22,25], net escape velocity [31], etc. Similar to indoor ventilation [32–36], the key point of applying these ventilation indices is based on the UCL ventilation processes as below: The surrounding external air is relatively clean and can be transported into urban areas to aid pollutant dilution with physical processes of horizontal dilution, vertical transport, recirculation of contaminants, and turbulent diffusion. In this sense, the tracer gas technique originated from indoor ventilation sciences has been used to predict outdoor ventilation by analyzing the final steady-state pollutant concentrations or the transient concentration decay history, which illustrates the processes of pollutant dilution and ventilation. For example, the concept of purging flow rate (PFR) was adopted to assess the net flow rate of flushing the urban domain [2] or the entire urban canopy layer [23] induced by the convection and turbulent diffusion. Moreover, the urban age of air [3–7] represents how long the external air can reach a place after it enters the UCL space. Hang et al. [3] first adopted the homogeneous emission method into CFD simulations [32] to calculate the urban age of air for quantifying the characteristics of wind supplying external air into UCL space for pollutant dilution.

In particular, the air change rate per hour is one of the most widely used indoor ventilation concepts [32–36], calculated by $ACH = 3600Q_T/Vol$, where Q_T is the total volumetric flow rate and Vol is the room volume. Later it is adopted to quantify the volumetric air exchange rate of the entire UCL volume per hour, including two-dimensional street canyon ventilation assessment [37,38] and three-dimensional UCL ventilation modeling [39–44]. Previous researchers usually calculated outdoor ACH indexes based on the volumetric flow rate obtained by integrating the mean value of the component of the mean velocity normal to the UCL boundaries and the effective flow rate due to turbulence based on integrating half the standard deviation (rms-value) of the velocity fluctuations on street roofs [37–44]. Specially, vertical turbulent exchange across a street roof has been verified to significantly influence UCL ventilation and pollutant removal in urban-like geometries [37–40] because street roofs are open at the top with a large area. However, the mean velocity often exhibits recirculation and the velocity fluctuations across the UCL open roof are bidirectional and therefore air or pollutant can return to the given UCL space several times. Thus, these two ACH indexes do not represent the entire UCL air volume that is really exchanged ACH times in an hour by external air, i.e., they are not the actual air change rate per hour in UCL models calculated by the actual flow rate flushing or leaving this space and never returning again. There are efficiency problems in UCL ventilation assessment [3,5].

The purpose of this paper is to predict the actual or net air change rate per hour in UCL models and how it is related to urban morphologies and atmospheric conditions. The net or actual air change rate is defined as the exchange of air flushing the space and never returning to the UCL space again. It depends on several time scales, i.e., the overall turn over time (air volume (Vol) divided by the total volumetric flow rate (Q_T)), purging time for air moving from entrances to exits, time constant for local vortices and turbulent mixing. The concentration decay method has been widely used to predict the actual ACH and age of air in indoor environments in which the concentration temporal profile accords with the exponential decay law and the concentration decay history shows the pollutant dilution and ventilation processes in room ventilation [32–36,45,46]. However, to date, this method has not been introduced to quantify the actual or net ACH index and age of air in UCL models.

Therefore, as a novelty, this paper aims to verify whether the temporal decay profile of concentration in UCL models accords with the exponential decay law and explore the effectiveness of applying concentration decay method for outdoor ventilation assessment. As a start, the typical medium-dense urban canopy layers ($H/W = 1$, $\lambda f = \lambda_p = 0.25$) are first studied. The effects of urban size, ambient wind directions, overall urban forms, and open space arrangements are evaluated under neutral atmospheric conditions.

2. Methodology

2.1. Turbulence Models for Urban Airflow Modeling

As outlined by the review papers, CFD simulations have been widely applied in urban flow/dispersion modeling outdoor in the last two decades [47–49]. Large eddy simulations (LES) are known to perform better in predicting turbulence than the Reynolds-Averaged Navier-Stokes (RANS) approaches. However, there are still some challenges involved with widely applying LES because of the strongly increased computational requirements, the development of advanced sub-grid scale models, and the difficulty in specifying appropriate time-dependent inlet and wall boundary conditions [16,17,20,26,38,39]. Therefore, in quantitative work one has to adopt time-averaged turbulence models (i.e., RANS approaches). Actually, in spite of their deficiencies in predicting turbulence, steady RANS models have become the most popular CFD approaches for urban flow modeling, and the standard *k*-ε model is one of the most widely adopted in the literature [2,3,5–7,9–11,21–23], with successful validation by wind tunnel measurements and good performance in predicting mean flows. This paper selects the standard *k*-ε model and the CFD validation case is conducted to evaluate the reliability of CFD methodologies by wind tunnel data.

2.2. Model Description in the CFD Validation Case

As shown in Figure 1a, Brown et al. [50] measured turbulent airflows in a seven-row and 11-column cubic building array with a parallel approaching wind. Building width (*B*), building height (*H*), and street width (*W*) are the same ($B = H = W = 15$ cm, $H/W = 1$, building packing densities $\lambda_p = \lambda_f = 0.25$, $Lx = 13H$, $Ly = 21H$). The scale ratio to full-scale models is 1:200. *x*, *y*, and *z* are the stream-wise, span-wise (lateral), and vertical directions. $x/H = 0$ represents the location of windward street opening. $y/H = 0$ is the vertical symmetric plane of the middle column. Point V*i* represents the center point of the secondary street No. *i*. The measured vertical profiles of time-averaged (or mean) velocity components (stream-wise velocity $\overline{u}(z)$, vertical velocity $\overline{w}(z)$), and turbulence kinetic energy $\kappa(z)$ at Points V*i* are used to evaluate CFD simulations.

In the CFD validation case, the full-scale seven-row building array is numerically investigated ($B = H = W = 30$ m, $\lambda_p = \lambda_f = 0.25$, $H/W = 1$; see Figure 1b). According to the literature [23,51–53], the lateral urban boundaries hardly affect airflows in the middle column as the wind tunnel model is sufficiently wide in the lateral (*y*) direction ($Ly = 21H$). Thus, it is effective to only consider half of this middle column (shaded area in Figure 1a) to reduce the computational requirement. Figure 1b shows model geometry, grid arrangements, CFD domain, and boundary conditions. In particular, the distances of UCL boundaries to the domain inlet, domain outlet, and domain top are 6.7*H*, 40.3*H*, and 9.0*H*, respectively. Zero normal gradient condition is used at the domain top, domain outlet, and two lateral domain symmetry boundaries. At the domain inlet, the measured power-law profile of time-averaged (or mean) velocity $U_0(z)$ in the upstream free flow is adopted (see Equation (1a)) (Brown et al. [51]). Moreover, the vertical profiles of turbulent kinetic energy $k(z)$ and its dissipation rate (ε) at the domain inlet are calculated by Equations (1b) and (1c) [23,51–53]:

$$U_0(z) = U_{ref} \times (z/H)^{0.16} \tag{1a}$$

$$k(z) = u_*^2 \Big/ \sqrt{C_\mu} \tag{1b}$$

$$\varepsilon(z) = C_\mu^{3/4} k^{3/2} / (\kappa_v z), \tag{1c}$$

where U_{ref} (=3.0 ms^{-1}) is the reference velocity at the building height (H = 30 m), C_μ is a constant (=0.09), u_* is the friction velocity (=0.24 ms^{-1}), and κ_v is von Karman's constant (=0.41).

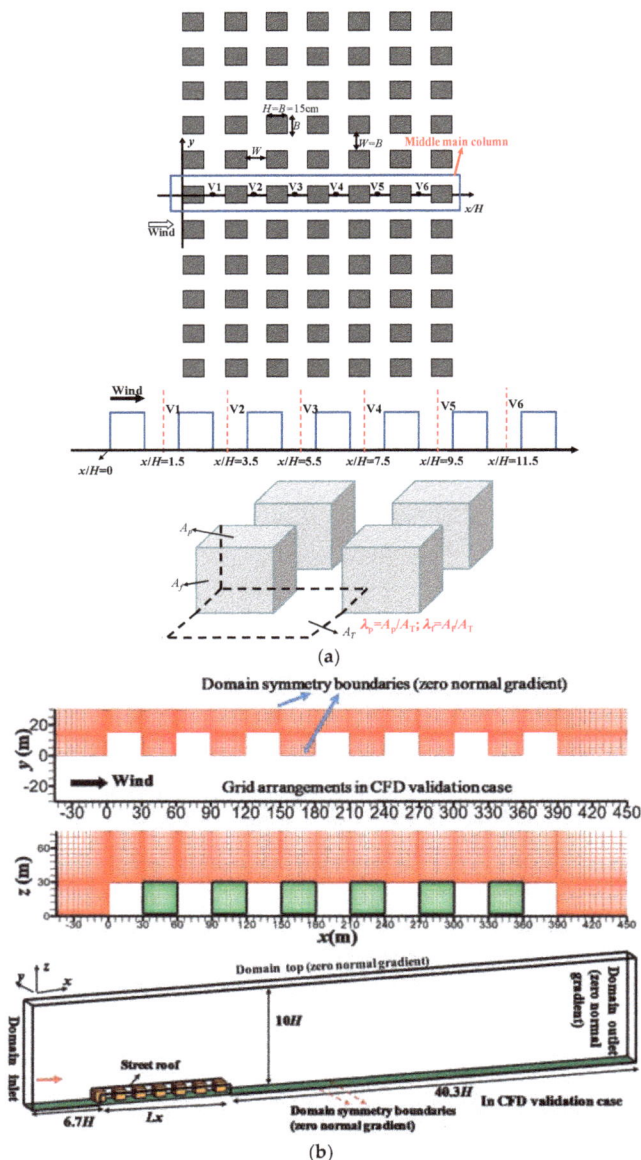

(a)

(b)

Figure 1. (**a**) Model description of wind tunnel test; (**b**) setups in CFD validation case.

For this CFD validation case, hexahedral cells of 531,657 and the minimum cell sizes of 0.5 m (H/60) are used (the medium grid). Following the CFD guidelines [54,55], four hexahedral cells exist below the pedestrian level (z = 0 m to 2 m). The grid expansion ratio from wall surfaces to the surrounding is 1.15. With such grid arrangements, the normalized distance from wall surfaces y^+ ($y^+ = yu_\tau/\nu$) ranges from 60 to 1000 at most regions of wall surfaces within UCL space.

No slip boundary condition with standard wall function [56] is used at wall surfaces. The practice guidelines for setting the upstream and downstream ground are followed to reproduce a horizontally homogeneous atmospheric boundary layer surrounding urban areas [46,57]: The roughness height k_S and the roughness constant C_S are correlated by the aerodynamic roughness length z_0:

$$k_S = \frac{9.793 z_0}{C_S}. \tag{2}$$

Note that the distance between the center point P of the ground adjacent cell and the ground surface is $Y_P = 0.25$ m. Fluent 6.3 does not allow k_S to be larger than Y_P [57]. Thus, according to van Hooff and Blocken [46], a user-defined function is used to set the roughness constant $C_S = 4$ because Fluent 6.3 does not allow it to be greater than 1. Then as $C_S = 4$ and $z_0 = 0.1$ m, $k_S = 0.245$ m is smaller than $Y_P = 0.25$ m.

ANSYS FLUENT with the finite volume method is used to predict the steady isothermal urban airflows [56]. The SIMPLE scheme is adopted to couple the pressure to the velocity field. The first-order upwind scheme is not appropriate for all transported quantities since the spatial gradients of the quantities tend to become diffusive due to a large viscosity. Thus, as recommended by the literature [54,55], all transport equations are discretized by the second-order upwind scheme for better numerical accuracy. The under-relaxation factors for pressure term, momentum term, k, and ε terms are 0.3, 0.7, 0.5, and 0.5, respectively. The solutions do not stop until all residuals become constant. The fine grid with a minimum size of 0.2 m ($H/150$) is also adopted to perform a grid independence study. The above CFD settings on domain sizes, grid arrangements and boundary conditions fulfill the major requirements given by CFD guidelines [54,55]. In particular, steady CFD simulations are first carried out about 4000 iterations with the first-order upwind scheme, then continued with the second-order upwind scheme until all residuals became constant. The residuals reached the following minimum values or less: 10^{-4} for the continuity equation, 0.5×10^{-5} for the velocity components and k, 0.5×10^{-5} and 0.5×10^{-4} for pollutant concentration and ε. After solving the steady-state airflow field, the uniform initial concentration $C(0)$ is defined in the entire urban canopy layer, then the unsteady concentration decay history is calculated and recorded for ventilation assessment as the solved flow field remains constant.

2.3. Model Description in All CFD Test Cases

As displayed in Table 1 and Figure 2a–d, two groups of medium-dense UCL models ($H = B = W = 30$ m, $\lambda_p = \lambda_f = 0.25$) are investigated in CFD simulations. The row and column numbers are referred to the numbers of buildings along the stream-wise direction (x, main streets) and span-wise direction (y, secondary streets). Wind directions are represented by the angles ($\theta = 0°$ to $90°$) between the approaching wind and the main streets.

Table 1. Test cases investigated ($\lambda_p = \lambda_f = 0.25$, $H/W = 1$).

Case Name	Number of Rows/Columns, Urban Sizes Lx and Ly	Wind Direction($\theta°$)
Group I		
[5-5, $\theta°$]	5 rows, 5 columns, $Lx = Ly = 270$ m	
[7-7, $\theta°$]	7 rows, 7 columns, $Lx = Ly = 390$ m	$0°, 15°, 30°, 45°$
[10-5, $\theta°$]	10 rows, 5 columns, $Lx = 570$ m, $Ly = 270$ m	$0°, 15°, 30°, 45°, 60°, 75°, 90°$
Group II		
[5-5, $\theta°$, Oij]	5 rows, 5 columns, $Lx = Ly = 270$ m, Oij = O21, O22, O23, O24, O33, O34	$0°, 15°, 30°, 45°$

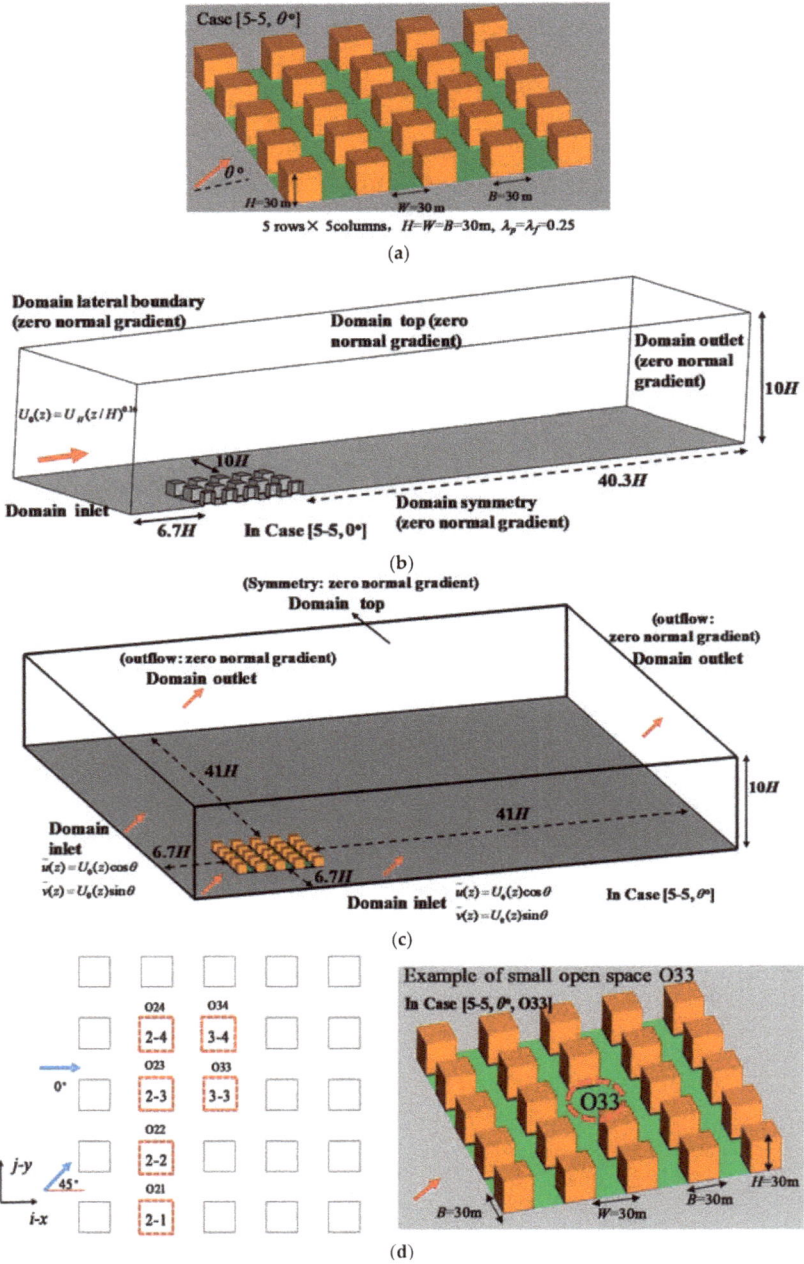

(a)

(b)

(c)

(d)

Figure 2. *Cont.*

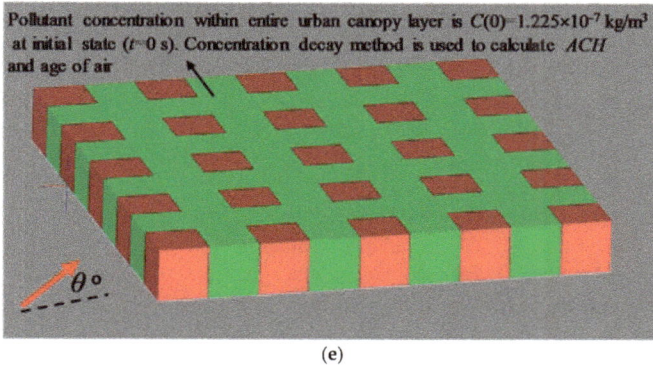

Figure 2. (**a**) Model description and (**b**) computational domain of Case [5-5, 0°] (θ = 0°); (**c**) Computational domain in Case [5-5, θ] (θ =15°, 30°, 45°); (**d**) Model description of Case [5-5, θ, Oij]; (**e**) Definition of initial-state concentration field in the entire UCL space.

In Group I (Table 1 and Figure 2a), test cases without open space are named as Case [row number-column number, wind direction θ°]. For cases with a parallel approaching wind (θ = 0°, Figure 2b), three cases are included (i.e., Case [5-5, 0°], Case [7-7, 0°], Case [10-5, 0°]) with only half of computational domain simulated. The distances from UCL boundaries to the domain top, domain outlet, domain inlet, and the domain lateral boundary are 9H, 40.3H, 6.7H, and 10H, respectively. Zero normal gradient boundary condition is used at the domain outlet, domain top, and the lateral domain boundary. At the domain inlet, Equation (1) is used to provide the boundary condition.

For test cases with oblique wind (θ = 15°, 30°, 45°, 60°, 75°, 90°, Figure 2c), the full CFD domains are used. There are two domain inlets and two domain outlets. The distances from UCL boundaries to the domain top, domain outlets, domain inlets are 9H, 41H and 6.7H. At two domain inlets, the vertical profiles of time-averaged (or mean) velocity components in Equations (3a)–(3c) and turbulent quantities defined in Equations (1b) and (1c) are used to define boundary conditions:

$$\bar{u} = U_0(z) \cos \theta \tag{3a}$$

$$\bar{v} = U_0(z) \sin \theta \tag{3b}$$

$$\bar{w}(z) = 0. \tag{3c}$$

At the domain outlets and domain top, the zero normal gradient boundaries are used.

In Group II (Table 1 and Figure 2d), for UCL models with five rows and five columns, open space arrangements are included. In total 24 test cases of Case [5-5, θ°, Oij] are investigated (θ = 0°, 15°, 30°, 45°). Here Oij represents only one building of position i-j (2-1, 2-2, 2-3, 2-4, 3-3, or 3-4) (see Figure 2d) is removed to attain an open space effect. Computational domain and boundary conditions are similar to Group I.

For all the test cases in Table 1, the grid arrangements are similar to the medium grid of the CFD validation case. The total number of hexahedral cells is from 531,657 to 3,360,096. The above CFD settings on domain sizes, grid arrangements, and boundary conditions fulfill the major requirements recommended by the best CFD guidelines.

2.4. ACH Indexes and Age of Air by Concentration Decay Method

2.4.1. Volumetric Flow Rates and the Corresponding ACH [23,37–42]

To analyze the flow field, the velocity components and wind speed are all normalized by the reference velocity in the upstream free flow at the same height, i.e., Equation (1a) at the domain inlet $U_0(z) = U_{ref} \times (z/H)^{0.16}$. For example, for velocity at $z = 15$ m = $0.5H$, the velocity is normalized by dividing it by U_0 ($z = 15$ m) = 2.685 ms^{-1}.

To quantify the ventilation capacity, the mean volumetric flow rates are calculated by integrating the normal air velocity across all UCL boundaries (Equation (4a)); moreover, the effective flow rate through street roofs due to turbulent exchange is defined by integrating the fluctuation velocity across open street roofs (Equation (4b)) [23,37–42]:

$$Q = \int_A \vec{V} \bullet \vec{n} \, dA \tag{4a}$$

$$Q_{roof}(turb) = \pm \int_{A_{roof}} 0.5\sigma_w dA, \tag{4b}$$

where, in Equation (4a), $\vec{V}$ is the velocity vector, consisting of three time-averaged (or mean) velocity components ($\vec{V} = (\overline{u}, \overline{v}, \overline{w})$ in x, y, z directions), and $\vec{n}$ and A are the normal direction and area, respectively, of UCL boundaries. In Equation (4b), A_{roof} is the area of street roofs, $\sigma_w = \sqrt{\overline{w'w'}} = \sqrt{2k/3}$ is the fluctuation velocity based on the approximation of isotropic turbulence in all k-ε turbulent models where u', v', w' are the stream-wise, span-wise, vertical velocity fluctuations ($u' = v' = w'$) and turbulent kinetic energy is $k = \frac{1}{2}(\overline{u'u'} + \overline{v'v'} + \overline{w'w'})$.

It is worth noting that the values of +0.5 and −0.5 in Equation (4b) represent the fact that turbulent fluctuations induce the same upward and downward air fluxes across street roofs [23,37–42]. As pollution sources exist within UCL space, the literature confirmed that turbulent fluctuations across street roofs significantly contribute to pollutant removal since the upward pollutant flux through street roofs usually exceeds the downward pollutant flux because the pollutant concentration below street roofs is higher than that above them [21,31]. Due to the much greater total area of street roofs, the effective flow rates by turbulent exchange across street roofs $Q_{roof}(turb)$ are also important for the urban canopy layer ventilation compared to mean flows across UCL boundaries (Equation (4a)). Thus, this paper mainly adopts $Q_{roof}(turb)$ to assess the effect of turbulent fluctuations on UCL ventilation [23,37–44].

Due to the flow balance, the total outflow rate induced by mean flows across UCL boundaries (Q_{out}) equals that entering UCL boundaries (Q_{in}). They are named the total mean volumetric flow rates Q_T (see Equation (5)) [23,37–44], which is the sum of all volumetric rates through all street openings and street roofs:

$$Q_T = |Q_{in}| = |Q_{out}|. \tag{5}$$

Then *ACH* by Q_T and Q_{roof}(turb) are used to quantify the volumetric air exchange rate per hour, as below [23,37–44]:

$$ACH_T = 3600Q_T / Vol \tag{6a}$$

$$ACH_{turb} = 3600Q_{roof}(turb) / Vol, \tag{6b}$$

where *Vol* is the entire UCL volume, and Q_{roof}(turb) is defined in Equation (4b).

2.4.2. Actual or Net ACH and Urban Age of Air by the Concentration Decay Method

In urban ventilation sciences, the flow rates across a control surface are usually defined and recorded by integrating the normal velocity through it. Generally the complicated circulating flow is generated in an urban area, hence there are fluid particles coming back into UCL space across UCL boundaries. Moreover turbulent fluctuations induce upward and downward fluxes across the open street roofs, representing a large amount of fluid particles leave and re-enter UCL space. Thus, similar to the indoor environment [35,36], fluid particles flowing across UCL boundaries can be divided into two groups, one is for air leaving UCL space for never returning again and the other for air returning or revisiting UCL space (it has been in the UCL space before). The first group is much more significant to UCL ventilation and pollutant dilution. Therefore, only a fraction of the flow rates defined in Equations (4a) and (4b) contributes to flushing UCL space by external air or diluting pollutants within it. Furthermore, the *ACH* indexes defined in Equation (6) do not represent the entire UCL air volume that is really exchanged *ACH* times in an hour by external air.

To attain and assess the actual and net *ACH* induced by mean flows (ACH_T) and turbulent exchange (ACH_{turb}) across UCL boundaries, the concentration decay method is introduced into CFD simulations of UCL ventilation modeling. Similarly with indoor *ACH* by concentration decay method [32,45,46], if the predicted temporal decay profile of spatial mean concentration in the entire UCL volume accords with the exponential decay law, this decay rate is correlated to air change rate per hour in UCL models.

After the steady-state flow field is solved, a uniform initial time-averaged concentration of tracer gas (CO, carbon monoxide) is defined in the entire UCL air volume at time of $t = 0$ s ($C(0) = 1.225 \times 10^{-7}$ kg/m^3, see Figure 2e). Then the transient concentration decay process $C(t)$ is numerically simulated and recorded while the flow field keeps steady.

The unsteady governing equation of time-averaged concentration $C(t)$ is:

$$\frac{\partial C}{\partial t} + \overline{u}_j \frac{\partial C}{\partial x_j} - \frac{\partial}{\partial x_j}\left((D_m + D_t)\frac{\partial C}{\partial x_j}\right) = 0, \tag{7}$$

where $\overline{u}_j$ is the time-averaged (or mean) velocity components ($\overline{u}$, $\overline{v}$, $\overline{w}$), and D_m and D_t are the molecular and turbulent diffusivity of pollutants or tracer gas. Here $D_t = \nu_t / Sc_t$, ν_t is the kinematic eddy viscosity and Sc_t is the turbulent Schmidt number ($Sc_t = 0.7$) [3–7,21–23].

The decay duration of all test cases is sufficiently long (400 s). The default time step is $dt = 1$ s. In the CFD validation case, time steps of $dt = 0.5$ s and 2 s are also tested. The second-order upwind scheme is used for Equation (7) with the under-relaxation factors of 1.0. For each time step, the residual of Equation (7) reaches a value below 10^{-10}. For Equation (7), the inflow pollutant concentration at the domain inlet is set as zero and zero normal flux condition is used at all wall surfaces; moreover, zero normal gradient condition is applied at the domain top, domain outlet, and domain lateral boundaries.

The net and actual *ACH* index is calculated by the decay rate of spatial mean concentration in the entire UCL space ($<C(t)>/C(0)$) (Equation (8)) [32–34]:

$$< C(t) > = \frac{\int C(t)dxdydz}{Vol} \tag{8a}$$

$$ACH = \frac{3600}{t}\left|\ln \frac{<C(t)>}{C(0)}\right|, \tag{8b}$$

where $C(0) = 1.225 \times 10^{-7}$ kg/m^3 is a uniform initial concentration, *Vol* is the entire volume of UCL space, $<C(t)>$ is the spatial mean concentration in the entire UCL space, and 3600 denotes one hour or 3600 s.

Moreover, the concentration decay method can be used to predict urban age of air (τ_p s) at a given point (see Equation (9)) [32,36]:

$$\tau_p = \frac{\int_0^\infty C(t)dt}{C(0)}. \tag{9}$$

Obviously, the physical meanings of Equations (8) and (9) are that the larger decay rate (K) of $<C(t)>/C(0)$ represents greater *ACH* and the pollutant dilution rate of the entire UCL volume; moreover, a larger K at a point denotes that it requires a shorter time for external air to reach this point (i.e., the age of air is smaller).

3. Results and Discussion

3.1. Evaluation of CFD Simulations Using Wind Tunnel Data

To validate CFD simulations, Figure 3 compares wind tunnel data and CFD results in the CFD validation case by using the fine and medium grid, including time-averaged stream-wise velocity $\bar{u}(z)$ and turbulent kinetic energy $k(z)$ at Point V2 and Point V5. Compared to wind tunnel data, the standard k-ε model with present grid arrangements can provide results of mean flows $\bar{u}(z)$ in good agreement with wind tunnel data, but does a little worse at predicting $k(z)$ and can only predict the shape of vertical profile well. Such findings are similar with the same CFD validation studies the literature [23,51–53]. In addition, CFD results with the fine grid are only slightly different from those with the present medium grid. Given the results from the validation tests, we conclude that the application of the standard k-ε model with present grid arrangement is acceptable for the purposes of our research. There are relatively large discrepancies of stream-wise velocity between the simulated and measured results at the higher area. The possible reason is the ratio of vertical grid size to the stream-wise grid size is relatively large at much higher levels above the building roofs (i.e., at the higher area). However, the simulation results below and near the building roofs are performed in good quality, which is more important for UCL ventilation assessment.

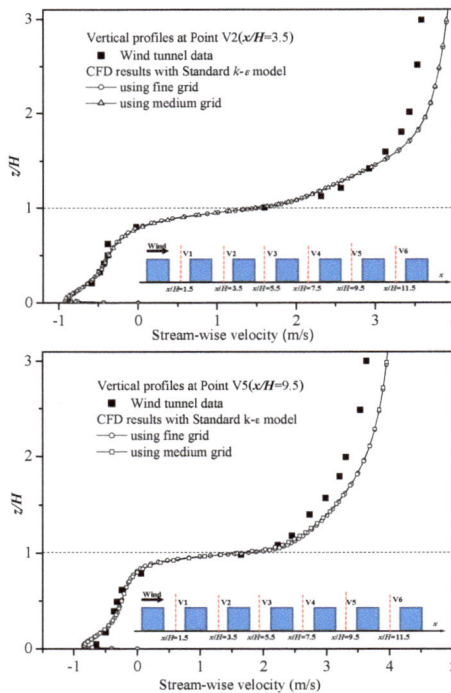

(a) Validation case: vertical profiles of $\bar{u}(z)$ at Point V2 and Point V5

Figure 3. *Cont.*

(**b**) Validation case: vertical profiles of $k(z)$ at Point V5

(**c**) $<C(t)>/C(0)$ with various time steps

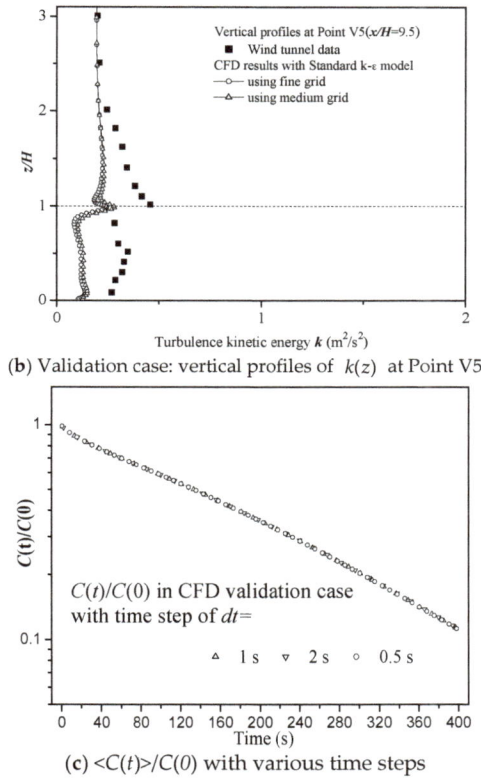

Figure 3. CFD results and experimental data in the CFD validation case: (**a,b**) Vertical profiles of $\bar{u}(z)$ and $k(z)$ at Point V2 and/or Point V5. (**c**) $<C(t)>/C(0)$ with various time steps.

To further quantitatively evaluate the CFD models with the present grid arrangement and standard *k-ε* model, several statistical performance metrics are calculated, including the mean value, the standard deviation (named *St dev.* here), the fraction of predictions (here it is the CFD result) within a factor of two of observations (wind tunnel experiment data) named *FAC2*, the normalized mean error (*NMSE*), the fraction bias (*FB*), and the correlation coefficient (*R*). Results of $\bar{u}(z)$ at point V2 and V5 and $k(z)$ at point V5 are shown in Table 2. According to the recommended reference criteria [52], for $\bar{u}(z)$ prediction, a good performing simulation model is supposed to meet the statistical metrics standards as below: FAC2 $\geq$ 0.5; NMSE $\leq$ 1.5; $-0.3 \leq$ FB ≤ 0.3. All the results satisfy the recommended criteria except the FB value of $k(z)$ at point V5, which exceeds the upper limit, while its FAC2 just reaches the threshold. Furthermore, $k(z)$ has a quite low *R* between wind tunnel data and CFD results. Results suggest a poorer quality of numerical prediction of $k(z)$ than $\bar{u}(z)$. As for $\bar{u}(z)$, although it is overestimated as the FR is positive, the value of *R* is particularly high (0.93) at both V2 and V5, which implies a credible prediction of $\bar{u}(z)$. Assessing models' acceptance requires considering all relevant performance metrics rather than one specific index, so the CFD models are considered to have quite a satisfactory performance on the whole.

Table 2. Statistical performance metrics in CFD validation case.

Variable (Position)	Cases	$k(z)$ (V5)	$\bar{u}(z)$ (V2)	$\bar{u}(z)$ (V5)
Average				
	Wind tunnel	0.31	1.61	1.58
	CFD	0.17	1.66	1.69
Standard deviation				
	Wind tunnel	0.07	1.76	1.70
	CFD	0.05	1.83	1.85
FAC2		0.50	0.93	0.93
NMSE		0.52	0.01	0.02
FB		0.61	−0.03	−0.07
R		0.13	0.93	0.93

Finally, with this CFD setup, Figure 3c shows the effects of time steps on the decay history of spatial mean concentration (<C(t)>/C(0)). The decay rates for three time steps (dt = 0.5, 1, 2 s) are almost the same, confirming that the present time step (dt = 1 s) is good enough for *ACH* prediction.

3.2. Flow and Concentration Decay in an Example Case [5-5, 0°] ($\theta = 0°$)

As an example to analyze the concentration decay history related to UCL ventilation, Figure 4 shows 3D streamline, normalized velocity ($V = \sqrt{\bar{u}^2 + \bar{v}^2 + \bar{w}^2}$), normalized lateral velocity ($\bar{v}$) in the plane of z = 0.05H, and normalized concentration ($C(t)/C(0)$) at a time of 10 s to 400 s in z = 0.5H in Case [5-5, 0°]. Here the velocity and lateral velocity are normalized by the freestream velocity at the same height of z = 0.05H (see Equation (1)), similar to the literature [17,18]. Then Figure 5 displays the concentration history ($C(t)/C(0)$) and the age of air (τ_p) at various points in z = 0.5H in Case [5-5, 0°]. In the main streets (Figure 4a), the flow is channeled toward downstream regions. In the secondary streets 3D vortices and helical flows exist in building wake regions. Across the lateral UCL boundary (at y = 135 m), there are lateral helical airflows leaving or re-entering UCL volume. Points with bigger concentration decay rates (K) experience better ventilation, a greater dilution rate, and smaller age of air (τ_p). Figures 4b and 5a,b show that the ventilation in upstream regions is better than in downstream regions (Figure 5a), and that in the main streets it is better than in the secondary streets (Figure 5a); those (Point S1b–S4b) near the lateral UCL boundary are better than those (Point S1o–S4o) in urban center regions (Figure 5b). Figure 5b displays that K at Point S4b is near to Point S3b, and that at Point S4b near the UCL lateral boundaries it is much bigger than at Point S4o because there is a strong helical inflow across lateral boundaries, bringing in external air to help with pollutant dilution at Point S4b (Figure 4a). Similarly, in Figure 5c, the τ_p at Points M1b to M4b in the main streets is much smaller than Points S1a to S4a and Points S1o to S4o in the secondary streets. Moreover, air at Points S1b to S4b near lateral UCL boundaries is younger than Points S1o to S4o far from lateral UCL boundaries. More importantly, the τ_p at Points S4b, S4a, and S4o (89.5 to 249.4 s) verifies that the lateral inflow across UCL lateral boundaries significantly improves the ventilation at Point S4b.

There are qualitative differences between the decay curves in Figure 5a,b. Analyzing the behavior of the decay curves in Figure 5a, for example, after some time the decay rates at Points S1a and M1b are similar to each other. This is a manifestation that there is a coupling interaction between the two locations (more explanation can be found by comparing Figure 11.2 in [32]). However, in Figure 5b all curves exhibit different decay rates, which shows that is not feedback (no backflow against the wind direction) that connects the different locations (see also Chapter 11 in [32]).

(a) 3D streamline and normalized velocity in z = 0.05H in Case [5-5, 0°] (θ = 0°)

(b) C(t)/C(0) at time of 10 s to 400 s at z = 0.5H in Case [5-5, 0°] (θ = 0°)

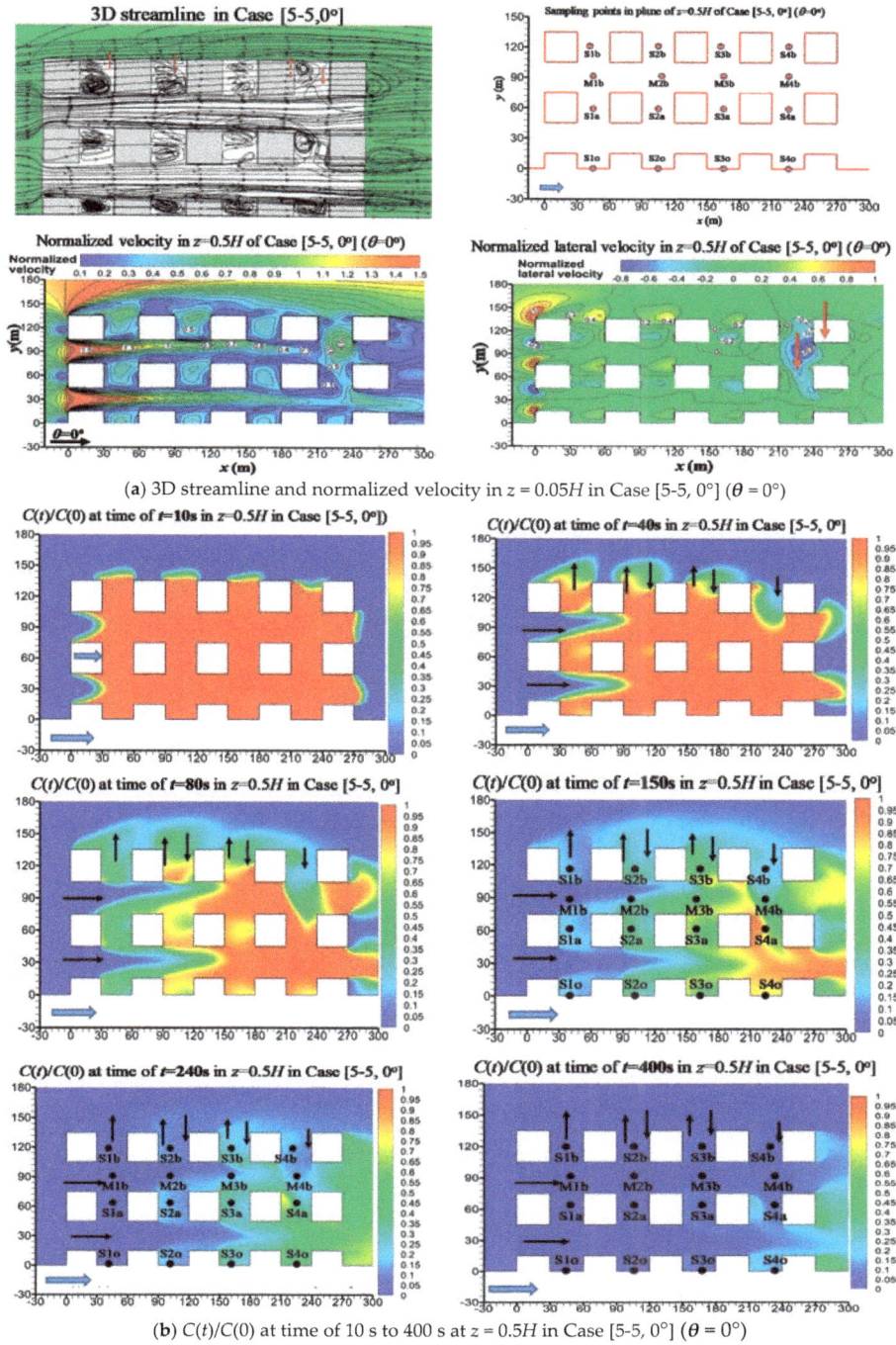

Figure 4. In Case [5-5, 0°] (θ = 0°): (a) 3D streamline and normalized velocity in z = 0.05H; (b) C(t)/C(0) at time of 10 s to 400 s in plane of z = 0.5H.

(**a**) $C(t)/C(0)$ at Points S1a–S4a and M1b–M4b at $z = 0.5H$ in Case [5-5, 0°]

(**b**) $C(t)/C(0)$ at Points S1o–S4o and S1b–S4b at $z = 0.5H$ in Case [5-5, 0°]

(**c**) Age of air at these points at $z = 0.5H$ in Case [5-5, 0°]

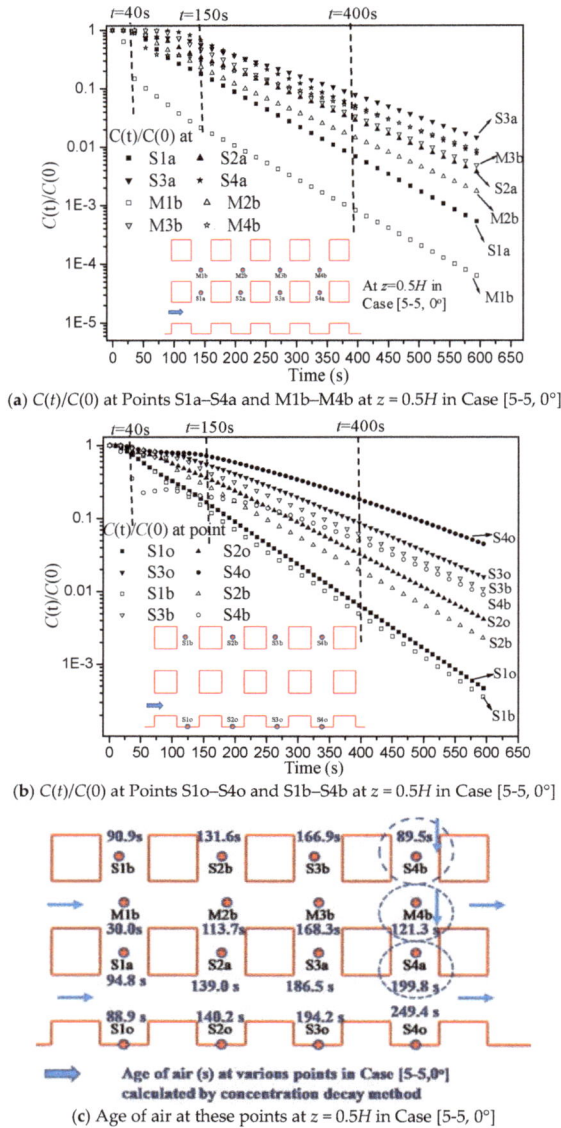

Figure 5. $C(t)/C(0)$ in Case [5-5, 0°]: (**a**) at Points S1a–S4a and M1b–M4b in $z = 0.5H$, (**b**) at Points S1o–S4o and S1b–S4b in $z = 0.5H$. (**c**) Age of air at these points.

3.3. Effect of Overall Urban Form and Ambient Wind Direction on UCL Ventilation

As an example, Figure 6 shows 3D streamline, normalized velocity, $C(t)/C(0)$, and τ_p at various points in $z = 0.5H$ in Case [5-5, 45°]. Here the velocity is normalized by the freestream velocity at the same height of $z = 0.5H$. Obviously wind brings clean air into an urban area downstream for pollutant dilution. There are recirculation regions with small wind speed (Figure 6a). Thus, the concentration decay rate (the age of air) is relatively small (large) in downstream regions and recirculation regions (Figure 6b,c).

Then Figure 7a,b further displays the normalized velocity in Case [7-7, $\theta°$] (square overall urban form) and Case [10-5, $\theta°$] (rectangular overall urban form) with similar total UCL air volume. Obviously ambient wind directions significantly influence the flow pattern for both overall urban forms (Figure 7a,b). As $\theta = 0°$, there are lateral flows across lateral UCL boundaries. With oblique winds, the flows enter UCL across two sides in upstream regions and leave across the other two toward downstream. The velocity is relatively small in recirculation regions and the presence of street crossings produces considerable momentum and scalar exchange between neighbor streets. In particular, Figure 7a confirms that, for square overall urban form, UCL models with $\theta = 45°$ experience more recirculation regions and quicker wind reduction than those with $\theta = 0°$. It is consistent with the literature that building arrays with $\theta = 45°$ produce greater flow resistances and worse UCL ventilation performance than $\theta = 0°$ [2,7,23,58,59]. Such characteristics with $\theta = 45°$ produce adverse effects on its ventilation performance in contrast to $\theta = 0°$.

Then Figure 7c,d displays the net *ACH* by concentration decay method, ACH_T and ACH_{turb} in cases of Group I. Since Case [5-5, $\theta°$] and Case [7-7, $\theta°$] are of square overall urban form, the flows with $\theta = 0°, 15°, 30°, 45°$ are the same with those with $\theta = 90°, 75°, 60°, 45°$. Obviously for Case [5-5, $\theta°$] ($Lx = Ly = 270$ m) and Case [7-7, $\theta°$] ($Lx = Ly = 390$ m), $\theta = 0°$ attains smaller ACH_T and ACH_{turb} but bigger net *ACH* than $\theta = 15°, 30°, 45°$. Therefore, the parallel wind ($\theta = 0°$) experiences the best overall UCL ventilation with the highest net ventilation efficiency. It can be explained that UCL models with non-parallel approaching wind (for example $\theta = 45°$ in Figure 7a) experience more recirculation regions and weaker wind due to the greater blockage induced by buildings, which can reduce the flow rates flushing through UCL space that never return or decrease the actual or net air change rate of the entire UCL space. Then for Case [10-5, $\theta°$] with a rectangular overall urban form ($Lx = 570$ m, $Ly = 270$ m), as θ varies from $0°$ to $90°$, ACH_T rises from 16.1 h^{-1} to 29.0 h^{-1}, ACH_{turb} first increases from $\theta = 0°$ to $\theta = 45°$ then decreases to $\theta = 90°$, and the net *ACH* rises from 14.3 h^{-1} to 28.7 h^{-1}. Thus $\theta = 90°$ obtains the best overall UCL ventilation in Case [10-5, $\theta°$], in which the approaching wind is parallel to the shorter urban size ($Ly = 270$ m). Finally, the net *ACH* by the concentration decay method are always much smaller than the sum of ACH_{turb} and ACH_T. This verifies that UCL ventilation efficiency is limited. As discussed in Section 2.4.2, the recirculation flows in UCL space and turbulent fluctuations across street roofs induce a significant fraction of fluid particles to return or revisit UCL space across UCL boundaries after they leave it. However, UCL ventilation mainly depends on the flow rates flushing UCL space and leaving it to never return. Thus only a fraction of ACH_{turb} and ACH_T contributes to UCL ventilation and the actual air change rate is limited due to the ventilation efficiency problems.

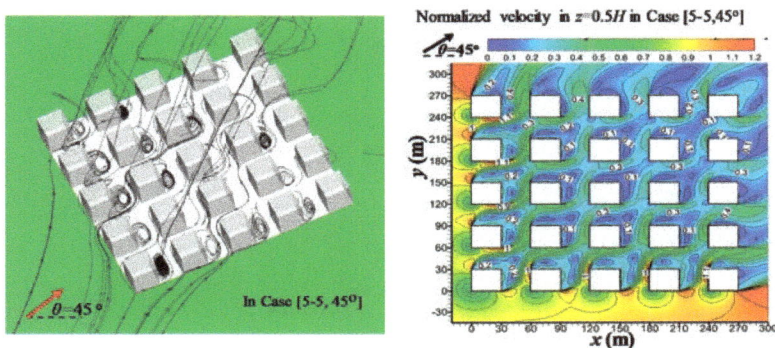

(a) 3D streamlines and the normal velocity at $z = 0.5H$ in Case [5-5, 45°]

Figure 6. *Cont.*

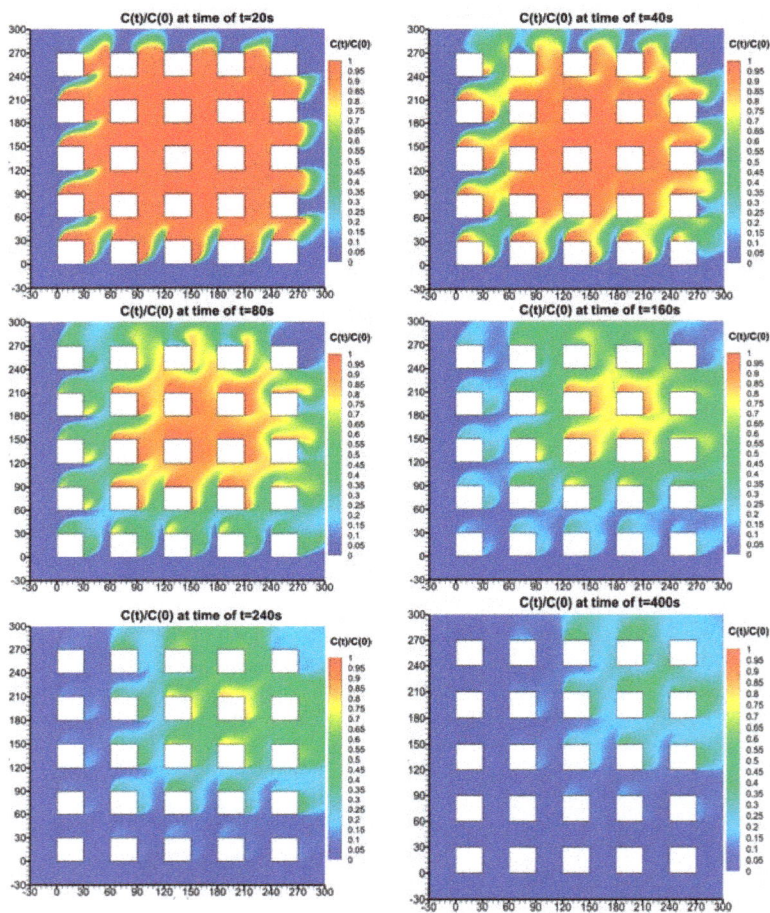

(**b**) $C(t)/C(0)$ at $z = 0.5H$ at time of 20 s to 400 s in Case [5-5, 45°]

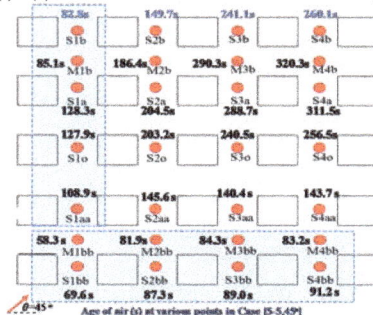

(**c**) Age of air at some example points at $z = 0.5H$ in Case [5-5, 45°]

Figure 6. In Case [5-5, 45°]: (**a**) 3D streamline and the normal velocity in $z = 0.5H$; (**b**) $C(t)/C(0)$ in $z = 0.5H$ at time of 20 s to 400 s; (**c**) age of air at some example points at $z = 0.5H$.

By comparing the net *ACH* in these three cases (Figure 7d), *ACH* in Case [5-5, $\theta°$] (~21.1–30.2 h^{-1}) are larger than those in Case [7-7, $\theta°$] (~16.8–18.6 h^{-1}) for all wind directions because Case [5-5, $\theta°$] has smaller urban size and requires shorter time for the approaching wind to flow through and will be exchanged more times by the external air within one hour. Rectangular urban form (Case [10-5, $\theta°$]) attains greater ACH (~21.1–28.7 h^{-1}) than the square urban form (Case [7-7, $\theta°$]~16.8–18.6 h^{-1}) for most ambient wind directions except θ = 0° and 15° (~14.3 and 16.9 h^{-1}). It can be explained by two example cases: for Case [10-5, 0°] there are 10 rows of buildings for the approaching wind to flush, much longer than Case [10-5, 90°], in which there are only five rows of buildings for wind to flow through.

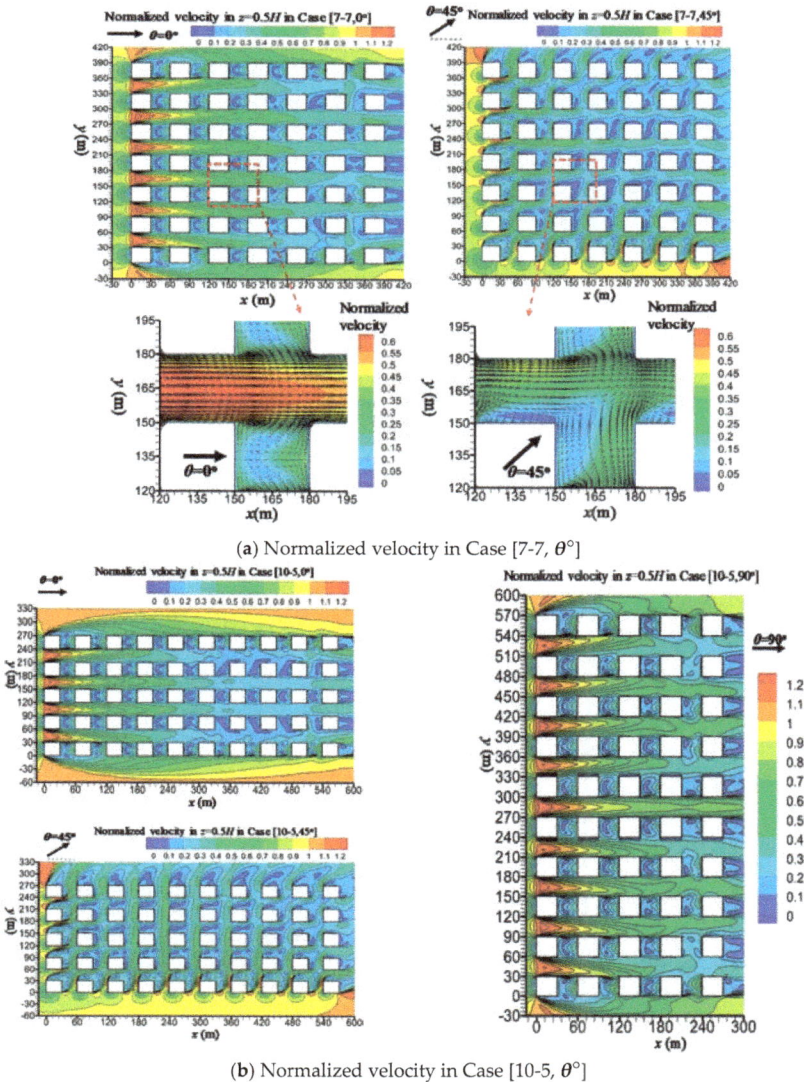

(**a**) Normalized velocity in Case [7-7, $\theta°$]

(**b**) Normalized velocity in Case [10-5, $\theta°$]

Figure 7. *Cont.*

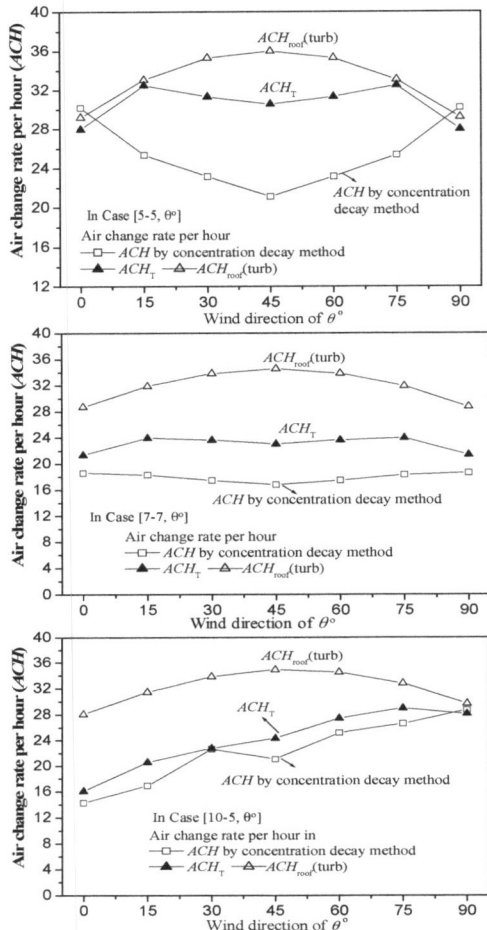

(c) Various *ACH* indexes in Case [5-5, $\theta°$], Case [7-7, $\theta°$], Case [10-5, $\theta°$]

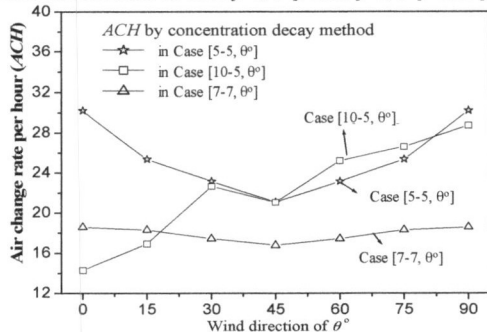

(d) Net *ACH* by concentration decay method in these three cases

Figure 7. Normalized velocity in (**a**) Case [7-7, $\theta°$]; (**b**) case [10-5, $\theta°$]; (**c**) Various *ACH* indexes in Case [5-5, $\theta°$], Case [7-7, $\theta°$], Case [10-5, $\theta°$]; (**d**) Net *ACH* by concentration decay method in these three cases.

3.4. Effect of Open Space Arrangements and Ambient Wind Direction

Open space arrangements have been regarded as one possible way to improve UCL ventilation. As shown in Table 1 (Group II), this subsection investigates 24 test cases with six kinds of open space arrangements under four wind directions (θ = 0°, 15°, 30°, 45°). Note that Oij represents the building of position i-j is removed for better ventilation. Figure 8a displays the 3D streamline in two example cases; Figure 8b further summarizes the net *ACH* by the concentration decay method in all 24 test cases.

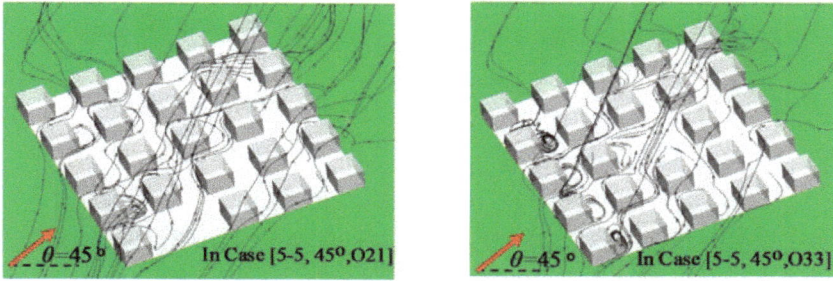

(a) 3D streamline in Case [5-5, 45°, O21] and Case [5-5, 45°, O33]

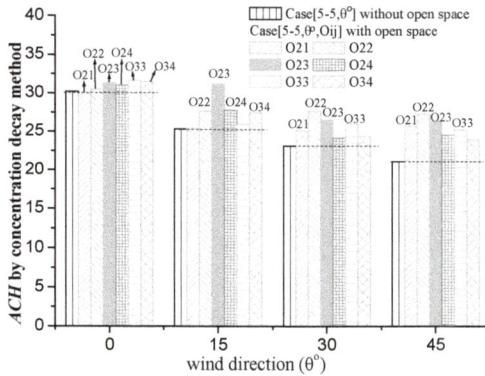

(b) The net *ACH* in all 24 test cases

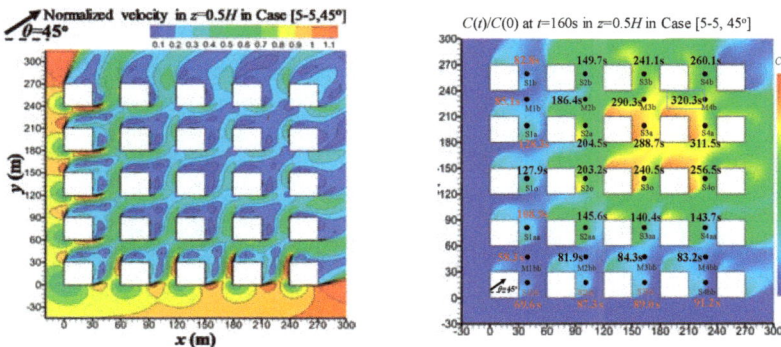

(c) $C(t)/C(0)$ at t = 160 s, age of air at various points and normalized velocity in z = 0.5H in Case [5-5,45°]

Figure 8. *Cont.*

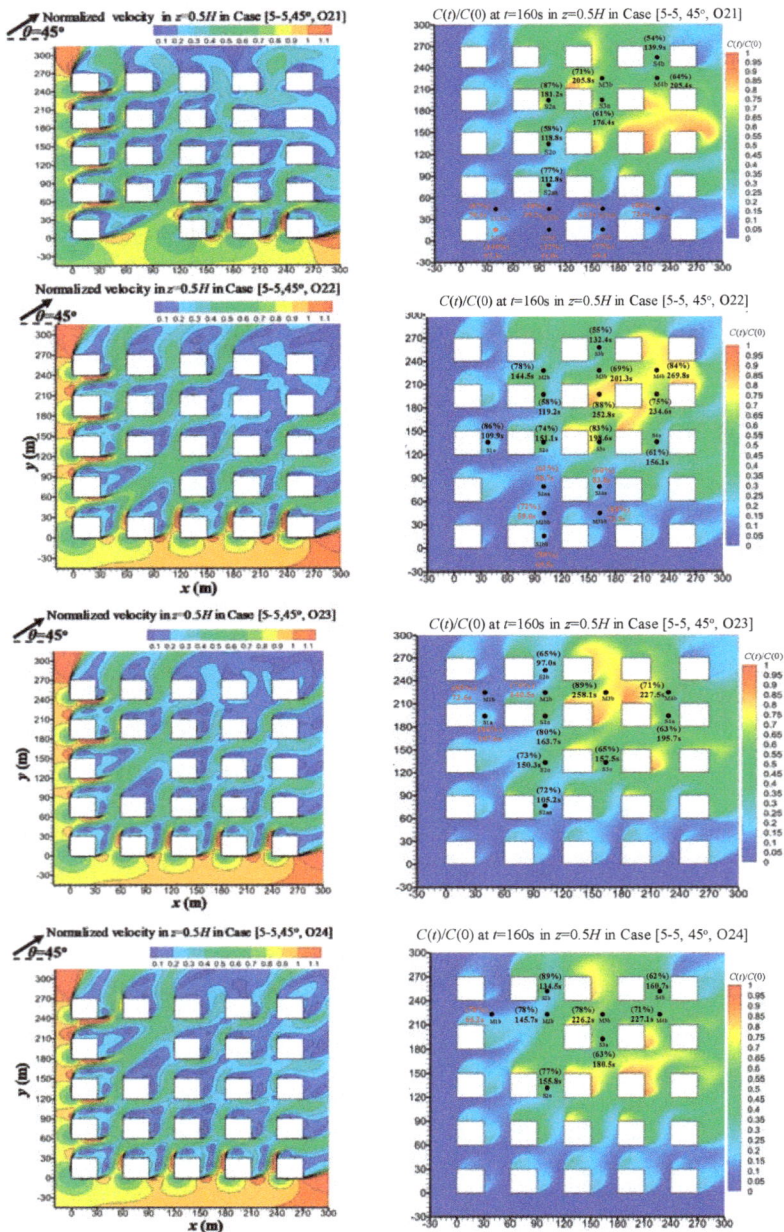

Figure 8. *Cont.*

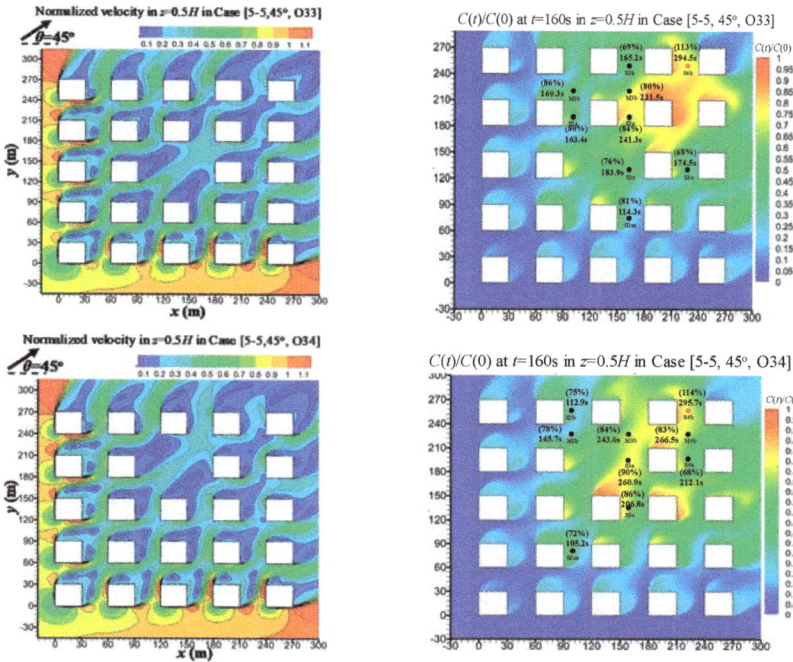

(**d**) C(t)/C(0) at t = 160 s, age of air and normalized velocity at z = 0.5H in Case [5-5, 45°, Oij]

Figure 8. (**a**) 3D streamline in Case [5-5, 45°, O21] and Case [5-5, 45°, O33]; (**b**) the net ACH in all 24 test cases. (**c,d**) C(t)/C(0) at t = 160 s, age of air at various points, and normalized velocity at z = 0.5H in Case [5-5, 45°] and Case [5-5, 45°, Oij].

Obviously, with oblique wind (θ = 15°, 30°, 45°) most open space arrangements improve UCL ventilation more than θ = 0° (see Figure 8b). The improvement of UCL ventilation is the best as θ = 45°. Thus θ = 45° is emphasized in the below analysis. Figure 8c,d shows C(t)/C(0) at t = 160 s, age of air (τ_p) at various points and normalized velocity in z = 0.5H in Case [5-5, 45°] and Case [5-5, 45°, Oij]. Because the open space arrangements act as a kind of ventilation corridor and reduce the total flow resistances induced by buildings, wind speed increases in regions near/surrounding open space and in its downstream regions; subsequently, the concentration decay processes speed up and the ventilation becomes better. The percentage data of τ_p in Figure 8d refer to the ratio between τ_p with open space (Case [5-5, 45°, Oij] in Figure 8d) and that without open space (Case [5-5, 45°] in Figure 8c). This percentage at some points can be from 12% to 65%, showing that τ_p significantly decreases in the regions near these points. Obviously, the effects of open space on flow pattern satisfy the variations of τ_p (Figure 8c,d) and overall *ACH* (Figure 8b).

4. Conclusions

As a novelty, this paper confirms the temporal decay profile of concentration in UCL models accords with the exponential decay law, and it is effective to introduce the concentration decay method into CFD simulations to predict the net air change rate (*ACH*) flushing UCL space and never returning. Street-scale (~100 m), medium-dense ($\lambda_f = \lambda_p = 0.25$, H/W = 1), urban-like geometries are studied under neutral atmospheric conditions. The standard k-ε model is first successfully evaluated by wind tunnel data. Then the flow pattern, the concentration decay rate and age of air (τ_p) at some points in UCL space, the net *ACH* are analyzed.

With a parallel approaching wind ($\theta = 0°$), UCL models with larger urban sizes ($Lx = Ly = 390$ m) experience smaller ACH (~16.8–18.6 h^{-1}) than smaller UCL models ($Lx = Ly = 270$ m, ACH~21.1–30.2 h^{-1}). The urban age of air τ_p at points near upstream/lateral UCL boundaries is smaller (or the air is younger) than far from them. For the square overall urban form, the parallel wind attains greater net ACH than non-parallel wind ($\theta = 15°, 30°, 45°$). For the rectangular overall urban form ($Lx = 570$ m, $Ly = 270$ m), ACH is greater than the square overall urban form ($Lx = Ly = 390$ m) under most wind directions (~21.1–28.7 h^{-1} as $\theta = 30°$ to $90°$), and the ventilation is the best as the approaching wind is parallel to the shorter urban size (~28.7 h^{-1} as $\theta = 90°$). With open space arrangements, the dilution capacity near/surrounding open space and in its downstream regions is enhanced; moreover, most open space arrangements are more effective at improving UCL ventilation under oblique wind directions ($\theta = 15°, 30°, 45°$) than $\theta = 0°$, and the best ventilation improvement by open space appears as $\theta = 45°$.

Similar to the purging flow rate [2,23], ACH calculated by the concentration decay approach has been proven effective to evaluate the effects of urban morphologies on the overall UCL ventilation capacity induced by mean flows and turbulent diffusions, which seems to be a better ventilation index than ACH calculated by the volumetric flow rates integrating the normal mean velocity or fluctuation velocity across UCL boundaries. Turbulent diffusions across open street roofs have been proven to significantly contribute to UCL ventilation, but further investigations are still required to analyze the net ventilation efficiency by mean flows and turbulence.

Acknowledgments: This study was financially supported by the National Natural Science Foundation of China (grant No. 51478486) and the National Natural Science Foundation—Outstanding Youth Foundation (grant No. 41622502) as well as the Science and Technology Program of Guangzhou, China (grant No. 201607010066) and the Fundamental Research Funds for the Central Universities (No. 161gzd01).

Author Contributions: Qun Wang did the major CFD simulations. Yuanyuan Lin performed the CFD validation study. Mats Sandberg and Shi Yin contributed to the research discussion and writing the introduction. Jian Hang finished the manuscript and acts as the supervisor of the first author and corresponding author.

Conflicts of Interest: The authors declare no conflict of interest.

Nomenclature

A	area of a surface (m^2)
ACH	air change rate per hour by concentration decay method
ACH_T	ACH calculated by Q_T for entire UCL volume
ACH_{turb}	ACH calculated by Q_{roof}(turb) for entire UCL volume
B, H, L, W	building width, building height, total length, street width
$C, <C>$	time-averaged pollutant concentration and its spatial mean value
K_c, ν_t	turbulent eddy diffusivity of pollutant and momentum $K_c = \nu_t/S_{ct}$
λ_p	building area density (or plan area index)
λ_f	frontal area density (or frontal area index)
k, ε	turbulent kinetic energy and its dissipation rate
$\vec{n}$	normal direction of street openings or canopy roofs
Q	flow rate through street openings or street roofs
Q_{in}, Q_{out}	total inflow and outflow rate by mean flows across UCL boundaries
Q_T	total ventilation flow rate by mean flows
Q_{roof}(turb)	effective flow rate across street roofs by turbulence
S_{ct}	turbulent Schmidt number
σ_w	fluctuation velocity on street roofs
τ_p	age of air (s)
$U_0(z)$	velocity profiles used at CFD domain inlet for ventilation cases
U_{ref}	reference velocity at $z = H$ at the domain inlet
$\bar{u}_j, x_j$	velocity and coordinate components

Atmosphere **2017**, *8*, 169

$\overline{V}_j$	velocity vector
Vol	control volume
x, y, z	stream-wise, span-wise, vertical directions
$\overline{u}, \overline{v}, \overline{w}$	stream-wise, lateral, vertical velocity components
u', v', w'	stream-wise, lateral, vertical velocity fluctuations

References

1. Oke, T.R. *Boundary Layer Climates*, 2nd ed.; Methuen Publishing: London, UK, 1987; p. 289.
2. Bady, M.; Kato, S.; Huang, H. Towards the application of indoor ventilation efficiency indices to evaluate the air quality of urban areas. *Build. Environ.* **2008**, *43*, 1991–2004. [CrossRef]
3. Hang, J.; Sandberg, M.; Li, Y.G. Age of air and air exchange efficiency in idealized city models. *Build. Environ.* **2009**, *44*, 1714–1723. [CrossRef]
4. Buccolieri, R.; Sandberg, M.; Di Sabatino, S. City breathability and its link to pollutant concentration distribution within urban-like geometries. *Atmos. Environ.* **2010**, *44*, 1894–1903. [CrossRef]
5. Hang, J.; Li, Y.G. Age of air and air exchange efficiency in high-rise urban areas. *Atmos. Environ.* **2011**, *45*, 5572–5585. [CrossRef]
6. Ramponi, R.; Blocken, B.; Laura, B.; Janssen, W.D. CFD simulation of outdoor ventilation of generic urban configurations with different urban densities and equal and unequal street widths. *Build. Environ.* **2015**, *92*, 152–166. [CrossRef]
7. Hang, J.; Luo, Z.W.; Sandberg, M.; Gong, J. Natural ventilation assessment in typical open and semi-open urban environments under various wind directions. *Build. Environ.* **2013**, *70*, 318–333. [CrossRef]
8. Kim, J.J. Assessment of observation environment for surface wind in urban areas using a CFD model. *Atmosphere* **2015**, *25*, 449–459. (In Korean)
9. Ashie, Y.; Kono, T. Urban-scale CFD analysis in support of a climate-sensitive design for the Tokyo Bay area. *Int. J. Climatol.* **2011**, *31*, 174–188. [CrossRef]
10. Yang, X.; Li, Y.G. The impact of building density and building height heterogeneity on average urban albedo and street surface temperature. *Build. Environ.* **2015**, *90*, 146–156. [CrossRef]
11. Hang, J.; Luo, Z.; Wang, X.; He, L.; Wang, B.; Zhu, W. The influence of street layouts and viaduct settings on daily CO exposure and intake fraction in idealized urban canyons. *Environ. Pollut.* **2017**, *220*, 72–86. [CrossRef] [PubMed]
12. Luo, Z.; Li, Y.G.; Nazaroff, W.W. Intake fraction of nonreactive motor vehicle exhaust in Hong Kong. *Atmos. Environ.* **2010**, *44*, 1913–1918. [CrossRef]
13. Ng, W.; Chau, C. A modeling investigation of the impact of street and building configurations on personal air pollutant exposure in isolated deep urban canyons. *Sci. Total Environ.* **2014**, *468*, 429–448. [CrossRef] [PubMed]
14. Meroney, R.N.; Pavegeau, M.; Rafailidis, S.; Schatzmann, M. Study of line source characteristics for 2-D physical modelling of pollutant dispersion in street canyons. *J. Wind Eng. Ind. Aerodyn.* **1996**, *62*, 37–56. [CrossRef]
15. Li, X.X.; Liu, C.H.; Leung, D.Y.C.; Lam, K.M. Recent progress in CFD modelling of wind field and pollutant transport in street canyons. *Atmos. Environ.* **2006**, *40*, 5640–5658. [CrossRef]
16. Li, X.X.; Liu, C.H.; Leung, D.Y.C. Numerical investigation of pollutant transport characteristics inside deep urban street canyons. *Atmos. Environ.* **2009**, *43*, 2410–2418. [CrossRef]
17. Zhang, Y.W.; Gu, Z.; Lee, S.C.; Fu, T.M.; Ho, K.F. Numerical simulation and in situ investigation of fine particle dispersion in an actual deep street canyon in Hong Kong. *Indoor Built Environ.* **2011**, *20*, 206–216. [CrossRef]
18. Lin, L.; Hang, J.; Wang, X.X.; Wang, X.M.; Fan, S.J.; Fan, Q.; Liu, Y.H. Integrated effects of street layouts and wall heating on vehicular pollutant dispersion and their reentry into downstream canyons. *Aerosol Air Qual. Res.* **2016**, *16*, 3142–3163. [CrossRef]
19. Dong, J.L.; Tan, Z.J.; Xiao, X.M.; Tu, J.Y. Seasonal changing effect on airflow and pollutant dispersion characteristics in urban street canyons. *Atmosphere* **2017**, *8*, 43. [CrossRef]
20. Gu, Z.; Zhang, Y.; Cheng, Y.; Lee, S.C. Effect of uneven building layout on air flow and pollutant dispersion in non-uniform street canyons. *Build. Environ.* **2011**, *46*, 2657–2665. [CrossRef]

21. Hang, J.; Li, Y.; Sandberg, M.; Buccolieri, R.; Di Sabatino, S. The influence of building height variability on pollutant dispersion and pedestrian ventilation in idealized high-rise urban areas. *Build. Environ.* **2012**, *56*, 346–360. [CrossRef]

22. Panagiotou, I.; Neophytou, M.K.A.; Hamlyn, D.; Britter, R.E. City breathability as quantified by the exchange velocity and its spatial variation in real inhomogeneous urban geometries: An example from central London urban area. *Sci. Total Environ.* **2013**, *442*, 466–477. [CrossRef] [PubMed]

23. Lin, M.; Hang, J.; Li, Y.G.; Luo, Z.W.; Sandberg, M. Quantitative ventilation assessments of idealized urban canopy layers with various urban layouts and the same building packing density. *Build. Environ.* **2014**, *79*, 152–167. [CrossRef]

24. Lee, D.; Kim, J.J. A study on the characteristics of flow and reactive pollutants' dispersion in step-up street canyons using a CFD model. *Atmosphere* **2015**, *25*, 473–482. (In Korean) [CrossRef]

25. Chen, L.; Hang, J.; Sandberg, M.; Claesson, L.; Di Sabatino, S.; Wigo, H. The impacts of building height variations and building packing densities on flow adjustment and city breathability in idealized urban models. *Build. Environ.* **2017**, *118*, 344–361. [CrossRef]

26. Li, X.X.; Britter, R.E.; Norford, L.K. Transport processes in and above two-dimensional urban street canyons under different stratification conditions: results from numerical simulation. *Environ. Fluid Mech.* **2015**, *15*, 399–417. [CrossRef]

27. Nazarian, N.; Kleissl, J. Realistic solar heating in urban areas: Air exchange and street-canyon ventilation. *Build. Environ.* **2016**, *95*, 75–93. [CrossRef]

28. Cui, P.Y.; Li, Z.; Tao, W.Q. Wind-tunnel measurements for thermal effects on the air flow and pollutant dispersion through different scale urban areas. *Build. Environ.* **2016**, *97*, 137–151. [CrossRef]

29. Wang, X.X.; Li, Y.G. Predicting urban heat island circulation using CFD. *Build. Environ.* **2016**, *99*, 82–97. [CrossRef]

30. Fan, Y.F.; Hunt, J.C.R.; Li, Y.G. Buoyancy and turbulence-driven atmospheric circulation over urban areas. *J. Environ. Sci.* **2017**. [CrossRef]

31. Hang, J.; Wang, Q.; Chen, X.Y.; Sandberg, M.; Zhu, W.; Buccolieri, R.; Di Sabatino, S. City breathability in medium density urban-like geometries evaluated through the pollutant transport rate and the net escape velocity. *Build. Environ.* **2015**, *94*, 166–182. [CrossRef]

32. Etheridge, D.; Sandberg, M. *Building Ventilation: Theory and Measurement*; John Wiley & Sons: New York, NY, USA, 1996; pp. 573–633.

33. Lim, E.S.; Ito, K.; Sandberg, M. New ventilation index for evaluating imperfect mixing conditions—Analysis of Net Escape Velocity based on RANS approach. *Build. Environ.* **2013**, *61*, 45–56. [CrossRef]

34. Li, J.Q.; Ward, I.C. Developing computational fluid dynamics conditions for urban natural ventilation study. In Proceedings of the Building Simulation, Beijing, China, 3–6 September 2007.

35. Zaki, S.A.; Hagishima, A.; Tanimoto, J. Experimental study of wind-induced ventilation in urban building of cube arrays with various layouts. *J. Wind Eng. Ind. Aerodyn.* **2012**, *103*, 31–40. [CrossRef]

36. Padilla-Marcos, M.Á.; Meiss, A.; Feijó-Muñoz, J. Proposal for a simplified CFD procedure for obtaining patterns of the age of air in outdoor spaces for the natural ventilation of buildings. *Energies* **2017**, *10*, 1252. [CrossRef]

37. Li, X.X.; Liu, C.H.; Leung, D.Y.C. Development of a *k*-ε model for the determination of air exchange rates for street canyons. *Atmos. Environ.* **2005**, *39*, 7285–7296. [CrossRef]

38. Liu, C.H.; Leung, D.Y.C.; Barth, M.C. On the prediction of air and pollutant exchange rates in street canyons of different aspect ratio using large-eddy simulation. *Atmos. Environ.* **2005**, *39*, 1567–1574. [CrossRef]

39. Moonen, P.; Dorer, D.; Carmeliet, J. Evaluation of the ventilation potential of courtyards and urban street canyons using RANS and LES. *J. Wind Eng. Ind. Aerodyn.* **2011**, *99*, 414–423. [CrossRef]

40. Hang, J.; Sandberg, M.; Li, Y. Effect of urban morphology on wind condition in idealized city models. *Atmos. Environ.* **2009**, *43*, 869–878. [CrossRef]

41. Hang, J.; Li, Y.G. Ventilation strategy and air change rates in idealized high-rise compact urban areas. *Build. Environ.* **2010**, *45*, 2754–2767. [CrossRef]

42. Liu, J.; Luo, Z.; Zhao, J.; Shui, T. Ventilation in a street canyon under diurnal heating conditions. *Int. J. Vent.* **2012**, *11*, 141–154. [CrossRef]

43. Yang, L.; Li, Y.G. City ventilation of Hong Kong at no-wind conditions. *Atmos. Environ.* **2009**, *43*, 3111–3121. [CrossRef]

44. Yang, L.; Li, Y.G. Thermal conditions and ventilation in an ideal city model of Hong Kong. *Energy Build.* **2011**, *43*, 1139–1148. [CrossRef]

45. Gao, N.P.; Niu, J.L.; Perino, M.; Heiselberg, P. The airborne transmission of infection between flats in high-rise residential buildings: Tracer gas simulation. *Build. Environ.* **2008**, *43*, 1805–1817. [CrossRef]

46. Hooff, V.T.; Blocken, B. CFD evaluation of natural ventilation of indoor environments by the concentration decay method: CO_2 gas dispersion from a semi-enclosed stadium. *Build. Environ.* **2013**, *61*, 1–17. [CrossRef]

47. Fernando, H.J.S.; Zajic, D.; Di Sabatino, S.; Dimitrova, R.; Hedquist, B.; Dallman, A. Flow, turbulence, and pollutant dispersion in urban atmospheres. *Phys. Fluids* **2010**, *22*, 051301. [CrossRef]

48. Di Sabatino, S.; Buccolieri, R.; Salizzoni, P. Recent advancements in numerical modelling of flow and dispersion in urban areas: a short review. *Int. J. Environ. Pollut.* **2013**, *52*, 172–191. [CrossRef]

49. Blocken, B. Computational fluid dynamics for urban physics: Importance, scales, possibilities, limitations and ten tips and tricks towards accurate and reliable simulations. *Build. Environ.* **2015**, *91*, 219–245. [CrossRef]

50. Brown, M.J.; Lawson, R.E.; DeCroix, D.S.; Lee, R.L. *Comparison of Centerline Velocity Measurements Obtained Around 2D and 3D Buildings Arrays in a Wind Tunnel*; Report LA-UR-01-4138; Los Alamos National Laboratory: Los Alamos, NM, USA, 2001; p. 7.

51. Lien, F.S.; Yee, E. Numerical modelling of the turbulent flow developing within and over a 3-D building array, part I: A high-resolution Reynolds-averaged Navier-Stokes approach. *Bound. Layer Meteorol.* **2004**, *112*, 427–466. [CrossRef]

52. Santiago, J.L.; Martilli, A.; Martin, F. CFD simulation of airflow over a regular array of cubes Part I: Three-dimensional simulation of the flow and validation with wind-tunnel measurements. *Bound. Layer Meteorol.* **2007**, *122*, 609–634. [CrossRef]

53. Hang, J.; Li, Y.G. Wind conditions in idealized building clusters: Macroscopic simulations by a porous turbulence model. *Bound. Layer Meteorol.* **2010**, *136*, 129–159. [CrossRef]

54. Tominaga, Y.; Mochida, A.; Yoshie, R.; Kataoka, H.; Nozu, T.; Yoshikawa, M.; Shirasawa, T. AIJ guidelines for practical applications of CFD to pedestrian wind environment around buildings. *J. Wind Eng. Ind. Aerodyn.* **2008**, *96*, 1749–1761. [CrossRef]

55. Franke, J.; Hellsten, A.; Schlunzen, H.; Carissimo, B. The COST732 Best Practice Guideline for CFD simulation of flows in the urban environment: A summary. *Inter. J. Environ. Pollut.* **2011**, *44*, 419–427. [CrossRef]

56. FLUENT V6.3. User's Manual. Available online: http://www.fluent.com (accessed on 20 August 2017).

57. Blocken, B.; Stathopoulos, T.; Carmeliet, J. CFD simulation of the atmospheric boundary layer: Wall function problems. *Atmos. Environ.* **2007**, *41*, 238–252. [CrossRef]

58. Kim, J.J.; Baik, J.J. A numerical study of the effects of ambient wind direction on flow and dispersion in urban street canyons using the RNG k-ε turbulence model. *Atmos. Environ.* **2004**, *38*, 3039–3048. [CrossRef]

59. Kanda, M. Large-eddy simulations on the effects of surface geometry of building arrays on turbulent organized structures. *Bound. Layer Meteorol.* **2006**, *18*, 151–168. [CrossRef]

atmosphere

MDPI

Article

Improving Residential Wind Environments by Understanding the Relationship between Building Arrangements and Outdoor Regional Ventilation

Wei You, Zhi Gao, Zhi Chen and Wowo Ding *

School of Architecture and Urban Planning, Nanjing University, Nanjing 210093, China;
youwei@nju.edu.cn (W.Y.); zhgao@nju.edu.cn (Z.G.); zhchentxy@163.com (Z.C.)
* Correspondence: dww@nju.edu.cn; Tel.: +96-25-8359-7205

Academic Editors: Riccardo Buccolieri and Jian Hang
Received: 23 April 2017; Accepted: 6 June 2017; Published: 9 June 2017

Abstract: This paper explores the method of assessing regional spatial ventilation performance for the design of residential building arrangements at an operational level. Three ventilation efficiency (VE) indices, Net Escape Velocity (NEV), Visitation Frequency (VF) and spatial-mean Velocity Magnitude (VM), are adopted to quantify the influence of design variation on VE within different regional spaces. Computational Fluid Dynamics (CFD) method is applied to calculate VE indices mentioned above. Several residential building arrangement cases are set to discuss the effect of different building length, lateral spacing and layouts on four typical space patterns under wind directions oblique or perpendicular to the main (long) building facade. The simulation results prove that NEV, VF and VM are useful VE indices, which can reflect different features of flow pattern in studied regional domains. Preliminary parametric studies indicate that wind direction might be the most important factor for improving spatial ventilation. When the angle between main building facade and wind direction is more than 30°, ventilation of different exterior spaces could improve evidently. When wind direction is perpendicular to main building façade, decreasing building length can increase NEV of the middle space by 50%, while decreasing lateral spacing would decrease NEV of the intersection space by 35%.

Keywords: residential wind environments; building arrangements; space pattern; ventilation efficiency; CFD simulation

---

1. Introduction

Exterior wind conditions are important in urban residential areas. Wind flow around buildings can dilute pollutants and remove excess heat, both of which are closely related to people's health and quality of life. Figure 1 shows examples of typical residential building arrangements in China. These patterns of building arrangements all reflect the considerations of architects regarding block shape, land use, design regulations, aesthetics and even solar access during the design process. Although many studies have revealed that wind environments of exterior spaces strongly depend on the arrangements of the buildings around them, and some studies have established the powerful influence of design variation on the wind environments of exterior spaces [1,2], during the designing process, designers generally adopt the trial-and-error method [3], since the co-relationship between the geometric exterior space and the wind conditions is still unclear. Thus, for the designing purpose, more studies are needed of the relationship between building arrangements and outdoor wind conditions at the micro-scale.

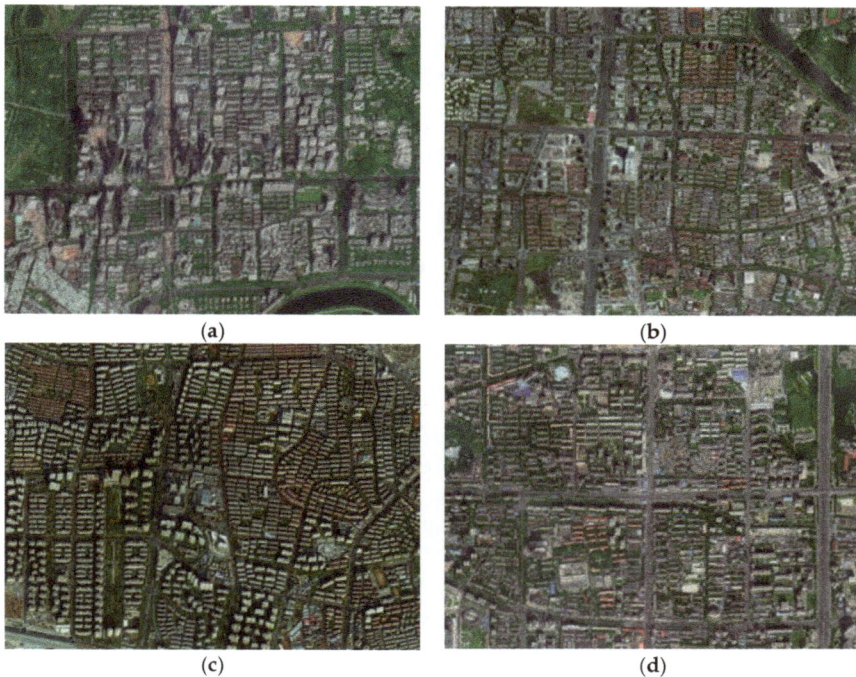

Figure 1. View of typical residential building arrangements in China: (**a**) Shenzhen; (**b**) Nanjing; (**c**) Shanghai; and (**d**) Beijing.

Based on measurements (field and wind tunnel tests) and/or numerical simulations, numerous studies have discussed the influence of building arrangements on exterior wind conditions. In Section 2 of this paper, a literature review reveals that most studies perform their analysis using generic urban geometries, such as aligned and/or staggered arrays of cubes (urban-like building groups) [4,5] or two-row or multi-row long strips (street canyons) [6–8]. By varying the distances between buildings, as well as building heights and wind directions, the influence of design changes under different wind directions can be assessed. These Computational Fluid Dynamics (CFD) simulation results have provided a large amount of valuable information. However, it should be noted that most of these studies, such as discussions of building packing densities and frontal area densities, focus on the characteristics of the whole city. These studies might provide overall guidance about urban planning, but, in further design process, more studies are needed to analyze the effects of specific design changes (building length, distance and layouts) on ventilation in different local spaces. It is very important for architects to design building arrangements for different living units. In addition to ventilation studies of urban-like building groups, several recent studies have also discussed the influence of building arrangements on exterior wind environments. These studies have examined building sizes and distances [9], layouts [10,11] and canyon configurations [12]. However, the geometrical models used in these studies simply combine several individual buildings and ignored the effects of surrounding buildings.

Some studies applied the concept of indoor ventilation into urban environments and assessed the ventilation efficiency (VE) of urban areas [13–17]. These studies provided a new perspective from which to assess and improve wind environments of exterior spaces. Our previous study had investigated influence of varying lateral spacing and lengths in residential buildings on VE in various typical outdoor spaces [18]. In that study, three VE indices—purging flow rate (PFR: the effective airflow rate required to purge pollutants from the domain), visitation frequency (VF: the number of times a pollutant enters the domain and passes through it) and air residence time (TP: the time elapsed between when a pollutant enters or is generated in the domain until it exits)—introduced by Bady et al. were adopted by performing calculations using the CFD method. However, the results showed that PFR and TP depend greatly on space volume, as Bady et al. noted [13]. These indices were limited to analyzing the effect of design variations on regional spatial ventilation.

This paper further explores the method of assessing regional spatial ventilation performance using net escape velocity (NEV) as an index for the design of residential building arrangements at an operational level. Furthermore, to fully discuss spatial ventilation performance, spatial mean velocity magnitude (VM) and visitation frequency (VF) are also employed as indices in this study, as they can reflect the air flow rates and recirculation phenomena of the calculated domains. To calculate these ventilation performance indices, CFD simulation with ANSYS-Fluent 13.0 is adopted. In this paper, the multi-residential building district is selected as an example for use in studying ventilation performance, as these districts account for the largest proportion of districts in China today. Five examples of building arrangements are chosen according to the designs typically found in real residential districts. They represent possible design changes in building length, lateral spacing and layout patterns. In consideration of the effect of surrounding buildings, all cases have similar surrounding conditions. The CFD simulations are performed for two directions, south wind (S) $\theta = 0°$ and southeast wind (SE) $\theta = 30°$, which means wind direction are perpendicular and non-perpendicular to the building's main facade.

2. Literature Review on Urban Wind Flow Prediction for Building Arrangement Design

2.1. Influence of Wind Conditions on Building Arrangements

The effects of design variations on urban ventilation have been discussed by many studies in the field of urban forms and street canyons. Among studies of urban form, for example, Mfula et al. [4], using wind tunnel tests, discussed effect of building spacing and density changing on pollutant dispersion. Buccolieri et al. [15] also studied the effect of spacing changes on air exchange rates using numerical modeling. These studies mostly focus on the influence of building density on overall urban wind flows. Therefore, arrays of square buildings are chosen as objects of CFD simulation, with equal longitudinal and lateral spacing used to investigate the influence of spacing variations on pollutant concentration distribution. Buccolieri et al. [19] also chose square arrays when investigating influence of longitudinal and lateral spacing; they showed that variations in spacing perpendicular to wind direction had a more noticeable influence on vertical exchanges of air flow. Similar studies have been performed by Di Sabatino et al. [20], Hang et al. [21–23], Ramponi et al. [24], Razak et al. [25] and Lee et al. [26] on building heights, layouts and street widths. In the field of street canyons, as early as 1988, Oke et al. [27] identified three types of characteristic flow based on the width/length (w/h) ratio of street sections. Then, Sini et al. [28], using numerical modeling, discussed different characteristics of wind fields with and without the heating of walls in street canyons of infinite length. They varied w/h ratios of street sections from 0.3 to about 10. CFD simulation results proved the conclusion of Oke et al. [27] and found that wind flow pattern changed radically when windward facade walls are heated. Similar studies were also carried out by Simoëns and Wallace on pollutant dispersion [6,7]. In addition to the w/h ratios of street sections, Chan et al. [29,30] also studied effects of different length/height (l/h) ratio of building facades and building height changes on dispersion of pollutants at different locations. The results showed that non-uniform building heights are beneficial to urban ventilation and that the maximum l/h ratio of building facades should be controlled within the range of 5. Most of these studies have been concerned with proposing a better evaluation method, but their guidance value for design practice is very limited.

In addition, studies related to residential building layouts have been performed in recent years. For example, Hong and Lin [10] compared six layout patterns of multilayer residential buildings under the same density and coverage, and the results showed that layout and orientation of buildings have significant effects on the outdoor wind environment at the pedestrian level. Yang et al. [11] analyzed effect of standard and staggered layouts of roadside multi-floor buildings on the pollutants dispersion. Ying [31] and Iqbal [32] also compared the layout patterns of high-rise residential buildings. However, most of these studies perform CFD simulations by simply combining several individual buildings, without considering the effect of surrounding buildings.

2.2. The CFD Approaches for Urban Wind Flow Modelling

The application of CFD simulation for urban wind flow modelling has developed rapidly in the last 20 years [33]. Compared to wind-tunnel or full-scale testing, CFD simulation method has some important advantages in predicting wind flow around buildings [33–35]. In the detailed review of 50 years of computational wind engineering [33], Blocken stated that "They can provide detailed information on the relevant flow variables in the whole calculation domain ('whole-flow field data'), under well-controlled conditions and without similarity constraints. However, the accuracy and reliability of CFD are of concern and solution verification and validation studies are imperative".

For wind flow predicting using CFD simulation, Reynolds-Averaged Navier-Stokes (RANS) and Large Eddy Simulation (LES) are two main approaches. RANS approaches include steady RANS and unsteady RANS (URANS). In addition, hybrid RANS/LES approaches also exist, although they are rarely used in urban physics and wind engineering [33]. Studies using these approaches for urban physics have been reviewed by Stathopoulos [34,36], Moonen et al. [37], Blocken et al. [35,38] and Blocken [33,39]. From these reviews, it can be stated URANS are relatively rare adopted in urban flow simulation, as it does not simulate the turbulence, but only its statistics. Moreover, URANS also requires a high-spatial resolution, so it is recommended that LES is directly used to model wind flow. The main limitation of RANS is that it cannot incorporate the transient behavior of separation and recirculation flow of windward edges. LES on the other hand can resolve the large and generally most important turbulent eddies. Thus, it is generally acknowledged that LES can provide more accurate result than steady RANS. However, if the urban wind prediction focused on mean wind speed rather than on an effective wind speed, RANS approach could be sufficient. Thus, Blocken [39] conclude that "It (steady RANS) is by far the most widely used approach in most urban physics focus areas. The reason for this is twofold: (i) the computational expense of LES and (ii) the increased model complexity of LES in combination with the absence of extensive best practice guidelines for LES".

Some guidelines, such as AIJ (Architectural Institute of Japan) guideline [40] and COSTA (European Cooperation in Science and Technology) action [41], have provided important recommendations of using the CFD technique for appropriate prediction of pedestrian wind environments. In these guidelines, the basic technique demands are provided, including computational domain size, grid and boundary conditions. These demands provide solid bases for our ventilation performance study.

2.3. Assessment of Urban Spatial Ventilation

To assess ventilation of exterior spaces, wind velocity and pollutant concentration are commonly used in previous studies. Ng et al. carry out a series of studies using wind velocity and pollutant concentration as indices to analyze building permeability, frontal area index and air path [42–46]. Some technical guidelines and policies for urban planning in high-density cities are also summarized [47]. Guidelines for residential neighborhoods are also provided by Chan et al. [29,30] and Kubota et al. [48] based on CFD simulation of pollutant concentration and wind velocity.

In addition to the traditional and commonly used indices, in the last ten years, some indoor ventilation indices have been developed to access VE in urban areas. For example, Bady et al. [13] and Kato and Huang [14] introduced some indoor ventilation parameters and developed a series of scales concept for evaluating VE in urban spaces. Through evaluation of several examples, the studies showed that these ventilation indices appear to be a promising tool for urban ventilation study. Hang et al. [16] used some simple idealized city models to explore the effect of urban morphology on the local mean age of the air (the time it takes for urban external "fresh" air to reach a given location). The air exchange efficiency (the frequency, in a certain area, with which air is replaced by outside "fresh" air) in such idealized city models was also studied [22]. Buccolieri et al. [15,19] also proposed a conceptual framework for city breathability. The overall flow rate across boundaries (sides and street top) and the local mean age of air are used to discuss influence of building packing density. Recently, Hang et al. [17] further discussed city breathability in medium-density urban-like geometries using pollutant transport rate and net escape velocity.

3. Residential Building Configurations and Typical Spaces Studied

3.1. Single Residential Building Sizes

At the level of single building (Figure 2a), Liu and Ding classified China's recently built living units into a number of types and provided the size range for each type [49]. As shown in Figure 2b, in consideration of different family demands, three unit types—two-bedroom (U1), three-bedroom (U2), and four-bedroom (U3) units—are selected in this study. The depth L of living unit equals that of residential building (L = 12 m). Building length W is determined by the combination of the living units, which can be divided into two and multi-unit combination, as shown in Figure 2c. It is also worth noting that, in consideration of economy, living units with larger areas (U3) are generally selected as side units in two-unit and multi-unit groups, which means the minimum W can be set as 30 m. The buildings each have six floors, with a uniform floor height of 2.8 m, for a total height H of 18 m including parapet above and interior-exterior height difference below. The undulation of balconies on southward façade is considered as a uniform single plane to simplify the computation.

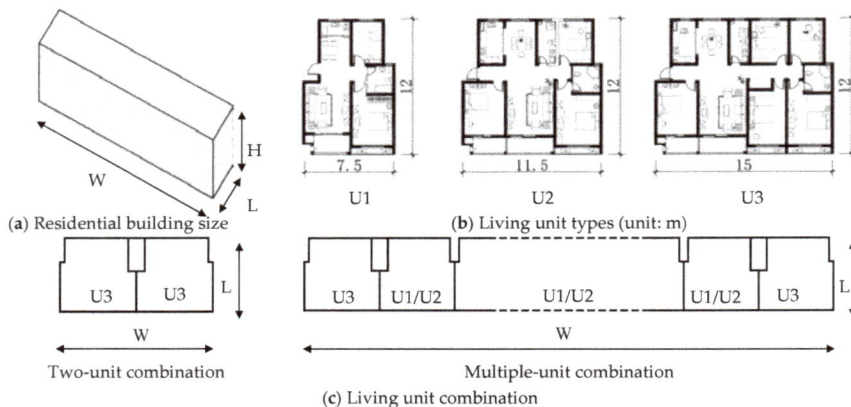

Figure 2. Living units and unit combination in a residential building: (**a**) Residential building size; (**b**) Living unit types (unit: m); (**c**) Living unit combination.

3.2. Residential Building Groups

Residential areas in China have common arrangement patterns that are easy to identify and that have resulted from demands for function, sunshine, fire protection and economical use of land resources, as shown in Figure 1. Building groups are uniform strip arrays, such as those in Nanjing City, where up to 65% of residential areas have strip-shaped spacing and slab buildings [50]. Based upon the patterns of existing residential areas, a model of 3 × 3 building groups is set up as the object for

CFD simulation, with the central group (CG) providing variation in the configuration analysis and the peripheral groups (PG) setting the environmental conditions, as shown in Figure 3a. The widths (B2 and D3) of the four streets (St1–St4) are determined to be 30 m according to current design code (Figure 3a). For CG, five types are established as study cases for comparison, as shown in Figure 3b, where Case A is set as the initial state and is also chosen as CG building. Building widths in Case A are all determined to be W1 = W2 = 72 m (six units). The longitudinal, namely south–north, spacing B1 is 24 m, which meets the requirement of sunlight spacing, while the east–west spacing is determined as 12 m according to living unit combination and design code. Case B and Case C are both variations developed upon Case A, with consideration of the influences of lateral spacing (D1 and D2) and length variation (W1–W4) on outdoor wind environment. Case D and Case E also follow Case A and discuss the influence of building width and staggering variation on spatial ventilation.

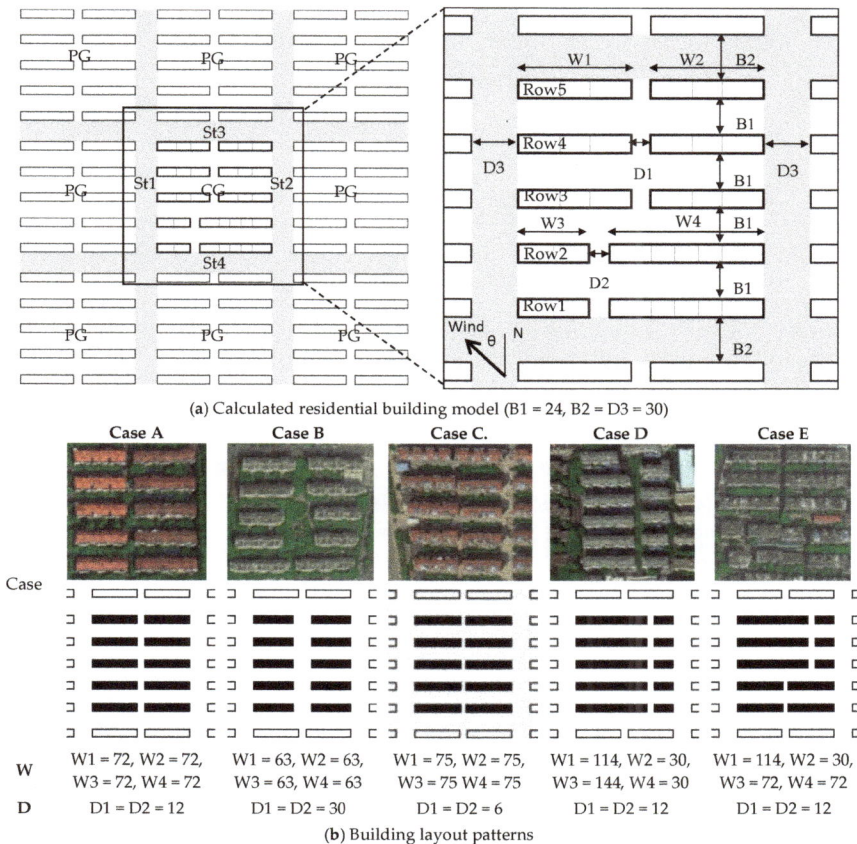

(a) Calculated residential building model (B1 = 24, B2 = D3 = 30)

	Case A	Case B	Case C	Case D	Case E
W	W1 = 72, W2 = 72, W3 = 72, W4 = 72	W1 = 63, W2 = 63, W3 = 63, W4 = 63	W1 = 75, W2 = 75, W3 = 75 W4 = 75	W1 = 114, W2 = 30, W3 = 144, W4 = 30	W1 = 114, W2 = 30, W3 = 72, W4 = 72
D	D1 = D2 = 12	D1 = D2 = 30	D1 = D2 = 6	D1 = D2 = 12	D1 = D2 = 12

(b) Building layout patterns

Figure 3. Setup of calculation cases (Units m): (**a**) Calculated residential building model (B1 = 24, B2 = D3 = 30); (**b**) Building layout patterns.

3.3. Ventilation Performance Area and Spatial Patterns

The area selected for data extraction is the space between the third (Row3) and fourth row (Row4) of buildings, as shown in Figure 3a. To compare the effect of variation in design parameters on VE in different areas, some typical domains are selected for comparison, as shown in Figure 4a. These domains could be classified into four spatial patterns, as shown in Figure 4b. RM space stands for the outdoor middle space of the investigated building, and RS space represents the outdoor intersection space; ROD space represents the outdoor outward-side space that adjoins streets St1 and St2; RID space refers to the outdoor inward-side space that adjoins the RS space. The VEs of the four spaces stand for four types of exterior wind environments of dwelling units. The domain volumes of RM (RM1 and RM2), ROD (ROD1 and ROD2) and RID (RID1 and RID2) are invariant, with a uniform size of 12 m wide and 24 m long. The volume of the RS space will vary as the building spacing changes. The domain volume height is from the ground to building height H.

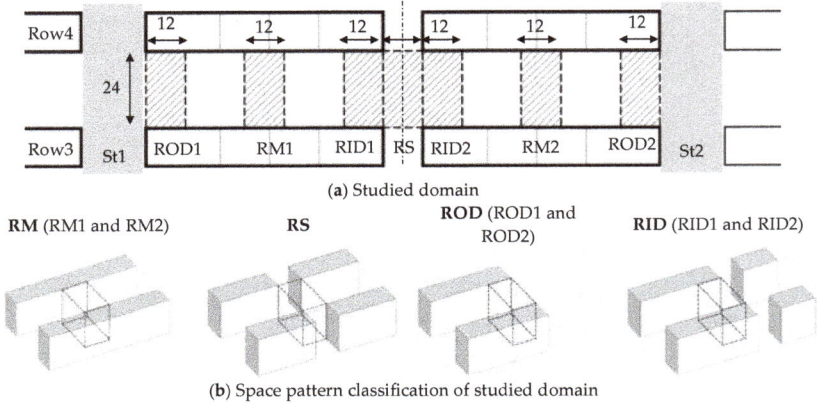

(a) Studied domain

(b) Space pattern classification of studied domain

Figure 4. Typical studied areas and space pattern classification (Unit: m): (a) Studied domain; (b) Space pattern classification of studied domain.

4. Computational Settings of Building Arrangements

4.1. Computational Domain and Boundary Conditions

Figure 5 shows the computational domain and boundary conditions in CFD simulation. The domain size refers to AIJ guidelines [40] and some CFD simulation studies of urban ventilation [51–53]. The lateral and inflow boundaries are set to 5 H away from the building groups, where H is the uniform building height. The outflow boundary is 20 H away from the building groups and the height of the computational domain is 11 H.

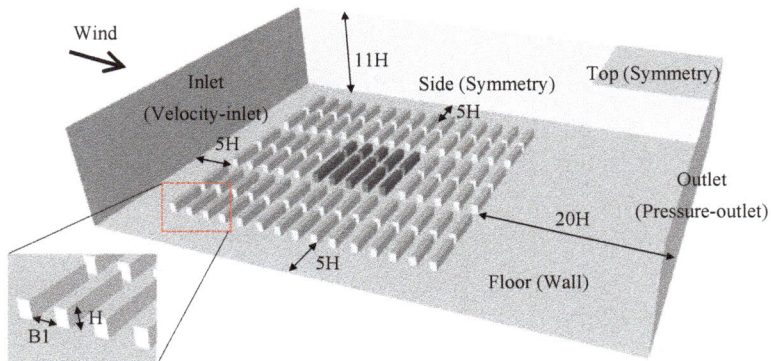

Figure 5. Computational domain and boundary conditions.

Symmetry boundary conditions, required to enforce a parallel flow, were imposed on the top and lateral sides of the domain. At the outlet boundary of the domain a pressure-outlet condition was used. No-slip wall boundary conditions were used on all solid surfaces. As for the inlet boundary condition, according to AIJ guideline, a power-law velocity profile was applied:

$$U(z) = U(s) \left(\frac{z}{z_s} \right)^{\alpha} \tag{1}$$

where $U(s)$ is the velocity at reference height, z_s, and α is the power-law exponent determined by terrain category. In the development of the AIJ guideline for wind environment prediction, the Working Group carried out some wind tunnel experiments [54]. In these experimental studies, some urban configurations are investigated. The inlet velocity profiles of these studies are adopted in our CFD simulation as they have similar features of urban configuration $\alpha = 0.25$, and the roughness length z_0 is 0.01 m. The thickness of the atmospheric boundary layer is 250 m. The reference wind speed $U(s)$ is 4 m/s at the reference building height $z_s = H$. The turbulent kinetic energy profile (k) and turbulent dissipation rate profile (ε) are calculated as:

$$k(z) = \frac{U^{*2}}{\sqrt{C_\mu}} \tag{2}$$

$$\varepsilon(z) = \frac{U^{*3}}{\kappa(z + z_0)} \tag{3}$$

where C_μ is a constant (0.09); the friction velocity $U^* = 0.33$ m/s; and κ is the von Karman constant, which is determined to be 0.4.

4.2. Computational Grid and Solver Settings

The computational domain was built using hexahedral elements (about 3.9–4.2 million for different Cases). The grid resolution meets the major computational requirements recommended by Tominaga et al. [40]. As shown in Figure 6, the minimum grid control in direction z is 0.028 H, the minimum grid control in the x and y directions are 0.056 H, and the maximum expansion factor between grids is below 1.25. The grid sensitivity analysis (see Supplementary Material) shows the errors caused by grid resolutions have an unnoticeable effect on the numerical results.

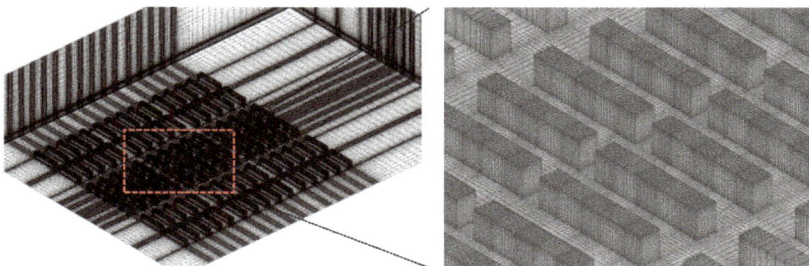

Figure 6. Gird resolution in the computational domain (Case A).

Based on the CFD approaches discussed in previous Section 2.2, the steady RANS approach with standard k-ε turbulence model is adopted in this study. The SIMPLE algorithm is utilized for pressure–velocity coupling. Pressure interpolation is in second order accuracy. For both the convection terms and the viscous terms of the governing equations, second-order discretization schemes are used. Validation study was performed by comparing the CFD simulation results with a wind-tunnel experiment of strip-type building groups (see Supplementary Material). As the experimental building group is different from that used in this study, we built an extra building model, with identical inlet boundary and geometry conditions to those in Zhang et al. [55]. The CFD simulation results, overall, have a good agreement with experimental results.

5. Ventilation Performance of Regional Space

When evaluating the ventilation performance of regional space, PFR and TR are some typical evaluating indices [13,14]. However, the values of these parameters depend greatly on space volume. Unlike indoor space, the boundary of exterior regional space is uncertain, and the determination of space volume is arbitrary. Thus, air flow patterns, i.e., wind velocity and flow recirculation, in the studied domains might be an important aspect of regional spatial ventilation performance. Generally speaking, greater wind velocity and less recirculation could be beneficial for space ventilation. Recently, Lim et al. [56] presented a new concept of the ventilation index, NEV, which is the velocity that corresponds to PFR. Compared to traditional parameters of wind velocity, NEV reflects the effective and net contaminant transport and dilution velocity, which relate not only to the wind speed but also to the flow reversal. Hang et al. [17] further adopted the parameter of net escape velocity to access the influence of city size, building height variations and wind direction for the entire pedestrian volume (throughout the $z = 0$–2 m volume). By assuming that the pollutant source is generated homogeneously in the studied regional space, the net escape velocity of this space can be calculated using Equation (1) [17]:

$$NEV = \frac{S_c \times Vol}{\langle C \rangle} / A_p = \frac{S_c \times Vol}{\int\limits_{Vol} C dxdydz / Vol} / A_p \tag{4}$$

where *Vol* is the studied volume (m³), and <C> is the spatially averaged concentration in the studied volume (kg/m³). S_c is the release rate of uniform pollutants (kg/m³-s). In Hang's study, A_p is defined as the entire area of boundaries for the entire studied (pedestrian) volume in urban areas. However, using the entire area of boundaries to calculate the NEV could lead to lower NEV values, especially when assessing spaces with large lateral areas, i.e., cross section spaces. From the NEV discussion in Lim et al. [56], we find that it might be more appropriate to define A_p as the normal outflow area of the space's boundary openings. In this study, we will explore the feasibility of this method to assess the ventilation efficiencies of different spaces.

In addition to NEV, we also employ wind VM and VF as the evaluating parameters to access the air flow pattern in the studied domain. VM and VF can be calculated using Equations (2) and (3) [13,16]:

$$VF = 1 + \frac{\Delta q_p}{S_c \times Vol} \tag{5}$$

$$VM = \frac{1}{Vol} \iiint\limits_{Vol} V(x, y, z) dxdydz \tag{6}$$

where Δq_p is the inflow flux of pollutants into the domain (kg/s). V is the velocity magnitude. By applying the above concepts, this paper quantifies the effects of design variations for different regional spaces and wind directions on the air flow patterns and the space's ventilation capacity.

6. Comparison of Ventilation Efficiency in Different Spaces

Figure 7 shows the simulation result of wind flow fields for the five cases under two wind directions ($\theta = S\,0°$ and $\theta = SE\,30°$). It reveals that under wind direction $\theta = S\,0°$, wind flows pass through the south–north streets and permeate into the spaces between long-strip buildings. The wind flow velocity and recirculation feature of the studied domains varies greatly. Wider streets (St1 and St2) could generally improve the wind velocity of space adjoined to the streets. However, under the wind direction $\theta = SE\,30°$, the air flow velocity of different domains improved evidently.

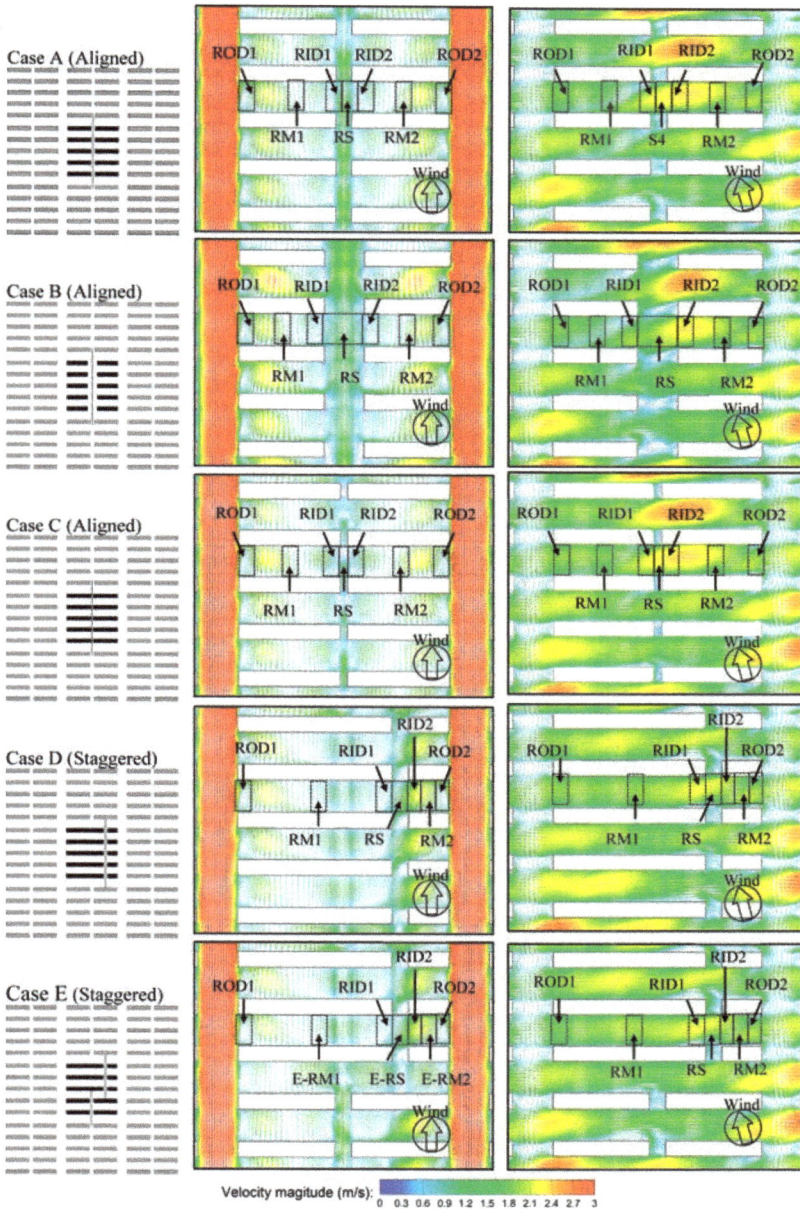

Figure 7. Wind flow patterns for five cases under two wind directions (θ = S 0° and θ = SE 30°) and studied domains for each case (z = 0.1 H).

Based on the calculated flow fields, NEV and VF of studied domains are calculated by setting pollutant source generated homogeneously within the study domains (Figure 4). Figure 8 shows the calculated NEV, VF and VM for different domains under wind directions θ = S 0° and θ = SE 30°. From the simulation results, it can be found that wind direction could influence VE of different outdoor spaces greatly. As wind direction θ varies from S 0° to SE 30°, VM of ROD1 and RID1 domains of case A, for example, increased by 33% and 106% respectively. The cause is the strip-shaped outdoor spaces, which forms a ventilation corridor between the southward oriented buildings, allowing increasing airflow into the spaces between, with the wind direction angle changing from perpendicular to parallel to the main building facades. Meanwhile, the VF descends remarkably the other way, especially in spaces near the lateral (ROD1 and ROD2, RID1 and RID2). For example, VF of ROD1 and RID1 domain drops to 1.04 and 1.05, respectively, from 1.57 and 1.63, as wind direction θ varies from S 0° to SE 30°. As a combined effect of the two results above, NEV increases considerably, which shows improvement in ventilation conditions resulting from the wind direction. Therefore, the local prevailing wind direction needs to be included in the design process to maintain a certain angle between the wind and the main building façade for effective improvement of outdoor ventilation.

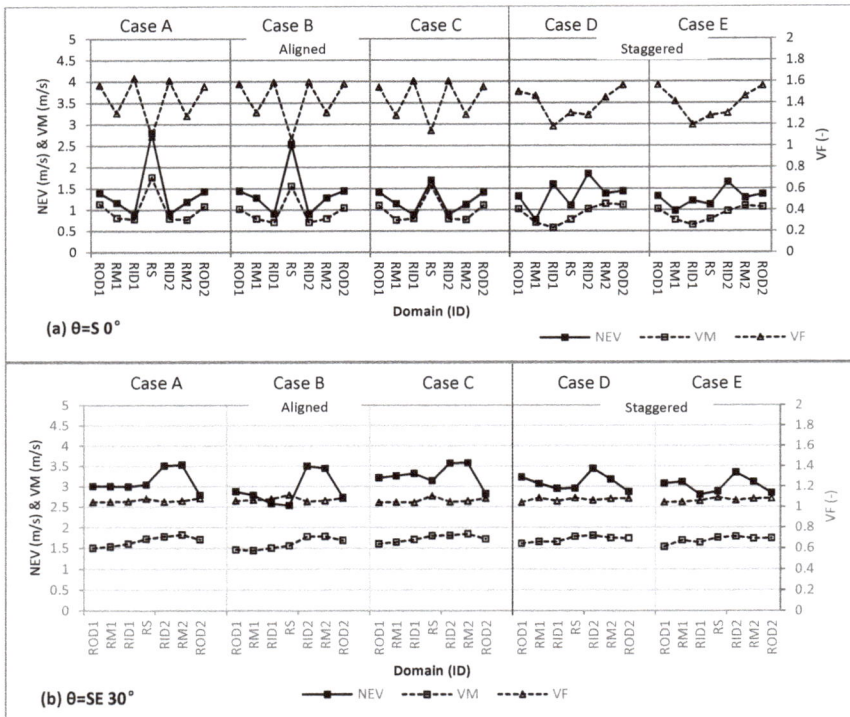

Figure 8. Ventilation performance of different studied domains for five cases under wind direction: (a) θ = S 0°; and (b) θ = SE 30°.

Under wind condition θ = S 0°, VE indices of different spaces differ noticeably from one another in both values and range. The ventilation efficiency shows a similar variation tendency under the aligned (Case A, Case B, and Case C) and staggered (Case D and Case E) conditions. Under the aligned condition, RS space shows relatively better VE. NEV could reach a more than 20% increase over that of the other three spaces (RM, ROD and RID) (Figure 8a). In these spaces, wind flows pass through the space directly, with relatively high VM and low VF. NEV of the ROD (ROD1 and ROD2) and RID (RID1 and RID2) spaces varies greatly, as shown in Figure 8a. Compared to ROD spaces, RID spaces show the worst VE. VFs of the two spaces are approximately 1.6. However, VM of ROD spaces (ROD1 and ROD2) is much larger than that of RID spaces (RID1 and RID2). This difference is mainly due to the widening of street width (St1 and St2) beside the space. Compared to the ROD and RID spaces, NEV of the RM space (RM1 and RM2) is in the middle, due to the combined effect of lower VF and low mean VM. Under the staggered conditions, NEV of the RID space decreased greatly compared to that under the aligned conditions, mainly due to the effect of staggered buildings, which leads to the great decrease in wind velocity.

Under wind direction θ = SE 30°, NEVs of different spaces do not change very much (Figure 8b). In general, the VE of spaces on the upwind side (RM2, ROD2, and RID2) is better than that of spaces on the downwind side (RM1, ROD1 and RID1). This is mainly because VMs of spaces on the upwind side (RID2–ROD2) are better than that of spaces on the downwind side, and VFs of these spaces do not change very much.

7. Effect of Design Change on Spatial Ventilation Efficiency and Its Related Pollutant Dilution

7.1. Effect of Design Change in Middle Space RM

Figure 9 shows the concentration fields within space RM (RM1 and RM2) for different building lengths, spacing distance and staggered conditions under wind directions θ = S 0° and θ = SE 30°. From the simulation results, it can be seen that comparing to building spacing and staggered condition, building lengths might be more relevant to the outdoor ventilation of the RM spaces under wind direction θ = S 0°. In Figure 9a, the pollutant concentrations in the study domains generally increase as building length W increases. As W increases, mean VF of the RM space decreases, and pollutant VFs are around 1.4. These values can be observed in Figure 10a. In this figure, when W = 30 (i.e., in Case D), VM of RM space in case D (D-RM2) increases by approximately 44.5%, 49.1% and 49.5%, respectively, from its values at RM2 in case B, Case A and Case C, where building lengths are 62 m, 73 m and 75 m. When W = 30, VF of space RM in case D (D-RM2) is also larger than that in case B (W = 63), Case A (W = 72) and Case C (W = 75). This reflects the strengthening of wind flow recirculation in these domains as the building length increases. However, when W = 30, VF of the RM space in case D (D-RM2) is also larger than that in Case A, Case B and Case C. This is mainly because building length is 30 m and the RM space (D-RM2) is close to ROD space, where VF is much larger (above 1.4). Under the combined effect, NEV of RM space in Case D (when W = 30) is obviously lower than that in Case A, Case B and Case C. NEV of the RM space in Case E is similar to that in Case D.

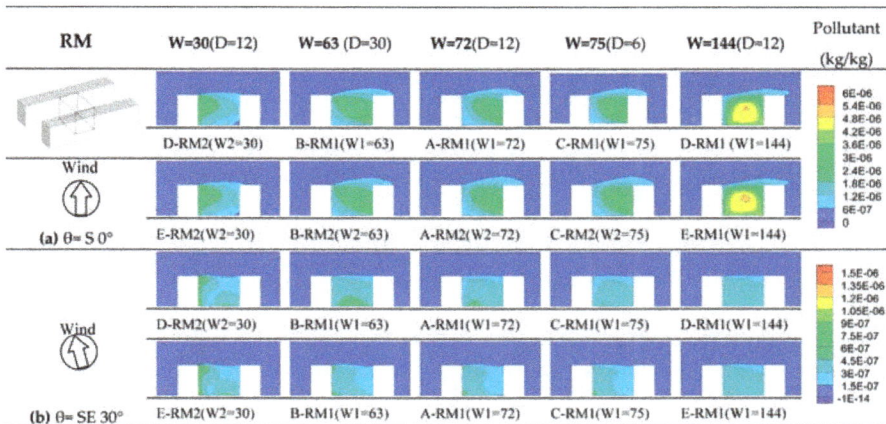

Figure 9. Concentration fields within the studied RM1 and RM2 domains (RM space) for different design variations under wind direction: (**a**) θ = S 0°; and (**b**) θ = SE 30° (Unit: m).

Under wind direction θ = SE 30°, the level of pollutant concentration is much lower than that under wind direction θ = S 0°, as shown in Figure 9b. It is mainly due to the improvement of air flow conditions in these studied domains. From Figure 10, it can be seen that when wind direction varies from south 0° to southeast 30°, VM of the RM spaces generally increases from 1 m/s to 1.8 m/s, and VF decreases from 1.4 to 1.04, which shows the great improvement in wind removal efficiency. Under these combined effect, the NEV of RM spaces improves evidently, increasing by about 180%. In addition, the distribution of pollutant concentration in the RM2 space of Case D and Case E is less uniform than that in the RM1 and RM2 spaces of other cases. The cause might be that the wind direction is form southeast, and when building length is 30 m, RM2 space is closely adjacent to the building's east corner space, at which more recirculation flow occurs.

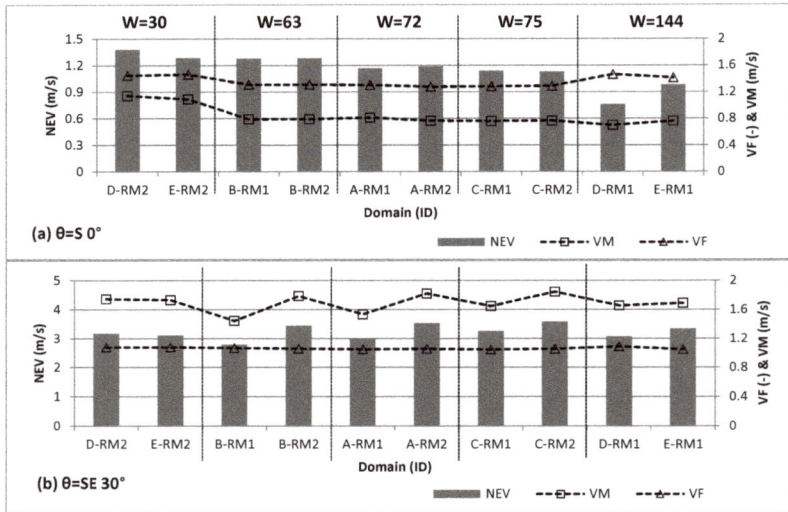

Figure 10. Effect of building length on ventilation efficiency indices within RM1 and RM2 domains (RM space) under wind direction: (**a**) θ = S 0°; and (**b**) θ = SE 30° (Unit: m).

7.2. Effect of Design Change in Intersection Space RS

Figure 11 shows the concentration fields within RS space of different cases of building spacing and staggering, including changes under wind directions θ = S 0° and θ = SE 30°. Under the south 0° wind direction, pollutant concentrations in the RS space decreases slightly as building spacing D (D1 and D2) increases. However, this variation is not obvious, even when D increases to 30 m. The variation is mainly due to the effects of surrounding buildings. The building spacing of the south up-wind building is set at 12 m. Although D increases to 30 m, mean VM in space RS does not increase obviously, which leads to the limited improvement of pollutant dispersion. When the building arrangement is staggered, the level of pollutant concentration obviously increases. This increase is mainly due to the effect of staggered building arrangement, which blocks the air flow path in studied area. Figure 12a shows the influence of changes in building spacing and staggering on NEV, VM and VF of RS spaces. From the figure, it can be clearly observed that NEV generally decreases slightly as building spacing increases. When D = 12 (i.e., A-RS), NEV of the RS space in case A increases by approximately 65% from its value in case C (D = 6). This increase is mainly due to the increase of wind speed within the studied domain, as D increases under the aligned condition. However, when D increases to 30 m, mean VM of this studied domain does not increase. This failure to increase might be due to the effects of surrounding buildings, as discussed above. VF does increase slightly, so NEV of this domain decreases slightly.

Under the staggered condition, it is clearly observed that the mean VM of this studied domain decreases, which induces the decrease of VE in the domain (35–60% decrease of NEV).

Under the wind direction θ = SE 30°, as the building length increases, the level of pollutant concentration in the study domains increases as D increases (Figure 11b). This is mainly due to the volume increase in the studied domain (increase in amount of pollutant released) and the limited improvement in ventilation efficiency. In Figure 12b, it can be observed that as D increases, although mean VM increases, VF also increases; and under the combined effect, NEV decreases slightly, which means the decreases of ventilation efficiency in this space.

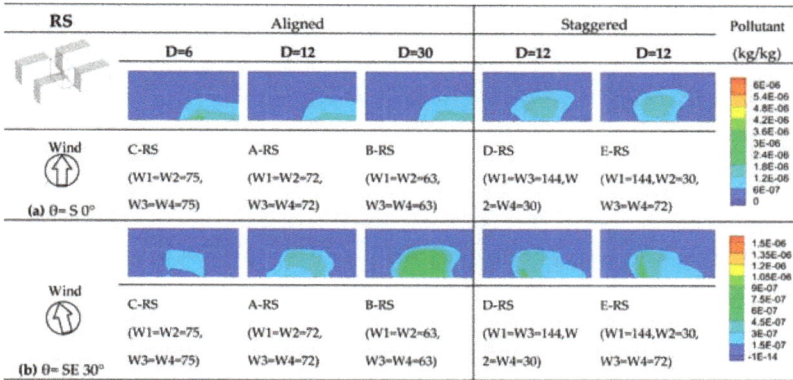

Figure 11. Concentration fields within studied RS space for different arrangements and building spacing under wind direction: (**a**) θ = S 0°; and (**b**) θ = SE 30° (Unit: m).

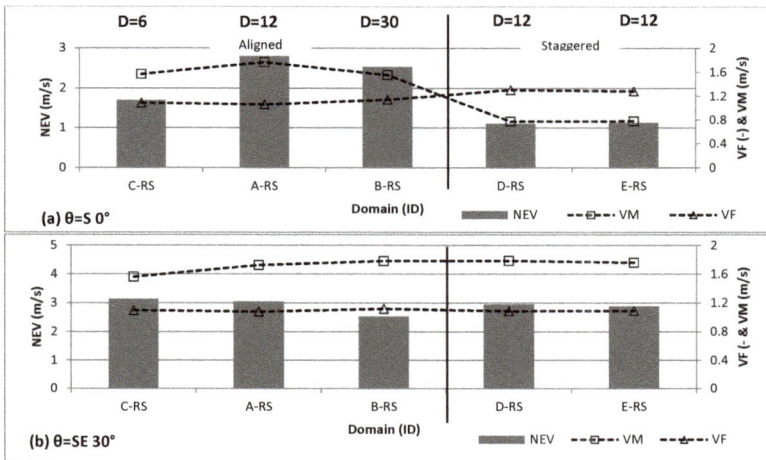

Figure 12. Effect of building spacing on ventilation efficiency indices within the RS space under wind direction: (**a**) θ = S 0°; and (**b**) θ = SE 30° (Unit: m).

7.3. Effect of Design Change in Outward-Side Space ROD

For the ROD space, these design variations nearly have negligible effect on spatial ventilation. Figure 13 shows the concentration fields within the ROD space (ROD1 and ROD2) for different design changes under wind directions θ = S 0° and θ = SE 30°. From Figure 13a, it can be observed that the level of pollutants within the domains show similar distribution patterns under the south 0° wind direction. This is mainly because the wind flow patterns within the studied domains are mainly influenced by the streets (St1 and St2) beside the space. As the street width does not change, air flow pattern in the studied domain does not change very much. Figure 14a shows the effects of design variations on VE indices within the ROD spaces. From this figure, it is obvious that VF, VM and their related NEV all do not change very much for different design cases. The change range is below 10%.

Under southeast 30° wind direction, distribution of pollutants within ROD2 domains is much less uniform than that of the ROD1 domains, and in some parts, the level of pollutants in the ROD2 domains is much higher (Figure 13b). The cause is similar to that in RM spaces. As the wind direction is from southeast, recirculation of flows within the ROD2 domains is more prominent than that of the RM1 domains, which induce the pollutants gathering in parts close to the buildings' corners. In Figure 14b, it can also be seen that NEVs within the ROD2 domains are slightly higher than that of the RM1 domains.

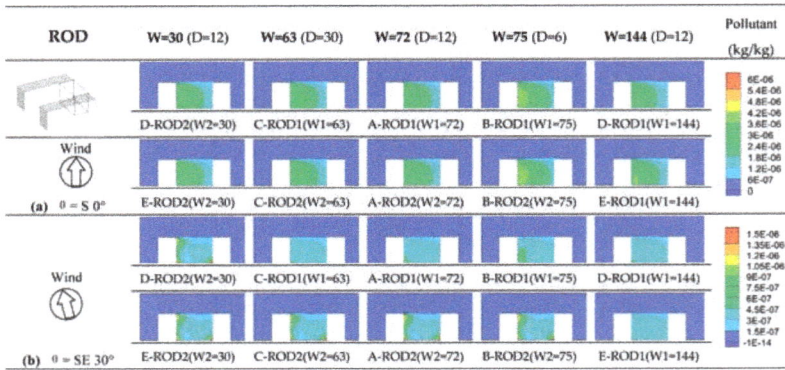

Figure 13. Concentration fields within studied ROD1 and ROD2 domains (ROD space) for different design variations under wind direction: (**a**) θ = S 0°; and (**b**) θ = SE 30° (Unit: m).

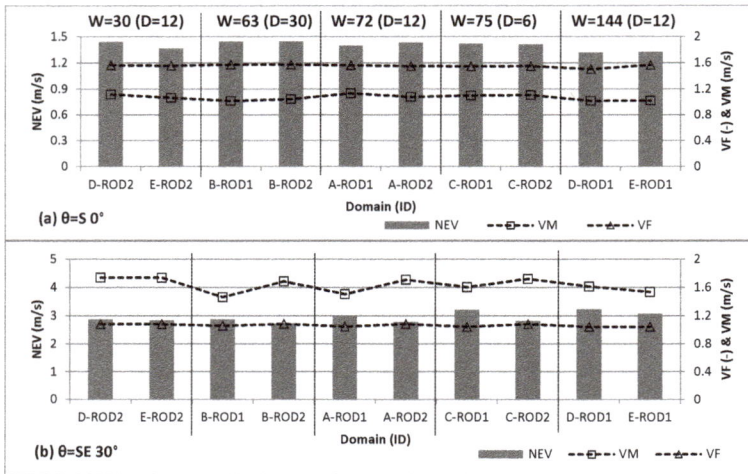

Figure 14. Effect of design variations on ventilation efficiency indices within ROD1 and ROD2 domains (ROD space) under wind direction: (**a**) θ = S 0°; and (**b**) θ = SE 30° (Unit: m).

7.4. Effect of Design Change in Inward-Side Space RID

Figure 15 shows the concentration fields within the RID space of different cases under wind directions θ = S 0° and θ = SE 30°. In Figure 15a, it can be observed that, under south 0° wind direction, pollutant concentrations in Case D and Case E are evidently less than that in Case A, Case B and Case C. It is mainly due to the decrease of building length. In case D and Case E, when building length W2 is 30 m, the RID space is close to streets St1. The air flow pattern improves evidently, which can be seen in Figure 16a. In case D and Case E, VF decreases in all the RID spaces and VM even increase in the RID2 spaces. These flow characteristics induce that the value of NEV increase greatly. Moreover, in Case A, Case B and Case C, pollutant concentrations evidently decrease as building spacing D increases. This is mainly due to the increase in pollutant dispersion in the RID space, which is closely related to the air exchange rate. However, from Figure 16a, it can be observed that under these wind conditions, increase of NEV is limited, which means that pollutant spreading speed does not improve greatly as building spacing widens. This variation is coincident with that of spatial VM and VF. As building spacing increases, VM does not improve evidently, and VF decreases slightly. With these combined effects, NEV is slightly improved. This is mainly due to the effect of up-wind surrounding buildings.

Under the southeast 30° wind direction, wider building spacing might decrease the RID space's VE. In Figure 15b, it can be observed that pollutant concentration in the RID1 space increases as building spacing D increases. This is mainly due to the effect of recirculation of flows, similar to the ROD space discussed above. However, under this wind direction, variation in building spacing has a nearly negligible effect on pollutant concentration of the RID2 space. The variation range of NEV in these spaces is below 2% (Figure 16b).

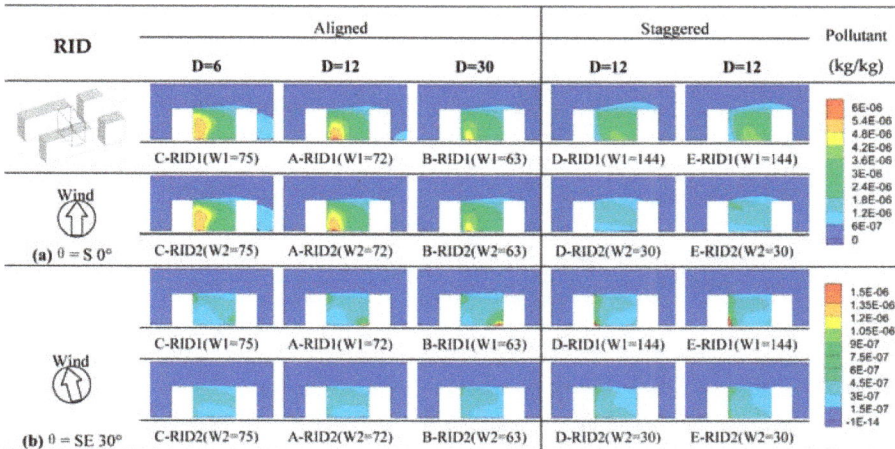

Figure 15. Concentration fields within the studied RID1 and RID2 domains (RID space) for different design variations under wind direction: (**a**) θ = S 0°; and (**b**) θ = SE 30° (unit: m).

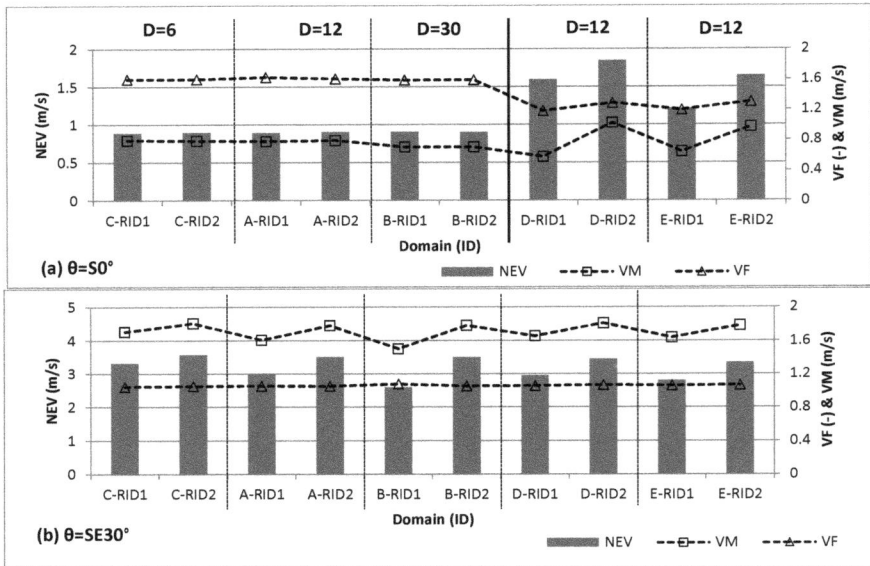

Figure 16. Effect of design variations on ventilation efficiency indices within the RID1 and RID2 domains (RID space) under wind direction: (**a**) θ = S 0°; and (**b**) θ = SE 30° (unit: m).

8. Conclusions

This study presents a preliminary investigation of the relevance of ventilation efficiency in outdoor space to various spatial patterns of residential buildings with the aid of CFD simulation techniques and the possibility of optimization through simulations at different spaces in typical cases. NEV, VM and VF were used to quantify the contributions of pollutant removal by mean flow and turbulent diffusion, as well as their ventilation capacities.

The simulation results indicate that NEV is a useful ventilation efficiency index that can comprehensively reflect the pollutant removal ability of wind velocity and flow recirculation in different regional spaces. VM and VF are also useful indicators that can reflect the influence on pollutant removal of wind velocity and flow recirculation, respectively. By combined applying these three indices, regional spatial ventilation performance can be effectively assessed.

From the cases studies, it can be determined that spatial patterns can strongly influence a space's ventilation efficiency, especially when wind direction is approximately perpendicular to the main building facade. For residential building arrangements, strip-shaped buildings form a ventilation corridor between the southward-oriented buildings due to solar access demands. Thus, when wind direction changes from parallel to perpendicular along the building's long facades, space ventilation generally decreases. In this study, when wind direction is perpendicular to buildings' main facades, intersection space has the best ventilation efficiency, in general, due to the air-path, which leads to high VM and lower VF; the outward-side space shows relatively better ventilation efficiency due to wind inducing the air-path of streets (St1 and St2) beside the outward-side spaces. Ventilation efficiency of the inward-side spaces and middle spaces depend heavily on design variations in building length and spacing. For example, in case A, NEV of the middle space is 31% higher than that of inward-side space, which means ventilation efficiency of the middle space is better than that of the inward-side space, but when building length increases, NEV of the middle space in case D is 14% lower than that of the inward-side space in case A.

Atmosphere **2017**, *8*, 102

Through preliminary analysis of the influence of design variation on ventilation efficiency of different spaces, it can be observed that building spacing evidently influences ventilation performance of the inward-side space and intersection space. As building spacing increases, ventilation efficiency of these spaces improves evidently, but the improvement is restricted greatly by the effect of up-wind surrounding buildings. Variation in building length mainly has a great influence on ventilation performance of the middle spaces. As building length increases, ventilation efficiency of the middle spaces decreases. When building length becomes very long (i.e., 144 m in Case D), NEV of the middle spaces can decrease by about 50% due to decreased wind velocity in this domain.

Thus, it can be concluded that local prevailing wind direction should be considered in design operations specifically dealing with building orientation; consideration of wind direction is especially important for residential building arrangements. When a certain degree (i.e., 30°) exists between prevailing wind direction and building's long facade, all ventilation efficiency indices can show sizable improvements in different regional spaces. Existing case studies show that building length, lateral spacing and staggered positioning can influence on ventilation efficiency of regional spaces. However, building length change is more relevant to outdoor ventilation of RM spaces, and residential buildings should be restricted to eight-living-unit combinations of dwellings. This might lead to more wind flow into the middle spaces, which can be beneficial for the middle living units. Lateral spacing and staggered positioning might have more of an effect on the intersection space and inward-side spaces. However, when considering lateral spacing and staggered design, surrounding building arrangements in the prevailing up-wind direction should be considered to organize air flow path, which could be beneficial for ventilation performance in these spaces.

This paper discusses the method of evaluating regional spatial ventilation performance for architects to improve residential building arrangements design. As a preliminary analysis of the relevance of spatial patterns to the wind environment in the city, this study uses a rather limited quantity and typology of spatial patterns and calculated data, which indicates a preliminary trend of variation in the influence of spatial patterns. Therefore, the need for more concrete and accurate evidence calls for further studies in the future.

Supplementary Materials: Supplementary materials can be found at www.mdpi.com/2073-4433/8/6/102/s1, Figure S1: Position P1, P2 and P3 of vertical line and central street line located in Case A, Figure S2: The wind velocity profiles along a vertical line at (a) position P1, (b) position P2, (c) position P3, (d) Distribution of wind velocity along the central street line (z = 0.1 H), Figure S3: Building configurations and arrangements (H = 18 m, W = 9 m, L1 = 48 m, L2 = 24 m), Figure S4: (a) Computational domain, Figure S5: Gird resolution in the computational domain, Table S1: The calculation conditions utilized in CFD simulation, Figure S6: Comparison of the numerical and experimental results at (a) position P1, (b) position P2, (c) position P3.

Acknowledgments: This study was financially supported by the National Natural Science Foundation of China (Grant No. 51508262 and No. 51538005).

Author Contributions: Wei You, Zhi Gao and Wowo Ding conceived and designed the experiments; Wei You performed the experiments and acquired the data; Wei You, Zhi Gao and Zhi Chen analyzed the data; and Wei You wrote the paper.

Conflicts of Interest: The authors declare no conflict of interest.

References

1. Erell, E.; Pearlmutter, D.; Williamson, T. *Urban Microclimate: Designing the Spaces between Buildings*; Earthsan: Oxon, UK; New York, NY, USA, 2011.
2. Givoni, B. *Climate Considerations in Building and Urban Design*; John Wiley & Sons: New York, NY, USA, 1998.
3. Shi, X.; Zhu, Y.; Duan, J.; Shao, R.; Wang, J. Assessment of pedestrian wind environment in urban planning design. *Landsc. Urban Plan.* **2015**, *140*, 17–28. [CrossRef]
4. Mfula, A.M.; Kukadia, V.; Griffithsa, R.F.; Hall, D.J. Wind tunnel modelling of urban building exposure to outdoor pollution. *Atmos. Environ.* **2005**, *39*, 2737–2745. [CrossRef]

5. Chen, L.; Hang, J.; Sandberg, M.; Claesson, L.; Di Sabatino, S.; Wigo, H. The impacts of building height variations and building packing densities on flow adjustment and city breathability in idealized urban models. *Build. Environ.* **2017**, *118*, 344–361. [CrossRef]

6. Simöens, S.; Ayraulta, M.; Wallace, J.M. The flow across a street canyon of variable width—Part 1: Kinematic description. *Atmos. Environ.* **2007**, *41*, 9002–9017. [CrossRef]

7. Simöens, S.; Wallace, J.M. The flow across a street canyon of variable width—Part 2: Scalar dispersion from a street level line source. *Atmos. Environ.* **2008**, *42*, 2489–2503. [CrossRef]

8. Dong, J.; Tan, Z.; Xiao, Y.; Tu, J. Seasonal Changing Effect on Airflow and Pollutant Dispersion Characteristics in Urban Street Canyons. *Atmosphere* **2017**, *8*, 43. [CrossRef]

9. Tsang, C.W.; Kwok, K.C.; Hitchcock, P.A. Wind tunnel study of pedestrian level wind environment around tall buildings: Effects of building dimensions, separation and podium. *Build. Environ.* **2012**, *49*, 167–181. [CrossRef]

10. Hong, B.; Lin, B. Numerical studies of the outdoor wind environment and thermal comfort at pedestrian level in housing blocks with different building layout patterns and trees arrangement. *Renew. Energy* **2015**, *73*, 18–27. [CrossRef]

11. Yang, F.; Gao, Y.; Zhong, K.; Kang, Y. Impacts of cross-ventilation on the air quality in street canyons with different building arrangements. *Build. Environ.* **2016**, *104*, 1–12. [CrossRef]

12. Ng, W.Y.; Chau, C.K. A modeling investigation of the impact of street and building configurations on personal air pollutant exposure in isolated deep urban canyons. *Sci. Total Environ.* **2014**, *468*, 429–448. [CrossRef] [PubMed]

13. Bady, M.; Katob, K.; Huang, H. Towards the application of indoor ventilation efficiency indices to evaluate the air quality of urban areas. *Build. Environ.* **2008**, *43*, 1991–2004. [CrossRef]

14. Kato, K.; Huang, H. Ventilation efficiency of void space surrounded by buildings with wind blowing over built-up urban area. *J. Wind Eng. Ind. Aerodyn.* **2009**, *97*, 358–367. [CrossRef]

15. Buccolieri, R.; Sandberg, M.; Di Sabatino, D. City breathability and its link to pollutant concentration distribution within urban-like geometries. *Atmos. Environ.* **2010**, *44*, 1894–1903. [CrossRef]

16. Hang, J.; Sandberg, M.; Li, Y. Age of air and air exchange efficiency in idealized city models. *Build. Environ.* **2009**, *44*, 1714–1723. [CrossRef]

17. Hang, J.; Wang, Q.; Chen, X.; Sandberg, M.; Zhu, W.; Buccolieri, R.; Di Sabatino, S. City breathability in medium density urban-like geometries evaluated through the pollutant transport rate and the net escape velocity. *Build. Environ.* **2015**, *94*, 166–182. [CrossRef]

18. You, W.; Ding, W.W. Assessment of outdoor space's ventilation efficiency around residential building: Effects of building dimension, separation and orientation. Proceedings of 50th International Conference of the Architectural Science Association (ASA), The University of Adelaide, Adelaide, Australia, 7–9 December 2016; pp. 219–228.

19. Buccolieri, R.; Salizzoni, P.; Soulhac, L.; Garbero, V.; Di Sabatino, S. The breathability of compact cities. *Urban Clim.* **2015**, *13*, 73–93. [CrossRef]

20. Di Sabatino, S.; Buccolieri, R.; Pulvirenti, B.; Britter, R. Simulations of pollutant dispersion within idealized urban-type geometries with CFD and integral models. *Atmos. Environ.* **2007**, *41*, 8316–8329. [CrossRef]

21. Hang, J.; Li, Y. Age of air and air exchange efficiency in high-rise urban areas and its link to pollutant dilution. *Atmos. Environ.* **2011**, *45*, 5572–5585.

22. Hang, J.; Li, Y.; Sandberg, M.; Buccolieri, R.; di Sabatino, S. The influence of building height variability on pollutant dispersion and pedestrian ventilation in idealized high-rise urban areas. *Build. Environ.* **2012**, *56*, 346–360. [CrossRef]

23. Lin, M.; Hang, J.; Li, Y.; Luo, Z.; Sandberg, M. Quantitative ventilation assessments of idealized urban canopy layers with various urban layouts and the same building packing density. *Build. Environ.* **2014**, *79*, 152–167.

24. Ramponi, R.; Blocken, B.; Laura, B.; Janssen, W.D. CFD simulation of outdoor ventilation of generic urban configurations with different urban densities and equal and unequal street widths. *Build. Environ.* **2015**, *92*, 152–166. [CrossRef]

25. Razak, A.A.; Hagishima, A.; Ikegaya, N.; Tanimoto, J. Analysis of airflow over building arrays for assessment of urban wind environment. *Build. Environ.* **2013**, *59*, 56–65. [CrossRef]

26. Lee, R.X.; Jusuf, S.K.; Wong, N.H. The study of height variation on outdoor ventilation for Singapore's high-rise residential housing estates. *Int. J. Low-Carbon Technol.* **2015**, *10*, 15–33. [CrossRef]

27. Oke, T.R. Street Design and Urban Canopy Layer Climate. *Energy Build.* **1988**, *11*, 103–113. [CrossRef]

28. Sini, J.F.; Anquetin, S.; Mestayer, P.G. Pollutant dispersion and thermal effects in urban street canyons. *Atmos. Environ.* **1996**, *30*, 2659–2677. [CrossRef]

29. Chan, A.T.; So, E.S.; Samad, S.C. Strategic guidelines for street canyon geometry to achieve sustainable street air quality. *Atmos. Environ.* **2001**, *35*, 4089–4098. [CrossRef]

30. Chan, A.T.; Au, W.T.; So, E.S. Strategic guidelines for street canyon geometry to achieve sustainable street air quality—Part II: Multiple canopies and canyons. *Atmos. Environ.* **2003**, *37*, 2761–2772. [CrossRef]

31. Ying, X.; Zhu, W.; Hokao, K.; Ge, J. Numerical research of layout effect on wind environment around high-rise buildings. *Archit. Sci. Rev.* **2013**, *56*, 272–278. [CrossRef]

32. Iqbal, Q.M.Z.; Chan, A.L.S. Pedestrian level wind environment assessment around group of high-rise cross-shaped buildings: Effect of building shape, separation and orientation. *Build. Environ.* **2016**, *101*, 45–63. [CrossRef]

33. Blocken, B. 50 years of Computational Wind Engineering: Past, present and future. *J. Wind Eng. Ind. Aerodyn.* **2014**, *129*, 69–102. [CrossRef]

34. Stathopoulos, T. Computational wind engineering: Past achievements and future challenges. *J. Wind Eng. Ind. Aerodyn.* **1997**, *67*, 509–532. [CrossRef]

35. Blocken, B.; Stathopoulos, T.; Carmeliet, J.; Hensen, J.L. Application of computational fluid dynamics in building performance simulation for the outdoor environment: An overview. *J. Build. Perform. Simul.* **2011**, *4*, 157–184. [CrossRef]

36. Stathopoulos, T. Pedestrian level winds and outdoor human comfort. *J. Wind Eng. Ind. Aerodyn.* **2006**, *94*, 769–780. [CrossRef]

37. Moonen, P.; Defraeye, T.; Dorer, V.; Blocken, B.; Carmeliet, J. Urban Physics: Effect of the micro-climate on comfort, health and energy demand. *Front. Archit. Res.* **2012**, *1*, 197–228. [CrossRef]

38. Blocken, B.; Stathopoulos, T.; Van Beeck, J.P.A.J. Pedestrian-level wind conditions around buildings: Review of wind-tunnel and CFD techniques and their accuracy for wind comfort assessment. *Build. Environ.* **2016**, *100*, 50–81. [CrossRef]

39. Blocken, B. Computational Fluid Dynamics for urban physics: Importance, scales, possibilities, limitations and ten tips and tricks towards accurate and reliable simulations. *Build. Environ.* **2015**, *91*, 219–245. [CrossRef]

40. Tominaga, Y.; Mochida, A.; Yoshie, R.; Kataoka, H.; Nozue, T.; Yoshikawa, M.; Shirasawa, T. AIJ guidelines for practical applications of CFD to pedestrian wind environment around buildings. *J. Wind Eng. Ind. Aerodyn.* **2008**, *96*, 1749–1761. [CrossRef]

41. Franke, J.; Hellsten, A.; Schlünzen, K.H.; Carissimo, B. The COST 732 Best Practice Guideline for CFD simulation of flows in the urban environment: A summary. *Int. J. Environ. Pollut.* **2011**, *44*, 419–427. [CrossRef]

42. Yuan, C.; Ng, E. Building porosity for better urban ventilation in high-density cities—A computational parametric study. *Build. Environ.* **2012**, *50*, 176–189. [CrossRef]

43. Yuan, C.; Ng, E.; Norford, L.K. Improving air quality in high-density cities by understanding the relationship between air pollutant dispersion and urban morphologies. *Build. Environ.* **2014**, *71*, 245–258. [CrossRef]

44. Ng, E.; Yuan, C.; Chen, L.; Ren, C.; Fung, J.C. Improving the wind environment in high-density cities by understanding urban morphology and surface roughness: A study in Hong Kong. *Landsc. Urban Plan.* **2011**, *101*, 59–74. [CrossRef]

45. Wong, M.S.; Nichol, J.; Ng, E. A study of the "wall effect" caused by proliferation of high-rise buildings using GIS techniques. *Landsc. Urban Plan.* **2011**, *102*, 245–253. [CrossRef]

46. Yuan, C.; Ng, E. Practical application of CFD on environmentally sensitive architectural design at high density cities: A case study in Hong Kong. *Urban Clim.* **2014**, *8*, 57–77. [CrossRef]

47. Ng, E. Policies and technical guidelines for urban planning of high-density cities—Air ventilation assessment (AVA) of Hong Kong. *Build. Environ.* **2009**, *44*, 1478–1488. [CrossRef]

48. Kubota, T.; Miura, M.; Tominaga, Y.; Mochid, A. Wind tunnel tests on the relationship between building density and pedestrian-level wind velocity: Development of guidelines for realizing acceptable wind environment in residential neighborhoods. *Build. Environ.* **2008**, *43*, 1699–1708. [CrossRef]

49. Liu, Q.; Ding, W. Morphological study on units of Fabric that constitute contemporary residential plot in the Yangtze River Delta, China. In Proceedings of the 19th International Seminar on Urban Form (ISUF), Delft, The Netherlands, 16–19 October 2012; pp. 689–694.

50. Zhao, Q.; Ding, W. Relevance study: Relationship of morphological characteristics between residential plot and building pattern in Nanjing, China. In Proceedings of the 21th International Seminar on Urban Form (ISUF), Porto, Portugal, 3–6 July 2014.

51. Franke, J.; Hellsten, A.; Schlünzen, H.; Carissimo, B. *Best Practice Guideline for the CFD Simulation of Flows in the Urban Environment*; COST Office: Brussels, Belgium, 2007; ISBN 3-00-018312-4.

52. Tominag, Y.; Stathopoulos, T. Numerical simulation of dispersion around an isolated cubic building: Model evaluation of RANS and LES. *Build. Environ.* **2010**, *45*, 2231–2239. [CrossRef]

53. Blocken, B.; van der Hout, A.; Dekker, J.; Weiler, O. CFD simulation of wind flow over natural complex terrain: Case study with validation by field measurements for Ria de Ferrol, Galicia, Spain. *J. Wind Eng. Ind. Aerodyn.* **2015**, *147*, 905–928. [CrossRef]

54. Tominaga, Y.; Mochida, A.; Shirasawa, T.; Yoshie, R.; Kataoka, H.; Harimoto, K.; Nozu, T. Cross Comparisons of CFD Results of Wind Environment at Pedestrian Level around a high-rise Building and within a Building Complex. *J. Asian Archit. Build. Eng.* **2004**, *3*, 63–70. [CrossRef]

55. Zhang, A.; Gao, C.; Zhang, L. Numerical simulation of the wind field around different building arrangements. *J. Wind Eng. Ind. Aerodyn.* **2005**, *93*, 891–904. [CrossRef]

56. Lim, E.; Ito, K.; Sandberg, M. New ventilation index for evaluating imperfect mixing conditions—Analysis of Net Escape Velocity based on RANS approach. *Build. Environ.* **2013**, *61*, 45–56. [CrossRef]

atmosphere

MDPI

Article

New Surrogate Model for Wind Pressure Coefficients in a Schematic Urban Environment with a Regular Pattern

Tam Nguyen Van [1,2,*] and Frank De Troyer [1]

1   Department of Architecture, Faculty of Engineering Science, KU Leuven, Kasteelpark Arenberg 1-box 2431, 3001 Heverlee, Belgium; frank.detroyer@kuleuven.be
2   Civil Engineering Department, College of Technology, Cantho University, 3/2 Street, 900000 Cantho City, Vietnam
*   Correspondence: tam.nguyenvan@kuleuven.be or ngvtam.ae@gmail.com

Received: 30 January 2018; Accepted: 16 March 2018; Published: 19 March 2018

Abstract: Natural ventilation and the use of fans are recognized as sustainable design strategies to reduce energy use while reaching thermal comfort. A big challenge for designers is to predict ventilation rates of buildings in dense urban areas. One significant factor for calculating the ventilation rate is the wind pressure coefficient (Cp). Cp values can be obtained at a high cost, via real measurements, wind tunnel experiments, or high computational effort via computational fluid dynamic (CFD) simulation. A fast surrogate model to predict Cp for a schematic urban environment is required for the integration in building performance simulations. There are well-known surrogate models for Cp. The average surface pressure coefficient model integrated in EnergyPlus considers only a box-shaped building, without surrounding buildings. CpCalc, a surrogate model for Cp, considers only one height of neighbouring buildings. The Toegepast Natuurwetenschappelijk Onderzoek (TNO) Cp Generator model was available via web interface, and could include several box-shaped buildings in the surrounding area. These models are complex for fast integration in a natural ventilation simulation. For optimization processes, with thousands of simulation runs, speed is even more essential. Our study proposes a new surrogate model for Cp estimation based on data obtained from the TNO CP Generator model. The new model considers the effect of different neighbouring buildings in a simplified urban configuration, with an orthogonal street pattern, box-shaped buildings, and repetitive dimensions. The developed surrogate model is fast, and can easily be integrated in a dynamic energy simulation tool like EnergyPlus for optimization of natural ventilation in the urban areas.

Keywords: wind pressure coefficient; airflow network; multiple linear regression; natural ventilation; urban layout; surrogate model; schematic urban environment

---

1. Introduction

Building energy consumption is responsible for about 40% of the global energy demand [1–3]. Due to increasing comfort requirements, global warming, and the urban heat island effect, the demand for air conditioning by individual households has risen [4–6]. To reduce energy consumption, "passive architecture" or "nearly zero energy buildings", together with sustainable occupant behavior, have become essential.

Natural ventilation is crucial for passive design in warm and hot humid climates, and also important in other conditions [7]. Studies in the context of Malaysia and Singapore have proven that thermal comfort can be reached for a large period of the year only with ventilation [8]. Airflow through building openings is a critical factor that influences heat and moisture exchange between thermal zones and the outdoor environment [9–12]. The wind pressure is the driving force for airflow and should be predicted carefully [13]. In the next section, an airflow network model and existing surrogate models for wind pressure coefficient (Cp) values are reviewed, in order to develop an improved model. The aim is to have a surrogate model that can easily be integrated into dynamic energy simulations in schematic urban environments.

2. Literature Review

2.1. Multi-Zone Airflow Network Model

Airflow networks have been applied in almost all building performance simulations. The airflow network model, embedded in the standard version of building performance simulation as EnergyPlus, provides the ability to simulate airflow, which is driven by wind pressure, through multiple zones. The air flow rate highly influences the predicted energy requirement [14]. In order to obtain the air flow rates, the building performance simulation takes the wind's velocity and direction from the hourly weather data, and predicts the wind pressure coefficients at different external nodes (Figure 1).

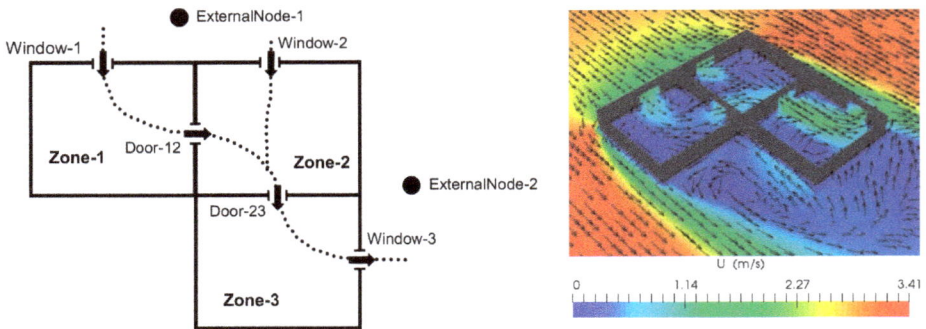

Figure 1. Plan view of a simple airflow network and computational fluid dynamic (CFD) results show possible airflow pattern.

2.2. Wind Pressure Coefficient Estimation Method

The wind pressure distribution on the envelope of a building is described by Cp, which is defined as the ratio of the dynamic pressure at a point on the surface over the dynamic pressure in the undisturbed flow pattern, measured at a reference height. Cp values, used to simulate multi-zone airflow network models for natural ventilation, can be obtained from many sources. The first source is full-scale measurements when an existing building is being studied. Precise pressures on a particular building in a specific environment can be measured. However, those pressures are only applicable to that specific building-layout and to that unique environment. Thus, they offer less relevant results for new designs. Moreover, these real scale measurements require a long measurement period and generate a high cost. Another source is wind tunnel tests, which can give more relevant results, because changes in building layout and urban layout are made easily. The limitations of wind tunnels include the requirement of special tools and large wind tunnels for investigating urban models. Finally, the computational fluid dynamic (CFD) approach has the same advantages as wind tunnel tests. CFD analysis is done on powerful computers, but is still time-consuming, requiring validation and user expertise.

The Cp Generator of TNO (Netherlands Organisation for Applied Scientific Research), a meta-model available via the web, was developed in the Netherlands [15]. This meta-model is based on finite element calculations, and has been verified through wind tunnel experiments [16] and measured data [17]. This approach offers a rather good correlation between measurements and meta-model predictions. The TNO Cp Generator approach was also used in another study, in order to obtain the Cps for a large urban fragment [18].

The CpCalc model was developed by analysing data from wind tunnel tests using a parametrical approach [19]. The CpCalc model considers only one height of neighbouring buildings. The model considers climate, environment, and building parameters. These independent parameters were varied to generate the data that were used to obtain polynomial functions via regression. Beside the climate parameters, the model allows variance in parameters, such as plan area density, relative building height, building layout, and relative position on façades. The model uses many regression functions. For each of them, the appropriate parameters have to be loaded from a data table. Using many data tables makes it complex to integrate the model into a whole-building energy simulation program.

Blocken and his group have compared the Cp values estimated through these different methods: CpCalc, Cp Generator, and Cp integrated in EnergyPlus [19]. Cp integrated in EnergyPlus is the average surface wind pressure coefficient based on the method developed by Swami and Chandra in 1988. Results of the three methods are similar [20].

To date, only a few studies consider surrogate models for the wind pressure coefficient of terraced buildings in urban areas, such as CpCalc, TNO Cp Generator, and free-standing average Cp [19,21]. A new parametric equation for the wind pressure coefficient for low-rise buildings fits the Tokyo database values with a goodness-of-fit value of $R^2 = 0.992$ [22]. Although existing surrogate models for Cp are good for analysing individual cases without obstructions, or considering average height of neighbouring box buildings, these models are not appropriate to be integrated into a dynamic simulation with parametric models or an optimisation process. Therefore, as a novelty in this research, a new surrogate model for Cp is developed. This model offers time saving and easy incorporation into the optimization processes, whereby a large number of simulations are required and both building geometry and urban layout can be changed.

3. Data Description and Methods

The method consists of five steps: (1) defining a schematic urban environment; (2) collecting Cp data from Cp Generator of TNO; (3) multiple linear regression; (4) CFD simulation; and (5) sensitivity analysis. Cp data for 36 orientations for each urban layout are collected to obtain regression functions for each orientation (in steps of 10°).

3.1. Schematic Urban Environment

The simplified urban layout model, as shown in Figure 2, can be considered as a representation for major urban areas with terraced housing around the world. This urban layout is described by the following key parameters: the height of each of the surrounding buildings (H1 to H14), road width in two directions (Rd, Rl), back garden (Bg), the building depth (D), and building length (L). Different urban layouts are generated based on the ranges and step size mentioned in Table 1. These ranges are defined based on case studies [23,24].

Table 1. Numerical variables and their design options for urban layout.

Design Parameters	Parameters	Range (m)	Step Size (m)
Height of surrounding buildings	H_1 to H_{14}	3 to 36	3
Width of road parallel with terraced row	R_w	12 to 24	4
Width of road perpendicular with terraced row	R_l	12 to 24	4
Back garden depth	B_g	8 to 24	4
House row depth	D	12 to 20	2
House row length	L	40 to 120	10

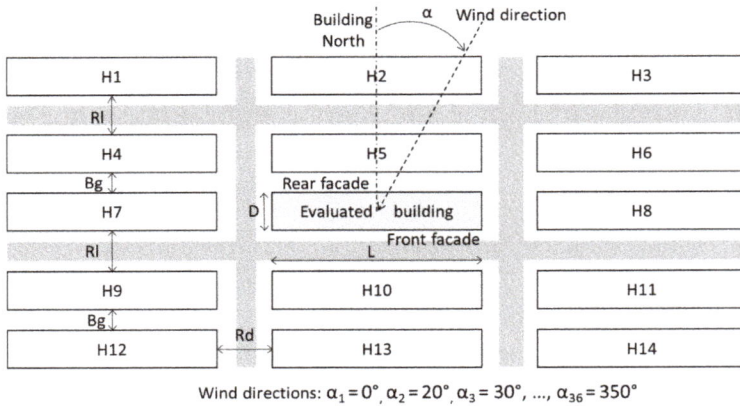

Figure 2. Schematic urban environment (for dimensions see Table 1).

3.2. Collecting Cp Data from the TNO Cp Generator

The "Latin Hypercube Sampling method" is applied to generate 200 combinations of 19 independent parameters (14 building heights and five urban parameters) of the building and urban layouts. Input data of those combinations for the Cp Generator consists of defining the terrain's roughness for the wind's different flow directions (kept constant as "urban") as well as "obstacles". In the TNO Cp Generator, each obstacle is referred to via a unique name. The following characteristics are associated: locations (x, y) of a corner point, orientation (β), and size of the buildings. Those data have to be transmitted to the programme as a text file with the correct formatting. In a next step, the text files are uploaded. The results of the Cp values are returned, and consist of a data table for each evaluated point on facades and roofs, and of graphical files, as can be seen in Figures 3 and 4. For our meta-model, seven points are considered: red points on the front facade, blue points on the rear façade, and the green point on the roof. Other positions on the facades will be considered in further research. Points on each facade are equally distributed over the height in the middle of the building. The point on the roof is in the middle.

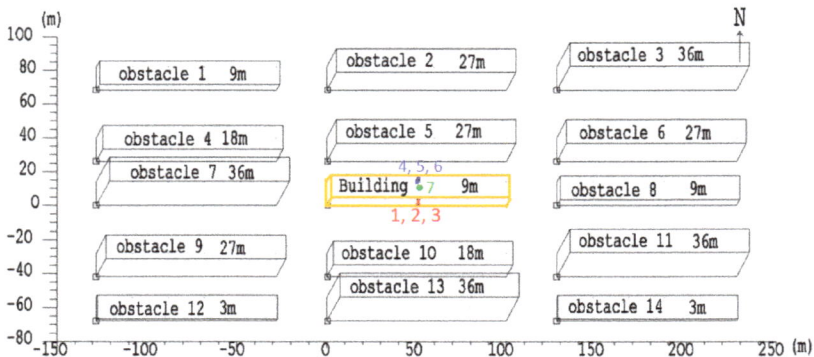

Figure 3. The geometry of the building and obstacles as submitted to the Toegepast Natuurwetenschappelijk Onderzoek (TNO) Cp Generator TNO. Results are obtained for points 1, 2, and 3 on the front façade; and points 4, 5, and 6 on the rear facade. with heights from the ground 1.5 m, 4.5 m, and 7.5 m, respectively. Point 7 is located in the middle on the roof.

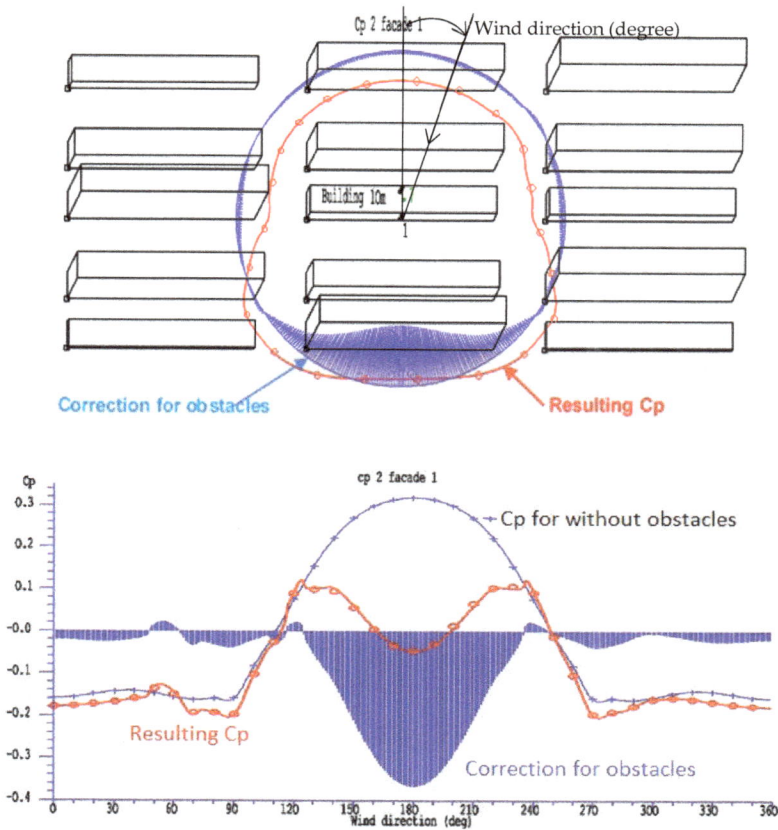

Figure 4. Wind pressure coefficient values of the middle point in the front façade of the terraced house, two buildings in front of the terraced building largely influence the wind pressure coefficient of direction 150° to 210°.

3.3. Surrogate Model: Multiple Linear Regression

The multiple linear regression approach is a well-known technique for modelling the relationship between two or more parameters. It was selected to develop the surrogate model of wind pressure coefficient. The general multiple linear regression model with parameter X is represented by the following equation.

$$Y = a_0 + a_1 X_1 + a_2 X_2 + \ldots + a_n X_n \tag{1}$$

where, Y is the predicted value; $a_0, a_1, \ldots a_n$ are constant coefficients of the regression function; $X_1, X_2, \ldots X_n$ are values of n parameters.

To develop the regression model for the wind pressure coefficients, 200 generated design cases were randomly separated into two sub-samples: the training samples (90% of the cases) and testing samples (10% of the cases). The training dataset was used for model development. The testing data was used to evaluate the model.

Initial trials showed a very low correlation between Cp values and urban parameters, including the orientations. However, Cp values had a very high correlation with the urban parameters when the regression function was developed for each orientation. Therefore, the surrogate model algorithms consisted of 36 functions for 36 orientations (in steps of 10°). The subdivision into 36 regression functions for 36 orientations was also based on the concept of the wind shadow model [25,26]. The surrogate model for the wind pressure coefficients is described by the following equations.

- Cps in middle of front facade:

$$Cp_{frontj} = a_{fr0j} + \sum(a_{frij} * H_i) + (a_{fr5j} * D) + (a_{fr6j} * L) + (a_{fr17j} * R_d) + (a_{fr18j} * R_L) + (a_{fr19j} * B_g)$$

- Cps in middle of rear facade:

$$Cp_{rearj} = a_{re0j} + \sum(a_{reij} * H_i) + (a_{re15j} * D) + (a_{re16j} * L) + (a_{re17j} * R_d) + (a_{re18j} * R_L) + (a_{re19j} * B_g)$$

- Cps in middle of the roof:

$$Cp_{roofj} = a_{ro0j} + \sum(a_{roij} * H_i) + (a_{ro5j} * D) + (a_{ro16j} * L) + (a_{ro17j} * R_d) + (a_{ro18j} * R_L) + (a_{ro19j} * B_g)$$

where j is the orientation, $j = 1, 2, \ldots, 36$ (wind orientations); i is the number of neighbouring buildings ($i = 1, 2, \ldots, 14$ buildings in Figure 2); Cp_{frontj}, Cp_{rearj} and Cp_{froofj} are wind pressure coefficients of front, rear and roof surfaces, respectively; H_i, L, D, Rw, RL and Bg were defined in the Table 1.

3.4. Computational Fluid Dynamic Simulation

In this section, the CFD is used to provide qualitative and quantitative results to understand the effect of the different wind directions and the urban layout on Cp values. The CFD simulations use OpenFoam with Reynolds-Averaged Navier-Stokes (RANS) equations and a simple algorithm. The computational domain is $(600 \times 600 \times 110)$ m³, full scale. All buildings are 9 m high and 24 m for Rl, Rd, and Bg. The depth and length of the terraced building are 24 m and 120 m, respectively. The basic square grids size is 5 m × 5 m × 5 m, but the mesh (grid) is resized near the buildings, in order to increase the quality of the calculation, Figure 5. The dimensions of the computational domain were chosen based on best practice guidelines by Franke et al. [27] and Tominaga et al. [28]. The inlet flow with fixed velocity and the outlet with free pressure were applied for this simulation.

Figure 5. The computational domain $(600 \times 600 \times 110)$ m³ full scale with the grid is resized near the buildings.

3.5. Sensitivity Analysis

In this study, the standardized regression coefficient (SRC) was used to identify the impact of each urban parameter on Cp values. SRC_i is calculated by dividing the standard deviation of each parameter (σ_{xi}) by the standard deviation of the Cp (σ_{Cp}), and multiplying the regression coefficient (a$_i$) of each parameter in the multiple linear regression model. The building and urban parameters, together with Cp results of 200 urban layouts, were used to obtain SRC. The parameters with a high absolute value of SRC have an important impact on the Cp values.

$$SRC_i = a_i \times \frac{\sigma_{xi}}{\sigma_{Cp}}$$

4. Results and Discussion

4.1. Cp Surrogate Model Based on the TNO Cp Generator

The model was developed by using 90% case data and 10% data for testing. The predicted Cp results of 20 cases from the testing data were good predictions compared with the TNO Cp Generator values, with R^2 = 0.93 as shown in Figure 6. The testing process is repeated with random data sets, and similar R^2 results were always obtained. The few cases with larger errors when predicting Cp are the cases whereby many neighbouring buildings reach the maximum height (36 m).

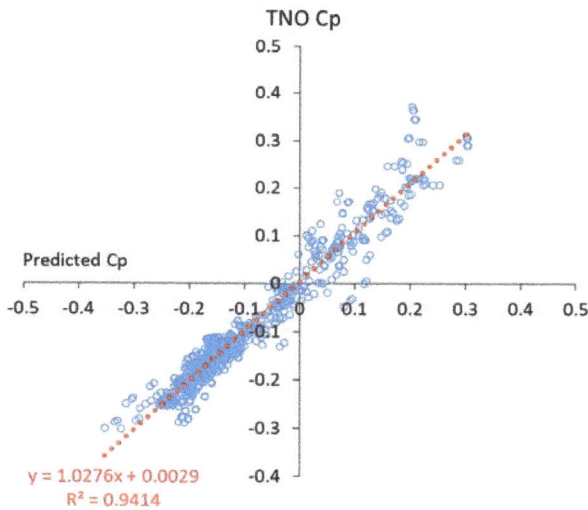

Figure 6. Predicted Cp and Cp obtained by the TNO Cp Generator of 20 urban forms of the testing data. A total of 720 points (36 orientations × 20 cases) are represented in the Scatter chart.

The factors of the multiple linear regression are reported in the Appendix 5. The regression functions show a high correlation (Figure 7). For each urban layout, the R^2 value for 36 Cp values is calculated, in order to obtain the correlation between Cp values obtained from the TNO Cp Generator and estimated Cp values. The R^2 values of 200 urban layouts are shown in Figure 8.

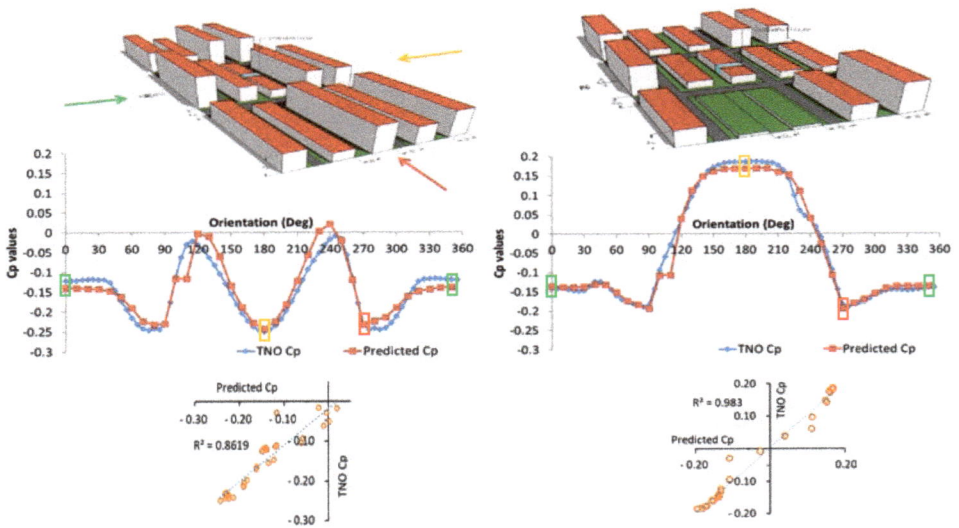

Figure 7. Cp value estimation for terraced house patterns by using regression approach, based on Cp results of 200 scenarios from the TNO Cp Generator web base. 3D models at the left and right show two examples of randomly-generated urban layouts. The analysed building is always in the centre.

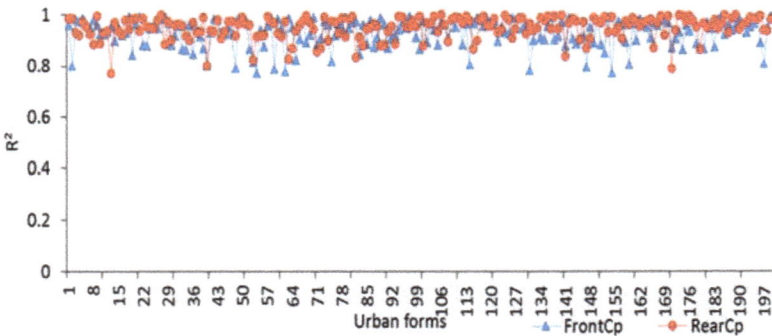

Figure 8. Goodness-of-fit value (R^2) values of the linear link between the predicted Cp values and Cp values of the TNO Cp Generator of 200 urban patterns; average R^2 of the front façade = 0.93 and average R^2 of the rear façade = 0.95.

4.2. Computational Fluid Dynamic Results and Sensitivity Analysis of Cp Values for the Schematic Urban Layout

The neighbouring buildings modify the external wind flow around the evaluated building. The velocity and pressure of the wind is related to the urban parameters and different angles of attack. The CFD result visualized in Figure 9 shows the effect of the urban layout on the wind pressure and wind direction.

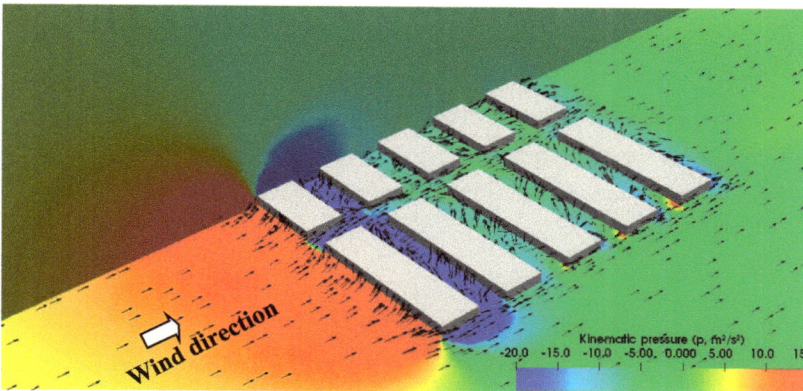

Figure 9. The external air pressure of the case when the wind angle is 0° (perpendicular the front façade).

The detailed graphical and numerical results are provided in Figures 10–13. The Cp values of the middle point on the front façade of 200 cases were used to calculate SRC (the right part of the Figures 10–13). The results shows the air pressure in a horizontal plane at half the building height above the ground, and in the Section A-A (Figure 10). On the right, the SRC of Cp results from the TNO model indicate the impact of each parameter on the Cp values for a certain direction.

- If the wind angle is 0° (Figure 10), the height of building in font (H5) has the strongest effect on the Cp values. The width (D), length of the building (L) and the back garden (Bg) also have a strong impact on the Cp. They impact the turbulence of the wind flow. Buildings H2 and H5 block the wind flow and generate negative Cp on the front façade of the evaluated building.
- If the wind angle is 30° (Figure 11), building H5 has also a high impact on the Cp value.
- If the wind angle is 60° (Figure 12), building H10 has a very high effect on the Cp value, because the back garden varies from 8 m to 24 m.
- When the wind flows parallel to the façade (Figure 13), building H10 and the width of road parallel the terraced row strongly impact on the Cp value.

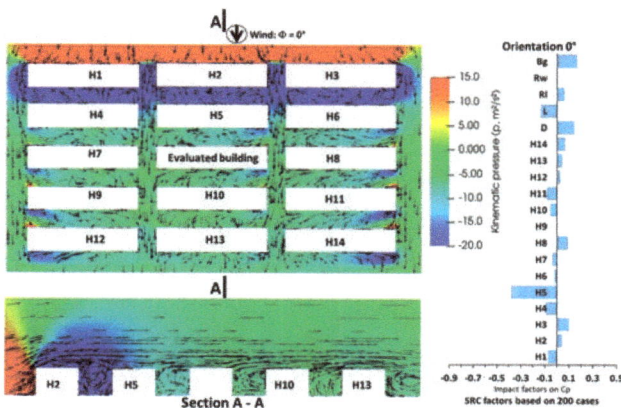

Figure 10. (**Left**) visualization of pressure (colour) and wind direction (arrows); (**Right**) Standardized regression coefficient (SRC) for orientation at 0° for the 19 variables, based on input data and Cp values of 200 cases.

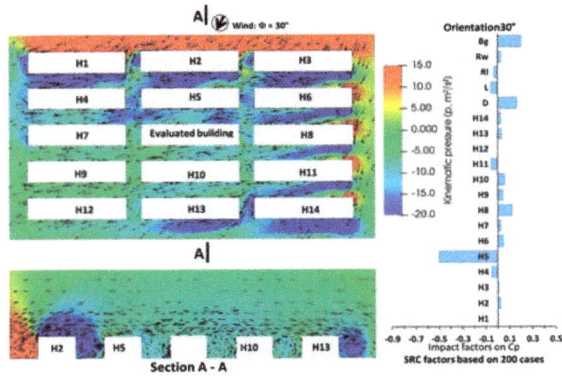

Figure 11. (**Left**) visualization of pressure (colour) and wind direction (arrow); (**Right**) SRC for orientation at 30° for the 19 variables, based on input data and Cp values of 200 cases.

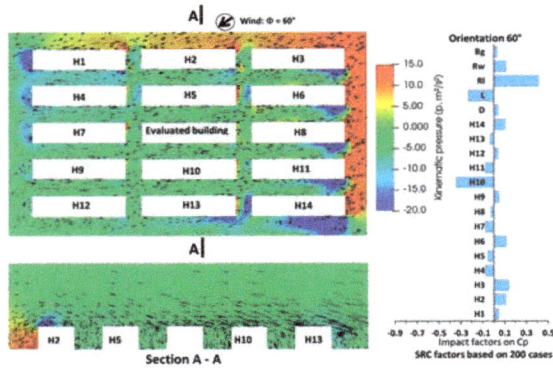

Figure 12. (**Left**) visualization of pressure (colour) and wind direction (arrows); (**Right**) SRC for orientation at 60° for the 19 variables, based on input data and Cp values of 200 cases.

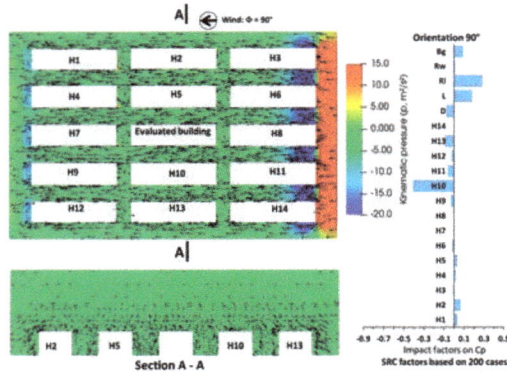

Figure 13. (**Left**) visualization of pressure (colour) and wind direction (arrows); (**Right**) SRC for orientation at 90° for the 19 variables, based on input data and Cp values of 200 cases.

The four cases illustrate that the dimensions of the neighbouring buildings and the width of the roads will generate various Cp values. This physical effect depends on the interplay of several parameters, such as wind direction, building layout, and obstructions. Figures 10–13 support the understanding for different wind directions, wind pressures, and the wind speed at different locations in 3D.

The Figure 14 shows that the height of the buildings in front (H10) and behind (H5) the evaluated building impact strongly the Cp of the front façade. The width of the road parallels the evaluated building (Rl) also has a strong effect on Cp values.

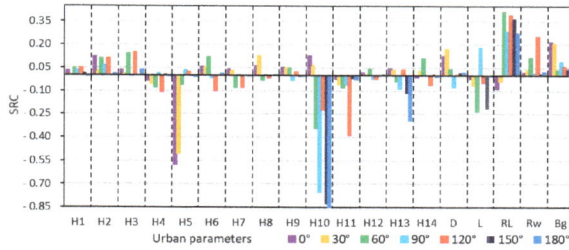

Figure 14. SRC results for different orientations: $0°$, $30°$, $60°$, $90°$, $120°$, $150°$, and $180°$ for the Cp values on front facade.

5. Conclusions

A computationally intensive step to calculate Cp values of various urban patterns and building geometries was solved by developing a surrogate model. This multiple linear regression model, with different functions for different orientations, is based on other surrogate models that were developed based on measurement in situ, wind tunnel measurement, and CFD simulations.

The developed meta-model of wind pressure coefficients allows for the inclusion of the effect of the schematic urban environment. The regression functions of this meta-model show a high correlation with the values of the existing surrogate model. This meta-model is able to predict the Cp values on different surfaces very quickly.

This surrogate model is very significant in a case where a huge number of simulations is required, like in an optimization process. However, the new Cp model considers the middle terraced unit of the terraced row. Estimating Cp of a house located closer to the edges of the terraced row requires further research. A more robust model will have to consider the Cp along the evaluated building, instead of only at the middle points of the building.

Author Contributions: This research was supported by KU Leuven University, Cantho University and the Ministry of Education and Training of Vietnam.

Appendix

Cp of the front façade:

$$Cp_{front} = a_{fr0} + a_{fr1} * H_1 + \ldots a_{fr14} * H_{14} + a_{fr15} * D + a_{fr16} * L + a_{fr17} * Rd + a_{fr18} * Rl + a_{fr19} * Bg$$

Orientation	afr0	afr1	afr2	afr3	afr4	afr5	afr6	afr7	afr8	afr9
0	-2.06×10^{-1}	1.24×10^{-4}	5.15×10^{-4}	1.44×10^{-4}	-1.49×10^{-4}	-2.39×10^{-3}	2.38×10^{-4}	1.77×10^{-4}	2.63×10^{-4}	2.22×10^{-4}
10	-2.02×10^{-1}	8.97×10^{-5}	4.03×10^{-4}	1.28×10^{-4}	-1.44×10^{-4}	-2.21×10^{-3}	2.35×10^{-4}	1.44×10^{-4}	2.93×10^{-4}	2.24×10^{-4}
20	-1.91×10^{-1}	7.09×10^{-5}	2.21×10^{-4}	7.09×10^{-5}	-1.61×10^{-4}	-1.76×10^{-3}	2.04×10^{-4}	9.45×10^{-5}	3.13×10^{-4}	1.92×10^{-4}
30	-1.76×10^{-1}	1.40×10^{-5}	6.68×10^{-5}	2.56×10^{-5}	-1.38×10^{-4}	-1.19×10^{-3}	1.28×10^{-4}	7.31×10^{-5}	2.97×10^{-4}	1.13×10^{-4}
40	-1.55×10^{-1}	-1.18×10^{-4}	6.73×10^{-4}	1.61×10^{-4}	-1.52×10^{-4}	-6.12×10^{-4}	-3.28×10^{-5}	-6.38×10^{-5}	1.45×10^{-4}	-4.14×10^{-7}
50	-1.69×10^{-1}	-4.51×10^{-5}	2.00×10^{-4}	3.70×10^{-4}	-4.57×10^{-5}	-1.81×10^{-4}	4.96×10^{-5}	-1.08×10^{-5}	-4.07×10^{-5}	6.48×10^{-5}
60	-2.10×10^{-1}	1.02×10^{-4}	2.41×10^{-4}	2.96×10^{-4}	-1.70×10^{-4}	-1.25×10^{-4}	2.50×10^{-4}	-1.69×10^{-4}	-6.61×10^{-4}	1.04×10^{-4}
70	-2.13×10^{-1}	1.21×10^{-4}	2.07×10^{-4}	1.08×10^{-4}	-1.77×10^{-4}	1.06×10^{-4}	1.50×10^{-4}	-1.84×10^{-4}	-8.77×10^{-5}	1.10×10^{-4}
80	-2.21×10^{-1}	1.17×10^{-4}	2.08×10^{-4}	4.54×10^{-5}	-4.06×10^{-5}	1.26×10^{-4}	3.68×10^{-5}	-3.14×10^{-5}	7.54×10^{-4}	8.87×10^{-4}
90	-2.12×10^{-1}	8.58×10^{-5}	1.73×10^{-4}	-1.55×10^{-5}	4.54×10^{-5}	8.34×10^{-5}	-4.14×10^{-5}	1.47×10^{-5}	5.76×10^{-6}	-7.38×10^{-5}
100	-1.28×10^{-1}	8.56×10^{-5}	9.53×10^{-5}	-6.54×10^{-5}	4.58×10^{-5}	8.97×10^{-5}	-5.95×10^{-5}	-1.58×10^{-5}	-4.42×10^{-5}	-1.34×10^{-4}
110	-7.11×10^{-2}	-5.57×10^{-5}	1.24×10^{-4}	1.79×10^{-4}	7.22×10^{-5}	1.06×10^{-4}	-1.45×10^{-5}	-6.48×10^{-5}	6.38×10^{-4}	-1.67×10^{-4}
120	-4.51×10^{-2}	2.13×10^{-4}	4.70×10^{-4}	6.22×10^{-4}	-4.50×10^{-4}	9.95×10^{-5}	-4.27×10^{-4}	-3.36×10^{-4}	-7.09×10^{-5}	1.05×10^{-4}
130	2.37×10^{-2}	1.20×10^{-4}	7.62×10^{-4}	5.18×10^{-4}	-3.38×10^{-4}	2.35×10^{-4}	-3.54×10^{-4}	-3.91×10^{-4}	-2.53×10^{-4}	2.37×10^{-4}
140	8.53×10^{-2}	4.50×10^{-5}	3.32×10^{-4}	3.49×10^{-4}	-9.37×10^{-5}	2.10×10^{-4}	-3.65×10^{-5}	-1.78×10^{-5}	1.18×10^{-4}	5.14×10^{-4}
150	1.19×10^{-1}	2.01×10^{-4}	3.69×10^{-5}	6.98×10^{-5}	1.43×10^{-4}	-8.99×10^{-5}	-4.03×10^{-5}	7.69×10^{-5}	7.24×10^{-5}	-1.34×10^{-4}
160	1.00×10^{-1}	1.09×10^{-4}	1.27×10^{-4}	3.10×10^{-4}	2.60×10^{-5}	-1.74×10^{-4}	1.86×10^{-4}	2.21×10^{-6}	5.42×10^{-5}	-1.36×10^{-4}
170	7.39×10^{-2}	-1.47×10^{-5}	1.78×10^{-4}	5.73×10^{-4}	-4.85×10^{-5}	-1.79×10^{-4}	2.37×10^{-4}	-8.12×10^{-5}	6.17×10^{-5}	-1.08×10^{-4}
180	6.49×10^{-2}	-7.62×10^{-5}	2.08×10^{-4}	6.76×10^{-4}	-7.07×10^{-5}	-1.58×10^{-4}	2.13×10^{-4}	-1.22×10^{-4}	3.59×10^{-5}	-1.16×10^{-4}
190	7.71×10^{-2}	-3.41×10^{-6}	1.68×10^{-4}	5.52×10^{-4}	-4.03×10^{-5}	-1.79×10^{-4}	2.39×10^{-4}	-7.62×10^{-5}	5.99×10^{-5}	-1.13×10^{-4}
200	1.06×10^{-1}	1.20×10^{-4}	1.12×10^{-4}	2.76×10^{-4}	2.63×10^{-5}	-1.62×10^{-4}	1.60×10^{-4}	1.86×10^{-5}	4.76×10^{-5}	-1.59×10^{-4}
210	1.21×10^{-1}	1.70×10^{-4}	5.39×10^{-5}	1.22×10^{-4}	2.62×10^{-4}	-1.21×10^{-4}	-2.45×10^{-5}	4.67×10^{-6}	1.48×10^{-4}	1.88×10^{-5}
220	8.66×10^{-2}	1.50×10^{-4}	4.25×10^{-4}	4.52×10^{-4}	-2.17×10^{-4}	2.04×10^{-4}	-2.91×10^{-4}	-1.43×10^{-4}	-7.18×10^{-5}	7.89×10^{-5}
230	-7.08×10^{-2}	-2.78×10^{-4}	5.10×10^{-4}	5.57×10^{-4}	-4.78×10^{-4}	1.74×10^{-4}	-1.78×10^{-4}	8.19×10^{-5}	-1.21×10^{-4}	-3.46×10^{-4}
240	-4.80×10^{-2}	-9.34×10^{-5}	1.59×10^{-4}	4.88×10^{-4}	-3.71×10^{-4}	-6.83×10^{-5}	-2.38×10^{-4}	1.62×10^{-4}	9.48×10^{-5}	-1.23×10^{-3}
250	-5.11×10^{-2}	1.06×10^{-4}	1.85×10^{-4}	1.35×10^{-4}	2.34×10^{-5}	-7.86×10^{-5}	-8.19×10^{-5}	-6.09×10^{-6}	-7.71×10^{-5}	-9.20×10^{-4}
260	-1.28×10^{-1}	8.60×10^{-5}	9.73×10^{-5}	-7.15×10^{-5}	3.68×10^{-5}	8.91×10^{-5}	-5.80×10^{-5}	-9.11×10^{-6}	-3.51×10^{-5}	-1.71×10^{-4}
270	-2.14×10^{-1}	9.09×10^{-5}	1.77×10^{-4}	-3.43×10^{-6}	4.20×10^{-5}	8.94×10^{-5}	-2.91×10^{-4}	7.76×10^{-6}	9.83×10^{-5}	-5.83×10^{-5}
280	-2.10×10^{-1}	-1.34×10^{-5}	1.97×10^{-4}	6.50×10^{-5}	-1.04×10^{-4}	-4.42×10^{-6}	-7.83×10^{-6}	8.26×10^{-5}	5.75×10^{-5}	5.32×10^{-5}
290	-2.23×10^{-1}	2.12×10^{-5}	2.79×10^{-4}	1.28×10^{-4}	-7.74×10^{-5}	-2.89×10^{-5}	8.59×10^{-5}	-1.44×10^{-4}	2.06×10^{-5}	4.53×10^{-5}
300	-1.92×10^{-1}	1.05×10^{-4}	1.63×10^{-4}	1.77×10^{-4}	3.90×10^{-4}	-7.21×10^{-5}	-1.90×10^{-4}	-9.17×10^{-5}	-8.87×10^{-5}	-8.35×10^{-5}
310	-1.60×10^{-1}	2.05×10^{-4}	1.21×10^{-5}	1.33×10^{-4}	-1.55×10^{-4}	-2.40×10^{-4}	-1.85×10^{-4}	-1.28×10^{-5}	4.54×10^{-5}	-7.32×10^{-5}
320	-1.65×10^{-1}	1.58×10^{-4}	-1.54×10^{-5}	7.50×10^{-6}	-1.61×10^{-4}	-6.58×10^{-4}	5.96×10^{-5}	8.94×10^{-5}	1.41×10^{-4}	4.68×10^{-5}
330	-1.76×10^{-1}	1.13×10^{-4}	5.76×10^{-5}	2.41×10^{-6}	-1.34×10^{-4}	-1.27×10^{-3}	1.60×10^{-4}	6.64×10^{-5}	2.52×10^{-4}	1.47×10^{-4}
340	-1.92×10^{-1}	7.08×10^{-5}	2.37×10^{-4}	7.94×10^{-5}	-1.64×10^{-4}	-1.82×10^{-3}	2.07×10^{-4}	1.04×10^{-4}	3.11×10^{-4}	1.97×10^{-4}
350	-2.02×10^{-1}	9.44×10^{-5}	4.21×10^{-4}	1.33×10^{-4}	-1.43×10^{-4}	-2.24×10^{-3}	2.36×10^{-4}	1.47×10^{-4}	2.88×10^{-4}	2.22×10^{-4}

Orientation	afr10	afr11	afr12	afr13	afr14	afr15	afr16	afr17	afr18	afr19
0	5.45×10^{-4}	-9.71×10^{-5}	8.15×10^{-5}	1.86×10^{-4}	-4.95×10^{-5}	2.18×10^{-3}	-3.28×10^{-5}	-7.51×10^{-4}	1.68×10^{-4}	1.89×10^{-3}
10	4.87×10^{-4}	-1.18×10^{-4}	6.62×10^{-5}	1.83×10^{-4}	-1.96×10^{-5}	2.04×10^{-3}	-2.44×10^{-5}	-6.67×10^{-4}	2.12×10^{-4}	1.76×10^{-3}
20	3.56×10^{-4}	-1.47×10^{-4}	4.70×10^{-5}	1.27×10^{-4}	2.71×10^{-4}	1.86×10^{-3}	-3.34×10^{-5}	-4.82×10^{-4}	2.19×10^{-4}	1.48×10^{-3}
30	1.52×10^{-4}	-1.46×10^{-4}	9.69×10^{-6}	8.48×10^{-5}	7.16×10^{-5}	1.67×10^{-3}	-6.23×10^{-5}	-1.92×10^{-4}	1.64×10^{-4}	1.03×10^{-3}
40	-8.81×10^{-5}	-1.36×10^{-4}	4.41×10^{-5}	6.93×10^{-5}	1.06×10^{-4}	9.64×10^{-4}	-9.34×10^{-5}	2.09×10^{-4}	-1.17×10^{-5}	5.60×10^{-4}
50	-3.72×10^{-4}	-5.14×10^{-5}	1.01×10^{-4}	-1.26×10^{-4}	1.91×10^{-4}	4.00×10^{-4}	-1.72×10^{-4}	7.35×10^{-4}	4.88×10^{-4}	4.40×10^{-4}
60	-7.25×10^{-4}	-1.70×10^{-4}	8.78×10^{-5}	-8.29×10^{-5}	2.36×10^{-4}	3.90×10^{-4}	-2.08×10^{-4}	1.85×10^{-3}	5.16×10^{-4}	1.57×10^{-4}
70	-1.28×10^{-3}	-3.21×10^{-4}	8.05×10^{-5}	-1.05×10^{-4}	2.40×10^{-4}	2.64×10^{-4}	-2.65×10^{-4}	2.15×10^{-3}	4.82×10^{-4}	1.46×10^{-5}
80	-1.76×10^{-3}	-2.72×10^{-4}	9.60×10^{-5}	-1.86×10^{-4}	2.20×10^{-4}	-9.46×10^{-5}	-1.68×10^{-4}	2.11×10^{-3}	4.93×10^{-4}	3.60×10^{-4}
90	-1.88×10^{-3}	-1.48×10^{-4}	-5.71×10^{-5}	-2.20×10^{-4}	-4.75×10^{-6}	-7.89×10^{-4}	1.91×10^{-4}	1.49×10^{-3}	7.32×10^{-5}	4.79×10^{-4}
100	-1.00×10^{-3}	-1.20×10^{-4}	-1.03×10^{-4}	-1.29×10^{-4}	-1.71×10^{-4}	-4.98×10^{-4}	4.38×10^{-4}	5.10×10^{-4}	4.65×10^{-5}	4.16×10^{-4}
110	-2.42×10^{-4}	-8.02×10^{-4}	-1.70×10^{-4}	7.35×10^{-5}	4.05×10^{-5}	-4.58×10^{-4}	3.71×10^{-4}	1.03×10^{-3}	6.43×10^{-4}	4.36×10^{-4}
120	-9.35×10^{-3}	-1.63×10^{-3}	-1.09×10^{-4}	1.61×10^{-4}	-2.51×10^{-4}	1.27×10^{-4}	-7.91×10^{-5}	3.45×10^{-3}	2.22×10^{-3}	5.17×10^{-4}
130	-3.86×10^{-3}	-1.01×10^{-3}	1.94×10^{-4}	-3.51×10^{-4}	-3.38×10^{-4}	7.29×10^{-5}	-6.92×10^{-4}	6.02×10^{-3}	2.46×10^{-3}	3.26×10^{-4}
140	-7.50×10^{-3}	-5.51×10^{-4}	1.05×10^{-4}	-8.56×10^{-4}	-1.27×10^{-4}	6.09×10^{-4}	-1.10×10^{-3}	8.62×10^{-3}	1.35×10^{-4}	7.03×10^{-4}
150	-1.10×10^{-2}	-2.49×10^{-4}	-6.36×10^{-5}	-1.53×10^{-3}	1.33×10^{-4}	1.21×10^{-3}	-1.21×10^{-3}	1.04×10^{-2}	3.61×10^{-4}	1.18×10^{-3}
160	-1.33×10^{-2}	-2.99×10^{-4}	-1.02×10^{-4}	-3.01×10^{-3}	6.53×10^{-5}	1.48×10^{-3}	-7.90×10^{-4}	1.05×10^{-2}	7.12×10^{-4}	1.48×10^{-3}
170	-1.45×10^{-2}	-3.84×10^{-4}	-9.71×10^{-6}	-4.52×10^{-3}	-9.06×10^{-5}	1.59×10^{-3}	-2.57×10^{-4}	1.03×10^{-2}	7.81×10^{-4}	1.92×10^{-3}
180	-1.49×10^{-2}	-4.52×10^{-4}	3.69×10^{-5}	-5.10×10^{-3}	-1.41×10^{-4}	1.50×10^{-3}	-3.86×10^{-5}	1.01×10^{-2}	8.23×10^{-4}	2.15×10^{-3}
190	-1.44×10^{-2}	-3.74×10^{-4}	-1.96×10^{-5}	-4.39×10^{-3}	-7.39×10^{-5}	1.60×10^{-3}	-3.05×10^{-4}	1.03×10^{-2}	7.76×10^{-4}	1.87×10^{-3}
200	-1.31×10^{-2}	-2.79×10^{-4}	-8.70×10^{-5}	-2.85×10^{-3}	6.57×10^{-5}	1.43×10^{-3}	-8.50×10^{-4}	1.06×10^{-2}	6.67×10^{-4}	1.48×10^{-3}
210	-1.06×10^{-2}	-3.49×10^{-4}	-1.64×10^{-4}	-1.40×10^{-3}	1.59×10^{-4}	9.98×10^{-4}	-1.19×10^{-3}	1.01×10^{-2}	5.66×10^{-4}	7.62×10^{-4}
220	-7.09×10^{-3}	-6.04×10^{-4}	-3.34×10^{-4}	-7.34×10^{-4}	3.09×10^{-4}	8.16×10^{-4}	-1.15×10^{-3}	8.36×10^{-3}	1.39×10^{-3}	6.99×10^{-4}
230	-3.58×10^{-3}	-2.01×10^{-4}	-4.77×10^{-4}	-2.44×10^{-4}	6.16×10^{-4}	1.22×10^{-3}	-8.12×10^{-4}	6.12×10^{-3}	2.65×10^{-3}	6.65×10^{-4}
240	-9.85×10^{-4}	-1.35×10^{-4}	-5.01×10^{-4}	9.42×10^{-5}	3.43×10^{-4}	3.21×10^{-4}	-1.92×10^{-5}	3.41×10^{-3}	2.14×10^{-3}	3.39×10^{-4}
250	-3.06×10^{-3}	-6.50×10^{-5}	-1.23×10^{-4}	2.32×10^{-4}	-1.44×10^{-4}	-8.08×10^{-4}	4.28×10^{-4}	8.16×10^{-4}	3.35×10^{-5}	1.76×10^{-4}
260	-1.10×10^{-3}	-7.73×10^{-5}	-9.45×10^{-5}	-1.40×10^{-4}	-1.63×10^{-4}	-4.76×10^{-4}	4.16×10^{-4}	6.09×10^{-4}	4.26×10^{-7}	4.35×10^{-4}
270	-1.90×10^{-3}	-1.71×10^{-4}	-5.55×10^{-5}	-2.15×10^{-4}	1.90×10^{-4}	-7.55×10^{-4}	1.56×10^{-4}	1.58×10^{-3}	1.02×10^{-4}	4.79×10^{-4}
280	-1.61×10^{-3}	-1.31×10^{-4}	9.98×10^{-4}	-2.56×10^{-4}	2.59×10^{-4}	-1.91×10^{-4}	-2.00×10^{-4}	1.71×10^{-3}	7.39×10^{-4}	9.80×10^{-5}
290	-1.17×10^{-3}	-2.11×10^{-4}	2.32×10^{-4}	-8.06×10^{-5}	2.69×10^{-4}	4.05×10^{-4}	-2.25×10^{-4}	2.05×10^{-3}	6.54×10^{-4}	-2.39×10^{-4}
300	-6.26×10^{-4}	-2.20×10^{-4}	2.47×10^{-4}	5.56×10^{-5}	2.21×10^{-4}	-2.52×10^{-4}	-2.34×10^{-4}	1.98×10^{-3}	4.96×10^{-4}	-3.18×10^{-5}
310	-1.43×10^{-4}	-9.77×10^{-5}	1.18×10^{-5}	2.18×10^{-4}	1.86×10^{-4}	9.03×10^{-4}	-1.52×10^{-4}	8.17×10^{-4}	-3.07×10^{-5}	4.10×10^{-4}
320	-8.61×10^{-6}	-1.02×10^{-4}	4.77×10^{-5}	5.21×10^{-5}	1.62×10^{-4}	1.08×10^{-3}	-9.62×10^{-5}	2.71×10^{-4}	-9.05×10^{-5}	6.31×10^{-4}
330	1.76×10^{-4}	-1.42×10^{-4}	2.31×10^{-4}	7.08×10^{-5}	4.66×10^{-5}	1.68×10^{-3}	-7.15×10^{-5}	-1.43×10^{-4}	5.53×10^{-5}	1.14×10^{-3}
340	3.70×10^{-4}	-1.45×10^{-4}	4.99×10^{-5}	1.37×10^{-4}	2.52×10^{-5}	1.86×10^{-3}	-3.10×10^{-5}	-4.94×10^{-4}	2.24×10^{-4}	1.52×10^{-3}
350	4.98×10^{-4}	-1.18×10^{-4}	7.01×10^{-5}	1.82×10^{-4}	-2.31×10^{-5}	2.06×10^{-3}	-2.58×10^{-5}	-6.87×10^{-4}	2.13×10^{-4}	1.77×10^{-3}

Cp of the rear façade:

$$Cp_{rear} = a_{re0} + a_{re1} * H_1 + \ldots a_{re14} * H_{14} + a_{re15} * D + a_{re16} * L + a_{re17} * Rd + a_{re18} * Rl + a_{re19} * Bg$$

Orientation	are0	are1	are2	are3	are4	are5	are6	are7	are8	are9
0	6.45×10^{-2}	-9.34×10^{-4}	-4.09×10^{-3}	-9.79×10^{-4}	-2.19×10^{-4}	-1.55×10^{-2}	2.42×10^{-4}	1.69×10^{-4}	-3.09×10^{-4}	-6.44×10^{-4}
10	7.46×10^{-2}	-9.25×10^{-4}	-3.40×10^{-3}	-9.47×10^{-4}	-1.92×10^{-4}	-1.51×10^{-2}	3.06×10^{-4}	2.17×10^{-4}	-3.25×10^{-4}	-5.90×10^{-4}
20	9.70×10^{-2}	-8.97×10^{-4}	-2.03×10^{-3}	-9.20×10^{-4}	-2.00×10^{-5}	-1.41×10^{-2}	3.57×10^{-4}	3.06×10^{-4}	-3.62×10^{-4}	-5.36×10^{-4}
30	1.10×10^{-1}	-8.43×10^{-4}	-8.58×10^{-4}	-1.03×10^{-3}	1.33×10^{-4}	-1.20×10^{-2}	3.39×10^{-4}	2.48×10^{-4}	-4.23×10^{-4}	-5.30×10^{-4}
40	1.00×10^{-1}	-6.42×10^{-4}	-1.55×10^{-4}	-1.24×10^{-3}	1.48×10^{-4}	-8.84×10^{-3}	2.83×10^{-4}	1.60×10^{-4}	-3.97×10^{-4}	-4.74×10^{-4}
50	6.36×10^{-2}	-3.22×10^{-4}	-1.06×10^{-5}	-1.24×10^{-3}	1.13×10^{-4}	-5.28×10^{-3}	9.55×10^{-5}	7.91×10^{-5}	-4.07×10^{-4}	-2.75×10^{-4}
60	-2.37×10^{-2}	-1.45×10^{-4}	-5.78×10^{-5}	-3.85×10^{-4}	-1.78×10^{-4}	-1.93×10^{-3}	-8.03×10^{-4}	2.36×10^{-4}	-2.25×10^{-4}	-4.29×10^{-4}
70	-7.49×10^{-2}	9.54×10^{-5}	1.20×10^{-5}	1.33×10^{-4}	-6.43×10^{-4}	-2.36×10^{-4}	-1.01×10^{-3}	1.95×10^{-5}	-1.51×10^{-4}	-1.25×10^{-4}
80	-1.32×10^{-1}	-6.89×10^{-6}	4.01×10^{-5}	-4.52×10^{-5}	-4.84×10^{-4}	-9.37×10^{-4}	-1.36×10^{-4}	-3.97×10^{-6}	-9.94×10^{-5}	-1.75×10^{-4}
90	-2.14×10^{-1}	-1.15×10^{-4}	8.61×10^{-5}	-1.83×10^{-4}	-6.61×10^{-5}	-2.14×10^{-3}	7.80×10^{-6}	-8.36×10^{-6}	-1.27×10^{-4}	-1.91×10^{-4}
100	-2.06×10^{-1}	-1.79×10^{-4}	5.94×10^{-5}	-3.20×10^{-4}	-8.33×10^{-4}	-2.17×10^{-3}	8.19×10^{-5}	-3.79×10^{-5}	-1.36×10^{-4}	-1.63×10^{-4}
110	-2.02×10^{-1}	2.94×10^{-6}	4.33×10^{-5}	-3.58×10^{-4}	-1.74×10^{-4}	-1.66×10^{-3}	1.45×10^{-4}	-1.90×10^{-5}	-2.39×10^{-4}	-8.05×10^{-5}
120	-1.93×10^{-1}	8.88×10^{-5}	8.27×10^{-5}	-2.35×10^{-4}	-7.33×10^{-5}	-1.15×10^{-3}	6.61×10^{-5}	7.50×10^{-5}	-2.01×10^{-4}	-9.19×10^{-5}
130	-1.75×10^{-1}	-4.16×10^{-5}	4.65×10^{-5}	-1.44×10^{-4}	-6.83×10^{-5}	-5.03×10^{-4}	-1.82×10^{-5}	-1.26×10^{-5}	-1.50×10^{-4}	6.73×10^{-5}
140	-1.62×10^{-1}	-7.40×10^{-5}	9.82×10^{-6}	-3.47×10^{-5}	3.16×10^{-5}	-4.33×10^{-5}	-8.71×10^{-5}	-2.68×10^{-5}	1.63×10^{-5}	8.74×10^{-5}
150	-1.50×10^{-1}	-4.81×10^{-5}	-5.11×10^{-5}	-7.03×10^{-5}	-5.44×10^{-5}	3.96×10^{-4}	-8.61×10^{-5}	-1.93×10^{-5}	3.75×10^{-5}	7.30×10^{-6}
160	-1.53×10^{-1}	-2.67×10^{-5}	-1.30×10^{-4}	-1.34×10^{-4}	-8.24×10^{-5}	7.68×10^{-4}	-1.28×10^{-4}	-3.43×10^{-5}	-1.79×10^{-5}	-2.91×10^{-5}
170	-1.58×10^{-1}	3.82×10^{-5}	-2.09×10^{-4}	-1.94×10^{-4}	-4.79×10^{-5}	1.01×10^{-3}	-1.61×10^{-4}	-2.31×10^{-5}	-4.36×10^{-5}	-6.68×10^{-5}
180	-1.60×10^{-1}	6.01×10^{-5}	-2.35×10^{-4}	-2.14×10^{-4}	-3.95×10^{-5}	1.09×10^{-3}	-1.77×10^{-4}	-2.54×10^{-5}	-6.85×10^{-5}	-9.24×10^{-5}
190	-1.58×10^{-1}	3.34×10^{-5}	-2.05×10^{-4}	-1.91×10^{-4}	-5.61×10^{-5}	9.87×10^{-4}	-1.58×10^{-4}	-2.74×10^{-5}	-3.93×10^{-5}	-6.18×10^{-5}
200	-1.53×10^{-1}	-2.68×10^{-5}	-1.21×10^{-4}	-1.24×10^{-4}	-8.39×10^{-5}	7.29×10^{-4}	-1.29×10^{-4}	-3.03×10^{-5}	-1.11×10^{-5}	-2.62×10^{-5}
210	-1.52×10^{-1}	-3.58×10^{-5}	-2.30×10^{-4}	-7.00×10^{-5}	-8.28×10^{-5}	3.49×10^{-4}	-6.67×10^{-5}	-2.62×10^{-5}	4.38×10^{-5}	5.41×10^{-7}
220	-1.52×10^{-1}	2.07×10^{-5}	9.80×10^{-6}	-5.69×10^{-5}	-6.32×10^{-5}	1.09×10^{-5}	-5.26×10^{-5}	-1.44×10^{-4}	-2.48×10^{-5}	-1.50×10^{-6}
230	-1.62×10^{-1}	-4.74×10^{-5}	6.88×10^{-6}	4.62×10^{-5}	-1.67×10^{-4}	-5.34×10^{-4}	-7.34×10^{-5}	-1.07×10^{-4}	-7.75×10^{-5}	4.03×10^{-5}
240	-1.97×10^{-1}	-9.77×10^{-5}	5.41×10^{-4}	-1.08×10^{-4}	-8.34×10^{-5}	-1.07×10^{-3}	5.13×10^{-5}	9.79×10^{-5}	-8.58×10^{-5}	1.62×10^{-4}
250	-2.04×10^{-1}	-1.58×10^{-4}	8.48×10^{-5}	-3.39×10^{-4}	-2.39×10^{-5}	-1.84×10^{-3}	1.90×10^{-4}	4.92×10^{-5}	-1.09×10^{-4}	-2.80×10^{-5}
260	-2.00×10^{-1}	-2.44×10^{-4}	9.02×10^{-5}	-3.33×10^{-4}	-3.81×10^{-4}	-2.32×10^{-3}	9.08×10^{-5}	1.14×10^{-4}	-8.63×10^{-5}	-1.38×10^{-4}
270	-2.14×10^{-1}	-1.27×10^{-4}	7.09×10^{-5}	-1.68×10^{-4}	-6.81×10^{-4}	-2.05×10^{-3}	1.32×10^{-6}	-7.80×10^{-6}	-1.24×10^{-4}	-2.02×10^{-4}
280	-1.31×10^{-1}	2.04×10^{-5}	1.03×10^{-7}	-3.93×10^{-4}	-1.61×10^{-4}	-8.10×10^{-4}	-6.32×10^{-5}	5.92×10^{-6}	-9.78×10^{-5}	-1.70×10^{-4}
290	-8.25×10^{-2}	-9.61×10^{-5}	9.10×10^{-5}	2.08×10^{-3}	-1.12×10^{-3}	-2.27×10^{-4}	-1.72×10^{-5}	7.93×10^{-5}	-3.14×10^{-4}	-2.05×10^{-4}
300	-4.34×10^{-2}	-6.36×10^{-4}	1.58×10^{-4}	-2.88×10^{-3}	-1.24×10^{-3}	-2.06×10^{-3}	1.09×10^{-4}	1.98×10^{-4}	-4.11×10^{-4}	-2.84×10^{-4}
310	5.10×10^{-2}	-1.22×10^{-3}	1.91×10^{-5}	-7.37×10^{-4}	-3.67×10^{-4}	-5.56×10^{-3}	3.60×10^{-4}	1.16×10^{-4}	-3.22×10^{-4}	-3.04×10^{-4}
320	7.89×10^{-2}	-1.32×10^{-3}	-2.78×10^{-4}	-7.82×10^{-4}	-9.92×10^{-5}	-9.17×10^{-3}	3.50×10^{-4}	2.97×10^{-4}	-1.62×10^{-4}	-5.93×10^{-4}
330	9.63×10^{-2}	-9.67×10^{-4}	-9.07×10^{-4}	-9.40×10^{-4}	1.16×10^{-4}	-1.22×10^{-2}	3.90×10^{-4}	3.34×10^{-4}	-3.14×10^{-4}	-5.27×10^{-4}
340	9.51×10^{-2}	-8.59×10^{-4}	-2.18×10^{-3}	-9.47×10^{-4}	-2.33×10^{-5}	-1.42×10^{-2}	3.43×10^{-4}	2.75×10^{-4}	-3.86×10^{-4}	-5.30×10^{-4}
350	7.17×10^{-2}	-9.28×10^{-4}	-3.52×10^{-3}	-9.46×10^{-4}	-2.01×10^{-4}	-1.52×10^{-2}	2.96×10^{-4}	2.09×10^{-4}	-3.25×10^{-4}	-5.99×10^{-4}

Orientation	are10	are11	are12	are13	are14	are15	are16	are17	are18	are19
0	-3.08×10^{-4}	2.32×10^{-4}	-6.08×10^{-5}	-3.95×10^{-4}	2.31×10^{-4}	2.67×10^{-3}	1.53×10^{-4}	2.34×10^{-3}	2.73×10^{-5}	1.12×10^{-2}
10	-2.18×10^{-4}	2.43×10^{-4}	-1.24×10^{-5}	-3.53×10^{-4}	3.38×10^{-4}	2.78×10^{-3}	-1.17×10^{-4}	2.28×10^{-3}	-9.36×10^{-5}	1.12×10^{-2}
20	-8.64×10^{-5}	2.33×10^{-4}	2.30×10^{-6}	-2.63×10^{-4}	4.96×10^{-4}	2.67×10^{-3}	-6.13×10^{-4}	2.08×10^{-3}	-2.15×10^{-4}	1.11×10^{-2}
30	-1.20×10^{-5}	1.89×10^{-4}	-6.09×10^{-5}	-2.55×10^{-4}	4.96×10^{-4}	2.41×10^{-3}	-8.92×10^{-4}	8.51×10^{-4}	1.22×10^{-4}	1.07×10^{-2}
40	3.34×10^{-5}	1.69×10^{-4}	-2.42×10^{-5}	-4.11×10^{-4}	5.47×10^{-4}	1.12×10^{-3}	-9.32×10^{-4}	1.85×10^{-4}	1.82×10^{-4}	1.04×10^{-2}
50	4.32×10^{-6}	8.81×10^{-5}	5.06×10^{-5}	-2.76×10^{-4}	3.46×10^{-4}	7.83×10^{-4}	-8.77×10^{-4}	-4.09×10^{-5}	3.57×10^{-4}	9.25×10^{-3}
60	1.85×10^{-5}	-1.46×10^{-4}	-6.98×10^{-6}	-2.03×10^{-4}	2.04×10^{-4}	1.57×10^{-3}	-3.03×10^{-4}	1.40×10^{-4}	1.26×10^{-3}	5.46×10^{-3}
70	-1.52×10^{-5}	-9.34×10^{-5}	5.30×10^{-6}	-1.82×10^{-4}	3.74×10^{-5}	-7.72×10^{-5}	3.72×10^{-4}	1.90×10^{-4}	7.72×10^{-4}	1.73×10^{-3}
80	1.80×10^{-5}	-1.12×10^{-4}	3.29×10^{-5}	-1.54×10^{-5}	-1.69×10^{-5}	-6.77×10^{-4}	5.15×10^{-4}	-6.71×10^{-5}	3.38×10^{-4}	3.99×10^{-4}
90	5.49×10^{-5}	-3.60×10^{-5}	5.96×10^{-5}	-9.72×10^{-5}	4.80×10^{-4}	-1.32×10^{-3}	4.13×10^{-4}	-1.36×10^{-4}	3.96×10^{-4}	1.58×10^{-3}
100	-2.98×10^{-5}	-1.05×10^{-4}	4.09×10^{-6}	-1.42×10^{-4}	1.06×10^{-4}	-8.64×10^{-4}	8.78×10^{-5}	2.41×10^{-4}	2.94×10^{-4}	2.82×10^{-3}
110	-1.02×10^{-4}	-4.72×10^{-5}	-5.65×10^{-5}	-1.39×10^{-4}	7.11×10^{-4}	1.84×10^{-4}	-1.80×10^{-4}	6.39×10^{-4}	3.43×10^{-6}	3.30×10^{-3}
120	-1.27×10^{-4}	2.27×10^{-4}	-7.14×10^{-5}	-1.38×10^{-4}	2.12×10^{-4}	6.23×10^{-4}	-2.93×10^{-4}	7.95×10^{-4}	2.62×10^{-5}	3.18×10^{-3}
130	-1.70×10^{-4}	1.33×10^{-4}	-1.57×10^{-5}	-7.17×10^{-5}	3.87×10^{-4}	7.55×10^{-4}	-2.76×10^{-4}	1.08×10^{-3}	-1.31×10^{-4}	2.09×10^{-3}
140	-3.54×10^{-4}	4.41×10^{-5}	3.81×10^{-5}	-3.75×10^{-5}	3.25×10^{-4}	5.23×10^{-4}	-7.73×10^{-5}	1.35×10^{-3}	-5.87×10^{-5}	3.39×10^{-4}
150	-7.53×10^{-4}	1.21×10^{-4}	1.16×10^{-4}	7.56×10^{-6}	2.20×10^{-4}	4.95×10^{-4}	3.56×10^{-5}	1.44×10^{-3}	9.13×10^{-5}	-9.51×10^{-4}
160	-1.20×10^{-3}	1.87×10^{-4}	7.79×10^{-5}	1.33×10^{-4}	1.50×10^{-4}	6.79×10^{-4}	1.18×10^{-4}	1.32×10^{-3}	8.39×10^{-5}	-1.56×10^{-3}
170	-1.52×10^{-3}	2.81×10^{-4}	3.60×10^{-5}	3.04×10^{-4}	9.12×10^{-5}	8.58×10^{-4}	1.53×10^{-4}	1.04×10^{-3}	-4.34×10^{-5}	-1.97×10^{-3}
180	-1.63×10^{-3}	3.28×10^{-4}	2.26×10^{-5}	3.88×10^{-4}	6.62×10^{-5}	9.84×10^{-4}	1.58×10^{-4}	7.02×10^{-4}	-7.77×10^{-5}	-2.16×10^{-3}
190	-1.50×10^{-3}	2.72×10^{-4}	3.89×10^{-5}	2.91×10^{-4}	9.58×10^{-5}	8.47×10^{-4}	1.52×10^{-4}	5.59×10^{-4}	-3.12×10^{-5}	-1.94×10^{-3}
200	-1.15×10^{-3}	1.83×10^{-4}	8.24×10^{-5}	1.18×10^{-4}	1.55×10^{-4}	6.78×10^{-4}	1.12×10^{-4}	9.97×10^{-4}	9.40×10^{-5}	-1.52×10^{-3}
210	-6.91×10^{-4}	1.63×10^{-4}	1.61×10^{-4}	-2.32×10^{-5}	2.30×10^{-4}	5.02×10^{-4}	2.58×10^{-5}	7.02×10^{-4}	1.46×10^{-4}	-8.92×10^{-4}
220	-2.49×10^{-4}	1.13×10^{-5}	3.26×10^{-4}	-1.72×10^{-4}	2.16×10^{-4}	2.67×10^{-4}	-1.41×10^{-4}	-1.56×10^{-4}	-2.09×10^{-5}	4.33×10^{-4}
230	-9.71×10^{-5}	-7.86×10^{-4}	2.01×10^{-4}	-1.57×10^{-4}	1.20×10^{-4}	4.51×10^{-4}	-2.87×10^{-4}	-2.31×10^{-4}	-8.01×10^{-5}	2.09×10^{-3}
240	-2.12×10^{-4}	-1.24×10^{-4}	4.37×10^{-5}	-1.59×10^{-4}	1.78×10^{-4}	5.83×10^{-4}	-3.06×10^{-4}	-1.77×10^{-4}	2.53×10^{-4}	3.21×10^{-3}
250	-1.04×10^{-4}	-5.61×10^{-5}	-3.01×10^{-5}	-1.18×10^{-4}	1.42×10^{-4}	2.60×10^{-4}	-1.70×10^{-4}	2.15×10^{-4}	2.25×10^{-4}	2.98×10^{-3}
260	6.56×10^{-5}	-6.98×10^{-6}	-2.74×10^{-5}	-1.79×10^{-4}	1.20×10^{-4}	-8.14×10^{-4}	9.37×10^{-5}	6.95×10^{-5}	3.64×10^{-4}	2.57×10^{-3}
270	5.90×10^{-5}	-5.88×10^{-5}	5.97×10^{-5}	-9.05×10^{-5}	3.92×10^{-5}	-1.35×10^{-3}	4.41×10^{-4}	-1.23×10^{-5}	4.05×10^{-4}	1.46×10^{-3}
280	4.22×10^{-5}	-1.11×10^{-4}	4.96×10^{-5}	1.65×10^{-5}	-2.88×10^{-5}	-7.18×10^{-4}	5.23×10^{-4}	2.15×10^{-4}	3.10×10^{-4}	3.80×10^{-4}
290	5.26×10^{-5}	-1.99×10^{-4}	4.83×10^{-5}	-2.26×10^{-4}	4.64×10^{-5}	-2.24×10^{-4}	3.90×10^{-4}	6.95×10^{-5}	9.92×10^{-4}	2.25×10^{-3}
300	-1.21×10^{-4}	-1.50×10^{-4}	7.98×10^{-5}	-3.55×10^{-4}	4.56×10^{-4}	1.81×10^{-3}	-2.88×10^{-4}	2.58×10^{-4}	1.43×10^{-3}	5.54×10^{-3}
310	-1.16×10^{-4}	1.34×10^{-4}	5.34×10^{-5}	-3.27×10^{-4}	5.45×10^{-4}	1.57×10^{-3}	-8.51×10^{-4}	3.18×10^{-4}	6.73×10^{-4}	8.85×10^{-3}
320	1.65×10^{-4}	1.28×10^{-4}	3.68×10^{-5}	-3.69×10^{-4}	5.81×10^{-4}	1.84×10^{-3}	-9.32×10^{-4}	6.69×10^{-4}	6.34×10^{-4}	1.07×10^{-2}
330	2.77×10^{-5}	2.00×10^{-4}	-3.97×10^{-5}	-2.45×10^{-4}	6.05×10^{-4}	2.56×10^{-3}	-8.83×10^{-4}	1.46×10^{-3}	1.30×10^{-4}	1.08×10^{-2}
340	-1.06×10^{-4}	2.20×10^{-4}	-3.12×10^{-6}	-2.66×10^{-4}	4.60×10^{-4}	2.67×10^{-3}	-5.64×10^{-4}	2.13×10^{-3}	-2.36×10^{-4}	1.11×10^{-2}
350	-2.33×10^{-4}	2.40×10^{-4}	-1.60×10^{-5}	-3.58×10^{-4}	3.23×10^{-4}	2.77×10^{-3}	-7.12×10^{-5}	2.29×10^{-3}	-7.57×10^{-5}	1.12×10^{-2}

Cp of the roof:

$$Cp_{roof} = a_{ro0} + a_{ro1} * H_1 + \dots a_{ro14} * H_{14} + a_{ro15} * D + a_{ro16} * L + a_{ro17} * Rd + a_{ro18} * Rl + a_{ro19} * Bg$$

Orientation	aro0	aro1	aro2	aro3	aro4	aro5	aro6	aro7	aro8	aro9
0	-4.12×10^{-1}	3.93×10^{-4}	2.87×10^{-3}	5.90×10^{-4}	-2.61×10^{-4}	1.44×10^{-3}	3.33×10^{-5}	4.55×10^{-4}	4.66×10^{-4}	6.89×10^{-4}
10	-4.13×10^{-1}	3.62×10^{-4}	2.33×10^{-3}	5.01×10^{-4}	-2.68×10^{-4}	1.43×10^{-3}	-1.85×10^{-5}	3.76×10^{-4}	5.04×10^{-4}	5.81×10^{-4}
20	-4.24×10^{-1}	3.26×10^{-4}	1.39×10^{-3}	4.31×10^{-4}	-3.60×10^{-4}	1.35×10^{-3}	-6.55×10^{-5}	2.50×10^{-4}	5.50×10^{-4}	4.33×10^{-4}
30	-4.37×10^{-1}	3.20×10^{-4}	6.40×10^{-4}	5.01×10^{-4}	-3.54×10^{-4}	1.03×10^{-3}	-1.66×10^{-4}	2.13×10^{-4}	4.86×10^{-4}	2.97×10^{-4}
40	-4.39×10^{-1}	3.54×10^{-4}	1.60×10^{-6}	7.64×10^{-4}	-3.28×10^{-4}	7.36×10^{-4}	-1.24×10^{-4}	1.87×10^{-5}	4.07×10^{-4}	1.22×10^{-4}
50	-4.13×10^{-1}	9.46×10^{-7}	9.64×10^{-6}	7.78×10^{-4}	6.74×10^{-5}	5.49×10^{-4}	2.32×10^{-4}	-2.47×10^{-4}	1.44×10^{-4}	1.40×10^{-4}
60	-3.81×10^{-1}	1.20×10^{-7}	-2.14×10^{-4}	3.09×10^{-4}	8.63×10^{-5}	-2.69×10^{-5}	9.21×10^{-4}	-1.39×10^{-4}	7.04×10^{-4}	-2.19×10^{-5}
70	-3.32×10^{-1}	-3.46×10^{-5}	-9.61×10^{-5}	5.17×10^{-5}	-1.09×10^{-4}	-6.93×10^{-5}	5.05×10^{-4}	-1.40×10^{-4}	-6.10×10^{-5}	-1.50×10^{-4}
80	-2.17×10^{-1}	5.84×10^{-5}	6.02×10^{-5}	-7.24×10^{-5}	-9.33×10^{-6}	2.55×10^{-5}	5.05×10^{-5}	-6.45×10^{-5}	-1.83×10^{-4}	-7.15×10^{-5}
90	-2.04×10^{-1}	6.49×10^{-5}	5.81×10^{-5}	-1.60×10^{-4}	5.31×10^{-5}	2.75×10^{-5}	1.11×10^{-4}	-1.18×10^{-4}	-1.24×10^{-4}	-2.33×10^{-5}
100	-2.13×10^{-1}	6.11×10^{-5}	8.72×10^{-5}	-8.22×10^{-5}	4.61×10^{-5}	7.29×10^{-5}	-1.02×10^{-5}	-4.61×10^{-5}	-1.25×10^{-4}	-3.62×10^{-5}
110	-3.00×10^{-1}	-3.76×10^{-5}	-5.70×10^{-5}	-4.51×10^{-5}	2.83×10^{-4}	4.04×10^{-5}	-1.77×10^{-4}	-9.06×10^{-5}	-1.89×10^{-4}	-9.09×10^{-5}
120	-3.42×10^{-1}	-1.55×10^{-4}	-1.34×10^{-4}	-1.93×10^{-4}	3.61×10^{-5}	2.91×10^{-5}	-8.67×10^{-5}	4.40×10^{-5}	-2.23×10^{-4}	1.10×10^{-4}
130	-3.94×10^{-1}	-4.15×10^{-4}	-2.88×10^{-4}	-2.23×10^{-4}	2.26×10^{-4}	1.35×10^{-5}	1.21×10^{-4}	1.91×10^{-4}	3.63×10^{-4}	2.80×10^{-4}
140	-4.11×10^{-1}	-2.31×10^{-4}	-3.41×10^{-4}	-2.23×10^{-4}	-4.81×10^{-5}	-5.58×10^{-5}	7.14×10^{-5}	3.28×10^{-4}	3.03×10^{-4}	5.39×10^{-5}
150	-4.27×10^{-1}	-3.04×10^{-4}	-2.30×10^{-4}	-1.07×10^{-4}	-3.26×10^{-4}	-8.45×10^{-5}	-1.20×10^{-4}	2.09×10^{-4}	1.33×10^{-4}	1.35×10^{-4}
160	-3.95×10^{-1}	-2.66×10^{-4}	-3.89×10^{-4}	-2.88×10^{-4}	-3.86×10^{-4}	-4.11×10^{-5}	-3.63×10^{-4}	2.66×10^{-4}	1.23×10^{-4}	1.60×10^{-4}
170	-3.71×10^{-1}	-2.56×10^{-4}	-4.40×10^{-4}	-4.30×10^{-4}	-4.06×10^{-4}	-9.59×10^{-5}	-3.42×10^{-4}	3.16×10^{-4}	1.38×10^{-4}	1.60×10^{-4}
180	-3.65×10^{-1}	-2.70×10^{-4}	-4.36×10^{-4}	-4.62×10^{-4}	-4.20×10^{-4}	-1.39×10^{-4}	-2.81×10^{-4}	3.30×10^{-4}	1.92×10^{-4}	1.66×10^{-4}
190	-3.74×10^{-1}	-2.55×10^{-4}	-4.43×10^{-4}	-4.22×10^{-4}	-4.07×10^{-4}	-8.91×10^{-5}	-3.47×10^{-4}	3.13×10^{-4}	1.35×10^{-4}	1.61×10^{-4}
200	-4.00×10^{-1}	-2.66×10^{-4}	-3.74×10^{-4}	-2.70×10^{-4}	-3.86×10^{-4}	-3.69×10^{-5}	-3.50×10^{-4}	2.52×10^{-4}	1.18×10^{-4}	1.68×10^{-4}
210	-4.32×10^{-1}	-3.15×10^{-4}	-2.32×10^{-4}	-1.61×10^{-4}	-3.87×10^{-4}	-7.08×10^{-5}	-1.13×10^{-4}	2.73×10^{-4}	1.60×10^{-4}	5.54×10^{-5}
220	-4.04×10^{-1}	-1.88×10^{-4}	-3.47×10^{-4}	-1.83×10^{-4}	7.84×10^{-5}	-2.38×10^{-5}	-2.67×10^{-4}	1.22×10^{-4}	2.67×10^{-4}	3.42×10^{-5}
230	-3.49×10^{-1}	1.29×10^{-5}	-1.48×10^{-4}	-8.90×10^{-5}	2.02×10^{-4}	-7.63×10^{-5}	5.59×10^{-6}	-3.92×10^{-4}	2.57×10^{-5}	3.94×10^{-4}
240	-3.75×10^{-1}	-8.53×10^{-5}	-2.02×10^{-4}	1.25×10^{-4}	1.82×10^{-4}	1.15×10^{-4}	-1.99×10^{-4}	-1.89×10^{-4}	-1.18×10^{-5}	8.76×10^{-4}
250	-3.07×10^{-1}	-1.64×10^{-4}	-6.81×10^{-5}	-9.13×10^{-5}	2.34×10^{-4}	1.63×10^{-4}	-1.02×10^{-4}	3.02×10^{-6}	-5.57×10^{-5}	3.88×10^{-4}
260	-2.11×10^{-1}	3.83×10^{-5}	9.71×10^{-5}	-8.07×10^{-5}	4.01×10^{-5}	4.76×10^{-5}	1.61×10^{-5}	-5.01×10^{-5}	-1.21×10^{-4}	-5.14×10^{-5}
270	-1.93×10^{-1}	-1.26×10^{-4}	3.52×10^{-4}	-8.72×10^{-5}	3.21×10^{-6}	-7.64×10^{-6}	2.67×10^{-5}	7.26×10^{-5}	-1.09×10^{-4}	1.15×10^{-5}
280	-2.22×10^{-1}	3.42×10^{-5}	7.72×10^{-5}	-6.31×10^{-5}	-3.78×10^{-5}	-4.68×10^{-6}	7.00×10^{-5}	-6.57×10^{-5}	-1.59×10^{-4}	-8.64×10^{-5}
290	-3.26×10^{-1}	-6.97×10^{-5}	-8.63×10^{-5}	-2.31×10^{-5}	4.31×10^{-4}	1.84×10^{-5}	-2.94×10^{-4}	-4.50×10^{-5}	-6.06×10^{-5}	-2.32×10^{-4}
300	-3.33×10^{-1}	4.37×10^{-4}	-4.87×10^{-4}	-1.88×10^{-4}	6.06×10^{-4}	-3.51×10^{-5}	-2.97×10^{-4}	1.05×10^{-4}	1.41×10^{-4}	-4.30×10^{-4}
310	-4.08×10^{-1}	8.48×10^{-4}	-1.94×10^{-4}	1.34×10^{-4}	7.81×10^{-4}	4.28×10^{-4}	-9.80×10^{-5}	1.58×10^{-4}	5.27×10^{-4}	-1.56×10^{-4}
320	-4.42×10^{-1}	7.71×10^{-4}	1.36×10^{-4}	2.86×10^{-4}	-9.78×10^{-5}	7.25×10^{-4}	1.08×10^{-5}	1.98×10^{-4}	4.99×10^{-4}	2.87×10^{-4}
330	-4.24×10^{-1}	4.47×10^{-4}	6.71×10^{-4}	4.01×10^{-4}	-3.36×10^{-4}	1.04×10^{-3}	-1.70×10^{-4}	1.46×10^{-4}	4.30×10^{-4}	3.30×10^{-4}
340	-4.21×10^{-1}	2.92×10^{-4}	1.48×10^{-4}	4.53×10^{-4}	-3.62×10^{-4}	1.37×10^{-3}	-5.27×10^{-5}	2.76×10^{-4}	5.48×10^{-4}	4.40×10^{-4}
350	-4.12×10^{-1}	3.69×10^{-4}	2.42×10^{-3}	5.14×10^{-4}	-2.60×10^{-4}	1.44×10^{-3}	-1.29×10^{-5}	3.88×10^{-4}	5.00×10^{-4}	5.99×10^{-4}

Orientation	aro10	aro11	aro12	aro13	aro14	aro15	aro16	aro17	aro18	aro19
0	1.50×10^{-4}	-3.34×10^{-4}	7.78×10^{-5}	-1.14×10^{-5}	-1.42×10^{-4}	3.95×10^{-3}	1.99×10^{-4}	-6.58×10^{-4}	1.04×10^{-4}	-2.26×10^{-3}
10	4.95×10^{-5}	-3.62×10^{-4}	1.05×10^{-5}	4.81×10^{-5}	-1.83×10^{-4}	4.15×10^{-3}	3.41×10^{-4}	-5.45×10^{-4}	1.72×10^{-4}	-2.39×10^{-3}
20	-5.67×10^{-5}	-3.47×10^{-4}	-1.80×10^{-5}	9.08×10^{-5}	-1.64×10^{-4}	4.84×10^{-3}	5.27×10^{-4}	-3.67×10^{-4}	2.59×10^{-4}	-2.47×10^{-3}
30	-3.36×10^{-5}	-2.56×10^{-4}	3.77×10^{-5}	6.83×10^{-5}	-4.15×10^{-5}	6.29×10^{-3}	4.88×10^{-4}	-1.13×10^{-5}	8.21×10^{-5}	-2.25×10^{-3}
40	-3.39×10^{-5}	-2.87×10^{-4}	6.31×10^{-5}	4.02×10^{-5}	7.51×10^{-5}	8.38×10^{-3}	3.32×10^{-4}	5.75×10^{-5}	-2.87×10^{-5}	-2.38×10^{-3}
50	-1.92×10^{-4}	-6.03×10^{-5}	5.22×10^{-5}	-9.15×10^{-5}	2.75×10^{-4}	8.35×10^{-3}	1.54×10^{-4}	4.17×10^{-4}	-6.50×10^{-4}	-1.27×10^{-3}
60	-1.27×10^{-4}	1.05×10^{-4}	-7.35×10^{-5}	-2.04×10^{-4}	9.70×10^{-5}	8.60×10^{-3}	-1.69×10^{-4}	2.61×10^{-4}	-4.23×10^{-4}	4.24×10^{-4}
70	-3.62×10^{-5}	-3.83×10^{-5}	-8.54×10^{-5}	-4.66×10^{-5}	1.43×10^{-4}	6.83×10^{-3}	1.42×10^{-4}	5.79×10^{-4}	2.71×10^{-4}	1.98×10^{-4}
80	-4.92×10^{-5}	-6.00×10^{-5}	-9.37×10^{-6}	-5.52×10^{-5}	-1.49×10^{-5}	5.63×10^{-4}	1.94×10^{-4}	2.41×10^{-4}	1.81×10^{-4}	3.98×10^{-4}
90	-1.76×10^{-4}	-1.75×10^{-4}	-1.00×10^{-4}	-1.05×10^{-4}	-2.62×10^{-5}	9.14×10^{-4}	6.63×10^{-5}	6.39×10^{-5}	6.11×10^{-4}	5.45×10^{-4}
100	-1.37×10^{-4}	-8.90×10^{-5}	-5.24×10^{-6}	-7.50×10^{-5}	-2.85×10^{-5}	6.37×10^{-4}	1.19×10^{-4}	-1.73×10^{-6}	2.11×10^{-4}	6.25×10^{-4}
110	-8.52×10^{-5}	4.62×10^{-4}	-6.68×10^{-5}	-1.34×10^{-4}	-5.16×10^{-5}	6.31×10^{-3}	-5.47×10^{-6}	-4.10×10^{-4}	-8.92×10^{-5}	9.15×10^{-4}
120	-3.93×10^{-4}	9.39×10^{-4}	8.46×10^{-5}	-2.96×10^{-4}	5.61×10^{-4}	8.52×10^{-3}	-2.13×10^{-4}	-5.39×10^{-4}	-7.49×10^{-4}	4.87×10^{-4}
130	-4.74×10^{-5}	5.51×10^{-4}	-1.35×10^{-6}	-4.91×10^{-4}	1.07×10^{-3}	9.05×10^{-3}	3.54×10^{-4}	-1.02×10^{-3}	-1.02×10^{-3}	6.86×10^{-5}
140	7.49×10^{-4}	2.51×10^{-4}	-3.01×10^{-5}	-1.39×10^{-4}	7.73×10^{-4}	8.16×10^{-3}	3.46×10^{-4}	-1.27×10^{-3}	-7.36×10^{-4}	-7.93×10^{-4}
150	1.31×10^{-3}	1.48×10^{-4}	-9.40×10^{-5}	4.47×10^{-4}	1.51×10^{-4}	7.20×10^{-3}	7.16×10^{-4}	-1.95×10^{-3}	-9.69×10^{-5}	-8.15×10^{-4}
160	1.73×10^{-3}	1.80×10^{-5}	-2.03×10^{-4}	1.30×10^{-4}	1.15×10^{-5}	5.73×10^{-3}	6.68×10^{-4}	-1.99×10^{-3}	-1.44×10^{-4}	-9.71×10^{-4}
170	1.92×10^{-3}	8.34×10^{-5}	-2.99×10^{-4}	2.27×10^{-3}	9.35×10^{-5}	5.28×10^{-3}	3.87×10^{-4}	-1.74×10^{-3}	-2.93×10^{-4}	-1.26×10^{-3}
180	1.98×10^{-3}	1.40×10^{-4}	-3.42×10^{-4}	2.69×10^{-3}	1.46×10^{-4}	5.28×10^{-3}	2.43×10^{-4}	-1.64×10^{-3}	-2.64×10^{-4}	-1.48×10^{-3}
190	1.92×10^{-3}	7.58×10^{-5}	-2.87×10^{-4}	2.19×10^{-3}	8.35×10^{-5}	5.33×10^{-3}	4.13×10^{-4}	-1.77×10^{-3}	-2.92×10^{-4}	-1.23×10^{-3}
200	1.71×10^{-3}	2.62×10^{-5}	-2.21×10^{-4}	1.19×10^{-3}	3.12×10^{-5}	5.89×10^{-3}	6.90×10^{-4}	-2.00×10^{-3}	-1.15×10^{-4}	-1.01×10^{-3}
210	1.08×10^{-3}	2.50×10^{-4}	-4.51×10^{-5}	4.03×10^{-4}	1.35×10^{-4}	7.38×10^{-3}	7.15×10^{-4}	-1.80×10^{-3}	-1.50×10^{-4}	-6.59×10^{-4}
220	7.46×10^{-4}	2.57×10^{-4}	6.10×10^{-4}	-1.98×10^{-4}	2.71×10^{-4}	7.94×10^{-3}	3.24×10^{-4}	-1.21×10^{-3}	-8.82×10^{-4}	-7.98×10^{-4}
230	2.46×10^{-4}	-6.82×10^{-5}	1.04×10^{-3}	-2.87×10^{-4}	3.52×10^{-5}	8.35×10^{-3}	-1.03×10^{-4}	-8.43×10^{-4}	-1.58×10^{-3}	-4.07×10^{-4}
240	-1.57×10^{-4}	3.87×10^{-5}	3.97×10^{-4}	-4.01×10^{-4}	1.33×10^{-4}	8.70×10^{-3}	-2.88×10^{-4}	-4.94×10^{-5}	-6.24×10^{-5}	6.33×10^{-4}
250	-2.00×10^{-4}	-6.59×10^{-5}	-8.26×10^{-4}	-1.92×10^{-4}	2.77×10^{-4}	6.38×10^{-3}	-6.12×10^{-5}	-2.45×10^{-4}	6.78×10^{-4}	8.10×10^{-4}
260	-1.30×10^{-4}	-6.55×10^{-5}	-9.94×10^{-6}	-5.47×10^{-5}	-1.82×10^{-5}	6.15×10^{-4}	1.18×10^{-4}	-3.87×10^{-5}	2.16×10^{-4}	5.61×10^{-4}
270	6.45×10^{-6}	-3.96×10^{-5}	-9.90×10^{-5}	-1.58×10^{-4}	6.16×10^{-5}	8.24×10^{-4}	6.67×10^{-5}	-4.79×10^{-4}	3.31×10^{-4}	2.59×10^{-4}
280	-3.21×10^{-5}	-4.52×10^{-5}	5.62×10^{-5}	-4.92×10^{-5}	8.14×10^{-7}	5.85×10^{-4}	2.20×10^{-4}	2.74×10^{-4}	2.31×10^{-4}	3.56×10^{-4}
290	6.42×10^{-5}	-7.02×10^{-5}	1.12×10^{-4}	-1.12×10^{-4}	8.79×10^{-5}	6.59×10^{-4}	1.05×10^{-5}	7.47×10^{-4}	2.01×10^{-5}	4.46×10^{-4}
300	-4.57×10^{-5}	5.33×10^{-5}	8.96×10^{-5}	1.42×10^{-4}	-3.51×10^{-5}	7.36×10^{-3}	-1.94×10^{-4}	3.69×10^{-4}	-4.80×10^{-4}	-1.62×10^{-4}
310	1.77×10^{-5}	-5.16×10^{-5}	2.30×10^{-5}	4.99×10^{-5}	2.36×10^{-4}	8.03×10^{-3}	1.80×10^{-4}	2.84×10^{-4}	-8.98×10^{-4}	-1.12×10^{-3}
320	1.11×10^{-5}	-2.25×10^{-4}	-3.68×10^{-5}	1.99×10^{-5}	1.59×10^{-4}	7.83×10^{-3}	3.06×10^{-5}	5.65×10^{-4}	-4.57×10^{-4}	-2.23×10^{-3}
330	-9.81×10^{-5}	-2.63×10^{-4}	-7.73×10^{-6}	5.63×10^{-5}	-1.50×10^{-4}	6.02×10^{-3}	5.00×10^{-4}	3.75×10^{-5}	-8.43×10^{-5}	-2.28×10^{-3}
340	-5.99×10^{-5}	-3.44×10^{-4}	-1.10×10^{-5}	9.77×10^{-5}	-1.67×10^{-5}	4.72×10^{-3}	5.15×10^{-4}	-3.94×10^{-4}	2.72×10^{-4}	-2.47×10^{-3}
350	6.14×10^{-5}	-3.56×10^{-4}	1.68×10^{-5}	4.03×10^{-5}	-1.76×10^{-4}	4.09×10^{-3}	3.19×10^{-4}	-5.76×10^{-4}	1.65×10^{-4}	-2.37×10^{-3}

References

1. Pérez-Lombard, L.; Ortiz, J.; Pout, C. A review on buildings energy consumption information. *Energy Build.* **2008**, *40*, 394–398. [CrossRef]
2. Fiaschi, D.; Bandinelli, R.; Conti, S. A case study for energy issues of public buildings and utilities in a small municipality: Investigation of possible improvements and integration with renewables. *Appl. Energy* **2012**, *97*, 101–114. [CrossRef]
3. Costa, A.; Keane, M.M.; Torrens, J.I.; Corry, E. Building operation and energy performance: Monitoring, analysis and optimisation toolkit. *Appl. Energy* **2013**, *101*, 310–316. [CrossRef]
4. Santamouris, M. On the energy impact of urban heat island and global warming on buildings. *Energy Build.* **2014**, *82*, 100–113. [CrossRef]
5. Santamouris, M.; Cartalis, C.; Synnefa, A.; Kolokotsa, D. On the impact of urban heat island and global warming on the power demand and electricity consumption of buildings—A review. *Energy Build.* **2015**, *98*, 119–124.
6. Oh, S.J.; Ng, K.C.; Thu, K.; Chun, W.; Chua, K.J.E. Forecasting long-term electricity demand for cooling of Singapore's buildings incorporating an innovative air-conditioning technology. *Energy Build.* **2016**, *127*, 183–193. [CrossRef]
7. Nguyen, A.T.; Reiter, S. Passive designs and strategies for low-cost housing using simulation-based optimization and different thermal comfort criteria. *J. Build. Perform. Simul.* **2013**, *7*, 1–14.
8. Liping, W.; Hien, W.N. Applying Natural Ventilation for Thermal Comfort in Residential Buildings in Singapore. *Archit. Sci. Rev.* **2007**, *50*, 224–233.
9. Hens, H.S.L.C. *Heat, Air and Moisture Transfer in Insulated Envelope Parts: Performance and Practice, International Energy Agency*; Final Report Annex 24; International Energy Agency: Paris, France, 2002.
10. Strømann-Andersen, J.; Sattrup, P.A. The urban canyon and building energy use: Urban density versus daylight and passive solar gains. *Energy Build.* **2011**, *43*, 2011–2020. [CrossRef]
11. Bady, M.; Kato, S.; Huang, H. Towards the application of indoor ventilation efficiency indices to evaluate the air quality of urban areas. *Build. Environ.* **2008**, *43*, 1991–2004. [CrossRef]
12. Bady, M.; Kato, S.; Takahashi, T.; Huang, H. Experimental investigations of the indoor natural ventilation for different building configurations and incidences. *Build. Environ.* **2011**, *46*, 65–74. [CrossRef]
13. Cóstola, D.; Blocken, B.; Hensen, J.L.M. Overview of pressure coefficient data in building energy simulation and airflow network programs. *Build. Environ.* **2009**, *44*, 2027–2036. [CrossRef]
14. Hopfe, C.J.; Hensen, J.L.M. Uncertainty analysis in building performance simulation for design support. *Energy Build.* **2011**, *43*, 2798–2805. [CrossRef]
15. Nederlandse Organisatie voor Toegepast Natuurwetenschappelijk Onderzoek (TNO). 2012. Available online: http://cpgen.bouw.tno.nl/cp (accessed in 19 March 2013).
16. Heijmans, N.; Wouters, P. *Impact of the Uncertainties on Wind Pressures on the Prediction of Thermal Comfort Performances*; IEA ECBCS Annex 35; International Energy Agency: Paris, France, 2003.
17. Knoll, B.; Phaff, J.C.; de Gids, W.F. Pressure Simulation Program. In Proceedings of the 16th AIVC Conference, Palm Springs, CA, USA, 18–22 September 1995.
18. Sun, Y.; Heo, Y.; Tan, M.; Xie, H.; Wu, C.F.J.; Augenbroe, G. Uncertainty quantification of microclimate variables in building energy models. *J. Build. Perform. Simul.* **2014**, *7*, 17–32. [CrossRef]
19. Grosso, M. Wind pressure distribution around buildings: A parametrical model. *Energy Build.* **1992**, *18*, 101–131. [CrossRef]
20. Ramponi, R.; Angelotti, A.; Blocken, B. Energy saving potential of night ventilation: Sensitivity to pressure coefficients for different European climates. *Appl. Energy* **2014**, *123*, 185–195. [CrossRef]
21. Swami, M.V.; Chandra, S. Correlations for pressure distribution on buildings and calculation of natural-ventilation airflow. *ASHRAE Trans.* **1988**, *94*, 243–266.
22. Muehleisen, R.T.; Patrizi, S. A new parametric equation for the wind pressure coefficient for low-rise buildings. *Energy Build.* **2013**, *57*, 245–249. [CrossRef]
23. Kesten, D.; Tereci, A.; Strzalka, A.M.; Eicker, U. A method to quantify the energy performance in urban quarters. *HVACR Res.* **2012**, *18*, 100–111.

24. Van, T.N.; de Troyer, F. Deriving Housing Preferences from advertising on the web for improving decision making by Economic and Social actors. In Proceedings of the "At Home in the Housing Market"—RC43 Conference 2013, Amsterdam University, Amsterdam, The Netherlands, 10–12 July 2013.

25. Walker, I.S.; Wilson, D.J.; Forest, T.W. Wind Shadow Model for Air Infiltration Sheltering by Upwind Obstacles. *HVACR Res.* **1996**, *2*, 265–282. [CrossRef]

26. Sawachi, T.; Maruta, E.; Takahashi, Y.; Ken-ichi, S. Wind Pressure Coefficients for Different Building Configurations with and without an Adjacent Building. *Int. J. Vent.* **2006**, *5*, 21–30. [CrossRef]

27. Franke, J.; Sturm, M.; Kalmbach, C. Validation of OpenFOAM 1.6.x with the German VDI guideline for obstacle resolving micro-scale models. *J. Wind Eng. Ind. Aerodyn* **2012**, *104–106*, 350–359. [CrossRef]

28. Tominaga, Y.; Mochida, A.; Yoshie, R.; Kataoka, H.; Nozu, T.; Yoshikawa, M.; Shirasawa, T. AIJ guidelines for practical applications of CFD to pedestrian wind environment around buildings. *J. Wind Eng. Ind. Aerodyn.* **2008**, *96*, 1749–1761. [CrossRef]

atmosphere

MDPI

Article

An Investigation of the Quantitative Correlation between Urban Morphology Parameters and Outdoor Ventilation Efficiency Indices

Yunlong Peng [1], Zhi Gao [1], Riccardo Buccolieri [2] and Wowo Ding [1,*]

[1]   School of Architecture and Urban Planning, Nanjing University, Nanjing 210093, China;
      ylpeng717@foxmail.com (Y.P.); zhgao@nju.edu.cn (Z.G.)
[2]   Dipartimento di Scienze e Tecnologie Biologiche ed Ambientali, University of Salento,
      S.P. 6 Lecce-Monteroni, 73100 Lecce, Italy; riccardo.buccolieri@unisalento.it
*   Correspondence: dww@nju.edu.cn; Tel.: +86-25-8359-7332

Received: 26 July 2018; Accepted: 12 January 2019; Published: 16 January 2019

Abstract: Urban outdoor ventilation and pollutant dispersion have important implications for urban design and planning. In this paper, two urban morphology parameters, i.e. the floor area ratio (FAR) and the building site coverage (BSC), are considered to investigate their quantitative correlation with urban ventilation indices. An idealized model, including nine basic units with FAR equal to 5, is considered and the BSC is increased from 11% to 77%, generating 101 non-repetitive asymmetric configurations, with attention to the influence of plan density, volume ratio, and building layout on ventilation performance within urban plot areas. Computational Fluid Dynamics (CFD) simulations are used to assess the ventilation efficiency at pedestrian level (2m above the ground) within each model central area. Six indices, including the air flow rate (Q), the mean age of air (τ_P), the net escape velocity (NEV), the purging flow rate (PFR), the visitation frequency (VF), and the resident time (TP) are used to assess the local ventilation performance. Results clearly show that, fixing the FAR, the local ventilation performance is not linearly related to BSC, but it also depends on buildings arrangement. Specifically, as the BSC increases, the ventilation in the central area does not keep reducing. On the contrary, some forms with low BSC have poor ventilation and some particular configurations with high BSC have better ventilation, which indicates that not all high-density configurations experience poor ventilation. The local ventilation performance can be effectively improved by rationally arranging the buildings. Even though the application of these results to real cities requires further research, the present findings suggest a preliminary way to build up a correlation between urban morphology parameters and ventilation efficiency tailored to develop a feasible framework for urban designers.

Keywords: outdoor ventilation; urban morphology; building site coverage; ventilation efficiency

---

1. Introduction

The urbanization process in the past few decades has accelerated the increase of urban density, which leads to significant differences between urban morphology and original form in natural conditions (such as number of streets, height of buildings, void spaces, etc.). Due to the compact urban spaces, air pollutant cannot be diluted and dispersed in time, causing a series of health problems [1]. With this rapid expansion of cities, especially for some developing countries, the relationship between the urban building density and the urban local wind environment has increasingly become the focus of attention. The density and morphological characteristics of buildings play an important role in the local ventilation performance and the dispersion of pollutants. Especially, in central space that is surrounded by buildings with a high density of pedestrians, the ventilation performance is very important for their comfort and health.

Urban outdoor ventilation related to urban morphology can be assessed by full-scale field measurement, reduced-scale wind-tunnel or water channel experiment, and Computational Fluid Dynamics (CFD). With regard to full-scale field measurement, it depends on the meteorological conditions and sufficient measurement points to cover the areas with pivotal turbulence characteristics, especially for a large district [2,3]. Reduced-scale experiment and CFD can provide a controllable initial and boundary conditions [4]. However, wind-tunnel and water channel experiments are limited by the numbers and density of selected positions in order to gain the whole characteristics of flow field. When compared to reduced-scale experiments, CFD allows for providing the whole flow field data within the cells of the computational domain. Especially in recent years, with the turbulence models, computational resources and the development of best practice guidelines continued to increase [5,6], CFD has gradually become a widely used method to evaluate urban ventilation and pollutant dispersion. However, validation and verification are critical concerns for assuring the accuracy and the reliability of CFD simulations [7,8].

In the past 20 years, a large substantial series of ventilation efficiency indices have been proposed, validated, and applied to assess the local outdoor wind environment of urban areas. Ventilation indices initially developed and applied for indoor air quality [9–12], such as local mean age of air, air change rate, air exchange efficiency, purging flow rate, visitation frequency, and the residential time, were extended to predict outdoor urban ventilation [13–17]. The local mean age of air (τ_P) implies the time that one air particle reaches the given point in the specific domain after entering the flow field. A low value of local mean age of air illustrates well pollutant removal capability and a lower pollutant concentration within urban canopy layer The air change rate (*ACH*) refers to the frequency with which a given canopy volume of air is completely replaced by "fresh" air, and the air exchange efficiency (ε_a) denotes the efficiency of flushing the street network with external air. A high level of the two indices implies that much fresh air flushes the given volume and a better wind ventilation performance. The purging flow rate (*PFR*) quantifies the capacity that the pollutant is completely removed from a specific volume (i.e. the canopy layer) and a lower value indicates poor ventilation within the canopy. The visitation frequency (*VF*) is the number of times that a pollutant enters the domain and passes through it [13]. A high visitation frequency indicates high level pollutant concentration. Finally, the residence time (*TP*) is the time that a pollutant takes from once entering or being generated in the domain until leaving the domain [18]. While the above mentioned indices have been originally proposed to evaluate indoor air quality, other indices proposed directly for the urban street canyon model and asymmetric real city model are the exchange velocity, the pollutant exchange velocity, the net escape velocity, and the air delay [19–26]. The exchange velocity (u_e) and the pollutant exchange velocity ($u_{e,pollutant}$) focus on the turbulence diffusion between the in-canopy and above-canopy flows. They parametrize the mean convective exchange fluxes at the rooftop level and are usually related to the so called "city breathability". Based on the concept of *PFR* in urban areas, the net escape velocity (*NEV*) was proposed to reflect the net pollutant removal capability from the specified volume by mean flow and turbulence diffusion. The air delay (τ_d) is a recent optimization measure that is independent of the initial values, as it presents the amount of time delay that is caused by the presence of the urban geometry in each point inside the domain [26].

By employing some or a combination of the above indices, several studies have investigated the relationship between geometric parameters and ventilation efficiency [21,27–31]. Various studies focused on idealized urban models and real-complexity urban models of different planar area density (λ_p) and frontal area density (λ_f) [16,24,32,33]. However, for urban designers and planners, key factors also involve economics, policy, and urban development strategies. In this perspective, in the early planning stage the floor area ratio (FAR) of urban plot is determined first, and the relationship between building site coverage (BSC) and ventilation performance in a specific area is evaluated [34]. Kubota et al. [35] collected the building coverage ratio (BCR, means BSC in this paper) and the FAR of 22 actual Japanese cities and used wind tunnel experiments to explore the relationship between BCR and wind velocity ratio. Results have shown that the average pedestrian-level wind velocity can be explained well by the BCR rather than by the FAR. Moreover, the FAR and the BSC have been employed by previous research to investigate the relationship between urban fabric and outdoor wind environment, especially in high-density urban areas [33,36,37].

Within this context, the two urban morphology parameters FAR and BSC are here considered to investigate their quantitative correlation with six urban ventilation indices. An idealized model, including nine basic units with FAR equal to 5 is considered; the BSC is increased from 11% to 77% generating 101 non-repetitive asymmetric configurations. Computational Fluid Dynamics (CFD) simulations are performed to assess the local ventilation efficiency at pedestrian level within each model central area. Here, the pedestrian level is taken at z = 2 m, as commonly employed in previous ventilation studies [13,15,24,26,38–41].

2. Methods

2.1. Description of Urban Blocks Morphology Characterists

As mentioned in the Introduction, during the early stage of urban planning and design, the floor area ratio (FAR) is usually fixed, being closely related to costs, benefits, development strategies, etc. The main factors affecting the morphology include the building site coverage (BSC) and the architectural arrangement. FAR represents the ratio between total floor areas and lot area, as indicated in Figure 1, which also shows the definition of BSC.

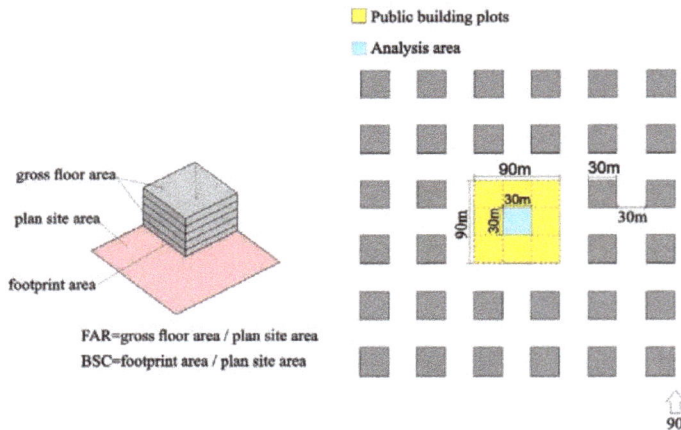

Figure 1. Description of floor area ratio (FAR) and building site coverage (BSC). The yellow area is the plot area considered in this paper (which can be covered by buildings), while the blue area is the central area which cannot be covered by buildings and where the ventilation indices are calculated.

Here, an idealized model consisting of nine basic units in the plot area is considered, where each unit size is a 30-meter square. The central square (unit) is the area that cannot be covered by buildings ("central area" hereinafter) and is the area where the ventilation indices are calculated, while the other eight units surrounding the central area may be covered by buildings. FAR is fixed to 5, and the initial BSC is 11%, meaning that there is only one high-rise building. Meanwhile, there are five possibilities for the position of building in the whole plot area (excluding symmetry conditions). The central area represents a typical square that is surrounded by buildings of different height and arrangement. Ventilation in this area is worth of investigation since people spend more time here than in streets.

Thus, in order to explore the correlation between morphology and ventilation performance of the central area, all possible configurations in the plot area are considered, as shown in Figure 2.

According to the architectural layout around the central area, the model is divided into "C form" and "B form". The C form means that the front and back directions of the central area are not blocked and a channeling effect is expected. The B1~B3 form means the number of buildings along the windward projection. The B(F1~F3) form means the number of building front (in the first two rows) along the windward projection. An approaching wind direction perpendicular to building façades is considered, as it is commonly associated with the worst scenario, leading to high pollution levels around the single building, in street intersections, and in the perpendicular streets [42,43]. As a consequence, the symmetrical configurations along the east-west direction are excluded. When BSC increases from 11% to 77%, 101 different asymmetrical configurations are generated in total. Please note that the building height diminishes moving from BSC=11% to 77%, specifically they are 135 m, 69 m, 45 m, 36 m, 27 m, 24 m, and 21 m, respectively.

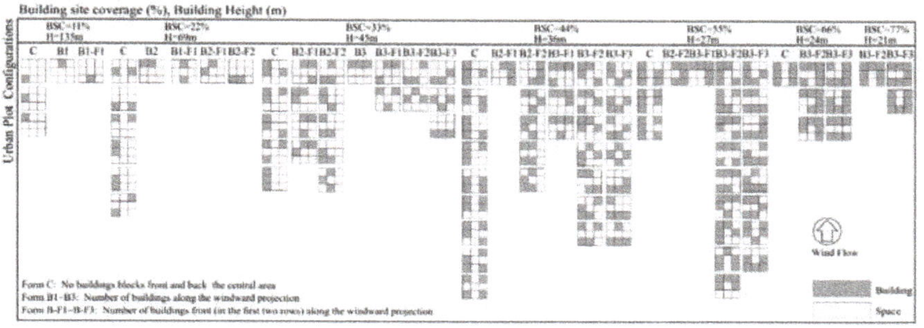

Figure 2. Idealized configurations with BSC increasing from 11% to 77% in the plot area, with indication of building heights and forms. Please note that the wind blows from below.

Conceptually, BSC is similar to λ_p. Here, BSC is employed in conjunction with FAR that limits the volume plot ratio, as typically done when a new building arrangement has to be built in China. Even though in the literature there are several studies relating λ_p to ventilation performance (see the Introduction), here we recognize that actually many possible combinations are available for the same λ_p (and thus BSC) for a given urban plot. Further, in Chinese cities, such as Nanjing, even in dense commercial plots, there are still some multi-storey or low-rise buildings surrounding the high-rise buildings. In this case, the corresponding urban texture becomes quite fuzzy. In this case, λ_p and λ_f may not clearly quantify the urban morphological characteristics of the fuzzy urban slice. With the morphological complexity, such as lift-up, road network inside the plot, etc., more parameters might more clearly describe the morphological characteristics of a specified area, such as FAR and BSC employed here. Finally, here we analyze the ventilation performance using six different ventilation indices (described below) and attempt to provide some suggestions about the indices that mostly correlate with urban morphological parameters.

The ventilation of the whole area, i.e. the city potential of removing pollutants and other scalars, is determined by the resistance generated by buildings, which usually implies that flow through the city downstream exit is lower than that entering the city upstream [16]. This is due to the drag force generated by the whole arrangement of buildings and cannot discriminate the single building [44]. From a practical and architectural perspective, the cases analyzed here mimic a real situation that an urban planner is faced with when a lot area of a given (fixed) size and FAR is allocated to the development of new built areas, where BSC and layout have to be chosen to optimize the ventilation.

We underline that idealized geometries have been here considered here to better isolate and elucidate the effect of changing BSC and building layout on ventilation and to evaluate the sensitivity of six different ventilation indices.

2.2. Description of the Outdoor Ventilation Efficiency Indices

The six ventilation indices employed to evaluate the ventilation in the central area are briefly described below.

The flow rate (Q) represents the capability of pollutant dilution through street openings and its roof by wind flow [15]. It is defined as:

$$Q = \int_{A_0} \overline{u} dA_0 \qquad (1)$$

Flow rates through street openings and roof are normalized by a reference flow rate (Q_∞) (the flow rate far upstream through the same area of the windward opening), as follows [15]:

$$Q_\infty = \int_{A_0} \overline{u}_\infty dA_0 \qquad (2)$$

$$Q^* = \int_{A_0} \overline{u} dA_0 / Q_\infty \qquad (3)$$

where Q^* is the normalized flow rate, $\overline{u}$ and $\overline{u}_\infty$ is the horizontal (stream-wise) velocity along the x direction and that far upstream, and A_0 is the area of street opening.

The purging flow rate (PFR) is the net flow rate by which the specified volume is purged out of pollutant [10]. If passive pollutants are uniformly released in the entire pedestrian volume (in our case, it is fixed and corresponds to the volume from the ground to the pedestrian height of the central area), the pedestrian purging flow rate (PFR_{ped}) is defined as:

$$PFR_{ped} = \frac{S_c \times Vol}{\langle C \rangle} = \frac{S_c \times Vol}{\int_{Vol} Cdxdydz / Vol} \qquad (4)$$

where S_c means the uniform pollutant generation rate (kg/m³·s), $\langle C \rangle$ is the spatially averaged concentration in the entire pedestrian volume (kg/m³), and Vol is the entire pedestrian volume (m³). The normalized PFR^*_{ped} is:

$$PFR^*_{ped} = PFR_{ped} / Q_\infty \qquad (5)$$

The net escape velocity NEV is based on the concept of PFR and has been proposed by Hang et al. [17]. NEV_{ped} represents the net capacity of removing/diluting pollutant from the entire pedestrian volume by both mean flow and turbulent diffusion and it is defined as:

$$NEV_{ped} = \frac{PFR_{ped}}{A_0} \qquad (6)$$

where A_0 is the area of boundaries for the entire pedestrian volume.

In addition to the level of the pollutant concentration, the pollutant behavior within the domains is also important for local ventilation assessment [13]. The visitation frequency (VF) is, in this sense, an index to measure the pollutant behavior. The VF can be calculated as:

$$VF = \frac{\Delta q_p}{S_c \times Vol} \qquad (7)$$

where Δq_p is the inflow flux of pollutants into the domain (kg/s). It is assumed that the inflow flux in the street canyon is almost exclusively affected by the value of the mean part of the flux [13].

The residence time (TP) is the time that it takes once a particle enters or is generated in the specified domain until it leaves there. Similar to VF, the residential time represents the time that pollutant enters or is generated in the street canyon. It is defined as:

$$TP = \frac{Vol}{PFR_{ped} \times VF} \qquad (8)$$

The age of air represents the mean time requiring the inflow air to reach a certain point in the specified volume. Currently, it is commonly employed to assess the ventilation efficiency of urban built-up areas [15,32,34,45]. The local mean age of air ($\bar{\tau}_p$) can be calculated by:

$$\bar{\tau}_p = \frac{\bar{c}}{S_c} \tag{9}$$

where $\bar{c}$ is the local pollutant concentration (kg/m^3). The local mean age of air can be normalized, as follows:

$$\bar{\tau}_p^* = \frac{\bar{\tau}_p \times Q_{ref}}{Vol_{ref}} \tag{10}$$

where Vol_{ref} (m^3) is the entire pedestrian volume.

2.3. CFD Simulations Set-Up

2.3.1. Computational Domain and Grid

The computational domain was built using hexahedral elements, with a finer resolution within the entire building area (the expansion rate between two consecutive cells was below 1.2). The influence on the predictions of the choice of grid size, using several refined meshes, was verified (see the Supplementary Material S1). Specifically, the minimum cell size were $\delta x = 0.8$ m, $\delta y = 0.8$ m, $\delta z = 0.5$ m in the plot area, which means that at least three cells were present up to pedestrian level and at least 10 cells per cube root of the building volume and 10 cells per building separation, as suggested by [5,6]. The lateral and inflow boundaries were set to 5H away from the building groups. In order to unify the computational domain size, we set H = 135 m, which corresponds to the highest building of BSC = 11%. The outflow boundary was 15H away from the building groups, and the height of the computational domain was 11H. Extra 30 m high buildings were considered around the central area. The grid resolution and the computational domain size fulfill the major microscale simulation requirements recommended by the AIJ (Architectural Institute of Japan) guidelines [6] and the COST action 732 [5]. The computational domain is shown in Figure 3.

We are aware that the surrounding buildings affect the flow field before reaching the central area. These buildings have been included to explicitly reproduce roughness elements and thus the real effects of urban elements (whose height is typical of eight to ten-story buildings in China), avoiding considering an isolated central area that does not represent a real situation. For this reason, we think that the results presented in Section 3 are representative of the effects of urban-like geometries on flow, ventilation, and pollutant concentration.

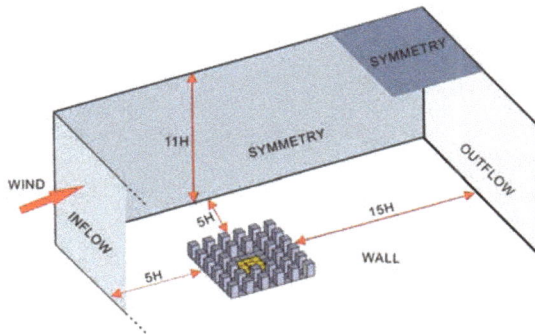

Figure 3. Schematic sketch of the computational domain and boundary conditions used in Computational Fluid Dynamics (CFD) simulations. The yellow area represents the plot area (see Figure 1).

2.3.2. Turbulence Model

For the choice of turbulence model, the standard Reynolds-averaged Navier–Stokes (RANS) and Large Eddy Simulation (LES) are most commonly compared and discussed in urban ventilation simulation. The accuracy of LES has been determined in adequate studies, especially in predicting the flow and turbulence characteristics in low wind speed region [25,46–48]. When compared to LES, the limitation of steady RANS method has been reported in many literature studies [49–52]. However, the computational cost of the LES model is several times higher than that of the RANS model. Meanwhile, RANS approach is still the most commonly approach of urban ventilation assessment, and it has shown good results for generic urban configurations [6,13,15–17,27,36]. For the purpose of this study, the employed RANS approach with the *k-ε* turbulence model is considered appropriate in comparison with other turbulence model, such as Large Eddy Simulations (LES), especially since the interest is on mean values ([53]). To further evaluate the standard *k-ε* turbulence model employed here, a validation study has been performed and results are presented in the Supplementary material S2.

2.3.3. Boundary Conditions and Solver Settings

For the inflow boundary conditions, the inlet profiles for $U(z)$, $k(i)$ and $\varepsilon(z)$ in the atmospheric boundary layer follows the COST recommendation presented by Richards and Hoxey [54].

The velocity profile (U) was:

$$U(z) = \frac{u^*_{ABL}}{\kappa} \ln\left(\frac{z + z_0}{z_0}\right) \tag{11}$$

where u^*_{ABL} is the atmospheric boundary layer friction velocity, z_0 is the aerodynamic roughness length, z is the height coordinate (m), and κ is the von Karman constant (0.4). Taking Nanjing as an example, we set $z_0 = 0.4$, which may represent woodlands within and surrounding the city [55], and a wind speed equal to about 2 m/s at 10 m, which roughly corresponds to the 10th percentile of the mean hourly wind speeds (https://weatherspark.com/y/132872/Average-Weather-in-Nanjing-China-Year-Round) and it is expected to lead to poorer ventilation conditions than larger wind speeds. These values lead to a friction velocity equal to 0.23 m/s, which is typical of isothermal conditions analyzed here. In this way, we represented a change of surface roughness (smooth–rough) mimicking the transition from the rural to the urban environment. The turbulent kinetic energy profile (k) and turbulent dissipation rate profile (ε) were:

$$k(z) = \frac{u^*_{ABL}{}^2}{\sqrt{C_\mu}} \tag{12}$$

$$\varepsilon(z) = \frac{u^*_{ABL}{}^3}{\kappa(z + z_0)} \tag{13}$$

where C_μ is a constant (0.09). Zero static gauge pressure condition was defined as the outflow boundary condition. At the top and lateral boundaries of the computational domain, the symmetry boundary conditions were applied to avoid the wall effect. Non-slip wall boundary conditions were set on the ground surface and all of the building surfaces.

The passive pollutant used is carbon monoxide (CO, $S_c = 10^{-5}$ kg m^{-3} s^{-1}) and the pollutant source is defined as a volume source from the ground to pedestrian height (i.e., from z = 0 to z = 2 m) of the central area using the uniform homogenous method that was widely employed in evaluating ventilation efficiency of urban canopy layers [56].

Fluent (version 15.06) was applied to solve three-dimensional (3D) steady RANS equations for incompressible and isothermal flow. Pressure interpolation was in second order accuracy. For both convection and viscous terms of the governing equations, second-order discretization schemes were used. The modeling settings are summarized in Table 1.

Table 1. CFD simulation condition settings.

Calculation conditions	Solver settings
Computational domain	(x)1080 m × (y)1920 m × (z)525 m
Grid resolution	about 2 million hexahedral cells
Turbulence model	Standard k–ε turbulence model
Algorithm for pressure-velocity	SIMPLE
Scheme for advection terms	Second-order discretization for convection terms and the viscous terms
Boundary conditions	Inflow: Boundary condition presented by Richards and Hoxey (1993) Outflow: Zero gradient condition Ground and block surfaces: Non-slip wall Top and lateral surfaces: Symmetry

3. Results and Discussion

3.1. Impact of Building Layout

As an example, Figure 4 shows the wind velocity of six cases at the pedestrian level. The flow feature in Figure 4a is consistent with simulation results of previous studies [13,32,46,57], i.e., when the BSC is 11%, the flow pattern is characterized by acceleration and flow separation, reattachment flow downstream and wake region. The velocity of the central area is obviously influenced by the high-rise buildings on the leeward side. By increasing the BSC to 44%, despite that the height (H = 36 m) is much lower than that of 11% (H = 135 m), the wind flow in the central area is still blocked by the leeward buildings (Figure 4b). Further, Figure 4c,d show that, still increasing the BSC, the flow patterns are very similar, which means that although the BSC increases and the building height decreases, this configurations have a similar indirect channeling effect. Finally, Figure 4e,f show that despite the geometries are similar (with building heights of 27 m and 21 m), the flow field characteristics are obviously different, which shows that, when the current row of buildings is completely blocked, the most significant factor affecting the central area is the building height but not the long space that is enclosed by the building.

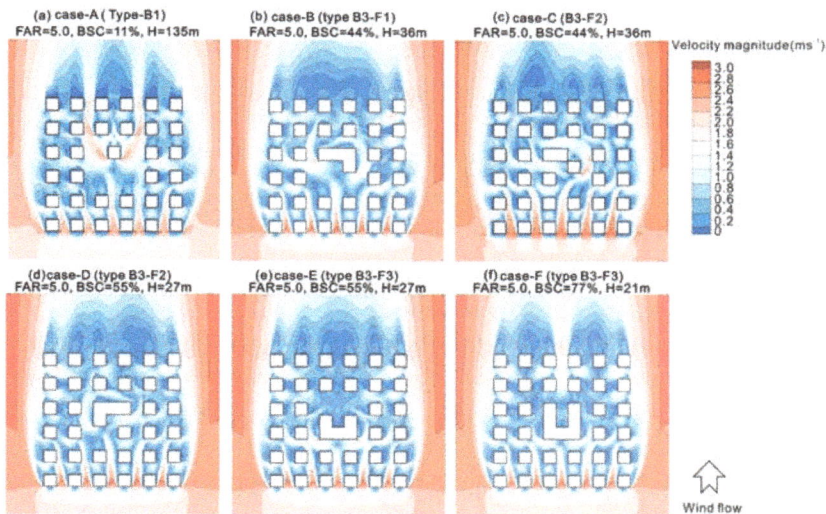

Figure 4. Velocity magnitude at pedestrian level for BSC equal to (**a**) 11%, (**b,c**) 44%, (**d,e**) 55%, (**f**) 77%.

The ventilation within the plot area is obviously related to the flow patterns that are shown in Figure 4. In order to have a complete picture of the ventilation, Figure 5 shows the indices *Q*, *PFR*, *NEV*, and *VF* for the different cases investigated. The ventilation performance of C forms is always better than other types for BSC of 11% to 66%. This is expected, since such C forms are not blocked by any front leeward obstacle. In addition, when BSC is 11% (H = 135 m), the corresponding C forms are made of single independent obstacle. Although there is no obstacle blocking the wind, no channeling effect occurs, so the air flow through the central area is less than that found for BSC of 22% (H = 69 m) to 55% (H = 27 m). When BSC is 66% (H = 24 m), the corresponding C form experiences an air channeling effect, but the lower building height makes part of the airflow exchanging through the street top and thus reduces the flow through the central area. When BSC is 77% (H = 21 m), no C type is possible, but B3–F2 and B3–F3 exists. Obviously, high BSC results in less air flow flushing into the central area by mean flow. However, the lower building height induces highly vertical turbulent diffusion exchange.

Figure 5 illustrates that B3–F3 forms also experience the poorest ventilation conditions within the central area. Contrary to other types, the local ventilation performance of B3–F3 forms central area that gradually increases with increasing BSC from 33% to 77%. Overall, this suggests that when FAR is fixed and the central area is blocked by surrounding obstacles, the local ventilation efficiency can be improved by reducing the average height of the building and increasing BSC within the plot area.

It should be also noted that when BSC increases from 11% to 77% in general the ventilation efficiency within the central area decreases due to the increase of buildings within the plot area. However, looking at *NEV* and *VF* (Figure 5c,d), it can be noted that, when BSC increases from 44% to 77%, *NEV* and *VF* slightly improves. Overall, there is no clear relationship between ventilation indices and the various BSC. Therefore, in order to achieve useful conclusions on such a relationship, the average value and the error have been calculated for each configuration, and the results are discussed in the next subsection.

The ventilation performance that is analyzed in Figure 5 directly affects the concentration distribution at the pedestrian level, as shown in Figure 6. When BSC is 11%, meaning only one high-rise building (H = 135 m) in the central area, i.e., case-A (type B1), as expected from flow pattern (Figure 4), the pollutant concentration is significantly affected by the isolated high-rise building. In particular, since the central area is on the leeward side of the building, less air flows into the central area by mean flow. Due to the high level aspect ratio (H/W > 2), third or fourth recirculation occurs and weakens the capability of pollutant dilution and removal from the pedestrian level, resulting in a high concentration level in this type of configurations. When the BSC increases to 44% (H = 36 m), looking at the concentration distribution of case-B (type B3–F1) and case-C (type B3–F2), an indirect wind channeling occurs in case-C. Therefore, the mean flow passing through the central area plays a major role in removing pollutant. However, the H/W corresponding to this BSC is about 1.0. From the vertical distribution of the pollutant concentration of case-B, it can be seen that, due to the obvious obstruction, turbulent diffusion begins to influence the central pedestrian level concentration through the roof, and part of the pollutant is removed by the turbulent diffusion effect, which implies that the building height is not the main influence factor of this BSC. When the BSC increases to 55% (H = 27 m) and 77% (H = 21 m), the case-E (type B3–F3) and case-F (type B3–F3) have a relative similar H/W and architectural layout. It can be seen that the turbulence diffusion through the roof now plays a significant role in diluting the pollutant at the pedestrian level. Meanwhile, when compared with the higher level of BSC, i.e., 55%, 66%, and 77%, very few configurations with wind channeling occur. Therefore, when the BSC exceeds 55%, the BSC becomes an important factor determining the ventilation efficiency within the central area rather than the building height.

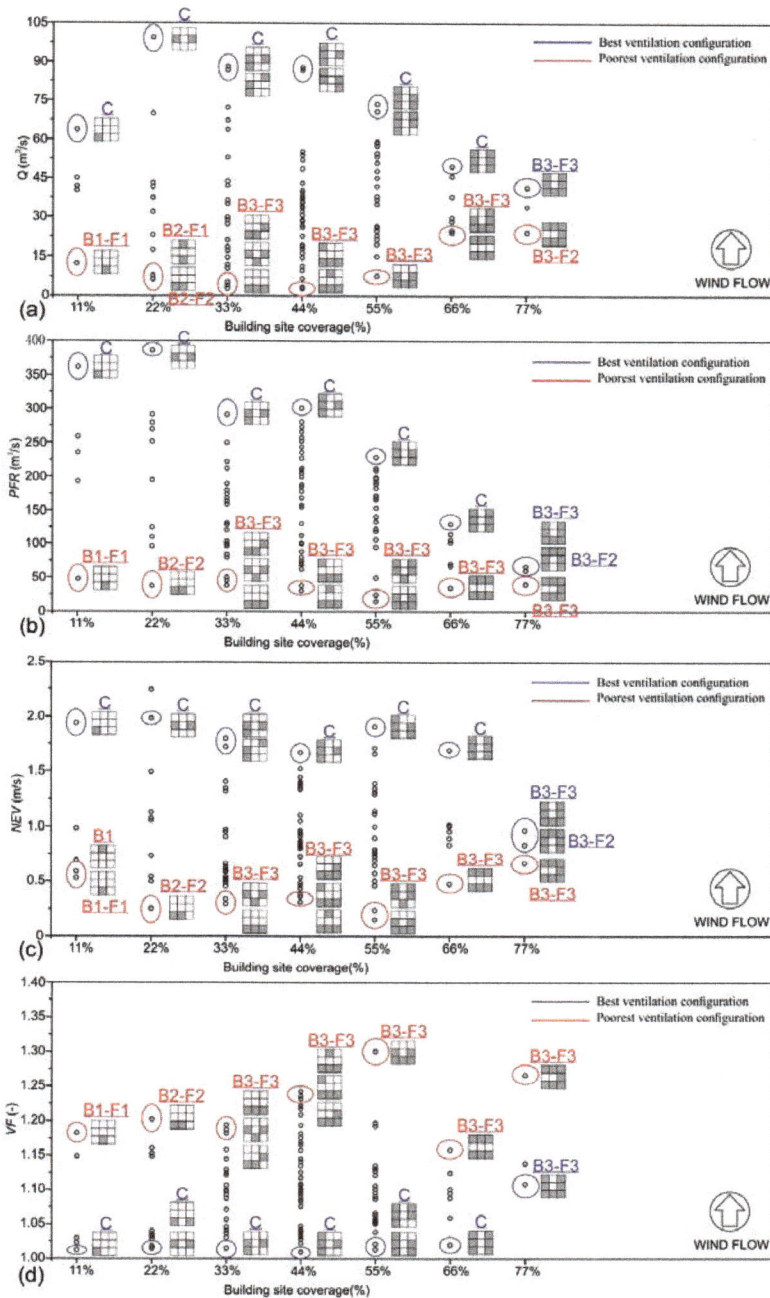

Figure 5. Values of ventilation indices (**a**) air flow rate (*Q*), (**b**) purging flow rate (*PFR*), (**c**) net escape velocity (*NEV*), and (**d**) visitation frequency (*VF*) for each configuration with BSC increasing from 11% to 77%. The best (blue) and poorest (red) configurations and the corresponding forms are explicitly indicated.

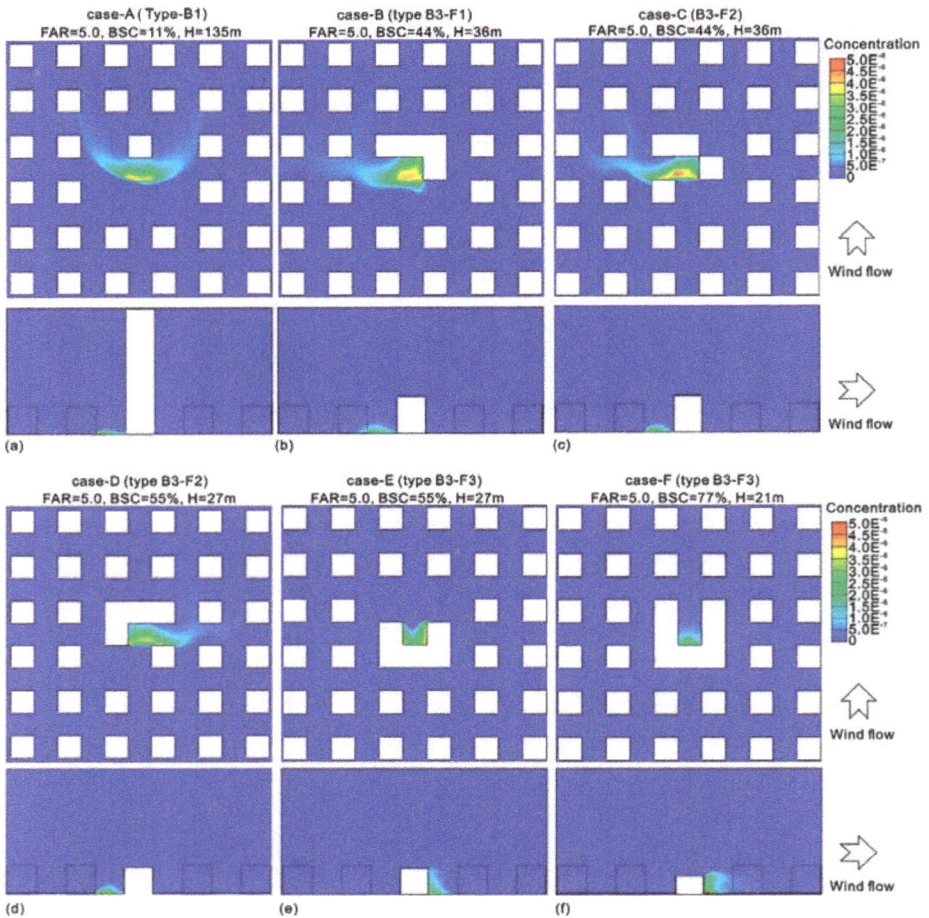

Figure 6. Pollutant concentration at pedestrian level and vertical distribution for BSC equal to (a) 11%, (b,c) 44%, (d,e) 55%, and (f) 77%.

3.2. Relationship between Local Ventilation Efficiency and BSC

Figure 7 shows the average and error values of the six ventilation indices (air flow rate (Q), τ_p, *NEV*, *PFR*, *VF*, and *TP*) for all the configurations investigated (see Figure 2). Q, *NEV*, and *PFR* (Figure 7a,c,d) decrease gradually with increasing BSC, which indicates that the ventilation performance within the central area becomes poorer. On the other hand, τ_p, *VF*, and *TP* gradually rise with increasing BSC, reflecting the decrease of pollutants removal capability in the central area (Figure 7b,e,f).

It is noted that the value of the aforementioned ventilation indices stopped decreasing when the BSC becomes 44%, but there is a slight increase with increasing BSC. This is due to the large impact of C forms, while the other forms have no direct wind channeling effect due to the blockage of the buildings. As BSC continues to increase from 44% (thus the building height decreases), it can be seen that the trend of the ventilation indices is no longer obvious as happens for BSC of 11% to 44%, although the increase of BSC causes the horizontal airflow channeling to gradually decrease. In fact, due to the reduction of building height, the turbulent diffusion effect within the building roof becomes remarkable. This is consistent with the research results of [24].

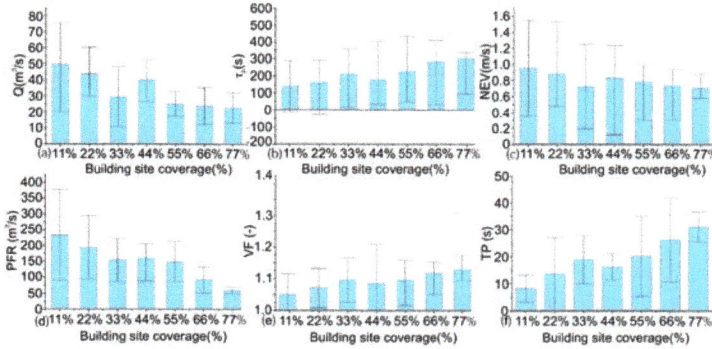

Figure 7. Average values and variances of ventilation indices (**a**) Q, (**b**) τ_p, (**c**) NEV, (**d**) PFR, (**e**) VF, and, (**f**) resident time (TP) for all the configurations with BSC increasing from 11% to 77%.

3.3. Sensitivity of the Six Ventialtion Indices to the BSC

Since the purpose of this paper is to try to establish a correlation between the morphological parameters and ventilation performance parameters, it is worth evaluating how the different indices change when BSC changes (sensitivity analysis). Figure 8 shows the average normalized values of the six indices (Q^*, VF, NEV^*, PFR^*, TP and $\overline{\tau p^*}$) of C, B2, and B3 forms.

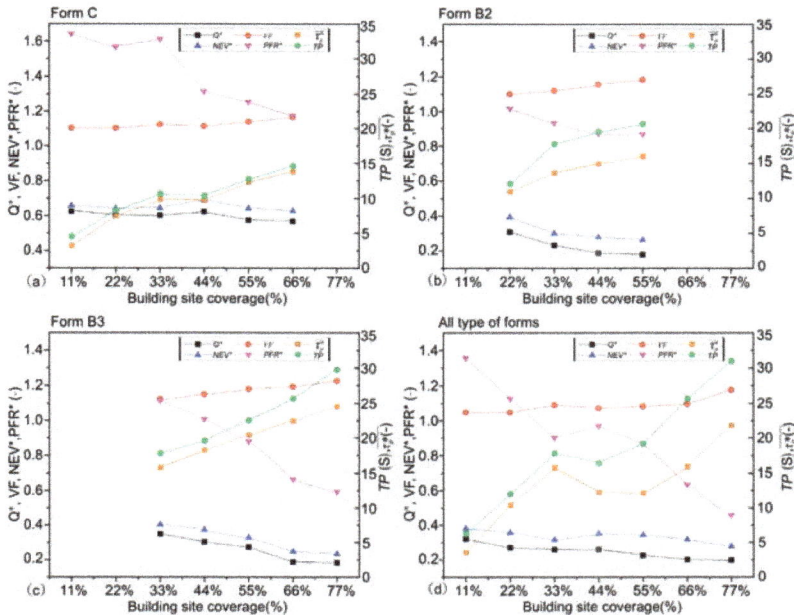

Figure 8. Average values of the six ventilation for all of the configurations with BSC increasing from 11% to 77%: (**a**) C forms, (**b**) B2 forms, (**c**) B3 forms, and (**d**) all type of forms.

As shown in Figure 8a (C forms), the values of each index do not change significantly with increasing BSC. This is due to the channeling effect that occurs in these cases, since, with increasing BSC, only the upwind airflow is affected and thus there is no significant effect on the air flowing in the horizontal direction downstream. For B2 forms (Figure 8b), even though BSC are lower or equal to C forms, the position change of buildings induces a significant impact on the ventilation performance within the central area. Finally, for B3 forms (Figure 8c), since the airflow is blocked upstream to the central area, the trend of each index is relatively obvious, especially for BSC of 33% to 55%. On the other hand, when BSC is 66% and 77%, the building arrangements are quite similar (i.e., there is less space to move buildings), resulting in a low change of each index.

To summarize, Figure 8d shows the trend of all the indices for all the forms (and thus for all the 101 configurations) investigated. The figure shows that *PFR*, $\overline{\tau_p}$, and *TP* are the most sensitive to BSC variation up to 44%, since the wind channeling effect is expected to dominate the exchange with the underlying atmosphere. With increasing BSC (and thus the density), the wind channeling effect is not the only and effective mechanism to dilute or remove air pollutants. In fact, since the average building height is relatively low, the effect of turbulence can effectively remove the pollutants at the pedestrian level and improve the local ventilation performance. At the same time, the results have shown that BSC is not the only factor that determines the ventilation performance within the urban plot, and the building patterns within the plot area also plays a significant role on flow and pollutant dispersion.

4. Conclusions

This paper aims to investigate a quantitative correlation between urban-like geometries of different building site coverage (BSC) and six ventilation indices that are commonly employed in the literature. The floor area ratio (FAR) is kept constant at 5.0, while BSC gradually increases from 11% to 77% (while building heights H decrease from 135 m to 21 m, respectively), resulting in a total of 101 asymmetrical idealized configurations. Results show that, among the indices investigated, *PFR*, *TP*, and $\overline{\tau_p}^*$ better correlate with urban morphological parameters. For improving ventilation and pollutant dispersion, the study has shown that:

- when BSC is 11% (H = 135 m), 22% (H = 69 m) and 33% (H = 45 m), the density of buildings in the central area is low, and the ventilation performance is greatly affected by the large building heights, regardless of the architectural arrangement with channeling effect;
- with increasing BSC to 44% (H = 36 m), 55% (H = 27 m), 66% (H = 24 m), and 77% (H = 21 m), the ventilation performance is slightly lower than before. However, due to the lower average building heights, more airflow flushes into the central area by turbulent flow and a vertical exchange through the roof occurs, diluting part of the pollutants; and,
- for the same BSC, the configurations experiencing a wind channeling effect are characterized by a better ventilation.

From an urban designing and planning perspective, results suggest that, when FAR is fixed in a plot area, intermediate BSCs (and building heights) may experience airflow channels by controlling the layout, which is an effective way to increase the overall ventilation and thus the ability of pollutant dispersion. These results show that by using BSC and FAR as morphological parameters, the ventilation performance of void spaces can be effectively improved in high-density cities by reducing the average height of buildings. Specifically, a BSC = 44% (H = 36 m) of C form (i.e., no buildings in front of the central area and with channeling effect) seems to simultaneously maximize the ventilation performance and the number of possible layouts that are also available to harmonize the new built area to the existing surrounding morphology.

This study investigated idealized configurations that were subjected to a wind perpendicular to building façades and thus further research is needed when considering more complex geometries (such as, for example, secondary roads within the central area) and oblique wind directions, as well as non-isothermal conditions with strong heat fluxes and low wind events, which may be critical with respect to ventilation and human thermo-physiological conditions. Further, the impact of passive methods [58], including vegetation [59–61], as well as lift-up design [62], which are becoming increasingly popular in China to improve the pedestrian level wind and thermal comfort.

Supplementary Materials: The supplementary materials are available online at http://www.mdpi.com/2073-4433/10/1/33/s1.

Author Contributions: W.D. and. Z.G were responsible for the study design and supervision. Y.P. wrote the manuscript, performed the CFD simulation and conducted data analysis; R.B. revised the paper and wrote sections of the manuscript. All authors contributed to the discussion of the results and have read and approved the final manuscript.

Funding: The research was supported by the key project funded by the National Science Foundation of China on "Urban form-microclimate coupling mechanism and control", Grant No. 51538005, and Supported by the Scientific Research Foundation of Graduate School of Nanjing University NO. 2017CL13 "Ventilation Performance and Pollutant Dispersion Mechanism under the Influence of Complex Urban Morphological Characteristics"

Conflicts of Interest: The authors declare no conflict of interest.

Nomenclature

FAR	floor area ratio (%)
BSC	building site coverage (%)
Q	flow rate (m^3 s^{-1})
Q_∞	reference flow rate far upstream (m^3 s^{-1})
Q^*	normalized flow rate through street openings or street roofs
PFR	pedestrian purging flow rate
PFR*	normalized pedestrian purging flow rate value
NEV_{ped}	pedestrian net escape velocity (m/s)
VF	visitation frequency
TP	pollutant residential time (s)
$\overline{\tau}_p, \overline{\tau}_p^*$	local mean age of air (s) and its normalized value
Sc	pollutant source rate (kg m^{-3} s^{-1})
$\langle C \rangle$	the spatially averaged concentration in the entire pedestrian domain (kg/m^3)
Vol	the pedestrian volume form ground up to 2m in the canopy (m^3)
k, ε	turbulent kinetic energy (m^2 s^{-2}) and its dissipation rate
u_{ABL}^*	atmospheric boundary layer friction velocity (m/s)
z_0	aerodynamic roughness length (m)
κ	the von Karman constant
z	height coordinate (m)
H	height of central area buildings

References

1. Raaschounielsen, O.; Andersen, Z.J.; Beelen, R.; Samoli, E.; Stafoggia, M.; Weinmayr, G.; Hoffmann, B.; Fischer, P.; Nieuwenhuijsen, M.J.; Brunekreef, B. Air pollution and lung cancer incidence in 17 european cohorts: Prospective analyses from the european study of cohorts for air pollution effects (ESCAPE). *Lancet Oncol.* **2013**, *14*, 813–822. [CrossRef]

2. Yoshie, R.; Mochida, A.; Tominaga, Y.; Kataoka, H.; Harimoto, K.; Nozu, T.; Shirasawa, T. Cooperative project for CFD prediction of pedestrian wind environment in the architectural institute of Japan. *J. Wind. Eng. Ind. Aerodyn.* **2007**, *95*, 1551–1578. [CrossRef]

3. Ketzel, M.; Berkowicz, R.; Lohmeyer, A. Comparison of numerical street dispersion models with results from wind tunnel and field measurements. *Environ. Monit. Assess.* **2000**, *65*, 363–370. [CrossRef]

4. Blocken, B.; Stathopoulos, T.; Carmeliet, J.; Hensen, J.L.M. Application of computational fluid dynamics in building performance simulation for the outdoor environment: An overview. *J. Build. Perform. Simul.* **2011**, *4*, 157–184. [CrossRef]

5. Franke, J.; Hellsten, A.; Schlünzen, H.; Carissimo, B. *Best Practice Guideline for the Cfd Simulation of Flows in the Urban Environment*; COST Office: Brussels, Belgium, 2007.

6. Tominaga, Y.; Mochida, A.; Yoshie, R.; Kataoka, H.; Nozu, T.; Yoshikawa, M.; Shirasawa, T. Aij guidelines for practical applications of cfd to pedestrian wind environment around buildings. *J. Wind. Eng. Ind. Aerodyn.* **2008**, *96*, 1749–1761. [CrossRef]

7. Blocken, B. Computational fluid dynamics for urban physics: Importance, scales, possibilities, limitations and ten tips and tricks towards accurate and reliable simulations. *Build. Environ.* **2015**, *91*, 219–245. [CrossRef]

8. Tominaga, Y.; Stathopoulos, T. Ten questions concerning modeling of near-field pollutant dispersion in the built environment. *Build. Environ.* **2016**, *105*, 390–402. [CrossRef]

9. Etheridge, D.W.; Sandberg, M. *Building Ventilation. Theory and Measurement*; John Wiley & Sons: Chichester, NY, USA, 1996.

10. Sandberg, M.; Sjöberg, M. The use of moments for assessing air quality in ventilated rooms. *Build. Environ.* **1983**, *18*, 181–197. [CrossRef]

11. Awbi, H.B. *Ventilation of Buildings*; E & FN Spon: London, UK, 1991.

12. Chen, Q. Ventilation performance prediction for buildings: A method overview and recent applications. *Build. Environ.* **2009**, *44*, 848–858. [CrossRef]

13. Bady, M.; Kato, S.; Huang, H. Towards the application of indoor ventilation efficiency indices to evaluate the air quality of urban areas. *Build. Environ.* **2008**, *43*, 1991–2004. [CrossRef]

14. Kato, S.; Huang, H. Ventilation efficiency of void space surrounded by buildings with wind blowing over built-up urban area. *J. Wind. Eng. Ind. Aerodyn.* **2009**, *97*, 358–367. [CrossRef]

15. Hang, J.; Sandberg, M.; Li, Y. Age of air and air exchange efficiency in idealized city models. *Build. Environ.* **2009**, *44*, 1714–1723. [CrossRef]

16. Buccolieri, R.; Sandberg, M.; Di Sabatino, S. City breathability and its link to pollutant concentration distribution within urban-like geometries. *Atmos. Environ.* **2010**, *44*, 1894–1903. [CrossRef]

17. Hang, J.; Li, Y.; Sandberg, M.; Buccolieri, R.; Di Sabatino, S. The influence of building height variability on pollutant dispersion and pedestrian ventilation in idealized high-rise urban areas. *Build. Environ.* **2012**, *56*, 346–360. [CrossRef]

18. Kato, S.; Ito, K.; Murakami, S. Analysis of visitation frequency through particle tracking method based on les and model experiment. *Indoor Air* **2010**, *13*, 182–193. [CrossRef]

19. Neophytou, M.; Britter, R. Modelling of atmospheric dispersion in complex urban topographies: A computational fluid dynamics study of the central London area. In Proceedings of the 5th GRACM International Congress on Computational Mechanics, Limassol, Cyprus, 29 June–1 July 2005.

20. Panagiotou, I.; Neophytou, K.A.; Hamlyn, D.; Britter, R.E. City breathability as quantified by the exchange velocity and its spatial variation in real inhomogeneous urban geometries: An example from central london urban area. *Sci. Total. Environ.* **2013**, *442*, 466–477. [CrossRef] [PubMed]

21. Liu, C.H.; Leung, D.Y.C.; Barth, M.C. On the prediction of air and pollutant exchange rates in street canyons of different aspect ratios using large-eddy simulation. *Atmos. Environ.* **2005**, *39*, 1567–1574. [CrossRef]

22. Kubilay, A.; Neophytou, K.A.; Matsentides, S.; Loizou, M.; Carmeliet, J. The pollutant removal capacity of an urban street canyon and its link to the breathability and exchange velocity. *Procedia Eng.* **2017**, *180*, 443–451. [CrossRef]

23. Lim, E.; Ito, K.; Sandberg, M. New ventilation index for evaluating imperfect mixing conditions—Analysis of net escape velocity based on rans approach. *Build. Environ.* **2013**, *61*, 45–56. [CrossRef]

24. Hang, J.; Wang, Q.; Chen, X.; Sandberg, M.; Zhu, W.; Buccolieri, R.; Di Sabatino, S. City breathability in medium density urban-like geometries evaluated through the pollutant transport rate and the net escape velocity. *Build. Environ.* **2015**, *94*, 166–182. [CrossRef]

25. Lim, E.; Ito, K.; Sandberg, M. Performance evaluation of contaminant removal and air quality control for local ventilation systems using the ventilation index net escape velocity. *Build. Environ.* **2014**, *79*, 78–89. [CrossRef]

26. Antoniou, N.; Montazeri, H.; Wigo, H.; Neophytou, M.; Blocken, B.; Sandberg, M. CFD and wind-tunnel analysis of outdoor ventilation in a real compact heterogeneous urban area: Evaluation using "air delay". *Build. Environ.* **2017**, *126*, 355–372. [CrossRef]

27. Hang, J.; Li, Y.; Sandberg, M.; Claesson, L. Wind conditions and ventilation in high-rise long street models. *Build. Environ.* **2010**, *45*, 1353–1365. [CrossRef]

28. Oke, T.R. Street design and urban canopy layer climate. *Energy Build.* **1988**, *11*, 103–113. [CrossRef]

29. Sini, J.F.; Anquetin, S.; Mestayer, P.G. Pollutant dispersion and thermal effects in urban street canyons. *Atmospheric Environ.* **1996**, *30*, 2659–2677. [CrossRef]

30. Salim, S.M.; Buccolieri, R.; Chan, A.; Di Sabatino, S. Numerical simulation of atmospheric pollutant dispersion in an urban street canyon: Comparison between rans and les. *J. Wind Eng. Ind. Aerodyn.* **2011**, *99*, 103–113. [CrossRef]

31. Thaker, P.; Gokhale, S. The impact of traffic-flow patterns on air quality in urban street canyons. *Environ. Pollut.* **2016**, *208*, 161–169. [CrossRef]

32. Ramponi, R.; Blocken, B.; Coo, L.B.D.; Janssen, W.D. CFD simulation of outdoor ventilation of generic urban configurations with different urban densities and equal and unequal street widths. *Build. Environ.* **2015**, *92*, 152–166. [CrossRef]

33. Yuan, C.; Ng, E.; Norford, L.K. Improving air quality in high-density cities by understanding the relationship between air pollutant dispersion and urban morphologies. *Build. Environ.* **2014**, *71*, 245–258. [CrossRef]

34. Peng, Y.; Gao, Z.; Ding, W. An approach on the correlation between urban morphological parameters and ventilation performance. *Energy Procedia* **2017**, *142*, 2884–2891. [CrossRef]

35. Kubota, T.; Miura, M.; Tominaga, Y.; Mochida, A. Wind tunnel tests on the relationship between building density and pedestrian-level wind velocity: Development of guidelines for realizing acceptable wind environment in residential neighborhoods. *Build. Environ.* **2008**, *43*, 1699–1708. [CrossRef]

36. Hang, J.; Sandberg, M.; Li, Y. Effect of urban morphology on wind condition in idealized city models. *Atmos. Environ.* **2009**, *43*, 869–878. [CrossRef]

37. Yoshie, R.; Tanaka, H.; Shirasawa, T.; Kobayashi, T. Experimental study on air ventilation in a built-up area with closely-packed high-rise buildings. *J. Environ. Eng.* **2008**, *73*, 661–667. [CrossRef]

38. Buccolieri, R.; Salizzoni, P.; Soulhac, L.; Garbero, V.; Di Sabatino, S. The breathability of compact cities. *Urban Clim.* **2015**, *13*, 73–93. [CrossRef]

39. Hang, J.; Li, Y.; Buccolieri, R.; Sandberg, M.; Di Sabatino, S. On the contribution of mean flow and turbulence to city breathability: The case of long streets with tall buildings. *Sci. Total. Environ.* **2012**, *416*, 362–373. [CrossRef]

40. Hu, T.; Yoshie, R. Indices to evaluate ventilation efficiency in newly-built urban area at pedestrian level. *J. Wind Eng. Ind. Aerodyn.* **2013**, *112*, 39–51. [CrossRef]

41. Shen, J.; Gao, Z.; Ding, W.; Yu, Y. An investigation on the effect of street morphology to ambient air quality using six real-world cases. *Atmos. Environ.* **2017**, *164*, 85–101. [CrossRef]

42. Yazid, A.W.M.; Sidik, N.A.C.; Salim, S.M.; Saqr, K.M. A review on the flow structure and pollutant dispersion in urban street canyons for urban planning strategies. *Simulation* **2014**, *90*, 892–916. [CrossRef]

43. Di Sabatino, S.; Buccolieri, R.; Kumar, P. Spatial distribution of air pollutants in cities. In *Clinical Handbook of Air Pollution-Related Diseases*; Springer: Berlin, Germany, 2018; pp. 75–95.

44. Buccolieri, R.; Wigö, H.; Sandberg, M.; Di Sabatino, S. Direct measurements of the drag force over aligned arrays of cubes exposed to boundary-layer flows. *Environ. Fluid Mech.* **2017**, *17*, 1–22. [CrossRef]

45. Tominaga, Y. Visualization of city breathability based on cfd technique: Case study for urban blocks in niigata city. *J. Vis.* **2012**, *15*, 269–276. [CrossRef]

46. Mittal, H.; Sharma, A.; Gairola, A. A review on the study of urban wind at the pedestrian level around buildings. *J. Build. Eng.* **2018**, *18*, 154–163. [CrossRef]

47. Tominaga, Y.; Stathopoulos, T. Numerical simulation of dispersion around an isolated cubic building: Model evaluation of rans and les. *Build. Environ.* **2010**, *45*, 2231–2239. [CrossRef]

48. Liu, J.; Niu, J. Cfd simulation of the wind environment around an isolated high-rise building: An evaluation of srans, les and des models. *Build. Environ.* **2016**, *96*, 91–106. [CrossRef]

49. Tominaga, Y.; Mochida, A.; Murakami, S.; Sawaki, S. Comparison of various revised k-ε models and les applied to flow around a high-rise building model with 1:1:2 shape placed within the surface boundary layer. *J. Wind Eng. Ind. Aerodyn.* **2008**, *96*, 389–411. [CrossRef]

50. Moonen, P.; Dorer, V.; Carmeliet, J. Evaluation of the Ventilation Potential of Countyards and Urban Street Canyons Using RANS and LES. *J. Wind Eng. Ind.* **2011**, *99*, 414–423. [CrossRef]

51. Gousseau, P.; Blocken, B.; Stathopoulos, T.; van Heijst, G.J.F. CFD simulation of near-field pollutant dispersion on a high-resolution grid: A case study by les and rans for a building group in downtown montreal. *Atmos. Environ.* **2011**, *45*, 428–438. [CrossRef]

52. Blocken, B.I. 50 years of computational wind engineering: Past, present and future. *J. Wind Eng. Ind. Aerodyn.* **2014**, *129*, 69–102. [CrossRef]

53. Blocken, B. LES over RANS in building simulation for outdoor and indoor applications: A foregone conclusion? *Build. Simul.* **2018**, *11*, 821–870. [CrossRef]

54. Richards, P.; Hoxey, R. Appropriate boundary conditions for computational wind engineering model using the k-ε turbulence model. In *Computational Wind Engineering 1*; Elsevier: Amsterdam, The Netherlands, 1993; pp. 145–153.

55. Yao, Y. *Nanjing: Historical Landscape and Its Planning from Geographical Perspective*; Springer Geography: Berlin, Germany, 2016; p. 221.

56. Hang, J.; Li, Y. Age of air and air exchange efficiency in high-rise urban areas and its link to pollutant dilution. *Atmos. Environ.* **2011**, *45*, 5572–5585. [CrossRef]

57. Princevac, M.; Baik, J.J.; Li, X.; Pan, H.; Park, S.B. Lateral channeling within rectangular arrays of cubical obstacles. *J. Wind. Eng. Ind. Aerodyn.* **2010**, *98*, 377–385. [CrossRef]

58. Gallagher, J.; Baldauf, R.; Fuller, C.H.; Kumar, P.; Gill, L.W.; McNabola, A. Passive methods for improving air quality in the built environment: A review of porous and solid barriers. *Atmos. Environ.* **2015**, *120*, 61–70. [CrossRef]

59. Buccolieri, R.; Santiago, J.L.; Rivas, E.; Sánchez, B. Review on urban tree modelling in cfd simulations: Aerodynamic, deposition and thermal effects. *Urban For. Urban Green.* **2018**, *31*, 212–220. [CrossRef]

60. Santamouris, M.; Ban-Weiss, G.; Osmond, P.; Paolini, R.; Synnefa, A.; Cartalis, C.; Muscio, A.; Zinzi, M.; Morakinyo, T.E.; Ng, E.; et al. Progress in urban greenery mitigation science–assessment methodologies advanced technologies and impact on cities. *J. Civ. Eng.* **2018**, *24*, 638–671. [CrossRef]

61. Tan, Z.; Lau, K.L.; Ng, E. Urban tree design approaches for mitigating daytime urban heat island effects in a high-density urban environment. *Energy Build.* **2016**, *114*, 265–274. [CrossRef]

62. Du, Y.; Mak, C.M.; Li, Y. Application of a multi-variable optimization method to determine lift-up design for optimum wind comfort. *Build. Environ.* **2018**, *131*, 242–254. [CrossRef]

atmosphere

MDPI

Review

Assessment of Indoor-Outdoor Particulate Matter Air Pollution: A Review

Matteo Bo [1], Pietro Salizzoni [2], Marina Clerico [1] and Riccardo Buccolieri [3,*]

[1] Politecnico di Torino, DIATI—Dipartimento di Ingegneria dell'Ambiente, del Territorio e delle Infrastrutture, corso Duca degli Abruzzi 24, Torino 10129, Italy; matteo.bo@polito.it (M.B.); marina.clerico@polito.it (M.C.)

[2] Laboratoire de Mécanique des Fluides et d'Acoustique, UMR CNRS 5509 University of Lyon, Ecole Centrale de Lyon, INSA Lyon, Université Claude Bernard Lyon I, 36, avenue Guy de Collongue, Ecully 69134, France; pietro.salizzoni@ec-lyon.fr

[3] Dipartimento di Scienze e Tecnologie Biologiche ed Ambientali, University of Salento, S.P. 6 Lecce-Monteroni, Lecce 73100, Italy

* Correspondence: riccardo.buccolieri@unisalento.it; Tel.: +39-0832-297-062

Received: 9 June 2017; Accepted: 20 July 2017; Published: 26 July 2017

Abstract: Background: Air pollution is a major global environmental risk factor. Since people spend most of their time indoors, the sole measure of outdoor concentrations is not sufficient to assess total exposure to air pollution. Therefore, the arising interest by the international community to indoor-outdoor relationships has led to the development of various techniques for the study of emission and exchange parameters among ambient and non-ambient pollutants. However, a standardised method is still lacking due to the complex release and dispersion of pollutants and the site conditions among studies. Methods: This review attempts to fill this gap to some extent by focusing on the analysis of the variety of site-specific approaches for the assessment of particulate matter in work and life environments. Results: First, the main analogies and differences between indoor and outdoor particles emerging from several studies are briefly described. Commonly-used indicators, sampling methods, and other approaches are compared. Second, recommendations for further studies based on recent results in order to improve the assessment and management of those issues are provided. Conclusions: This review is a step towards a comprehensive understanding of indoor and outdoor exposures which may stimulate the development of innovative tools for further epidemiological and multidisciplinary research.

Keywords: indoor-outdoor; mass concentration; nanoparticles; particle number concentration (PNC); PM_{10}; $PM_{2.5}$; sampling; Total Suspended Particles (TSP); ultrafine particles (UFP)

1. Introduction

In many countries, the persistence or the increasing of air pollution represents a major environmental and health issue [1], which largely depends on the amount of chemical energy used in our society (i.e., fossil, biomass). The relapses of the anthropic activities cannot be related exclusively to local emissions in urban and metropolitan areas, but also to the diffuse pollution involving entire territories or mega-city regions [2]. This is the case, for example, of the Po Valley in Italy, where the urban emission contribution is overlapped with the critical state of pollution at the regional scale, in particular during wintertime periods.

The assessment of source emissions and the measuring and modelling of outdoor concentrations is, therefore, fundamental to obtain a framework of pollution conditions of an area at different temporal and spatial scales. International and national legislation and policies are mainly based on these approaches. Despite a general improvement of observation and measurement techniques of outdoor pollutant concentrations in the last 20 years, due to technological developments and the adoption of some normative restrictions, the implemented policies and actions have shown limitations in reducing personal exposure [3]. While the policy-makers at various public entities' scales are challenged for the introduction of innovative actions, the scientific community is called to make a step forward in the assessment of air pollution and its relapses to different targets and in different environments.

The correlation among outdoor air pollution and health diseases, affecting in particular respiratory and cardiovascular systems, has been widely demonstrated [2,4–6]. However, an approach based exclusively on the assessment of outdoor air pollution has shown its limited effectiveness. People spend, in fact, most of their time indoors [7,8] and the correlation among personal exposure and outdoor concentrations of particulate matter is still weak in the literature [9,10]. For this reason, despite formerly and recent epidemiological studies referring mainly to outdoor particulate concentrations [11], the assessment of indoor and personal concentrations in work and life environments is necessary to evaluate the total exposure to air pollution.

Furthermore, direct health diseases are primary in a wide list of relapses of air pollution which also include disturbances to the population and the loss in quality and in the use of territories and indoor environments. This is even more serious especially in highly-populated areas, where the synergy of air pollution with other hazard factors (i.e., noise, vibrations, odours) may lead to increasing damages and disturbance to human health and land use [12,13]. Even though in most epidemiological studies the assessment of finer particulate sizes prevails, the employ of other indicators to understand the whole phenomena affecting human health and the use of environments is required.

Within this context, the present review analyses the existing works on the assessment of indoor-outdoor (I-O) particulate matter concentrations and relationships. Differently from other existing reviews on I-O particulate matter pollution [14–16], this work is not limited to investigate one specific parameter or approach, but studies which considered more indicators (i.e., Total Suspended Particles (TSP), PM_{10}, $PM_{2.5}$, $PM_{10-2.5}$, ultrafine particles (UFP), Nanoparticles (NP), and Indoor/Outdoor Ratio (I/O ratio), air exchange rate, infiltration factors) in residential and working environments are preferred to find potentials and weaknesses in the framework of I-O PM research methodologies. Such recent studies, in fact, proposed sampling or modelling approaches for the assessment of site-specific cases. The lack of standardised methods, due to the complex phenomenology of air pollution release and dispersion and the boundary conditions, is evident. This review attempts to partially fill such a gap.

2. Materials and Methods

Starting from the definition of the main objectives, this review collects a large dataset of papers based on several main searches of key criteria which include: large-scale international studies on indoor-outdoor air pollution issues, on-site assessment of I-O concentrations in specific life environments and work sectors, and on-site and experimental studies of particulate matter with different size between indoor and outdoor environments.

The dataset is then reduced by specific exclusion criteria. The main focus is toward studies that analysed more than one indicator; however, studies focusing on one specific indicator are also considered. Likewise, studies involving different indoor and outdoor environments, such as residences and workplaces, are considered. Furthermore, recent studies are generally preferred and, in particular, those proposing innovative approaches and new points of views on methodologies and results; studies published prior to 2005 are also considered for their significant contribution to following studies or that focused on "atypical" case studies. Large-scale exposure assessment studies developed at the end of the last century, such as EXPOLIS, PTEAM, and THEES [9,17–19], are intended as starting points for the purpose of this review.

Using the SciVal tool [20], a qualitative analysis for investigating the main tendencies in I-O studies is also performed. For the searching criteria, indicator terms associated to indoor and outdoor keywords was compiled in order to observe their usage in recent years (2011–2016 published papers). The analysis was developed both by considering the overall results from the SciVal DB and then by filtering to journal categories (JC) expressing subject areas (SA), such as: "Environmental Science"; "Earth and Planetary Sciences"; "Engineering"; and "Medicine" (each journal could be characterised by more than one subject area). The research was developed in May 2017.

3. Brief Summary of the Main Characteristics of PM in Indoor and Outdoor Environments

The main sources of outdoor PM pollution in most developed countries are commonly identified to be road traffic (including exhaust and non-exhaust emissions of vehicle combustion, tire wearing, and resuspension), power generation plants, industries, agriculture, and domestic heating systems [2,21]. While natural sources, which represent a consistent fraction of aerosols in many regions, contribute mainly on coarser particles, anthropic sources are well-known for the generation of primary and secondary fine, ultrafine, and nano-scale particulates [22–25]. The definition and adoption by normative frameworks of size-depending indicators to fix mass concentration limit values for outdoor air quality considered both the penetration in the human respiratory tract and the need to distinguish anthropogenic and natural emissions [26].

Indoor sources are associated to anthropic activities and the intended use of spaces. In life environments a significant role is played by smoking and cooking, followed by heating systems, cleaning, and resuspension due to the presence of humans [10,27–30]. Combustion processes and cleaning contribute significantly to fine, ultrafine, and nanoparticles emissions, while coarse fractions of PM are principally evidenced due to resuspension [22,24,27,31,32]. In working environments, PM size distribution, concentrations, and chemical properties are even more site-specific than in residential ones as these depend on the used materials, productive methods, and working typologies. Extensive literature on school environments, partially for assessing children's exposure (i.e., the tendency for health impairments, and the large percentage of daytime spent in those spaces) and as a major working sector for the number of employees, is found [14,33–35].

A large amount of works described the consistent contribution of outdoor PM to indoor concentrations. The heterogeneity in the estimation of such contributions found for different particulate size ranges is strictly linked to the different pathways of infiltration and aerodynamic behaviours of finer and coarse particles. A general trend describing a decreased penetration for coarser particles is found in the literature [27]. Other factors influencing the contribution of outdoor pollution on indoor environments are constituted by the type of ambient ventilation (i.e., natural or mechanical), distance to the sources, meteorological conditions, and by the building age and architectural characteristics [28,31].

Indoor versus outdoor levels are found to be heterogenic in the literature. In the absence of intense indoor sources, studies show a general trend of higher outdoor concentrations rather than indoor values [8,36,37]. Furthermore, spatial and temporal variability of outdoor PM could significantly affect the relation between I-O concentrations [38]. In urban areas, as an example, the source proximity and the primary and secondary particulate pathways lead to the high variability of the observed PM size distributions and concentrations. An even more uniform spatial distribution is generally observed for the finer particles, and frequent exceptions are reported in the literature due to the interference of local sources [26,39]. Moreover, the seasonal variability of I-O relationships may be referred mainly to the outdoor contribution, the influence of ventilation types, and occupant behaviours [26].

A widely-used indicator for evaluating the indoor-outdoor exchange is the ratio between the measured concentrations in those environments [40,41]. The results of I/O ratios varied in the literature from values tending to zero in modern mechanically-ventilated buildings with an absence of indoor activities to values over 10 in the occupational sector or for residential buildings with intense indoor activity and smoking in territories with relatively low outdoor concentrations [10,37,42]. This wide range is consistently influenced by resuspension, air exchange, and the deposition velocity of particles. Therefore, this indicator does not permit reaching a complete explanation of I-O relationships.

In the literature [15,40,43,44] various sets of parameters and models which consider the mechanisms of generation (i.e., the contribution of indoor sources), transport (i.e., the air exchange rate, infiltration factor, penetration factor, change in indoor concentration per unit change in outdoor concentration) ,and deposition (i.e., the decay rate) of particulate are presented. Evidence from the literature frameworks reinforce the review hypothesis of the need for methods which consider different indicators and parameters, rather than limiting the assessment to the ones required by the normative framework or for the assessment of specific patterns.

4. Indoor-Outdoor Particulate Matter Sampling and Assessment

4.1. Indicators

Within the reviewed papers, the analysis of the main indicators used to describe particulate matter mass and particle number concentrations (PNC) is developed. Both standard and non-standard indicators found in I-O studies are considered.

4.1.1. Total Suspended Particles (TSP)

Total Suspended Particles (TSP) is a historically-used indicator for the assessment of outdoor air pollution and its relapses. However, a limited employ emerges in the literature for the purposes of I-O assessment. This outcome follows the general trend of substitution of such indicators, due to the availability of consolidated technologies and techniques, with others taking into account the size distribution of airborne particulate (i.e., PM_{10}, $PM_{2.5}$) in outdoor applications. Moreover, even if the interest on assessing TSP for intense natural and human dust emissions persists, such an indicator has a limited effectiveness for observing outdoor particle infiltration phenomena into indoor ambient environments.

The few recent uses of TSP are mainly related to the determination of specific components of particulates. Nazir et al. [45] assessed I-O distributions of trace metal in the TSP of outdoor origins (i.e., industry, vehicles, soils sources). As expected, higher values of TSP are found outdoors, rather than indoors, with moderate correlation among the two environments (R = 0.415). Similar results are found in other studies [46,47] which investigate I-O concentrations of particle phase PAHs in total dust. Other applications are found in researches considering large sets of indicators including TSP, PM_{10}, PM_4, $PM_{2.5}$, PM_1, and UFP at different environments, such as museums, offices, industries, schools, and an Antarctic research station [48–52].

4.1.2. PM_{10} and $PM_{2.5}$

In the last two decades there has been an extensive and increasing use of PM_{10} and $PM_{2.5}$. These indicators are adopted in a variety of different I-O studies and are considered as main objects of the samplings or as references for the comparison within other indicators or with other studies.

Ranges of PM_{10} and $PM_{2.5}$ mass concentrations vary by orders of magnitude between, and within, indoor and outdoor environments. In the month of June in two modern offices with mechanical ventilation, Chatoutsidou et al. [53] report low indoor daily averaged PM_{10} concentrations (<3.5 µg/m^3) while simultaneously-collected outdoor measures ranged between 11 µg/m^3 and 21 µg/m^3. Higher outdoor than indoor concentrations are also reported by Diapouli et al. [54] at three residences with air-conditioning in the Athens urban area. However, they found approximately ten times higher 24 h-averaged PM_{10} indoor values (≈25–47 µg/m^3) compared to the results of Chatoutsidou et al. [53]. Additionally, in two naturally-ventilated commercial activities, higher mean PM_{10} indoor values (≈50–55 µg/m^3) rather than outdoor (≈25–45 µg/m^3) are found by Vicente et al. [55] over the sampling period. Same authors also found higher indoor concentrations of PM_{10} during working hours rather than non-working hours. A similar temporal variation between occupancy and non-occupancy is also found in a previous study by Branis et al. [56] which, however, developed the study using different sampling strategies and time references as discussed later in this review. High variability of $PM_{2.5}$ I-O concentrations is also described by other analysed studies [40,50,57–61]. In particular, the findings of Liu et al. [58] through different residential and commercial buildings in Beijing, clearly show the wide variability within indoor PM_{10} and $PM_{2.5}$ concentrations which are, resultantly, higher in restaurants, dormitories, and classrooms, rather than in supermarkets, computer rooms, offices, and libraries (PM$_{10}$ and PM$_{2.5}$ ranging, respectively, from 373.8 µg/m^3 and 136.6 µg/m^3 in restaurants to 33.8 µg/m^3 and 5.6 µg/m^3 in libraries).

In the literature, less frequently-adopted PM classes of indicators (PM$_5$, PM$_4$, PM$_2$, PM$_1$) are likewise investigated. The use of such indicators is, in many cases, related to the cut-off of the availability of instrumentation, the purposes of the occupational field investigations, or for evidence of specific size fraction emissions. In a former study, Monn et al. [8] investigated I-O and personal relationships by comparing indoor and outdoor PM$_{10}$ and PM$_{2.5}$ measurements with PM$_5$ personal samplings. The use of PM$_4$ is found in Weichenthal et al. [62] where the measure of indoor PM$_4$ in relation to I-O measurements of ultrafine particles (UFP) is reported. More recently, Diapouli et al. [54] monitored I-O mass and PN concentrations of UFP, black smoke, PM$_{10}$, and PM$_2$, the latter using a custom-made impactor (with a cut-off point at 2.1 µm and at 23 L/min). The use of PM$_1$ is documented in three studies at schools and universities of Central Europe and in studies assessing multiple indicators (including TSP, PM$_{10}$, and PM$_{2.5}$) [51,56–58,63,64].

A summary of the observed 24-h mean concentrations in indoor and outdoor environments is reported in Figure 1, which shows the box plot for PM$_{10}$ and PM$_{2.5}$. The selection of such indicators is ascribed to the availability of extensive data in the reviewed studies.

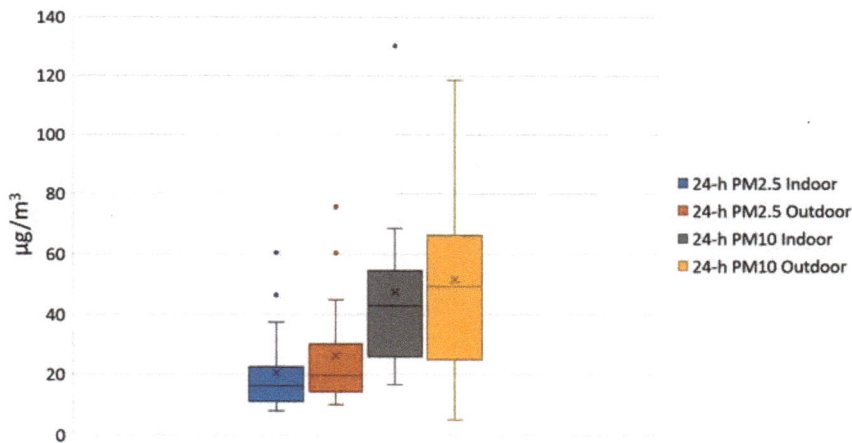

Figure 1. Box plot of 24 h-mean values of PM$_{10}$ and PM$_{2.5}$ [µg/m^3] in indoor and outdoor environments.

4.1.3. Ultrafine Particles (UFP) and Nanoparticles (NP)

Despite the limited amount of paper reporting the use of UFP compared to PM$_{10}$ and PM$_{2.5}$, the recent interest in these types of particles in I-O studies is well documented in the literature. The main content of the studies on UFP particles is related to the investigation of PNC and relapses to human health. The dominant contribution of such particles to number concentrations is, in fact, contraposed to the incidence of larger size particles to mass concentrations [26].

Weichental et al. [62] investigated the contribution to UFP from heating systems in life environments (in over 36 houses) taking into account outdoor concentrations, the age of buildings, types of cooking systems, and the presence of smokers. Results of the study describe the contribution of these sources in UFP indoor concentration, in particular in the residential buildings using electric baseboard heaters and wood stoves. However, as also reported by the authors, the ambient conditions and the potential dominant contribution of other sources (outdoor, cooking, and smoking) need to be taken into account in the comparison of the measured UFP concentrations in buildings with different heating systems. Increasing interest in the use of the UFP indicator is reported for specific workplaces, such as offices and copy centres, where significant indoor sources of fine and ultrafine particles are represented by laser printers [55,58,65,66].

Concentrations vary from low values (<10^3 particles/cm^3) in particularly clean indoor environments [37] to high values (>10^6 particles/cm^3) in the presence of intense indoor sources of such particles [67]. Diapouli et al. [68] report higher UFP mean concentrations outdoors (32,000 ± 14,200 particles/cm^3) rather than indoors (24,000 ± 17,900 particles/cm^3), with a maximum indoor mean value found in a library with a carpet-covered floor and a smoking office (both ≈52,000 particles/cm^3). Similar results are also found in other studies [53,60,61]. In particular, Zauli Sajani et al. [61] investigate I-O concentrations in the front and back of a building along a high traffic street: the highest UFP 1 h-mean value is found for the outdoor front sampling (≈25,500 particles/cm^3) and the lowest in the indoor located in the back of the building (≈3500 particles/cm^3). Comparable values are found indoors at the front and outdoors at the back (7635 particles/cm^3 and 7444 particles/cm^3, respectively) leading the authors to suggest similarities between the gradients of front-back (I-O) and of high-low traffic areas (I-O). Seasonal variability investigated by Wheeler et al. [60] show both indoor and outdoor 24 h-averaged values are higher during winter rather than summer. Furthermore, the difference between indoor and outdoor concentrations is narrow during one of the sampled periods (the second summer campaign) due to a significantly lower outdoor averaged value compared to an another considered period with the same awaited meteorological conditions (first summer campaign).

Related to nanoparticles, relatively few studies focused directly on the assessment of I-O relationships. However, a growing interest is found in occupational indoor environments due to the increase of productive activities employing innovative nanomaterials [69]. Dahm et al. [70], for example, investigate the exposure to carbon nanotubes and nanofibers in six productive sites (handling materials with diameter ranges between 1.1 nm and 140 nm). In this research, no evident trends are described for mass and PN concentrations among the different factories and within the I-O samplings by using three different real-time optical instruments and a filter-based method.

4.1.4. Miscellaneous

In the reviewed articles, a heterogeneity in the use of multiple indicators is observed: the joint use of two indicators is found in different studies with PM_{10} and $PM_{2.5}$ or $PM_{2.5}$ and UFP, while studies considering more than two indicators employed PM_{10} and $PM_{2.5}$ with UFP or TSP [38,60,61]. The recurrence of the use non-typical PM classes (i.e., PM_5, PM_4, PM_1) is also found in many studies, frequently joint with the most-used indicators [49,50,54,57,62,71].

One of the main direct results of the studies assessing more indicators is the investigation of correlations and contributions among different particle size ranges. Mass concentration of finer sizes appear to contribute variously to the values of greater size classes. Liu et al. [58] showed a significant correlation and contribution of PM_{10} to TSP (R^2 = 0.674–0.996, range 47–69%) and discrete for $PM_{2.5}$ to PM_{10} (R^2 = 0.144–0.894, range 16–45%) and PM_1 to PM_{10} (R^2 = 0.149–0.879, range 6–23%). In another study, Branis et al. [56] show a higher correlation of $PM_{2.5}$ and PM_{10} by comparing indoor $PM_{2.5}/PM_{10}$ ratio during workdays (R = 0.872 daytime; R = 0.912 nighttime) and weekends (R = 0.918 daytime; R = 0.991 nighttime). High correlation between indoor PM_1/PM_{10} and $PM_1/PM_{2.5}$ with a slighter increase during the weekend is also observed. The presence of a source of coarse particles during workdays, which does not appear during the other observed periods and can be associated to a good correlation among $PM_{10-2.5}$ and student presence (R = 0.683), can be attributable to the resuspension. Similar results are provided by Vicente et al. [55], who show ratios between different size classes ranging from 0.62 to 0.78 for $PM_{2.5}/PM_{10}$ and PM_4/PM_{10} during working hours and tending to unity during the hours of non-occupancy, as an effect of coarse particle decay. The same study reports ≈slight unity values for $PM_1/PM_{2.5}$ and PM_{10}/TSP, both for working and non-working periods. A correlation among indoor concentrations of PM_4 and UFP (R^2 = 0.53) is reported in the I-O research of Weichental et al. [62].

4.2. Sampling Methods

Based on the available systematic reviews on particulate air sampling technologies [72,73], it is worth examining instrumentation and approaches commonly employed in the reviewed I-O investigations.

Related to concentration samplings, gravimetric methods are the most used (Figure 2). The reason is due to the historical development of this technology, the normative references, and the need to collect mass on filters for further analysis (i.e., the composition of PM). In particular, gravimetric cyclones and impactors, such as the widely-employed Harvard Impactor [74], are the most used technologies for collecting mass. The microbalances (i.e., TEOM, QCM) appear marginally considered in I-O studies, with applications restricted in time sampling, assessed environments, used indicators, or limited to the validation of other instrumentation [75–77]. Relating to the material of membranes, Teflon filters are the most used, followed by quartz and glass filters. The diameter of membranes are typically represented by 25 mm, 37 mm, and 47 mm filters, with the smaller membranes associated with personal samplers as, for example, personal environmental monitors (PEM) reported by Meng et al. [40]. Pumping flow rates in the studies range between 2.3 L/min and 38.3 L/min (with a high frequency of 10 L/min and 16.7 L/min) due to instrument design and compliance with legislation [78].

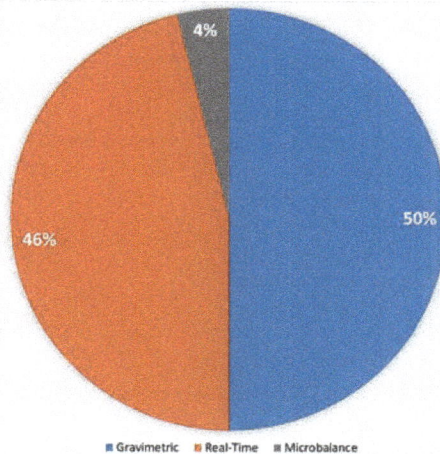

Figure 2. The employed instrument technologies for the assessment of mass and particle number concentrations in the reviewed studies.

An alternative approach to obtain approximate mass concentrations is represented by the use of real-time techniques, such as optically-based systems. The availability of "instantaneous" results and higher time resolution than gravimetric methods represents an important key factor for the sampling campaign design. Advantages comprise the possibility to develop short-term and spot approaches at different locations and extract trends and behaviours over long-term monitoring. The use of such instrumentations should be subjected to dedicated site-specific verifications in order to investigate the need of correction factors to reduce bias with concentrations obtained using gravimetric methods [73,79]. Additionally, the main use of real-time instruments in I-O studies is related to the collection of PNC and size distribution. Chatoutsidou et al. [53] developed measurements of mass (PM_{10}) and number concentrations in recently-built offices using real-time instruments: I/O simultaneously measures are collected by light scattering photometric instrumentation (PM_{10}), scanning mobility particle sizer spectrometers (SMPS, for particles under 0.7 μm), and aerodynamic particle sizer spectrometers (APS, for particles amid 0.5–18 μm). Results show both mass and PN concentrations to be higher outdoors rather than indoors (I/O < 0.3 for all the measures), with PNC dominated by finer particles (the concentrations of particles with a diameter lower than 0.5 μm over two orders of magnitude higher than the range 0.5–18 μm) and mass by coarser particles (related, in particular, to the human presence in working hours leading to resuspension). Despite air conditioning determined to cause a significant reduction of PM_{10} concentrations, temporal fluctuations indoor are found comparable to outdoor PM_{10} variability over time.

Most of the measurements are made by fixed instruments. However, personal samplers are used to investigate the personal exposure with on-board direct measurements. These studies describe the need to follow dedicated design criteria to ensure the validity of such approaches as, for example, measuring at fixed positions in bedrooms during nighttime and keeping the instrumentation far from high-humidity sources [80]. Moreover, such sampling methods should be easy to carry and should integrate noise control and sound insulation systems in order to not interfere with personal daytime activities and, therefore, the representativeness of the collected data. Technological developments improved such issues compared to heavy personal samplers used in former studies as Koistinen et al. [81]. The use of personal samplers for developing measurements at fixed position is also reported. The good correlation of this sampling solution with traditional fixed monitoring samplers found by Meng [40] suggests a sufficient accuracy of such instruments for describing I/O relationships. The feasibility of the methodologies requires, however, a site-specific testing activity in parallel with traditional approaches.

The characteristic time interval of sampling is found to be predominantly of 24 h or 48 h (Figure 3). The daily interval is widely adopted to follow and compare the results with outdoor normative prescriptions which are frequently based on daily concentrations from midnight to midnight. Custom 24-h samplings with starting and ending points at different hours of the day are reported due to the campaign design requirements. The sampling interval of 48 h is typically related to epidemiological studies. Similar percentages can be found on the review on $PM_{2.5}$ by Mohammed et al. [14]. Among the other intervals, the 8 h sampling period is associated with the typical working time and, therefore, is adopted on I-O studies developed at workplaces rather than in life environments. Furthermore, spot measurements (<8 h for each sampling position) are reported to investigate general trends of pollutants or due to limited available time to spend for in-site sampling. Moreover, a former study by Branis et al. [56] adopted a custom 12 h time interval in order to collect mass concentrations on filters describing different periods of use of a university hall (lecture hours vs. nighttime). Among long-term measurement campaigns, semi-continuous approaches are also reported [60]. Such methodologies do not permit the collection of representative data over the time series, but merely to have a general picture of the phenomena over the sampling period.

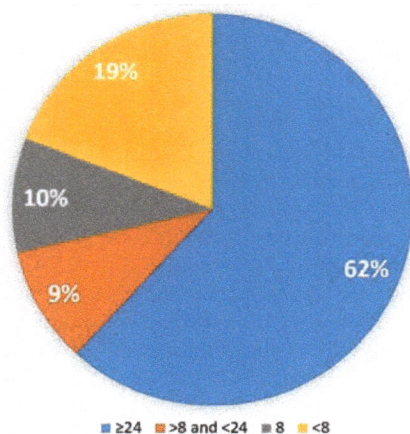

Figure 3. Sampling intervals adopted in the reviewed studies.

Most of the reviewed studies are developed using simultaneous sampling intervals between indoor and outdoor measurements. The design of contemporary measurement campaigns is adopted in order to neglect the time variability of PM among I-O environments. Few exceptions are reported in relation to limited parts of the studies (i.e., additional samplings, instrumentation failures) or for the unfeasibility of achieving simultaneous samplings as, for example, due to extreme weather conditions reported in the Antarctica study by Pagel et al. [49].

In accordance with good-practice and normative frameworks, all the studies set the instrument chains far from every potentially interfering source. For this reason, indoor sampling is typically made at the centre of the environment and at a height between 1 and 2 m (to simulate the breathing height of occupants). Distances from walls and heating sources are also identified as main design factors. Related to outdoor air samplings, in studies the height ranges from 1–2 m (referred to the front door or ground level) to the height of the floor corresponding to the indoor sampling (5–10 m from the ground). Measures on building roofs are also reported [57,79]. The horizontal distance from the external wall is frequently reported as another key parameter for the reduction of unwanted interference. Personal samplers are typically positioned in the breathing zone of the carrying person.

In some studies mainly related to epidemiological purposes [36,50,56] the outdoor concentrations are totally, or partially, obtained from central-site stations, instead of proximity samplings. The adoption of concentrations from the public monitoring stations or other samplings represents an opportunity to save resources. However, this requires both the detailed examination of representativeness of the station for describing outdoor values at the indoor sampling and the comparison of measurement methods. When possible, it is strongly recommended to perform measurements of outdoor concentrations in proximity to the indoor measure. The choice of the samplers' disposition is strictly related to the goals of the survey and requires an accurate aprioristic design, as underlined by Zauli Sajani et al. [61].

4.3. Studies of Site-Specific and Environmental Characteristics

As expected, different sites and ambient characteristics emerge through the reviewed literature. First, the type of analysed environments: a comparable number of case-studies is found between workplaces (productive, non-productive) and life environments (Figure 4). Residential homes appear as the most represented, followed by schools and universities, offices, and commercial buildings. Only a few I-O studies considered industries [50,70]. The choice to investigate different typologies of buildings is also reported [50,58] and clearly indicates the interest of the international community to develop I-O studies on a wide range of life and occupational environments.

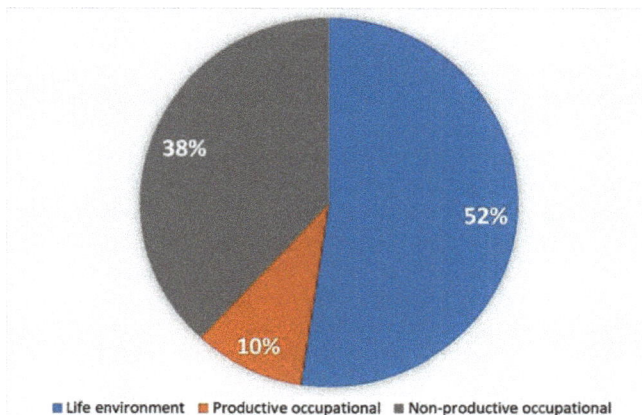

38%

52%

10%

■ Life environment ■ Productive occupational ■ Non-productive occupational

Figure 4. Synthesis of the environment typologies assessed in the reviewed studies.

Related to the spatial variability, urban and suburban areas prevail over rural areas. Most studied indoor life environments are kitchens and living rooms due to the long time spent by occupants and to the presence of the major residential indoor sources (cooking, smoking). In the occupational sector, offices and ambient conditions potentially subjected to intense indoor sources (i.e., printers in commercial copy centres) are investigated. The design of indoor measurements in unoccupied residential environments is also found due to epidemiological purposes [61]. Outdoor environments are frequently chosen as the front door and front windows of indoor ambient environments, followed by courtyards and roofs.

Another difference among the studies is represented by the seasonal and duration variability of assessment campaigns. The choice of seasonal periods is based principally on outdoor sampling design both for meteorological constraints (i.e., rainfall, temperature) and for the need of measuring during high- or low-pollution periods. Many studies analyse more different periods of the year covering cold and hot seasons. While a predominance of a specific month or season do not emerge, frequently multi-seasonal durations are represented by autumn-winter-spring campaigns for the assessment of I-O relationships during such a particularly-sensible period for outdoor PM pollution. Spring-autumn and winter-summer investigations are also frequently adopted as representative of antipode meteorological conditions. These observations are based on the campaign duration reported by the authors of the original works: the effective sampling intervals could involve only a part of the entire period of study. As a brief observation, samplings are generally conducted over the entire campaign period, or at least on multiple weeks/months, rather than limited to single days or weeks.

No prevalence of mechanically- or naturally-ventilated buildings emerges from the studies. The type of ventilation assessed by the studies is, in fact, mainly related to the available sampling areas, with the exception to studies on the effects of a specific ventilation design to the infiltration of outdoor particles in indoor environments [53,79,82]. Significant differences are shown, in fact, in relation to ventilation systems, as described in Section 5.

5. Exchange Factors

5.1. Indoor/Outdoor Ratio

For the purposes of this review, the term "exchange factors" denotes the indicators describing the I-O relationship's characteristics. The I/O ratio is widely used, even though it presents some limitations, as described in Section 3. In particular, its high variability also emerges among the reviewed studies and no consistent global trends can be stated a priori.

A general relationship with indoor activities and I/O ratio higher than the unity is found by Diapouli et al. [68] which report PM_{10} and $PM_{2.5}$ ratios higher than 1 in gym, offices, and classes, and lower than 1 for a library (used only for a limited part of the day by students). Similar results are obtained for PM_{10} by Vicente et al. [55] by observing I/O ratio at copy centres in workdays (mostly > 1, with a maximum of 2.38) and in weekends (≈0.7–0.8). Significantly high I/O ratios for TSP, PM_{10}, $PM_{2.5}$, and PM_1 (ranging from values approx. between 2 and 18, except for PM_1 between 0.98 and 8.9) are found at an Antarctic research station [49]. In this case study, the extreme environmental conditions and limited outdoor sources significantly affect the increase of the observed I/O ratios, which should depend on the indoor activities and the emissions of the research station (i.e., vehicles, incinerator). The maximum observed I/O ratio (30.40), reached in an air conditioned classroom during cleaning hours and in the presence of rainfall, is reported by Guo et al. [82]. I/O ratios less than unity is found in the published studies of either occupied or unoccupied indoor environments [55,57,58,61,64,68,82].

A significant example of the indicator variability can be clearly observed in the results of Challoner et al. [79]. The two highest I/O values (9.18 and 8.18), compared to other values between 1 and 3, are found for two offices (A and B) with consistent differences among them and some "contradictory" results from the expected trends: naturally (A) and mechanically (B) ventilated (no significant difference found in this study by building ventilation types); at the fifth (A) and ground (B) floors (expected different outdoor concentration incidence); small office (A) and large open plan (B) (incidence of air volume and ventilation pathways to indoor contribution); both during non-working hours (even if during working hours higher values are generally expected rather than non-working, in this case they are inferior and correspond to values of 4.68 and 2.87, respectively); and ventilation intake in front of a high-traffic road for (B) (with the expected effect of consistent outdoor concentrations on reducing the ratio). The study also developed two different campaigns for (B) considering a different disposition of the outdoor sampling: an I/O ratio four times higher is obtained with the outdoor station at the ground floor (8.18) than on the roof (2.04), where outdoor concentrations are expected to be lower. Considering the need of detailed correlation analysis with the indoor sources and the characteristic meteorological conditions of the area (i.e., temporal variation, well-noted high humidity, and frequent rainfalls in the Dublin region), these results confirm the extreme susceptibility of the I/O ratio to boundary conditions. Such variability among the reviewed studies is synthetized in Figure 5, which shows the box plot reporting mean, maximum, and minimum values of I/O ratios.

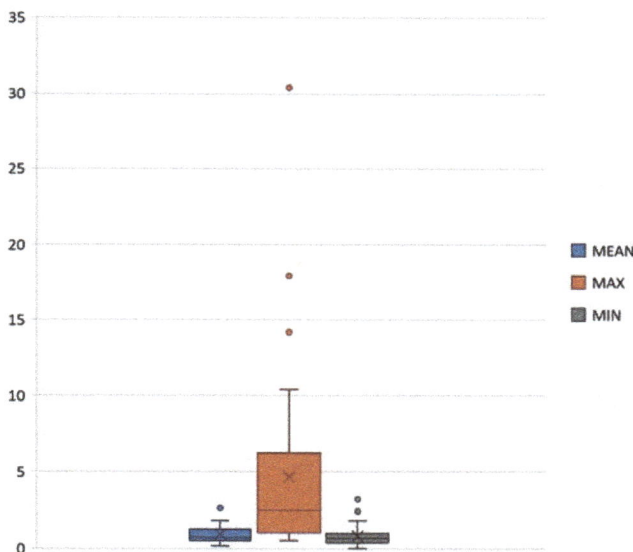

Figure 5. Variability of mean, maximum and minimum I/O ratios in the reviewed papers.

5.2. Air Exchange Rate (AER)

Evaluation of building ventilation is needed for the comprehension of influence of outdoor air pollution into indoor environments. Using the air exchange rate (AER), expressed as the air flow rate per volume of the indoor environment, a minor exchange of air is reported for mechanically-ventilated buildings rather than naturally-ventilated [40,60,62] ones. Volume estimation is obtained in the studies by direct and indirect approaches as measures of indoor spaces, cadastre data, and reported data by occupants. Measures of the AER are carried using tracer gasses or the pressurized blower method. Uses of perfluorocarbon, carbon dioxide, and perfluorinated methyl-cyclohexane are reported [40,60,61,82]. The distance of tracer sources from ventilation or heating sources is needed in order to reduce their interference. Different results for mechanical and natural ventilations are found due to the age of the building, occupants' behaviours, and the efficiency of filters. The greater seal in mechanically-ventilated buildings is generally associated to a lower penetration of PM of outdoor origin than for naturally-ventilated environments [54].

5.3. Other Approaches

Among the I-O reviewed studies, considerations on some exchange factors or modelling approaches, rather than the sole analysis of the I/O ratio, are reported. Meng et al. [40] adopted a single compartment mass balance model and a random component superposition statistical model (RCS), with the support of AER and PM direct measurements and values obtained from the literature, to evaluate the contribution of ambient and non-ambient sources to indoor and personal concentrations. Results showed an average contribution of outdoor pollution to $PM_{2.5}$ indoor concentrations over 60%. Furthermore, the increasing trend of the infiltration factor (IF) at the growth of AER values reported in the research shows the general reduction of outdoor contributions due to the decrease of air exchange. Using a two-compartment mass balance model, Chatoutsidou et al. [66] have also estimate the incidence of printer emissions in different rooms in a mechanically-ventilated building.

Within the results of the study published by Hoek et al. [38] emerges the role of particle size and composition to their infiltration and penetration in indoor environments: using a single mass balance model and comparing the results with other studies, the incidence of both dimensions and chemical components is confirmed by the authors from the variability of the average IF among sulphate, soot, and mass and particulate number concentrations (whole range between 0.06 and 0.87). The assessment of sulphate in parallel with the other I-O multiple samplings, is also found in other reviewed studies [50,68,83]. Actually, indoor sulphate concentration is a good estimator of IF due to the absence of significant indoor sources of such inorganic pollutants and constitutes, considering some limitations, a surrogate for describing the contribution of outdoor sources to indoor environments, as described in the literature [84].

As previously stated, the description in detail of infiltration and penetration factors, as well as modelling approaches, is intentionally considered marginal for the purposes of this review. Indicators and models for describing the dynamics of the unwanted entrance of outdoor pollution into indoor environments are systematically evaluated in other reviews [15,22,31,84]. We suggest to the readers who aim to obtain a complete framework of the subject to refer to such, and to other works in the existing literature.

6. Discussion

The framework emerging from the review of the existing literature shows a growing interest on PM I-O assessment. Most of recent studies focus on the evaluation of fine and ultrafine particulates due to the relevance of the relapses of potentially highly-penetrating particles to human health. Additionally, the use of a standard indicator, such as the PM_{10}, is widely documented and the adoption of innovative or non-standard indicators is also observed in recent studies. The observed outline from the analysis of indicators is representative of the variety of approaches adopted within the I-O studies.

With the aim of finding potential standard approaches for the design and execution of I-O assessment, the need to associate a common indicator to the ones adopted for the specific objective of the studies is strengthened. This may represent an opportunity to make comparisons among studies, considering prior research. The feasibility of such additional measurement with a standard indicator must be subjected to the projects' available resources and technologies. For this reason, the adoption of PM_{10} or $PM_{2.5}$ additional sampling and the description of the I-O ratio could represent a good compromise between the use of resources and having a general comparison of indoor and outdoor environments among studies. Furthermore, the use of other indicators can also generally lead to a better comprehension of indirect relapses of PM as an annoyance and disturbance or to evaluate a specific contribution to mass concentrations, as a resuspension of dust, by assessing coarse particles (i.e., $PM_{10-2.5}$).

In order to design a measurement campaign, the site-specific conditions of indoor and outdoor environments are not of secondary importance. The potential role of different sources should be evaluated: for outdoor sources the distance within the receptors, the emissions over time, and the dispersion pathways should be considered while, for indoor environments, human activities and behaviours (i.e., smoking, cooking, presence of pets) and internal and external architectural characteristics (i.e., age of the building, carpets, doors) represent crucial parameters for the definition of the assessment procedures [15,36,68]. In addition, sampling disposition should take into account the interference with unwanted sources of heat, ventilation, and other pollutants. Simultaneous measurements and outdoor sampling in the proximity of indoor environments should also be included as design criteria. The aim is to reduce spatial and temporal variability, which is widely described for indoor and outdoor PM. Contemporaneity is found to be generally considered in the reviewed studies, while a persistence of studies adopting outdoor values from other samplings is observed.

Related to human exposure, which is one of the main goals of the reviewed studies, the weak potential of using the sole outdoor concentrations to represent personal exposure to PM is confirmed by the literature. The good correlation found among indoor workplaces and personal samplings during working hours or, likewise, indoor residential and leisure time personal concentrations [85], show the feasibility of a simplified approach-based fixed monitoring in different I-O environments to obtain personal exposures [60]. Furthermore, the choice to associate I-O measurement with personal samplings represents a major opportunity: a well-designed personal measurement can describe unintended individual behaviours and might cover specific exposures to PM which can be difficult to assess by adopting only fixed monitoring. An example is represented by in-vehicle exposures which can consistently contribute to the "personal cloud" [9] for people spending a large amount of daytime moving from one indoor environment to another (i.e., the workplace and home) or working in transportation sectors. The recent developments of low-cost, light, and real-time samplers, used in association with GPS technologies, represent an opportunity for further studies employing personal measurements as shown by Steinle et al. [77]. The effectiveness of personal concentration measurements, however, requires as a critical parameter of design the verification of the aptitude of participants to carry the sampler, as also reported by Meng et al. [40]. Definitely, the design of both fixed and personal approaches is suggested.

The good precision and the correlation with traditional fixed monitoring stations, demonstrated in the literature [40], suggest the feasibility of using personal samplers in fixed positions for some specific purposes. In fact, the use of personal samplers might represent a good alternative to assess indoor and outdoor concentrations instead of traditional approaches, as described by Snyder et al. [86]. Different scenarios comprehend limitations in instrumentation and resources, difficulties of carrying heavy samplers and reaching the investigated site, the need of short-time or spot sampling, and the unavailability of connection to the electricity grid. It is, however, important to underline that the good correlation founded in some studies [40] does not necessarily imply a complete correspondence of the obtained concentrations and requires a reliable choice of sampling parameters considering the site-specific boundary conditions.

As a final consideration, the available literature showed the wide interest in I/O studies on the assessment of particulate matter chemical components between indoor and outdoor environments, in different contexts, especially schools [34,45,49,61,68,83]. In particular, a broad investigation of PM-bound PAHs is widely reported by recently-published papers [87–90] and a dedicated review on such topics is under consideration by the authors, given the increasing interest of the scientific community.

7. Conclusions

This paper reviews recent studies on the assessment of I-O particulate matter air pollution. An increasing interest by the international scientific community to indoor-outdoor relationships is evidenced by the development of several techniques for the study of emission and exchange parameters among ambient and non-ambient pollutants. The present review emphasises the importance of reducing divergences among I-O studies, derived from differences in measuring approaches, site characteristics, and campaign periods, and by the definition and adoption of standard approaches. This statement is joint with the need of efforts for the definition of dedicated normative frameworks, with the implementation of international common policies and specific value limits for indoor environments.

The opportunity given by the recent technological developments and the need to assess the human exposure to recently-introduced materials and in less studied environments are continuously leading to a growing interest in the development of innovative epidemiological and multidisciplinary studies in the field of I-O assessment.

Author Contributions: Matteo Bo, Pietro Salizzoni, Marina Clerico, and Riccardo Buccolieri conceived and designed the structure of the paper. Matteo Bo conducted the literature research, analysed the data, and wrote the main part of the paper. All authors contributed to the discussion of the results and have read and approved the final manuscript.

Conflicts of Interest: The authors declare no conflict of interest.

Abbreviations

I-O	indoor-outdoor
Mass concentration	concentration expressed in micrograms per cube meters [$\mu g/m^3$]
NP	nanoparticles
PM	(airborne) particulate matter
PN	Particle number
PNC	Particle number concentration = concentration expressed in number of particles per cubic centimetre [particles/cm^3]
UFP	ultrafine particles

References

1. World Health Organization (WHO). Ambient Air Pollution: A Global Assessment of Exposure and Burden of Disease. Available online: http://www.who.int/phe/publications/air-pollution-global-assessment/en/ (accessed on 11 May 2017).
2. International Agency for Research on Cancer (IARC). *International Agency for Research on Cancer (IARC) Monographs on the Evaluation of Carcinogenic Risks to Humans, Volume 109*; Outdoor Air Pollution; IARC: Lyon, France, 2015; Available online: http://monographs.iarc.fr/ENG/Monographs/vol109/index.php (accessed on 12 July 2016).
3. World Health Organization (WHO). WHO Guidelines for Indoor Air Quality: Selected Pollutants. 2010. Available online: http://www.euro.who.int/en/health-topics/environment-and-health/air-quality/publications/2010/who-guidelines-for-indoor-air-quality-selected-pollutants (accessed on 12 May 2017).
4. Brunekreef, B.; Holgate, S.T. Air pollution and health. *Lancet* **2002**, *360*, 1233–1242. [CrossRef]
5. Raaschou-Nielsen, O.; Andersen, Z.J.; Beelen, R.; Samoli, E.; Stafoggia, M.; Weinmayr, G.; Hoffmann, B.; Fischer, P.; Nieuwenhuijsen, M.; Brunekreef, B.; et al. Air pollution and lung cancer incidence in 17 European cohorts: Prospective analyses from the European Study of Cohorts for Air Pollution Effects (ESCAPE). *Lancet Oncol.* **2013**, *14*, 813–822. [CrossRef]
6. Seaton, A.; Godden, D.; MacNee, W.; Donaldson, K. Particulate air pollution and acute health effects. *Lancet* **1995**, *345*, 176–178. [CrossRef]
7. Brasche, S.; Bischof, W. Daily time spent indoors in German homes—Baseline data for the assessment of indoor exposure of German occupants. *Int. J. Hyg. Environ. Health* **2005**, *208*, 247–253. [CrossRef] [PubMed]
8. Monn, C.; Fuchs, A.; Högger, D.; Junker, M.; Kogelschatz, D.; Roth, N.; Wanner, H.U. Particulate matter less than 10 μm (PM_{10}) and fine particles less than 2.5 μm ($PM_{2.5}$): Relationships between indoor, outdoor and personal concentrations. *Sci. Total Environ.* **1997**, *208*, 15–21. [CrossRef]
9. Ozkaynak, H.; Xue, J.; Spengler, J.; Wallace, L.; Pellizzari, E.; Jenkins, P. Personal exposure to airborne particles and metals: Results from the Particle TEAM study in Riverside, California. *J. Expo. Anal. Environ. Epidemiol.* **1996**, *6*, 57–78. [PubMed]
10. Wallace, L. Indoor Particles: A Review. *J. Air Waste Manag. Assoc.* **1996**, *46*, 98–126. [CrossRef] [PubMed]
11. Chen, R.; Kan, H.; Chen, B.; Huang, W.; Bai, Z.; Song, G.; Pan, G. Association of Particulate Air Pollution with Daily Mortality: The China Air Pollution and Health Effects Study. *Am. J. Epidemiol.* **2012**, *175*, 1173–1181. [CrossRef] [PubMed]

12. Bo, M.; Clerico, M.; Pognant, F. Annoyance and disturbance hazard factors related to work and life environments: A review. *Geam-Geoing. Ambient. E Mineraria-Geam-Geoengin. Environ. Min.* **2016**, *149*, 27–34.

13. Kephalopoulos, S.; Koistinen, K.; Paviotti, M.; Schwela, D.; Kotzias, D. Proceedings of the International Workshop on "Combined Environmental Exposure: Noise, Air Pollutants and Chemicals". Available online: https://ec.europa.eu/jrc/en/publication/eur-scientific-and-technical-research-reports/proceedings-international-workshop-combined-environmental-exposure-noise-air-pollutants-and (accessed on 10 May 2017).

14. Mohammed, M.O.A.; Song, W.-W.; Ma, W.-L.; Li, W.L.; Ambuchi, J.J.; Thabit, M.; Li, Y.-F. Trends in indoor–outdoor PM$_{2.5}$ research: A systematic review of studies conducted during the last decade (2003–2013). *Atmos. Pollut. Res.* **2015**, *6*, 893–903. [CrossRef]

15. Chen, C.; Zhao, B. Review of relationship between indoor and outdoor particles: I/O ratio, infiltration factor and penetration factor. *Atmos. Environ.* **2011**, *45*, 275–288. [CrossRef]

16. Lin, C.-C.; Peng, C.-K. Characterization of Indoor PM$_{10}$, PM$_{2.5}$, and Ultrafine Particles in Elementary School Classrooms: A Review. *Environ. Eng. Sci.* **2010**, *27*, 915–922. [CrossRef]

17. Clayton, C.A.; Perritt, R.L.; Pellizzari, E.D.; Thomas, K.W.; Whitmore, R.W.; Ozkaynak, H.; Spengler, J.D. Particle Total Exposure Assessment Methodology (PTEAM) study: Distributions of aerosol and elemental concentrations in personal, indoor, and outdoor air samples in a southern California community. *J. Expo. Anal. Environ. Epidemiol.* **1993**, *3*, 227–250. [PubMed]

18. Jantunen, M.J.; Hänninen, O.; Katsouyanni, K.; Knöppel, H.; Kuenzli, N.; Lebret, E.; Maroni, M.; Saarela, K.; Srám, R.; Zmirou, D. Air pollution exposure in European cities: The "EXPOLIS" study. *J. Expo. Anal. Environ. Epidemiol.* **1998**, *8*, 495–518.

19. Lioy, P.J.; Waldman, J.M.; Buckley, T.; Butler, J.; Pietarinen, C. The personal, indoor and outdoor concentrations of PM-10 measured in an industrial community during the winter. *Atmos. Environ. Part B Urban Atmos.* **1990**, *24*, 57–66. [CrossRef]

20. SciVal. Available online: https://www.scival.com/ (accessed on 7 May 2017).

21. Karagulian, F.; Belis, C.A.; Dora, C.F.C.; Prüss-Ustün, M.A.; Bonjour, S.; Adair-Rohani, H.; Amann, M. Contributions to cities' ambient particulate matter (PM): A systematic review of local source contributions at global level. *Atmos. Environ.* **2015**, *120*, 475–483. [CrossRef]

22. Biswas, P.; Wu, C.-Y. Nanoparticles and the Environment. *J. Air Waste Manag. Assoc.* **2005**, *55*, 708–746. [CrossRef] [PubMed]

23. Brauer, M.; Amann, M.; Burnett, R.T.; Cohen, A.; Dentener, F.; Ezzati, M.; Henderson, S.B.; Krzyzanowski, M.; Martin, R.V.; Dingenen, R.V.; et al. Exposure Assessment for Estimation of the Global Burden of Disease Attributable to Outdoor Air Pollution. *Environ. Sci. Technol.* **2012**, *46*, 652–660. [CrossRef] [PubMed]

24. Murr, L.E.; Garza, K.M. Natural and anthropogenic environmental nanoparticulates: Their microstructural characterization and respiratory health implications. *Atmos. Environ.* **2009**, *43*, 2683–2692. [CrossRef]

25. Wangchuk, T.; He, C.; Dudzinska, M.R.; Morawska, L. Seasonal variations of outdoor air pollution and factors driving them in the school environment in rural Bhutan. *Atmos. Environ.* **2015**, *113*, 151–158. [CrossRef]

26. Monn, C. Exposure assessment of air pollutants: A review on spatial heterogeneity and indoor/outdoor/personal exposure to suspended particulate matter, nitrogen dioxide and ozone. *Atmos. Environ.* **2001**, *35*, 1–32. [CrossRef]

27. Abt, E.; Suh, H.H.; Allen, G.; Koutrakis, P. Characterization of indoor particle sources: A study conducted in the metropolitan Boston area. *Environ. Health Perspect.* **2000**, *108*, 35–44. [CrossRef] [PubMed]

28. Jones, N.C.; Thornton, C.A.; Mark, D.; Harrison, R.M. Indoor/outdoor relationships of particulate matter in domestic homes with roadside, urban and rural locations. *Atmos. Environ.* **2000**, *34*, 2603–2612. [CrossRef]

29. Lai, H.K.; Kendall, M.; Ferrier, H.; Lindup, I.; Alm, S.; Hänninen, O.; Jantunen, M.; Mathys, P.; Colvile, R.; Ashmore, M.R.; et al. Personal exposures and microenvironment concentrations of PM$_{2.5}$, VOC, NO$_2$ and CO in Oxford, UK. *Atmos. Environ.* **2004**, *38*, 6399–6410. [CrossRef]

30. Yocom, J.E. A Critical Review. *J. Air Pollut. Control Assoc.* **1982**, *32*, 500–520. [CrossRef]

31. Nazaroff, W.W. Indoor particle dynamics. *Indoor Air* **2004**, *14*, 175–183. [CrossRef] [PubMed]

32. Fuoco, F.C.; Stabile, L.; Buonanno, G.; Trassiera, C.V.; Massimo, A.; Russi, A.; Mazaheri, M.; Morawska, L.; Andrade, A. Indoor Air Quality in Naturally Ventilated Italian Classrooms. *Atmosphere* **2015**, *6*, 1652–1675. [CrossRef]

33. Rivas, I.; Viana, N.; Morento, T.; Pandolfi, M.; Amato, F.; Reche, C.; Bouso, L.; Alvarez-Pedrerol, M.; Alastuey, A.; Sunyer, J.; et al. Child exposure to indoor and outdoor air pollutants in schools in Barcelona, Spain. *Environ. Int.* **2014**, *69*, 200–212. [CrossRef] [PubMed]

34. Tofful, L.; Perrino, C. Chemical Composition of Indoor and Outdoor PM$_{2.5}$ in Three Schools in the City of Rome. *Atmosphere* **2015**, *6*, 1422–1443. [CrossRef]

35. Mainka, A.; Zajusz-Zubek, E.; Kaczmarek, K. PM$_{2.5}$ in Urban and Rural Nursery Schools in Upper Silesia, Poland: Trace Elements Analysis. *Int. J. Environ. Res. Public Health* **2015**, *12*, 7990–8008. [CrossRef] [PubMed]

36. Lachenmyer, C. Urban Measurements of Outdoor-Indoor PM$_{2.5}$ Concentrations and Personal Exposure in the Deep South. Part I. Pilot Study of Mass Concentrations for Nonsmoking Subjects. *Aerosol Sci. Technol.* **2000**, *32*, 34–51. [CrossRef]

37. Riesenfeld, E.; Chalupa, D.; Gibb, F.R.; Oberdörster, G.; Gelein, R.; Morrow, P.E.; Utell, M.J.; Frampton, M.W. Ultrafine Particle Concentrations in a Hospital. *Inhal. Toxicol.* **2000**, *12*, 83–94. [CrossRef] [PubMed]

38. Hoek, G.; Kos, G.; Harrison, R.; de Hartog, J.; Meliefste, K.; ten Brink, H.; Katsouyanni, K.; Karakatsani, A.; Lianou, M.; Kotronarou, A.; et al. Indoor–outdoor relationships of particle number and mass in four European cities. *Atmos. Environ.* **2008**, *42*, 156–169. [CrossRef]

39. Hussein, T.; Hämeri, K.; Aalto, P.P.; Paatero, P.; Kulmala, M. Modal structure and spatial–temporal variations of urban and suburban aerosols in Helsinki—Finland. *Atmos. Environ.* **2005**, *39*, 1655–1668. [CrossRef]

40. Meng, Q.Y.; Turpin, B.J.; Korn, L.; Weisel, C.P.; Morandi, M.; Colome, S.; Zhang, J.; Stock, T.; Spektor, D.; Winer, A. Influence of ambient (outdoor) sources on residential indoor and personal PM$_{2.5}$ concentrations: Analyses of RIOPA data. *J. Expo. Sci. Environ. Epidemiol.* **2005**, *15*, 17–28. [CrossRef] [PubMed]

41. Brunekreef, B.; Janssen, N.A.H.; de Hartog, J.J.; Oldenwening, M.; Meliefste, K.; Hoek, G.; Lanki, T.; Timonen, K.L.; Vallius, M.; Pekkanen, J.; et al. Personal, indoor, and outdoor exposures to PM$_{2.5}$ and its components for groups of cardiovascular patients in Amsterdam and Helsinki. *Res. Rep. Health Eff. Inst.* **2005**, *127*, 1–70.

42. Shilton, V.; Giess, P.; Mitchell, D.; Williams, C. The Relationships between Indoor and Outdoor Respirable Particulate Matter: Meteorology, Chemistry and Personal Exposure. *Indoor Built Environ.* **2002**, *11*, 266–274. [CrossRef]

43. Zhou, B.; Zhao, B.; Guo, X.; Chen, R.; Kan, H. Investigating the geographical heterogeneity in PM$_{10}$-mortality associations in the China Air Pollution and Health Effects Study (CAPES): A potential role of indoor exposure to PM$_{10}$ of outdoor origin. *Atmos. Environ.* **2013**, *75*, 217–223. [CrossRef]

44. Zhao, B.; Wu, J. Particle deposition in indoor environments: Analysis of influencing factors. *J. Hazard. Mater.* **2007**, *147*, 439–448. [CrossRef] [PubMed]

45. Nazir, R.; Shaheen, N.; Shah, M.H. Indoor/outdoor relationship of trace metals in the atmospheric particulate matter of an industrial area. *Atmos. Res.* **2011**, *101*, 765–772. [CrossRef]

46. Krugly, E.; Martuzevicius, D.; Sidaraviciute, R.; Ciuzas, D.; Prasauskas, T.; Kauneliene, V.; Stasiulaitiene, I.; Kliucininkas, L. Characterization of particulate and vapor phase polycyclic aromatic hydrocarbons in indoor and outdoor air of primary schools. *Atmos. Environ.* **2014**, *82*, 298–306. [CrossRef]

47. Halsall, C.J.; Maher, B.A.; Karloukovski, V.V.; Shah, P.; Watkins, S.J. A novel approach to investigating indoor/outdoor pollution links: Combined magnetic and PAH measurements. *Atmos. Environ.* **2008**, *42*, 8902–8909. [CrossRef]

48. Alves, C.; Duarte, M.; Ferreira, M.; Alves, A.; Almeida, A.; Cunha, Â. Air quality in a school with dampness and mould problems. *Air Qual. Atmos. Health* **2016**, *9*, 107–115. [CrossRef]

49. Pagel, É.C.; Reis, N.C.; Alvarez, C.E.; Santos, J.M.; Conti, M.M.; Boldrini, R.S.; Kerr, A.S. Characterization of the indoor particles and their sources in an Antarctic research station. *Environ. Monit. Assess.* **2016**, *188*, 167. [CrossRef] [PubMed]

50. Saraga, D.; Pateraki, S.; Papadopoulos, A.; Vasilakos, C.; Maggos, T. Studying the indoor air quality in three non-residential environments of different use: A museum, a printery industry and an office. *Build. Environ.* **2011**, *46*, 2333–2341. [CrossRef]

51. Polednik, B. Particulate matter and student exposure in school classrooms in Lublin, Poland. *Environ. Res.* **2013**, *120*, 134–139. [CrossRef] [PubMed]

52. Worobiec, A.; Samek, L.; Krata, A.; van Meel, K.; Krupinska, B.; Stefaniak, E.A.; Karaszkiewicz, P.; van Grieken, R. Transport and deposition of airborne pollutants in exhibition areas located in historical buildings–study in Wawel Castle Museum in Cracow, Poland. *J. Cult. Herit.* **2010**, *11*, 354–359. [CrossRef]

53. Chatoutsidou, S.E.; Ondráček, J.; Tesar, O.; Tørseth, K.; Ždímal, V.; Lazaridis, M. Indoor/outdoor particulate matter number and mass concentration in modern offices. *Build. Environ.* **2015**, *92*, 462–474. [CrossRef]

54. Diapouli, E.; Eleftheriadis, K.; Karanasiou, A.A.; Vratolis, S.; Hermansen, O.; Colbeck, I.; Lazaridis, M. Indoor and Outdoor Particle Number and Mass Concentrations in Athens. Sources, Sinks and Variability of Aerosol Parameters. *Aerosol Air Qual. Res.* **2011**, *11*, 632–642. [CrossRef]

55. Vicente, E.D.; Ribeiro, J.P.; Custódio, D.; Alves, C.A. Assessment of the indoor air quality in copy centres at Aveiro, Portugal. *Air Qual. Atmos. Health* **2017**, *10*, 117–127. [CrossRef]

56. Braniš, M.; Řezáčová, P.; Domasová, M. The effect of outdoor air and indoor human activity on mass concentrations of PM_{10}, $PM_{2.5}$, and PM_1 in a classroom. *Environ. Res.* **2005**, *99*, 143–149. [CrossRef] [PubMed]

57. Goyal, R.; Kumar, P. Indoor-outdoor concentrations of particulate matter in nine microenvironments of a mix-use commercial building in megacity Delhi. *Air Qual. Atmos. Health* **2013**, *6*, 747–757. [CrossRef]

58. Liu, Y.; Chen, R.; Shen, X.; Mao, X. Wintertime indoor air levels of PM_{10}, $PM_{2.5}$ and PM_1 at public places and their contributions to TSP. *Environ. Int.* **2004**, *30*, 189–197. [CrossRef]

59. Schembari, A.; Triguero-Mas, M.; de Nazelle, A.; Davdand, P.; Vrijheid, M.; Cirach, M.; Martinez, D.; Figueras, F.; Querol, X.; Basagaña, X. Personal, indoor and outdoor air pollution levels among pregnant women. *Atmos. Environ.* **2013**, *64*, 287–295. [CrossRef]

60. Wheeler, A.J.; Wallace, L.A.; Kearney, J.; van Ryswyk, K.; You, H.; Kulka, R.; Brook, J.R.; Xu, X. Personal, Indoor, and Outdoor Concentrations of Fine and Ultrafine Particles Using Continuous Monitors in Multiple Residences. *Aerosol Sci. Technol.* **2011**, *45*, 1078–1089. [CrossRef]

61. Sajani, S.Z.; Ricciardelli, I.; Trentini, A.; Bacco, D.; Maccone, C.; Castellazzi, S.; Lauriola, P.; Poluzzi, V.; Harrison, R.M. Is particulate air pollution at the front door a good proxy of residential exposure? *Environ. Pollut.* **2016**, *213*, 347–358. [CrossRef] [PubMed]

62. Weichenthal, S.; Dufresne, A.; Infante-Rivard, C.; Joseph, L. Indoor ultrafine particle exposures and home heating systems: A cross-sectional survey of Canadian homes during the winter months. *J. Expo. Sci. Environ. Epidemiol.* **2006**, *17*, 288–297. [CrossRef] [PubMed]

63. Colbeck, I.; Nasir, Z.A.; Ali, Z. Characteristics of indoor/outdoor particulate pollution in urban and rural residential environment of Pakistan. *Indoor Air* **2010**, *20*, 40–51. [CrossRef] [PubMed]

64. Majewski, G.; Kociszewska, K.; Rogula-Kozłowska, W.; Pyta, H.; Rogula-Kopiec, P.; Mucha, W.; Pastuszka, J. Submicron Particle-Bound Mercury in University Teaching Rooms: A Summer Study from Two Polish Cities. *Atmosphere* **2016**, *7*, 117. [CrossRef]

65. Barthel, M.; Pedan, V.; Hahn, O.; Rothhardt, M.; Bresch, H.; Jann, O.; Seeger, S. XRF-Analysis of Fine and Ultrafine Particles Emitted from Laser Printing Devices. *Environ. Sci. Technol.* **2011**, *45*, 7819–7825. [CrossRef] [PubMed]

66. Chatoutsidou, S.E.; Serfozo, N.; Glytsos, T.; Lazaridis, M. Multi-zone measurement of particle concentrations in a HVAC building with massive printer emissions: Influence of human occupation and particle transport indoors. *Air Qual. Atmos. Health* **2017**, 1–15. [CrossRef]

67. Lee, C.-W.; Hsu, D.-J. Measurements of fine and ultrafine particles formation in photocopy centers in Taiwan. *Atmos. Environ.* **2007**, *41*, 6598–6609. [CrossRef]

68. Diapouli, E.; Chaloulakou, A.; Mihalopoulos, N.; Spyrellis, N. Indoor and outdoor PM mass and number concentrations at schools in the Athens area. *Environ. Monit. Assess.* **2008**, *136*, 13–20. [CrossRef] [PubMed]

69. Kuhlbusch, T.A.; Asbach, C.; Fissan, H.; Göhler, D.; Stintz, M. Nanoparticle exposure at nanotechnology workplaces: A review. *Part. Fibre Toxicol.* **2011**, *8*, 22. [CrossRef] [PubMed]

70. Dahm, M.M.; Evans, D.E.; Schubauer-Berigan, M.K.; Birch, M.E.; Deddens, J.A. Occupational Exposure Assessment in Carbon Nanotube and Nanofiber Primary and Secondary Manufacturers: Mobile Direct-Reading Sampling. *Ann. Occup. Hyg.* **2013**, *57*, 328–344. [PubMed]

71. Xu, H.; Guinot, B.; Shen, Z.; Ho, K.F.; Niu, X.; Xiao, S.; Huang, R.J.; Cao, J. Characteristics of Organic and Elemental Carbon in $PM_{2.5}$ and $PM_{0.25}$ in Indoor and Outdoor Environments of a Middle School: Secondary Formation of Organic Carbon and Sources Identification. *Atmosphere* **2015**, *6*, 361–379. [CrossRef]

72. Amaral, S.S.; De Carvalho, J.A.; Costa, M.A.M.; Pinheiro, C. An Overview of Particulate Matter Measurement Instruments. *Atmosphere* **2015**, *6*, 1327–1345. [CrossRef]

73. Chow, J.C.; Doraiswamy, P.; Watson, J.G.; Chen, L.-W.A.; Ho, S.S.H.; Sodeman, D.A. Advances in Integrated and Continuous Measurements for Particle Mass and Chemical Composition. *J. Air Waste Manag. Assoc.* **2008**, *58*, 141–163. [CrossRef] [PubMed]

74. Marple, V.A.; Rubow, K.L.; Turner, W.; Spengler, J.D. Low Flow Rate Sharp Cut Impactors for Indoor Air Sampling: Design and Calibration. *JAPCA* **1987**, *37*, 1303–1307. [CrossRef] [PubMed]

75. Crist, K.C.; Liu, B.; Kim, M.; Deshpande, S.R.; John, K. Characterization of fine particulate matter in Ohio: Indoor, outdoor, and personal exposures. *Environ. Res.* **2008**, *106*, 62–71. [CrossRef] [PubMed]

76. Kuo, H.-W.; Shen, H.-Y. Indoor and outdoor $PM_{2.5}$ and PM_{10} concentrations in the air during a dust storm. *Build. Environ.* **2010**, *45*, 610–614. [CrossRef]

77. Steinle, S.; Reis, S.; Sabel, C.E.; Semple, S.; Twigg, M.M.; Braban, C.F.; Leeson, S.R.; Heal, M.R.; Harrison, D.; Lin, C.; et al. Personal exposure monitoring of $PM_{2.5}$ in indoor and outdoor microenvironments. *Sci. Total Environ.* **2015**, *508*, 383–394. [CrossRef] [PubMed]

78. United States Environmental Protection Agency (US EPA). 2012 National Ambient Air Quality Standards (NAAQS) for Particulate Matter (PM). Available online: https://www.epa.gov/pm-pollution/2012-national-ambient-air-quality-standards-naaqs-particulate-matter-pm (assessed on 16 May 2017).

79. Challoner, A.; Gill, L. Indoor/outdoor air pollution relationships in ten commercial buildings: $PM_{2.5}$ and NO_2. *Build. Environ.* **2014**, *80*, 159–173. [CrossRef]

80. Weisel, C.P.; Zhang, J.; Turpin, B.J.; Morandi, M.T.; Colome, S.; Stock, T.H.; Spektor, D.M.; Korn, L.; Winer, A.; Alimokhtari, S. Relationship of Indoor, Outdoor and Personal Air (RIOPA) study: Study design, methods and quality assurance/control results. *J. Expo. Sci. Environ. Epidemiol.* **2005**, *15*, 123–137. [CrossRef] [PubMed]

81. Koistinen, K.J.; Kousa, A.; Tenhola, V.; Hänninen, O.; Jantunen, M.J.; Oglesby, L.; Kuenzli, N.; Georgoulis, L. Fine Particle (PM_{25}) Measurement Methodology, Quality Assurance Procedures, and Pilot Results of the EXPOLIS Study. *J. Air Waste Manag. Assoc.* **1999**, *49*, 1212–1220. [CrossRef] [PubMed]

82. Guo, H.; Morawska, L.; He, C.; Gilbert, D. Impact of ventilation scenario on air exchange rates and on indoor particle number concentrations in an air-conditioned classroom. *Atmos. Environ.* **2008**, *42*, 757–768. [CrossRef]

83. Fromme, H.; Diemer, J.; Dietrich, S.; Cyrys, J.; Heinrich, J.; Lang, W.; Kiranoglu, M.; Twardella, D. Chemical and morphological properties of particulate matter (PM_{10}, $PM_{2.5}$) in school classrooms and outdoor air. *Atmos. Environ.* **2008**, *42*, 6597–6605. [CrossRef]

84. Diapouli, E.; Chaloulakou, A.; Koutrakis, P. Estimating the concentration of indoor particles of outdoor origin: A review. *J. Air Waste Manag. Assoc.* **2013**, *63*, 1113–1129. [CrossRef] [PubMed]

85. Kousa, A.; Oglesby, L.; Koistinen, K.; Künzli, N.; Jantunen, M. Exposure chain of urban air $PM_{2.5}$—Associations between ambient fixed site, residential outdoor, indoor, workplace and personal exposures in four European cities in the EXPOLIS-study. *Atmos. Environ.* **2002**, *36*, 3031–3039. [CrossRef]

86. Snyder, E.G.; Watkins, T.H.; Solomon, P.A.; Thomas, E.D.; Williams, R.W.; Hagler, G.S.W.; Shelow, D.; Hindin, D.A.; Kilaru, V.J.; Preuss, P.W. The Changing Paradigm of Air Pollution Monitoring. *Environ. Sci. Technol.* **2013**, *47*, 11369–11377. [CrossRef] [PubMed]

87. Oliveira, M.; Slezakova, K.; Madureira, J.; de Oliveira Fernandes, E.; Delerue-Matos, C.; Morais, S.; do Carmo Pereira, M. Polycyclic aromatic hydrocarbons in primary school environments: Levels and potential risks. *Sci. Total Environ.* **2017**, *575*, 1156–1167. [CrossRef] [PubMed]

88. Romagnoli, P.; Balducci, C.; Perilli, M.; Vichi, F.; Imperiali, A.; Cecinato, A. Indoor air quality at life and work environments in Rome, Italy. *Environ. Sci. Pollut. Res.* **2016**, *23*, 3503–3516. [CrossRef] [PubMed]

89. Błaszczyk, E.; Rogula-Kozłowska, W.; Klejnowski, K.; Fulara, I.; Mielżyńska-Švach, D. Polycyclic aromatic hydrocarbons bound to outdoor and indoor airborne particles ($PM_{2.5}$) and their mutagenicity and carcinogenicity in Silesian kindergartens, Poland. *Air Qual. Atmos. Health* **2017**, *10*, 389–400. [CrossRef] [PubMed]

90. Rogula-Kopiec, P.; Rogula-Kozłowska, W.; Kozielska, B.; Sówka, I. PAH Concentrations Inside a Wood Processing Plant and the Indoor Effects of Outdoor Industrial Emissions. *Pol. J. Environ. Stud.* **2015**, *24*, 11–17. [CrossRef]

![atmosphere logo] *atmosphere*

MDPI

Article

Prediction of Wind Environment and Indoor/Outdoor Relationships for PM$_{2.5}$ in Different Building–Tree Grouping Patterns

Bo Hong [1], Hongqiao Qin [1] and Borong Lin [2,*]

[1] College of Landscape Architecture & Arts, Northwest A&F University, Yangling 712100, China;
   hongbo@nwsuaf.edu.cn (B.H.); qinhq88@163.com (H.Q.)
[2] Department of Building Science, School of Architecture, Tsinghua University, Beijing 100084, China
[*] Correspondence: linbr@tsinghua.edu.cn; Tel.: +86-010-6278-5691

Received: 26 October 2017; Accepted: 22 January 2018; Published: 24 January 2018

Abstract: Airflow behavior and indoor/outdoor PM$_{2.5}$ dispersion in different building–tree grouping patterns depend significantly on the building–tree layouts and orientation towards the prevailing wind. By using a standard k-ε model and a revised generalized drift flux model, this study evaluated airflow fields and indoor/outdoor relationships for PM$_{2.5}$ resulting from partly wind-induced natural ventilation in four hypothetical building–tree grouping patterns. Results showed that: (1) Patterns provide a variety of natural ventilation potential that relies on the wind influence, and buildings that deflect wind on the windward facade and separate airflow on the leeward facade have better ventilation potential; (2) Patterns where buildings and trees form a central space and a windward opening side towards the prevailing wind offer the best ventilation conditions; (3) Under the assumption that transported pollution sources are diluted through the inlet, the aerodynamics and deposition effects of trees cause the lower floors of a multi-storey building to be exposed to lower PM$_{2.5}$ compared with upper floors, and lower indoor PM$_{2.5}$ values were found close to the tree canopy; (4) Wind pressure differences across each flat showed a poor correlation ($R^2 = 0.059$), with indoor PM$_{2.5}$ concentrations; and (5) Patterns with the long facade of buildings and trees perpendicular to the prevailing wind have the lowest indoor PM$_{2.5}$ concentrations.

Keywords: wind environment; Natural Ventilation Potential (NVP); PM$_{2.5}$; building–tree grouping patterns; Computational Fluid Dynamics (CFD)

---

1. Introduction

With the rapid development of urbanization, particulate matter (PM) pollution, especially types with an aerodynamic diameter of less than 2.5 μm (PM$_{2.5}$), has led to a dramatic decline in urban air quality. PM$_{2.5}$ pollution has been confirmed to have a close relationship with the human respiratory system, resulting in cardiopulmonary system damage and high incidences of cancer [1]. In addition, a nationally representative survey of communities, families, and individuals in China showed that air pollution reduced hedonic happiness and raised the rate of depression symptoms [2]. In built environments, outdoor-generated particles are major contributors to indoor pollution, without strong internal pollution sources [3–6]. Since most people spend approximately 85–90% of their time indoors, determining the relationships between outdoor particle sources and the corresponding indoor concentrations are especially significant when measuring particulate concentrations in occupied residences [7].

Natural ventilation is extensively used to provide better indoor air quality without using electricity [8]. Some knowledge of the relationships between wind velocity, wind direction, and ventilation characteristics is needed to ventilate naturally [9]. In addition to the types of natural vents, the outdoor wind environment acts as a significant factor that affects the natural ventilation potential [10]. A comparison of Computational Fluid Dynamics (CFD) simulation results has been carried out to examine the accuracy of wind environment measurements around a single building and building complexes in urban areas [11]. Wind fields strongly depend on the building layout and prevailing wind, and staggered arrangements could provide a comfortable wind environment and sufficient natural ventilation potential with S–N and SE–NW wind directions [12,13]. Additionally, CFD simulations indicated that the outdoor wind environment relies on building geometry, spacing, grouping patterns, and orientation towards the prevailing wind [14–18]. Furthermore, the aerodynamic effects of trees on airflows have also attracted attention. The accuracy of the aerodynamic effects of trees has been examined for a pedestrian-level wind environment assessment [19]. The optimum arrangement of trees for creating a comfortable outdoor environment and sufficient natural ventilation was tested with numerical simulations and field experiments in a residential neighborhood [20–22]. The influence of different building–tree arrangements on natural ventilation potential and outdoor thermal comfort was investigated via a simulation platform for an outdoor thermal environment [23].

Considerable research has also examined the behavior of indoor/outdoor air pollutant dispersion in different types of buildings. The indoor/outdoor (I/O) ratio, penetration, and infiltration factors are considered the main factors expressing the connection between indoor and outdoor particle concentrations [24]. The measured data showed that indoor particles were highly correlated with outdoor particles, but negatively coordinated with wind velocity [25–28]. The infiltration rate was affected by wind velocity, but little by temperature [29,30]. A field investigation indicated that $PM_{2.5}$ and PM_{10} were approximately equal when measured inside and outside of a building next to roads with high traffic emissions. The I/O ratio also differs whether building windows were open and closed [31]. Field experiments on indoor/outdoor and seasonal variations have indicated that particle concentrations are highest in winter due to low wind speed and high outdoor humidity compared with other seasons [32]. More pollutants from outdoors entered naturally ventilated rooms in winter than in other seasons [33].

Several numerical studies have also investigated indoor particle concentrations generated from outside pollutants [34]. Simulations have confirmed that particle deposition was mainly affected by the ventilation conditions, and that the deposition rate under displacement ventilation was lower than mixed ventilation, but the escaped particle mass is larger, and the average pollutants concentration is higher in the former than the latter [35]. In addition, a combined empirical and simulated method was used to appraise the effectiveness of deep bag and electrostatic filters to reduce PM. The results revealed that an air filter can significantly reduce the indoor PM, and particle dynamics and movement in office buildings under different ventilation were quantified [36]. However, few studies have examined the effect of trees on indoor and outdoor particle dispersion, and the study of indoor $PM_{2.5}$ removal by natural ventilation exerted by building–tree arrangements is one of fundamental and practical significances in the design stage.

Overall, many full-scale field experiments and simulation studies have illustrated that indoor particles originate largely outdoors, and that building ventilation has a significant influence on the particle diffusion process [25,37–40]. Some studies have illustrated airflow fields around different building grouping patterns [17], or various building–tree arrangements on airflows and outdoor particle dispersion [41]. Moreover, according to the Annual Report on Chinese Building Energy Conservation Development 2016, most residents in cold regions still opened the windows in the cold winter days for ventilation over 1 h accumulatively to reduce indoor CO_2 concentration, due to a lack of air conditioning systems [42]. Based on the monitored data, the probability of window opening would be randomly high or low over a short period during the winter in Beijing. For example, when outdoor $PM_{2.5}$ concentration was 245 μg/m^3 in Beijing during the monitoring periods, the corresponding probability of window opening was 100% [43]. This suggests that partly wind-driven natural ventilation in cold seasons could affect the indoor $PM_{2.5}$ dispersions. Under the circumstance of high $PM_{2.5}$ concentrations during the winter in Beijing, it may be possible to analyze the influence of building–tree arrangements on indoor and outdoor relationships for PM$_{2.5}$ dispersion in winter resulting from partly wind-driven natural ventilation, to uncover optimal patterns that improve indoor air quality.

This study performed numerical simulations with CFD code, coupled with the standard *k-ε* model and the generalized drift flux model, to determine: (1) The effects of four building–tree grouping patterns on the outdoor wind environment and the indoor and outdoor relationships for PM$_{2.5}$ as a result of partly wind-induced natural ventilation in Beijing; (2) The relationship between the resulting wind pressure differences and indoor PM$_{2.5}$ concentrations; and (3) Which configuration can provide the best ventilation potential or provide the lowest indoor PM$_{2.5}$ concentrations.

2. Climate of Beijing

Beijing's climate is characterized as typical of a warm, continental, monsoon zone, with daily average temperatures in summer and winter of 26.5 °C and −3.8 °C, respectively. The daily relative humidity ranges from 56.6 to 74.1% in summer, and from 39.4 to 45.8% in winter. The prevailing winds are southerly and southeasterly in summer, and the average wind velocity reaches 2.0 m/s. Prevailing winds during the winter are northerly and northwesterly, with an average velocity of 2.6 m/s. Based on weekly data, the highest level of PM$_{2.5}$ pollution is between mid-November and December [44]. Data reported by the Beijing Meteorological Service indicated that the average wind velocity in November and December 2015 was 2.3 m/s, and the dominant wind directions were northerly, northwesterly, and westerly. The average temperature ranged from −0.5 °C to 5.4 °C, and relative humidity ranged from 45.4 to 55.6%. The average PM$_{2.5}$ during this time period was 134 μg/m^3 [45], which is almost four times higher than the World Health Organization's interim target-1 (IT-1) level. The annual mean concentration of IT-1 is 35 μg/m^3, and this level is associated with approximately a 15% higher long-term mortality risk relative to the air quality guideline (AQG) level [46].

3. Simulation Case Descriptions

The airflow and PM$_{2.5}$ dispersion around the grouped buildings and trees are affected by several factors, including: building geometry, grouping patterns, orientation with respect to the prevailing wind, and tree-building distance. Based on Beijing's climate in November and December 2015, four common buildings configurations with a reference wind velocity of 2.3 m/s and two prevailing wind directions (northerly and westerly) were selected for testing.

The first configuration is two parallel rows of housing blocks. The second block contains a T-shaped central space that adopts a semi-open array architectural layout. The third is a staggered pattern with the buildings turned 90° to one another. The architectural layout of the fourth is known as a peripheral array in which each building block measures 30 m × 15 m × 18 m and rises six storeys. The building blocks are regarded as solid blocks with no wind permeability considered. The area of the target site is 6400 m^2 (80 m × 80 m), with coverage ratios of 71% to allow space for outdoor activity. Cypress (*Platycladus orientalis*) trees, with conical crown morphology, are planted along the building 2.5 m from the building base. The leaf area density of cypress is 2.3 m^2/m^3, and the deposition velocity for PM$_{2.5}$ is 0.0458 m/s [47]. The tree height, crown diameter, and clear bole height are 5 m, 3 m, and 2 m, respectively (Figure 1).

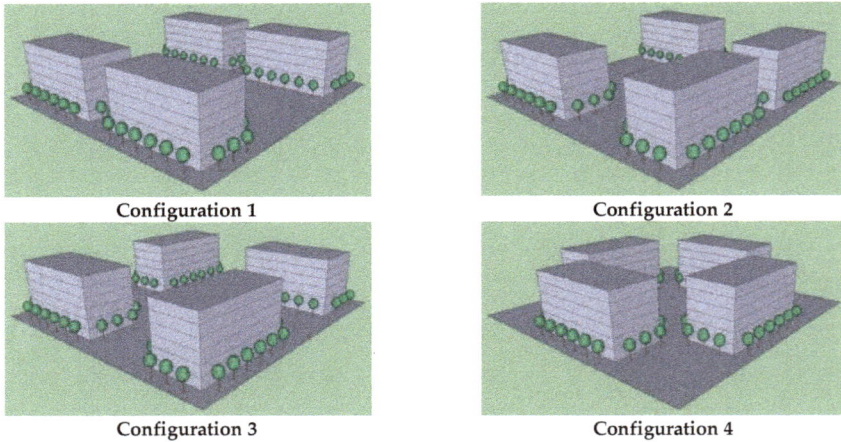

Configuration 1

Configuration 2

Configuration 3

Configuration 4

Figure 1. Illustration of building–tree grouping patterns modeled in this study. **Configuration 1** is the parallel rows; **Configuration 2** is the internal T; **Configuration 3** is staggered; **Configuration 4** is the peripheral array.

4. The CFD Code

4.1. Simulation Models

Parabolic Hyperbolic or Elliptic Numerical Integration Code Series (PHOENICS) was used in this study as a three-dimensional simulation tool, and the standard k-ε model was selected. The Semi-Implicit Method for Pressure-Linked Equations (SIMPLE) algorithm with the Quadratic Upstream Interpolation for Convective Kinematics (QUICK) discretization scheme is applied to all governing equations except the concentration equation [48], for which a Sharp and Monotonic Algorithm for Realistic Transport (SMART) scheme is adopted to avoid negative concentrations near sources. The simulation process was facilitated by the use of an i7 2.67 GHz processor, while, the PHOENICS 2009 program was used to process the solution.

In this model, vegetation is described as a porous medium with branches and truck comprising the canopy. The influence of vegetation on the flow field was to decrease air velocity by exerting drag forces and pressure, and the canopy elements that reduced air velocity created additional turbulence levels [49]. The consequent effects of vegetation on turbulent flow fields were modeled, and drag forces were included in the momentum equations. Additional source terms were also included in transport equations for turbulent kinetic energy, k, and its dissipation rate, ε. Turbulence production and accelerated turbulence dissipation within the canopy were accounted for by the additional turbulence source terms [50].

Flow resistance, which is generated by turbulent flow through the plant canopy, is represented by introducing the following sink term into the momentum equations:

$$S_{d,i} = -C_d \alpha(z) |U| u_i \tag{1}$$

where C_d is the drag coefficient, $\alpha(z)$ is the leaf area density (LAD) (m^2/m^3), z is the vertical space coordinate, $|U|$ is the magnitude of the superficial velocity vector, and u_i is the superficial value of the Cartesian velocity in the direction i (m/s).

Vegetation density is characterized by the integral value of leaf area density. The integral value is determined by the leaf area index (*LAI*), which is defined by the following equation:

$$LAI = \int_0^h \alpha(z)dz \tag{2}$$

where h is the average height of the canopy.

The turbulent interactions between the plant canopy and airflow, including the following additional source terms in the transport equations for k and ε, are described as:

$$S_k = C_d\alpha(z)\left(\beta_p|U|^3 - \beta_d|U|k\right) \tag{3}$$

$$S_\varepsilon = C_d\alpha(z)\left(C_{4\varepsilon}\beta_p|U|^3\frac{\varepsilon}{k} - C_{5\varepsilon}\beta_d|U|\varepsilon\right) \tag{4}$$

where the constant β_p is the fraction of mean-flow kinetic energy converted to wake-generated k by canopy drag, and β_d is the fraction of k dissipated by the Kolmogorov energy cascade short-circuiting. In this study, the constants β_p and β_d, and the closure constants $C_{4\varepsilon}$ and $C_{5\varepsilon}$ are given the values 1.0, 3.0, 1.5, and 1.5, respectively [51–53].

To model the particle dispersion procedure, the revised generalized drift flux model, which was modified from the Eulerian model by adding additional terms for particle absorption and the resuspension by trees, was used [54]. In the model, particles are treated as a continuum, and particle movements are assumed to not affect turbulence. The convective velocity of the particle phase is assumed to be the same as the airflow, which has been widely used in recent studies [55–57], and greatly reduced the complexity of the two-phase flow simulation. The model is expressed as follows:

$$\frac{\partial\left[\left(V_j + V_{slip,j}\right)C\right]}{\partial x_j} = \frac{\partial}{\partial x_j}\left[\varepsilon_p\frac{\partial C}{\partial x_j}\right] + S_c - S_{\sin k} + S_{resuspension} \tag{5}$$

The additional terms S_{sink} and $S_{resuspension}$ are calculated as:

$$S_{sink} = V_d \times C \times \alpha \tag{6}$$

$$S_{resuspension} = V_r \times C_{sink} \times \alpha \tag{7}$$

where V_j is the spatial mean fluid (air) velocity (m/s); $V_{slip,j}$ is the gravitational settling velocities of particlesin jdirection (m/s); C is the inlet concentration ($\mu g/m^3$); ε_p is the eddy diffusivity (m^2/s); S_c is the generated rate of the pollution source ($kg/(m^3{\cdot}s)$); S_{sink} is the mass of particles absorbed per unit vegetation volume per unit time ($\mu g/m^3$); $S_{resuspension}$ is the secondary pollutants from foliage per unit vegetation volume per unit time ($\mu g/m^3$) [58]; V_d is the particle deposition velocity on plant foliage (m/s); V_r is the particle resuspension velocity from plant foliage (m/s); C_{sink} is the particle concentration deposited on plant foliage ($\mu g/m^3$); and α is LAD (m^2/m^3).

4.2. Model Validation

To validate models against the wind tunnel dataset issued by the Laboratory of Building and Environmental Aerodynamics at the Karlsruhe Institute of Technology (KIT) [59], models of street canyons with two rows of avenue trees were built and set according to the wind tunnel model represented. The height, width, and length of street canyons in the model were H, $2H$, and $10H$, respectively. H is equal to 18 m. The tree canopy was 7.56 m wide ($0.42H$), 12 m high ($2/3H$), and 180 m ($10H$) long. They were planted 5.22 m ($0.29H$) away from street canyons. Traffic exhaust sources were placed in two rows along the street. Based on literature sources [60], the distance from inlet to windward building, from downwind building to outlet, and from buildings to both boundaries were set to $5H$, $20H$, and $9H$, respectively. The height of the domain was $11H$. The grid in the target area consisted of elements with $X_{min} = Y_{min} = Z_{min} = 0.028H$, and grid independence evaluated by the grid convergence index (GCI) showed that the GCI value was 2.53% (less than 5%), demonstrating that the used grid arrangement was fine enough [61]. Boundary conditions were set according to wind tunnel experiments [62].

Since the models should be verified against experimental data, the simulated parameters should be selected considering the Concentration Data of Street Canyons (CODASC) wind tunnel experiments. In the wind tunnel tests, trees were modeled with porous media with different porosities, and the deposition effect was not considered. The different porosity was described by the pressure loss coefficient C_x. The C_x in the CODASC wind tunnel models were 80 and 200 m^{-1} with trees, representing 0.53 and 1.33 m^{-1} at a scale of 1:150, respectively. Correspondingly, simulation models set additional terms (S_{sink} and $S_{resuspension}$) to zero, and the $C_x = 1.33$ m^{-1} was selected to compare the simulated data with scenarios in CODASC wind tunnel experiments. To obtain comparable results, vertical concentrations 0.75 m from both walls in the street canyon were measured in the simulation. Figure 2 shows the scatter plots of $C_x = 1.33$ m^{-1}. It can be seen that there were strong correlations between the simulated and wind tunnel data ($R^2 = 0.926$ in wall A and $R^2 = 0.991$ in wall B). According to Hanna and Chang [63], a set of acceptance criteria: $-0.3 < FB < 0.3$, $NMSE < 1.5$, $0.5 < FAC2 < 2$, $NAD < 0.3$, and $R > 0.8$ was recommended for further analysis. All of the metrics are within acceptance ranges, indicating the numerical models were suitable for predicting the airflow and pollutant dispersion within the street canyon (Table 1).

Additionally, the implemented models have been validated by the author in a previous study that compared simulated and measured data from a residential district in Beijing [64]. Parameters including wind velocity and $PM_{2.5}$ concentrations were compared, and the results showed high correlations ($R^2 = 0.910$ for wind velocity, and $R^2 = 0.906$ for $PM_{2.5}$ concentrations), indicating that the models accurately represented the real environment.

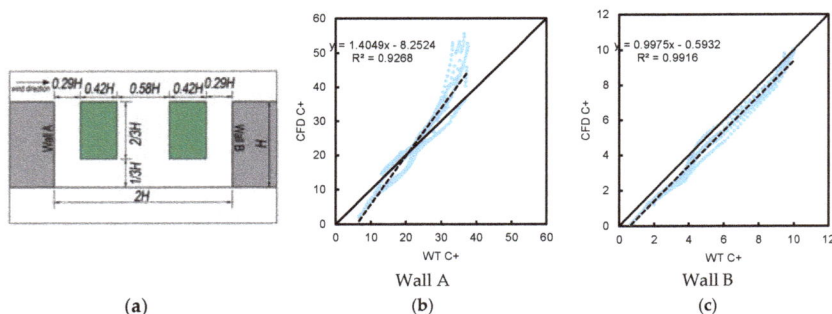

Figure 2. Model lateral view and dimensions (Wall A is the leeward wall of the upwind building, wall B is the windward wall of the downwind building) (**a**), and comparisons between normalized concentration results in wind tunnel experiments (WT C^+) and simulations (CFD C^+) (**b**,**c**). C^+ is normalized by $C^+ = C_m H U_H / Q_l$ (C_m is measured concentration, H is building height (m), U_H is wind velocity at height H (m/s), and Q_l is the emission rate of a line source (m^2/s)). Dashed lines are linear fits, with linear equations and R^2-values alongside. Diagonals depict perfect matches.

Table 1. Mean concentration results and statistical analysis metrics for $C_x = 1.33$ m^{-1}.

Wall	Mean Concentration*			Statistical Analysis Metrics				
	WT C^+	CFD C^+	RD	FB	NMSE	FAC2	NAD	R
A	20.77	20.92	6.6%	−0.007	0.052	1.008	0.083	0.961
B	3.90	3.30	−15.4%	0.1674	0.033	0.846	0.084	0.994

* RD is relative difference; FB is fractional bias; NMSE is normalized mean square error; FAC2 is the fraction of predictions within a factor of two of observations; NAD is normalized absolute difference; and R is relation coefficient.

4.3. Domain Size and Grid Independence Testing

To match published recommendations [14,48], the distance of the computational domain inlet to the target area was set to 5H, and the outlet boundary was set 15H distant from the target area. A symmetry condition was imposed at the left and right lateral sides of 5H, and the height from the ground to the top plane was set to 11H (Figure 3). A mesh size was set using a horizontal and vertical hierarchy to ensure reasonable file sizes and computing time. To achieve this, a tetrahedral meshing scheme was used. Based on European Cooperation in Science & Technology (COST) recommendations, more than 10 cells per cube root of the building volume must be set as the mesh size [48]. Considering the inputs of this study, a mesh size larger than 1.5×10^6 has been maintained to ensure a higher resolution, and three different grid sizes were tested by grid convergence index (GCI) to validate the grid independency of the numerical simulation, following the procedures proposed by Hefny and Ooka [65] (Table 2). Tests indicated that the GCI (u) for coarse and fine meshes differed by 2.72%, whereas those for fine and finest mesh differed by 2.61%. The GCIs (u) differences were all less than 5%, which indicated that fine meshes were sufficient. Element dimensions were selected as $X_{min} = Y_{min} = Z_{min} = 0.05H$ to assure rational file sizes and save computation time.

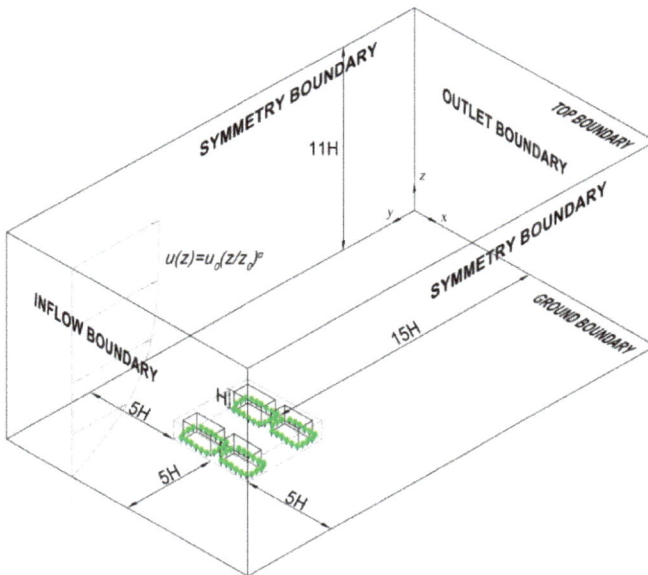

Figure 3. The computational domain.

Table 2. Different grid size for the computational domain.

Mesh	Minimum Grid Dimension	Total Cellnumber
coarse mesh	$X_{min} = Y_{min} = Z_{min} = 0.1H$	1,566,234
fine mesh	$X_{min} = Y_{min} = Z_{min} = 0.05H$	3,001,458
finest mesh	$X_{min} = Y_{min} = Z_{min} = 0.02H$	5,863,544

4.4. Boundary Conditions and Turbulence Model

The boundary conditions implemented in this study were simply velocity inlet and outflow. As no internal pollution sources in the target area was assumed, $PM_{2.5}$ sources were transported by wind from outside of the target area [64]. Thus, the source of $PM_{2.5}$ was added through the inlet, and the concentration was evenly distributed throughout the entire inflow boundary. The gradient wind was set as the velocity inlet boundary, and expressed as:

$$u(z) = u_0(z/z_0)^\alpha \qquad (8)$$

where $u(z)$ is the horizontal air velocity at the height z; and u_0 is the horizontal air velocity at a reference height of z_0. In the model, $u_0 = 2.3$ m/s, $z = 1.5$ m, and $z_0 = 10$ m. α is set to 0.25 [66].

To model the outflow boundary, a fixed pressure and zero gradients were specified. The lateral planes were regarded as symmetry boundaries. For the ground boundary, a fully rough wall function with a surface roughness of 0.495 m (0.0275H) was implemented. Moreover, the constant horizontal velocity and turbulent kinetic energy were fixed for the top boundary according to the inflow profiles.

The turbulent kinetic energy, k (m^2/s^2), and its dissipation rate, ε (m^2/s^3), were determined as follows:

$$k = \frac{u_*^2}{\sqrt{C_\mu}}\left(1 - \frac{z}{\delta}\right) \qquad (9)$$

$$\varepsilon = \frac{u_*^3}{kz}\left(1 - \frac{z}{\delta}\right) \qquad (10)$$

where u_* is the friction velocity, which is set as 0.52 m/s, δ is the boundary layer depth, and k is the von Kàrmàn constant, which is set as 0.4. The constant C_μ is set as 0.09.

There are several specification methods for the turbulence parameters, including turbulence intensity and length scale, which can be calculated independently from the transport equations. Based on the Reynolds number of the case, the turbulence intensity is determined to be 10%, and the length scale simply equals $0.06l$, where l is the inlet height in this study.

4.5. Convergence Criteria

Based on the recommendation of COST Action 732 [48], a convergence of scaled residuals down to 10^{-5} was adopted as criteria to examine the results with the main output parameters monitored. In this study, the average wind pressure on building facades, and airflow velocity at the pedestrian level (1.5 m height), were selected for soughting. These results were monitored in different convergence criteria, starting with larger residuals convergence criteria 10^{-3} up to 10^{-6}. There was no significant difference in results in the case of using 10^{-6} convergence criteria compared to 10^{-5}, indicating that the solution converged.

4.6. Simulation Target

Wind-induced ventilation depends strongly on the pressure distribution on building facades, and it is significantly affected by the upstream building characteristics and the prevailing wind [67]. Wind pressure is the main natural ventilation driving force in Beijing. It is normally positive on the windward facades, but negative on the roof and leeward facade, due to boundary layer separation occurring by a building's edges. Based on Bernoulli's principle, the wind pressure is defined by:

$$P_w = 0.5K\rho V^2 \qquad (11)$$

where P_w is the wind pressure (Pa); K is the static pressure coefficient; ρ is the reference air density (Kg/m^3); and V is the time-mean wind speed at datum level (m/s).

In this study, each building was six storeys high, and each storey was divided into four flats (*a*, *b*, *c*, and *d*). The wind pressure differences across flat facades were calculated, and the natural ventilation potential across each flat was evaluated by the resulting average wind pressure difference across each flat. These were achieved by the following procedures for each case:

(1)   Average wind pressure of each building façade was recorded.

(2)   Pressure differences across flats *a* to *d* in each building were computed (Tables 3–6).

(3)   Contours of wind velocity and streamlines at the pedestrian level, and contours of wind velocity and streamlines in sections *A-A'* and *B-B'* were also presented for better understanding the surrounding wind environment.

The indoor $PM_{2.5}$ concentration can be calculated as:

$$C_{indoor} = A * \left(\frac{U_i}{\exp \ln(U_i)} \right)^B * \left(\frac{RH_i}{\exp \ln(RH_i)} \right)^C * C_{outdoor} \tag{12}$$

where C_{indoor} and $C_{outdoor}$ are the mean concentrations indoors and outdoors; and U_i and RH_i are the outdoor wind velocity (m/s) and relative humidity (%) under the steady state *i*, respectively. *A* is the permeability coefficient of an exterior window crack. *B* and *C* represent the correction coefficients for outdoor wind velocity and relative humidity. Based on a previous study, the factors *A*, *B*, and *C* are set as 0.5719, −0.1825, and 0.1622, respectively, given weather conditions in Beijing [38]. Equation (12) has been validated by a monthly measured data in February 2014 in Beijing, and the predicted indoor $PM_{2.5}$ agreed well with the monitored indoor data ($R^2 = 0.95$) [38]. Hence, this equation proved to be an effective and feasible mathematical model to predict indoor $PM_{2.5}$ in a naturally ventilated building with windows closed.

To achieve the prediction of outdoor and indoor $PM_{2.5}$ concentrations, the following procedures were implemented for each examined case:

(1)   Outdoor vertical $PM_{2.5}$ concentrations were measured 0.5 m from the building facade. This was done for each of the four buildings.

(2)   Indoor concentrations across flats *a* to *d* in each building can be computed by Equation (12), as presented in Tables 7–10.

(3)   Contours of pedestrian level $PM_{2.5}$ concentrations, and $PM_{2.5}$ dispersion in vertical level were also presented to understand the vertical $PM_{2.5}$ concentrations (Figures 8–11).

5. Results and Discussion

5.1. Analysis of Outdoor Wind Environment

5.1.1. Configuration 1

The velocity contours and streamline distributions at the pedestrian level ($Z = 1.5$ m), average wind pressure on facades, and wind pressure difference across flats in configuration 1 are presented in Table 3. In the 360° wind direction scenario, the outdoor airflow largely passed through the narrow space between two gable walls where the greatest wind speed occurred. Velocities suddenly increase at the leading sharp edge. The central space between two rows generated large vortexes and significant reversed-airflows. Vortexes at the two flanks were stronger than in the center. Due to the shading of windward buildings, the central space experienced low airflow movement. Meanwhile, streamlines converged at the space between two gables of leeward buildings, and the velocity increased in the convergence region. Wind speed decreased noticeably within trees, and the minimum speed occurred at the point where the trees were planted and vortexes arose (Figure 4a).In the 270° wind direction scenario, the airflow crossed over simulation fields easily from short facades without impediment. It appeared that the high wind speed in the wind corridor and the windward sharp edge had a rapidly increasing velocity. Vortexes arose mainly within trees with decreasing wind speed, and the minimum velocity occurred where the trees were planted. In addition, the 270° direction had more vortexes that were smaller than with north wind scenarios, and exerted more reversed airflows (Figure 4b).

Table 3. Results of configuration 1 showing velocity contours and streamlines around blocks at 1.5 m height (m/s), average wind pressure on facades, $P_{av.}$, (Pa), and pressure differences across each flat, dP, (Pa).

V	Case	*V* Contours at 1.5 m	$P_{av.}$ (Pa)	*dP* across Flats *a* to *d* (Pa)				
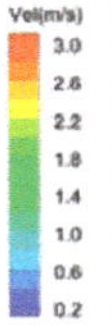	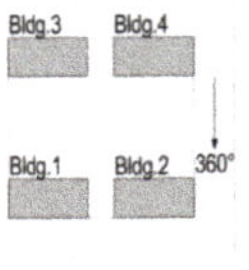	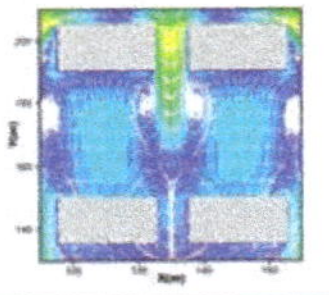	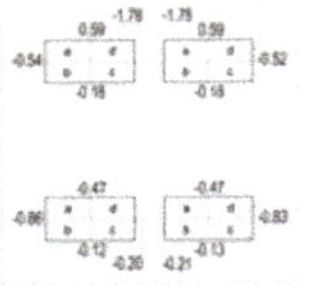		Block No.			
					Bldg.1	*Bldg.2*	*Bldg.3*	*Bldg.4*
				a	−0.39	0.26	1.13	2.37
				b	0.74	0.08	−0.37	−1.60
				c	0.08	0.70	−1.16	−0.34
				d	0.27	−0.36	2.38	1.10
	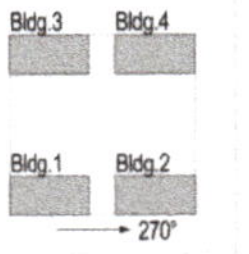	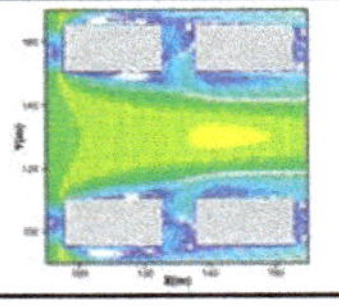	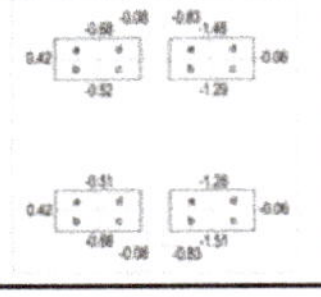		Block No.			
					Bldg.1	*Bldg.2*	*Bldg.3*	*Bldg.4*
				a	0.93	−0.43	1.10	−0.65
				b	1.11	−0.68	0.95	−0.45
				c	−0.61	1.45	−0.45	1.23
				d	−0.43	1.19	−0.60	1.42

An analysis of the wind pressure difference at 360° showed that the absolute values of windward building (Buildings 3 and 4) were higher than leeward buildings (Buildings 1 and 2), indicating the front blocks have better ventilation potential. Table 3 shows that Flats *a*, *b*, *c* and *d* of Building 3 are respectively equivalent to Flats *d*, *c*, *b*, and *a* of Building 4. Meanwhile, Buildings 1 and 2 perform the same corresponding relationship, and their corresponding flats have the same wind pressure difference. Thus, the resulting average wind pressure difference of Buildings 1 and 3 should be analyzed in this scenario. An analysis of Building 1 showed that the wind pressure difference for Flat *b* increased with building height, while Flat *c* had poor ventilation potential because of a small wind pressure difference (0.08 Pa).There is a higher negative pressure difference for the upper storey of Flats *a* and *d*. Wind pressure differences in Building 3 show those airflows through the building from the windward to leeward flats. In general, Flats *c* and *d* have better potential ventilation compared with Flats *a* and *b*. In the 270° wind direction scenario, the pressure differences among the buildings were relatively uniform, which was due to the airflow pattern that traveled from a wide space with few obstacles created by two parallel buildings exerting similar airflow surrounded by leeward and windward buildings. Meanwhile, wind pressure differences for leeward buildings are slightly higher than windward buildings. For this situation, the pressure differences of Buildings 1 and 2 are telling. An analysis of Building 1 showed that the absolute values for Flats *a* and *b* were higher than those of Flats *c* and *d*. An analysis of Building 2 showed that wind pressure differences were always positive and increased with building height for Flats *c* and *d*, with values of 1.45 and 1.19 Pa, respectively. The results from all of the buildings indicated that the highest absolute values appeared at the top storey, and the higher storeys had better ventilation potential.

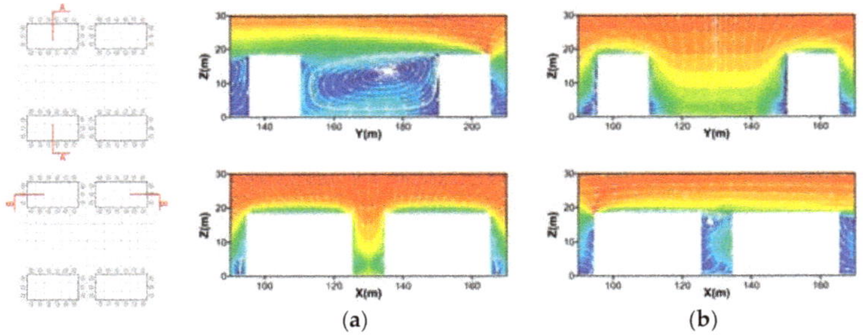

(a) (b)

Figure 4. Velocity contours and streamlines presented at vertical sections of configuration 1 when the wind angle was 360° (**a**) and 270° (**b**), respectively.

5.1.2. Configuration 2

Table 4 depicts contours of velocity and streamline distributions for Z = 1.5 m, the average wind pressure on facades, and the wind pressure difference across flats in configuration 2. In the 360° wind direction scenario, the central space between Buildings 1 and 3, where airflow reversed significantly and exerted obvious vortexes, was wind shaded. The left vortex was stronger than the right vortex. An analysis of the 270° wind direction indicated that airflow passes through the large area between windward buildings, and encountered leeward buildings, resulting in a partial separation. The wind converged between two gables of leeward buildings and reached its maximum velocity in the convergence area. There are more vortexes revealed by comparing these, in the case of the 360° wind direction (Figure 5a,b).

By analyzing the results of different wind directions, it was found that the highest absolute values of wind pressure difference all occurred in Building 2. In the case of the 360° wind direction, wind pressure increased with height for Flats *b* and *c* of Building 1. For Building 2, Flats *a* and *d* experienced low wind pressure differences, with values of −0.34 and −0.79 Pa, respectively. Flats *b* and *c* in Building 3 had negative values, but positive values in Flats *a* and *d*. In the case of the 270° wind direction, it was only necessary to analyze Buildings 1 and 2. For Building 1, the absolute value of wind pressure differences ranged from −0.40 to 1.06 Pa. Flats *a* and *b* had positive values of 0.90 and 1.06 Pa, respectively. Flat *a* had the highest values compared with other flats in Building 2. The total absolute wind pressure difference of Building 2 was higher than that of Building 1, with a difference of 4.16 Pa.

Table 4. Results of configuration 2 showing velocity contours and streamlines around blocks at 1.5 m height (m/s), average wind pressure on facades, $P_{av.}$, (Pa), and pressure differences across each flat, dP, (Pa).

V	Case	V Contours at 1.5 m	$P_{av.}$ (Pa)	dP across Flats *a* to *d* (Pa)			
				Block No.			
				Bldg.1	*Bldg.2*	*Bldg.3*	*Bldg.4*
				a	−0.38	−0.34	1.14	1.52
				b	0.71	1.07	−0.38	−0.98
				c	0.59	1.52	−0.23	−0.63
				d	−0.26	−0.79	0.99	1.16
				Block No.			
				Bldg.1	*Bldg.2*	*Bldg.3*	*Bldg.4*
				a	0.90	3.05	1.05	1.59
				b	1.06	1.58	0.92	3.05
				c	−0.55	−0.49	−0.41	−1.95
				d	−0.40	−1.95	−0.55	−0.50

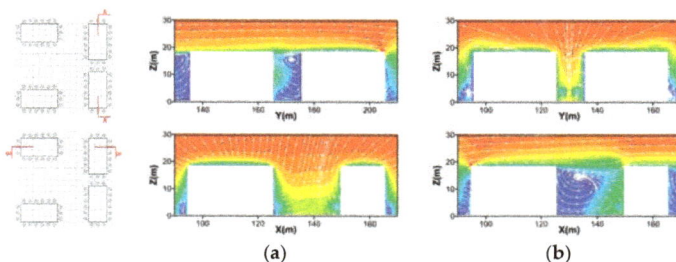

(a)   **(b)**

Figure 5. Velocity contours and streamlines presented at vertical sections of configuration 2 when the wind angle was 360° (**a**), and 270° (**b**), respectively.

5.1.3. Configuration 3

Table 5 shows the contours of velocity and streamline distributions at the pedestrian level, the average wind pressure on facades, and the wind pressure difference across flats in configuration 3. The results of the leeward and windward aspects in the 360° wind direction were the same for those when the wind direction was 270°. Buildings 3, 4, 1, and 2 in the case of the 360° wind were equivalent to Buildings 1, 3, 2, and 4 at the 270° wind direction. Thus, it is only necessary to analyze the contours of velocities and *dP* across each flat for the 360° wind direction.

An analysis of streamline distributions indicated that airflow penetrated the central space between windward buildings and converged near the long facade of Building 2. The area between Buildings 4 and 2 with trees created wind deflection, resulting in obvious reversed flow and small vortexes (Figure 6a,b). The resulting wind pressure difference ranged from −0.68 to 1.51 Pa of Building 1. The special tendency for Building 2 is that airflow travels from indoors to outdoors at the lower height and oppositely at the upper height. There is a sharp velocity increase in the top of Buildings 3 and 4, and the two blocks have the same tendency for having a higher absolute value of pressure difference, with values of 3.71 Pa and 2.72 Pa, respectively.

Table 5. Results of configuration 3 showing velocity contours and streamlines around blocks at 1.5 m height (m/s), average wind pressure on facades, $P_{av.}$, (Pa), and pressure differences across each flat, dP, (Pa).

V	Case	V Contours at 1.5 m	$P_{av.}$ (Pa)	dP across Flats *a* to *d* (Pa)				
					Block No.			
Vel(m/s) 3.0 2.6 2.2 1.8 1.4 1.0 0.6 0.2	Bldg 3 Bldg 4 Bldg 2 Bldg 1 360°				*Bldg.1*	*Bldg.2*	*Bldg.3*	*Bldg.4*
				a	1.51	−0.38	1.09	1.03
				b	−0.68	0.49	−0.54	−0.18
				c	−0.15	1.15	−0.78	−0.33
				d	0.97	0.29	1.33	1.18
					Block No.			
	Bldg 3 Bldg 4 Bldg 2 Bldg 1 270°				*Bldg.1*	*Bldg.2*	*Bldg.3*	*Bldg.4*
				a	1.29	1.00	1.21	0.31
				b	1.09	1.53	1.03	−0.36
				c	−0.55	−0.71	−0.19	0.51
				d	−0.75	−0.19	−0.37	1.18

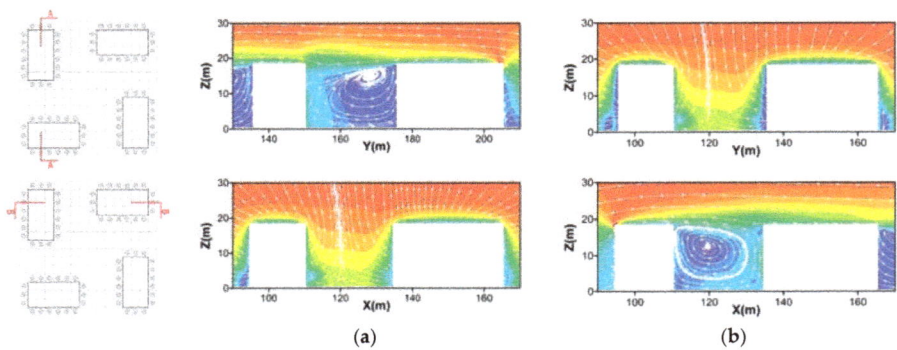

(a)   (b)

Figure 6. Velocity contours and streamlines presented at vertical sections of configuration 3 when the wind angle was 360° (**a**), and 270° (**b**), respectively.

5.1.4. Configuration 4

Table 6 shows the results of the velocity contours around blocks and streamlines at Z = 1.5 m (m/s), average wind pressure on facades, and pressure differences across each flat for configuration 4. The wind speed accelerated at the edge of Buildings 2 and 3 at the pedestrian level. Due to the narrow space for air to flow through, and the wind shading of the windward building, few airflows exerted slow movement in the central space created by the buildings envelope, and vortexes were small in each case (Figure 7a,b).

The absolute wind pressure differences were always near 0 Pa in the area surrounding Building 1, due to the wind-shading effect. Therefore, the flats in Building 1 had poor natural ventilation potential, and people may feel uncomfortable in interior space in this scenario. Flat *c* of Building 2 and Flat *b* of Building 3 both had slight improvements compared with other flats. In general, configuration 4 produced the poorest ventilation potential; as a result, it is the worst layout for airflow transportation.

Table 6. Results of configuration 4 showing velocity contours and streamlines around blocks at 1.5 m height (m/s), average wind pressure on facades, $P_{av.}$, (Pa), and pressure differences across each flat, *dP*, (Pa).

V	Case	V Contours at 1.5 m	$P_{av.}$ (Pa)	dP across Flats *a* to *d* (Pa)			
					Block No.		
				Bldg.1	Bldg.2	Bldg.3	Bldg.4
				a	−0.02	1.09	0.60	0.81
				b	−0.09	−0.63	−0.14	−0.03
				c	−0.09	−0.15	−0.63	−0.02
				d	−0.01	0.61	1.09	0.81
					Block No.		
				Bldg.1	Bldg.2	Bldg.3	Bldg.4
				a	0.59	0.80	0.01	1.09
				b	1.11	0.81	0.01	0.59
				c	−0.66	−0.03	−0.08	−0.14
				d	−0.14	−0.02	−0.08	−0.64

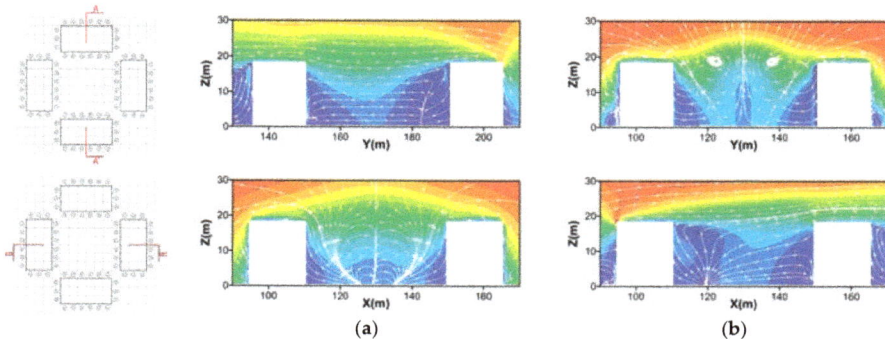

Figure 7. Velocity contours and streamlines presented at vertical sections of configuration 4 when the wind angle was 360° (**a**) and 270° (**b**), respectively.

5.2. Analysis of Indoor/Outdoor PM$_{2.5}$

5.2.1. Configuration 1

Table 7 depicts the PM$_{2.5}$ concentrations around blocks at 1.5 m height, the average vertical PM$_{2.5}$ concentrations, and the indoor concentration across each flat for configuration 1. At a 360° wind direction, long building facades, except the windward facade of leeward buildings, remained the lowest PM$_{2.5}$ within tree crown height. The leeward facades of Buildings 1 and 2 had low particle concentrations on both lateral sides. On the leeward facades of buildings, particles were significantly affected by trees, and the deposition effect continued to the tops of buildings; however, its influence appeared only near the tree crown for windward building facades (Figure 8a). In the case of a 270° wind direction, the concentration distributions of lower layer for leeward facades of Buildings 1 and 3 were uniform. Thus, Buildings 1 and 2 need analysis. The leeward facades of Building 2 had lower concentrations when Z < 3 m, and the concentrations declined when 135 m < X < 140 m. The leeward facade of Building 1 had a lower concentration within the tree crown, and the deposition effects of trees on particulate matters reached to Z = 6 m (Figure 8b).

With a 360° wind direction, the PM$_{2.5}$ distributions around Buildings 1 and 3 were symmetrical to Buildings 2 and 4. Therefore, the indoor concentrations of Buildings 1 and 3 need to be analyzed. The concentrations of the leeward building (Building 1) were significantly lower than that of the windward building (Building 3), with a difference of 59.6 µg/m^3. In Building 1, the PM$_{2.5}$ of Flats *b* and *c* were lower, with the values of 26.5 and 28.8 µg/m^3, which meant 20% of the inlet value. The concentrations of Flats *b* and *c* in Building 3 were the lowest, and Flat *d* was the highest, with the value of 76.0 µg/m^3. The results from a 270° wind direction showed that the concentrations of each flat were similar with a difference of 3.3 µg/m^3, and Flats *a* and *b* in Building 1 had maximum concentrations, with 74.7 and 75.9 µg/m^3, respectively. The concentrations for each flat of Building 2 ranked lowest to highest in order are: Flat *c* < Flat *d* < Flat *b* < Flat *a*.

Table 7. Results of configuration 1 showing PM$_{2.5}$ concentrations around blocks at 1.5 m height (µg/m^3), average vertical PM$_{2.5}$ concentrations (0.5 m away from facades), (µg/m^3), and indoor $C_{av.}$ across flats *a* to *d*, (µg/m^3).

C	Case	*C* Contours at 1.5 m	Outdoor Vertical $C_{av.}$ (µg/m³)	Indoor $C_{av.}$ across Flats *a* to *d* (µg/m³)			
				Block No.			
				Bldg.1	*Bldg.2*	*Bldg.3*	*Bldg.4*
				a	64.7	64.8	68.4	76.0
				b	26.5	28.7	50.6	49.2
				c	28.8	26.2	49.4	50.9
				d	64.8	64.6	76.0	68.4
				Block No.			
				Bldg.1	*Bldg.2*	*Bldg.3*	*Bldg.4*
				a	74.7	71.1	73.6	69.5
				b	75.9	69.4	74.8	70.9
				c	74.1	66.5	73.1	68.1
				d	72.6	68.1	75.3	66.4

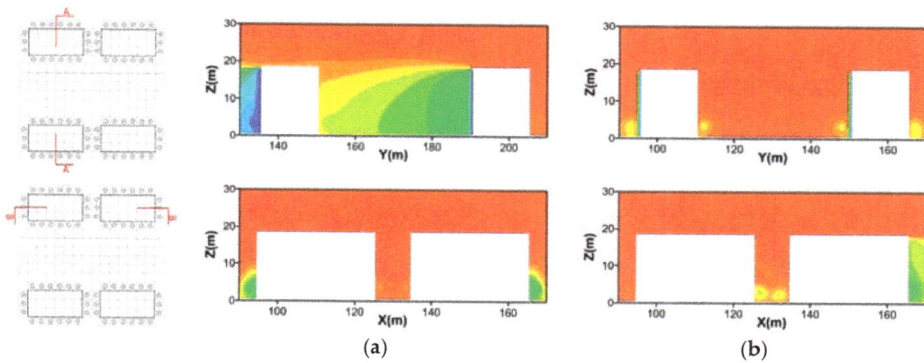

Figure 8. PM$_{2.5}$ concentrations presented at vertical sections of configuration 1 when the wind angle is 360° (**a**), and 270° (**b**), respectively.

5.2.2. Configuration 2

Table 8 depicts PM$_{2.5}$ concentrations around blocks at 1.5 m height, the average vertical PM$_{2.5}$ concentrations, and the indoor concentration across each flat for configuration 2. In the case of a 360° wind direction, the concentrations of the leeward façade of Building 1 were lower than that of other facades, and showed significant decreases within the tree crown. The concentration distributions of the vertical facades of Building 3 were similar to those of Building 1. Meanwhile, the concentrations of windward and leeward facades of Building 2 were the lowest within the height of the tree crown. The indoor particle concentrations of Buildings 1 and 3 were lower than Buildings 2 and 4. Flats *b* and *c* had lower concentrations compared with Flats *a* and *d* for Building 1, and the lowest concentrations appeared at the height within tree crown height (from 2 m to 5 m). For Building 2, the PM$_{2.5}$ had no obvious change, and the indoor average values ranged from 64.3 to 69.8 µg/m³. In Building 3, the concentrations of Flats *b* and *c* varied in the same way, and the influence of the surrounding trees continued to the top of the buildings. In Building 4, the concentrations of Flats *b* and *c* were similar from the bottom to the top of the building (Figure 9a).

Table 8. Results of configuration 2 showing PM$_{2.5}$ concentrations around blocks at 1.5 m height (µg/m³), average vertical PM$_{2.5}$ concentrations (0.5 m away from facades), (µg/m³), and indoor $C_{av.}$ across flats *a* to *d*, (µg/m³).

C	Case	C Contours at 1.5 m	Outdoor Vertical $C_{av.}$ (µg/m³)	Indoor $C_{av.}$ across Flats *a* to *d* (µg/m³)			
				Block No.			
	Bldg.3 Bldg.4 / Bldg.2 / Bldg.1 360°		131.2 131.2; 131.4 · 126.6; 84.1 84.1; 128.0 · 128.5; 119.5 119.3; 125.2 · 124.3; 124.1 · 125.0; 73.9 76.5		*Bldg.1*	*Bldg.2*	*Bldg.3*	*Bldg.4*
				a	69.6	69.8	74.6	68.2
				b	46.7	66.0	54.2	69.6
				c	49.6	64.3	53.7	74.3
				d	69.8	66.8	76.9	72.1
				Block No.			
	Bldg.3 Bldg.4 / Bldg.2 / Bldg.1 270°		127.0 125.6; 129.6 · 82.7; 131.1 118.0; 130.8 · 86.4; 131.0 118.7; 130.9 · 85.7; 129.6 · 81.9; 127.4 126.3		*Bldg.1*	*Bldg.2*	*Bldg.3*	*Bldg.4*
				a	71.8	77.1	70.1	65.5
				b	73.0	65.5	71.9	77.1
				c	70.4	50.9	70.3	55.4
				d	70.1	55.1	72.6	51.5

PM (µg/m³): 130, 120, 110, 100, 90, 80, 70, 60, 50, 40, 30, 20, 10

With a 270° wind direction, the patterns of Buildings 1 and 3 were symmetrical to Buildings 2 and 4. So, the results of Buildings 1 and 2 must be analyzed. The concentrations appeared lower on the leeward facades of back buildings, and decreased within the tree canopy. For Building 1, concentrations of Flat *a* gradually decreased with the building height. Flats *c* and *d* had lower concentrations, with a value of approximately 70 μg/m³. The concentrations for the flats of Building 2, ranked lowest to highest, in order, are: Flat *c* < Flat *d* < Flat *b* < Flat *a*. Flats *c* and *d* have the lowest concentrations, with the values of 50.9 and 55.1 μg/m³, respectively (Figure 9b).

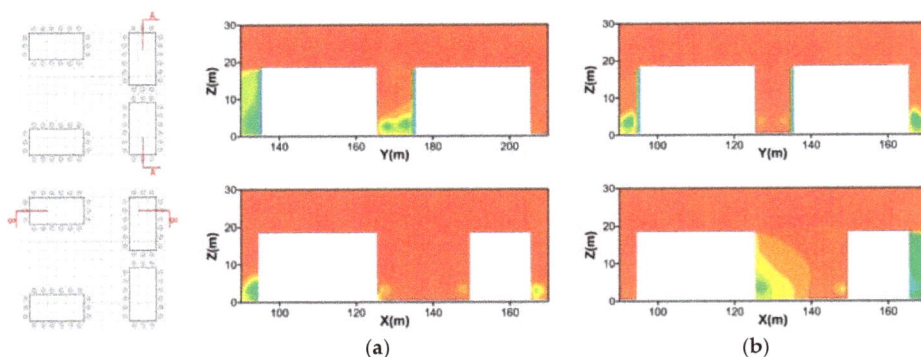

Figure 9. $PM_{2.5}$ concentrations presented at vertical sections of configuration 2 when the wind anglewas 360° (**a**) and 270° (**b**), respectively.

5.2.3. Configuration 3

Table 9 depicts $PM_{2.5}$ concentrations around blocks at 1.5 m height, the average vertical $PM_{2.5}$ concentrations, and the indoor concentration across each flat for configuration 3. The building layout in configuration 3 is centrally symmetrical; hence, only the indoor and outdoor particle concentrations distribution at 360° wind needs to be analyzed. The concentrations of the leeward facade of Building 1 were significantly lower than the other facades, and reached their lowest values within the tree canopy. The low concentrations of the long facades of Buildings 3 and 4 also occurred within the tree canopy, while the concentrations of the long facades of Building 2 increased with the building height (Figure 10). An analysis of indoor $PM_{2.5}$ concentrations showed that the concentrations of Flats *b* and *c* were low in Building 1, with the values of 45.1 μg/m³ and 48.3 μg/m³, respectively. In Building 2, the concentrations of Flats *b* and *c* differed indistinctly. In Building 3, the concentrations of Flat *b* changed obviously between each storey, with the maximum concentration occurring at the 5 m height. The concentrations of Flats *b* and *c* represented lower values in Building 4, with values of 64.8 μg/m³ and 64.9 μg/m³.

5.2.4. Configuration 4

Table 10 shows $PM_{2.5}$ concentrations around blocks at 1.5 m height, the average vertical concentration, and the indoor concentrations across Flats *a* to *d* in configuration 4. This layout is also centrally symmetrical. In the scenario with an angle of 360°, the concentrations of leeward facade of Building 1 were lower within the tree crown, and the deposition effects reached to 15 m on both lateral sides of the building facades. These deposition effects on the windward facade of Building 2 reached to 12 m (the fourth storey). The concentrations of the leeward facade of Building 2 also obtained their minimum values within the tree canopy. Buildings 2 and 3 were symmetrical for all four flats. The concentrations of the leeward facade of Building 4 gradually increased from the middle to the top of the building (Figure 11). An analysis of indoor $PM_{2.5}$ indicated that Flats *b* and *c* have the same change tendency, with a lower value of 24.7 μg/m³ and 24.6 μg/m³, respectively. In Building 2, the concentrations of Flat *d* reached the lowest value compared with other flats. Building 3 was symmetrical with Building 2, and their changes were similar. In Building 4, the concentrations of Flats *b* and *c* were similar. Flats *a* and *d* have the same changes that occur elsewhere along the vertical facades.

Table 9. Results of configuration 3 showing PM$_{2.5}$ concentrations around blocks at 1.5 m height (μg/m³), average vertical PM$_{2.5}$ concentrations (0.5 m away from facades), (μg/m³), and indoor $C_{av.}$ across flats *a* to *d*, (μg/m³).

C	Case	C Contours at 1.5 m	Outdoor Vertical $C_{av.}$ (μg/m³)	Indoor $C_{av.}$ across Flats *a* to *d* (μg/m³)			
				Block No.			
	Bldg.3 Bldg.4 Bldg.1 Bldg.2 360°			Bldg.1	Bldg.2	Bldg.3	Bldg.4
				a 68.9	74.4	71.7	77.0
				b 45.1	67.0	74.5	64.8
				c 48.3	65.7	71.2	64.9
				d 75.3	69.2	68.3	73.8
	Bldg.3 Bldg.4 Bldg.1 Bldg.2 270°			Block No.			
				Bldg.1	Bldg.2	Bldg.3	Bldg.4
				a 68.5	75.1	73.8	69.0
				b 71.9	69.0	77.0	74.3
				c 74.7	45.3	65.5	66.8
				d 71.8	49.1	65.4	65.7

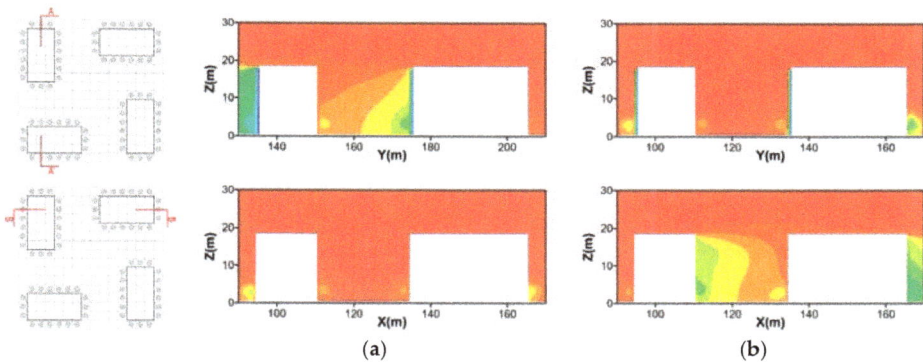

Figure 10. PM$_{2.5}$ concentrations presented at vertical sections of configuration 3 when the wind angle was 360° (**a**), and 270° (**b**), respectively.

The above-mentioned discussion aimed at exploring the resulting ventilation potential and outdoor and indoor PM$_{2.5}$ concentrations of each configuration. However, the question of which configuration can give the best ventilation potential, which can provide the lowest indoor PM$_{2.5}$ concentrations, and the relationship between wind pressure difference and indoor PM$_{2.5}$ concentrations remain outstanding. Thus, further assessments of each scenario considering the actual architectural and tree planting design were conducted. A general assessment can be given by summing up the pressure difference values and indoor PM$_{2.5}$ across the sixteen flats of each configuration. Since positive or negative pressure differences only indicate the flow direction, pressure differences should be converted to their absolute values when summed.

Table 10. Results of configuration 4 showing $PM_{2.5}$ concentrations around blocks at 1.5 m height ($\mu g/m^3$), average vertical $PM_{2.5}$ concentrations (0.5 m away from facades), ($\mu g/m^3$), and indoor $C_{av.}$ across flats *a* to *d*, ($\mu g/m^3$).

C	Case	C Contours at 1.5 m	Outdoor Vertical $C_{av.}$ ($\mu g/m^3$)	Indoor $C_{av.}$ across Flats a to d ($\mu g/m^3$)				
					Block No.			
$PM_{2.5}$ ($\mu g/m^3$) 130 120 110 100 90 80 70 60 50 40 30 20 10	Bldg 4 / Bldg 2 Bldg 3 / Bldg 1 — 360°		131.2 131.2 / 100.7 109.1 / 127.0 111.8 112.4 126.4 / 126.4 116.6 116.9 128.2 / 121.7 121.5 / 37.6 37.4		Bldg.1	Bldg.2	Bldg.3	Bldg.4
				a	73.6	68.1	61.1	71.4
				b	24.7	69.4	75.8	64.4
				c	24.6	75.6	69.3	64.2
				d	73.6	60.7	67.8	71.4
					Block No.			
	Bldg 4 / Bldg 2 Bldg 3 / Bldg 1 — 270°		126.9 128.0 / 113.2 116.8 / 131.2 108.4 121.7 41.8 / 131.2 109.2 121.8 41.8 / 128.7 116.7 / 127.6 128.4		Bldg.1	Bldg.2	Bldg.3	Bldg.4
				a	61.2	71.4	73.3	67.7
				b	68.0	71.4	73.3	61.5
				c	69.1	64.1	27.6	75.9
				d	75.9	63.7	27.6	69.1

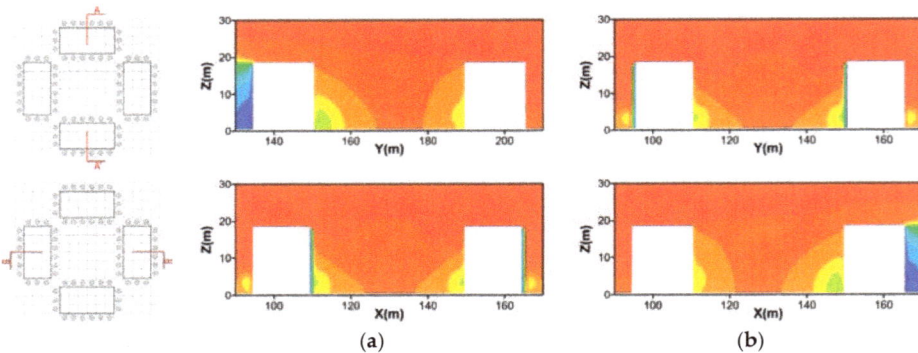

(a) (b)

Figure 11. $PM_{2.5}$ concentrations presented at a vertical section of configuration 4 when the wind angle was 360° (**a**) and 270° (**b**), respectively.

Figure 12 depicts the total absolute values for pressure differences and interior $PM_{2.5}$ concentrations with the wind directions of 360° and 270°. The high pressure difference in configuration 2 with a 270° wind direction results in the best ventilation potential. The poorest ventilation potential is found in configuration 4, in two wind directions with similar wind pressure differences because of the wind-shading effect. This scenario showed 7.51 Pa and 2.70 Pa, respectively, which means a difference of 178% (Figure 12a).Configuration 1 had the lowest indoor $PM_{2.5}$ with the wind angle of 360°, and highest concentration with a 270° wind direction (Figure 12b). The overall values were similar among the other three configurations at different wind angles. A scatter plot of indoor $PM_{2.5}$ concentration versus wind pressure difference had an R^2 of 0.059 for $PM_{2.5}$ (Figure 12c).

The results of this study suggest that wind pressure difference has little effect on indoor $PM_{2.5}$ concentration. This is similar to the results concluded by Chithra and Shiva Nagendra [30]. It differs, however, from the simulation study using a CONTAMW simulation, as our study shows the lower floors of a multi-storey building having lower $PM_{2.5}$ compared with the upper floors. The CONTAMW simulation drew the opposite conclusion [68]. However, this was only because our study considered the aerodynamic and deposition effects of trees on particles dispersion, and airflow fields among building complexes is different from single office buildings.

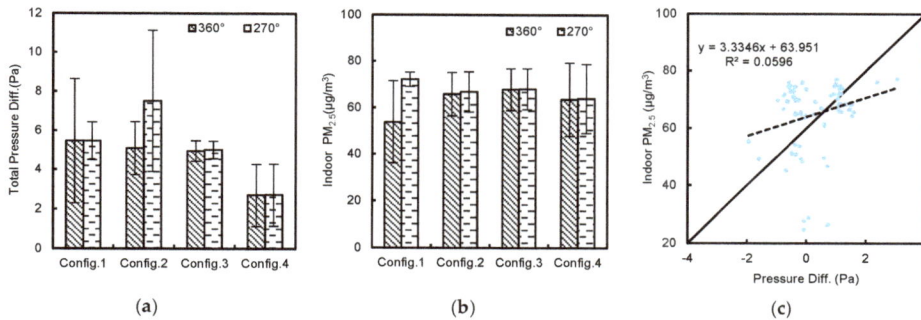

Figure 12. The resulting gross absolute pressure difference for each configuration (**a**), indoor $PM_{2.5}$ for each configuration (**b**), and correlation between indoor $PM_{2.5}$ and wind pressure difference (**c**). Dashed lines are linear fits, with linear equations and R^2-values given alongside. Diagonals depict perfect matches.

Our study found that lower floors are subject to less $PM_{2.5}$ than upper floors, because we assume that the pollution source was introduced through the inlet from a representative, dispersed upstream source. Based on regional statistics [45], the predominant source of $PM_{2.5}$ pollution in Beijing in 2015 was derived from atmospheric transmission, which accounted for 38.6% of all sources (e.g., traffic, industry, and burning). Thus, this study added the pollution source through the inlet. However, in many urban environments, the pollution source is at street level in the vicinity of the buildings. An indepth study should be carried out to consider this alternative scenario.

This research qualitatively and quantitatively assessed the airflow field and indoor/outdoor relationships for $PM_{2.5}$ within partly wind-driven natural ventilation in four typical building–tree configurations. Our study had some limitations that should be addressed. First, the airflow and particle dispersion simulation was only conducted for some building–tree patterns, and trees around the buildings were simulated as evergreen trees and planted aligned to the road. Additionally, weather parameters that were only typical during November to December in 2015 were used in this simulation study. Thus, more building and tree arrangements in serious climate conditions should be considered. Second, field experiments measuring the pressure difference and particle transportation indoor and outdoor in multi-storey buildings are needed in subsequent study to verify our conclusions. Third, particles are regarded as a continuum, and particle movements are assumed to not affect turbulence. The convective velocity of the particle phase is assumed to be the same as the airflow in this study. However, wind flow in the real environments is transient, and even a steady wind condition gives transient vortex shedding in wind-tunnel/field experiments under certain configurations [69]. These transient flows are important, both for pollutant mixing and building ventilation. Thus, an experimental validation of the models is still needed. Fourth, the findings of this study are only valid for wind-driven ventilation. There may be other mechanisms that are important under thermally-driven conditions, i.e., solar heat, anthropogenic heat, heat absorption/emission of building masses, etc. More complicated scenarios, such as non-isothermal cases with solar radiation, and convective heat transfer inside buildings and tree clusters, would add to our findings. Finally, this study assumed that the pollution sources were diluted by wind transportation, and the indoor generation of particles was not considered. This is because the objective of this research was to explore the effects of building–tree patterns on indoor particle concentrations generated outdoors. In the real built environment, indoor air pollution is affected not only by outdoor pollution, but also by indoor originated particles. A comprehensive analysis linking various factors of I/O ratio, particle penetration, and infiltration, and indoor generated could contribute to a thorough understanding of actual indoor/outdoor particles dispersion in multi-storey buildings with different building–tree arrangements.

6. Conclusions

In this study, the effects of four typical building–tree configurations on the pedestrian-level wind environment, the wind pressure on building facades, and the indoor/outdoor relationships for $PM_{2.5}$ due to partly wind-driven natural ventilation were studied by the standard k-ε model and the revised generalized drift flux model. The results clearly indicated that airflow and indoor/outdoor $PM_{2.5}$ dispersion strongly depended on the relationships of building layouts, tree arrangements, and orientation towards the prevailing wind. The following conclusions can be drawn from the simulated results: (1) Building and tree configurations can provide natural ventilation potentials relying on wind influence, and buildings that experience wind deflection on their windward facade and airflow separation on the leeward side have better ventilation potential; (2) Patterns with buildings and trees forming a central space and an opening side towards the prevailing wind could offer the best ventilation conditions, that is to say, configuration 2 with the prevailing wind direction of 270° can offer the best ventilation potential, and a great difference of 178% existed when comparing it with the poorest configuration; (3) Under the conditions of the particle sources introduced to the interior through the inlet, the aerodynamics and deposition effects of trees allowed the lower floors of a multi-storey building to be exposed to lower $PM_{2.5}$ compared with the upper storeys. As well, lower indoor $PM_{2.5}$ were found close to the tree canopy, leading to a significant reduction on $PM_{2.5}$ with the height of tree canopy; (4) Wind pressure differences across each flat showed poor correlation ($R^2 = 0.059$) with indoor $PM_{2.5}$. This is because pressure change usually causes global pollutant dispersion changes, rather than producing effects on microscopic pollutants dispersion [70]; and (5) Configurations with long building facades and trees perpendicular to prevailing wind had the lowest indoor $PM_{2.5}$. The highest $PM_{2.5}$ was observed in configuration 1 with a 270° wind direction, while the lowest was also observed in configuration 1, with a wind angle of 360°. This emphasized the importance of prevailing wind in the transportation of pollution.

Acknowledgments: This study is supported by the National Natural Science Foundation of China (51708451, 51561135001 and 51638003) and the Innovative Research Groups of the National Natural Science Foundation of China (51521005).

Author Contributions: Bo Hong and Borong Lin conceived and designed the experiments; Bo Hong and Hongqiao Qin performed the experiments and analyzed the data; Bo Hong wrote the paper.

Conflicts of Interest: The authors declare no conflict of interest.

References

1. Brunekreef, B.; Holgate, S.T. Air pollution and health. *Lancet* **2002**, *360*, 1233–1242. [CrossRef]
2. Zhang, X.; Zhang, X.; Chen, X. Happiness in the air: How does a dirty sky affect mental health and subjective well-being? *J. Environ. Econ. Manag.* **2017**, *85*, 81–94. [CrossRef] [PubMed]
3. Abt, E.; Suh, H.H.; Catalano, P.; Koutrakis, P. Relative contribution of outdoor and indoor particle sources to indoor concentrations. *Environ. Sci. Technol.* **2000**, *34*, 3579–3587. [CrossRef]
4. Ramachandran, G.; Adgate, J.L.; Pratt, G.C.; Sexton, K. Characterizing indoor and outdoor 15 minute average $PM_{2.5}$ concentrations in urban neighborhoods. *Aerosol Sci. Technol.* **2003**, *37*, 33–45. [CrossRef]
5. Schneider, T.; Jensen, K.A.; Clausen, P.A.; Afshari, A.; Gunnarsen, L.; Wåhlin, P.; Glasius, M.; Palmgren, F.; Nielsen, O.J.; Fogh, C.L. Prediction of indoor concentration of 0.5–4 μm particles of outdoor origin in an uninhabited apartment. *Atmos. Environ.* **2004**, *38*, 6349–6359. [CrossRef]
6. Matson, U. Indoor and outdoor concentrations of ultrafine particles in some Scandinavian rural and urban areas. *Sci. Total Environ.* **2005**, *343*, 169–176. [CrossRef] [PubMed]
7. Jones, A.P. Indoor air quality and health. *Atmos. Environ.* **1999**, *33*, 4535–4564. [CrossRef]

8. Liao, C.M.; Huang, S.J.; Yu, H. Size-dependent particulate matter indoor/outdoor relationships for a wind-induced naturally ventilated airspace. *Build. Environ.* **2004**, *39*, 411–420. [CrossRef]

9. De Jong, T.; Bot, G.P.A. Air exchange caused by wind effects through (window) openings distributed evenly on a quasi-infinite surface. *Energy Build.* **1992**, *19*, 93–103. [CrossRef]

10. Miguel, A.F.; Van de Braak, N.J.; Silva, A.M.; Bot, G.P.A. Wind-induced airflow through permeable materials part I: Air infiltration in enclosure. *J. Wind Eng. Ind. Aerodyn.* **2001**, *89*, 59–72. [CrossRef]

11. Tominaga, Y.; Mochida, A.; Shirasawa, T.; Yoshie, R.; Kataoka, H.; Harimoto, K.; Nozu, T. Cross comparisons of CFD results of wind environment at pedestrian level around a high-rise building and with a building complex. *J. Asian Archit. Build.* **2004**, *3*, 63–70. [CrossRef]

12. Zhang, A.; Gao, C.; Zhang, L. Numerical simulation of the wind field around different building arrangements. *J. Wind Eng. Ind. Aerodyn.* **2005**, *93*, 891–904. [CrossRef]

13. Asfour, O.S.; Gadi, M.B. A comparison between CFD and Network models for predicting wind-driven ventilation in buildings. *Build. Environ.* **2007**, *42*, 4079–4085. [CrossRef]

14. Tominaga, Y.; Mochida, A.; Yoshie, R.; Kataoka, H.; Nozu, T.; Yoshikawa, M.; Shirasawa, T. AIJ guidelines for practical applications of CFD to pedestrian wind environment around buildings. *J. Wind Eng. Ind. Aerodyn.* **2008**, *96*, 1749–1761. [CrossRef]

15. Kubota, T.; Miura, M.; Tominaga, Y.; Mochida, A. Wind tunnel tests on the relationship between building density and pedestrian-level wind velocity: Development of guidelines for realizing acceptable wind environment in residential neighborhoods. *Build. Environ.* **2008**, *43*, 1699–1708. [CrossRef]

16. Mochida, A.; Lun, I.Y.F. Prediction of wind environment and thermal comfort at pedestrian level in urban area. *J. Wind Eng. Ind. Aerodyn.* **2008**, *96*, 1498–1527. [CrossRef]

17. Asfour, O.S. Prediction of wind environment in different grouping patterns of housing blocks. *Energy Build.* **2010**, *42*, 2061–2069. [CrossRef]

18. You, W.; Gao, Z.; Chen, Z.; Ding, W. Improving residential wind environments by understanding the relationship between building arrangements and outdoor regional ventilation. *Atmosphere* **2017**, *8*, 102. [CrossRef]

19. Mochida, A.; Tabata, Y.; Iwata, T.; Yoshino, H. Examining tree canopy models for CFD prediction of wind environment at pedestrian level. *J. Wind Eng. Ind. Aerodyn.* **2008**, *96*, 1667–1677. [CrossRef]

20. Chen, H.; Ooka, R.; Kato, S. Study on optimum design method for pleasant outdoor thermal environment using genetic algorithms (GA) and coupled simulation of convection, radiation and conduction. *Build. Environ.* **2008**, *43*, 18–30. [CrossRef]

21. Hong, B.; Lin, B.; Hu, L.; Li, S. Optimal tree design for sunshine and ventilation in residential district using geometrical models and numerical simulation. *Build. Simul.* **2011**, *4*, 351–363. [CrossRef]

22. Hong, B.; Lin, B.; Wang, B.; Li, S. Optimal design of vegetation in residential district with numerical simulation and field experiment. *J. Cent. South Univ.* **2012**, *19*, 688–695. [CrossRef]

23. Hong, B.; Lin, B. Numerical studies of the outdoor wind environment and thermal comfort at pedestrian level in housing blocks with different building layout patterns and trees arrangement. *Renew. Energy* **2015**, *73*, 18–27. [CrossRef]

24. Chen, C.; Zhao, B. Review of relationship between indoor and outdoor particles: I/O ratio, infiltration factor and penetration factor. *Atmos. Environ.* **2011**, *45*, 275–288. [CrossRef]

25. Braniš, M.; Řezáčová, P.; Domasová, M. The effect of outdoor air and indoor human activity on mass concentrations of PM_{10}, $PM_{2.5}$, and PM_1 in a classroom. *Environ. Res.* **2005**, *99*, 143–149. [CrossRef] [PubMed]

26. Massey, D.; Masih, J.; Kulshrestha, A.; Habil, M.; Taneja, A. Indoor/outdoor relationship of fine particles less than 2.5 μm ($PM_{2.5}$) in residential homes locations in central Indian region. *Build. Environ.* **2009**, *44*, 2037–2045. [CrossRef]

27. Chithra, V.S.; Nagendra, S.M.S. Indoor air quality investigations in a naturally ventilated school building located close to an urban roadway in Chennai, India. *Build. Environ.* **2012**, *54*, 159–167. [CrossRef]

28. Mohammadyan, M.; Ghoochani, M.; Kloog, I.; Abdul-Wahab, S.A.; Yetilmezsoy, K.; Heibati, B.; Pollitt, K.J.G. Assessment of indoor and outdoor particulate air pollution at an urban background site in Iran. *Environ. Monit. Assess.* **2017**, *189*, 235. [CrossRef] [PubMed]

29. Hahn, I.; Brixey, L.A.; Wiener, R.W.; Henkle, S.W. Parameterization of meteorological variables in the process of infiltration of outdoor ultrafine particles into a residential building. *J. Environ. Monit.* **2009**, *11*, 2192–2200. [CrossRef] [PubMed]

30. Chithra, V.S.; Nagendra, S.M.S. Impact of outdoor meteorology on indoor PM_{10}, $PM_{2.5}$ and PM_1 concentrations in a naturally ventilated classroom. *Urban Clim.* **2014**, *10*, 77–91. [CrossRef]

31. Riain, C.M.N.; Mark, D.; Davies, M.; Harrison, R.M.; Byrne, M.A. Averaging periods for indoor-outdoor ratios of pollution in naturally ventilated non-domestic buildings near a busy road. *Atmos. Environ.* **2003**, *37*, 4121–4132. [CrossRef]

32. Massey, D.; Kulshrestha, A.; Masih, J.; Taneja, A. Seasonal trends of PM_{10}, $PM_{5.0}$, $PM_{2.5}$& $PM_{1.0}$ in indoor and outdoor environments of residential homes located in North-Central India. *Build. Environ.* **2012**, *47*, 223–231.

33. Zhao, X.; Zhang, X.; Xu, X.; Xu, J.; Meng, W.; Pu, W. Seasonal and diurnal variations of ambient $PM_{2.5}$ concentration in urban and rural environments in Beijing. *Atmos. Environ.* **2009**, *43*, 2893–2900. [CrossRef]

34. Yang, F.; Kang, Y.; Gao, Y.; Zhong, K. Numerical simulations of the effect of outdoor pollutants on indoor air quality of buildings next to a street canyon. *Build. Environ.* **2015**, *87*, 10–22. [CrossRef]

35. Zhao, B.; Zhang, Y.; Li, X.; Yang, X.; Huang, D. Comparison of indoor aerosol particle concentration and deposition in different ventilated rooms by numerical method. *Build. Environ.* **2004**, *39*, 1–8. [CrossRef]

36. Quang, T.N.; He, C.; Morawska, L.; Knibbs, L.D. Influence of ventilation and filtration on indoor particle concentrations in urban office buildings. *Atmos. Environ.* **2013**, *79*, 41–52. [CrossRef]

37. Kopperud, R.J.; Ferro, A.R.; Hildemann, L.M. Outdoor versus indoor contributions to indoor particulate matter (PM) determined by mass balance methods. *J. Air Waste Manag. Assoc.* **2004**, *54*, 1188–1196. [CrossRef] [PubMed]

38. Zhao, L.; Chen, C.; Wang, P.; Chen, Z.; Cao, S.; Wang, Q.; Xie, G.; Wan, Y.; Wang, Y.; Lu, B. Influence of atmospheric fine particulate matter ($PM_{2.5}$) pollution on indoor environment during winter in Beijing. *Build. Environ.* **2015**, *87*, 283–291. [CrossRef]

39. Hussein, T. Indoor-to-outdoor relationship of aerosol particles inside a naturally ventilated apartment-A comparison between single-parameter analysis and indoor aerosol model simulation. *Sci. Total Environ.* **2017**, *596*, 321–330. [CrossRef] [PubMed]

40. Bo, M.; Salizzoni, P.; Clerico, M.; Buccolieri, R. Assessment of Indoor-Outdoor Particulate Matter Air Pollution: A Review. *Atmosphere* **2017**, *8*, 136. [CrossRef]

41. Hong, B.; Lin, B.; Qin, H. Numerical investigation on the coupled effects of building-tree arrangements on fine particulate matter ($PM_{2.5}$) dispersion in housing blocks. *Sustain. Cities Soc.* **2017**, *34*, 358–370. [CrossRef]

42. Building Energy Conservation Research Center, Tsinghua University (BECRC). *Annual Report on Chinese Building Energy Conservation Development 2016*; China Architecture & Building Press: Beijing, China, 2016.

43. Shi, S.; Zhao, B. Occupants' interactions with windows in 8 residential apartments in Beijing and Nanjing, China. *Build. Simul.* **2016**, *9*, 221–231. [CrossRef]

44. Ye, B.; Ji, X.; Yang, H.; Yao, X.; Chan, C.K.; Cadle, S.H.; Chan, T.; Mulawa, P.A. Concentration and chemical composition of $PM_{2.5}$ in Shanghai for a 1-year period. *Atmos. Environ.* **2003**, *37*, 499–510. [CrossRef]

45. Beijing Municipal Environmental Monitoring Center (BMEMC). Real-Time Air Quality in Beijing. Available online: http://www.bjmemc.com.cn (accessed on 1 January 2016).

46. World Health Organization (WHO). WHO Air Quality Guidelines for Particulate Matter, Ozone, Nitrogen Dioxide and Sulfur Dioxide. Available online: http://apps.who.int/iris/bitstream/10665/69477/1/WHO_SDE_PHE_OEH_06.02_eng.pdf (accessed on 2 June 2005).

47. Freer-Smith, P.H.; Beckett, K.P.; Taylor, G. Deposition velocities to *Sorbus aria, Acer campestre, Populus deltoids×trichocarpa* 'Beaupré', *Pinus nigra* and× *Cupressocyparis leylandii* for coarse, fine and ultra-fine particles in the urban environment. *Environ. Pollut.* **2005**, *133*, 157–167. [CrossRef] [PubMed]

48. Franke, J.; Hellsten, A.; Schlunzen, K.H.; Carissimo, B. The COST 732 best practice guideline for CFD simulation of flows in the urban environment: A summary. *Int. J. Environ. Pollut.* **2011**, *44*, 419–427. [CrossRef]

49. Lin, B.; Li, X.; Zhu, Y.; Qin, Y. Numerical simulation studies of the different vegetation patterns' effects on outdoor pedestrian thermal comfort. *J. Wind Eng. Ind. Aerodyn.* **2008**, *96*, 1707–1718. [CrossRef]

50. Finnigan, J. Turbulence in plant canopies. *Annu. Rev. Fluid Mech.* **2000**, *32*, 519–571. [CrossRef]

51. Sanz, C. A note on k-epsilon modelling of vegetation canopy air-flows. *Bound.-Layer Meteorol.* **2003**, *108*, 191–197. [CrossRef]

52. Katul, G.G.; Mahrt, L.; Poggi, D.; Sanz, C. One-and two-equation models for canopy turbulence. *Bound.-Layer Meteorol.* **2004**, *113*, 81–109. [CrossRef]

53. Endalew, A.M.; Hertog, M.; Delele, M.A.; Baetens, K.; Persoons, T.; Baelmans, M.; Ramon, H.; Nicolai, B.M.; Verboven, P. CFD modelling and wind tunnel validation of airflow through plant canopies using 3D canopy architecture. *Int. J. Heat Fluid Flow* **2009**, *30*, 356–368. [CrossRef]

54. Ji, W.; Zhao, B. Numerical study of the effects of trees on outdoor particle concentration distributions. *Build. Simul.* **2014**, *7*, 417–427. [CrossRef]

55. Vranckx, S.; Vos, P.; Maiheu, B.; Janssen, S. Impact of trees on pollutant dispersion in street canyons: A numerical study of the annual average effects in Antwerp, Belgium. *Sci. Total Environ.* **2015**, *532*, 474–483. [CrossRef] [PubMed]

56. Jeanjean, A.P.R.; Monks, P.S.; Leigh, R.J. Modelling the effectiveness of urban trees and grass on $PM_{2.5}$ reduction via dispersion and deposition at a city scale. *Atmos. Environ.* **2016**, *147*, 1–10. [CrossRef]

57. Santiago, J.L.; Martilli, A.; Martin, F. On dry deposition modelling of atmospheric pollutants on vegetation at the microscale: Application to the impact of street vegetation on air quality. *Bound.-Layer Meteorol.* **2016**, *162*, 1–24. [CrossRef]

58. Bell, J.N.B.; Treshow, M. (Eds.) *Air Pollution and Plant Life*, 2nd ed.; John Willey & Sons Ltd.: Chichester, UK, 2003.

59. Concentration Data of Street Canyons (CODASC). Available online: http://www.windforschung.de/CODASC.htm (accessed on 1 January 2008).

60. Gromke, C.; Blocken, B. Influence of avenue-trees on air quality at the urban neighborhood scale. Part I: Quality assurance studies and turbulent Schmidt number analysis for RANS CFD simulations. *Environ. Pollut.* **2015**, *196*, 214–223. [CrossRef] [PubMed]

61. Roache, P.J. Perspective: A method for uniform reporting of grid refinement studies. *J. Fluids Eng.* **1994**, *116*, 405–413. [CrossRef]

62. Gromke, C.; Ruck, B. Pollutant concentrations in street canyons of different aspect ratio with avenues of trees for various wind directions. *Bound.-Layer Meteorol.* **2012**, *144*, 41–64. [CrossRef]

63. Hanna, S.; Chang, J. Acceptance criteria for urban dispersion model evaluation. *Meteorol. Atmos. Phys.* **2012**, *116*, 133–146. [CrossRef]

64. Hong, B.; Lin, B.; Qin, H. Numerical investigation on the effect of avenue trees on $PM_{2.5}$ dispersion in urban street canyons. *Atmosphere* **2017**, *8*, 129. [CrossRef]

65. Hefny, M.M.; Ooka, R. CFD analysis of pollutant dispersion around buildings: Effect of cell geometry. *Build. Environ.* **2009**, *44*, 1699–1706. [CrossRef]

66. Barratt, R. *Atmospheric Dispersion Modeling: An Introduction to Practical Applications*; Earthscan: London, UK, 2001.

67. Jin, R.; Hang, J.; Liu, S.; Wei, J.; Liu, Y.; Xie, J.; Sandberg, M. Numerical investigation of wind-driven natural ventilation performance in a multi-storey hospital by coupling indoor and outdoor airflow. *Indoor Built Environ.* **2016**, *25*, 1226–1247. [CrossRef]

68. Lee, B.H.; Yee, S.W.; Kang, D.H.; Yeo, M.S.; Kim, K.W. Multi-zone simulation of outdoor particle penetration and transport in a multi-storey building. *Build. Simul.* **2017**, *10*, 525–534. [CrossRef]

69. King, M.; Gough, H.L.; Halios, C.; Barlow, J.F.; Robertson, A.; Hoxey, R.; Noakes, C.J. Investigating the influence of neighbouring structures on natural ventilation potential of a full-scale cubical building using time-dependent CFD. *J. Wind Eng. Ind. Aerodyn.* **2017**, *169*, 265–279. [CrossRef]

70. Chan, A.T. Indoor–outdoor relationships of particulate matter and nitrogen oxides under different outdoor meteorological conditions. *Atmos. Environ.* **2002**, *36*, 1543–1551. [CrossRef]

atmosphere

MDPI

Article

Optimisation of Heat Loss through Ventilation for Residential Buildings

Dariusz Suszanowicz

Faculty of Natural Sciences and Technology, University of Opole, 45-040 Opole, Poland;
d.suszanowicz@uni.opole.pl; Tel.: +48-77-401-6690

Received: 28 January 2018; Accepted: 6 March 2018; Published: 8 March 2018

Abstract: This study presents the results of research on heat loss from various types of residential buildings through ventilation systems. Experimental research was done to analyse the effectiveness of ventilation systems of different types and determine the parameters of air discharged via the ventilation ducts. A model of heat loss from the discharge of exhaust air outside through air ducts has since been developed. Experiments were conducted on three experimental systems of building ventilation: gravitational, mechanical, and supply-exhaust ventilation systems with heat recovery. The proposed model dependencies were used to chart the daily fluctuations of the optimum multiplicity of air exchange for precise control of the parameters of mechanical ventilation systems in residential buildings. This study proves that natural ventilation in residential buildings fulfils its function only by increasing the air flow into the building, and that this incurs significant heat loss from buildings during the heating season.

Keywords: heat loss; optimisation; residential building; air quality; carbon dioxide concentration; ventilation system

1. Introduction

In the context of the debate on climate change and the need to increase the energy efficiency of buildings, viable means of reducing heat loss from the ventilation of residential buildings that have to be heated must be found. In some climate zones, the use of natural ventilation brings the expected effects [1,2]. However, in the climatic conditions throughout Poland, as well as most other countries of Central and Northern Europe, this ventilation system does not guarantee high energy efficiency. About 80% of single-family and multi-family low-rise residential buildings (up to four stories) in Poland have natural ventilation (Polish standards do not permit natural ventilation in multi-family high-rise buildings). It is therefore necessary to undertake research to determine the performance of the different types of ventilation systems used in residential buildings in Central and Northern Europe to reduce energy loss through ventilation systems.

Ensuring comfort in residential buildings requires selecting the right air parameters, i.e., temperature, humidity, and concentration of gas, particulate pollutants, and microorganisms. These factors significantly affect the well-being and health of people living in such spaces [3–5]. Failure to provide adequate air quality causes residents to experience a number of symptoms of sick building syndrome (SBS) [6,7] including headache, dizziness, dry linings, drowsiness, shortness of breath, and fainting.

It should be noted, however, that the effects of exposure of a human being to a harmful substance are assessed by its dose, and not the concentration or distribution. Reducing residents' exposure to harmful substances requires reducing the concentration of harmful substances or the time they spend in a polluted room. Shortening their time in living spaces is not generally feasible (e.g., at night): the only solution is to reduce the concentrations of pollutants. This can be achieved by reducing emissions or by bringing in more fresh air to dilute pollutants, and this can be achieved by a ventilation system [8,9].

Ventilation of a residential space is the exchange of air throughout the whole space or parts of it in order to remove the used, contaminated air and introduce fresh air from outside. In other words, ventilation is an organised exchange of air in a given space in order to refresh it and drive the contaminants which are produced in that space outside [10–12].

The concept of indoor air distribution must be developed in such a way as to obtain the required ventilation effects. The efficiency of the ventilation system should be defined via measurable flow parameters, such as the local distributions of air velocity and temperature, concentrations of pollutants, etc. In the zone where people are continually present, the analysis of air parameters allows specification of the optimum distribution of indoor air and achievement of the most favourable parameters [13,14].

It should also be emphasised that the ventilation air stream, when properly adjusted for living spaces, can reduce the pollution of suspended particles emitted by internal sources, e.g., household appliances, computer printers, cigarettes, or electronic cigarettes. This contributes significantly to the comfort of people in rooms where high air quality is maintained [15–18].

In Poland, the regulations concerning ventilation air streams and the performance of ventilation systems are specified in the Polish Standards and the following legal acts:

- The Act of 7 July 1994, Construction Law (as amended) [19];
- Regulation of the Minister of Infrastructure of 10 December 2010, on the technical requirements for buildings and their locations [20];
- Polish Standards:

  - PN-B-03421:1978 "Ventilation and air conditioning—parameters for indoor air in the habitats designated for permanent presence of people" [21],
  - PN-B-03430:1983/Az3:2000 "Ventilation in residential, common living, and public buildings—requirements" [22],
  - PN-EN 13779:2007 "Ventilation for non-residential buildings—performance requirements for ventilation and room-conditioning systems" [23].

This work does not refer to European standards or standards in other countries of Central and Northern Europe, and the requirements set out in the above regulations vary significantly. The research was carried out in facilities located in Poland, so only Polish regulation is referenced. Unfortunately, Polish regulation does not always align with European Union regulation.

Meeting the requirements set forth in the above regulations does not ensure appropriate levels of air pollutants in buildings. The air that enters the building through the ventilation system is, in ventilation terminology, called fresh air, and it of course does not always meet the quality parameters required of the air in living spaces. External air in city centres can carry both exhaust and dust pollution [24], and substantial flows of external air into a building can reduce the air pollution in ventilated spaces only slightly or in extreme situations actually increase it. In most ventilation systems, fresh air is supplied by infiltration, i.e., the spontaneous flow of air through the leaks in doors, windows, and walls.

Standard PN-EN 13779:2007 [23] indicates three classes of outside air based on purity (as presented in Table 1).

Table 1. Outside air classification based on PN-EN 13779 [23].

Category	Description
ODA 1 (ZEW 1)	Clean air, which can be dusty only periodically (e.g., with pollen—in accordance with the WHO 1999 recommendations)
ODA 2 (ZEW 2)	Outside air with high-level concentration of pollutants: particulates and gas
ODA 3 (ZEW 3)	Outside air with high-level concentration of pollutants: particulates and gas (WHO standards exceeded more than 1.5 times)

The classification of the outdoor air according to the categories listed in the Polish Standard PN-EN 13779:2007 is not, however, precise enough for use in the optimisation of ventilation systems in residential spaces. Therefore, studies and analyses of outdoor air were carried out at measuring points in the provinces of Silesia, Opole, and Lower Silesia. The research was conducted by the author during field studies, which involved university students, in areas where buildings selected as research objects in this paper were located. This made it possible to determine the parameters of outside air supplied to the building by the ventilation systems in different climatic conditions. The author recorded the average concentrations of pollutants in the outdoor air at measuring stations located in areas of varying character throughout the region of analysis in Poland, and the figures are shown in Table 2.

Table 2. Sample average concentrations of pollutants in outdoor air.

Measuring Stations	Concentrations of Pollutants					
	CO_2	CO	SO_2	NO_2	PM_{10}	Total Suspended Particulates
	(ppm)	$(mg \cdot m^{-3})$	$(\mu g \cdot m^{-3})$	$(\mu g \cdot m^{-3})$	$(\mu g \cdot m^{-3})$	$(mg \cdot m^{-3})$
Rural areas with low population density	320	1	4	5–35	do 15	0.5
Small towns with no large production plants	390	1.5–4.0	6–14	15–40	20–40	0.5–1.0
Central districts of large urban developments or cities with production plants	580	3.5–8.0	20–60	30–80	30–70	1.0–2.0

As can be inferred from the data summarised in Tables 1 and 2, large streams of fresh air do not always ensure good air quality in living spaces. On the other hand, excessive ventilation of living spaces during the heating season causes greater heat loss in the building as the ventilation system sends heated air outside [25,26]. Substantial heat losses by ventilation systems translate directly into higher costs of heating of the building, and ways should be found to reduce them.

An initiative was undertaken to create a model for optimisation of the performance of the ventilation system in a building that would ensure the required indoor air parameters recommended by Polish Standards, as well as by WHO (concentration of CO_2 below 1000 ppm) [27], with minimal building heat loss.

2. Experiments

As confirmed by the analyses carried out by a number of authors [11,28–31], 30 to 40% of the heat loss from residential buildings in various climatic conditions is discharged into the atmosphere together with the used air via the natural ventilation systems. In order to optimise the performance parameters of the ventilation system and simultaneously minimise heat losses via ventilation in a residential building, it is necessary to determine the unit heat flow driven out of the building together with the used air over one hour. Unit heat flow can be determined by Equation (1):

$$\dot{Q}_w = \frac{1}{3600} \cdot \dot{V} \cdot \rho \cdot c_p \cdot (T_w - T_o) \tag{1}$$

where $\dot{Q}_w$ is the heat loss-unit heat flow driven out of the building together with the used air $(W \cdot h^{-1})$; $\dot{V}$ is the ventilation airflow $(m^3 \cdot h^{-1})$; ρ is the air density $(kg \cdot m^{-3})$; c_p is the specific heat of the air at constant pressure $(J \cdot (kg \cdot K)^{-1})$; T_w is the indoor air temperature (K); and T_o is the temperature of fresh outside air (K).

In Equation (1), the air temperature inside the building, T_w, is a constant. In the heating season, it is 293 K (20 °C); out of the heating season, it is 296 K (23 °C). Therefore, it can be assumed that the values of density ρ and specific heat c_p of the air are also constant. The outside temperature value is read from the climate databases (measured during field tests at the specific location of the building, the average temperature values are given for the selected months). In the case of natural ventilation, where fresh air is supplied into the building from outside in an uncontrolled manner (via leaks in construction elements), no changes in temperature are possible. Raising the temperature of fresh air is possible for mechanical ventilation with heat recovery by means of a recuperator or a ground heat exchanger. To a large extent, however, it is possible to control the value of the ventilation airflow $\dot{V}$. Carbon dioxide concentration is one of the important indicators of indoor air quality. Particulate pollutants were not taken into account during the analysis because they are nearly completely eliminated by the inlet filters regardless of whether air is taken in by mechanical ventilation or natural ventilation. Therefore, when defining the minimum ventilation airflow, carbon dioxide concentration and the relative humidity of the air are the primary indicators of air pollution inside a living space [32]. Air humidity affects only the flow of ventilation air being removed from bathrooms or kitchens. For the total living space, it is sufficient to correlate the flow of ventilation air with the concentration of carbon dioxide. Therefore, Equation model (2) (based on the established balance sheet equations, extended by empirical factors determined in the course of this analysis) was selected for determining the optimum ventilation flow for residential buildings and living spaces:

$$\dot{V} = \alpha \cdot \frac{\varepsilon_M}{C_D - C_Z} \cdot V_B \tag{2}$$

where α is the correction coefficient of the standard ventilation rate (no unit); ε_M is the indoor carbon dioxide emissions from human sources (ppm·h^{-1}); C_D is the permissible level of carbon dioxide in indoor air in residential buildings (ppm); C_Z is the concentration of carbon dioxide in the outdoor air (ppm); and V_B is the volume of heated space in the building (m^3).

The correction coefficient of the standard ventilation rate, α, is an empirical coefficient that takes into account the number of people in the facility and their activities, as well as other sources of carbon dioxide, e.g., a gas stove or a fireplace. The database of the values of α, which is necessary for optimisation calculations, covering most of the circumstances occurring in residential premises and buildings, was prepared on the basis of long-term studies of ventilation systems in residential buildings, having analysed the correlation between the number and activity of the residents and the concentration of carbon dioxide.

The values of the carbon dioxide emission coefficient for individuals with different physical characteristics and their activity profiles were determined by the author during metabolic tests. Metabolic tests were carried out in apartment 2 and they involved recording changes in the concentration of carbon dioxide in the selected room (with a specific cubic capacity) used by one person. In that room air exchange was stopped, and in a subsequent series of tests selected individuals performed different activities (e.g., 4 h of sleep, computer work, or simple physical exercises).

In the case of mechanical ventilation with heat recovery by means of a recuperator or ground heat exchanger, the temperature of the fresh air entering the building through the ventilation system needs to be specified. Therefore, Equation (3) is proposed to specify the temperature of the fresh air entering the building, T_o:

$$T_o = T_Z + _R \left[T_W - T_Z + _G \left(T_Z + T_G \right) \right] \tag{3}$$

where T_Z is the outdoor air temperature (K); T_G is the ground temperature in the ground heat exchanger (K); η_R is the recuperator heat recovery efficiency (no unit); and η_G is the thermal efficiency of ground heat exchanger (no unit).

The modified classical balance Equations (1)–(3), the climate database for southwestern Poland, and the empirical factors determined on the basis of the conducted studies and analyses resulted in a model for selecting optimal performance parameters for a residential building ventilation system. In order to verify the proposed model dependencies and analyse the effectiveness of ventilation systems, as well as to specify the parameters of the exhaust air driven outside via the ventilation ducts, experimental research was conducted and theoretical calculations were made for seven ventilation systems in residential buildings and apartments equipped with various ventilation systems. The following objects of study were selected:

- three apartments:

  - Apartment 1 with a floorspace of 48 m^2 and mechanical exhaust ventilation;
  - Apartment 2 with a floorspace of 37 m^2 and natural ventilation;
  - Apartment 3 with a floorspace of 69 m^2 and natural ventilation, after an energy efficient upgrade involving the improvement of the building insulation and replacement of window joinery for better air tightness;

- four single-family/detached houses:

  - House 1 with a floorspace of 170 m^2 and natural ventilation;
  - House 2 with a floorspace of 117 m^2 and natural ventilation, after an energy efficient upgrade;
  - House 3 with a floorspace of 158 m^2 and mechanical ventilation—intake and exhaust with heat recovery;
  - House 4 with a floorspace of 204 m^2 and mechanical ventilation—intake and exhaust with heat recovery and ground exchanger;

In order to compare the results of the research for individual research objects, residential apartments inhabited by three people were selected, and single-family buildings were selected with four inhabitants.

The research objects were selected specifically because they have ventilation systems of the three types of air flow distribution, presented graphically in Figure 1.

The tests were carried out in the selected research objects from October to April, i.e., in the period when the rooms were heated. Measurements were carried out in five-day cycles, 24 h a day, recording the measured air parameters every 5 min at selected points in the tested objects. The tests were conducted only on weekdays, when the daily cycles of residents' activity were nearly identical. This approach allowed for a comparison of the results of tests conducted in different objects and in different weather conditions. All windows in the research objects were closed at all times. The fresh air was introduced into the spaces only through micro ventilation openings and wall diffusers. According to research conducted by numerous authors, ventilation by opening windows in buildings with natural ventilation can reduce the CO_2 concentration but can simultaneously increase dust pollution and heat demand for the building [33].

Measurements of the concentration of carbon dioxide, relative humidity, and air temperature were taken in kitchens, bathrooms, and other rooms in the objects; an average value for each entire object was then determined. Air streams in individual ventilation ducts were calculated on the basis of the velocity of air flow at various points of entry into the ventilation ducts. The velocity was measured by a Kestrel 2000 anemometer, which allows for measurement definition up to 0.1 m/s and measuring accuracy of ±3%. Measurements of the air pressure, temperature, and relative humidity were taken by a Commeter C4130 thermo-hygro-barometer. It recorded temperature to an accuracy of 0.4 °C, relative humidity within ±2.5%, and atmospheric pressure within ±2 hPa. Carbon dioxide concentration measurements were taken using the multifunction carbon dioxide meter AZ 77535, accurate to 10 ppm ±5% of the reading. Tests were only carried out on weekdays, when the daily activity cycles of the inhabitants were nearly identical. The same procedure was applied for all tested objects. This approach made it possible to compare the results of tests carried out at different times of the year and in different weather conditions.

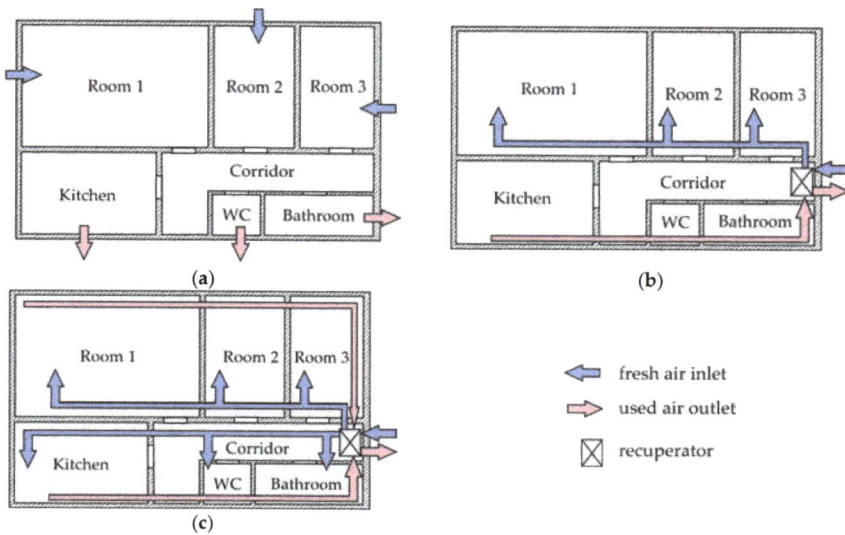

Figure 1. Example air flow distributions in the ventilation systems of residential buildings: (**a**) natural ventilation; (**b**,**c**) mechanical ventilation with heat recovery.

3. Results and Discussion

The test results were used to draw up graphs of variability of the carbon dioxide concentrations, relative humidity, and air temperature in the analysed objects. Having compared the graphs of air parameters variability, it was immediately apparent that the study objects with natural ventilation, i.e., apartments 2 and 3 and residential buildings 1 and 2, cannot keep concentrations of carbon dioxide below the recommended level of 1000 ppm. Also, the relative air humidity in the bathrooms and kitchens in the buildings with natural ventilation exceeded the level of 65%; such indoor humidity persists for long periods of time and may cause the growth of fungi or mould in the construction elements of the building.

Sample comparison of the variability in the concentration of carbon dioxide in apartment 1, with mechanical ventilation, and apartment 2, with natural ventilation is shown in Figure 2.

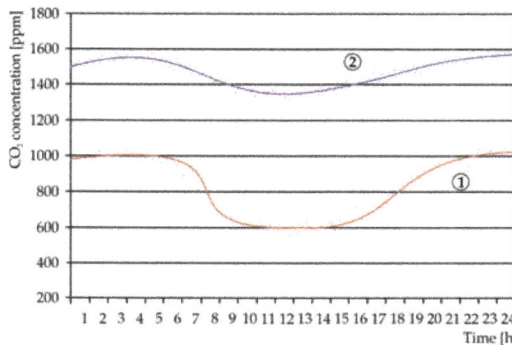

Figure 2. Comparison of the variability in the concentration of carbon dioxide in apartment 1, with mechanical ventilation, apartment 2, with natural ventilation.

Mechanical ventilation systems used in the other buildings and apartments ensured greater stability of the parameters of the air inside the analysed objects.

Table 3 presents the ventilation air streams (in column 2, determined according to Reference [22], taking into account the number of inhabitants and the number and use of rooms), averaged streams of heat lost via the ventilation system, and averaged unit heat losses via ventilation (with reference to 1 m^2 of the floorspace of the building) for the analysed research objects. These are all the data defined during the tests, calculated in accordance with the standards and by means of model equations.

Table 3. Defined ventilation air streams and heat loss for the buildings analysed.

Analysed Buildings	Ventilation Air Stream Calculated in Accordance with the Polish Standard	Ventilation Air Stream Calculated with Results of Measurements	Ventilation Air Stream Calculated by Model Equations	Averaged Stream of Heat Lost via the Ventilation System	Averaged Unit Heat Losses via Ventilation
	$(m^3 \cdot h^{-1})$	$(m^3 \cdot h^{-1})$	$(m^3 \cdot h^{-1})$	$(W \cdot h^{-1})$	$(W \cdot (m^2 \cdot h)^{-1})$
Apartment 1	120.0	106.1	67.5	1071.0	22.3
Apartment 2	120.0	84.2	46.9	850.0	22.9
Apartment 3	140.0	71.3	63.2	720.0	10.4
House 1	185.0	120.7	169.0	1234.0	8.2
House 2	165.0	92.6	131.2	935.0	7.7
House 3	180.0	191.3	151.4	644.0	4.2
House 4	204.0	211.8	196.4	427.0	2.1

The calculated ventilation air streams and the air parameters measured during the tests were used to create graphs for each research object to identify the optimal work parameters for the ventilation system of the building. An example graph that was used to optimise the work parameters of the ventilation system of the building (drawn for apartment 1) is shown in Figure 3.

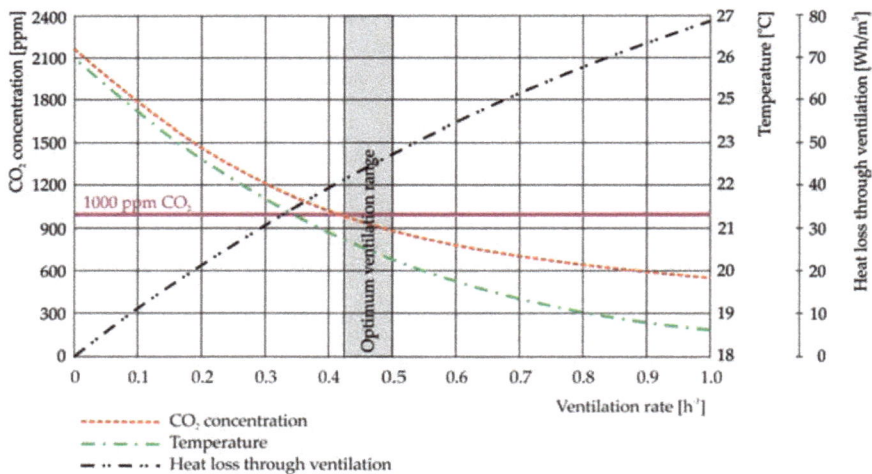

Figure 3. Example graph of performance parameters optimisation for a ventilation system in an apartment.

The measurements of carbon dioxide concentration and relative humidity in the residential buildings in the analysis indicate that there are changes in pollutant concentrations and relative humidity that are specific to weekdays, and that they affect the optimum multiplicity of air exchange in the living spaces. The optimal ventilation air stream with respect to the cubic capacity of a residential building, determined by Equation (2), allows determination of the optimal air exchange rate through the ventilation system. Figure 4 shows an example of the daily variation of the optimum rate of air exchange defined by means of the model dependencies throughout a standard weekday for residential building 2.

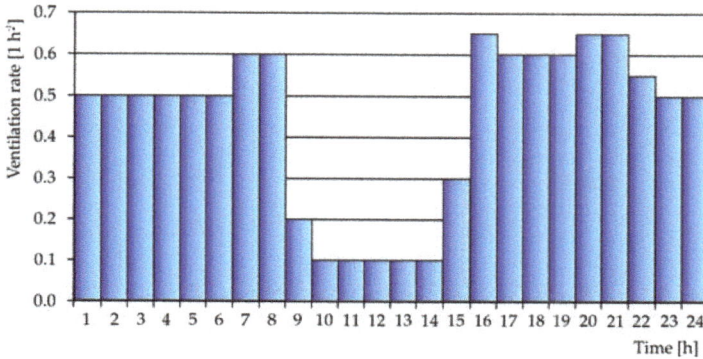

Figure 4. Optimum multiplicity of air exchange n_k throughout a standard weekday.

Graphs of the daily fluctuation of the optimum rate of air exchange were drawn for each research object, and the graphs are the basis for precise control of the parameters of mechanical ventilation systems in the objects.

It is apparent that the smallest heat loss via the ventilation system occurs with the use of mechanical ventilation—intake and exhaust with heat recovery and a ground heat exchanger. Ventilation systems with heat recovery may be implemented not only in single-family/detached houses, but also in apartment buildings. In existing apartment buildings, heat may be recovered from ventilation systems by the use of recuperators installed on the walls.

4. Conclusions

The tests show that natural ventilation should not be used in residential buildings or in communal residential buildings, because natural ventilation does not permit precise control of the ventilation airflow.

As research object 3 shows, the energy efficiency upgrade of a building (the most common type of residential building in Poland) involving only the installation of wall insulation and replacement of window joinery increases the airtightness of a building, which reduces the effectiveness of natural ventilation as well as the danger of SBS—sick building syndrome.

Mechanical ventilation by an exhaust fan meets the requirements of Polish Standards; however, due to the intermittent mode of its operation, it does not keep the concentration of carbon dioxide below the required level nor remove the excess water vapour from the air in the bathroom or kitchen.

Only the mechanical intake-exhaust ventilation system with heat recovery meets the requirements of the air quality in residential spaces, as set out in the Polish Standard PN-B-03430: 1983/Az3 2000, and at the same time minimises the loss of heat from the ventilated rooms.

This research also indicates the need to make changes in Polish regulations on construction to require the owners of residential buildings to replace natural ventilation systems with mechanical ventilation with heat recovery systems.

As can be seen in the example diagram of the variability of the optimum air exchange rate (Figure 4), it is only the control of the operation parameters of the ventilation system by changing the ventilation flow rate during the daily lifecycle of a residential building that can minimise heat loss through ventilation while maintaining the required air quality inside the building.

By determining the optimum air exchange rate for the analysed buildings by means of the suggested dependencies, it is possible to reduce the heat demand for heating of buildings by 9–12%.

Acknowledgments: This work was financed by The Faculty of Natural Sciences and Technology, Chair of Process Engineering basic (statutory) research projects.

Conflicts of Interest: The author declares no conflict of interest.

References

1. Tong, Z.; Chen, Y.; Malkawi, A. Estimating natural ventilation potential for high-rise buildings considering boundary layer meteorology. *Appl. Energy* **2017**, *193*, 276–286. [CrossRef]
2. Chen, Y.; Tong, Z.; Malkawi, A. Investigating natural ventilation potentials across the globe: Regional and climatic variations. *Build. Environ.* **2017**, *122*, 386–396. [CrossRef]
3. Ickiewicz, I. Building Thermomodernization and Reducing Air Pollution. *Ecol. Chem. Eng. S* **2013**, *20*, 805–816. [CrossRef]
4. Webb, A.L. Energy retrofits in historic and traditional buildings: A review of problems and methods. *Renew. Sustain. Energy Rev.* **2017**, *77*, 748–759. [CrossRef]
5. Majewski, G.; Kociszewska, K.; Rogula-Kozłowska, W.; Pyta, H.; Rogula-Kopiec, P.; Mucha, W.; Pastuszka, J.S. Submicron Particle-Bound Mercury in University Teaching Rooms: A Summer Study from Two Polish Cities. *Atmosphere (Basel)* **2016**, *7*, 117. [CrossRef]
6. Antoniadou, P.; Papadopoulos, A.M. Occupants' thermal comfort: State of the art and the prospects ofpersonalized assessment in office buildings. *Energy Build.* **2017**, *153*, 136–149. [CrossRef]
7. Zender-Swiercz, E.; Telejko, M. Impact of insulation Building on the Work of Ventilation. *Proc. Eng.* **2016**, *161*, 1731–1737. [CrossRef]
8. Fisk, W.J.; Mirer, A.G.; Mendell, M.J. Quantitative relationship of sick building syndrome symptoms with ventilation rates. *Indoor Air* **2009**, *19*, 159–165. [CrossRef] [PubMed]
9. Gaczoł, T. Gravitational ventilation in residential premises-chosen questions. *J. Civ. Eng. Environ. Archit.* **2015**, *32*, 81–88. [CrossRef]
10. Pinto, M.; Viegas, J.; Freitas, V. Performance sensitivity study of mixed ventilation systems in multifamily residential buildings in Portugal. *Energy Build.* **2017**, *152*, 534–546. [CrossRef]
11. Słodczyk, E.; Suszanowicz, D. Optimization of carbon dioxide concentration in the didactic rooms by the regulation of ventilation. *Ecol. Chem. Eng. A* **2016**, *23*, 275–286. [CrossRef]
12. Cao, G.; Awbi, H.; Yao, R.; Fan, Y.; Sirén, K.; Kosonen, R.; Zhang, J. A review of the performance of different ventilation and airflow distribution systems in buildings. *Build. Environ.* **2014**, *73*, 171–186. [CrossRef]
13. Chludzińska, M.; Bogdan, A. The role of the front pattern shape in modelling personalized airflow and its capacity to affect human thermal comfort. *Build. Environ.* **2017**, *126*, 373–381. [CrossRef]
14. Chen, Q. Ventilation performance prediction for buildings: A method overview and recent applications. *Build. Environ.* **2009**, *44*, 848–858. [CrossRef]
15. Fuoco, F.C.; Stabile, L.; Buonanno, G.; Scungio, M.; Manigrasso, M.; Frattolillo, A. Tracheobronchial and alveolar particle surface area doses in smokers. *Atmosphere (Basel)* **2017**, *8*, 19. [CrossRef]
16. Scungio, M.; Stabile, L.; Buonanno, G. Measurements of electronic cigarette-generated particles for the evaluation of lung cancer risk of active and passive users. *J. Aerosol. Sci.* **2018**, *115*, 1–11. [CrossRef]
17. Stabile, L.; Buonanno, G.; Ficco, G.; Scungio, M. Smokers' lung cancer risk related to the cigarette-generated mainstream particles. *J. Aerosol. Sci.* **2017**, *107*, 41–54. [CrossRef]
18. Scungio, M.; Vitanza, T.; Stabile, L.; Buonanno, G.; Morawska, L. Characterization of particle emission from laser printers. *Sci. Total Environ.* **2017**, *586*, 623–630. [CrossRef] [PubMed]
19. Act of 7 July 1994 Construction Law. Journal of Laws (2016) Pos. 290, 961, 1165, 1250. Available online: http://isap.sejm.gov.pl/DetailsServlet?id=WDU20160000290&min=1 (accessed on 17 October 2017).
20. Regulation of Ministry of Infrastructure on Technical Requirements to Be Fulfilled by Buildings and Their Localization. Journal of Laws (2010) No. 239 Pos. 1597. Available online: http://isap.sejm.gov.pl/DetailsServlet?id=WDU20102391597 (accessed on 17 October 2017).

21. Polish StandardPN-B-03421:1978, Ventilation and Air Conditioning—Calculated Parameters for Indoor Air in the Habitats Destinated for Permanent Presence of People. Available online: http://sklep.pkn.pl/pn-en-13779-2008p.html (accessed on 17 October 2017).

22. Polish Standard-PN-B-03430:1983/Az3:2000 Ventilation in Residential, Common Living and Public Buildings-Requirements. Available online: http://sklep.pkn.pl/pn-en-13779-2008p.html (accessed on 17 October 2017).

23. Polish Standard-PN-EN 13779:2007 Ventilation for Non-Residential Buildings-Performance Requirements for Ventilation and Room-Conditioning Systems. Available online: http://sklep.pkn.pl/pn-en-13779-2008p.html (accessed on 17 October 2017).

24. Olszowski, T. Comparison of PM_{10} washout on urban and rural areas. *Ecol. Chem. Eng. S* **2017**, *24*, 381–395. [CrossRef]

25. Cosar-Jorda, P.; Buswell, R.A. Estimating the air change rates in dwellings using a heat balance approach. *Energy Procedia* **2015**, *78*, 573–578. [CrossRef]

26. Liddamen, M.W. A Review of Ventilation and the Quality of Ventilation Air. *Indoor Air* **2000**, *10*, 193–199. [CrossRef]

27. Theakston, F. *Air Quality Guidelines for Europe*; European Series; WHO Regional Publications, Regional Office for Europe Copenhagen: Copenhagen, Denmark, 2000; p. 91, ISBN 92-890-1358-3.

28. Schlueter, A.; Thesseling, F. Building information model based energy/exergy performance assessment in early design stages. *Autom. Constr.* **2009**, *18*, 153–163. [CrossRef]

29. Harvey, D.D.L. Reducing energy use in the buildings sector: Measures, costs, and examples. *Energy Effic.* **2009**, *2*, 139–163. [CrossRef]

30. Akbari, K.; Oman, R. Impacts of Heat Recovery Ventilators on Energy Savings and Indoor Radon in a Swedish Detached House. *WSEAS Trans. Environ. Dev.* **2013**, *9*, 24–34.

31. Ng, L.C.; Payne, W.V. Energy Use Consequences of Ventilating a Net-Zero Energy House. *Appl. Therm. Eng.* **2016**, *96*, 151–160. [CrossRef] [PubMed]

32. Stabile, L.; Dell'Isola, M.; Frattolillo, A.; Massimo, A.; Russi, A. Effect of natural ventilation and manual airing on indoor air quality in naturally ventilated Italian classrooms. *Build. Environ.* **2016**, *98*, 180–189. [CrossRef]

33. Stabile, L.; Dell'Isola, M.; Russi, A.; Massimo, A.; Buonanno, G. The effect of natural ventilation strategy on indoor air quality in schools. *Sci. Total Environ.* **2017**, *595*, 894–902. [CrossRef] [PubMed]

atmosphere

MDPI

Article

Pedestrian-Level Urban Wind Flow Enhancement with Wind Catchers

Lup Wai Chew [1,*], Negin Nazarian [2] and Leslie Norford [3]

[1] Department of Mechanical Engineering, Massachusetts Institute of Technology, Cambridge, MA 02139, USA
[2] Center for Environmental Sensing and Modeling, Singapore-MIT Alliance for Research and Technology, Singapore 138602, Singapore; negin@smart.mit.edu
[3] Department of Architecture, Massachusetts Institute of Technology, Cambridge, MA 02139, USA; lnorford@mit.edu
* Correspondence: lupwai@mit.edu

Received: 10 July 2017; Accepted: 22 August 2017; Published: 25 August 2017

Abstract: Dense urban areas restrict air movement, causing airflow in urban street canyons to be much lower than the flow above buildings. Boosting near-ground wind speed can enhance thermal comfort in warm climates by increasing skin convective heat transfer. We explored the potential of a wind catcher to direct atmospheric wind into urban street canyons. We arranged scaled-down models of buildings with a wind catcher prototype in a water channel to simulate flow across two-dimensional urban street canyons. Velocity profiles were measured with Acoustic Doppler Velocimeters. Experiments showed that a wind catcher enhances pedestrian-level wind speed in the target canyon by 2.5 times. The flow enhancement is local to the target canyon with little effect in other canyons. With reversed flow direction, a "reversed wind catcher" has no effect in the target canyon but reduces the flow in the immediate downstream canyon. The reversed wind catcher exhibits a similar blockage effect of a tall building amid an array of lower buildings. Next, we validated Computational Fluid Dynamics (CFD) simulations of all cases with experiments and extended the study to reveal impacts on three-dimensional ensembles of buildings. A wind catcher with closed sidewalls enhances maximum pedestrian-level wind speed in three-dimensional canyons by four times. Our results encourage better designs of wind catchers to increase wind speed in targeted areas.

Keywords: urban street canyon; wind enhancement; architectural intervention; water channel experiment; CFD simulation; passive ventilation

1. Introduction

In the process of urbanization, natural land covers are replaced with built materials and, consequently, the land surface roughness is significantly modified [1]. These alterations further impact the airflow in urban areas, as tall buildings obstruct and separate the wind [2–4] and canyon vortices are formed in urban street canyons [5–7], often resulting in a low wind speed near the ground or at the pedestrian level.

The importance of pedestrian-level ventilation in an urban environment is manifold. First, urban areas are severely subjected to the urban heat island (UHI) phenomena [8–12], which is in part due to the decreased momentum and heat exchange from the land surface to the atmosphere in the street canyons [13]. Subsequently, poor pedestrian-level ventilation can exacerbate UHI [5,13] and exposes urban dwellers to a higher air temperature. This factor, together with the decreased wind speed at the pedestrian level, cause a significant threat to human thermal comfort in urban areas [14]. Second, building energy consumption is closely tied with the pedestrian-level ventilation and the canyon air temperature [15]. Accordingly, improving ventilation in street canyons is instrumental for achieving a low-energy urban design [16]. Lastly, with the increased rate of emissions in urban areas, air quality in street canyons and city breathability for urban dwellers are closely tied to the efficiency of the city to ventilate itself [17–21]. Therefore, it is paramount that we evaluate the methods of enhancing wind speed in urban street canyons.

The role of urban morphology and architectural elements on urban flow and thermal field is indisputable. For instance, canyon aspect (height-to-width) ratio has been identified as one of the most important parameters in categorizing the flow field inside an urban street canyon [5–7]. Additionally, the drag coefficient and flow field in the urban canopy layer are strongly dominated by the urban packing density [18,22,23] and the layout of building arrays [24,25]. These studies also demonstrated that the geometrical inhomogeneity in urban design (e.g., building configurations and variability of street width) significantly modifies the local ventilation capacity, and Hang et al. [7] showed that height variability determines the momentum flux as well as the removal of pedestrian-level pollutants in urban street canyons.

In addition to the street-to-meso scale features of urban morphology, small-scale features of the street architecture also influence urban airflow [26]. Several studies provided evidence that roof shapes significantly modify the urban airflow and dispersion [27,28]. Huang et al. [29] demonstrated that different orientations of wedge-shaped roofs significantly alter the structure of circulation vortices induced in the canyons. Abohela et al. [30] evaluated the effects of roof shape on above-roof flow acceleration and found that a vaulted roof produces the largest flow acceleration in an aligned wind direction, while a dome roof accelerates the flow consistently in all wind directions. Additionally, Aliabadi et al. [31] showed that an active roof-level roughness design can improve thermal comfort and air quality in the canyon for specific times of the day. These studies point to the role of architectural elements on urban ventilation, and draw attention to urban architecture as an adaptation method to urban environmental concerns.

As architectural elements, wind towers and wind catchers have been prevalent as historical designs in the Middle East and North Africa [32,33]. They are effective as a passive cooling and natural ventilation method for the indoor/outdoor interface, although the airflow rate is strongly influenced by the geometry of the wind catchers and the wind direction [34,35]. In modern architecture, however, such architectural interventions are uncommon, and their performance in enhancing pedestrian-level wind speed has not been fully evaluated. We aim to address this gap and study the effect of a wind catcher on urban ventilation in two-dimensional (2D) and three-dimensional (3D) urban street canyons.

The structure of the paper is illustrated as a flowchart in Figure 1. First, we introduce a simplified form of a wind catcher as an architectural intervention and present a robust assessment of this element in a water channel (Sections 2 and 3). The impact of a wind catcher on flow enhancement is compared to idealized 2D urban street canyons, as well as other architectural elements such as step-up/step-down canyons. Second, we use the experimental results to evaluate and validate a CFD model (Section 4), which is then used for visualizing the detailed flow fields. The CFD model is used to extend the analysis to 3D urban street canyons and improve the design of a wind catcher. Conclusions, limitations, and future work are discussed in Section 5.

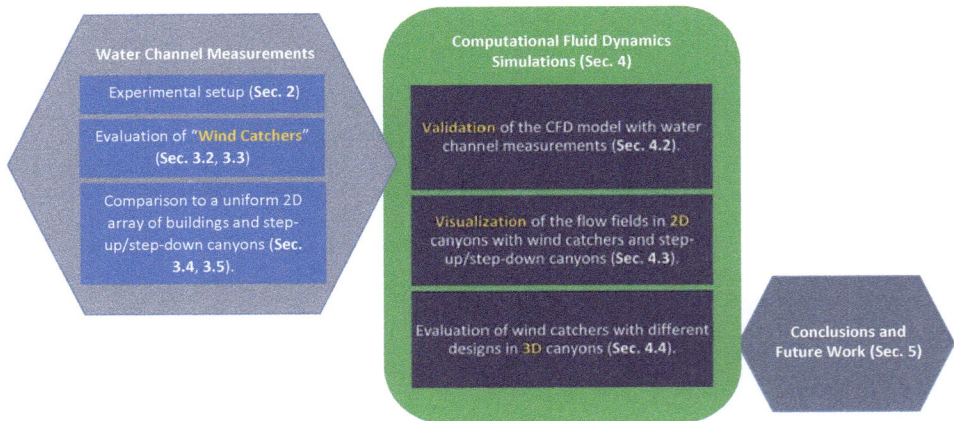

Figure 1. The flow chart and methodology of the current study.

2. Experimental Setup

A recirculating water channel in the Hydraulic Engineering Laboratory at the Department of Civil and Environmental Engineering, National University of Singapore, was used for all experiments. The sketch in Figure 2 shows the side view of the water channel. The water channel is 15 m long, 0.6 m tall and 0.6 m wide. High flow velocity at the test section was needed in some of the experiments, thus a partition made of marine plywood was installed to reduce the effective width from 0.6 m to 0.3 m at the test section. The test section measured 3.6 m long, 0.6 m tall and 0.3 m wide. The models of buildings spanned across the whole width of the test section, simulating flow across 2D urban street canyons ("canyons" hereafter), as shown in Figure 3. Flow straighteners made with a combination of plastic tubes, wire mesh and honeycomb minimize span-wise and vertical velocity components at the inlet. Two layers of ceramic marbles (0.5 inch or 1.27 cm diameter) accelerate the flow profile development such that the flow profile is fully developed at the test section. Far downstream of the water channel, an adjustable floodgate controls the water level. The water is circulated back to the tank with a pump. The flow rate was controlled by turning the valve and measured by a digital flowmeter. The maximum flow rate is 50 L/s. Acoustic Doppler Velocimeters, ADV (Vectrino by Nortek AS, Oslo, Norway), were used throughout the experiments to measure flow velocities. The ADV can measure all three components (stream-wise, vertical and span-wise) of velocities and velocity fluctuations up to 200 Hz. The accuracy is $\pm 0.5\%$ of measured value ± 1 mm/s. The ADV was mounted on metal frames with its vertical position adjustable. Although ADV is categorized as an intrusive measurement device, comparison with Laser Doppler Velocimeter (LDV, which is non-intrusive) measurements in Li et al. [36] verifies that our ADV probes did not disrupt the flow. Measurements taken by ADV and LDV are discussed in detail in Section 3.1.

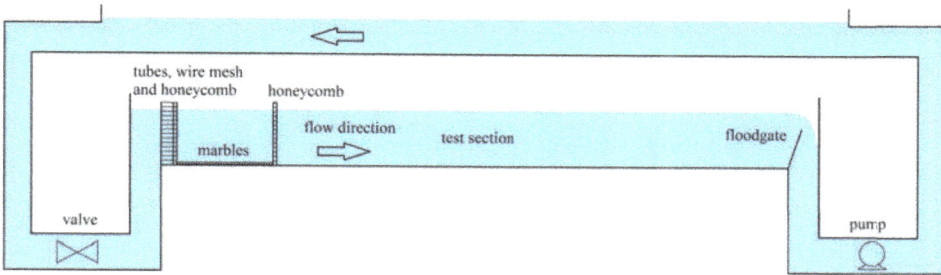

Figure 2. Side view of the recirculating water channel with an adjustable floodgate and a control valve to vary the flow rate and water level. The model blocks are placed at the test section, which is at the middle of the water channel.

Figure 3. Arrangement of model blocks in the test section to form seven canyons of aspect ratio 1 with height, $H = 12$ cm.

Table 1 summarizes four sets of experiments conducted: (i) the reference case, (ii) wind catcher, (iii) wind catcher in a reversed flow direction, and (iv) step-up and step-down canyons. In case (i), eight pieces of 12 cm tall $\times$ 10 cm wide $\times$ 30 cm long wooden blocks were spaced 12 cm apart from each other to simulate an array of 12 cm $\times$ 12 cm canyons of aspect ratio 1, as shown in Figure 3. For case (ii) and case (iii), the dimension of the scaled-down prototype of the wind catcher is shown in Figure 4. The fourth canyon, or canyon 4, is the target canyon, for which we aim to enhance the near-ground flow. Therefore, the wind catcher (and the reversed wind catcher) is installed above canyon 4. For case (iv), the fifth model block is replaced by a taller block (44% taller) to simulate a step-up (i.e., taller downwind building) canyon 4 and a step-down (i.e., shorter downwind building) canyon 5. The Reynolds number, Re, is defined as $Re = U_{ref}H/\nu$, where U_{ref} is a reference velocity, H the canyon height and ν the kinematic viscosity. In all cases, the Reynolds numbers were above 10,000 to achieve fully turbulent regime [37–39]. The water depth was 3.4H for case (i) and 3.8H for cases (ii)–(iv).

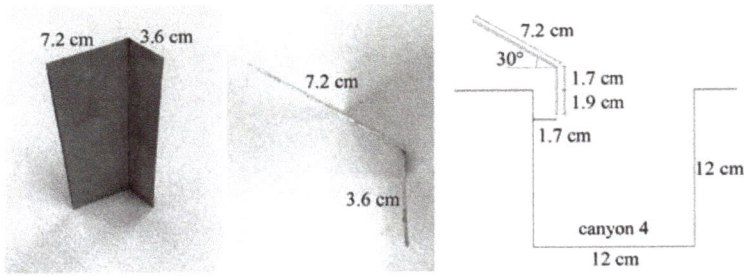

Figure 4. Photos and the dimensions of the wind catcher prototype.

Table 1. Summary of cases studied in the water channel. Case (i) represents the reference case with canyons of aspect ratio 1, cases (ii)–(iv) demonstrate the architectural interventions evaluated in this analysis. The middle-canyon profiles of all cases were measured in canyons 3–6 indicated by the dashed line shown in case (i) (not shown in cases (ii)–(iv) for clarity).

3. Experimental Results

This section discusses the experimental results for all four cases in Table 1. All three components of velocity were measured, but the mean span-wise velocity is zero, so only the mean stream-wise velocity, u, the mean vertical velocity, w, the stream-wise fluctuation, u', and the vertical fluctuation, w', are plotted. The mean velocities are time-averaged, while the velocity fluctuations are the root-mean-squared deviations from the mean (i.e., the standard deviation). All velocity components are normalized by a reference velocity, U_{ref}. In the literature, U_{ref} is either taken as the velocity at the roof level or the free-stream velocity. The velocity at the roof level is conveniently available but is very sensitive to the uncertainty in measurement position due to a large velocity gradient. On the other hand, the free-stream velocity remains relatively constant. For example, in a canyon with height H, the velocity at $0.1H$ above the roof level ($z/H = 1.1$, where z is the vertical distance from the ground) differs by about 50% compared to the velocity at the roof level ($z/H = 1.0$). In contrast, the difference between the velocity at $z/H = 2$ and $z/H = 3$ is less than 2% [36]. Kastner-Klein et al. [40] and

Li et al. [36] proposed taking the velocity at $z/H = 2$ or higher as the reference velocity, since it is less sensitive to the measurement position. We adopted their recommendation in this study by taking the free-stream velocity as U_{ref}.

3.1. Reference Case with Canyons of Aspect Ratio 1

Canyons with a unity aspect ratio (case (i) in Table 1) have been studied extensively in the literature, both experimentally [36,38,39,41] and numerically [31,42–44]. We repeated the measurements of this reference case to verify that our experimental setup can reproduce the results in the literature. We measured the middle-line velocity profiles of canyons 3–6 (indicated as the dashed lines in Table 1). The measurement frequency was 50 Hz and the measurement period was 60 s. Two separate sensitivity tests were conducted (one with double measurement frequency and the other with double measurement period) and verified that 50 Hz and a 60 s measurement period were sufficient. The eddy turnover time can be estimated as $H/U_{ref} \approx 0.7$ s, where $H = 0.12$ m and $U_{ref} = 0.17$ m/s. Therefore, the averaging period was about two orders of magnitude larger than the eddy turnover time. The flow was let to settle for 20 min (>1500 eddy turnover time) to ensure stationary before taking measurements. To confirm the repeatability of the experimental setup, three sets of measurements were taken on three separate days. The pump was turned off and the water was allowed to drain off after each set of measurement, i.e., each set of experiment was a new start by turning on the pump and allowing the flow to settle for 20 min before measurements. The three runs produced negligibly small run-to-run standard deviations: 0.001 m/s for the mean velocities and 0.0005 m/s for the velocity fluctuations, confirming the repeatability of the experimental setup.

The in-canyon profiles of canyons 3–6 are identical. This means that the flow has developed into the urban roughness flow at canyon 3, in agreement with Brown et al. [41] and Meroney [38]. For this reference case, since the velocity profiles in canyons 3–6 are similar, we plot only the profiles of canyon 4 in Figure 5. All profiles are normalized by U_{ref}. The result in Li et al. [36] serves as a benchmark comparison. This study used Acoustic Doppler Velocimeter (ADV) and had an $Re = 19,000$, while Li et al. [36] used Laser Doppler Velocimeter (LDV) and had an $Re = 11,000$. The good agreement between both studies verifies that the ADV in our experiments did not disrupt the flow. In addition, Li et al. [36] only measured up to $z/H = 1.2$, and we extended the measurement to $z/H = 2.5$ to check at what elevation the flow recovers to the free-stream velocity. Figure 5a shows that at z/H about 2, the stream-wise velocity approaches the free-stream velocity. Near the ground level, u/U_{ref} has a magnitude up to 30% in the negative x-direction. The magnitude of u/U_{ref} decreases almost linearly with increasing z/H to zero at the mid-canyon height ($z/H = 0.5$), then increases to 30% at the roof level ($z/H = 1.0$). On the other hand, Figure 5b shows that w/U_{ref} has a magnitude up to 10% at the mid-canyon height. Near the ground and above the canyon, w/U_{ref} are negligible. For turbulence, Figure 5c,d show that both u'/U_{ref} and w'/U_{ref} are the highest near the roof level and decay with increasing distance from the roof. This is expected, as most turbulence is generated at the roof level, where the velocity gradient is the highest. Inside the canyon ($z/H < 1$), u'/U_{ref} and w'/U_{ref} stay relatively constant at about 5%. The mean span-wise velocity (not shown) was zero throughout, indicating the 2D nature of the flow.

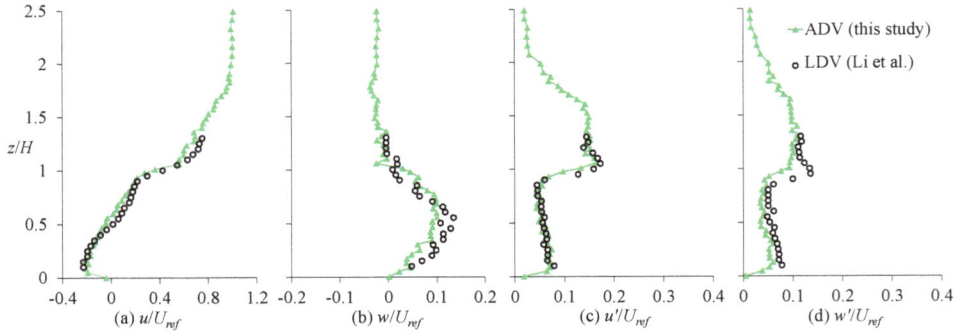

Figure 5. Normalized middle-line velocity profiles at canyon 4 measured with Acoustic Doppler Velocimeter (ADV, this study) and Laser Doppler Velocimeter (LDV, from Li et al. [36]). (**a**) mean stream-wise velocity, (**b**) mean vertical velocity, (**c**) stream-wise velocity fluctuation, and (**d**) vertical velocity fluctuation.

3.2. Canyons with a Wind Catcher

Next, we added a wind catcher above canyon 4, without changing the arrangement of the model blocks (case (ii) in Table 1). Similar to the reference case, the measurement frequency was 50 Hz and the measurement period was 60 s. To speed up the measurement process, U_{ref} was increased by about two times and, correspondingly, the measurement frequency was doubled to 100 Hz and the measurement period was halved to 30 s. Both sets of experiments produced similar results, so only the latter (100 Hz and 30 s measurements) is reported. We are interested in quantifying the effect of the wind catcher in the target canyon 4, the upstream canyon 3, and the downstream canyons 5 and 6.

Figure 6a shows that u/U_{ref} is distinctive in canyon 4 compared to the other canyons. First, u/U_{ref} in canyon 4 is positive below the mid-canyon height ($z/H < 0.5$) while the other canyons have negative u/U_{ref}. Second, near the ground level, u/U_{ref} in canyon 4 is about 2.5 times larger in magnitude compared to the other canyons. Third, u/U_{ref} approaches zero near $z/H = 0.7$, and becomes negative at $z/H > 0.7$. This suggests a strong counter-clockwise vortex in canyon 4, with the vortex core located near $z/H = 0.7$. This is caused by the protrusion of the wind catcher above the roof level, which induces additional blockage to the flow upstream. Based on the geometry in Figure 4, the protrusion of the wind catcher is 5.3 cm above the roof level. The canyon height is 12 cm, so the effective blockage to the flow is 12 cm + 5.3 cm = 17.3 cm. In contrast to a counter-clockwise vortex in canyon 4, negative u/U_{ref} below $z/H = 0.5$ implies that clockwise vortices form in canyons 3, 5, and 6. The normalized vertical velocity, w/U_{ref}, remains small (<10%) in canyons 3, 5, and 6, as shown in Figure 6b. Between $z/H = 0.7$ and $z/H = 1.5$, canyon 4 has higher negative w/U_{ref} of up to 10%, possibly induced by the wake of the wind catcher. Concerning turbulence, Figure 6c,d show that overall, u'/U_{ref} is larger than w'/U_{ref}. Near the ground level ($z/H < 0.2$), u'/U_{ref} in canyon 4 is twice as high compared to the other canyons. Above canyon 4, the maximum u'/U_{ref} and w'/U_{ref} are recorded near $z/H = 1.4$, which is the protrusion height of the wind catcher. These high velocity fluctuations travel downstream so canyons 5 and 6 record higher u'/U_{ref} and w'/U_{ref} compared to canyon 3, which is upstream of the wind catcher.

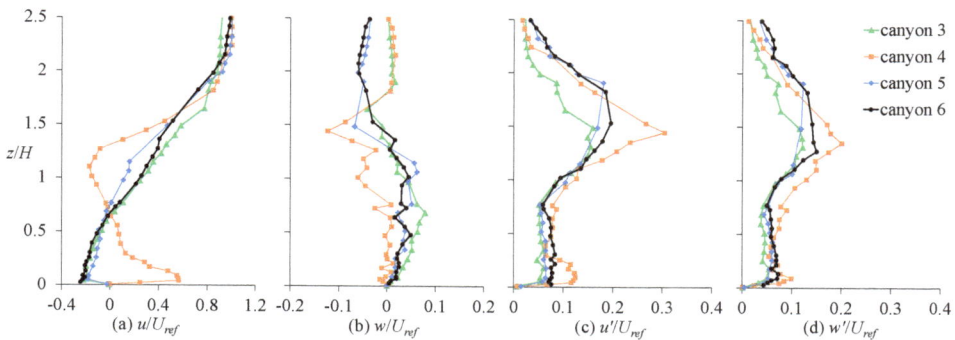

Figure 6. Normalized middle-line velocity profiles of canyons 3–6 with a wind catcher above canyon 4. (**a**) mean stream-wise velocity, (**b**) mean vertical velocity, (**c**) stream-wise velocity fluctuation, and (**d**) vertical velocity fluctuation.

3.3. Wind Catcher in a Reversed Flow Direction

A wind catcher can be installed with its inlet oriented towards the prevailing wind direction to maximize the flow enhancement in a canyon. If the wind comes at an opposite direction (e.g., diurnal wind where the wind changes direction at night), a wind catcher becomes a "reversed wind catcher", depicted as case (iii) in Table 1. In this setting, canyon 4 is still the target canyon, as the wind catcher is nearest to canyon 4. The measurement frequency and period were 100 Hz and 30 s. Figure 7 summarizes the middle-line velocity profiles of canyons 3–6 with a reversed wind catcher. The flow profiles of canyons 3 and 4 are similar to those in the reference case, showing that the reversed wind catcher has little effect on the upstream canyon 3 and the target canyon 4. In canyon 5, u/U_{ref} drops to nearly zero throughout the canyon. The wake induced by the protrusion of the reversed wind catcher above the roof level is apparent at canyon 5, as the magnitude of u/U_{ref} remains small up to $z/H = 1.4$. Above $z/H = 1.4$, u/U_{ref} increases rapidly to recover to the free-stream velocity. In canyon 6, near-ground u/U_{ref} recovers to about 20%, slightly lower compared to canyon 4. Concerning turbulence, Figure 7c,d show that u'/U_{ref} is larger than w'/U_{ref}, consistent with all previous cases. In canyons 3, 4, and 6, both u'/U_{ref} and w'/U_{ref} are between 5% and 10%, except near the ground level, where they decay to zero. Canyon 5 records the smallest u'/U_{ref} and w'/U_{ref} at about 2% inside the canyon.

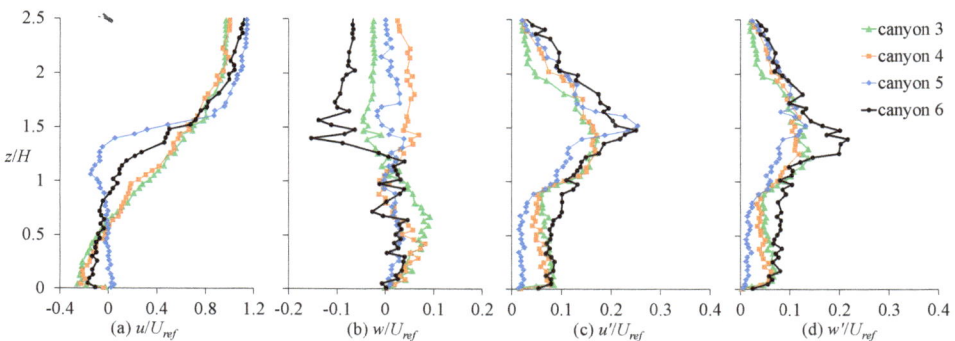

Figure 7. Normalized middle-line velocity profiles of canyons 3–6 with a reversed wind catcher above canyon 4. (**a**) Mean stream-wise velocity, (**b**) mean vertical velocity, (**c**) stream-wise velocity fluctuation, and (**d**) vertical velocity fluctuation.

3.4. Step-Up/Step-Down Canyons

The wind catcher captures atmospheric wind with an inlet protruding above the roof level, where the wind speed is higher due to less obstruction to flow. The wind catcher in case (ii) protrudes 5.3 cm above the roof level of a 12 cm tall canyon. Does a taller building with an equivalent height induce similar flow enhancement effect in canyon 4? We attempt to answer this question by studying flows across step-up and step-down canyons (case (iv) in Table 1). In this experiment, the fifth building model was 17.3 cm tall (to match the effective blockage height of the wind catcher in case (ii)), while all other building models remained at 12 cm tall. Canyon 4 is a step-up canyon, while canyon 5 is a step-down canyon. The measurement frequency and period were 100 Hz and 30s. Figure 8 summarizes the experimental results.

Figure 8a shows that canyon 5 records almost zero flow, while canyons 3, 4, and 6 have about the same u/U_{ref} profiles inside the canyons. These profiles are similar to the u/U_{ref} profiles in the case of a reversed wind catcher in Figure 7a. This means that the tall building obstructs the flow in a similar fashion as the reversed wind catcher. The profiles of w/U_{ref} are also comparable across different canyons, except above canyon 4. Compared to the reversed wind catcher (Figure 7b), Figure 8b shows that w/U_{ref} above canyon 4 is slightly higher in this case. This may be due to the geometry of the blockage. In the case with a reversed wind catcher, the flow above canyon 4 has a milder turn of 30 degrees when it approaches the sloped plate of the reversed wind catcher. In the case with a tall building downwind of canyon 4, the flow approaching building 5 (the downwind building of canyon 4) turns 90 degrees to align with the vertical windward wall of building 5. The abrupt turning of 90 degrees induces a higher w/U_{ref} compared to a milder turn of 30 degrees in the case with a reversed wind catcher. Concerning turbulence, Figure 8c,d are similar to Figure 7c,d, further suggesting that a taller building obstructs the flow similarly to a reversed wind catcher.

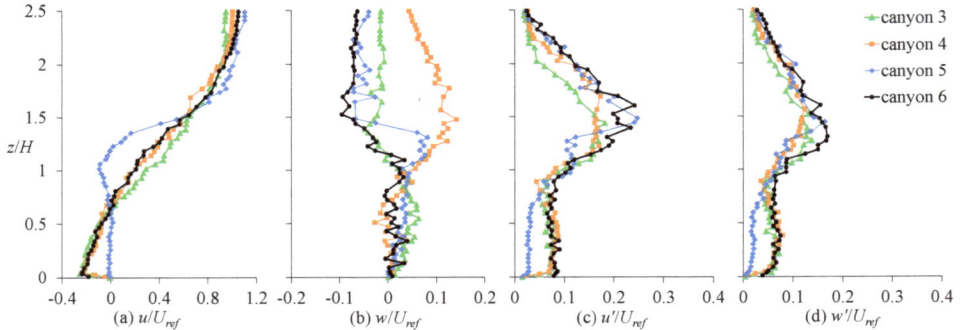

Figure 8. Normalized middle-line velocity profiles of canyons 3–6 with a step-up canyon 4 and a step-down canyon 5. (**a**) mean stream-wise velocity, (**b**) mean vertical velocity, (**c**) stream-wise velocity fluctuation, and (**d**) vertical velocity fluctuation.

3.5. Comparison between Wind Catcher, Reversed Wind Catcher, and Step-Up/Step-Down Canyons

Figure 9 plots the velocity profiles in canyons 3–6, comparing the three types of architectural interventions to the reference case with canyons of aspect ratio 1. Since w/U_{ref} is small near the ground level, only $|u|/U_{ref}$ is plotted (the absolute values are taken to ease comparison, since the reference case has negative u/U_{ref} near the ground level). To focus on the pedestrian-level wind speed, the y-axis ranges up to the mid-canyon height at $z/H = 0.5$ (instead of $z/H = 2.5$). Figure 9a shows that none of the three architectural interventions affects near-ground wind speed in the upstream canyon 3. Figure 9b shows that in the target canyon 4, the reversed wind catcher and the tall building have no effect, while the wind catcher enhances near-ground wind speed by more than 2.5 times.

Figure 9c shows that the tall building and the reversed wind catcher reduce the flow to nearly zero in the downstream canyon 5, while the wind catcher has negligible effect. Lastly, Figure 9d shows that in the further downstream canyon 6, all cases have about the same near-ground wind speed, except the case with a reversed wind catcher, which has a slightly lower speed.

In summary, the wind catcher is shown to be effective in channeling atmospheric wind into a target canyon. A reversed wind catcher does not enhance flow in the target canyon and reduces flows into downstream canyons. This drawback of wind speed reduction in an unfavorable wind direction can be overcome with improved designs such as incorporating a rotatable inlet. Similar to a reversed wind catcher, a tall building amid an array of lower buildings does not enhance flow in the target canyon and reduces flow in the immediate downstream canyon.

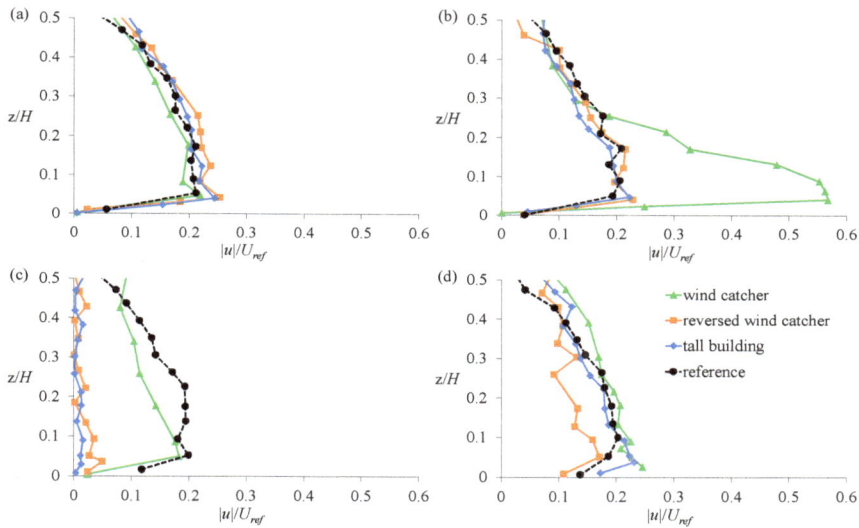

Figure 9. Normalized (absolute) stream-wise velocity comparing the cases with a wind catcher, a reversed wind catcher, and step-up/step-down canyons with reference to canyons of aspect ratio 1 in: (**a**) canyon 3, (**b**) canyon 4, (**c**) canyon 5, and (**d**) canyon 6.

4. CFD Simulations

We repeated the studies of all experimental cases with CFD simulations. Both the experimental and computational approaches have their strengths and weaknesses [45,46]. We could not perform flow visualization with our experimental setup because the partition walls were not transparent. A validated CFD model serves as a valuable tool to complement this drawback of experiment, as the CFD result can reveal the holistic flow fields. We also need CFD simulations to study the flow across a real-scale built environment, since full-scale building models will not fit even in the largest wind tunnel and water channel. We conducted CFD simulations at both the experimental scale (i.e., reduced scale) and full scale to justify the scale reduction in our water channel experiments.

4.1. Numerical Model

We used ANSYS DesignModeler (Release 17.2) for geometrical modelling and ANSYS Meshing (ANSYS Inc., Canonsburg, PA, USA) [47] for mesh generation. The CFD domain had similar geometrical dimensions as the experimental setup, except the span-wise length was reduced to H with a periodic boundary condition in the span-wise direction to reduce the total number of grids

[31,48–50]. The top face had a symmetric boundary condition. The domain height was 3.4H or 3.8H, which corresponded to the water depth in the experiments. A sensitivity study with 10H domain height showed that the mean stream-wise velocity was smaller for $1 < z/H < 2$, but had no effect in the canyon ($z/H < 1$). The inlet, located 4H from the first model block, was prescribed with the velocity profile measured in the experiments at the same location. The outlet, located 15H from the last block [51], had a zero-gradient boundary condition. All walls including the bottom had a no-slip boundary condition. In the experiments, all model blocks were coated with epoxy, which smoothed the surfaces and all walls in the CFD domain were set to be smooth. Figure 10 shows an example of the mesh for the reference case. The model had 2.5 million grids in total. The canyons had a fine mesh resolution with uniform grid size $H/60$. The same resolution was maintained above the roof level up to 1.8H. Above 1.8H, the mesh was coarsened. Similarly, the mesh was coarsened upstream of the first building and downstream of the last building (not shown). The maximum mesh expansion ratio was 1.2. The span-wise direction had 25 uniform grids.

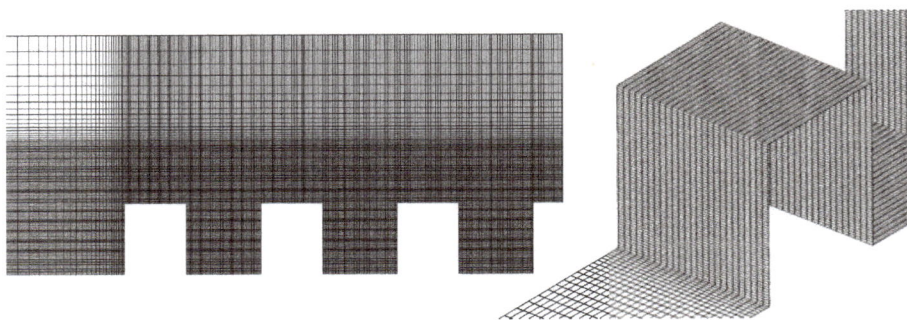

Figure 10. Side view of the central plane (only the first three canyons shown) and 3D view (only the first model block shown) of the Computational Fluid Dynamics domain.

The open-source, finite volume solver Open Field Operation and Manipulation (OpenFOAM), was adopted for CFD simulations. All simulations were run with OpenFOAM version 3.0.1(OpenCFD Ltd., Bracknell, UK) [52] in the Linux platform Ubuntu 15.10, (Canonical Ltd., London, UK) in a Dell Precision Tower 7910 Workstation (Dell Inc., Round Rock, TX, USA) with 48 processors. Up to 46 processors were used for a single run, depending on the number of grids in the mesh. We used the built-in Reynolds-averaged Navier–Stokes (RANS) solver, "simpleFOAM", which is a steady-state solver for incompressible turbulent flow with SIMPLE (Semi-Implicit Method for Pressure Linked Equations) pressure-velocity coupling and k-ϵ turbulence closure. There are three commonly used k-ϵ closure schemes: standard k-ϵ, Re-Normalization Group (RNG) k-ϵ, and realizable k-ϵ. Hang et al. [7] recommends the standard k-ϵ scheme over the RNG k-ϵ scheme, as the former agrees better with their experimental results. We tested the standard k-ϵ and realizable k-ϵ schemes and found that the former agrees better with our experimental results. Therefore, the standard k-ϵ was adopted for all our simulations. Second order Gaussian integration with linear interpolation was used for all gradient schemes and divergence schemes. All Laplacian schemes were based on Gaussian integration with linear interpolation and non-orthogonal correction. The standard wall function was employed to reduce computational cost by allowing a coarser mesh. The tolerance of residuals was set at 10^{-5} for all parameters, and iterations were continued until all residuals reach a plateau (do not change with further iterations) [53]. Post-processing was done with the open-source software ParaView version 5.3.0 (Kitware Inc., Clifton Park, NY, USA) [54].

4.2. Simulation of the Reference Case

We first look at simulation results of the reference case with canyons of aspect ratio 1. The reduced-scale experiment had a canyon height, $H = 12$ cm, and a corresponding Reynolds number, $Re = 19,000$. In a real-scale built environment, Re is much higher. For example, a 3 m/s wind flow across a 20 m tall canyon has an $Re = 3.8 \times 10^6$. Such a high Re cannot be achieved in reduced-scale experiments so we need to use CFD simulations for full-scale studies. Two CFD simulations were run: one with $H = 12$ cm with reference velocity, $U_{ref} = 0.52$ m/s in water ($Re = 62,000$); another one with $H = 20$ m with $U_{ref} = 3$ m/s in air ($Re = 3.8 \times 10^6$) to simulate a real-scale built environment. In addition, a full-scale simulation with a refined mesh (with twice the resolution in all three directions) was run to check for mesh independence. Figure 11 compares numerical and experimental results. Since canyons 3–6 have similar velocity profiles, only the velocity profiles at canyon 4 are plotted.

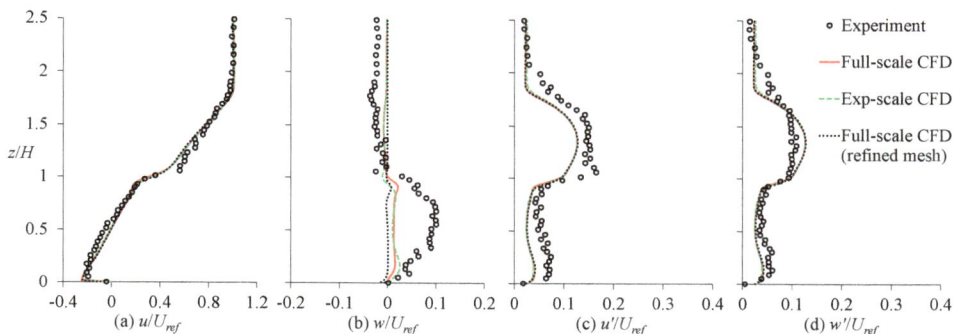

Figure 11. Normalized middle-line velocity profiles at canyon 4 comparing experimental and CFD results for the reference case. (**a**) mean stream-wise velocity, (**b**) mean vertical velocity, (**c**) stream-wise velocity fluctuation, and (**d**) vertical velocity fluctuation. "Exp-scale" and "Full-scale" are experimental scale and real scale in a built environment.

Overall, the full-scale CFD simulation produced results similar to the reduced-scale CFD simulation, verifying that the reduced-scale CFD simulation (and the reduced-scale experiment) is representative of flows across full-scale built environments. The flow pattern does not change when increasing Re from 19,000 to 62,000 and further to 3.8×10^6, suggesting that Reynolds independence is achieved at $Re > 10,000$ [37–39]. Compared to experimental data, Figure 11a shows a satisfactory agreement of u/U_{ref} at both scales. Nevertheless, Figure 11b shows that the CFD simulations under-predicted w/U_{ref} at both scales. Figure 11c,d show that above the roof level ($z/H > 1$), both u'/U_{ref} and w'/U_{ref} agree well with the experiment. Inside the canyon ($z/H < 1$), CFD simulations predicted lower turbulence. Note that the k-ϵ turbulence closure scheme produces isotropic turbulence so u'/U_{ref} equals w'/U_{ref} in the CFD simulations. Lastly, refining the mesh produced similar results, hence mesh independence is achieved. Quantitative evaluations of model validation and mesh independence are provided in Appendix A.

4.3. Simulations of Cases with Different Types of Architectural Interventions

This section discusses CFD model validation of the cases with the three types of architectural interventions: a wind catcher, a reversed wind catcher, and step-up/step-down canyons. Figure 12 plots the u/U_{ref} profiles of canyons 3–6 for the three cases. The experimental results are plotted as filled circles, while the CFD results are plotted as solid lines. For brevity, only u/U_{ref} profiles are plotted (the profiles of w/U_{ref}, u'/U_{ref} and w'/U_{ref} in selected canyons are plotted and discussed later in this section). All three cases show good agreement between experimental and simulation results.

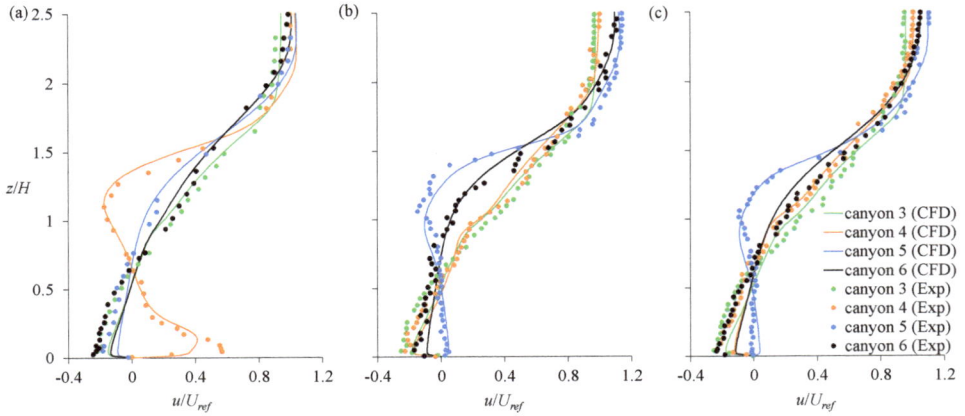

Figure 12. CFD model validation with experiments for the cases of (**a**) wind catcher, (**b**) reversed wind catcher, and (**c**) step-up/step-down canyons.

Similar to the reference case, we conducted CFD simulations at reduced scale (Re in the order of 10^4) and full scale (Re in the order of 10^6). Figure 13 plots the profiles of u/U_{ref}, w/U_{ref}, u'/U_{ref}, and w'/U_{ref} in canyon 4 for the case with a wind catcher. The CFD simulations are able to predict the trends of both the mean velocities and the velocity fluctuations observed in the experiment. Although Re differs by two orders of magnitude, the normalized velocity profiles at full scale coincide with the profiles at experimental scale, justifying that the reduced-scale experiment can reproduce the flow patterns with a wind catcher in a full-scale built environment. Similarly, Figure 14 plots the velocity profiles in canyon 5 for the case with a reversed wind catcher at both the experimental scale and full scale. Note that canyon 5 is selected because the flow profiles in canyon 5 are the most distinctive in the case with a reversed wind catcher. The CFD simulations are able to predict the trends of both the mean velocities and the velocity fluctuations observed in the experiment. The velocity profiles for the case of step-up/step-down canyons are not shown here, as they are quite similar to the profiles in Figure 14. Quantitative evaluations of model validation are provided in Appendix A.

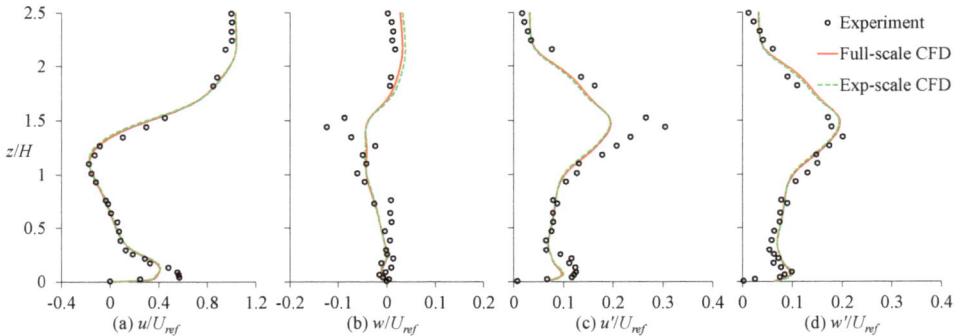

Figure 13. Normalized middle-line velocity profiles at canyon 4 comparing experimental and CFD results for the case with a wind catcher. (**a**) mean stream-wise velocity, (**b**) mean vertical velocity, (**c**) stream-wise velocity fluctuation, and (**d**) vertical velocity fluctuation. "Exp-scale" and "Full-scale" are experimental scale and real scale in a built environment.

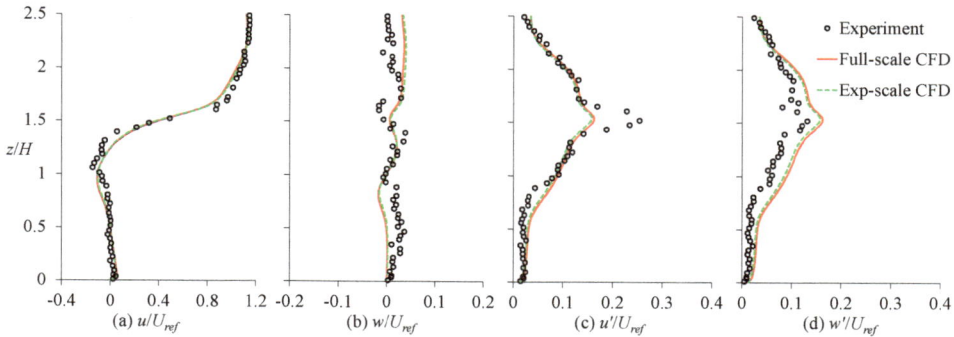

Figure 14. Normalized middle-line velocity profiles at canyon 5 comparing experimental and CFD results for the case with a reversed wind catcher. (**a**) Mean stream-wise velocity, (**b**) mean vertical velocity, (**c**) stream-wise velocity fluctuation, and (**d**) vertical velocity fluctuation. "Exp-scale" and "Full-scale" are experimental scale and real scale in a built environment.

To visualize the flow patterns in all simulated cases, Figure 15 plots the middle-plane normalized velocity magnitude (U_{mag}/U_{ref}) contours and vectors of canyons 3–6. For the reference case, Figure 15a shows that all canyons exhibit identical flow fields with a large clockwise-rotating vortex in each canyon. Each vortex has its core near the center of the canyon. The maximum near-ground velocity is about 25% of the free-stream velocity.

For the case with a wind catcher, Figure 15b reveals how the wind catcher channels atmospheric wind into the target canyon 4. Atmospheric wind captured at the inlet increases its momentum while squeezing through the narrowing channel between the roof and the top plate of the wind catcher. The high-speed jet exiting the outlet of the wind catcher travels vertically downward and turns into the stream-wise direction upon impinging the ground, boosting wind speed at the pedestrian level. This jet then turns again and moves up along the windward wall, before exiting to the atmosphere. The positively upward flow along the windward wall and the negatively downward flow along the leeward wall induces a strong counter-clockwise vortex, as opposed to clockwise vortices formed in canyons 3, 5, and 6. The cores of the vortices in canyon 4 and canyon 5 are located at around $z/H = 0.7$.

For the case with a reversed wind catcher, Figure 15c shows that the flow field in canyon 4 is not altered by the reversed wind catcher, except near the top right corner, where flow escapes through the "outlet" of the reversed wind catcher. The reversed wind catcher is installed near the windward wall so it has little effect on the velocity profiles at the middle line of the canyon. This explains the similarity between the velocity profiles of canyon 3 and canyon 4 measured in the experiment. The protrusion of the reversed wind catcher above the roof level induces a horizontally-elongated separation bubble, which spans above canyon 5. This separation bubble inhibits atmospheric wind from flowing into canyon 5, causing canyon 5 to be quiescent. Further downstream at canyon 6, part of the atmospheric flow is able to enter canyon 6, but the velocity magnitude is lower compared to canyon 3.

Lastly, for the case of step-up/step-down canyons, Figure 15d confirms that a tall building exhibits blockage effects similarly to a reversed wind catcher. The flow fields in canyons 3, 4, and 6 remain relatively unchanged compared to the reference case. In canyon 5, which is a step-down canyon, there is almost no flow at the pedestrian level due to blockage of the upstream tall building.

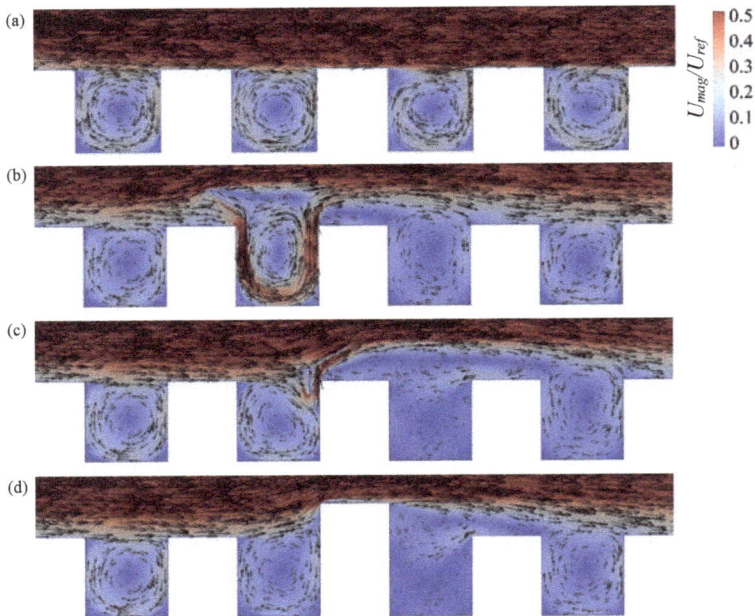

Figure 15. Normalized velocity magnitude contours and vectors in canyons 3–6 for cases: (**a**) reference case, (**b**) wind catcher, (**c**) reversed wind catcher, and (**d**) step-up/step-down canyons. The scale ranges from 0 to 0.5 (not 1.0) to emphasize the in-canyon flow.

4.4. Wind Catcher in 3D Canyons

All analyses discussed thus far are for 2D canyons. This section extends the study of wind catchers to 3D canyons with a finite building length. As our water channel has a limited width (30 cm), it is not suitable to study flow across 3D canyons. The wind tunnel experiment across nine arrays of twelve rectangular model blocks in Hang et al. [7] was used to validate our CFD model with 3D canyons. RANS with standard k-ϵ turbulence closure was adopted, with boundary conditions similar to that described in Section 4.1. Figure 16 compares CFD simulations with experimental results. The RANS simulation result in Hang et al. [7] is also included for comparison. Note that the CFD model in Hang et al. [7] employed a symmetric span-wise boundary condition while our CFD model employed a periodic span-wise boundary condition (we repeated the simulation with a symmetric span-wise boundary condition and obtained identical results so both boundary conditions are applicable in the span-wise direction). Overall, Figure 16 shows that both CFD simulations agree well with experiments, except the location of the peak turbulence kinetic energy. The CFD simulations predicted the peak near $x/B = 0$ while the experiment observed the peak near $x/B = 4$. For $x/B > 4$, the turbulence kinetic energy profiles from both CFD and experiment show a decreasing trend. Quantitative evaluations of model validation are provided in Appendix A.

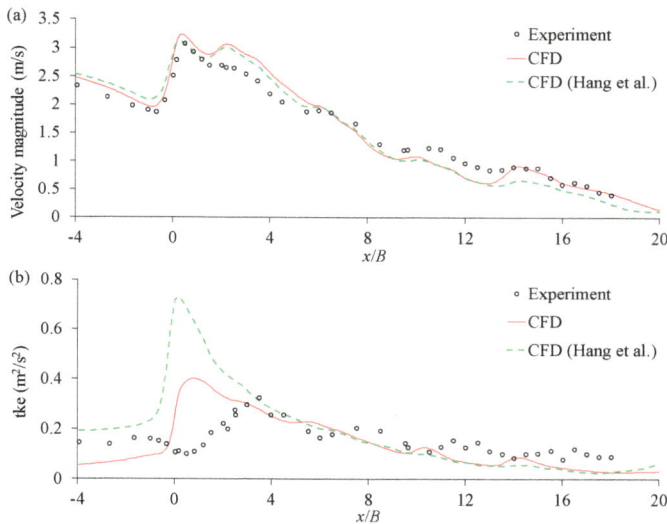

Figure 16. Along-street horizontal profiles of (**a**) velocity magnitude and (**b**) turbulence kinetic energy at elevation $z/B = 0.5$, where $B = 30$ mm is the building width. The x-axis is normalized by B, and $x/B = 0$ corresponds to the first building. Solid lines are from our CFD simulation, both the experimental data and the dashed-lines CFD are from Hang et al. [7].

Having validated the CFD model with 3D canyons, we simulated flow across wind catchers in 3D canyons, as depicted in Figure 17a. Both the height and the width of the canyons remain at H, while the span-wise building length is $5H/12$, representing narrow buildings commonly found in urban areas. Both the wind catcher span-wise length and the street width equal $5H/12$. As shown in Figure 17a, we modelled only one array of buildings. To represent an infinite array of buildings, we employed a periodic/symmetric boundary condition in the span-wise faces of the computational domain (i.e., the array repeats itself span-wise infinitely). The domain height and length, as well as the boundary conditions at the inlet, outlet, top, and walls are the same as outlined in Section 4.1. The canyons had a fine mesh resolution with uniform grid size $H/60$. The mesh was coarsened with a mesh expansion ratio (<1.2) at three regions: above $1.8H$, upstream of the first building, and downstream of the last building. The span-wise direction had uniform grid size $H/60$. Since the flow across 3D canyons behaves differently than the flow across 2D canyons, a new reference case with full-scale 3D canyons is simulated. The Re based on H is 2.0×10^6. Figure 18 plots the magnitude of u/U_{ref} up to the mid-canyon height ($z/H = 0.5$), as we are interested in the pedestrian-level flows. In a 3D canyon, near-ground wind speed is small (<10% of the free-stream velocity). Installing a wind catcher boosts the near-ground wind speed to about 15%. Recall that a wind catcher in a 2D canyon boosts near-ground wind speed by 2.5 times. This means that the wind catcher is less effective in terms of pedestrian-level wind speed enhancement in 3D canyons. The velocity contour in Figure 17c reveals that the wind catcher captures high-speed atmospheric wind via its inlet, but the high-speed downward jet loses momentum before reaching the ground level. To understand what happens between the leeward wall and the vertical plate of the wind catcher, we plot the velocity contours of section A-A in Figure 17e. Note that, in Figure 17e, we plot the magnitude of span-wise velocity, v, and vertical velocity, w, but exclude the stream-wise velocity, u, to emphasize the span-wise flow. The magnitudes of v and w are normalized with U_{ref}. Right below the top plate of the wind catcher (the white line), part of the flow turns around the side edges of the plate. This flow near the side edges then either flows upward or turns around further to hit the top surface of the plate. This flow pattern is induced by the pressure difference across the top plate of the wind catcher. The top surface of the plate has a low pressure due to flow separation, while the bottom surface has a high pressure due to impinging high-speed flow captured at the inlet. This pressure difference drives the span-wise leakage, analogous to an induced vortex at the wingtip of an airfoil. By sealing the sides of the wind catcher with sidewalls as shown in Figure 17b (as opposed to no sidewalls in Figure 17a, Figure 18 shows that near-ground u/U_{ref} increases to over 30%. The velocity contours in Figure 17d,f confirm that, by preventing the span-wise leakage with sidewalls, the high-speed jet captured at the inlet of the wind catcher travels downward with little momentum loss to the span-wise direction until it reaches the ground. It then turns into both the span-wise and the stream-wise directions. The flow that turns into the stream-wise direction is observed as the high-speed jet in Figure 18.

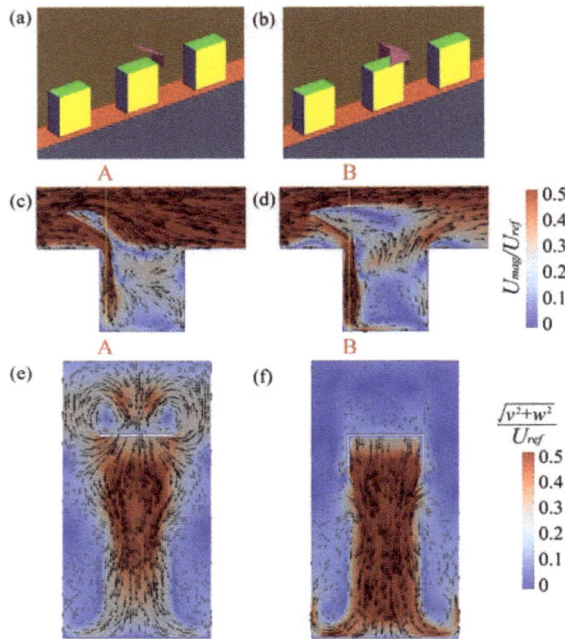

Figure 17. (**a**) Wind catcher and (**b**) wind catcher with sidewalls. Normalized velocity magnitude contours and vectors in canyon 4 for (**c**) wind catcher and (**d**) wind catcher with sidewalls. Normalized span-wise and vertical velocity contours and vectors of (**e**) section A-A and (**f**) section B-B.

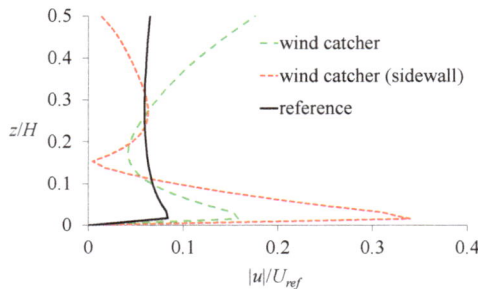

Figure 18. Comparison of normalized stream-wise velocity magnitude at the middle of canyon 4.

5. Conclusions

Urban ventilation is a key element in the microclimate of cities, as it alleviates several major environmental challenges such as air quality, thermal comfort, and urban heat island effect. Urban morphology and design are the major contributing factors to urban ventilation and should be taken into great consideration for new development or extensions of cities. In existing urban areas facing climate challenges, however, alternative solutions are needed. Architectural interventions that improve the natural ventilation of urban street canyons can be an example of such solutions.

The focus of this study is on the pedestrian-level ventilation in urban environments, and the role of architectural interventions such as wind catchers was evaluated. The strength of the current study is the comprehensive approach that (a) combines the experimental measurements with Computational Fluid Dynamics simulations of flow field with wind catchers, (b) extends the analyses to both 2D and 3D canyons, and (c) further compares the results with step-up/step-down canyons, as a representative of a tall building with a height equivalent to the wind catcher. The summary of our findings is as follows:

- We employed water channel measurements over an idealized array of 2D street canyons with an aspect ratio of 1 and evaluated the addition of a wind catcher in the aligned and reversed direction of the approaching wind. We found that a wind catcher significantly enhances pedestrian-level ventilation by increasing the local wind speed by 2.5 times. When installed in a reversed wind direction, however, the wind catcher acts similarly to a tall building with an equivalent height, such that the airflow in the downstream canyon is decreased. Therefore, further engineering analysis is required for the design of wind catchers that adapt to the wind direction.
- Using the validated CFD model, we visualized the flow field in the presence of a wind catcher, and demonstrated that a counter-clockwise vortex larger than the size of the canyon is formed when the wind catcher is aligned with the wind direction. This may result in a slight velocity decrease in the immediate downstream canyon; therefore, it is important that the deployment of wind catchers in real environments includes a holistic evaluation including the surrounding canyons.
- We extended the CFD simulations to 3D canyons and found that the characteristics of the canyon vortices are significantly different than in 2D canyons. An improved design of wind catcher with closed sidewalls enhances maximum near-ground wind speed by four times.

The main limitations of the current study and proposed future work are as follows:

- The cases evaluated here are limited in the representation of urban configuration, where only a homogeneous urban area with canyons of aspect ratio 1 is examined. Future work should evaluate the effectiveness of wind catchers in both 2D and 3D canyons with different aspect ratios, and possibly other building arrangements.
- In the present work, only two wind directions with respect to the wind catcher inlet are considered, and the effect of wind direction is not fully included. Accordingly, the results of the reversed wind catcher demonstrate the need for a comprehensive analysis on wind directions that can further inform an effective design of a wind catcher adaptable to the incoming wind direction.
- Future research should incorporate the structural and economical feasibility analyses for the installment of wind catchers in existing urban environments.

Acknowledgments: This research is supported by the National Research Foundation Singapore under its Campus for Research Excellence and Technological Enterprise programme. The Center for Environmental Sensing and Modeling is an interdisciplinary research group of the Singapore-MIT Alliance for Research and Technology. The authors appreciate the discussions with Prof. Rajasekhar Balasubramanian and Dr. Kian Yew Lim on the experimental setup and results analyses. Help from the technical staff in the Hydraulic Engineering Laboratory is acknowledged.

Author Contributions: Lup Wai Chew and Leslie Norford conceived and designed the experiments. Lup Wai Chew conducted the experiments and simulations. Lup Wai Chew and Negin Nazarian analyzed the data and wrote the paper.

Conflicts of Interest: The authors declare no conflict of interest.

Appendix A

CFD model validation should be evaluated in a quantitative way [53]. We adopted two quantitative performance measures: the fractional bias (FB) and the normalized mean-square error (NMSE) in Hanna and Chang [55]:

$$FB = 2(\overline{C_o} - \overline{C_p})/(\overline{C_o} + \overline{C_p}), \tag{A1}$$

$$NMSE = \overline{(C_o - C_p)^2}/(\overline{C_o} * \overline{C_p}). \tag{A2}$$

C can be any variable, subscript o represents observed (or measured), subscript p represents predicted (or simulated), and the over-bar represents average (the spatial average of the line profiles in our cases). We calculated FB and NMSE of the line profiles plotted in Figures 11, 13, 14 and 16. Hanna and Chang [55] suggest $|FB| < 0.67$ and NMSE < 6 as acceptable criteria for simulations of urban areas. For the mesh independence study (dashed green line and dotted black line) in Figure 11, Table A1 shows that all velocity components, except w/U_{ref}, are well within the acceptable criteria, confirming that mesh independence is achieved. High values of FB and NMSE for w/U_{ref} are due to very small means of w/U_{ref}, as shown in Figure 11b. Since the means of variables appear in the denominators, small values of means will amplify FB and NMSE (in fact, both FB and NMSE approach infinity in the limit of zero means). For CFD model validation, Tables A2–A4 summarize FB and NMSE comparing simulated results with experimental measurements for the reference case (Figure 11), the case with a wind catcher (Figure 13), and the case with a reversed wind catcher (Figure 14), respectively. All three models satisfied the acceptable criteria, except for w/U_{ref} due to the same reason discussed above. Lastly, small FB and NMSE in Table A5 show that both the velocity magnitude and turbulence kinetic energy were well predicted by the model with 3D canyons (Figure 16).

Table A1. Fractional bias (FB) and normalized mean-square error (NMSE) between the simulations with default mesh and refined mesh for the mesh independence study in Section 4.2.

	u/U_{ref}	w/U_{ref}	u'/U_{ref}	w'/U_{ref}
FB	−0.0069	−1.9170	0.0103	0.0103
NMSE	0.0003	679.4798	0.0030	0.0030

Table A2. FB and NMSE between experiment and simulation for the reference case with canyons of aspect ratio 1 in Section 4.2.

	u/U_{ref}	w/U_{ref}	u'/U_{ref}	w'/U_{ref}
FB	0.0080	−1.3691	−0.3332	−0.0118
NMSE	0.0048	33.4916	0.1712	0.0832

Table A3. FB and NMSE between experiment and simulation for the case with a wind catcher in Section 4.3.

	u/U_{ref}	w/U_{ref}	u'/U_{ref}	w'/U_{ref}
FB	0.0032	−4.6399	0.0001	0.0001
NMSE	0.0002	−12.4428	0.0012	0.0012

Table A4. FB and NMSE between experiment and simulation for the case with a reversed wind catcher in Section 4.3.

	u/U_{ref}	w/U_{ref}	u'/U_{ref}	w'/U_{ref}
FB	−0.0145	0.4035	-0.0392	0.3203
NMSE	0.0082	3.7988	0.1221	0.1599

Table A5. FB and NMSE between experiment and simulation for the case with 3D canyons in Section 4.4.

	Velocity Magnitude	**tke**
FB	0.0323	0.0642
NMSE	0.0146	0.4109

References

1. Roth, M. Review of atmospheric turbulence over cities. *Q. J. R. Meteorol. Soc.* **2000**, *126*, 941–990.
2. Ng, E. Policies and technical guidelines for urban planning of high-density cities–air ventilation assessment (AVA) of Hong Kong. *Build. Environ.* **2009**, *44*, 1478–1488.
3. Gu, Z.L.; Zhang, Y.W.; Cheng, Y.; Lee, S.C. Effect of uneven building layout on air flow and pollutant dispersion in non-uniform street canyons. *Build. Environ.* **2011**, *46*, 2657–2665.
4. Zaki, S.A.; Hagishima, A.; Tanimoto, J.; Ikegaya, N. Aerodynamic parameters of urban building arrays with random geometries. *Bound. Layer Meteorol.* **2011**, *138*, 99–120.
5. Oke, T.R. Street design and urban canopy layer climate. *Energy Build.* **1988**, *11*, 103–113.
6. Li, X.X.; Liu, C.H.; Leung, D.Y. Numerical investigation of pollutant transport characteristics inside deep urban street canyons. *Atmos. Environ.* **2009**, *43*, 2410–2418.
7. Hang, J.; Li, Y.; Sandberg, M.; Buccolieri, R.; Di Sabatino, S. The influence of building height variability on pollutant dispersion and pedestrian ventilation in idealized high-rise urban areas. *Build. Environ.* **2012**, *56*, 346–360.
8. Oke, T.R. The energetic basis of the urban heat island. *Q. J. R. Meteorol. Soc.* **1982**, *108*, 1–24.
9. Oke, T.R. City size and the urban heat island. *Atmos. Environ.* **1973**, *7*, 769–779.
10. Arnfield, A.J. Two decades of urban climate research: A review of turbulence, exchanges of energy and water, and the urban heat island. *Int. J. Climatol.* **2003**, *23*, 1–26.
11. Rizwan, A.M.; Dennis, L.Y.; Chunho, L. A review on the generation, determination and mitigation of Urban Heat Island. *J. Environ. Sci.* **2008**, *20*, 120–128.
12. Santamouris, M.; Cartalis, C.; Synnefa, A.; Kolokotsa, D. On the impact of urban heat island and global warming on the power demand and electricity consumption of buildings—A review. *Energy Build.* **2015**, *98*, 119–124.
13. Bueno, B.; Roth, M.; Norford, L.; Li, R. Computationally efficient prediction of canopy level urban air temperature at the neighbourhood scale. *Urban Clim.* **2014**, *9*, 35–53.
14. Nazarian, N.; Fan, J.; Sin, T.; Norford, L.; Kleissl, J. Predicting outdoor thermal comfort in urban environments: A 3D numerical model for standard effective temperature. *Urban Clim.* **2017**, *20*, 251–267.
15. Santamouris, M.; Papanikolaou, N.; Livada, I.; Koronakis, I.; Georgakis, C.; Argiriou, A.; Assimakopoulos, D. On the impact of urban climate on the energy consumption of buildings. *Sol. Energy* **2001**, *70*, 201–216.
16. Hui, S.C. Low energy building design in high density urban cities. *Renew. Energy* **2001**, *24*, 627–640.
17. Neophytou, M.K.; Britter, R.E. Modelling the wind flow in complex urban topographies: A Computational-Fluid-Dynamics simulation of the central London area. In Proceedings of the Fifth GRACM International Congress on Computational Mechanics, Limassol, Cyprus, 29 June–1 July 2005; Volume 29.
18. Buccolieri, R.; Sandberg, M.; Di Sabatino, S. City breathability and its link to pollutant concentration distribution within urban-like geometries. *Atmos. Environ.* **2010**, *44*, 1894–1903.
19. Hang, J.; Li, Y.; Buccolieri, R.; Sandberg, M.; Di Sabatino, S. On the contribution of mean flow and turbulence to city breathability: The case of long streets with tall buildings. *Sci. Total Environ.* **2012**, *416*, 362–373.

20. Panagiotou, I.; Neophytou, M.K.A.; Hamlyn, D.; Britter, R.E. City breathability as quantified by the exchange velocity and its spatial variation in real inhomogeneous urban geometries: An example from central London urban area. *Sci. Total Environ.* **2013**, *442*, 466–477.

21. Di Sabatino, S.; Buccolieri, R.; Salizzoni, P. Recent advancements in numerical modelling of flow and dispersion in urban areas: A short review. *Int. J. Environ. Pollut.* **2013**, *52*, 172–191.

22. Grimmond, C.; Oke, T.R. Aerodynamic properties of urban areas derived from analysis of surface form. *J. Appl. Meteorol.* **1999**, *38*, 1262–1292.

23. Ng, E.; Yuan, C.; Chen, L.; Ren, C.; Fung, J.C. Improving the wind environment in high-density cities by understanding urban morphology and surface roughness: A study in Hong Kong. *Landsc. Urban Plan.* **2011**, *101*, 59–74.

24. Lin, M.; Hang, J.; Li, Y.; Luo, Z.; Sandberg, M. Quantitative ventilation assessments of idealized urban canopy layers with various urban layouts and the same building packing density. *Build. Environ.* **2014**, *79*, 152–167.

25. Ramponi, R.; Blocken, B.; Laura, B.; Janssen, W.D. CFD simulation of outdoor ventilation of generic urban configurations with different urban densities and equal and unequal street widths. *Build. Environ.* **2015**, *92*, 152–166.

26. Kastner-Klein, P.; Rotach, M.W. Mean flow and turbulence characteristics in an urban roughness sublayer. *Bound. Layer Meteorol.* **2004**, *111*, 55–84.

27. Xie, X.; Huang, Z.; Wang, J.S. Impact of building configuration on air quality in street canyon. *Atmos. Environ.* **2005**, *39*, 4519–4530.

28. Hosseini, S.H.; Ghobadi, P.; Ahmadi, T.; Calautit, J.K. Numerical investigation of roof heating impacts on thermal comfort and air quality in urban canyons. *Appl. Therm. Eng.* **2017**, *123*, 310–326.

29. Huang, Y.; Hu, X.; Zeng, N. Impact of wedge-shaped roofs on airflow and pollutant dispersion inside urban street canyons. *Build. Environ.* **2009**, *44*, 2335–2347.

30. Abohela, I.; Hamza, N.; Dudek, S. Effect of roof shape, wind direction, building height and urban configuration on the energy yield and positioning of roof mounted wind turbines. *Renew. Energy* **2013**, *50*, 1106–1118.

31. Aliabadi, A.A.; Krayenhoff, E.S.; Nazarian, N.; Chew, L.W.; Armstrong, P.R.; Afshari, A.; Norford, L.K. Effects of Roof-Edge Roughness on Air Temperature and Pollutant Concentration in Urban Canyons. *Bound. Layer Meteorol.* **2017**, *164*, 149–179.

32. Sharples, S.; Bensalem, R. Airflow in courtyard and atrium buildings in the urban environment: A wind tunnel study. *Sol. Energy* **2001**, *70*, 237–244.

33. Montazeri, H. Experimental and numerical study on natural ventilation performance of various multi-opening wind catchers. *Build. Environ.* **2011**, *46*, 370–378.

34. Montazeri, H.; Azizian, R. Experimental study on natural ventilation performance of one-sided wind catcher. *Build. Environ.* **2008**, *43*, 2193–2202.

35. Dehghan, A.; Esfeh, M.K.; Manshadi, M.D. Natural ventilation characteristics of one-sided wind catchers: Experimental and analytical evaluation. *Energy Build.* **2013**, *61*, 366–377.

36. Li, X.X.; Leung, D.Y.; Liu, C.H.; Lam, K. Physical modeling of flow field inside urban street canyons. *J. Appl. Meteorol. Clim.* **2008**, *47*, 2058–2067.

37. Snyder, W.H. *Guideline for Fluid Modeling of Atmospheric Diffusion*; Technical Report; Environmental Protection Agency: Research Triangle Park, NC, USA, 1981.

38. Meroney, R.N.; Pavageau, M.; Rafailidis, S.; Schatzmann, M. Study of line source characteristics for 2-D physical modelling of pollutant dispersion in street canyons. *J. Wind Eng. Ind. Aerodyn.* **1996**, *62*, 37–56.

39. Baik, J.J.; Park, R.S.; Chun, H.Y.; Kim, J.J. A laboratory model of urban street-canyon flows. *J. Appl. Meteorol.* **2000**, *39*, 1592–1600.

40. Kastner-Klein, P.; Fedorovich, E.; Rotach, M. A wind tunnel study of organised and turbulent air motions in urban street canyons. *J. Wind Eng. Ind. Aerodyn.* **2001**, *89*, 849–861.

41. Brown, M.J.; Lawson, R.; Decroix, D.S.; Lee, R.L. Mean flow and turbulence measurements around a 2-D array of buildings in a wind tunnel. In Proceedings of the 11th joint AMS/AWMA conference on the applications of air pollution meteorology, Long Beach, CA, USA, 9–14 January 2000.

42. Cui, Z.; Cai, X.; J Baker, C. Large-eddy simulation of turbulent flow in a street canyon. *Q. J. R. Meteorol. Soc.* **2004**, *130*, 1373–1394.

43. Salim, S.M.; Buccolieri, R.; Chan, A.; Di Sabatino, S. Numerical simulation of atmospheric pollutant dispersion in an urban street canyon: Comparison between RANS and LES. *J. Wind Eng. Ind. Aerodyn.* **2011**, *99*, 103–113.

44. Dong, J.; Tan, Z.; Xiao, Y.; Tu, J. Seasonal Changing Effect on Airflow and Pollutant Dispersion Characteristics in Urban Street Canyons. *Atmosphere* **2017**, *8*, 43.

45. Memon, R.A.; Leung, D.Y.; Liu, C.H. Effects of building aspect ratio and wind speed on air temperatures in urban-like street canyons. *Build. Environ.* **2010**, *45*, 176–188.

46. Blocken, B.; Stathopoulos, T.; Van Beeck, J. Pedestrian-level wind conditions around buildings: Review of wind-tunnel and CFD techniques and their accuracy for wind comfort assessment. *Build. Environ.* **2016**, *100*, 50–81.

47. ANSYS. Available online: http://www.ansys.com (accessed on 16 June 2017).

48. Santiago, J.L.; Martilli, A.; Martín, F. CFD simulation of airflow over a regular array of cubes. Part I: Three-dimensional simulation of the flow and validation with wind-tunnel measurements. *Bound. Layer Meteorol.* **2007**, *122*, 609–634.

49. Li, X.X.; Liu, C.H.; Leung, D.Y. Large-eddy simulation of flow and pollutant dispersion in high-aspect-ratio urban street canyons with wall model. *Bound. Layer Meteorol.* **2008**, *129*, 249–268.

50. Hang, J.; Li, Y.; Sandberg, M. Experimental and numerical studies of flows through and within high-rise building arrays and their link to ventilation strategy. *J. Wind Eng. Ind. Aerodyn.* **2011**, *99*, 1036–1055.

51. Franke, J.; Hirsch, C.; Jensen, A.; Krüs, H.; Schatzmann, M.; Westbury, P.; Miles, S.; Wisse, J.; Wright, N. Recommendations on the use of CFD in wind engineering. In Proceedings of the International Conference on Urban Wind Engineering and Building Aerodynamics, COST Action C14, von Karman Institute, Sint-Genesius-Rode, Belgium, 5–7 May 2004.

52. OpenFOAM. Available online: http://www.openfoam.com (accessed on 16 June 2017).

53. Blocken, B. Computational Fluid Dynamics for urban physics: Importance, scales, possibilities, limitations and ten tips and tricks towards accurate and reliable simulations. *Build. Environ.* **2015**, *91*, 219–245.

54. ParaView. Available online: https://www.paraview.org (accessed on 16 June 2017).

55. Hanna, S.; Chang, J. Acceptance criteria for urban dispersion model evaluation. *Meteorol. Atmos. Phys.* **2012**, *116*, 133–146.

![atmosphere logo] *atmosphere*

MDPI

Article

Seasonal Changing Effect on Airflow and Pollutant Dispersion Characteristics in Urban Street Canyons

Jingliang Dong [1,2], Zijing Tan [1,3], Yimin Xiao [1,*] and Jiyuan Tu [2]

[1] College of Urban Construction and Environmental Engineering, Chongqing University, Chongqing 400030, China; jingliang.dong@gmail.com (J.D.); tanzijing01@163.com (Z.T.)
[2] School of Engineering, RMIT University, Bundoora, VIC 3083, Australia; jiyuan.tu@rmit.edu.au
[3] School of Civil Engineering, Chang'an University, Chang'an 710064, China
* Correspondence: xiaoyimin@cqu.edu.cn; Tel.: +86-23-6512-0756

Academic Editors: Riccardo Buccolieri and Jian Hang
Received: 8 January 2017; Accepted: 17 February 2017; Published: 23 February 2017

Abstract: In this study, the effect of seasonal variation on air flow and pollutant dispersion characteristics was numerically investigated. A three-dimensional urban canopy model with unit aspect ratio ($H/D = 1$) was used to calculate surface temperature distribution in the street canyon. Four representative time events (1000 LST, 1300 LST, 1600 LST and 2000 LST) during typical clear summer and winter days were selected to examine the air flow diurnal variation. The results revealed the seasonal variation significantly altered the street canyon microclimate. Compared with the street canyon surface temperature distribution in summer, the winter case showed a more evenly distributed surface temperature. In addition, the summer case showed greater daily temperature fluctuation than that of the winter case. Consequently, distinct pollutant dispersion patterns were observed between summer and winter scenarios, especially for the afternoon (1600 LST) and night (2000 LST) events. Among all studied time events, the pollutant removal performance of the morning (1000 LST) and the night (2000 LST) events were more sensitive to the seasonal variation. Lastly, limited natural ventilation performance was found during the summer morning and the winter night, which induced relatively high pollutant concentration along the pedestrian height level.

Keywords: street canyon; seasonal variation; air flow; pollutant dispersion; pollutant removal; natural ventilation

---

1. Introduction

According to the latest statistics from the World Health Organization (WHO), the world's urban population now stands at 3.7 billion, implying more than half of the global population resides in cities. This rapid urbanization poses challenges for sustainable development and public health. In urban environments, especially in those areas where population and traffic density are relatively high, human exposure to hazardous substances is significantly increased due to continuous traffic emissions and poor natural ventilation in street canyons.

The ventilation and pollutant dispersion process in street canyons have been extensively investigated [1–3]. Canyon geometry, ambient wind condition and thermal stratification are found to be the main determinants of air flow regimes in urban street canyons [4,5]. Influential factors such as building morphology, canyon aspect ratio, ambient wind speed and direction were intensively studied. Hang et al. [6–8] investigated the effect of geometry morphology on street canyon ventilation by varying aspect ratio, length and building packing density. They found that lowering aspect ratios or increasing street lengths may enhance the pollutant removal. As driven by ambient wind, street canyon flow is significantly influenced by external wind direction [9]. Soulhac et al. [10] built a theoretical two-region street canyon model and revealed that the mean longitudinal velocity is proportional to the cosine of the angle of incident for any wind direction. Ryu et al. [11] presented four regression equations of canyon-averaged wind speed as the function of canyon aspect ratio for different ambient wind directions. Wind tunnel experimental results reported by Gromke and Ruck [12] indicated that the trends of pollutant concentration at side walls with increasing crown porosity or tree-stand density may be altered by ambient wind direction.

Numerous research studies were also conducted on air flow and pollutant dispersion changes due to the street canyon surface heating effect [13–15]. Measurement results demonstrated that the surface temperature varies with the solar radiation and the surface albedo [16–18], and the temperature difference between sunlit and shadow walls in summer may exceed 10 °C [19,20]. Heating effect may cause unstable stratification, and significantly influent flow and pollutant transport in urban street canyons [21,22]. Studies conducted by Cheng et al. [23], Cheng and Liu [24] revealed the pollutant removal performance was improved when the stability decreases with ground heating. Cai [25] found the differential wall heating significantly affected street canyon ventilation, and the venting velocity and exchange velocity linearly increased with wall heating. Bottillo et al. [26,27] investigated the impact of solar radiation on the wind flow field and heat transfer within a street canyon. They found apparent vortex structure and heat transfer coefficient changes occurred along the length of the canyon for all studied wind conditions.

Recently, studies focused on the differential surface heating due to atmospheric instability and solar tilt becomes an emerging research topic [28]. Kwak and Baik et al. [29] demonstrated non-uniform wall heating played crucial roles in the pollutant removal process, and its exchange amount can be comparable with that by turbulent flow during afternoon. Nazarian and Kleissl [30] performed unsteady simulations of a street-scale urban environment based on idealized geometries, in which, non-uniform surface heating caused by solar insolation and building shadowing were dynamically coupled with the airflow field. In their following study [31], they found the highest convective heat transfer coefficient occurred at the windward wall throughout the day. Our previous studies [32,33] have identified obvious variation of air flow and pollutant dispersion due to surface heating diurnal variation under clear and haze-fog weather conditions.

Despite the numerous studies previously conducted, researches focusing on the seasonal difference considering the solar radiation and the anthropogenic heating from the building interiors remain limited, and their effects on air flow and pollutant transport are not clear [34]. To reveal the seasonal differences of flow patterns and pollutant dispersion in street canyons between summer and winter, numerical simulations considering exterior solar radiation and interior anthropogenic heating were performed in this paper.

2. Methodology

2.1. Numerical Model

In this study, a three-dimensional hypothetical street canyon model with an aspect ratio of unity was used (Figure 1). The street axis was aligned with the x-direction (north-south). Both the building height (H) and street canyon width (W) were set as 20 m. The fluid domain is $5H$ (x-direction) $\times$ $3H$ (y-direction) $\times$ $4H$ (z-direction).

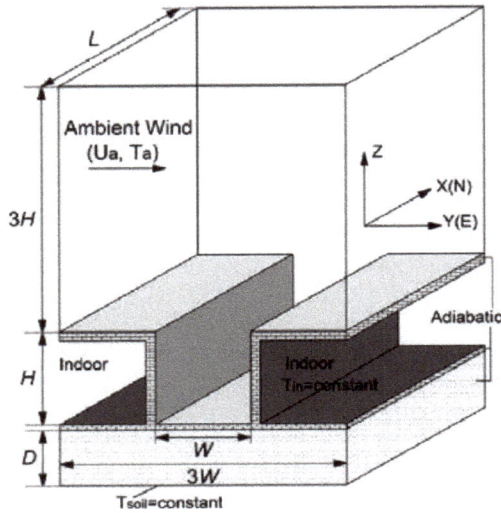

Figure 1. Schematic diagram of the computational domain ($H = 20$ m; $W = 20$ m; $D = 2$ m; $L = 100$ m; U_a: ambient wind velocity; T_a: ambient air temperature).

Symmetry boundary condition was imposed at two side planes, while all other solid facets were set as no slip boundaries. A prescribed logarithmic profile [35] was adopted as the prevalent wind condition at the upstream inlet.

$$U(z) = \frac{u_{ABL}^*}{\kappa} \ln\left(\frac{z + z_0}{z_0}\right) \tag{1}$$

$$k = \frac{u_{ABL}^{*}{}^2}{\sqrt{C_\mu}} \tag{2}$$

$$\varepsilon = \frac{u_{ABL}^{*}{}^3}{\kappa(z + z_0)} \tag{3}$$

$$u_{ABL}^* = \kappa U_{10} \ln^{-1}\left(\frac{10 + z_0}{z_0}\right) \tag{4}$$

where u_{ABL}^* is the atmospheric boundary layer (ABL) friction velocity, z is the height above the street, z_0 is the aerodynamic roughness length, set as 0.03 m, κ is von Karman's constant, set as 0.41, U_{10} is the reference horizontal velocity at 10 m high, set as 1 m/s [36].

Figure 2 illustrates the sunlight angle variation through four representative time events, where the leeward wall receives direct solar radiation during the morning event (Figure 2a), the ground receives direct solar radiation during the noon event (Figure 2b), the windward wall receives direct solar radiation during the afternoon event (Figure 2c), and no solar radiation exists during the night (Figure 2d).

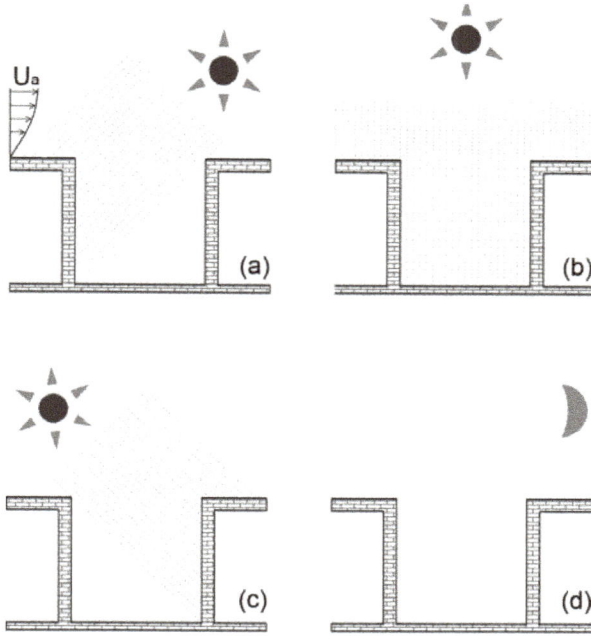

Figure 2. Diurnal variation of sunlight angles for four typical time events: (**a**) Morning (1000 LST); (**b**) Noon (1300 LST); (**c**) Afternoon (1600 LST); (**d**) Night (2000 LST).

The Typical Meteorology Year weather data file of Designer's Simulation Toolkit (DeST) software was used as the meteorological database [37,38]. Two sets of meteorological data selected from clear days in summer (18 July) and winter (17 January) of Beijing (39.92° N, 116.46° E), China were chosen and plotted in Figure 3.

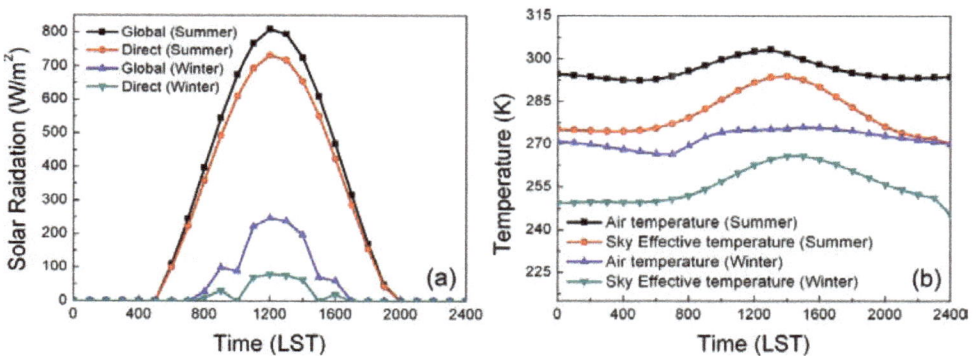

Figure 3. Meteorological conditions: (**a**) solar radiation; and (**b**) air temperature, dew point and sky effective temperature.

The ambient air temperature was imposed at the inlet and the outlet of the computational domain, while the thermal stratification was set as neutral. Both indoor and underground surfaces were specified as constant temperature conditions. The internal building temperature was set as 299 K [26] for summer and 293 K [39] for winter, respectively. The daily-averaged air temperature (296 K for summer, and 275 K for winter) was assumed to be the soil temperature at the depth of 2 m. In reality, the soil temperature fluctuates annually and daily due to variations in air temperature and solar radiation, and it varies at different depths due to the soil texture, soil water content changes. In this study, these influential factors were not considered.

The solar ray tracing algorithm and the discrete ordinates (DO) model were chosen to include the diurnal solar radiation effect and radiation transfer between street canyon facets [30]. The transmissivity of the inlet, outlet and the top boundary was set as 1.0, and the effective sky temperature was used as external radiation condition [40]. The physical properties of all facets were referred from Idczak's study [41], and model details are listed in Table 1.

Table 1. Physical properties of street canyon materials.

Location	α	ε	C_p (J·kg^{-1}·K^{-1})	ρ (kg·m^{-3})
street	0.2	0.95	880	2600
walls	0.6	0.95	800	2000
soil	-	-	1500	1440

α, ε, C_p and ρ represent absorption coefficient, emission coefficient, specific heat capacity and density of the solid materials, respectively.

Tracing particles representing vehicular emissions were released from a line source located along the center-line of the street for the street canyon ventilation assessment purpose [42,43]. Inert ash particles with the diameter of 2×10^{-6} m and density of 1 kg/m^3 were used to represent airborne PM$_{2.5}$ pollutant. The total mass of the tracing particles was set as 1×10^{-2} kg [32]. Discrete Phase Model (DPM) was used to predict the particle dispersion characteristic, and all street canyon facets were set as reflect boundary for the particles. The governing equation is given below [44]:

$$\frac{du_p}{dt} = F_D(u - u_p) + \frac{g_x(\rho_p - \rho)}{\rho_p} + F_x \tag{5}$$

where F_x is an additional acceleration (force/unit particle mass) term, $F_D(u - u_p)$ is the drag force per unit particle mass and

$$F_D = \frac{18\mu}{\rho_p d_p^2} \frac{C_D \text{Re}}{24} \tag{6}$$

Here, u is the fluid phase velocity, u_p is the particle velocity, μ is the molecular viscosity of the fluid, ρ is the fluid density, ρ_p is the density of the particle, and ρ_p is the particle diameter. Re is the relative Reynolds number, which is defined as

$$\text{Re} \equiv \frac{\rho d_p |u_p - u|}{\mu} \tag{7}$$

The Discrete Random Walk (DRW) model used in ANSYS Fluent was used to simulate dispersed particles due to turbulent dispersion.

A mesh independency study was conducted over six grid scales. To address the buoyancy force in near wall regions, six different first layer grid thicknesses varied from 0.05 m to 0.4 m were chosen, while the inflation layer expansion ratio was kept as 1.1 in the whole domain. The mesh elements number is 2,070,145 and 560,583 for the finest and coarsest grids, respectively. As listed in Table 2, the air flow velocity became stable once the mesh elements number exceeded 780,318. Taking the computational efficiency into consideration, the case having 1,026,221 mesh elements (the first layer thickness is 0.15 m, 0.75% of the street's width) was chosen for numerical analysis in this study. The scalable wall function was adopted to solve the near-wall convection in this model. y+ values of street canyon surfaces were around 100, which is within the desirable range of the scalable wall function.

Table 2. Mesh independence study.

Elements Number	First Layer Thickness (m)	Grid Expansion Ratio	Maximum Element Size (m)	U (m/s) *
560,583	0.4	1.1	0.6	0.104
645,480	0.3	1.1	0.5	0.09
780,318	0.3	1.1	0.4	0.083
1,026,221	0.15	1.1	0.3	0.083
1,237,698	0.1	1.1	0.2	0.083
2,070,145	0.05	1.1	0.2	0.083

* Velocity at the reference location (x = 50 m, y = 0 m, z = 5 m) at 1600 LST.

The CFD model of this study is based on fluid flow and transport principles for incompressible turbulence flow in terms of mass, momentum and energy conservation equations. Reynolds number in all cases of this study is around 1×10^6. Since many previous [45,46] and recent studies [47–49] indicate that the RNG k-epsilon model has good performance on both street canyon flow pattern and pollutant dispersion predictions, equations for turbulent kinetic energy and turbulent dissipation rate were solved with RNG k-eplison closure scheme. Boussinesq approximation was employed to address temperature induced density variation [45,47]. All governing equations were solved by the commercial CFD code ANSYS Fluent (ANSYS, Canonsburg, PA, USA) with finite volume method. The SIMPLE scheme was used for the pressure and velocity coupling. For all transport equations, the second-order scheme was used to provide better numerical accuracy, and the residual criterion for convergence was set as 1×10^{-4}. The energy conservation solved in Fluent is based on the equation given below [50]:

$$\frac{\Delta}{\Delta t}(\rho E) + \nabla \cdot (\vec{v}(\rho E + p)) = \nabla \cdot \left(k_{eff} \nabla T - \sum_j h_j \vec{J}_j + \left(\overline{\overline{\tau}}_{ef} \cdot \vec{v} \right) \right) + S_h \tag{8}$$

where k_{eff} is the effective conductivity, and $\vec{J}_j$ is the diffusion flux of species j.

2.2. Model Validation

Despite many field measurements and wind tunnel experiments have been conducted, almost none of them combined surface heating and air flow measurement together. In the current study, the numerical model was validated against a field measurement [41] for surface heating comparison and a wind tunnel experiment [51] for air flow velocity comparison, separately.

Figure 4 compares the numerical prediction of surface temperature with a field measurement conducted by Idczak et al. [41]. It should be noted that this field measurement records thermal environment data including solar radiation, surface and air temperature based on a street canyon with the aspect ratio of 2.48, and the street direction was kept E–W, which is different from the benchmark model used in this study. Thus, an extra street canyon model was built for numerical validation purpose, which applied identical meteorology and geometry conditions with Idczak et al.'s [41] measurement. RNG k- epsilon turbulence model with scalable wall function was adopted to solve the governing equations. 235,985 structured cells were built in the domain with the minimum size of 0.018 m (0.75% of the street's width) near the surfaces, and an expansion factor equals to 1.1 was used. The results showed $y+$ values of street canyon surfaces were around 100. The surface temperature distributions of northern and southern walls were compared. Despite minor over predictions can be found around noon for the southern wall, the simulated surface temperature well agreed with the recorded data. This minor discrepancy can be attributed to the underestimated wall thermal storage, and similar discrepancy can be founded in other relevant numerical studies [30,41].

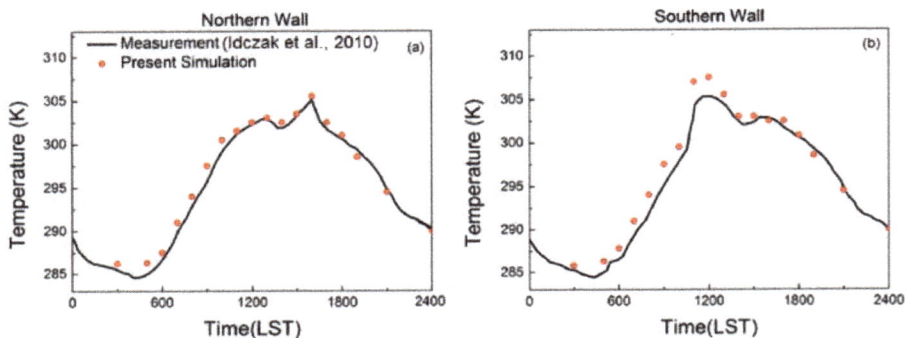

Figure 4. Surface temperature comparison between numerical simulations with field measurements: (**a**) Northern wall; (**b**) Southern wall [41].

Figure 5 compares the normalized horizontal velocity along the centerline of the street canyon with wind tunnel data referred from Uehara et al.'s work [51]. In the wind tunnel experiment, isolated cubic building blocks with the aspect ratio of 1 were used, and constant vertical temperature difference was kept by heating the bottom surface. Numerical simulations using the current street canyon model (shown in Figure 1) were compared with the wind tunnel data under the same bulk Richardson number ($Rb = -0.21$); the Rb calculation method is shown as below:

$$Rb = \frac{gH(T_{in} - T_g)}{(T_a + 273) \cdot U_{2H}^2} \tag{9}$$

where, H is the building height, T_a is the average air temperature inside the street canyon, and U_{2H} is the horizontal velocity at $Z = 2H$, T_{in} is the inlet air temperature (equals to the ambient air temperature), and T_g is the ground temperature.

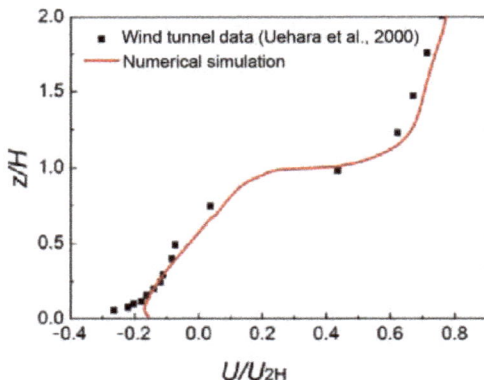

Figure 5. Comparison of normalized horizontal velocity along the centerline of the street canyon between numerical simulation and wind tunnel data (Uehara et al., 2000).

The data comparison showed a good agreement. However, the current numerical model was slightly under predicted the near ground wind. This under prediction could be attributed to the geometric difference: the street canyon model used in the wind tunnel experiment has a shorter street length than that of the current numerical model, which allows additional airflow enters the street canyon from building sidewalls.

Based on the good agreements demonstrated through above two case studies, the prediction accuracy of the current numerical model is validated, which confirms the proposed numerical modelling approach is capable to provide wind and thermal environment prediction analysis for street canyons.

3. Results and Discussion

In this section, four typical time events, morning (1000 LST), noon (1300 LST), afternoon (1600 LST) and night (2000 LST) were used for results analysis. The diurnal variation of surface temperature, air flow patterns and pollutant dispersion were analyzed and compared between summer and winter seasons.

3.1. Surface Temperature

Figure 6 illustrates diurnal surface temperatures of leeward wall, ground and windward wall under summer and winter conditions. For the ground temperature, the same diurnal variation trend was observed both in winter and summer cases, which peaks at 1300 LST for both cases. In contrast, considerable seasonal differences were found for the windward and leeward walls. For the summer condition, the maximum temperature of leeward and windward wall occurred at 1000 LST (319 K) and 1600 LST (313.2 K). For the winter condition, almost identical diurnal varying trend was found between the windward and leeward walls, which peaks at 1300 LST with the value of 288.5 K. This is mainly because the different anthropogenic heating condition inside buildings. During the summer season, indoor temperature keeps lower than ambient air temperature during daytime due to the cooling effect of air conditioning systems. The solar radiation effect dominates building surface temperature variation. However, for the winter season, combining heating effects from the solar radiation and indoor heating systems contributes to the building facets temperature changing. Compared with the anthropologic heating, the heat energy sourced from the solar radiation is minor. Therefore, almost identical temperature distributions were observed between windward and leeward walls. Consequently, the leeward wall and windward wall temperature fluctuates more wildly in summer with a maximum daily temperature difference of 24 K and 18.2 K, respectively. For the winter condition, the maximum daily temperature difference remains 11 K for both leeward and windward walls, which accounts half of that in summer. Additionally, the ground temperature exceeds other building walls at 1300 LST in summer, but remains the lowest for the whole day in winter.

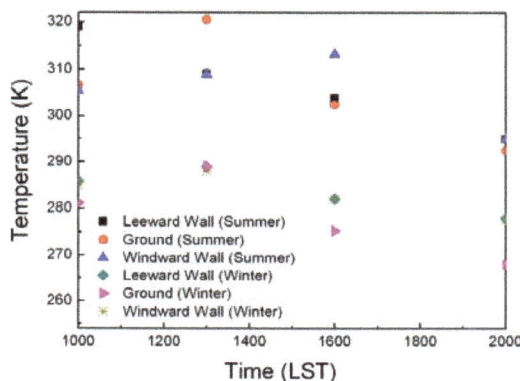

Figure 6. Diurnal variation of canyon facets temperatures between summer and winter conditions.

Figure 7 compares absolute temperature differences among all facets, namely leeward wall and ground (Figure 7a), leeward wall and windward wall (Figure 7b), the most heated facet and ambient air (Figure 7c) under summer and winter. In general, greater temperature difference occurred in summer during most of the day than that of winter. For the morning case (1000 LST), all three temperature differences under summer condition were found greater than those in winter. As the only one sunlight surface during the morning, the leeward wall was found to be the most heated surface for both seasons. The largest temperature difference between the leeward wall and ambient was 19.3 K in summer and 11.6 K in winter (Figure 7c). Despite notable surface temperature was established on the leeward wall for both seasons in the morning, the surface heating patterns were very different among all canyon facets. For the noon case (1300 LST), the temperature differences among all facets (namely leeward wall and ground, leeward wall and windward wall) were fairly small (<1 K) in winter, which indicates all three street canyon facets were evenly heated. For the noon case in summer, similar surface temperatures were established between the leeward wall and the windward wall. Only the ground received direct solar radiation, and its surface temperature exceeded the rest two side walls about 11.5 K.

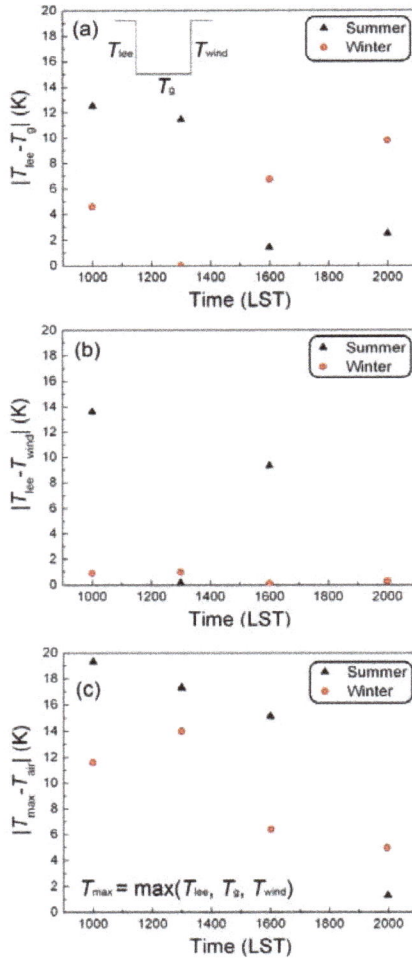

Figure 7. Temperature comparison for all time events: (**a**) leeward wall and ground; (**b**) leeward wall and windward wall; (**c**) the most heated wall and the ambient air. T_{lee} represents leeward wall temperature; T_g represents ground temperature; T_{wind} represents windward wall temperature; T_{max} represents maximum temperature among all canyon facets; T_{air}: ambient air temperature.

For the afternoon event (1600 LST), the most heated facet was the windward wall as the solar radiation directly acted on the windward wall. The temperature difference between the leeward wall and ground was only 1.4 K in summer, but nearly 8 K in winter. The temperature difference between the two side walls was 9.3 K in summer and 0.1 K in winter. This can be attributed to the significant indoor heating effect during the winter season. Due to the combined effect of solar radiation and anthropologic heating, the temperature difference between the windward wall and ambient approached to 15.1 K in summer and 6.3 K in winter, respectively.

For the night event (2000 LST) in summer, temperature differences among all facets was significantly dropped (less than 2.5 K). However, for the winter condition, a 9.9 K temperature difference between the leeward wall and ground was established. Both leeward and windward walls showed similar temperature, which exceeds the ambient temperature about 4.9 K.

Therefore, significant seasonal and diurnal variations were found for the canyon facets heating. Especially for the winter condition, more evenly distributed facets temperature with weak diurnal fluctuation is expected due to the combined effect of solar radiation and anthropologic heating.

3.2. Air Flow Pattern

In this section, the along-canyon averaged streamline fields [52] were calculated. Figure 8 details the wind-buoyancy-driven flow behavior within the street canyon under different seasonal conditions. Streamlines emanating from the core of the vortex is observed in Figure 8b,f, which can be attributed to the presence of axial flow. Since the streamlines were averaged along street canyon from $x/L = 0.25$ to $x/L = 0.75$, slight axial flow caused by the canyon end effect might be captured in the calculated results. For the morning event (1000 LST) as shown in Figure 8a,b, a clockwise rotating vortex centering close to the windward wall was observed under both seasons. However, a small secondary vortex was established at the bottom corner of the windward wall under winter condition (Figure 8b). Since the winter case does not show significant temperature variation between the windward wall and the leeward wall (Figure 7), and both of them hold higher temperature value than that of either ground or ambient air, a secondary vortex imposing an opposite force on the main wind-driven flow was observed due to the buoyancy effect close to the windward wall.

Similar flow patterns with two counter-rotating vortices were found for the noon event (1300 LST) for both seasons (Figure 8c,d). For the summer case, the main airflow vortex was compressed towards its vortex core, and the secondary vortex size is slightly larger than that under the winter condition. In contrast, the air flow was intensified along near wall regions in the winter case, and the size and intensity of the secondary vortex were greatly reduced. This is mainly attributed to different surface heating between summer and winter. In general, the ground was the most heated surface (Figure 7), and no significant temperature difference can be found between the leeward and windward walls. In summer, the ground temperature exceeded approximately 12 K compared with the other two side walls, and 17 K compared with the ambient temperature. This temperature difference induced strong buoyancy, which pushes the main vortex $0.05H$ upwards and compresses the wind-driven flow (Figure 8c). In contrast, all canyon facets were evenly heated in winter, but higher than the ambient about 14 K. Thus, the airflow in the near wall region was intensified due to the buoyancy effect.

For the afternoon case (1600 LST), completely different airflow patterns were found. For the summer condition, the direct solar radiation on the windward wall during afternoon induced a 15 K temperature increase when compared with the ambient. This heated windward wall generated significant upward buoyancy against the main wind-driven airflow, which breaks the previous single main vortex structure into multiple disturbed patterns. Thus, a complex air flow structure with multi-vortex was observed in summer (Figure 8e). While for winter, a single vortex with intensified airflow along canyon facets was established (Figure 8f). Based on the surface heating showed in Figure 7, the two side walls hold almost identical surface temperatures in winter, but only exceeded around 6 K when compared with the ambient. Thus, the circumferential velocity of the main vortex was reduced. Comparable flow patterns are also reported by Xie et al. [53] under similar stratification condition.

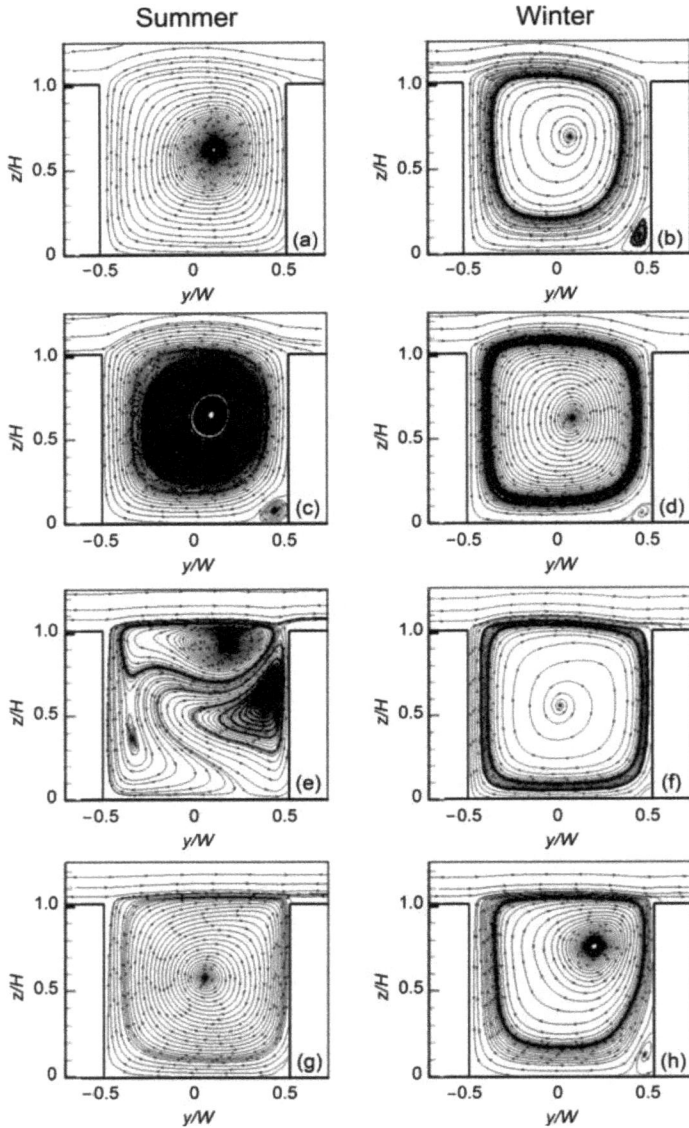

Figure 8. Along-canyon averaged streamlines for all time events: (**a,b**) represents 1000 LST; (**c,d**) represents 1300 LST; (**e,f**) represents 1600 LST; (**g,h**) represents 2000 LST. The averaging calculation was taken from $x/L = 0.25$ to $x/L = 0.75$.

For the night event (2000 LST), a clockwise rotating vortex with slight disturbance of the ambient wind was observed under both seasons. Similar with the morning event, the primary vortex occupies the whole street canyon in summer, and a small secondary vortex was observed at the windward corner in winter. As disclosed by Figure 7, for the summer condition, the temperature differences between canyon facets and ambient air were small during the night. Thus, the ambient wind dominated the vortex formation in summer. For the winter condition, the primary vortex center was moved upwards towards the windward wall, and a secondary vortex was established at the windward wall bottom corner. This is mainly due to the considerable temperature differences between the side walls and the ground (10 K), the side walls and the ambient air (5 K).

Figure 9 illustrates along-canyon averaged horizontal velocity profiles at the canyon center ($y = 0$). For the morning event (Figure 9a), the airflow velocity gradually changed from downwind direction at the top of the canyon to upwind direction at the bottom of the canyon for both seasons, which indicates the existence of the main vortex rotating in clockwise direction. Despite the velocity profile during the noon (Figure 9b) showed a similar overall trend with the morning case, the velocity magnitude under summer weather condition, especially for the lower part of the canyon ($Z < 10$ m), slightly exceeds that of winter condition, and the near ground peak wind velocity ($Z = 2.2$ m) increases from 0.72 m/s (1000 LST) to 0.95 m/s (1300 LST). Furthermore, the near ground peak velocities during the noon for both seasons were found to be the quickest compared with all other time events. The afternoon event showed complete difference horizontal velocity profiles (Figure 9c). Despite similar velocity pattern with previous events persisted in winter, but both the near ground and top wind speeds were significantly reduced. However, for the summer situation, due to the multiple vortex structure (Figure 8e), no significant air flow movement were observed at the lower half of the vertical center canyon region ($Z < 10$ m). For the night event (2000 LST), small air flow velocity ($< \pm 0.1$ m/s) was found in the central part of the street canyon (5 m $< Z < 15$ m) for both seasons. While, for the top and near ground level regions, the air flow velocity was increased in summer due to the bulk motion of ambient wind.

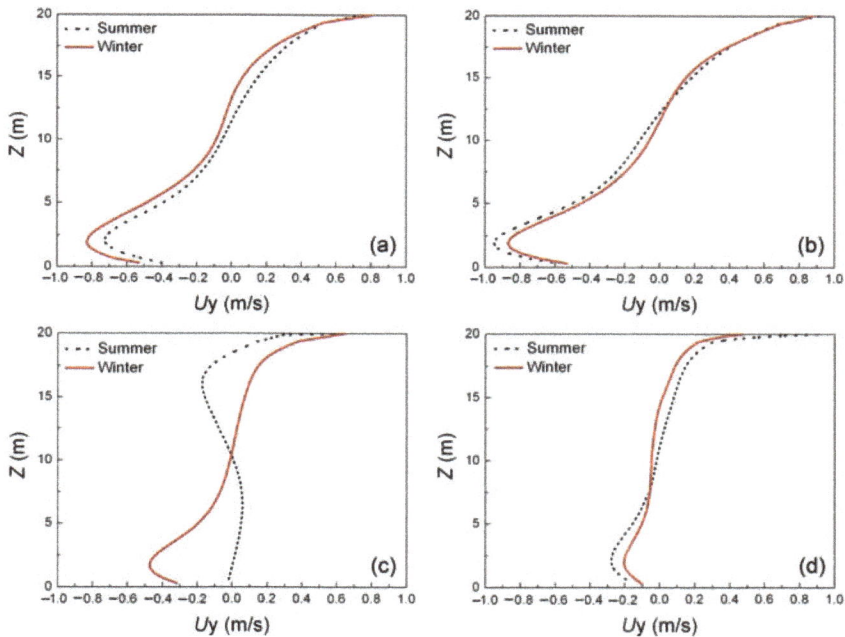

Figure 9. Averaged horizontal velocity profile on the symmetry plane ($y = 0$): (**a**) represents 1000 LST; (**b**) represents 1300 LST; (**c**) represents 1600 LST; (**d**) represents 2000 LST.

3.3. Pollutant Dispersion Characteristics

Figure 10 depicts the normalized along-canyon averaged tracing pollutants concentrations ($C/\overline{C}$). These results are obtained by linear interpolation and averaging methods. Firstly, these results were extracted from $x/L = 0.25$ to $x/L = 0.75$, with an interval of $x/L = 0.05$; Then, all field data was averaged based on eleven Y-Z slices. In general, distinct differences were observed for all case studies. In summer, the morning event showed significant pollutant accumulations in the windward half of the canyon. In winter, the night event showed a notable near-ground contaminants build-up in the leeward half of the canyon.

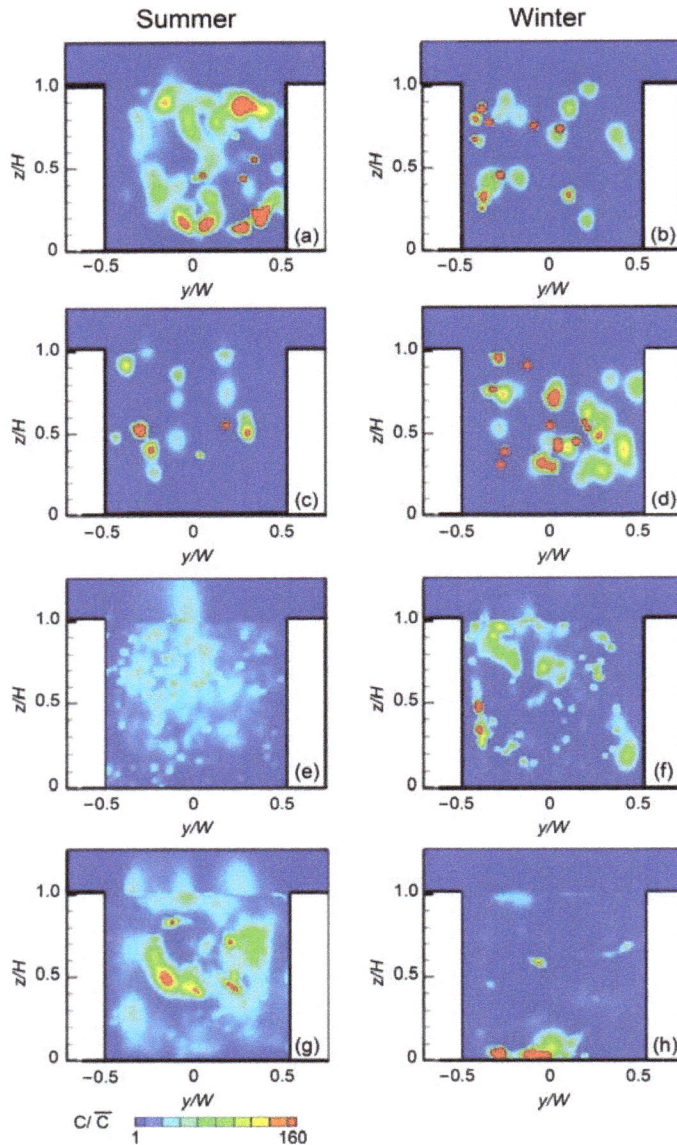

Figure 10. Along-canyon averaged pollutant concentration: (**a,b**) represents 1000 LST; (**c,d**) represents 1300 LST; (**e,f**) represents 1600 LST; (**g,h**) represents 2000 LST. Please note, pollutant concentration C was normalized by averaged pollutant concentration $\overline{C}$.

For the morning event (Figure 10a), a lower pollutant concentration was observed in the leeward half of the canyon in summer. In contrast, the winter case showed a reversed pattern with relatively more pollutant accumulating within the leeward half of the canyon (Figure 10b). For the noon event (Figure 10c,d), both cases showed a good air quality in the near-ground-zone ($z \leq 0.3H$) due to the strong lifting force imposed by the buoyancy. For the afternoon event, the summer case showed an evenly distributed pollutants concentration pattern in most of the canyon space (Figure 10e), while the winter case showed two elevated accumulation spots along the middle region of the leeward wall (Figure 10f). For the night event, despite the fact that notable pollutant accumulation can be found in summer (Figure 10g), the majority of the pollutants were centered in the upper half of the canyon space. Although majority of the pollutants have been removed out of the canyon, the night event for the winter case was the worst time slot with serious pollutant build-up along the pedestrian's height level (Figure 10h).

To further quantify the pollutant removal performance, air exchange rate (*AER*), defined as the volumetric air exchange rate per unit time, is calculated by the following equations [13]:

$$AER = \iint_{\Sigma} \overline{w_+}|_{roof} \, ds + \frac{1}{2} \iint_{\Sigma} \overline{w'' w''}^{1/2}|_{roof} \, ds \tag{10}$$

$$\overline{w'' w''}^{1/2}|_{roof} = \sqrt{\frac{2}{3} k}|_{roof} \tag{11}$$

where $\overline{w_+}$ is the mean positive vertical velocity, w'' is the mean positive velocity fluctuation, Σ is the roof area of the street canyon, and k is the kinetic energy.

Figure 11 compares the normalized *AER* between summer and winter. In general, for both seasons, the normalized *AER* peaks at the noon event, and the lowest air exchange rate occurs at the night event. For the morning and night cases, greater normalized *AER* was found in winter. While, for the noon and afternoon cases, greater normalized *AER* value occurred in summer.

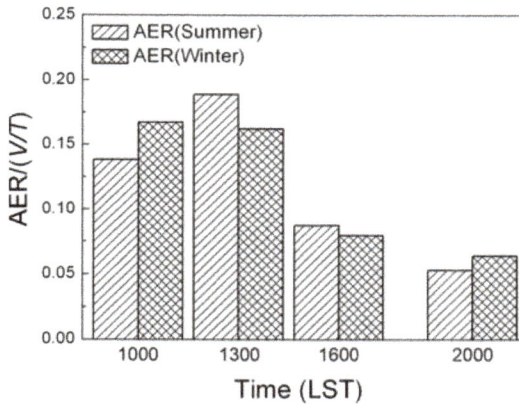

Figure 11. Normalized air exchange rate (AER) of the street canyon. Please note, V stands for the volume of street canyon of unity aspect ratio; T stands for the reference time scale H/U_a.

Lastly, the relative residual pollutant percentages in the near-wall-zone, near-ground-zone, and inhabitant-zone were calculated to quantify the air quality in human occupied zones (Figure 12). The near-wall-zone, near-ground-zone and inhabitant-zone was defined as the adjacent layer close to building side walls within 0.2W, the bottom layer in the street canyon ($z < 0.3H$), and the combination of near-ground-zone and near-wall-zone, respectively. Figure 12a summarizes the percentage of pollutants concentrated within all inhabitant zones, including near-ground zone, and near-wall zone.

The results showed a great increase for the morning event in summer with almost 90% of residual pollutant concentrating in these regions. In contrast, the winter condition showed a worst air quality with 75% of residual pollutant persisting in all inhabitant zones during the night event. These peaks indicate the natural ventilation during summer morning and the winter night are poor, and they are worst time slots for residents.

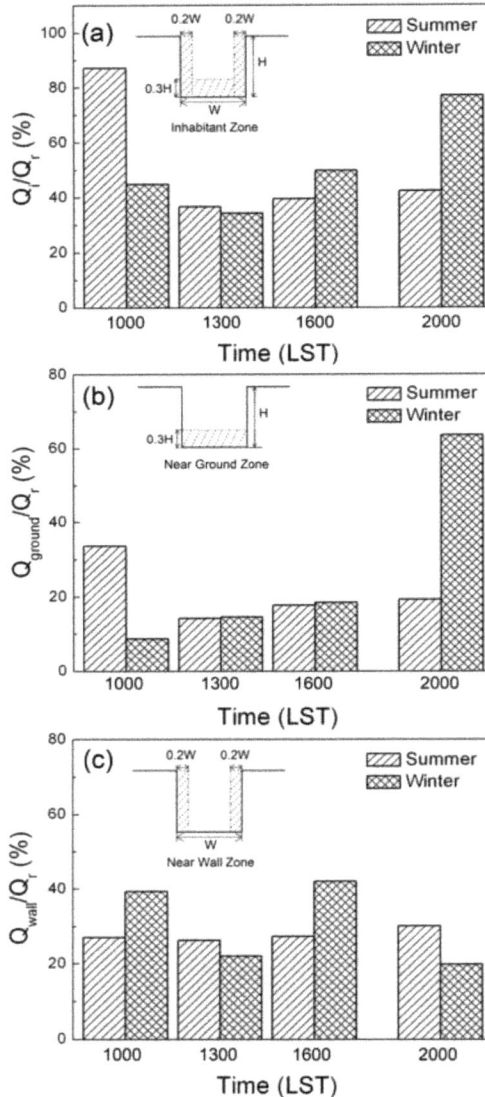

Figure 12. Residual pollutant percentage within inhabitant zones: (**a**) represents the combined results, (**b**) and (**c**) represent the localized concentration. (Q_{wall}: residual pollutant mass in the near-wall-zone; Q_{ground}: residual pollutant mass in the near-ground-zone; Q_i: residual pollutant mass in the inhabitant-zone, Q_r: total residual pollutant mass).

Through comparing the localized pollutant concentration (Figure 12b,c) with the combined results (Figure 12a), it is found majority of the residual pollutant in winter night (65%) were accumulated in the near-ground-zone due to weak ground ventilation. However, no apparent pollutant concentration differences were observed in near-wall-zones for the summer condition.

4. Conclusions

In this study, the impact of seasonal changes, including solar radiation and anthropogenic heating effects, on urban canyon ventilation as well as pollutant dispersion characteristics was investigated. The numerical model was first validated against experimental data [41,51], then the seasonal changing effects were investigated in detail. Our results showed a more evenly distributed surface temperature with relatively weak diurnal fluctuation was found in winter. The summer afternoon case showed a very different flow pattern with multi-vortex flow structure due to strong buoyance disturbance generated by heated adjacent building walls. The near-ground velocity during the summer afternoon was significantly reduced (almost zero), which may even slower than the night event. Consequently, poor pollutant removal performance was observed for the summer afternoon with a scattered pollutant concentration. Lastly, the regional residual pollutant percentage analysis showed the summer morning and the winter night are two worst time slots with 87% and 77% of the residual pollutant being trapped in the inhabitant-zone, respectively.

Due to the steady state assumption, heat storage effects of building walls and ground soil were not considered in the present study, which may contribute over-predictions of surface temperature. Besides, real street canyon geometries are much more complicated. Despite some limitations of this study, such as the unit aspect ratio and constant ambient wind assumptions, research findings presented by the current paper can contribute to fully understandings of street canyon environment under different seasonal conditions. The proposed research method offers an effective solution for urban ventilation and wind path designs.

Acknowledgments: The financial supports of the Fundamental Research Funds for the Central Universities (Project ID: 106112016CDJCR211221), National Natural Science Foundation of China (Grant No.21277080 and No.51678088), Open-end Research Funds of Chongqing Meteorological Administration (Project ID: Kfjj-201202) are gratefully acknowledged.

Author Contributions: J.D., Z.T., Y.X. and J.T. conceived and designed the experiments; J.D. and Z.T. performed the experiments; J.D. and Z.T. analyzed the data; J.D. and Z.T. wrote the paper.

Conflicts of Interest: The authors declare no conflict of interest.

References

1. Krayenhoff, E.S.; Voogt, J.A. A microscale three-dimensional urban energy balance model for studying surface temperatures. *Bound.-Layer Meteorol.* **2007**, *123*, 433–461. [CrossRef]
2. Solazzo, E.; Britter, R. Transfer processes in a simulated urban street canyon. *Bound.-Layer Meteorol.* **2007**, *124*, 43–60. [CrossRef]
3. Lee, S.-H.; Park, S.-U. A vegetated urban canopy model for meteorological and environmental modelling. *Bound.-Layer Meteorol.* **2008**, *126*, 73–102. [CrossRef]
4. Vardoulakis, S.; Fisherb, B.E.A.; Pericleousa, K.; Gonzalez-Flescac, N. Modelling air quality in street canyons: A review. *Atmos. Environ.* **2003**, *37*, 155–182. [CrossRef]
5. Georgakis, C.; Santamouris, M. On the air flow in urban canyons for ventilation purposes. *Int. J. Vent.* **2004**, *3*, 53–65. [CrossRef]
6. Hang, J.; Li, Y. Wind conditions in idealized building clusters: Macroscopic simulations using a porous turbulence model. *Bound.-Layer Meteorol.* **2010**, *136*, 129–159. [CrossRef]
7. Hang, J.; Li, Y.; Sandberg, M. Experimental and numerical studies of flows through and within high-rise building arrays and their link to ventilation strategy. *J. Wind Eng. Ind. Aerodyn.* **2011**, *99*, 1036–1055. [CrossRef]

8. Hang, J.; Li, Y.; Sandberg, M.; Buccolieri, R.; di Sabatino, S. The influence of building height variability on pollutant dispersion and pedestrian ventilation in idealized high-rise urban areas. *Build. Environ.* **2012**, *56*, 346–360. [CrossRef]

9. Santamouris, M.; Georgakis, C.; Niachou, A. On the estimation of wind speed in urban canyons for ventilation purposes—Part 2: Using of data driven techniques to calculate the more probable wind speed in urban canyons for low ambient wind speeds. *Build. Environ.* **2008**, *43*, 1411–1418. [CrossRef]

10. Soulhac, L.; Perkins, R.J.; Salizzoni, P. Flow in a street canyon for any external wind direction. *Bound.-Layer Meteorol.* **2008**, *126*, 365–388. [CrossRef]

11. Ryu, Y.-H.; Baik, J.-J.; Lee, S.-H. A new single-layer urban canopy model for use in mesoscale atmospheric models. *J. Appl. Meteorol. Climatol.* **2011**, *50*, 1773–1794. [CrossRef]

12. Gromke, C.; Ruck, B. Pollutant concentrations in street canyons of different aspect ratio with avenues of trees for various wind directions. *Bound.-Layer Meteorol.* **2012**, *144*, 41–64. [CrossRef]

13. Xie, X.; Liu, C.H.; Leung, D.Y.C.; Leung, M.K.H. Characteristics of air exchange in a street canyon with ground heating. *Atmos. Environ.* **2006**, *40*, 6396–6409. [CrossRef]

14. Xie, X.; Liu, C.H.; Leung, D.Y.C. Impact of building facades and ground heating on wind flow and pollutant transport in street canyons. *Atmos. Environ.* **2007**, *41*, 9030–9049. [CrossRef]

15. Allegrini, J.; Dorer, V.; Carmeliet, J. Buoyant flows in street canyons: Validation of CFD simulations with wind tunnel measurements. *Build. Environ.* **2014**, *72*, 63–74. [CrossRef]

16. Yang, L.; Li, Y. City ventilation of Hong Kong at no-wind conditions. *Atmos. Environ.* **2009**, *43*, 3111–3121. [CrossRef]

17. Kantzioura, A.; Kosmopoulos, P.; Zoras, S. Urban surface temperature and microclimate measurements in Thessaloniki. *Energy Build.* **2012**, *44*, 63–72. [CrossRef]

18. Ahmed, A.Q.; Ossen, D.R.; Jamei, E.; Ahmad, M.H. Urban surface temperature behaviour and heat island effect in a tropical planned city. *Theoret. Appl. Climatol.* **2014**, *119*, 493–514. [CrossRef]

19. Huang, H.; Ooka, R.; Kato, S. Urban thermal environment measurements and numerical simulation for an actual complex urban area covering a large district heating and cooling system in summer. *Atmos. Environ.* **2005**, *39*, 6362–6375. [CrossRef]

20. Offerle, B.; Eliasson, I.; Grimmond, C.S.B.; Holmer, B. Surface heating in relation to air temperature, wind and turbulence in an urban street canyon. *Bound.-Layer Meteorol.* **2007**, *122*, 273–292. [CrossRef]

21. Li, X.-X.; Britter, R.E.; Koh, T.-Y.; Norford, L.K.; Liu, C.H.; Entekhabi, D.; Leung, D.Y.C. Large-eddy simulation of flow and pollutant transport in urban street canyons with ground heating. *Bound.-Layer Meteorol.* **2010**, *137*, 187–204. [CrossRef]

22. Li, X.-X.; Britter, R.E.; Norford, L.K.; Koh, T.-Y.; Entekhabi, D. Flow and pollutant transport in urban street canyons of different aspect ratios with ground heating: Large-eddy simulation. *Bound.-Layer Meteorol.* **2012**, *142*, 289–304. [CrossRef]

23. Cheng, W.; Liu, C.-H.; Leung, D.Y. On the correlation of air and pollutant exchange for street canyons in combined wind-buoyancy-driven flow. *Atmos. Environ.* **2009**, *43*, 3682–3690. [CrossRef]

24. Cheng, W.; Liu, C.-H. Large-eddy simulation of turbulent transports in urban street canyons in different thermal stabilities. *J. Wind Eng. Ind. Aerodyn.* **2011**, *99*, 434–442. [CrossRef]

25. Cai, X. Effects of differential wall heating in street canyons on dispersion and ventilation characteristics of a passive scalar. *Atmos. Environ.* **2012**, *51*, 268–277. [CrossRef]

26. Bottillo, S.; De Lieto Vollaro, A.; Galli, G.; Vallati, A. CFD modeling of the impact of solar radiation in a tridimensional urban canyon at different wind conditions. *Sol. Energy* **2014**, *102*, 212–222. [CrossRef]

27. Bottillo, S.; De Lieto Vollaro, A.; Galli, G.; Vallati, A. Fluid dynamic and heat transfer parameters in an urban canyon. *Sol. Energy* **2014**, *99*, 1–10. [CrossRef]

28. Yazid, A.W.M.; Sidik, N.A.C.; Salim, S.M.; Saqr, K.M. A review on the flow structure and pollutant dispersion in urban street canyons for urban planning strategies. *Simulation* **2014**, *90*, 892–916. [CrossRef]

29. Kwak, K.-H.; Baik, J.-J. Diurnal variation of NO x and ozone exchange between a street canyon and the overlying air. *Atmos. Environ.* **2014**, *86*, 120–128. [CrossRef]

30. Nazarian, N.; Kleissl, J. CFD simulation of an idealized urban environment: Thermal effects of geometrical characteristics and surface materials. *Urban Clim.* **2015**, *12*, 141–159. [CrossRef]

31. Nazarian, N.; Kleissl, J. Realistic Solar Heating in Urban Areas: Air Exchange and Street-Canyon Ventilation. *Build. Environ.* **2015**, *95*, 79–93. [CrossRef]

32. Tan, Z.; Dong, J.; Xiao, Y.; Tu, J. Numerical simulation of diurnally varying thermal environment in a street canyon under haze-fog conditions. *Atmos. Environ.* **2015**, *119*, 95–106. [CrossRef]

33. Tan, Z.; Dong, J.; Xiao, Y.; Tu, J. A numerical study of diurnally varying surface temperature on flow patterns and pollutant dispersion in street canyons. *Atmos. Environ.* **2015**, *104*, 217–227. [CrossRef]

34. Yaghoobian, N.; Kleissl, J. An Improved Three-Dimensional Simulation of the Diurnally Varying Street-Canyon Flow. *Bound.-Layer Meteorol.* **2014**, *153*, 251–276. [CrossRef]

35. Richards, P.; Hoxey, R. Appropriate boundary conditions for computational wind engineering models using the k-ε turbulence model. *J. Wind Eng. Ind. Aerodyn.* **1993**, *46*, 145–153. [CrossRef]

36. Qu, Y.; Milliez, M.; Musson-Genon, L.; Carissimo, B. Numerical study of the thermal effects of buildings on low-speed airflow taking into account 3D atmospheric radiation in urban canopy. *J. Wind Eng. Ind. Aerodyn.* **2012**, *104*, 474–483. [CrossRef]

37. Yan, D.; Xia, J.; Tang, W.; Song, F.; Zhang, X.; Jiang, Y. DeST-An Integrated Building Simulation Toolkit Part I: Fundamentals. *Build. Simul.* **2008**, *1*, 95–110. [CrossRef]

38. Zhang, X.L.; Xia, J.; Jiang, Z.; Huang, J.; Qin, R.; Zhang, Y.; Liu, Y.; Jiang, Y. DeST-An Integrated Building Simulation Toolkit Part: Applications. *Build. Simul.* **2008**, *1*, 193–209. [CrossRef]

39. Stazi, F.; Mastrucci, A.; di Perna, C. The behaviour of solar walls in residential buildings with different insulation levels: An experimental and numerical study. *Energy Build.* **2012**, *47*, 217–229. [CrossRef]

40. Jones, A.; Underwood, C. A thermal model for photovoltaic systems. *Sol. Energy* **2001**, *70*, 349–359. [CrossRef]

41. Idczak, M.; Groleau, D.; Mestayer, P.; Rosant, J.-M.; Sini, J.-F. An application of the thermo-radiative model SOLENE for the evaluation of street canyon energy balance. *Build. Environ.* **2010**, *45*, 1262–1275. [CrossRef]

42. Michioka, T.; Takimoto, H.; Sato, A. Large-Eddy Simulation of Pollutant Removal from a Three-Dimensional Street Canyon. *Bound.-Layer Meteorol.* **2014**, *150*, 259–275. [CrossRef]

43. Salizzoni, P.; Soulhac, L.; Mejean, P. Street canyon ventilation and atmospheric turbulence. *Atmos. Environ.* **2009**, *43*, 5056–5067. [CrossRef]

44. Fluent, A. *ANSYS Fluent 14.0 User's Guide*; ANSYS Inc.: Canonsburg, PA, USA, 2011.

45. Kang, Y.S.; Baik, J.J.; Kim, J.J. Further studies of flow and reactive pollutant dispersion in a street canyon with bottom heating. *Atmos. Environ.* **2008**, *42*, 4964–4975. [CrossRef]

46. Kim, J.J.; Baik, J.J. Urban street-canyon flows with bottom heating. *Atmos. Environ.* **2001**, *35*, 3395–3404. [CrossRef]

47. Memon, R.A.; Leung, D.Y.C.; Liu, C.H. Effects of building aspect ratio and wind speed on air temperatures in urban-like street canyons. *Build. Environ.* **2010**, *45*, 176–188. [CrossRef]

48. Hang, J.; Luo, Z.; Wang, X.; He, L.; Wang, B.; Zhu, W. The influence of street layouts and viaduct settings on daily carbon monoxide exposure and intake fraction in idealized urban canyons. *Environ. Pollut.* **2016**, *220*, 72–86. [CrossRef] [PubMed]

49. Blocken, B. Computational Fluid Dynamics for urban physics: Importance, scales, possibilities, limitations and ten tips and tricks towards accurate and reliable simulations. *Build. Environ.* **2015**, *91*, 219–245. [CrossRef]

50. Fluent, I. *ANSYS FLUENT 14: Theory Guide*; Fluent Inc.: Canonsburg, PA, USA, 2012.

51. Uehara, K.; Murakami, S.; Oikawa, S.; Wakamatsu, S. Wind tunnel experiments on how thermal stratification affects flow in and above urban street canyons. *Atmos. Environ.* **2000**, *34*, 1553–1562. [CrossRef]

52. Kwak, K.H.; Baik, J.J.; Lee, S.H.; Ryu, Y.H. Computational Fluid Dynamics Modelling of the Diurnal Variation of Flow in a Street Canyon. *Bound.-Layer Meteorol.* **2011**, *141*, 77–92. [CrossRef]

53. Xie, X.; Huang, Z.; Wang, J.; Xie, Z. The impact of solar radiation and street layout on pollutant dispersion in street canyon. *Build. Environ.* **2005**, *40*, 201–212. [CrossRef]

atmosphere

MDPI

Article

On Street-Canyon Flow Dynamics: Advanced Validation of LES by Time-Resolved PIV

Radka Kellnerová [1,†,*], **Vladimír Fuka** [2,‡], **Václav Uruba** [1,‡], **Klára Jurčáková** [1,‡], **Štěpán Nosek** [1,‡], **Hana Chaloupecká** [1,2,‡] **and Zbyněk Jaňour** [1,‡]

[1] Institute of Thermomechanics CAS, v.v.i., Dolejskova 1402/5, 182 00 Prague 8, Czech Republic; uruba@it.cas.cz (V.U.); klara.jurcakova@it.cas.cz (K.J.); nosek@it.cas.cz (S.N.); hanach@it.cas.cz (H.CH.); janour@it.cas.cz (Z.J.)

[2] Faculty of Mathematics and Physics, Department of Atmospheric Physics, Charles University in Prague, V Holesovickach 2, 180 00 Prague 8, Czech Republic; vladimir.fuka@mff.cuni.cz

* Correspondence: radka.kellnerova@it.cas.cz; Tel.: +420-266-053-202

† Current address: Institute of Thermomechanics CAS, v.v.i., Dolejskova 1402/5, 182 00 Pragu 8, Czech Republic.

‡ These authors contributed equally to this work.

Received: 29 January 2018; Accepted: 21 April 2018; Published: 25 April 2018

Abstract: The advanced statistical techniques for qualitative and quantitative validation of Large Eddy Simulation (LES) of turbulent flow within and above a two-dimensional street canyon are presented. Time-resolved data from 3D LES are compared with those obtained from time-resolved 2D Particle Image Velocimetry (PIV) measurements. We have extended a standard validation approach based solely on time-mean statistics by a novel approach based on analyses of the intermittent flow dynamics. While the standard Hit rate validation metric indicates not so good agreement between compared values of both the streamwise and vertical velocity within the canyon canopy, the Fourier, quadrant and Proper Orthogonal Decomposition (POD) analyses demonstrate very good LES prediction of highly energetic and characteristic features in the flow. Using the quadrant analysis, we demonstrated similarity between the model and the experiment with respect to the typical shape of intensive sweep and ejection events and their frequency of appearance. These findings indicate that although the mean values predicted by the LES do not meet the criteria of all the standard validation metrics, the dominant coherent structures are simulated well.

Keywords: wind tunnel; LES; validation; street canyon; coherent structures

1. Introduction

Numerical codes for computational fluid dynamics (CFD) have been developed and used in industrial CFD applications where a variety of practical problems are predicted and tested (e.g., [1]). Owing to the enormous complexity of turbulence and extremely variable boundary conditions, the modelling of the micro-meteorological scale has been delayed with respect to their practical implementations. After years of intensive development, the numerical codes for near-surface atmospheric flow have attained a level of sufficient precision in mathematical description and spatial resolution. Time-resolving frameworks, such as Large-Eddy Simulation (LES) and Direct Numerical Simulation (DNS), have the potential now to become truly credible calculation tools for solving air quality issues since these models are capable of capturing the time-dependent behaviour of turbulence.

The CFD model needs to undergo a thorough validation procedure before its practical implementation. The validation determines to what extent the model is in agreement with real physics. An extensive review of the evaluation methodologies as well as the definition of the nomenclature used can be found in Oberkampf and Trucano [2].

For turbulent boundary-layer flow problems, the best data sources are field measurements. Unfortunately, field measurements are very rarely available and, as Schatzmann and Leitl [3] pointed out, the micro-meteorological flow found in field measurements exhibits much larger scatter than the data from a closely-controlled wind-tunnel experiment. Therefore, field data represent a greater challenge in terms of post-processing and preparation for a validation. Thus, the CFD results are often compared with those from wind-tunnel experiments [4–9]. That said, the CFD validations against data from street-canyon field experiments were performed as well (e.g., [10–12]).

A validation procedure for atmospheric boundary layer dispersion or velocity distribution predictions by CFD was compiled within the frame of COST 732 [13–15], COST C14 [16] and AIJ Tominaga et al. [17]. These guidelines addressed various types of models including Gaussian models, and RANS and LES models. The latter, LES model, represents an affordable combination of direct simulation of large turbulent structures while modelling small unresolved scales by means of an embedded sub-grid model [18].

Since common validation techniques target just the variables that are available from all the discussed models, only temporally-averaged values are usually retrieved for comparison. Illustrative applications of the validation of temporally-averaged values from LES against various experiments can be found in Jimenez and Moser [18]. The most suitable experimental data currently available are those obtained from Particle Image Velocimetry (PIV) measurement techniques as they can provide multi-points time-resolved synchronised data. The use of PIV for the validations is, however, still rare [4,19,20]. Since there may be a lack of a sufficiently long time period from LES or from PIV to achieve legitimate statistical averages, a feasible strategy for the validation of LES is, according to Hertwig et al. [20], not only the comparison of mean values but also a multi-point time-series analysis by means of advanced statistical tools, which allows us to detect transient structures, their shape and also frequency distribution.

In this paper, we present an extension of a validation procedure introduced by the research of [20–22] and Hertwig [23], in which a three-level validation hierarchy, consisting of global statistics, eddy statistics and flow structure statistics was employed. The extension comprises time-series analyses such as the spectral and spatial modification of the quadrant analysis, and the temporally-spatial analyses, namely the spatial correlation or the Proper Orthogonal Decomposition (POD). By this extension, we demonstrate that spatial data from the time-resolved particle image velocimetry (TR-PIV) improve the validation procedure. As suggested in Jimenez and Moser [18], the Reynolds stress is also a very sensitive quantity for testing the LES sub-grid model. Hence we present the investigation of the momentum flux by an additional spatial quadrant analysis as well.

This strategy respects the time-dependent nature of the coherent features and reveals the similarity between the predicted and measured turbulent flow in a more detailed manner. Our goal is to validate the capability of the LES to simulate both the larger-scale coherent features above the canyon, known to be crucial for street canyon ventilation (e.g., [7,24–26]), and to model the small-scale swirls, responsible for the intense mixing process inside the street canyon itself (e.g., [6]). Strengths and shortcomings of the well-established validation metrics are thoroughly discussed and the benefits of the innovative methods are introduced.

2. Methods and Data Pre-Processing

2.1. Experimental Method

The experiment was performed in a pressure-driven open-circuit wind tunnel with a test section of cross-sectional dimensions 0.25×0.25 m and a length of 3 m. A fan, followed by a 3 m long tunnel duct and a contraction pipe, was installed upstream of the test section. The street-canyon model covered the entire test section. The model was built from 30 identical parallel street canyons, spanned laterally across the width of the tunnel (0.25 m). The building height and the roof height was 0.03 m and 0.02 m, respectively. The aspect ratio of the street canyon (street-canyon height $H = 0.05$ m to street-canyon width W) was equal to one $H/W = 1$ (Figure 1).

The triangular shape of the street-building roofs was chosen according to roofs typical for European city centres. Kellnerova et al. [24] pointed out that triangle roofs generate more turbulent dynamics than flat ones and to successfully validate the flow behind the triangle roofs is therefore more challenging. The presented experimental set-up was a part of the measurement campaign comparing a skimming flow and its dynamics between street canyons with the flat and pitched roofs in a neutrally stratified boundary layer flow. A detailed description of the campaign can be found in Kellnerova et al. [24] or Kellnerova [27].

The PIV measurements were taken downstream behind the 20th street canyon and only at the tunnel axis in order to guarantee a fully developed flow and to avoid a wall effect. Based on an empty wind-tunnel measurement, the wall affects the mean velocity in distances up to 40–60 mm from the walls. Although we do not consider the boundary layer above the street canyons as fully representative of the atmospheric flow due to the excessive aerodynamical blocking caused by buildings (achieving value of 20%), the lowest part of the boundary layer shows significant similarity with a true atmospheric layer in terms of normalised lower and higher moments and cross-moments of the velocity [27]. The numerical large-eddy simulation copies the cross-sectional test section dimension and simulates the aforementioned effects of the walls.

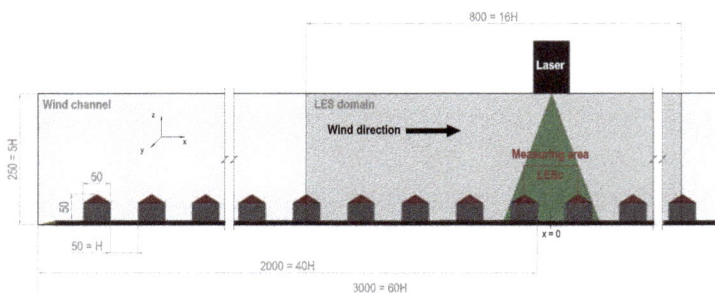

Figure 1. Isometric view of the model positioned in the tunnel test section. The measured area by TR-PIV system is indicated by the red rectangle within the green laser sheet. The dimensions are in mm.

The wind-tunnel coordinate system (x, y, z) corresponds to the streamwise (u), lateral (v) and vertical (w) instantaneous velocity, where the mean velocity components are labelled U, V and W, and their fluctuations according to Reynolds decomposition, as u', v' and w'. The mean wind speed measured above the street-canyon centre ($x/H = 0$, $z/H = 2$) was $U_{2H} = 5.7$ ms^{-1}. Thus, the Reynolds building number became $Re_{2H} = U_{2H}H/\nu = 19\,000$, where ν is the kinematic viscosity of the air. The velocity measurements were conducted with the TR-PIV system, providing 2-D snapshots of the instantaneous velocity vectors in a single streamwise-vertical plane (xz) placed 2000 mm (40 H) downstream from the test section entrance (see Figure 1). One run of PIV measurements consisted of 1634 image-pairs. An overview of the PIV parameters is listed in Table A1 in Appendix A.

Additional measurements above the height of $z/H = 1$ were performed by means of CTA hot-wire anemometry (HWA), with a single-wire probe DANTEC 55P01 (tungsten wire with a diameter of 0.005 mm and a length of 1.25 mm). The HWA provided a sampling frequency of 25 kHz for the 1-dimensional streamwise velocity component u. Prior to the validation procedure, the HWA velocity data were compared with the PIV data in terms of the mean velocity, turbulence intensity, skewness, flatness, histogram and spectra for the streamwise velocity component. Good agreement was found in all cases [27]. The spanwise homogeneity of the mean velocity was 2.3%. The HWA system reached a very good accuracy of 1% for the streamwise velocity component.

2.1.1. Numerical Method

For this study, we used the LES model called the Charles University Large-eddy Microscale Model (CLMM) developed by Fuka [28]. CLMM is an in-house finite difference solver of the incompressible Navier-Stokes equations that uses LES to model the smallest scales in the flow. The code uses uniform staggered Cartesian grids. Solid obstacles are treated by using the immersed boundary method [29,30].

CLMM uses implicit filtering, i.e., it is assumed that the variables are filtered by using approximate numerical schemes on a finite grid. The solution of the discrete Poisson system is performed using the open source library PoisFFT [28]. The spatial derivatives are computed using the second-order central differences with the exception of the momentum advection, which is computed using the fourth-order central differences [31] that reduce to second order differences in cells closest to the wall. The discrete system is integrated usign the 3rd order Runge-Kutta method with variable time step keeping the Courant number below 0.9.

The subgrid scale eddy viscosity is computed by the σ-model by Nicoud et al. [32] and the shear stresses on solid walls are computed using a wall model with a logarithmic wall function when the wall coordinate is $y^+ > 11.2$. The wall model uses the instantaneous velocity in the closest gridpoint to compute the shear stress. The wall coordinate y^+ was lower than 30 in this simulation.

For this study the model was run on a uniform grid with a constant resolution of 2 mm in each direction. The boundary conditions of the model follow the boundary conditions of the wind-tunnel test section as much as possible. The vertical and lateral dimensions of the wind tunnel are fully solved by LES over the whole extent (5 H of both vertical and lateral direction) with the no-slip boundary conditions on the tunnel walls. The streamwise dimension covers the length of 16 H (0.26 of the full test-section length). Cyclic boundary conditions were specified only in the x-direction for the upstream and downstream boundaries, and therefore the model represents an infinite number of idealised street canyons along the streamwise direction. A detailed description and the principal equations can be found in Appendix B.

3. Data Pre-Processing

For the LES validation procedure, we performed two levels of data analyses. The first level comprises a qualitative and quantitative comparison between the temporally and spatially averaged quantities of the wind-tunnel experiment and the LES model by means of profiles and standard validation metrics. The second level represents comparisons of intermittent flow dynamics between the experiment and the LES model by means of well-established statistical methods for detection of the turbulent coherent structures (spectral and quadrant analysis, correlation and POD). For each analytical approach, we performed a specific data pre-processing procedure.

3.1. Analyses of Time-Averaged Flow

All of the velocity data were normalized by the reference velocity, U_{ref}, which was temporally and horizontally averaged along the streamwise direction at the reference height ($x/H \in [-0.5, 0.5]$, $H_{ref} = z/H = 1.5$), where the LES data as well as the PIV data were available. The HWA measurement was conducted at the canyon centre and the reference velocity was achieved at a single point ($x/H = 0$, $H_{ref} = z/H = 1.5$). The temporal scale was converted into the dimensionless time, t^*, based on the formula [33]

$$t^* = t\frac{U_{ref}}{H_{ref}} \tag{1}$$

where t is the real or simulated time during the experiment or LES simulation, respectively.

For the time-mean analysis, all the experimental and numerical data available were gathered and averaged. The data from three PIV runs, each containing 1634 snapshots, were combined into one ensemble with 4902 snapshots in total, labelled PIVa. The total dimensionless time corresponds to $t^*_{PIVa} = 1295$. All LES computational periods were equal to 18 s and steady-state results were obtained from the last 10 s of the computations. Since the cyclic boundary conditions were implemented at the inlet and outlet of the domain boundaries, the streamwise position of the individual canyons did not matter. The overall statistics of the LES data were derived from the temporal and spatial averaging across all eight canyons and are labelled LESa. By assembling this large database, the effective simulation time increased to a value of t_a =80 s and the number of snapshots reached 80,000. The dimensionless total time became $t^*_{LESa} = 1528$.

3.2. Analyses of Intermittent Flow

For validation of the simulated time-dependent flow dynamics, we used the continuous time-resolved measurements from the PIV run (labelled PIVc) with the following parameters: 1634 snapshots, $t_c = 3.2$ s, $t^*_{PIVc} = 181$. The continuous HWA time-series (the streamwise velocity only, labelled HWAc) were recorded over $t_c = 30$ s, and corresponded to $t^*_{HWAc} = 1704$, respectively. For the continuous time-resolved LES data (labelled LESc), the time-series lasting $t_c = 10$ s and $t^*_{LESc} = 191$ from one canyon were employed in the validation procedures concerning the flow dynamics.

The grid resolution of the PIV data is finer (1.2 mm) than the LES resolution (2 mm). For the purpose of the validation, we started to prepare a systematic comparison between the different grid resolutions of the LES. Based on the very preliminary, not-yet-published results from the LES with half resolution (1 mm), the number of grid points, P, is considered by the authors of this paper to have a crucial impact on the LES performance. An overview of the methodology regarding the set-ups for PIV and LES is presented in Table 1.

Table 1. Parameters of the methodologies HWA, PIV and LES.

Measurement Methodology	HWA	PIV	LES
Number of samples/snapshots	750,000	4902	10,000
Spatial resolution [mm]	5	1.2	2
Sampling frequency [kHz]	25	0.5	1
Acquisition time t_a (averaged) [s]	30	22.7	80
Dimensionless time t^*_a (averaged) [-]	1704	1295	1528
Acquisition time t_c (intermittent) [s]	-	3.2	10
Dimensionless time t^*_c (intermittent) [-]	-	181	191

3.3. Measurement Uncertainty

It is common practice to determine the *uncertainty* (also referred to as the measurement error) by means of the standard deviation (STD) of many repeated measurements at reference points. Unfortunately, obtaining a high number of repetitions with PIV requires large data storage. Therefore, only a few runs are usually executed with the PIV system. We repeated the PIV measurement three times. Two measurement runs lasted 3.2 s with a sampling frequency of 500 Hz in order to capture the transient flow dynamics. A third run was performed in order to acquire turbulence statistics and lasted 16.3 s with a low sampling frequency of 100 Hz. The detailed specification of the variation observed between the individual runs is inserted in Appendix C.

Due to the occurrence of spurious vectors just next to the solid walls and edges of the area measured by PIV, the outmost grid points were omitted in the validation metric procedures, which will be discussed later in Section 4.3. We determined the measurement error from the PIV measurements as the spatial average of the STD within the remaining *restricted rectangular area* inside the street canyon. The spatially-averaged STD of the given dimensionless quantities from three PIV samples, the streamwise and vertical velocity component, U/U_{ref} and W/U_{ref}, the momentum flux, $< u'w' > /U^2_{ref}$, and the reduced turbulent kinetic energy, $TKE/U^2_{ref} = 0.5(< u'^2 > + < w'^2 >)/U^2_{ref}$, are listed in Table 2. These values also serve as the absolute deviation criterion for the Hit rate validation procedure discussed in Section 4.3.1.

Table 2. Measurement error expressed as spatially-averaged standard deviation of the given dimensionless quantity (used as the Hit rate absolute deviation criterion A in Section 4.3.1).

Quantity X	Measurement Error (Used as A)
U/U_{ref}	0.008
W/U_{ref}	0.007
$<u'w'>/U_{ref}^2$	0.002
TKE/U_{ref}^2	0.003

4. Results: Model Performance Assessment for Time-Averaged Flow

4.1. Profiles

The vertical profiles of the dimensionless mean streamwise velocity, U/U_{ref}, at the centre of the street canyon are compared in Figure 2a, where the PIVa data are displayed with the horizontal bars representing the measurement error derived from the STD at each elevation. The PIVa (*black squares*) and HWA (*grey squares*) data match remarkably well with the output from LESa for the upper part of the streamwise velocity profile, i.e., within the interval $0.9 < z/H < 1.5$. The lower part is predicted less successfully since the LES model fails to capture properly the exact shape of the recirculation zone inside the canyon, owing to an underestimation of the negative streamwise velocities near the canyon bottom. These discrepancies in the velocity and the corresponding momentum flux deviation plays a significant role in the recirculation vortex pattern, as will be shown later.

The simulated vertical velocity, W/U_{ref}, deviates from the measured one inside the street canyon ($0 < z/H < 0.8$), but agrees well at the roof level (Figure 2b). Likewise, the simulated dimensionless reduced turbulent kinetic energy, $TKE/U_{ref}^2 = 0.5(<u'^2> + <w'^2>)/U_{ref}^2$, and the vertical momentum flux, $<u'w'>/U_{ref}^2$, of LESa exhibited very good agreement with those from PIV (Figure 2c,d) within the upper canyon ($0.5 < z/H < 1.5$). The peak of reduced TKE at $z/H = 1.1$ in Figure 2d is predicted extremely well. The peak of momentum flux was slightly overestimated by LESa in terms of magnitude, but mostly within the range of the PIV measurement error. A noticeable discrepancy appears in the lower part of the momentum flux profile ($0.3 < z/H < 0.4$ in Figure 2c). As will be explained later, this positive increase in the total momentum flux in Figure 2c relates to the formation of secondary vortices inside the street.

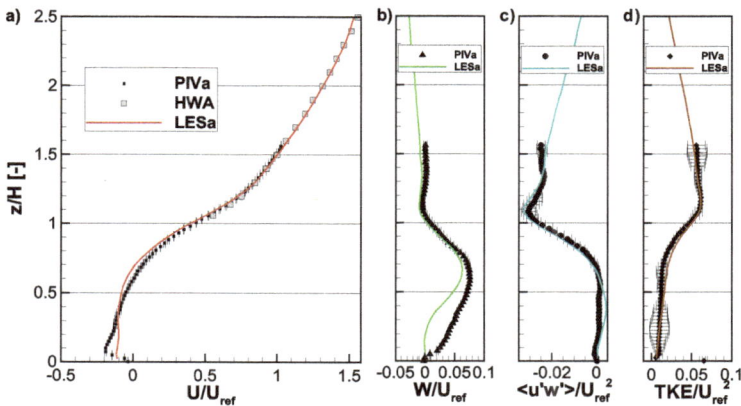

Figure 2. Comparison of the profiles of the mean dimensionless (**a**) streamwise and (**b**) vertical velocities, (**c**) momentum flux and (**d**) reduced turbulent kinetic energy at the centre of the street canyon ($x/H = 0$). The solid lines represent the LES simulation; the symbols denote the wind-tunnel experiment.

4.2. Velocity and Momentum Flux Fields

The mean velocity xz fields for both the streamwise and the vertical velocity, and for the corresponding momentum flux, are depicted in Figure 3a,b. The first line shows the results from the PIVa, and the second line represents the LESa performance. All quantities are normalised by the reference velocity U_{ref}.

Generally, the streamline pattern exhibits a single vortex with the core slightly shifted upward and downstream, compared to the flat roof case [27,34]. This matches previous wind-tunnel studies (e.g., [35,36]) and CFD studies (e.g., [34]) since the exact position depends strongly on either the specific roof geometry [37] or the roof aspect ratio (the roof height to the building width) (e.g., [34]).

A comparison between the dimensionless fields of PIVa (Figure 3a) and LESa (Figure 3b) shows a very good agreement in the region above the street canyon, but a less successful agreement in the region inside the canyon. The streamlines presented in the first column of Figure 3a,b show that LESa predicts a single primary vortex of similar shape as the observed one. The position of the primary vortex centre is predicted by LESa slightly higher compared to PIVa. Further, the secondary windward bottom-corner counter-rotating vortex in the street canyon is over-predicted by the LES model whereas the leeward bottom-corner vortex is completely missing in the simulation (see right and left bottom of the street canyon in Figure 3aI,bI).

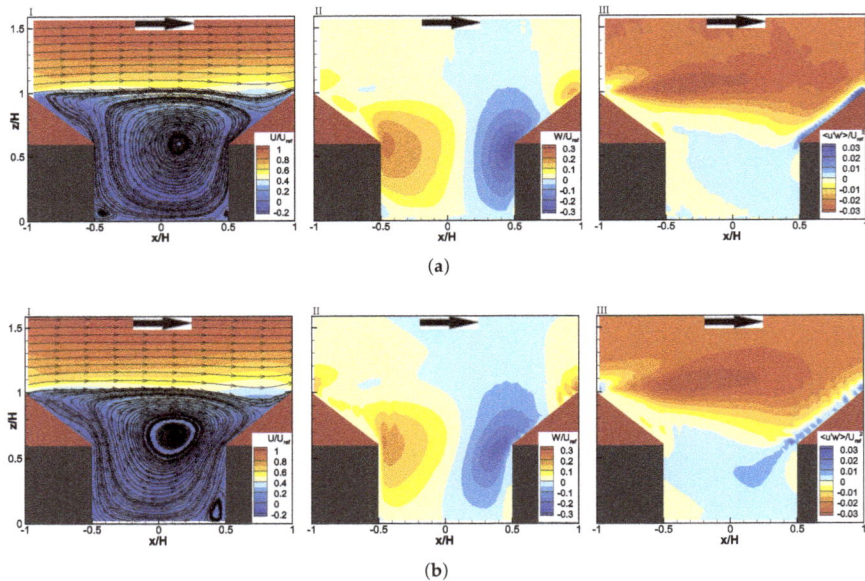

(a)

(b)

Figure 3. The mean dimensionless streamwise velocity with streamlines (first column) and vertical (second column) velocity, and momentum flux (third column) for (**a**) PIVa; and (**b**) LESa.

This corresponds with the notably streamwisely tilted pattern of the simulated vertical velocity field in the second column of Figure 3b, compared to the experiment. Another important observation can be interpreted from the fields of the momentum flux (third column in Figure 3a,b), namely the LES very well predicts the vortex shedding in the roof-top region but it over-predicts the positive streamwise momentum flux within the canyon. As will be shown later in Section 5.3, the vortex in the roof-top region is part of a rather large-scale and smooth up-roof flow, which is well simulated by the LES in general. The zone inside the canyon, on the other hand, consists of many small vortices and shear patterns under the spatial resolution of the LES grid.

4.3. Validation Metrics

A set of general performance metrics to quantitatively assess the agreement between a model and an observation was established by the U.S. Environmental Protection Agency [38] to provide a common framework for various air dispersion model comparisons. The metrics were designed to compare the concentration values by means of their mean or maximum values; they specifically target statistical quantiles and percentiles in order to assess the performance of the model. Only some of the metrics were selected for this paper. Since we work solely with velocity-based data, we modified the following formulae with respect to the used quantities X ($X = U, W, < u'w' >, TKE$):

Fractional bias (FB):

$$\textbf{FB} = 2\frac{\overline{X_e} - \overline{X_m}}{\overline{X_e} + \overline{X_m}} \tag{2}$$

Root normalised mean square error (RNMSE):

$$\textbf{RNMSE} = \sqrt{\frac{\overline{(X_e - X_m)^2}}{\overline{X_e}\,\overline{X_m}}} \tag{3}$$

Geometric mean bias (MG) and geometric variance (VG):

$$\textbf{MG} = exp\left(\overline{\ln X_e} - \overline{\ln X_m}\right), \textbf{VG} = exp\left[\overline{(\ln X_e - \ln X_m)^2}\right] \tag{4}$$

Pearson correlation (r):

$$\textbf{r} = \frac{\overline{X_e - \overline{X_e}}\ \overline{X_m - \overline{X_m}}}{\sigma_e \qquad \sigma_m} \tag{5}$$

where the subscripts e and m refer to the *experiment* and *modelled* data, respectively, $\overline{X}$ denotes the space-averaged value of the time-mean quantity over the restricted rectangular area and the angle bracket $X = \langle x \rangle$ denotes the time-averaged value at a point.

The application of FB, RNMSE, MG and VG to values achieving both positive and negative values can lead to a non-physically-justified experiment-simulation agreement since those values can compensate each other in Equations (2)–(5) as pointed out by Schatzmann et al. [15]. The visual comparison of the data (Figure 3) shows that there is no area where the experimental and modelled data differ in the sign and have a non-dimensionless magnitude higher than 0.1. Therefore, the authors applied those metrics to the absolute values of the investigated quantities with the awareness that the results of the metrics are affected by this phenomenon, but not significantly. An overview of these general metrics is listed in Table 3. The bold values denote the satisfaction of the particular criterion listed in the third column.

The general validation metrics are successfully satisfied in all cases. In comparison with the qualitative validation previously done on the velocity fields, the quantitative validation by means of the general metrics shows a surprisingly good agreement between PIVa and LESa. The main explanation could be that the general metrics work with fixed threshold values, which are not based on the uncertainty of the experimental data, and which were originally proposed for the concentration values. The velocity data more easily comply with these not-so-stringent criterion thresholds. We propose, at least for the Pearson correlation coefficient, that the acceptance criterion should be 0.9 instead of 0.8. The uncertainty itself can be introduced via the *Hit rate* metric, specifically designed by VDI's guidelines for validations of predicted mean velocity fields.

Table 3. General metrics applied to PIVa and LESa.

Metric	Ideal	Accepted	U/U_{ref}	W/U_{ref}	$<u'w'>/U_{ref}^2$	TKE/U_{ref}^2
FB	0	$\in(-0.3,0.3)$	**0.029**	0.209	**0.028**	**0.023**
RNMSE	0	$<\sqrt{1.5}$	**0.005**	**0.175**	**0.010**	**0.012**
MG	1	$\in(0.7,1.3)$	**1.097**	**1.194**	0.824	**1.023**
VG	1	<4	**2.001**	3.707	**2.657**	**1.067**
r	1	>0.8	**0.996**	**0.940**	**0.990**	**0.983**

The bold values denote the satisfaction of the particular criterion listed in the third column.

4.3.1. Hit Rate

The Hit rate metric applied in guideline VDI [39] for the validation of prognostic micro-scale wind-field models was used in order to quantitatively compare the experimental and modelled results. The Hit rate metric, q, is calculated by the equation using the normalised numerical model data M_i (modelled) and the normalised observed data E_i (experiment) as

$$q = \frac{1}{P}\sum_{i=1}^{P} q_i \quad \text{with} \quad q_i = \begin{cases} 1 & \text{for} \quad |\frac{M_i-E_i}{E_i}| \leq D \quad \text{or} \quad |M_i - E_i| \leq A \\ 0 & \text{else} \end{cases} \tag{6}$$

where D and A represent the relative and absolute deviation, respectively, according to the guidelines [15,39], with P being the number of grid points within the restricted rectangular area. The Hit rate absolute deviation value, A, was attributed to the normalised measurement error obtained from the given quantity (see Table 2). The relative deviation limit $D = 0.25$ was adopted from Schatzmann et al. [15]. Both acceptance levels A and D are presented in the scatter plots in Figure 4 along with each quantity.

The algorithm (6) is applied to a limited region $(-0.5 < x/H < 0.5; 0 < z/H < 1.5)$. The LES computed grid (1120 spatial points) was fitted to the PIV measured grid (2400 points) by a bilinear interpolation in order to provide one unified spatial domain. The interpolation error was evaluated from a mutual comparison of the linear, cubical and spline version of the interpolating MATLAB procedure applied to typical comparative quantities U, W, $<u'w'>$ and reduced TKE. During the interpolation the values at the points located within the buildings were artificially set to zero. This resulted in lower interpolated values just next to the walls in comparison with those measured by the PIV. The points next to the wall were, however, excluded from Hit rate and other metrics. The difference between the type of interpolations achieved a normalised value of 0.0001. Therefore, the interpolation error is not involved in the absolute value of A.

The final Hit rate values between the LES simulation of the inter-canyons averaged data (labelled LESa) and the averaged values of the repeated experiments (PIVa) are listed in Table 4. Again, the bold values meet the acceptance criterion of **0.66** as defined in VDI [39] for each velocity component. We adopted this criterion in agreement with other studies (e.g., [15,20]) for all the investigated quantities.

Based on the numbers in Table 4, our LES simulation fails to simulate the mean velocity field for U and W, according to the established Hit Rate metric. On the other hand, the higher moments are seemingly calculated with a sufficiently high precision. However, without a proper simulation of the velocity gradient, the moments can not be captured well by LES. To understand the nature of the problem, the scatter plots of measured and simulated values from LESa are shown in Figure 4.

The scatter plot between the PIVa and the LESa values in Figure 4a is located along the 1:1 line, though the absolute deviation A (dashed and dotted lines) for the streamwise velocity component is so small that most of the points do not fall in the prescribed region. Most of them fulfil only the relative deviation criterion D (dashed lines) in Figure 4a. This leads to a preliminary conclusion that the LES model fails in the prediction of the very low velocity values inside the canyon where small-scale vortices and intense shear exists, but it succeeds in the prediction of the fast velocity flow above the canyon.

The Hit rate scatter plot illustrates that the streamwise and vertical velocity components are subtly and moderately underestimated by the LESa according to Figure 4a,b, respectively. The momentum flux (Figure 4c) and the reduced *TKE* (Figure 4d) are predicted very well and reach a high Hit rate score. The additional discussion on the Hit rate results can be found in Section 6.

Table 4. Hit rate metric *q*.

Quantity	Full Street Canyon
U	0.59
W	0.36
$< u'w' >$	**0.83**
TKE	**0.83**

The bold values meet the acceptance criterion of **0.66** as defined in VDI [39] for each velocity component.

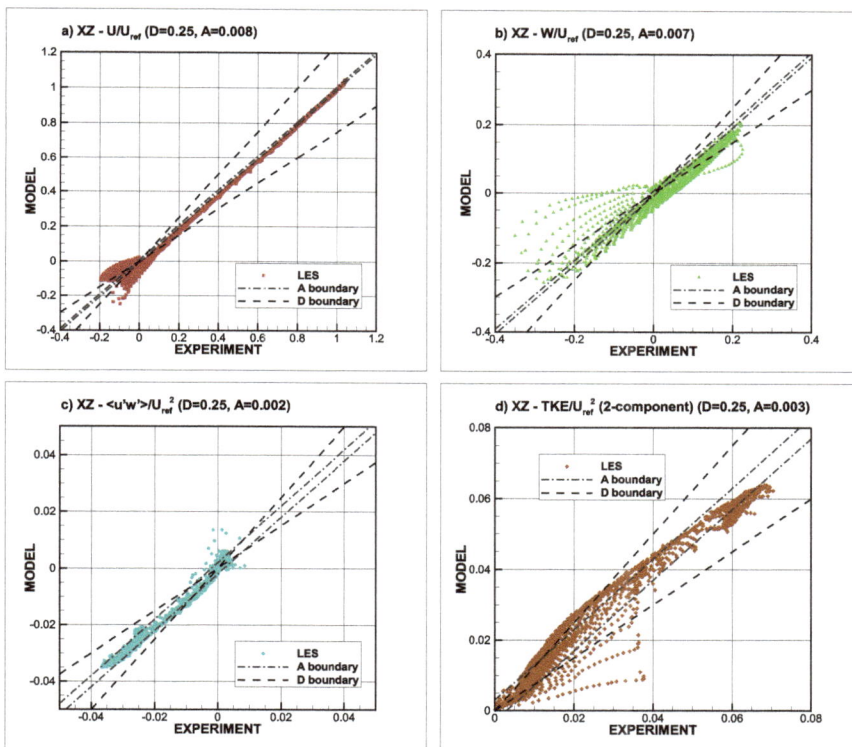

Figure 4. The Scatter plots of the measured (PIVa) and simulated (LESa) values obtained within the street canyon for the dimensionless (**a**) streamwise velocity; (**b**) vertical velocity; (**c**) momentum flux and (**d**) turbulent kinetic energy. Dashed lines denote the relative deviation *D*; dashed-and-dotted lines denote the absolute deviation *A*. 2400 points are plotted for each run.

4.3.2. Spatial Hit Rate

The STD of each quantity significantly differs from position to position in the street canyon due to the influence of long-term, large-scale flow structures on the stationarity of the data. For example, in our case, the values of the absolute deviation *A*, as derived from the STD of the streamwise velocity, exhibit a spatial scatter of $A(\mathbf{x}) \in (0.001 - 0.040)$ inside the rectangular area. The area where the long-term structures occur (e.g., in the up-roof region) supposedly has a high STD, while the area of an intense small-scale mixing (e.g., canyon centre) exhibits low STD. If the measurement error is determined from a one-point measurement only, this reference point should be chosen with the utmost care.

To test the role of the space-dependent differences in the deviation criterion $A(\mathbf{x}) = A(x_i, z_i)$ for the validation, we included a particular measurement error into the Hit rate formula at each location separately as

$$q = \frac{1}{P} \sum_{i=1}^{P} q_i \quad \text{with} \quad q_i = \begin{cases} 1 & \text{for} \quad \left|\frac{M_i - E_i}{E_i}\right| \leq D \quad \text{or} \quad |M_i - E_i| \leq A(x_i, z_i) \\ 0 & \text{else} \end{cases} \tag{7}$$

The final *spatially sensitive* Hit rate agreement is shown in Table 5. The spatially sensitive Hit rate is supposed to be more precise since it reflects the spatial inhomogeneity of the flow. In our case, the spatial Hit rate on LESa is consistent with the standard Hit rate metric output. The larger $A(\mathbf{x})$ occurs mostly around the roof level, where velocities were predicted sufficiently well, and $A(\mathbf{x})$ remains small inside the canyon, where velocities were predicted poorly. Considering the fact that the rounding-off upward effect of the spatially averaged error $A = \overline{A(\mathbf{x})}$ poses less stringency on the Hit rate validation procedure in the inside-canyon area than a local $A(\mathbf{x})$, we found the standard Hit rate metric was resulting in slightly more optimistic results than the spatially sensitive Hit rate. Although the spatially sensitive Hit rate is considered by the authors of this paper as more precise, the standard single Hit rate validation metric does not deviate significantly in terms of results, and it is much more easily-applied. Hence, single Hit rate criterion is regarded as a suitable comparable tool for the comparison of a simulation with an experiment.

Table 5. Hit rate metric q without and with spatial sensitivity.

Quantity	LESa Standard	LESa Spatial
U	0.59	0.58
W	0.36	0.47
$< u'w' >$	**0.83**	**0.68**
TKE	**0.83**	**0.77**

The bold values meet the acceptance criterion of **0.66** as defined in VDI [39] for each velocity component.

4.4. Spectral Analysis

Both the TR-PIV experiment and the LES simulation have an advantage over the RANS and Gaussian models in providing the time-resolved data with a satisfactorily high temporal resolution, to which the spectral analysis is worth being applied. The power spectral densities were calculated as the square of the fast Fourier transformation of the streamwise velocity fluctuations time series. The resulting spectra were smoothed by non-overlapping rectangular blocks. The length of the blocks was exponentially increasing with the increasing frequency to get equidistant points on the logarithmic axis (about eight estimates per frequency decade). The comparison of the power spectral density obtained from the HWAc and the LESc time series at height $z/H = 2$ is shown in Figure 5. Contrary to the locations inside the street canyon, at this elevation Taylor's hypothesis on frozen turbulence can be considered. Above the reduced frequency of $n = fz/U = 0.1$, where f is the frequency, z is the elevation and U is the mean streamwise velocity component, there is agreement between the numerical and experimental data.

Since the LES is inherently limited for a correct simulation of the whole turbulent spectral range, it is, by definition, capable of simulating only *low-frequency ranges* and the *inertial subrange*, provided the grid resolution is high enough. The sub-grid vortices are only modelled ones, leading to the energy deficiency in the *high-frequency tail* of the spectral density plot ($n > 10$). Still, the LESc spectra follow the well-known Kolmogorov $-2/3$ law ($-5/3$ in non-weighted representation), as depicted by the solid line in Figure 5.

On the *low-frequency tail* of the LESc spectra, the sudden drop in spectral energy at approximately $n = fz/U = 0.1$ is clearly notable in comparison with the dashed Karman's theoretical curve. The drop was detectable in all the LES canyons. With a LES periodic domain length of $16 H$, only structures up to $8 H$ can be safely computed. However, structures of $16 H$ can exist and can in fact be very strong. The $8 H$ wave length corresponds with the dimensionless frequency $n = fz/U \approx 0.07$.

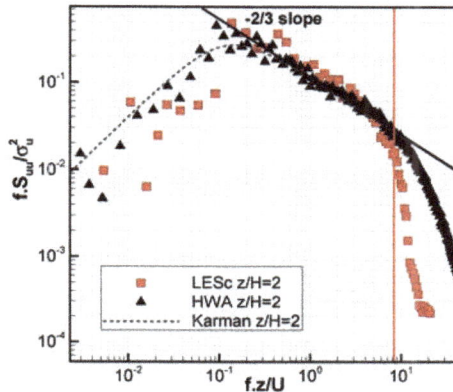

Figure 5. Comparison of the energy spectral density of the streamwise velocity at position $x/H = 0$, $z/H = 2$ for HWA (black triangles) and LESc (red squares). The solid black line represents the slope of the Kolmogorov $-2/3$ law, the dashed grey line represents the Karman's Theoretical curve. The solid vertical red line indicates the smallest scale not strongly affected by the sub-grid model and discretization errors with $\Lambda_{LES} \approx 0.24H = 12$ mm.

5. Results: Model Performance Assessment for Intermittent Flow

5.1. Quadrant Analysis

The quadrant analysis allows for a deeper insight into the dynamical behaviour of the flow, since it groups certain events based on their specific contribution to the total momentum flux, and provides their statistical overview. To be more specific, we obtain information about a particular contribution of the momentum flux $< u'w' >$ from the particular flux direction (i.e., quadrant) with the quadrant analysis. The definition of the quadrants, derived from the scatter plot of the streamwise and vertical velocity component fluctuations, was inspired by Willmarth and Lu [40]:

- $u' > 0, w' > 0$ - outward interaction (τ_1),
- $u' > 0, w' < 0$ - sweep (τ_2).
- $u' < 0, w' < 0$ - inward interaction (τ_3),
- $u' < 0, w' > 0$ - ejection (τ_4),

The particular contribution of the i-th quadrant τ_i to the total momentum flux $\tau_{xz} = < u'w' >$ is obtained from a formula of weighted average:

$$\tau_i = \frac{< u'w' >_i \cdot n_i}{N_{total}} \tag{8}$$

where $< u'w' >_i$ means the time-averaged momentum flux within the i-th quadrant, n_i is the number of events belonging to the i-th quadrant and N_{total} is the global number of events recorded during the time period $t^*_{PIVa} = 1295$ and $t_{LESa}* = 1528$. The vertical profiles of the total momentum flux, $< u'w' >$, with each of the event contributions are plotted in Figure 6a. The total momentum flux is a simple sum of all the contributions.

First, it is worth pointing out that the peak shape (at $z/H = 1.1$) is quite different from the peak shape of the street canyons with the same aspect ratio, but with flat roofs (e.g., [7,25,41]). While the peak of the profile of the total momentum flux is narrow and located just above the roof-top of the street canyons with flat roofs, the pitched roofs generate a vertically more extensive shear layer and consequently produce a stronger exchange between the canyon cavity and the boundary layer aloft (for more details see [24]).

The quadrant analysis confirms that the sweep events (orange triangles in Figure 6a) and ejection events (green triangles in Figure 6a) clearly contribute the most to the total momentum flux, while the inward (light blue triangles in Figure 6a) and outward (dark blue triangles in Figure 6a) interactions are negligible. As the contribution of the sweep events is forecasted by LESa (solid lines) extraordinarily well in both the peak elevation and the magnitude above the roof top, the moderate over-estimation of the ejections leads to a slight over-prediction of the total momentum flux at the peak level. Again, the general tendencies of strong sweep and ejection events are predicted extremely well.

The momentum flux exhibits a significant diversion from the measured data in the lower part of the canyon ($z/H < 0.5$). The over-prediction by LESa is caused by a strong contribution from the inward interaction quadrant compared to the completely missing inward interaction in the case of PIVa. The LES with a grid resolution of 2 mm is probably unable to resolve the small-scale shear motion properly. The preliminary results from a finer grid resolution of the LES (1 mm) indicates much better agreement (not shown here).

Considering the hypothesis that a model, primarily, has to correctly simulate the momentum flux in order to predict velocity properly [18], the over-prediction of the inward interaction provides a possible explanation why the LESa fails to correctly predict the mean velocity profiles and 2-D fields, as presented in previous sections in Figures 2 and 3. Figure 6b reveals that the increase of the inward interaction significantly correlates with the flow pattern of the second POD mode calculated from LESa data, indicated by the red streamlines (the details are explained later in Section 5.3).

The sweep and ejection events often travel in a compact shape across the street canyon. For a better demonstration of these phenomena, it is convenient to plot their conditionally averaged shape as derived from several strong events. An example of a typical sweep from PIVc and a typical sweep from LESc, in the canyon, is presented in Figure 7a,b, respectively. The negative (orange) values denote the region where a strong sweep takes place whereas the positive (green) values denote the existence of a dominant ejection. Further, we evaluated the spatial distribution of each quadrant in the planar vector field xz by means of a spatially relative cumulative contribution $\varepsilon_i(t^*)$ from particular event $\tau_i(t^*)$ to the total momentum flux $\tau_{total}(t^*)$ according to the formula:

$$\varepsilon_i(t^*) = \frac{\tau_i(t^*)}{|\tau_{total}(t^*)|} \cdot 100\% \tag{9}$$

$$\tau_i(t^*) = \sum_{k=1}^{K} \sum_{l=1}^{L} \tau_{kl,i}(t^*) \tag{10}$$

$$\tau_{total}(t^*) = \sum_{k=1}^{K} \sum_{l=1}^{L} |\tau_{kl}(t^*)| \tag{11}$$

where $k = 1, 2...K$ and $l = 1, 2...L$ are the row and column numbers in the vector field, $|...|$ representing the absolute values of the inner product. The total momentum flux $\tau_{total}(t^*)$ in Equation (11) is the sum of absolute values from each quadrant at each time.

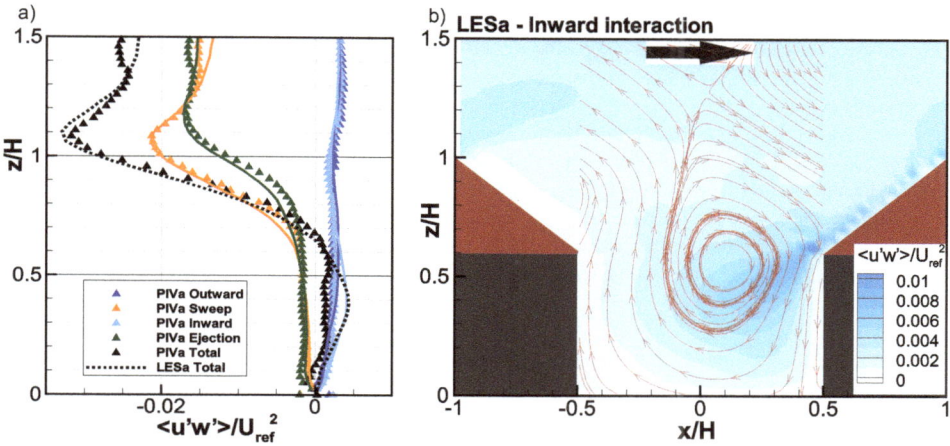

Figure 6. (**a**) The vertical profiles of the total vertical momentum flux and the particular event contributions to the total momentum flux from LESa (lines) and PIVa (triangles) at position $x/H = 0$. The colours correspond to the schematic diagram of the quadrant analysis in the bottom left corner in Figure 7; (**b**) 2D plane of the inward interaction from LESa (light blue contours) with the red streamlines of the second POD mode from LESa.

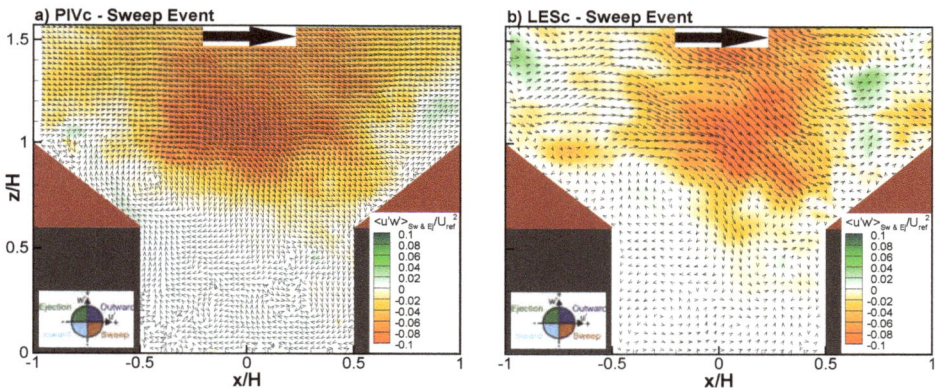

Figure 7. The characteristic examples of the sweep event (orange) at $y/H = 0$ for (**a**) PIVc and (**b**) LESc. A definition of the intensive events is explained in the following procedure and Figure 8.

The calculation of the contribution of the particular event $\varepsilon_i(t^*)$ from Equation (9) revealed that the sweep and ejection events occasionally represent up to $\varepsilon_{2,4}(t^*) > 80\%$ of the total momentum flux (see the red and green lines in Figure 8). Such a high contribution agrees with the many experimental and numerical studies dealing with the role of sweep and ejection in turbulent boundary layers (e.g., [40,42–44]). The specific time instants with $\varepsilon_2(t^*) > 80\%$ from the sweep quadrant serve as the data input for the conditional averaging of the momentum flux field in order to achieve a typical intensive sweep event in the canyon as shown in Figure 7.

The clear tendency for both events to pass the street canyon in an alternating fashion can be seen also in Figure 8a,b. This is supported by the large and negative value of the correlation coefficient between the relative contributions of sweeps and ejections, $R_{sw,ej} = -0.90$. When the sweep enters the canyon, the ejection is suppressed and vice versa. The frequency of such events occurring was analysed from the time series of their relative contribution $\varepsilon_i(t^*)$ defined by Equation (9). The mean value derived from all instantaneous values $\varepsilon_i(t^*)$ was subtracted from the time series and the spectral function was obtained by the same algorithm used in the case of velocity fluctuations described as in Section 4.4. The spectra revealed that the characteristic frequency of this pseudo-wavy pattern occurs approximately at $fz/U = 0.2 \approx 6H$, taking into account an estimated convective velocity of the quadrant events [27] for both the PIVc and the LESc (Figure 9). The sweep and ejection events pass the canyon or are induced by the canyon geometry with the same pseudo-frequency and with the same relative intensity in both the experiment and the simulation. It is necessary to note that sweep and ejection events are not part of any vortex since their wavelength is larger than either the vertical or lateral dimension of the wind tunnel. In conclusion, the quadrant analysis proves that the LES is capable of reliably modelling large intermittent and organised structures in the flow above the obstacles.

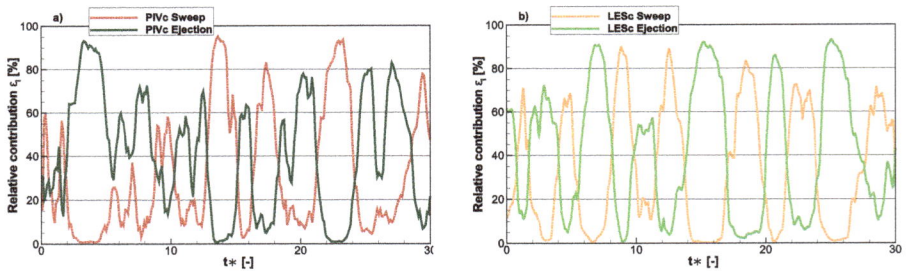

Figure 8. The segment with the time evolution of the sweep (red lines) and ejection (green line) relative contribution to the absolute total momentum flux for (**a**) PIVc and (**b**) LESc.

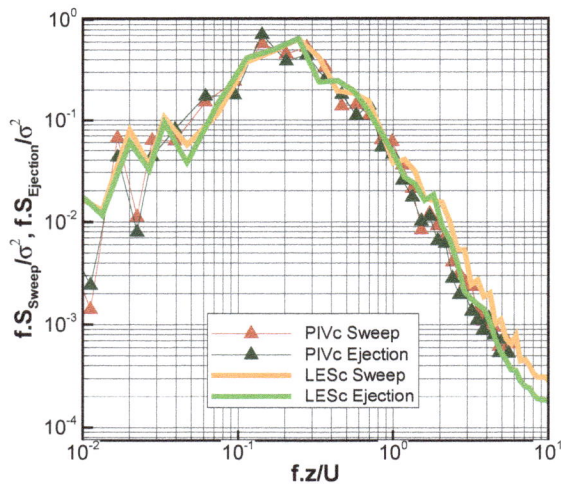

Figure 9. The spectral density plot of the relative contribution of the sweep and ejection events to the total momentum flux for PIVc (triangles) and LESc (lines).

5.2. Spatial Correlation

Spatial correlation helps to explore the link between transient flow dynamics across the field in a statistical sense. The normalised correlation coefficient at the point $\mathbf{x} + d\mathbf{x}$, where $\mathbf{x}$ is the reference point, was calculated according to formula

$$r_{uu}(\mathbf{x}, d\mathbf{x}) = \frac{1}{N-1} \sum_{i=1}^{N} \frac{u'(\mathbf{x}, t_i) u'(\mathbf{x} + d\mathbf{x}, t_i)}{\sigma_u(\mathbf{x}) \sigma_u(\mathbf{x} + d\mathbf{x})} \tag{12}$$

where N is the number of time-steps (snapshots) obtained from the continuous PIVc measurement and LESc simulation, and σ_u is the standard velocity deviation. For both PIVc and LESc vector fields, we calculated the correlation between the time series at the chosen reference point $\mathbf{x} = (x/H = 0, z/H = 1.25)$ and all other spatial locations. The correlation coefficients were calculated for velocity fluctuation u' and w', and for the momentum flux fluctuation $(u'w')'$. Since the spatial quadrant analysis (Section 5.1) reveals that appreciable regions of high momentum fluxes (i.e., the sweep and ejection events) pass the canyon, we aim to investigate the correlation between the time-series of these momentum fluxes $u'w'$ instead of a simple correlation between the quantity u' and w'. The contour-plots in Figure 10 display typical patterns of the spatial correlation above the street canyon similar to the one obtained in, e.g., Michioka et al. [7]. Regarding the qualitative comparison, the plots confirm a good agreement between the experimental (Figure 10a) and the numerical (Figure 10b) results.

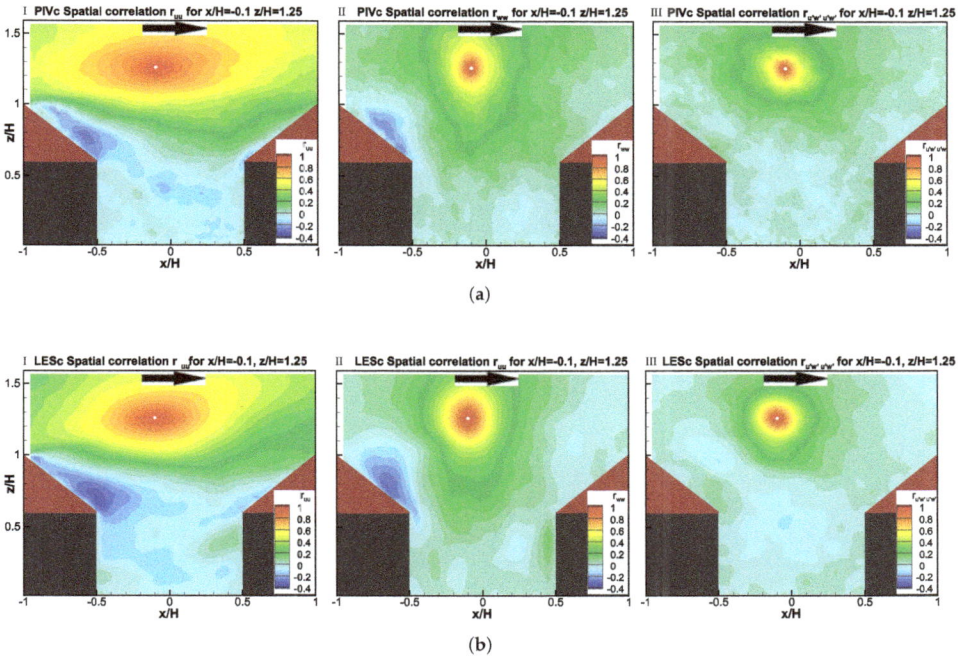

Figure 10. Spatial correlation of the normalised streamwise velocity fluctuation r_{uu} (first column), of the vertical velocity r_{ww} (second column) and of the fluctuation of the momentum flux $r_{u'w'u'w'}$ (third column) for the chosen point at $x/H = 0$, $z/H = 1.25$ for (**a**) PIVc and (**b**) LESc. The LES plot is interpolated onto the PIV grid (4977 points).

The contour-plots allow us to directly compare the correlation coefficient fields from the PIVc and LESc. To express the degree of similarity between the plots in Figure 10a,b, we calculated the *final correlation coefficient* via the following formula

$$R_{jj}(\mathbf{x}) = \frac{1}{P-1} \sum_{i=1}^{P} \frac{r_{jj,PIV}(\mathbf{x},d\mathbf{x_i})r_{jj,LES}(\mathbf{x},d\mathbf{x_i})}{\sigma_{r_{jj,PIV}}\sigma_{r_{jj,LES}}} \tag{13}$$

where $P = 4977$ is the number of locations $\mathbf{x}$ in the measurement area and $j = u, w$ or $u'w'$.

The results of R_{uu}, R_{ww} and $R_{u'w'u'w'}$ obtained for the shared grid points of both PIVc and LESc (together 132 points) are plotted in Figure 11. The appropriate final correlation coefficient R_{jj} is presented by means of the coloured scale and is assigned to a specific location (square) within the investigated area. Figure 11 suggests that the greatest deviation in the LES predictions, compared with the experimental results, is located mainly inside the street canyon, close to the vicinity of the walls and near the canyon bottom, especially in the case of the momentum flux, $R_{u'w'u'w'}$. Again, the LESc apparently fails to predict the recirculation zone together with the small-scale structures within the canyon cavity, probably due to the coarseness of the grid resolution. Other lower correlations can be seen in the vicinity of the upstream building roof, where small-vortex shedding occurs. This, in accordance with previous results of the Hit rate validation metric, confirms that the transient dynamics, and consequently the ventilation processes, at these critical areas will not be predicted by the presented LES model in the proper manner.

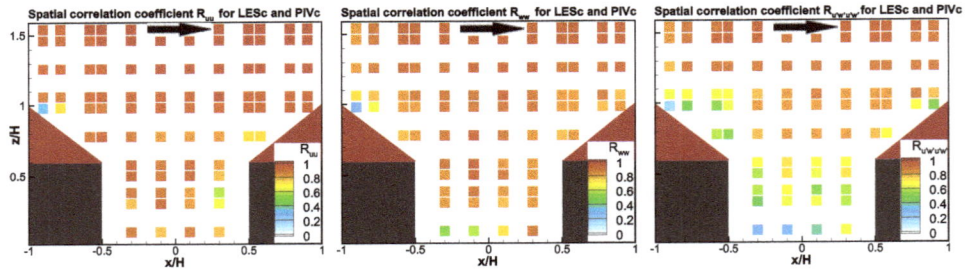

Figure 11. The final correlation coefficients between the PIVc and the LESc spatial correlation fields of the streamwise velocity R_{uu} (first column), of the vertical velocity R_{ww} (second column) and of the momentum flux $R_{u'w'u'w'}$ (third column). Each point shows the comparison based on the spatial correlation of the correlation coefficient fields between PIVc and LESc at one of the 132 shared points of the PIV and LES grids.

The range of the mean values of the correlation coefficient that corresponds to the individual canyons, and which is spatially-averaged over the entire investigated area within each canyon (always 132 shared points between the PIV and LES grids) for u, w and $u'w'$, are listed in Table 6. The relatively high unified values in Table 6 indicate that the correlation method for a single velocity component u or w is rather tolerant in measuring the degree of similarity between the predicted and observed short-time intermittent motions within the shear layer at the canyon bottom. In other words, this spatially-average single number from each canyon will not reveal the subtle differences between the measured and forecasted transient dynamics and will yield a rather high percentage of agreement. We presume that validations based solely on the spatially-averaged correlation of u and w would lead to an overestimation of the LES model with respect to the prediction of intermittent motions. However, the detailed picture of locally dependent final correlations, as depicted in Figure 11, is a suitable tool for the detection of problematic areas in the LES prediction.

Table 6. Final correlation coefficient spatially-averaged over the each canyon.

Quantity	LESc1-8
R_{uu}	0.92–0.94
R_{ww}	0.90–0.92
R_{uwuw}	0.82–0.85

5.3. Proper Orthogonal Decomposition

Contrary to a one-point spatial correlation, which simply carries out the statistical behaviour of the fluctuations with respect to one arbitrarily chosen reference point, the POD groups the correlated motions into contextual merit-based ensembles and thus provides a certain insight into the coherent intermittent dynamics [45]. Simply stated, for fluid mechanic applications, POD assembles the intermittent events together by finding their most appropriate representation (basal vectors). This representation expresses the highest TKE content in each vector from a statistical point of view. The POD may be applied for both the experimental data (e.g., [20,24,25,46]) and the data obtained from CFD simulations (e.g., [47]). Hence, it provides a valuable tool for comparing the results from numerical modelling with those from a physical experiment.

The POD was proposed by Lumley [48] as a tool for the detection of large coherent structures in the flow. This method is based on the assumption that we can extract special functions (alias basal vectors $\{\boldsymbol{\varphi}_m\}$) from the chaotic turbulent flow, those which possess the mathematical description of the coherent structures as the most probable flow features. It is thus possible to describe every value of the instantaneous velocity fluctuation at every instant of time and space $u'(\mathbf{x}, t)$ by a set of new basal vectors $\boldsymbol{\varphi}_p$ and their corresponding expansion coefficients a_p, where $p = 1, 2, ...2P$ is equal to double the number of grid points P, as

$$u'(\mathbf{x}, t) = \sum_{p=1}^{2P} a_p(t)\boldsymbol{\varphi}_p(\mathbf{x}) \tag{14}$$

$$\boldsymbol{\varphi}_p^T \boldsymbol{\varphi}_q = \delta_{pq} \tag{15}$$

$$corr(a_p, a_q) = \delta_{pq} \tag{16}$$

where $corr$ represents the correlation coefficient and δ_{pq} the Kronecker delta achieving value of 1 for $p = q$ and 0 otherwise.

The basis $\{\boldsymbol{\varphi}_p\}$ meets the orthogonality and normality criteria, so that the vectors $\boldsymbol{\varphi}_p$ (i.e., POD modes) are perpendicular to each other and are properly normalised. In the case of POD, this basis is not chosen *a priori* as in the Fourier or the wavelet analysis, but according to the input data. It is important to note that the POD modes are functions of space, not of time. The *p-th* eigenvalue λ_p corresponds to the *p-th* POD mode and contains information about the contribution of the *p-th* mode to the total turbulent kinetic energy (TKE). A reordering of the eigenvalues based on the descending TKE (and the ascending order of p)

$$\lambda_1 > \lambda_2 > ... > \lambda_{2P} \tag{17}$$

reveals the most dominant modes in the flow according to their relative contribution Π to the TKE [49]:

$$\Pi = \frac{\lambda_p}{\sum \lambda_p} * 100\% \tag{18}$$

We applied the POD to the horizontal and vertical velocity components inside a limited rectangular area of the street canyon. Data from both the LES and PIV were normalised by the reference velocity. Since the number of grid points obtained from PIV post-processing (2400 points) differs from the number of grid points in the LES simulation (1120 points), and also because the slightly different dimensions of the investigated areas (due to the different outer boundaries of the grids) might cause certain differences in the POD results, we tested the sensitivity of the POD to these aspects first. We observed that it is better to keep the dimension of the region identical, irrespective of the differences in spatial resolution. It is also better to interpolate only the POD output, rather than the input data. Thus, every POD mode was calculated for the original grid, and then each LES mode φ_p was interpolated to the PIV grid.

In order to illustrate the spatial distribution of the modes φ_p, the four most dominant POD modes for PIVa and LESa in terms of *TKE* are depicted in Figure 12a–d. It has to be emphasized that the black lines with the arrows, the so-called streamlines, were painted manually, hence their density and precise location serve for display purposes only.

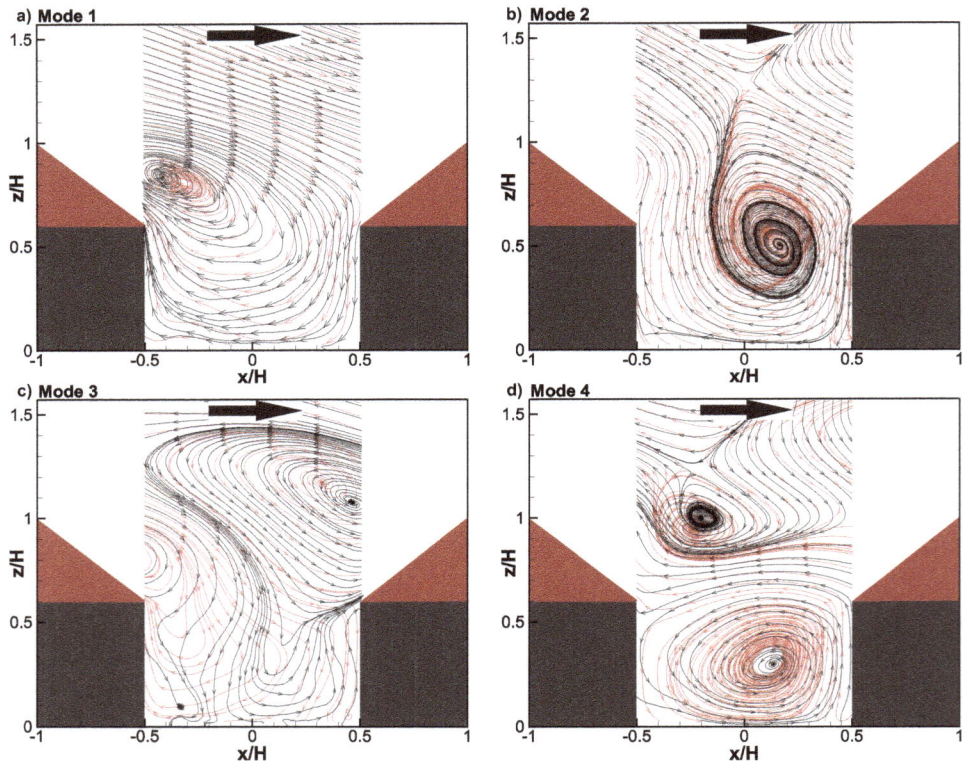

Figure 12. The POD modes for the restricted rectangular area within the street canyon for PIVa and LESa. (**a–d**) correspond to Mode 1–4. The black streamlines denote PIVa, the red streamlines denote LESa. The LESa modes are interpolated on the PIV grid.

The first mode in Figure 12a displays the vortex behind the roof, which is well pronounced and dominant in the case of the intermittent flow. Kellnerova et al. [24] proved that this vortex accurately captures the dominant street-canyon flow dynamics. The LESa version, involving the data from all of the simulated canyons, shows a good agreement with the PIV results with respect to the core of the vortex. This is a very important finding, since the first mode contributes the most (by 32%) to the total turbulent energy budget. The deviation of this roof-vortex core of PIVa from that of LESa ($\Delta x/H = 0.1$) can be considered small enough, resulting in an overall consistency of the modal shape and leading to a satisfactory validation.

Regarding the second mode (Figure 12b), the inflection point at the area above the roof is predicted precisely. The LES seems to push the center of the vortex between the walls. The recirculation vortex, containing 7% of total TKE, is the main dynamical pattern responsible for the acceleration and deceleration of the mean recirculation vortex. The shape of the recirculation zone also moderately deviates from the experimental results since the LES enhances the backward flow near the downstream edge of the vortex. This backward flow contributes to the inward momentum flux (see Figure 6b) leading to an elevation and an inclination of the primary recirculation zone (see Figure 3a,b).

The third mode (Figure 12c) shows definite deviation from the experimental results concerning the curved trajectory of wind in the whole lower canyon. The LES's third mode further contributes to the inward interaction momentum flux in this area. Meanwhile, the other flow patterns, such as the vortices above the windward and leeward roof, are appropriately predicted. The fourth mode (Figure 12d), again, is very well forecasted. The fourth mode represents the late phase of the recirculation vortex, when the vortex is moving toward the ground. Although the latter two modes contribute only 6% and 4%, respectively, to the total TKE, Kellnerova et al. [24] showed that they occasionally play a significant role in street-canyon flow dynamics.

To evaluate the agreement between the simulated POD modes and the measured ones more clearly, the scatter plots for the first four dominant modes are displayed in Figure 13. The excellent similarity between the experiment and model for the streamwise and vertical velocity fluctuations, u'/U_{ref} and w'/U_{ref}, respectively, is further confirmed quantitatively by the Hit rate metrics results, and are listed in Table 7. The Hit rate score is evaluated in a similar way as in Section 4.3.1, i.e., the absolute deviation boundary A is taken as $A = 0.01$ and $A = 0.005$ for streamwise fluctuation and vertical fluctuation, respectively. These values of A are calculated as the mean standard deviation of corresponding velocity fluctuation. The only difference is that here, the member q_i in Equation (6) achieves a value of one when both the streamwise and vertical fluctuations satisfy the criteria for the streamwise and vertical deviation boundaries simultaneously.

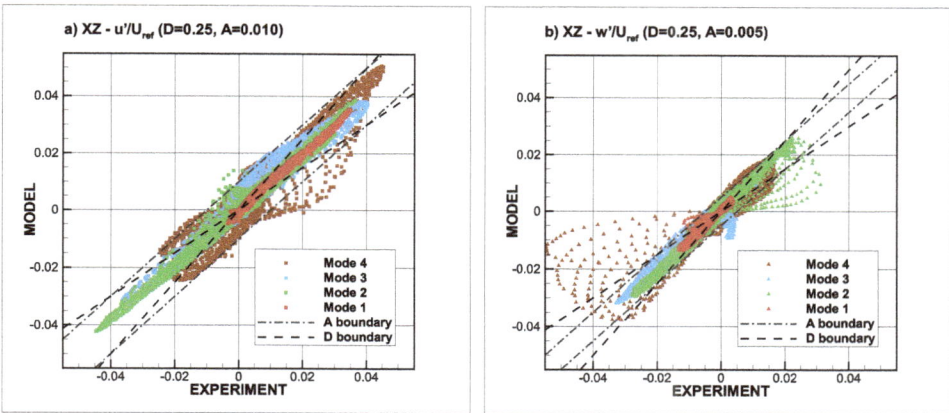

Figure 13. Scatter plots for modes 1-4 between the PIVa (experiment) and LESa (model) for (a) u'/U_{ref} and (b) w'/U_{ref}.

Table 7. Hit rate metric *q* for POD modes.

Quantity	Mode 1	Mode 2	Mode 3	Mode 4
u'/U_{ref} and w'/U_{ref}	**0.99**	**0.96**	**0.90**	**0.79**

The bold values meet the acceptance criterion of **0.66** as defined in VDI [39] for each velocity component.

Figure 14 displays the relative and cumulative contribution of each POD mode to the *TKE*. The relative contribution of the first mode calculated from all-the-canyons data (LESa), matched well with the experiment. Generally, in the case of higher modes, which mostly represent the vortices inside the canyon, the LES predicts their systematically higher percentages of the contribution in terms of the turbulent kinetic energy. The LES forecast therefore exhibits a higher degree of coherency in correspondence with the faster rate of the cumulative contribution convergence. This indicates that the LES generates either a higher number of vortices or a higher rotational speed of the vortices within the recirculation zone compared to the experiment, especially in connection with the vortex associated with the second and third POD mode. Incidentally, Figure 6b illustrates the close relationship of the second POD mode with the inward interaction. The stronger second and third modes in LES consequently enhance the inward interaction of the momentum flux at the lower part of the canyon, thereby leading to a higher positive total momentum flux (the third column in Figure 3) and a streamwise inclined recirculation pattern (the first and the second column in Figure 3) that modifies the velocity profiles in the lower canyon part (Figure 2a).

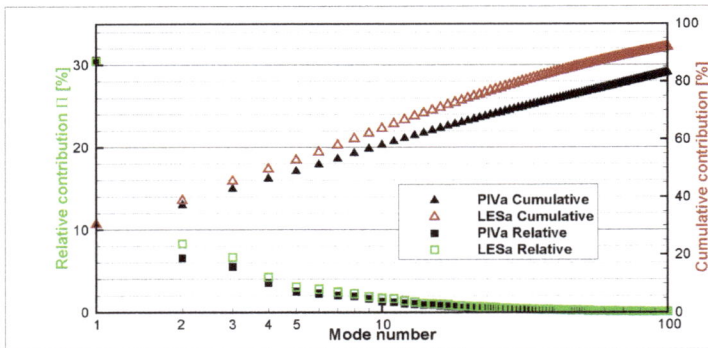

Figure 14. The comparison of the relative (squares) and cumulative (triangles) contributions of the first 100 individual POD modes of PIVa and LESa to the TKE.

6. Discussion on Validation Metrics

Since the general validation metrics Section 4.3 are originally proposed for the concentration data, the velocity data easily comply with their criterion thresholds. As mentioned previously, we therefore propose for the Pearson correlation coefficient to increase the acceptance criterion to 0.9, instead of 0.8.

The Hit rate metric points out issues encountered while using a PIV measurement technique for a street-canyon model topology. Firstly, the Hit rate benefit lies in a simple implementation of the measurement error, as obtained from repeated measurements at a well-chosen reference position or in the investigated region. With the PIV and other multi-points techniques, however, it might be difficult to achieve a sufficiently high number of repetitive runs, while maintaining a sufficiently long acquisition time. Hence, the results from the Hit rate method should not be regarded as the "benchmark" validation metric before the PIV runs provide enough repetitive data. Near-wall areas, where the PIV post-processing is likely to substitute a large portion of spurious vectors, have to be cautiously dealt with.

Secondly, the Hit rate is extremely sensitive to the magnitudes of the preselected data. In our case, the relative deviation boundary, $D = 0.25$, is a much less stringent criterion than the absolute deviation boundary, as determined from a very low experimental uncertainty. If the data are selected from locations containing only large values (above the roof area, for example), the majority of the data fits into the relative deviation limits. This factor can be effectively demonstrated for the above-roof region $(-0.5 < x/H < 0.5, 1 < z/H < 1.5)$ with large values of the streamwise velocity, U, the momentum flux and reduced turbulent kinetic energy, and with almost negligible vertical velocity, W. The final Hit rate score achieves a value of 1 for U/U_{ref}, TKE/U_{ref}^2 and $< u'w' > /U_{ref}^2$, seemingly suggesting the ideal model situation. On the other hand, the vertical velocity component, W/U_{ref}, achieves an extremely low value of 0.20, indicating a rather poor LES performance (see the left column in Table 8).

Table 8. Hit rate metric q.

Quantity	Above-Roof Region	Windward Region
U	**1.00**	0.52
W	0.20	**0.75**
$< u'w' >$	**1.00**	**0.68**
TKE	**1.00**	**0.85**

The bold values meet the acceptance criterion of **0.66** as defined in VDI [39] for each velocity component.

In the region containing a small streamwise velocity component, U, and the large quantities, W, TKE, and $< u'w' >$, as in the windward region of the canyon ($0.25 < x/H < 0.5, 0.25 < z/H < 1$), the results would be different. The Hit rate score would achieve high values for W/U_{ref}, TKE/U_{ref}^2 and $< u'w' > /U_{ref}^2$, while the streamwise component U/U_{ref} would not satisfy the threshold criterion (see the right column in Table 8). The result suggests that the validation technique should employ data from all the possible locations available from both the experimental and numerical output.

Hypothetically, when using precise measurement techniques in a well-controlled experiment (ensuring a low uncertainty), a fast-motion flow region would always comfortably satisfy the Hit rate relative criterion, whereas a slow-motion flow region would hardly meet the low experimental uncertainty. An absolute deviation criterion depending on the precision of the measurement technique that varies from study to study may make the inter-studies comparison of the Hit rate scores problematic. It might be more practical to pose one unified absolute deviation criterion A in the form of a percentage (e.g., 2% of the nominal velocity, $A = 0.02U_{ref}$).

7. Conclusions

The advanced validation procedures were performed, and others introduced, in order to compare the time-resolved LES simulations with the time-resolved PIV measurements on street-canyon topology. The validation procedure followed the general procedures recommended by Schatzmann et al. [15], Cox and Tikvart [38], which employ time-space averaged statistics. Further, an innovative approach based on coherent detection methods was implemented to compare the predicted and observed values of both the intermittent large- and small-scale features in the flow.

The general metrics, encompassing the time-space averaged data, were fulfilled for all the tested quantities. The LES performance failed to satisfy the minimum Hit rate criterion (0.66) in both the mean streamwise and vertical velocity components, while the quantities based on the fluctuations provided a satisfactory agreement. The strengths and weaknesses of the Hit rate metric were thoroughly discussed.

The large-scale structures in the up-roof area were predicted accurately, in contrast to the small-scale vortices near the street-canyon bottom-area. The discrepancies between the modelled and the measured mean velocities at the bottom area, where a strong shear takes place, are apparently associated with the deviation in the momentum flux.

The advanced validation techniques, the quadrant analysis together with POD, revealed that the enhanced inward interaction in the case of the LES simulation is associated with a more intensive vortex formation inside the canyon. Further, the quadrant analysis shows that the dominant flow structures in the up-roof region - the strong sweep and ejection events - have similar compact shapes and pass through the street canyon with identical frequency and intensity in both the simulation and the experiment. The detailed Fourier spectral analysis revealed an undefined drop of energy in LES in the energy-containing range across the entire time domain. Nevertheless, the inertial subrange was modelled extraordinarily well. Finally, the POD demonstrated that the shape of the coherent structures from the LES is strikingly similar to the shape of the observed structures. The Hit rate metrics applied to the POD modes showed an excellent agreement. The turbulent kinetic energy contained in the most dominant LES structure was identical with the measured one, while the less dominant LES structures exhibited a slightly more turbulent kinetic energy. This suggests that the simulated flow is more organised and less noisy than the flow observed in the experiment.

In an ideal case, the tested model should pass all the available validation metrics. However, in the real situation we might need to prioritize some metrics over others to prove satisfactory model performance in given applications. In our study we focused on the flow within an urban roughness layer, where the dynamical exchange of pollution and heat is a crucial factor. The dominant momentum events, which are correlated with heat and emission transport [50,51], can be quite easily described by the quadrant analysis in terms of contribution to TKE and their spectral characteristics, and we recommend to use it. The quadrant analysis can be executed either as one-point method or spatial multi-point analysis, as drafted in this paper.

If the synchronised multi-points data are at disposal, the POD would be a quick and useful tool to compare the large structures, both visually and numerically. On the other hand, using a spatial correlation might overestimate the LES performance with respect to the simulation of intermittent motions.

The validation methods presented, including the coherent detection methods, proved to be an indispensable tool for the proper validation of the time-resolved numerical codes. The analysis of the coherent features helped us to easily validate, qualitatively and quantitatively, the prediction against the observation, as well as to explain their possible contradictions. We demonstrated that although the modelling of sub-grid, small-scale structures inside a street canyon did not meet the Hit rate validation criteria, the dominant dynamics of the larger scale can be modelled in a satisfactory manner.

Author Contributions: R.K. researched the literature, designed the experiment, developed the code for coherent structured detection methods, executed the calculations and wrote the paper. V.F. performed the LES simulation. V.U. installed the PIV system and performed the experiment. K.J., Š.N. and H.C. suggested the type of the utilised methods, discussed the results and provided feedback for both the velocity and concentrations data. Z.J. supervised the project and corrected the paper.

Acknowledgments: The authors kindly thank the Czech Science Foundation GA CR-GA15-18964S for its financial support. Access to the CERIT-SC computing and storage facilities provided under the program Center CERIT Scientific Cloud, part of the Operational Program Research and Development for Innovations, reg. No. CZ. 1.05/3.2.00/08.0144, is greatly appreciated.

Conflicts of Interest: The authors declare no conflict of interest. The founding sponsors had no role in the design of the study; in the collection, analyses, or interpretation of data; in the writing of the manuscript, and in the decision to publish the results.

Abbreviations

The following abbreviations are used in this manuscript:

AIJ	Architectural Institute of Japan
CFD	Computational Fluid Dynamics
COST	European Cooperation in Science and Technology
CTA	Constant Temperature Anemometry
DNS	Direct Numerical Simulation
EPA	Environmental Protection Agency
HWA	Hot-Wire Anemometry
LES	Large Eddy Simulation
PIV	Particle Image Velocimetry
POD	Proper Orthogonal Decomposition
RANS	Reynolds-averaged Navier–Stokes
STD	Standard Deviation
TKE	Turbulent Kinetic Energy
VDI	Verein Deutscher Ingenieure (The Association of German Engineers)

Appendix A. TR-PIV System Description

An overview of the PIV parameters is listed in Table A1.

Table A1. Parameters of the TR-PIV.

Laser	Peagus-PIV-Diode-Pumped (Dual Head) Nd:YLF
Beam diameter	1.5 mm
Repetition rate	100 & 2 × 500 Hz
Illuminated Area	100 × 100 mm
Energy per cavity (at 1 kHz)	10 mJ
Camera sensor resolution	1280 × 1024 pxs
Interrogation area	32 × 32 pxs
Overlapping	50% (80 × 64 = 5120 vectors)
Final spatial resolution	1.2 × 1.2 mm
Acquisition time	16.3 s & 2 × 3.27 s = 22.7 s
Number of snapshots N	3 × 1634 = 4902
Grid points/reduced grid points P	4977/2400

Appendix B. LES Model Description

For the LES model CLMM, the filtered Navier-Stokes equations, with an eddy-viscosity sub-grid, and the continuity equations are considered in the present study as

$$\frac{\partial \bar{u}}{\partial t} + \nabla \cdot (\bar{u}\bar{u}) = -\nabla \bar{P} + \nabla \cdot (\nu_{\text{eff}} \nabla \bar{u}) + f_i, \tag{A1}$$

$$\nabla \cdot \bar{u} = 0, \tag{A2}$$

where the overbar represents the filtered quantity, u is the velocity vector, $\bar{P} = \bar{p}/\rho + \frac{2}{3}k_r$ is the modified pressure, $\bar{p}$ is the pressure and $k_r = \frac{1}{2}\text{Tr}(\tau) = \frac{1}{2}\text{Tr}(\overline{uu} - \bar{u}\bar{u})$, defined using the trace of the subgrid stress tensor τ, is the subgrid kinetic energy. As a knowledge of p and k_r is not needed, only the modified pressure is computed during the solution. The effective viscosity ν_{eff} in Equation (A1) is the sum of the kinematic viscosity ν and the sub-grid scale (SGS) eddy viscosity ν_{sgs}. The last term in Equation (A1), f_i, is the constant volume force representing the pressure gradient driving the flow. The value of f_i was determined by previous coarser resolution tests so that the reference time-averaged velocity is close to the experiment. The sub-grid scale eddy viscosity is computed by the σ-model from Nicoud et al. [32]

$$\nu_{\text{sgs}} = (C_\sigma \Delta)^2 \frac{\sigma_3 (\sigma_1 - \sigma_2)(\sigma_2 - \sigma_3)}{\sigma_1^2}, \tag{A3}$$

where $\sigma_1 \geq \sigma_2 \geq \sigma_3 \geq 0$ are the singular values of tensor $\nabla \bar{u}$, Δ is the filter width and C_σ is an empirical model constant. The value $C_\sigma = 1.35$ was used for the present computations.

The coupling of the velocity and pressure fields is performed using the fractional step (pressure correction) method from Brown et al. [52]. The time integration is based on the 3-stage Wray's Runge-Kutta method [53].

Appendix C. Measurement Uncertainty Assessment

Firstly, we attested the representativeness of each PIV run by means of a convergence plot (not shown). The convergence plots of the streamwise and vertical velocity were calculated as the relative deviation of the time-averaged velocity over the gradually increasing time interval from the time-averaged velocity over the entire measurement duration. We evaluated the maximum deviation from the mean velocity value after the inclusion of 75% of the snapshots.

The convergence plots verified that 1634 snapshots are a reasonable, but still minimum, number of snapshots for a reliable PIV performance with a maximum deviation from the mean velocity magnitude value of approx. 2% for both velocity components, U/U_{ref} and W/U_{ref}, 8% for the momentum flux,

$< u'w' > /U_{ref}^2$, and 7% for the reduced turbulent kinetic energy, TKE/U_{ref}^2. Thus, we can conclude that the PIVc runs lasting for 3.2 s contain a sufficiently high number of representative flow events for the intermittent flow analysis. The same holds true for the LES performance, as the maximum deviation from the mean velocity magnitude value of both velocity components quickly achieves a low error percentage of 2%, the momentum flux of 13% and TKE of 8% after reaching 10,000 samples. The performance of LES with 10,000 snapshots for each canyon is therefore considered to be adequate for the validation of flow dynamics.

We also performed the sensitivity test to an acquisition time by comparing the mean values point-by-point between the individual runs. Using the Frobenius norm, we compared the mean velocities, the standard deviations and reduced TKE. The averaged deviation of the mean velocity between all three runs was 2.6% for the streamwise component and 5.5% for the vertical component. The standard deviation had a scatter of 11% and 6% for the U-component and W-component, respectively. Also, the difference for reduced TKE was only 3%. All results can be found in [27].

Further, the POD modes from individual runs were compared by means of the Frobenius norm. The first POD modes derived from the individual runs did not deviate more than 4.2% from each other. The results for other modes are published in [27].

These values were considered to be sufficiently low enough to assume that even a short measurement, lasting for 3.2 s and assembling 1634 snapshots, captures a representative number of the dominant transient features in the flow. Therefore, we collected all three runs together and performed the statistics introduced in this paper. Some advanced statistics are not sensitive to the sampling frequency, for example the POD, when the essential dynamics in the flow is captured by the snapshots. For the spectral characteristics, where either the sampling frequency or continuity of the data are crucial, only the continuous data from the 500 Hz run were used.

Generally, the repetitions of the measurements involve all the possible random variations in an experimental performance—the switching off of the facility (e.g., a wind tunnel) and the devices (e.g., a laser and a particle seeding generator). The measurement also involves a bias error caused by both the PIV system and the post-processing procedure (the presence of spurious vectors, a strong gradient velocity field, a sub-pixel interpolation, a pixel-locking effect). The authors considered the bias error to be notably greater than the random error. Since we were unable to estimate the type of hypothesised distribution of the velocity mean values, we just simply calculated the standard deviation as a true error uncertainty, which presumably involves both types of error.

References

1. Casey, M.; Wintergerste, T. *Best Practice Guidelines—ERCOFTAC Special Interest Group on Quality and Trust in Industrial CFD*; COST: Brussels, Belgium, 2000.
2. Oberkampf, W.L.; Trucano, T.G. Verification and validation in computational fluid dynamics. *Prog. Aerosp. Sci.* **2002**, *38*, 209–272. [CrossRef]
3. Schatzmann, M.; Leitl, B. Issues with validation of urban flow and dispersion CFD models. *J. Wind Eng. Ind. Aerodyn.* **2011**, *99*, 169–186. [CrossRef]
4. Moonen, P.; Gromke, C.; Dorer, V. Performance assessment of Large Eddy Simulation (LES) for modeling dispersion in an urban street canyon with tree planting. *Atmos. Environ.* **2013**, *75*, 66–76. [CrossRef]
5. Soulhac, L.; Perkins, R.J.; Salizzoni, P. Flow in a Street Canyon for any External Wind Direction. *Bound.-Lay. Meteorol.* **2008**, *126*, 365–388. [CrossRef]
6. Salizzoni, P.; Soulhac, L.; Mejean, P. Street canyon ventilation and atmospheric turbulence. *Atmos. Environ.* **2009**, *43*, 5056–5067. [CrossRef]
7. Michioka, T.; Sato, A.; Takimoto, H.; Kanda, M. Large-Eddy Simulation for the Mechanism of Pollutant Removal from a Two-Dimensional Street Canyon. *Bound.-Lay. Meteorol.* **2011**, *138*, 195–213. [CrossRef]
8. Soulhac, L.; Salizzoni, P.; Mejean, P.; Perkins, R. Parametric laws to model urban pollutant dispersion with a street network approach. *Atmos. Environ.* **2013**, *67*, 229–241. [CrossRef]
9. Mirzaei, P.A.; Carmeliet, J. Dynamical computational fluid dynamics modeling of the stochastic wind for application of urban studies. *Build. Environ.* **2013**, *70*, 161–170. [CrossRef]

10. Soulhac, L.; Salizzoni, P.; Mejean, P.; Didier, D.; Rios, I. The model {SIRANE} for atmospheric urban pollutant dispersion; {PART} II, validation of the model on a real case study. *Atmos. Environ.* **2012**, *49*, 320–337. [CrossRef]

11. Hofman, J.; Samson, R. Biomagnetic monitoring as a validation tool for local air quality models: A case study for an urban street canyon. *Environ. Int.* **2014**, *70*, 50–61. [CrossRef] [PubMed]

12. Hofman, J.; Lefebvre, W.; Janssen, S.; Nackaerts, R.; Nuyts, S.; Mattheyses, L.; Samson, R. Increasing the spatial resolution of air quality assessments in urban areas: A comparison of biomagnetic monitoring and urban scale modelling. *Atmos. Environ.* **2014**, *92*, 130–140. [CrossRef]

13. Franke, J.; Hellsten, A.; Schlunzen, H.; Carissimo, B. The Best Practise Guideline for the CFD simulation of flows in the urban environment: Quality Assurance of Microscale Meteorological Models. In Proceedings of the Fifth International Symposium on Computational Wind Engineering. Chapel Hill, NC, USA, 23–27 May 2010.

14. Franke, J.; Hellsten, A.; Schlünzen, H.; Carissimo, B. *COST 732 Quality Assurance and Improvement of Microscale Meteorological Models—The Best Practise Guideline for the CFD Simulation of Flows in the Urban Environment*; COST Action 732; COST: Brussels, Belgium, 2007.

15. Schatzmann, M.; Olesen, H.; Franke, J. *COST 732 Model Evaluation Case Studies: Approach and Results—Quality Assurance of Microscale Meteorological Models*; COST Action 732; COST: Brussels, Belgium, 2010.

16. Franke, J.; Hirsch, C.; Jensen, A.; Krus, H.; Schatzmann, M.; Westbury, P.; Miles, S.; Wisse, J.; Wright, N. Recommendations on the use of CFD in Wind Engineering. In Proceedings of the International Conference on Urban Wind Engineering and Building Aerodynamics: COST C14—Impact of Wind and Storm on City life and Built Environment. Rhode-Saint-Genese, Belgium, 5–7 May 2004.

17. Tominaga, Y.; Mochida, A.; Yoshie, R.; Kataoka, H.; Nozu, T.; Yoshikawa, M.; Shirasawa, T. AIJ guidelines for practical applications of CFD to pedestrian wind environment around buildings. *J. Wind Eng. Ind. Aerodyn.* **2008**, *96*, 1749–1761. [CrossRef]

18. Jimenez, J.; Moser, R.D. *A Selection of Test Cases for the Validation of Large-Eddy Simulations of Turbulent Flows*; North Atlantic Treaty Organization; AGARD: Neuilly sur Seine, France, 1998.

19. Westerweel, J.; Elsinga, G.E.; Adrian, R.J. Particle Image Velocimetry for Complex and Turbulent Flows. *Ann. Rev. Fluid Mech.* **2013**, *45*, 409–436. [CrossRef]

20. Hertwig, D.; Leitl, B.; Schatzmann, M. Organized turbulent structures: Link between experimental data and LES. *J. Wind Eng. Ind. Aerodyn.* **2011**, *99*, 296–307. [CrossRef]

21. Hertwig, D.; Patnaik, G.; Leitl, B. LES validation of urban flow, part I: Flow statistics and frequency distributions. *Environ. Fluid Mech.* **2017**, *17*, 521–550. [CrossRef]

22. Hertwig, D.; Patnaik, G.; Leitl, B. LES validation of urban flow, part II: Eddy statistics and flow structures. *Environ. Fluid Mech.* **2017**, *17*, 551–578. [CrossRef]

23. Hertwig, D. On Aspects of Large-Eddy Simulation Validation for Near-Surface Atmospheric Flows. Ph.D. Thesis, University of Hamburg, Hamburg, Germany, 2013.

24. Kellnerova, R.; Kukacka, L.; Jurckova, K.; Uruba, V.; Janour, Z. PIV measurement of turbulent flow within a street canyon: Detection of coherent motion. *J. Wind Eng. Ind. Aerodyn.* **2012**, *104–106*, 302–313. [CrossRef]

25. Perret, L.; Savory, E. Large-Scale Structures over a Single Street Canyon Immersed in an Urban-Type Boundary Layer. *Bound.-Lay. Meteorol.* **2013**, *148*, 111–131. [CrossRef]

26. Blackman, K.; Perret, L.; Savory, E. Effect of upstream flow regime on street canyon flow mean turbulence statistics. *Environ. Fluid Mech.* **2015**, *15*, 823–849. [CrossRef]

27. Kellnerova, R. Wind-tunnel Modelling of Turbulent Flow Inside the Street Canyon. Ph.D. Thesis, Charles University in Prague Faculty of Mathematics and Physics, Prague, Czech Republic, 2014.

28. Fuka, V. PoisFFT—A free parallel fast Poisson solver. *Appl. Math. Comput.* **2015**, *267*, 356–364. [CrossRef]

29. Kim, J.; Kim, D.; Choi, H. An Immersed-Boundary Finite-Volume Method for Simulations of Flow in Complex Geometries. *J. Comput. Phys.* **2001**, *171*, 132–150. [CrossRef]

30. Peller, N.; Duc, A.L.; Tremblay, F.; Manhart, M. High-order stable interpolations for immersed boundary methods. *Int. J. Numer. Methods Fluids* **2006**, *52*, 1175–1193. [CrossRef]

31. Morinishi, Y.; Lund, T.; Vasilyev, O.; Moin, P. Fully conservative higher order finite difference schemes for incompressible flow. *J. Comput. Phys.* **1998**, *143*, 90–124. [CrossRef]

32. Nicoud, F.; Toda, H.B.; Cabrit, O.; Bose, S.; Lee, J. Using singular values to build a subgrid-scale model for large eddy simulations. *Phys. Fluids* **2011**, *23*, 085106. [CrossRef]

33. Verein Deutcher Ingenieure. *Physical Modelling of Flow and Dispersion Processes in the Atmospheric Boundary Layer—Application of Wind Tunnels*; VDI 3783, Part 12; Beuth Verlag: Berlin, Germany, 2000.

34. Xie, X.; Huang, Z.; Wang, J.S. Impact of building configuration on air quality in street canyon. *Atmos. Environ.* **2005**, *39*, 4519–4530. [CrossRef]

35. Kastner-Klein, P.; Berkowicz, R.; Britter, R. The influence of street architecture on flow and dispersion in street canyons. *Meteorol. Atmos. Phys.* **2004**, *87*, 121–131. [CrossRef]

36. Pascheke, F.; Barlow, J.F.; Robins, A. Wind-tunnel Modelling of Dispersion from a Scalar Area Source in Urban-Like Roughness. *Bound.-Lay. Meteorol.* **2008**, *126*, 103–124. [CrossRef]

37. Takano, Y.; Moonen, P. On the influence of roof shape on flow and dispersion in an urban street canyon on the influence of roof shape on flow and dispersion in an urban street canyon. *J. Wind Eng. Ind. Aerodyn.* **2013**, *123*, 107–120. [CrossRef]

38. Cox, W.M.; Tikvart, J.A. A statistical procedure for determining the best performing air quality simulation model. *Atmos. Environ. Part A Gen. Top.* **1990**, *24*, 2387–2395. [CrossRef]

39. Verein Deutcher Ingenieure. *Environmental Meteorology—Prognostic Microscale Windfield Models—Evaluation for Flow around Buildings and Obstacles*; VDI 3783, Part 9; Beuth Verlag: Berlin, Germany, 2005.

40. Willmarth, W.W.; Lu, S.S. Structure of the {R}eynolds stress near the wall. *J. Fluid Mech.* **1972**, *55*, 65–92. [CrossRef]

41. Cui, Z.; Cai, X.; Baker, C.J. Large-eddy simulation of turbulent flow in a street canyon. *Q. J. R. Meteorol. Soc.* **2004**, *130*, 1373–1394. [CrossRef]

42. Robinson, S.K. Coherent Motions in the Turbulent Boundary Layer. *Ann. Rev. Fluid Mech.* **1991**, *23*, 601–639. [CrossRef]

43. Kanda, M.; Moriwaki, R.; Kasamatsu, F. Large-Eddy simulation of turbulent organized structures within and above explicitly resolved cube arrays. *Bound.-Lay. Meteorol.* **2004**, *112*, 343–368. [CrossRef]

44. Coceal, O.; Dobre, A.; Thomas, T.G.; Belcher, S.E. Structure of Turbulent Flow Over Regular Arrays of Cubical Roughness. *J. Fluid Mech.* **2007**, *589*, 375–409. [CrossRef]

45. Aubry, N.; Guyonnet, R.; Lima, R. Spatiotemporal Analysis of Complex Signals: Theory and Applications. *J. Stat. Phys.* **1991**, *64*, 683–739. [CrossRef]

46. Hilberg, D.; Lazik, W.; Fiedler, H.E. The Application of Classical POD and Snapshot POD in a Turbulent Shear Layer with Periodic Structures. *Appl. Sci. Res.* **1994**, *53*, 283–290. [CrossRef]

47. Sen, M.; Bhaganagar, K.; Juttijudata, V. Application of proper orthogonal decomposition (POD) to investigate a turbulent boundary layer in a tunnel with rough walls. *J. Turbul.* **2007**, *8*, 1–21. [CrossRef]

48. Lumley, J.L. *The Structure of Inhomogeneous Turbulent Flows*; Yaglom, A.M., Tatarski, V.I., Eds.; Nauka: Moscow, Russia, 1967; pp. 166–178.

49. Nobach, H.; Tropea, C.; Cordier, L.; Bonnet, J.P.; Delville, J.; Lewalle, J.; Farge, M.; Schneider, M.K.; Adrian, R.J. *Springer Handbook of Experimental Fluid Mechanics*; Tropea, C., Yarin, A., Foss, J., Eds.; Springer: New York, NY, USA, 2007; pp 1337–1398.

50. Christen, A.; van Gorsel, E.; Vogt, R. Coherent structures in urban roughness sublayer turbulence. *Int. J. Climatol.* **2007**, *27*, 1955–1968. [CrossRef]

51. Nosek, Š.; Kukačka, L.; Kellnerová, R.; Jurčáková, K.; Jaňour, Z. Ventilation Processes in a Three-Dimensional Street Canyon. *Bound.-Lay. Meteorol.* **2016**, *159*, 259–284. [CrossRef]

52. Brown, D.L.; Cortez, R.; Minion, M.L. Accurate Projection Methods for the Incompressible Navier–Stokes Equations. *J. Comput. Phys.* **2001**, *168*, 464–499. [CrossRef]

53. Spalart, P.R.; Moser, R.D.; Rogers, M.M. Spectral methods for the Navier-Stokes equations with one infinite and two periodic directions. *J. Comput. Phys.* **1991**, *96*, 297–324. [CrossRef]

atmosphere

MDPI

Article

Application of GPU-Based Large Eddy Simulation in Urban Dispersion Studies

Gergely Kristóf * and Bálint Papp *

Department of Fluid Mechanics, Faculty of Mechanical Engineering, Budapest University of Technology and Economics, Műegyetem rkp. 3, 1111 Budapest, Hungary
* Correspondence: kristof@ara.bme.hu (G.K.); p.balint14@gmail.com (B.P.)

Received: 20 September 2018; Accepted: 7 November 2018; Published: 13 November 2018

Abstract: While large eddy simulation has several advantages in microscale air pollutant dispersion modelling, the parametric investigation of geometries is not yet feasible because of its relatively high computational cost. By assuming an analogy between heat and mass transport processes, we utilize a Graphics Processing Unit based software—originally developed for mechanical engineering applications—to model urban dispersion. The software allows for the modification of the geometry as well as the visualization of the transient flow and concentration fields during the simulation, thus supporting the analysis and comparison of different design concepts. By placing passive turbulence generators near the inlet, a numerical wind tunnel was created, capable of producing the characteristic velocity and turbulence intensity profiles of the urban boundary layer. The model results show a satisfactory agreement with wind tunnel experiments examining single street canyons. The effect of low boundary walls placed in the middle of the road and adjacent to the walkways was investigated in a wide parameter range, along with the impact made by the roof slope angle. The presented approach can be beneficially used in the early phase of simulation driven urban design, by screening the concepts to be experimentally tested or simulated with high accuracy models.

Keywords: street canyon; CFD; Large Eddy Simulation (LES); urban air quality; pedestrian exposure; concentration fluctuation

1. Introduction

The increase in the population density of many modern cities is limited by technical issues related to the construction of a healthy and effective living environment. In the short term, the citizens of such densely populated urban areas benefit from the scientific results of air quality research in the form of improved health conditions. Technological improvements may also allow for further increases in population density, therefore long-term benefits might be expected from the urbanization itself (e.g., by creating more job opportunities for individuals, faster development for local enterprises, and a more competitive economy at the national level).

Urban air quality displays large fluctuations in both space and time; therefore, the mitigation of extremely high local pollution conditions is of particular importance. The street canyon as a minimal model of a city is studied extensively from the perspective of flow structures and transport phenomena. The character of the flow is principally determined by the H/W ratio of the building height and street width [1,2]. The literature on street canyons was comprehensively reviewed by Vardoulakis et al. [3] and Ahmad et al. [4].

As can be seen later, the flow generated by the roof-level turbulent stresses can be characterized by one large vortex occupying almost the entire cross-section of the street canyon. The canyon vortex conveys airborne pollutants towards the leeward wall at a pedestrian level. At the same time, fresh air is transported to the sidewalk by the descending flow near the windward wall, causing a local concentration decrease. As can be seen from Figure 1, there is a substantial concentration difference between the leeward and windward sides; the leeward side is critical regarding the pedestrian exposure to traffic induced pollution.

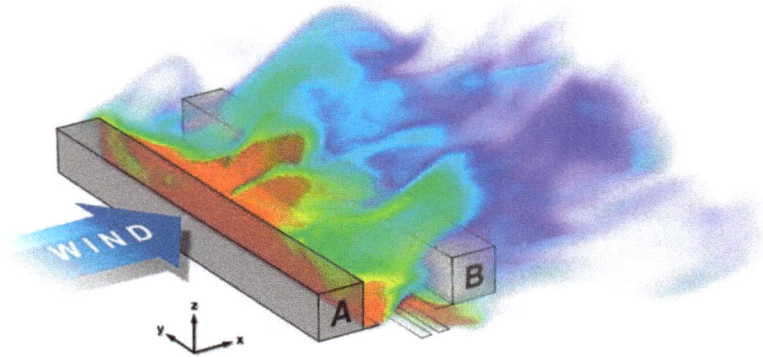

Figure 1. Instantaneous air pollutant concentration field in the vicinity of an isolated street canyon with a height-to-width ratio of 0.5, computed by ANSYS Discovery Live. Warmer colors indicate higher concentrations of air pollutants.

In practice, street-level air pollution can be mitigated by restricting emissions or by controlling the transmission process via geometrical changes. Buildings and other flow obstructions, such as urban vegetation, can be considered in the latter case. So far, most studies emphasizing the positive impact of urban vegetation are focused on the enhanced settling and filtering effects [5–7], although only minor improvements of around 1% relative to the average street level concentration can be achieved in this way, because the hydraulic resistance of the vegetation limits the air flux through the volume occupied by the plants. Numerous other studies [8–14] confirm that trees planted in the usual patterns decelerate the flow, thereby increasing the pollutant concentrations in critical locations, and this negative effect is at least one order of magnitude stronger than the gain from enhanced settling and filtering. The ambiguous impact of urban vegetation on air quality is comprehensively reviewed by Litschke and Kuttler [15], and Janhaell [16].

Based on a large eddy simulation (LES) model, McNabola et al. [17] showed that low boundary walls may be used effectively for improving air quality at the street level. Even a relatively low wall in the middle of the street substantially modifies the flow; the canyon vortex splits into two counter-rotating vortices, and consequently, the street level flow is directed towards the middle of the street on both sides, thus decreasing the concentration at the leeward sidewalk. The favorable effect can also be observed if two low boundary walls are placed between the traffic lanes and sidewalks on the opposite sides of the street. Using a similar LES modelling method, Gallagher et al. [18] showed that in asymmetric street canyons, the favorable effect of low boundary walls might be decreased or even reversed, therefore, a deeper investigation of the parametric dependencies is necessary.

Gromke et al. [19] proved the favorable effect of central boundary walls with wind tunnel experiments on a finite-length street canyon model, and also pointed out that the improvement in air quality can also be realized by using hedgerows of limited permeability instead of boundary walls. The unfavorable effects of the permeability and discontinuity of the hedgerows were also investigated in a realistic parameter range.

Most microscale (building level) numerical dispersion models utilize the Reynolds averaged turbulence modelling approach, in which only the average field properties are computed, and the enhanced mixing—caused by unresolved unsteady flow structures—is modelled depending on the average flow properties. The computational cost can be substantially reduced with the help of the

Reynolds averaged models, as the time dependency can be eliminated, and meaningful results can be obtained on relatively coarse meshes.

The Reynolds averaged approach has a number of shortcomings in urban air quality investigations [20] (e.g., the lateral diffusion is underestimated), because such models cannot capture the effect of large-scale unsteady flow structures in the wake of buildings. Dispersion models are affected by turbulence modelling uncertainties in two ways. Firstly, by impacting the advection of pollutants through the inaccuracies of the velocity field, and secondly, by directly affecting the diffusive transport, therefore the concentration field is predicted with much greater uncertainty than the velocity field. Another shortcoming is that Reynolds averaged models do not provide sufficient information about concentration fluctuations, which limits the model's applicability for assessing the health impact of toxic substances.

Globally unstable flows, attributed to the urban atmosphere, are usually more accurately predicted by scale-resolving turbulence models such as large eddy simulation (LES), detached eddy simulation (DES), or scale adaptive simulation (SAS) than by Reynolds averaged models. Best practice guidelines for scale resolving models are provided by Menter [21]. The ability to resolve the three-dimensional unsteady flow structures stemming from flow instabilities is a common feature of these models. In LES models, the effect of small-scale unresolved turbulence is taken into account by using algebraic sub-grid-scale stress (SGS) models, depending on the local shear rate and mesh size. The mesh needs to be locally adapted to the turbulent scales in LES models, in order to keep the ratio of resolved turbulent kinetic energy at a sufficiently high level. This requirement generally leads to very fine meshes in the vicinity of solid boundaries, which also places a limitation on the time step size, therefore LES is usually 20 to 100 times more expensive than RANS in terms of computational cost.

The application of LES models in urban atmospheric flow simulations is supported by the fact that the flow is governed by large scale turbulence originating from the free shear layers, and additionally, because the intensity of the emission does not depend on the flow. Therefore, the results are grossly insensitive for boundary layer resolution, unlike in many other engineering applications (e.g., in the case of heat transfer studies) [22]. Owing to the advantages mentioned above, a number of examples exist for the application of LES in urban atmospheric research. The flow and the passive scalar transport process in the street canyons of different H/W ratios was investigated by Liu et al. [23–25]. Moreover, So et al. [26] studied flow structures in asymmetric canyons characterized by different building heights on leeward and windward sides, and Michioka et al. [27] analyzed the evolution of the urban boundary layer above an evenly spaced row of street canyons. Furthermore, Baker et al. [28] and Kikumoto and Ooka [29] used the LES method for analyzing the reactive flows in street canyons. Despite its higher accuracy and predictive power, LES could not become widely adopted in urban air quality modelling because of its high computational cost.

Powered by the huge market for computer games, graphics processing units have developed rapidly in recent years. The migration of numerical models from CPUs to GPUs offers a remarkable improvement in terms of available computing power in an inexpensive way. The CUDA application programming interface, introduced by the NVIDIA Corporation in 2007, made GPUs available for general purpose processing, which made the graphics processing unit ideal for executing large-scale numerical computations. Presently, a top of the line GPU features several thousand processor cores, making parallel scalability the primary concern for programmers. Consequently, the use of numeric methods other than CPU applications is required in the field of computational fluid dynamics. Furthermore, the amount of graphical memory available at reasonable prices on graphics cards has significantly increased recently, which is also a key change in terms of Computational Fluid Dynamics (CFD) applications.

The perspectives of GPU technology in the fields of wind engineering and micrometeorology are indicated by some models of outstanding size and complexity. Schalkwijk et al. [30] investigated the cloud formation process using a LES-based micrometeorological model adapted for a GPU. The NVIDIA GTX 580 video card (with 512 cores) installed in the PC for this simulation was able to accelerate computation by a factor of nine, when compared to using a CPU alone. Onodera et al. [31] investigated the flow in a 10 km by 10 km area of Tokyo, with the help of a one-meter resolution LES simulation on a GPU cluster, using a self-developed lattice Boltzmann model that revealed unseen details of the complex urban flow field.

In February 2018, ANSYS Inc. released a new GPU-based software called Discovery Live, which aims to support creative mechanical engineering design. By utilizing the finite volume method, the program is capable of an almost real-time large eddy simulation of thermally coupled flows. The user can modify the geometry and observe changes in the results almost instantaneously, while the simulation is running, thus the process of simulation-driven product development speeds up considerably.

One of the aims of this study is to explore the applicability of this new GPU-based LES model in computational wind engineering. Firstly, a virtual wind tunnel is created in such a way that the resulting velocity and turbulence profiles would correspond to that of the wind tunnel experiments used for urban dispersion studies. Secondly, Discovery Live is validated by comparing its results with known wind tunnel experiments, and then, based on the analogy between heat and pollutant transport, the new software is used to analyze the air quality improvement effect of boundary walls in street canyons. The effects of the boundary wall height and the roof slope are examined in a wider range compared with earlier models. Finally, user experiences are summarized and recommendations are given to the software developers.

2. Methods

2.1. Street Canyon Geometry and Low Boundary Wall Configurations

One of the main purposes of this study is to investigate the applicability of the ANSYS Discovery Live simulation tool in urban dispersion studies, by reproducing some of the wind tunnel experiments of Gromke et al. [19], using the numerical model. In the investigated layout, illustrated in Figures 2 and 3, a generic street canyon model was formed by two parallel blocks, which consist of the leeward (A) and windward (B) buildings. Both the building height and breadth are H = B = 120 mm on both sides, and the street width is W = 2H = 240 mm. The building length (parallel to the street canyon length axis, *y*) is L = 10H = 1200 mm, that is, the street canyon's aspect ratios are W/H = 2 and L/H = 10. The geometrical parameters meet the dimensions of the reduced scale (*M* = 1:150) wind tunnel model, that is, the full-scale dimensions would be the following: H* = 18 m, W* = 36 m, and L* = 180 m. Additionally, there is an option of placing sloped roofs with a roof angle α on the top of each building. The changes in the flow structure and dispersion field as a function of the roof angle has been investigated in a parameter study in the 0–50° range.

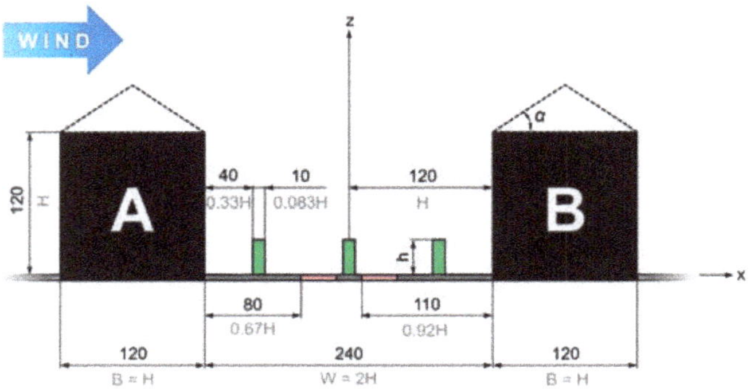

Figure 2. Side view of the street canyon. Low boundary walls are indicated with green and pollution sources are marked with pink.

Presently, Discovery Live is not able to handle porous zones, which limits the scope of the model validation, as the effect of the hedgerows of various permeability were studied in many of Gromke's experiments. In our investigation, low boundary walls (LBWs) were introduced in the arrangements described below.

The central boundary wall configuration (CBW) contains one low boundary wall located in the middle of the canyon, along the *y* axis (Figure 3). The sidewise boundary walls arrangement (SBW) consists of two longitudinal walls located at the opposite sides of the street canyon adjacent to the sidewalks, separating the traffic lanes from the pedestrian zones. In the case of the sidewise boundary walls, the distance between the LBW and the closest building is 40 mm. The low boundary walls have the same L = 1200 mm length as the buildings, and a width of 10 mm in all scenarios. The height of the boundary walls (*h*) ranges from 0 to 45 mm (up to 0.375H) in both arrangements, including a reference scenario with no boundary walls (NBW).

As stated before, along the entire length of the buildings, two 40 mm wide areas are separated by the sidewise boundary walls on the opposite sides next to the leeward and windward walls. This width, which equals 6 m when scaled up to full-size dimensions, enables the construction of additional reduced traffic zones—such as bikeways or parking spaces—separated from the heavy traffic areas. The aforementioned areas are used mostly by pedestrians; therefore, this will be the focus area of our pollutant dispersion study, similarly to earlier investigations [17,19].

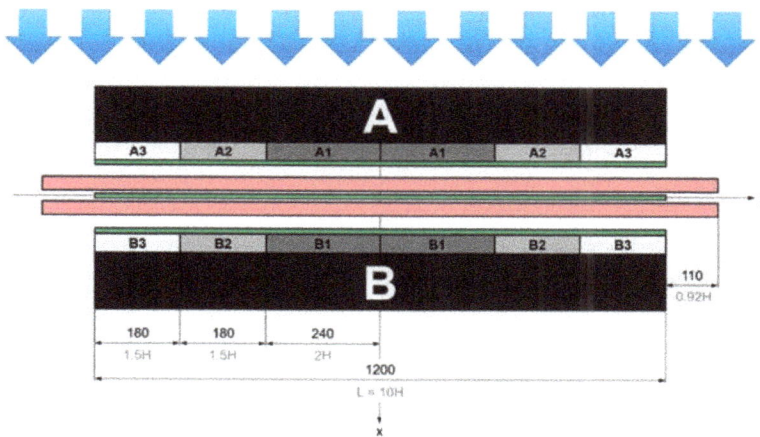

Figure 3. Top view of the street canyon. Location of the pre-defined concentration measurement areas (A1–A3 and B1–B3).

In order to validate the velocity field of our model, the geometrical parameters of a different street canyon defined by Gromke and Ruck [9]—with the measured velocity data most relevant to the air exchange phenomena—were adopted. This setup is similar in size to the previously described one, except for the lack of emission zones, low boundary walls, and pitched rooftops. Furthermore, the aspect ratio of this street canyon is H/W = 1.

2.2. Computational Domain and Boundary Conditions

In order to reproduce the physical conditions of the wind tunnel measurements, the street canyon geometry was placed into a virtual wind tunnel using the Fluids module in ANSYS Discovery Live 19.2. The length of the computational domain is 30H, the width is 16H, and the height is 5H. The first building is located at a distance of 20H from the upstream end of the enclosure. The boundary zones of different types are indicated in Figure 4, and are explained further in the following chapters.

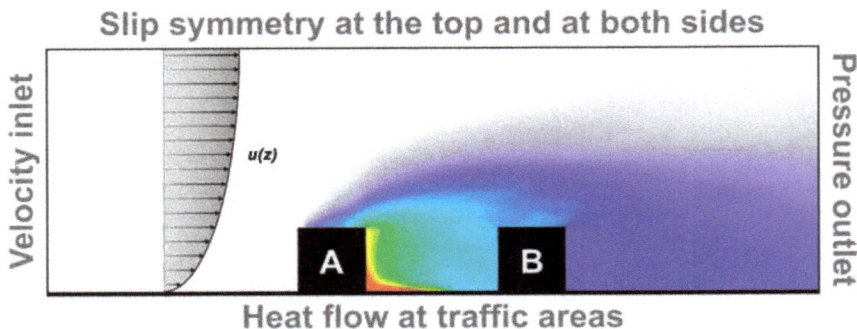

Figure 4. Boundary conditions in the virtual wind tunnel (geometric dimensions are not proportionate).

2.3. Simulation of the Atmospheric Boundary Layer Approach Flow

The flow approaching the street canyon in the wind tunnel [32] is characterized by the following vertical profiles, regarding time averaged horizontal velocity (u) and turbulence intensity (I_u):

$$u(z) = u_{ref}\left(z_{ref}\right)\left(\frac{z}{z_{ref}}\right)^{\alpha_u} = 4.39\left(\frac{z}{0.1\ (\mathrm{m})}\right)^{0.30} (\mathrm{m/s}) \tag{1}$$

$$I_u(z) = I_{u,ref}\left(z_{ref}\right)\left(\frac{z}{z_{ref}}\right)^{\alpha_{Iu}} = 22.4\left(\frac{z}{0.1\ (\mathrm{m})}\right)^{-0.36} (\%) \tag{2}$$

At the time that this paper is being written, Discovery Live does not allow for the specification of continuous inlet profiles; only spatially uniform, temporally constant velocity inlets can be used. To achieve the above-described characteristics of the approaching flow, several rectangle-shaped inlets containing passive turbulence generators (circular flow obstructions) are defined on the inlet plane; as shown in Figure 5. The height of the rectangular inlets, as well as the diameter of the circular turbulence generators grow with the distance from the ground in order to create the increasing turbulent length scale that characterizes the fully developed turbulent urban boundary layer. The number of circular cutouts in each rectangle and the inlet velocity magnitudes were fine-tuned in order to obtain the best fit both for the mean velocity and turbulence intensity in the vicinity of the street canyon. Details of the inlet configuration are given below.

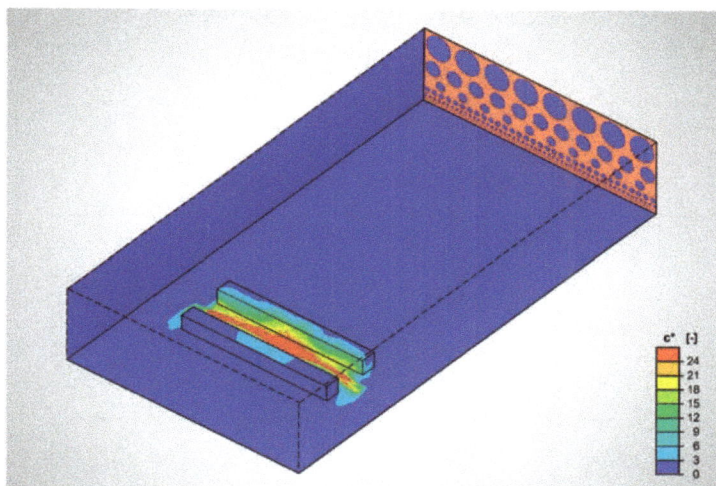

Figure 5. Overview of the virtual wind tunnel. Surfaces are colored by normalized concentration. The velocity inlets are marked with orange.

The comparison between the desired and computed velocity and turbulence intensity profiles is shown in Figure 6. The profiles realized in Discovery Live were obtained by averaging three time-averaged profiles taken in the mid-planes of three equal sectors along the y axis (at $y/H = \{-16/3; 0; 16/3\}$), each at H distance upstream from Building "A" ($x/H = -3$). The direct output of the desired profiles is not yet available in Discovery Live, therefore 20 virtual probes were defined in the heights corresponding to the experimental setup. The probes were used to export the time series of the velocity values for post-processing.

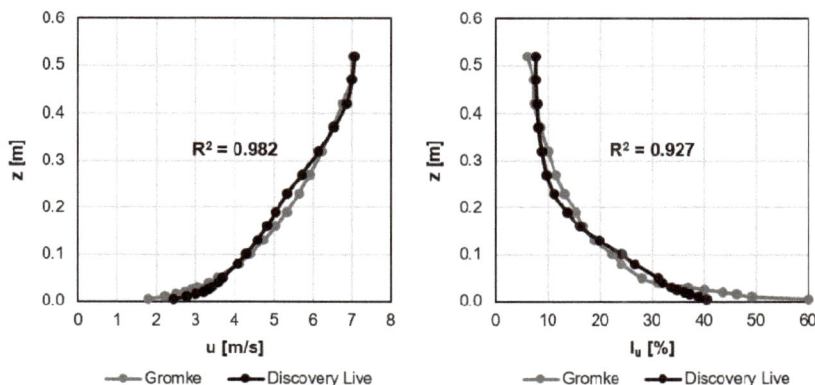

Figure 6. Comparison of the wind tunnel measurement and the CFD results regarding horizontal mean velocity (**left**) and turbulence intensity (**right**) profiles.

The flow velocity at a roof level is $u_H = 4.636$ m/s, based on Equation (1), and the reference height is H, therefore the Reynolds number is $Re_H = 37{,}000$—similarly to the original wind tunnel experiment, which can be considered sufficiently high to achieve a Reynolds number independent of flow and dispersion.

It must be noted, that the parameters of the optimum inlet configuration depend on the mesh resolution, thus the values presented in Table 1 may become sub-optimum even when the mesh is further refined. A mesh independent optimum inlet configuration would be desirable, although its development requires the perfect resolution of turbulence (DNS resolution), which is not yet feasible at the given Reynolds number.

Table 1. Summary of the velocity inlets.

Inlet No.	Rectangle Height (mm)	Circle Diameter (mm)	Circle Count (–)	Velocity (m/s)
1	13.93	2.86	137	3.4
2	22.29	10.28	86	5.3
3	35.66	22.16	53	9.0
4	57.06	41.17	33	10.0
5	91.29	71.58	10	8.5
6	146.07	120.24	11	10.6
7	233.70	198.09	8	12.0

2.4. Air Pollutant Dispersion Simulation Using Heat Transport Analogy

Presently, an arbitrary user-defined scalar transport cannot be modelled in Discovery Live; only a heat transport model is available. Therefore, a diffusion–heat transfer analogy has to be applied for modelling the pollutant dispersion process.

The analogy is based on the identical forms of the diffusive (Equation (3)) and thermal (Equation (4)) transport equations of constant property fluids supplemented with identical boundary conditions.

$$\frac{dc}{dt} = \nabla \cdot (D\nabla c) \tag{3}$$

$$\frac{dT}{dt} = \nabla \cdot (a\nabla T) \tag{4}$$

In Equations (3) and (4), c denotes the mass concentration of non-settling passive pollutants in the air (kg/m^3), t is time measured in seconds, T (K) is the absolute temperature, while D and a are the diffusivity and thermal diffusivity coefficients expressed in the same kinematic unit (m^2/s), respectively. The thermal analogy requires the heat conductivity of the fluid to be chosen in a way that the Lewis number ($Le = a/D$) is in unity, therefore the fluid heat conductivity (W/(m·K)) in the numerical model was calculated according to the following formula:

$$\lambda = D\rho c_p \tag{5}$$

where ρ (kg/m^3) is the density, and c_p (J/(kg·K)) is the specific heat of air at constant pressure. Note that both expansion work and viscous dissipation need to be neglected in the energy balance.

In the heat transport analogy, the temperature distribution takes the place of the air pollutant concentration field (with 0 °C indicating clear air), and the heat introduced into the model represents the rate of pollutants produced by a source.

Gromke et al. [19] presented non-dimensional concentrations using the following definition:

$$c^+ = \frac{cU_H H}{Q_l} \tag{6}$$

where U_H (m/s) is the characteristic velocity defined at roof level, H (m) is the building height, and Q_l (kg/(m·s)) is the line source intensity. Correspondingly, the normalized concentration c^* was computed from the numerical model results according to Equation (7):

$$c^* = \frac{U_H H L_Q \rho c_p}{Q} T = KT \tag{7}$$

in which L_Q (m) is the source length and Q (W) is the source intensity. For convenience, Q was chosen in a way that the value of K (1/°C) is unity, thus the Discovery Live temperature results (T, displayed in °C units) could be directly compared with the known normalized concentration data.

In order to assess the effect of the low boundary wall (LBW) configurations on air quality in comparison to the reference scenario not containing any boundary walls (NBW), the following formula of relative concentration change was used, similarly to Gromke's definition.

$$\Delta c^* = \frac{c^*_{LBW} - c^*_{NBW}}{c^*_{NBW}} \tag{8}$$

The line sources of the wind tunnel experiment—placed over the inner traffic lanes through the entire length of the street canyon, with a 0.92H long extension at each end to model the effect of the intersections—were represented in the present model by two rectangular emission zones of the same length on the inner sides of the road (indicated by the pink color in Figures 2 and 3).

The time dependent concentration field over the pedestrian areas was monitored in two different ways. Firstly, for direct comparison, local probes were placed in the 136 sampling points of the wind tunnel measurement, and secondly, to establish the tendencies in the change of the concentration field from the effect produced by the low boundary walls, area-averaged concentration values were tracked on pre-defined surfaces during the simulation process. Following Gromke's footsteps, the canyon floor was divided into the following areas (illustrated in Figure 3):

1. Center area ($|y/H| < 2.0$): the area of the street canyon where the dominant flow structure is the canyon vortex (characterized by a horizontal rotational axis);
2. Transitional area ($2.0 < |y/H| < 3.5$): the area of the street canyon where the effects of both the canyon vortex and the corner eddy can be observed;
3. Corner/end area ($3.5 < |y/H| < 5.0$): the area dominated by the corner eddy (with a vertical rotational axis).

These regions were defined on the ground surface of both the leeward and windward side pedestrian zones—as averaging in arbitrary inner surfaces, for example, at pedestrian head height, is not yet possible in Discovery Live. Hence, altogether, there are six different areas to monitor using output charts, from which the time series of the area-averaged concentration can be exported. The simulation covered 20 s in physical time, and to avoid errors originating from the transient effects, only the data from 5 to 20 s was used for the statistical analysis.

2.5. The Numerical Mesh and Mesh Convergence

In ANSYS Discovery Live, the user does not have full control over the mesh properties. The software uses finite volume method with large eddy simulation, and it is capable of generating an equidistant Cartesian mesh of an automatically determined resolution. The details of the discretization schemes are not yet disclosed, however, it is known that the spatial resolution depends on the amount of on-board graphical memory, on the Speed/Fidelity ratio set by the user within Discovery Live, and on the domain size. Consequently, a tradeoff needs to be made between the discretization error, which increases with the domain size, and the model errors due to boundary influences, which decrease with the domain size.

It must be noted, that in the current version of the software, no direct information is available about the mesh size, although it can be estimated from the model outputs with acceptable accuracy, by using the below described grid interference method. Apart from the mesh resolution, the time step size is also automatically determined by the software. The model outputs presented in the Results section can be reproduced by using the same type of graphics card, which was a single NVIDIA GTX 1080Ti, or by another graphics card with at least the same amount of video memory with a properly chosen speed/fidelity setting. To the best of our knowledge, other hardware components do not influence the software's performance.

To explore the effect of the speed/fidelity parameter, the NBW reference case was run using speed/fidelity ratios of 25%, 50%, and 100%, with the latter being the finest discretization available. The normalized air pollutant concentration results from all of the six focus areas of the street canyon against the different speed/fidelity values are compiled in Figure 7.

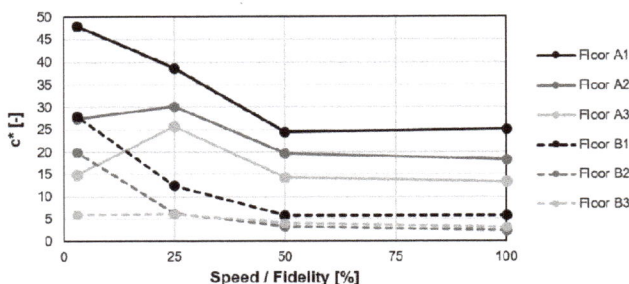

Figure 7. Comparison of computed normalized concentration values using different speed/ fidelity settings.

For estimating the mesh resolution, a staircase shaped test object was placed into the computational domain. The staircase consisted of 16 steps, each with 5 mm height and 40 mm length. As the stairs are lower than the mesh resolution, not every one of them is resolved by the mesh. By examining the streamlines while the flow simulation is running, it is possible to identify which steps are resolved (R) and which ones are skipped (S) by the mesh. The grid interference pattern can be represented by a 16-character string of R and S letters. Only a small range of mesh resolution produces exactly the same grid interference pattern, that is, by calculating the interference pattern with several hypothetic mesh resolutions, the valid range of the actual mesh resolution can be determined. Using this grid interference method, the mesh resolution could be calculated with ±0.58% accuracy at a 100% speed/fidelity ratio. The mesh details of the examined speed/fidelity cases are summarized in Table 2.

Table 2. Discretization parameters corresponding to the different speed/fidelity ratios.

Speed/Fidelity (%)	Estimated Mesh Resolution (mm)	Mesh Resolution Uncertainty (%)	Total Cell Count (Million)	Average Time Step Size (s)
100	7.74	±0.58	9.0	5.2×10^{-4}
50	9.71	±3.04	4.5	6.9×10^{-4}
25	15.52	±1.72	2.1	9.4×10^{-4}
3 [1]	38.33	±4.36	0.074	3.0×10^{-3}

[1] Note, that using the 0% speed/fidelity setting, the geometry has not been captured to a sufficient detail by Discovery Live.

The results of the scale resolving models change with the mesh resolution, because of both the change of the discretization error and the change of the portion of resolved turbulence; therefore, a regular mesh convergence cannot be expected. As can be seen in Figure 7, the 50% fidelity results are reasonably close to the results produced with maximum fidelity. In Discovery Live (using the aforementioned GPU), a large eddy simulation with a 20 s long time interval can be completed in approximately an hour at 100% fidelity, and therefore the use of the finest discretization can be considered reasonable, and thus all of the dispersion cases were computed at the maximum resolution. From the model outputs, the finest spatial resolution was estimated to 7.74 mm, thus in the flow direction, the computational domain is divided into approximately 465 voxels, the height of the buildings is around 16, and the street width can be estimated for 31 cells. The average time step size was 5.2×10^{-4} s. Based on these data and the maximum velocity (the greatest value measured in the velocity profile, see Figure 6), the Courant number is around 0.5.

3. Results and Discussion

In our study, we focused exclusively on cases with the street canyon subjected to a perpendicular wind direction (parallel to the x axis). First, we analyzed the flow field and the air pollutant distribution of the reference scenario with no boundary walls. Furthermore—in order to validate the Discovery Live model—the concentration results of this case were compared to Gromke's measurement results extensively, along with the comparison of the velocity field of the narrower street canyon (H/W = 1). Finally, the results of the parameter studies will be presented.

3.1. Flow Field Observed in the Reference Scenario without Boundary Walls

Being an isolated group of buildings, the flow regime around our street canyon can be described by the attributes of isolated roughness flow documented by Oke [1]. Between the buildings, however, the characteristics of wake interference flow can be observed.

As the flow passes the leeward building, its major part is directed above the roofs, but a substantial amount also swerves around the sides of the buildings. As can be seen in Figure 8b, a horseshoe vortex with a large recirculation zone is formed upstream from the leeward building near the ground, with the vortex axis parallel to the street canyon in front of Building "A".

Again, as a consequence of being a sole complex of buildings, the flow above the roof of the leeward building is not entirely parallel to the ground, therefore the evolution of an approximately H long, however shallow clockwise (forward) rotating vortex, can be observed above the roof of Building "A". These vortices described above have no direct effect on the propagation of the traffic induced pollutants in the pedestrian areas (although, the leeward side building-top vortex is able to transport pollutants upstream above the roof); the near-ground air pollutant distribution is mainly governed by the flow regime inside the street canyon.

In between the leeward and windward buildings, a clockwise-rotating vortex can be observed (Figure 8b), driven by the shear forces of the flow over the street canyon. This phenomenon was described earlier by Kastner-Klein et al. [33], Gromke and Ruck [9], and Gromke et al. [19] as canyon vortex, but it can be referred to as primary vortex [1,34], horizontal vortex [35], lee vortex cell [1,36], vortex circulation [36], or mixing circulation [37]. It can be observed in Discovery Live, that the vortex core is a strongly curved line throughout the length of the street canyon, similarly to the description by Kellnerova and Janour [38], and it also loses its continuity in the instantaneous flow field. It can be seen in Figure 8b, that the top of the canyon vortex is significantly raised above the roof of the buildings, which is one of the attributes of the wake interference flow. In the case of multiple street canyons placed one after another in an upstream and downstream flow direction, one can expect a skimming flow to develop over the buildings [1], which could lead to a flatter and straighter stream of air in the proximity of the rooftops.

Apart from the primary vortex, the flow field near the lateral ends of the street canyon is dominated by the effects of the air stream swerving around the sides of the buildings, which results in the formation of the corner eddies (illustrated in Figure 8a), which have a vertical axis of rotation in contrast to the horizontal rotational axis of the main canyon vortex. The corner vortices of roughly W diameter were previously observed by Hoydysh and Dabbert [35], Belcher [37], Gromke and Ruck [9], and Kellnerova and Janour [38].

The corner eddies direct air into the canyon on the windward side and out of the canyon on the leeward side. This movement, superposed with the effect of the primary canyon vortex, which moves the air to the leeward side in the vicinity of the street surface, results in a diagonal motion pointing from both windward ends towards the center of the street canyon on the leeward side. This stream can be clearly identified in Figure 8a.

Over the windward building, the flow attaches to the rooftop, and a large recirculation zone develops downstream from Building "B" along the entire length of the street canyon complex, and the wake of the objects can be observed even at the outlet of the virtual wind tunnel, 6H distance downstream from the actual geometry.

The phenomena described above also correspond to the CFD results by Buccolieri et al. [14] for the same geometry.

It can be observed from Figure 8a that the flow pattern displays some asymmetries, because the averaging time of the streamline visualization is limited in the present version of Discovery Live, although a long averaging would be necessary because of the slowly evolving vortex structure in the building wakes.

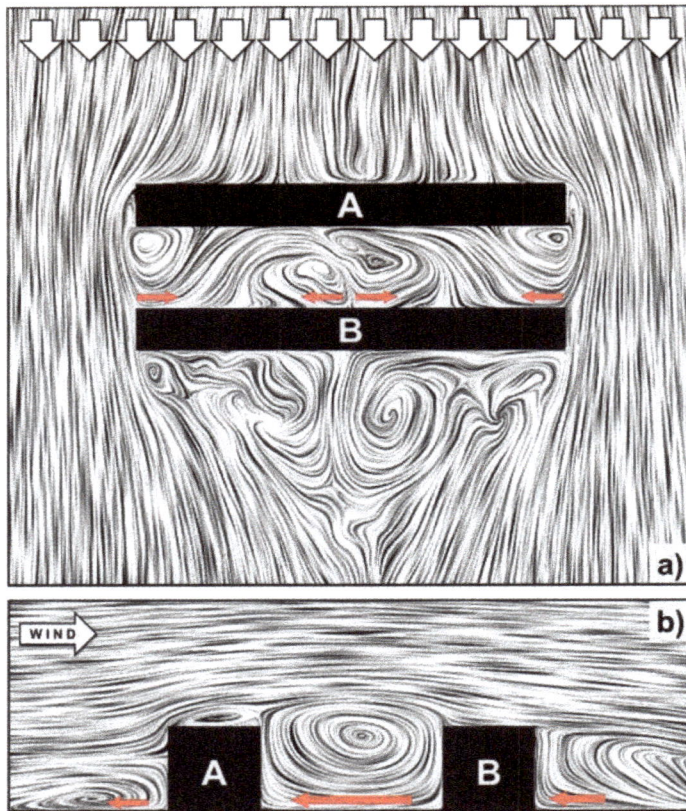

Figure 8. Time-averaged streamlines around the buildings (**a**) from top view, at $z/H = 0.5$; and (**b**) from side view, at $y/H = 4$.

3.2. Air Pollutant Dispersion in the Reference Scenario without Boundary Walls

The typical distribution of the traffic induced air pollutants in the NBW reference case is illustrated by Figure 5 (and later in Figure 13a,c).

The clockwise-rotating canyon vortex—depicted in Figure 8b—transports the pollutants from the near-ground emission zones to the leeward direction. Consequently, a significant difference (with a factor of nearly five) can be observed in terms of the normalized concentration between the full area of the leeward and windward pedestrian zones (19.6 and 4.0, respectively). Based on this asymmetry, it can be concluded that the crucial area regarding the pedestrian exposure to traffic induced air pollutants is the leeward side walkway.

The corner eddies—characterized by a vertical axis of rotation—are capable of sweeping out a substantial amount of the air pollutants on both ends of the leeward footpath, thus creating a relatively clear corner section on this side. On the other hand, over the windward walkway, their effect is inverted; the vortices transport air pollutants into the canyon, resulting in a relatively contaminated corner area—compared to the transition zone on the windward side.

As a consequence of the continuous flow towards the middle section from the corners, the most heavily polluted regions are located in the center areas on both the leeward and windward pedestrian zones, with mean normalized concentrations of 25.2 and 5.7, respectively. On the leeward side, the air quality improves over the footpath, as one arrives closer to the lateral ends of the canyon, with mean normalized concentrations of 18.3 in the transitional area, and 13.3 in the end area. As stated before, on the windward side, the center area has the highest normalized concentrations as well, but it must be noted that this concentration value is lower than that of the clearest area of the leeward pedestrian zone. As discussed above, the corner eddy has an unfavorable effect on the windward side's pollutant concentrations in the corner regions (with the mean value of 3.2), therefore the less contaminated region on the windward footpath is the transitional area, with 2.4 of mean normalized concentration.

As shown by the graphs of Figure 9, the normalized concentration displays drastic fluctuations at each point; the RMS (root mean square) value approaches the mean, and some of the peak values are three times higher than the average.

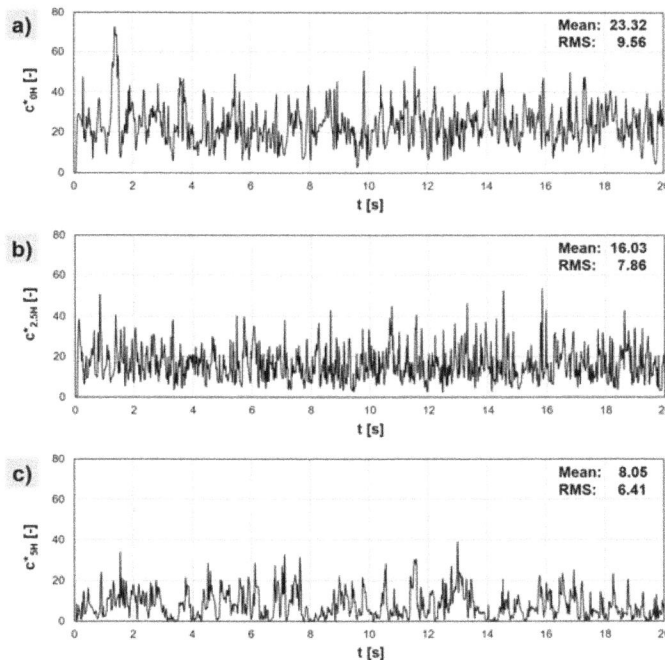

Figure 9. Time series of pointwise normalized concentrations over the leeward walkway, in $z/H = 1/12$ height and $x/H = 1/6$ distance from Building "A". Lateral positions: (a) $y/H = 0$; (b) $y/H = -2.5$; (c) $y/H = -5$.

3.3. Model Validation

In the course of the validation process, the following simulation results from the Discovery Live CFD model were compared to the wind tunnel measurements:

- The velocity and turbulence intensity profiles of the approaching flow of Gromke et al. [19];
- The pointwise normalized concentrations of Gromke (2016) at the Wall A, Wall B, Floor A, and Floor B measurement locations of the H/W = 0.5 aspect ratio street canyon, without the hedgerows or low boundary walls of Gromke et al. [19];
- The pointwise velocity data at the end planes and at the roof height of the H/W = 1 aspect ratio street canyon of Gromke and Ruck [9].

The comparison of the above-described measurement data and the simulation results is shown in Figures 6, 10 and 11.

As shown by the graphs of Figure 6, with the help of the inlet configuration detailed in Section 2.2, the desired velocity and turbulence intensity profiles (Equations (1) and (2)) could be reproduced with $R^2 = 0.982$ and 0.927 coefficients of determination, respectively. The deviation of the boundary layer characteristics in the wind tunnel experiment from the desired profiles is not known.

In Figure 10, the model results are compared with the measured surface normal velocity components at the lateral and upper boundaries of the street canyon. As can be seen from the graphs, the model captures the basic flow structures, and the maximum velocities are reproduced with remarkable accuracy. Most of the deviation from the reference data is caused by the slight displacement of the maximum inflow location on the lateral sides. The coefficient of determination of the velocity dataset is $R^2 = 0.697$.

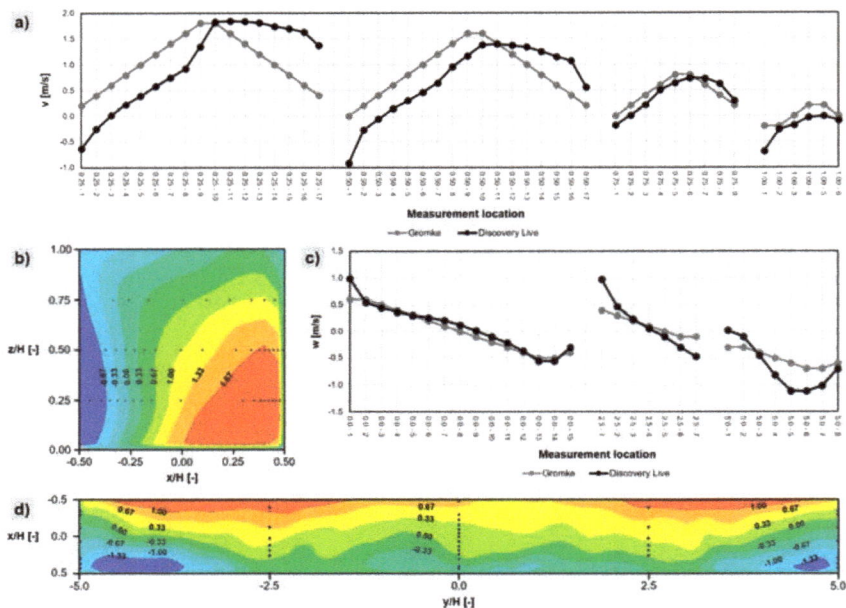

Figure 10. Visualization and comparison of time-averaged velocity component results from the wind tunnel measurement of Gromke and Ruck [9], and the Discovery Live simulation data. (**a,b**) show the inflow (+) and outflow (−) at the lateral ends of the street canyon in y direction. (**c,d**) show the upwards (+) and downwards (−) flow at roof level in z direction. The measurement locations are indicated with black dots.

According to the validation criteria established by COST Action 732 [39], the quality metrics for the normalized concentration results were calculated and listed in Table 3.

Table 3. Validation metrics of the normalized concentration results.

Validation Metric	Model Result	Recommended Range of Acceptance [1]	Aim Value
Correlation coefficient (R)	0.9551	>0.8	1
Hit rate (q)	0.6667	>0.66	1
Factor of two of observations (FAC2)	0.7917	>0.5	1
Fractional bias (FB)	−0.1238	$\in[-0.3; 0.3]$	0
Geometric mean bias (MG)	1.1579	$\in[0.7; 1.3]$	1
Normalized mean square error (NMSE)	0.0750	<4	0
Geometric variance (VG)	1.2323	<1.6	1

[1] Proposed by Balczó et al. [13] for the comparison of CFD results to a similar wind tunnel experiment.

The superior match between both the velocity and concentration results with the measurement data is partly due to the fact that the buildings being tested are sharp-cornered bodies with straight walls, which can be captured well by the Cartesian mesh of Discovery Live. Furthermore, as a consequence of the cuboid-shaped buildings, most of the flow field is not influenced by the boundary layers that develop on the surface of solid objects, which cannot be resolved by the current version of the software. The model errors can be considerably larger if the buildings to be inspected include curved surfaces.

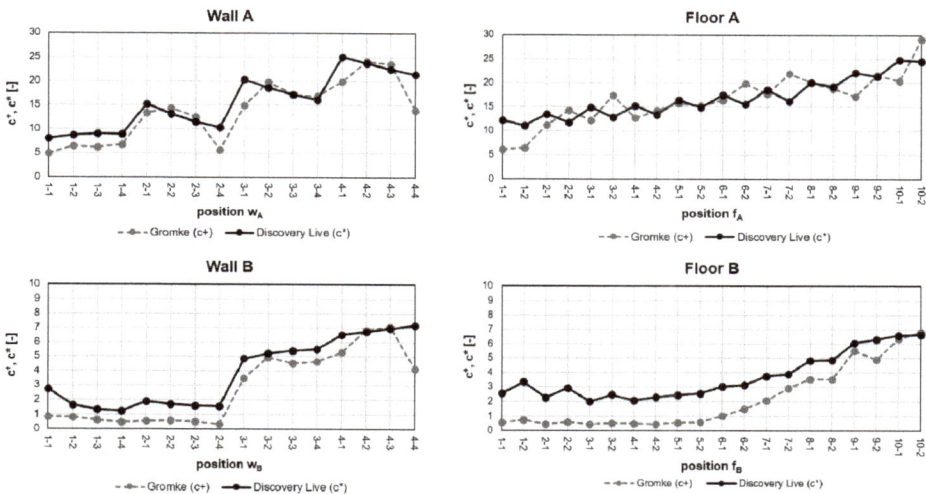

Figure 11. Comparison of the normalized concentration results from the wind tunnel experiment and the Discovery Live simulation data. Sampling positions are as defined by Gromke et al. [19].

The model shows some errors exceeding the absolute measurement error in the windward sidewall (Wall B 1.1 ... 6.2), but the measured concentration values are relatively small in these locations. Exploring the cause of the deviation requires further investigation. Possible causes may be (A) the numerical diffusion introduced by the solution method, (B) the imperfect resolution of the turbulence spectrum, (C) the inaccuracies in the approaching wind profiles, or (D) the errors due to the reduced domain size. Unfortunately, the details of the sub-grid-scale stress model and the flux scheme are not yet disclosed by the developers; therefore, a meaningful investigation of the above causes cannot be carried out.

3.4. Air Quality Improvements by the Low Boundary Walls

As a part of the parameter studies, different configurations of central and sidewise longitudinal low boundary walls were placed in the street canyon. The height of the walls in both types of arrangements were multiples of the edge length of the equidistant Cartesian mesh (7.74 mm). Including the wall-free reference case, the dispersion analysis was carried out with 13 different LBW configurations, with a maximum full-scale LBW size of 6.97 m (0.39H). The normalized concentration results for all of the scenarios regarding the six focus areas are compiled in Figure 12.

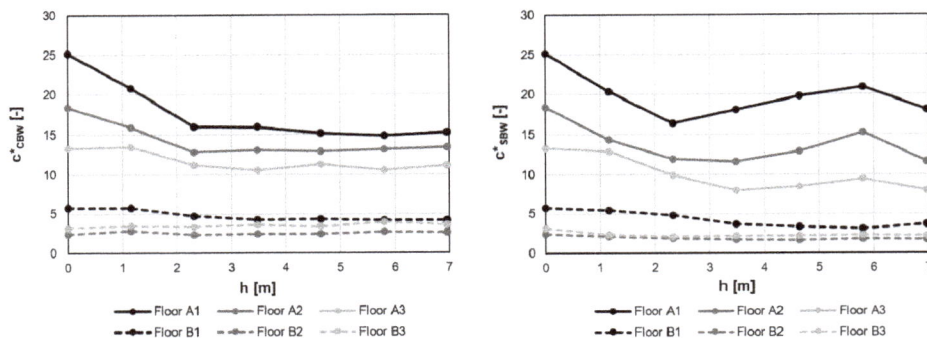

Figure 12. Area averaged normalized concentrations at the focus areas as a function of low boundary wall (LBW) height in the case of central (**left**) and sidewise (**right**) boundary walls.

The central boundary walls—even at a height of 1.2 m—are capable of deflecting the flow headed towards the leeward sidewalk from the street surface over the emission zones. At around three meters of CBW height (or higher), the flow is routed vertically upwards, therefore enabling the formation of a counter-rotating secondary canyon vortex, of approximately the same size as the primary canyon vortex (Figure 13e).

The consequence of the above-described phenomena is that on both sides of the central boundary wall, the near-ground flow is directed towards the middle lanes, then vertically upwards, transporting the majority of the air pollutants from the source areas to the mixing layer at the upper boundary of the street canyon. As a result, the traffic induced air pollutants are substantially diluted before reaching the crucial pedestrian areas.

This fundamental change to the flow field causes significant differences in the concentration field. The most heavily contaminated area shifts from the leeward pedestrian zone to the inner lanes, resulting in the improvement of air quality, viz. a negative concentration change of 12.4% to 33.4% at the leeward side walkway, as shown by the graphs of Figure 14. The concentration decrease is nearly proportional to the LBW height at the lower boundary walls; however, above two meters, the extra height results in only marginal improvements (between 30.5 and 33.4).

A minor air quality improvement is detected over the windward walkway, where the pedestrian exposure to the traffic induced air pollutants is inherently small, resulting in normalized concentration values between 3.5 and 4.2.

It is important to note, that by redirecting the flow, the CBW arrangements improve the air quality in the vicinity of both leeward and windward walls, providing the residents of the buildings with significantly cleaner air.

The sidewise boundary walls—contrary to the central boundary walls—do not restructure the flow within the street canyon. Although they modify the shape of the canyon vortex and generate two smaller eddies in their wake (shown in Figure 13h), their main role instead is sheltering the pedestrian zones from the heavy traffic areas.

As can be seen in Figure 14, a significant relative decrease in the normalized concentration (down to 13.1 on the leeward side) can be achieved using the SBWs of about three meters in height, reaching a relative change of −33.2% using SBWs, exceeding even the effect of the CBW with the same height. More importantly, the SBWs cannot decrease the concentration of air pollutants near the building walls as much as the CBW with the same height. Furthermore, it is worth noting that SBWs also prevent air exchange over the traffic lanes, thus increasing the average ground level concentration when the total area is taken into account. Therefore, SBWs may increase the exposure of vehicle drivers to traffic induced air pollutants, which prompts further investigation.

In conclusion, according to the present model, boundary walls with a medium height (around two to three meters) are capable of efficiently reducing pedestrian exposure to traffic induced air pollutants by one third at the critical location of the street canyon. Among the analyzed configurations, central boundary walls seem to be favorable, as they are able to decrease pollutant concentrations at the building walls, as well as at pedestrian head height on the leeward side.

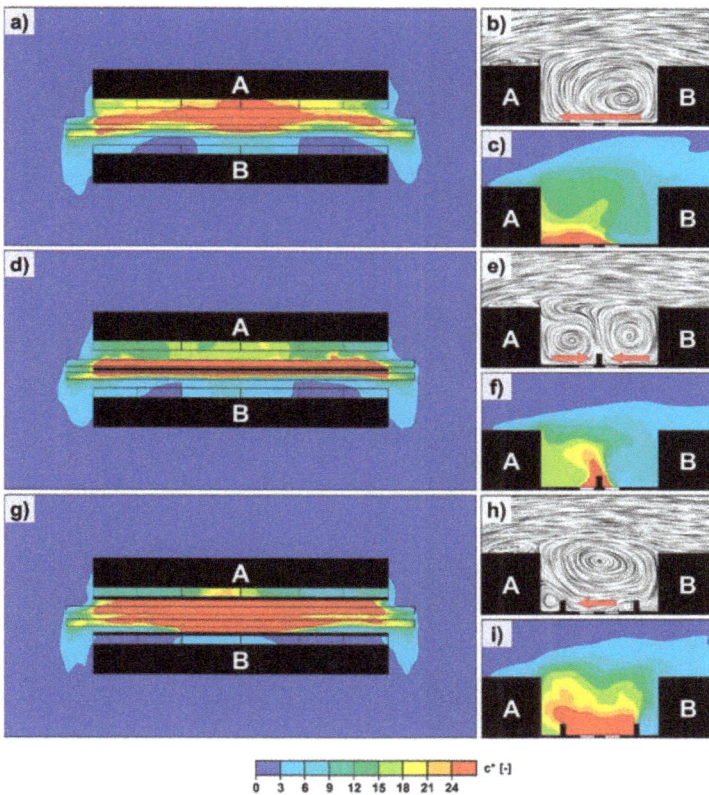

Figure 13. Time averaged streamlines and concentration contour plots. Left (**a,d,g**): typical distributions of the air pollutants at street level ($z/H = 0$). Right: typical flow fields (**b,e,h**) and normalized air pollutant concentration fields (**c,f,i**) in the center area around $y/H = 0$. The investigated cases from top to bottom: (**a–c**) no boundary walls (NBW; reference scenario); (**d–f**) central boundary walls (CBW); (**g–i**) sidewise boundary walls (SBW). The walls are 3.48 m (0.19H) high in the latter cases.

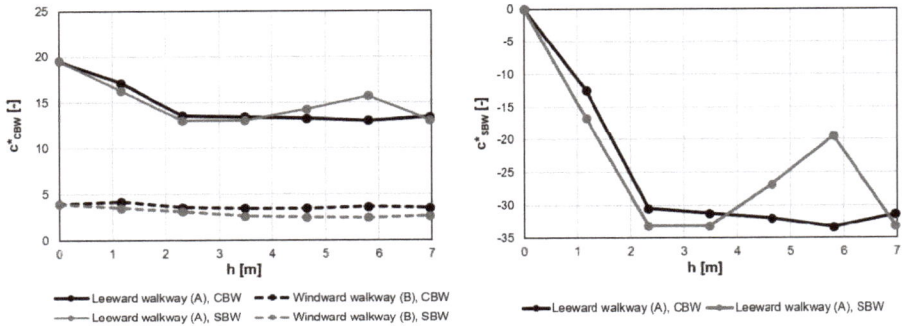

Figure 14. (**Left**) area-weighted averages of normalized concentration on the leeward and windward walkways (6 m wide surfaces along the entire building length), as functions of LBW height. (**Right**) normalized concentration changes at the leeward walkway as functions of LBW heights.

3.5. The Impact of the Roof Angle

The pitched roofs placed on top of the buildings (examined in the range of slope angle α, between 20° and 50° in four separate cases) alter the flow field by shifting the core of the canyon vortex, thus generating an intense, medium-sized recirculation zone in the vicinity of the roof of the windward building. Large-scale turbulence prevailed deeper in the street canyon. The flow structure (illustrated in Figure 15a–c) changes both spatially and temporally in the street canyon, although the advection caused by the shifted primary canyon vortex was observed along the whole length of the canyon during the entirety of the simulation.

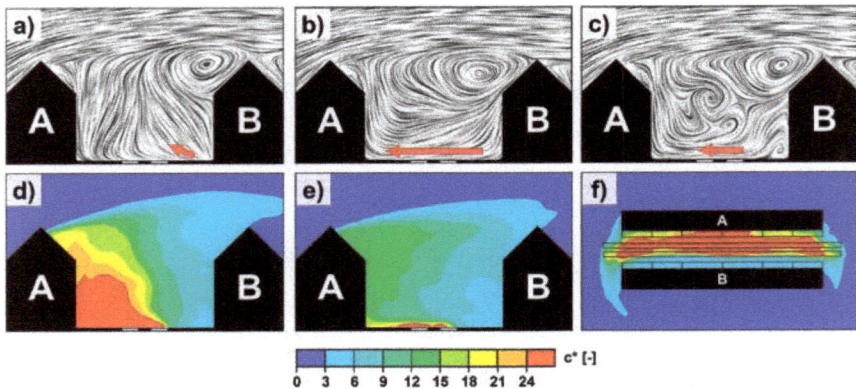

Figure 15. Typical flow structures of the street canyon completed with pitched roofs that have a 45° slope angle: (**a**) in the center area, at $y/H = 0$; (**b**) in the transitional area, at $y/H = 2.5$. (**c**) with a complex vortex structure in the center area. Typical normalized concentration contour plots: (**d**) in the center area, at $y/H = 0$; (**e**) in the transitional area, at $y/H = 2.5$ (**e**); (**f**) the typical distribution of air pollutants at ground level.

The shift of the canyon vortex in the case of buildings with pitched roofs (Figure 15) may depend heavily on model features such as the arrangement of crossroads or the number of successive buildings, which is confirmed by the fact that this phenomenon does not occur in the recent LES results of Kellnerova et al. [40] obtained in a periodic computational domain featuring infinitely wide pitched roofed buildings without crossroads.

The concentration distribution within the street canyon displays a pattern similar to that of the reference case characterized by flat roofs, although the more intensive turbulent mixing tends to balance leeward and windward concentrations in the case of steeper roofs. It was observed that the leeward side mean concentration is still higher than the windward side concentration by at least a factor of three in all cases.

Figure 16 shows that the introduction of the sloped roofs, which elevates the mixing layer, causes a rise in normalized concentration over both walkways as well. However, it can be observed, that at a 50° roof angle, the leeward concentration decreases and the windward concentration further rises, that is, the extremely high pollution in the middle of the leeward sidewalk is mitigated by the enhanced mixing.

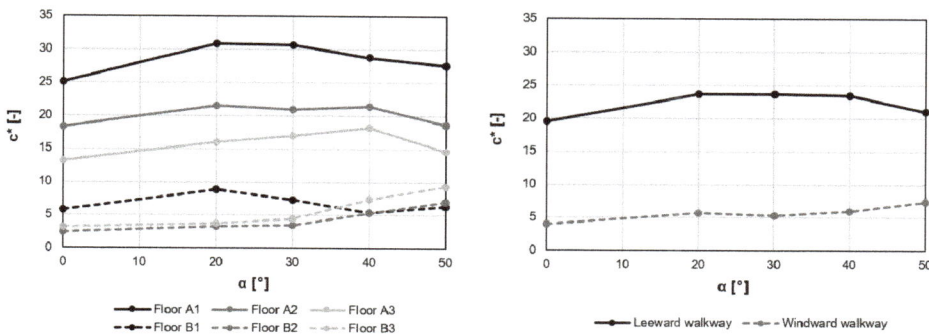

Figure 16. **Left**: normalized concentration at the focus areas as a function of the roof angle. **Right**: normalized concentration averaged on the leeward and windward walkways (6 m wide surfaces along the entire building length) as a function of the roof angle.

4. Conclusions and Outlook

ANSYS Discovery Live 19.2, which is primarily developed for mechanical engineering applications, seems to be a useful tool for examining the dispersion of urban air pollutants by utilizing the analogy between heat and mass transport processes. Owing to the GPU-based calculation method and automatic meshing, the large eddy simulation (LES) models can be significantly accelerated on commonly used PCs. Discovery Live can be used beneficially in the early phases of the design process for the selection of favorable concepts, which can be further examined with established CFD tools or wind tunnel experiments. With the help of this software, it was possible to produce statistically converged LES results on a PC using a NVIDIA GTX 1080Ti graphics card, in approximately one hour of computing time.

Using the inlet configuration and computational domain presented in this paper, the LES provided satisfactory agreement with the time averaged approaching velocity and turbulence profiles used in the earlier wind tunnel experiments [19], as well as the velocity distribution of the airflow entering and leaving the finite length street canyon. The concentration distribution of the LES model showed an excellent agreement with experimental results [19], according to the common validation metrics [39], which suggests that Discovery Live is applicable in urban dispersion studies subject to the following conditions:

- the investigated flow needs to be insensitive to boundary layer effects (e.g., only sharp-edged buildings are assumed);
- the unresolved geometrical features do not substantially impact the flow; thus, the use of porous media models is not necessary;
- the investigated pollutant can be well represented by a passive scalar;

- the flow is isothermal;
- the source intensity is pre-defined and does not depend on flow characteristics;
- the investigated geometrical configuration is simple enough to allow for the resolution of every important flow feature.

In this study, the favorable effect of low boundary walls (LBWs) on the air quality of the street canyon was analyzed in a wider parameter range compared to previous studies, investigating the relationship between flow structure and concentration distribution. In the reference scenario without any boundary walls, the highest concentration of pollutants is generated at the center of the leeward pedestrian zone in the case of the street canyon subjected to wind with a perpendicular approach. Based on our model results, pedestrian exposure to traffic induced air pollution can be reduced by LBWs—or dense hedgerows—placed either in the middle of the street or between the sidewalks and the traffic lanes. The maximum achievable concentration decrease using the LBWs was about 41% at the critical point, however, over the leeward walkway, the concentration can be mitigated by one third of the inherent mean concentration, according to the present model.

From the point of view of the air quality of the sidewalks, the LBWs placed on both sides result in a similar degree of concentration reduction as that of the central LBW; however, in the latter case, the concentration increases over the traffic lanes. Both solutions require a substantial LBW height; roughly $h = 0.15H$ is necessary in a street canyon with an $H/W = 0.5$ aspect ratio.

The building models used in the wind tunnel experiments of Gromke et al. [19] were completed with pitched roofs, and the effect of the roof's angle on air pollutant concentration over the pedestrian sidewalks was investigated. Although the lowest concentration was found in the case of flat top buildings, only a moderate concentration increase was found with a $50°$ roof angle, which suggests that the construction of slightly shorter buildings with pitched roofs, such as those utilized by loft apartments, could further improve air quality on the sidewalks, while retaining useful built-in volume. To explore the effects of such geometrical modifications, further investigations are planned by using the modelling approach presented.

In the authors' opinion, the relatively simple developments listed below could significantly improve the reliability and efficiency of Discovery Live in urban dispersion simulations:

- Providing direct information about mesh resolution and time step size.
- Detailed descriptions of the governing equations, the turbulence model, and the numerical schemes.
- The optional solution of a passive scalar transport equation.
- More control of the averaging period, and the ability to set the interval length or the start and end points, which would make the analysis easier and more accurate.
- Averaging on arbitrarily defined surfaces and volumes to enable the direct output of, for example, the pedestrian head height mean concentration, instead of the pointwise export.
- Porous media models for taking into account the effect of vegetation or similar finely structured flow obstructions.

As the impact of urban architecture on the propagation of air pollutants is still largely unexplored, it is necessary to study a wide variety of different geometrical configurations. We believe that, due to its superior cost–benefit ratio, the proposed GPU-based simulation method can be an effective component for such investigations.

Author Contributions: Conceptualization, G.K.; methodology, G.K. and B.P.; validation, B.P.; writing, B.P and G.K.; visualization, B.P.

Funding: This study was funded by grant no. K 124439, NKFIH from the National Research, Development, and Innovation Office, Hungary. The research reported in this paper was supported by the Higher Education Excellence Program of the Ministry of Human Capacities, within the framework of the Water Science and Disaster Prevention Research Area of the Budapest University of Technology and Economics (BME FIKP-VÍZ).

Acknowledgments: The authors acknowledge the support of CFD.HU Ltd., who provided access to the necessary hardware and software.

Conflicts of Interest: The authors declare no conflict of interest. The founding sponsors had no role in the design of the study; in the collection, analyses, or interpretation of data; in the writing of the manuscript; and in the decision to publish the results.

References

1. Oke, T.R. Street Design and Urban Canopy Layer Climate. *Energy Build.* **1988**, *11*, 103–113. [CrossRef]
2. Hang, J.; Li, Y.; Sandberg, M.; Claesson, L. Wind Conditions and Ventilation in High-Rise Long Street Models. *Build. Environ.* **2010**, *45*, 1353–1365. [CrossRef]
3. Vardoulakis, S.; Fisher, B.E.A.; Pericleous, K.; Gonzalez-Flesca, N. Modelling Air Quality in Street Canyons: A Review. *Atmos. Environ.* **2003**, *37*, 155–182. [CrossRef]
4. Ahmad, K.; Khare, M.; Chaudrhy, K.K. Wind Tunnel Simulation Studies on Dispersion at Urban Street Canyons and Intersections—A Review. *J. Wind Eng. Ind. Aerodyn.* **2005**, *93*, 697–717. [CrossRef]
5. Vogt, J.; Lauerbach, H.; Meurer, M.; Langner, M. The Influence of Urban Vegetation on Air Flow. In Proceedings of the Fifth International Conference on Urban Climate (ICUC-5), Lodz, Poland, 1–5 September 2003; pp. 471–474.
6. Freer-Smith, P.H.; El-Khatib, A.A.; Taylor, G. Capture of Particulate Pollution by Trees: A Comparison of Species Typical of Semi-Arid areas (Ficus Nitida and Eucalyptus Globulus with European and North American Species. *Water Air Soil Pollut.* **2004**, *155*, 173–187. [CrossRef]
7. Nowak, D.J.; Crane, D.E.; Stevens, J.C. Air Pollution Removal by Urban Trees and Shrubs in the United States. *Urban For. Urban Green.* **2006**, *4*, 115–123. [CrossRef]
8. Ries, K.; Eichhorn, J. Simulation of Effects of Vegetation on the Dispersion of Pollutants in Street Canyons. *Meteorol. Z.* **2001**, *10*, 229–233. [CrossRef]
9. Gromke, C.; Ruck, B. Influence of Trees on the Dispersion of Pollutants in an Urban Street Canyon— Experimental Investigation of the Flow and Concentration Field. *Atmos. Environ.* **2007**, *41*, 3287–3302. [CrossRef]
10. Gromke, C.; Denev, J.; Ruck, B. Dispersion of Traffic Exhausts in Urban Street Canyons with Tree Plantings—Experimental and Numerical Investigations. In Proceedings of the International Workshop on Physical Modelling of Flow and Dispersion Phenomena (PHYSMOD 2007), Orléans, France, 23–25 August 2007; pp. 121–128.
11. Gromke, C.; Buccolieri, R.; Di Sabatino, S.; Ruck, B. Dispersion Study in a Street Canyon with Tree Planting by Means of Wind Tunnel and Numerical Investigations—Evaluation of CFD Data with Experimental Data. *Atmos. Environ.* **2008**, *42*, 8640–8650. [CrossRef]
12. Gromke, C.; Ruck, B. On the Impact of Trees on Dispersion Processes of Traffic Emissions in Street Canyons. *Bound. Layer Meteorol.* **2009**, *131*, 19–34. [CrossRef]
13. Balczó, M.; Gromke, C.; Ruck, B. Numerical Modeling of Flow and Pollutant Dispersion in Street Canyons with Tree Planting. *Meteorol. Z.* **2009**, *18*, 197–206. [CrossRef]
14. Buccolieri, R.; Gromke, C.; Di Sabatino, S.; Ruck, B. Aerodynamic Effects of Trees on Pollutant Concentration in Street Canyons. *Sci. Total Environ.* **2009**, *407*, 5247–5256. [CrossRef] [PubMed]
15. Litschke, T.; Kuttler, W. On the Reduction of Urban Particle Concentration by Vegetation—A Review. *Meteorol. Z.* **2008**, *17*, 229–240. [CrossRef]
16. Janhaell, S. Review on Urban Vegetation and Particle Air Pollution—Deposition and Dispersion. *Atmos. Environ.* **2015**, *105*, 130–137. [CrossRef]
17. McNabola, A.; Broderick, B.M.; Gill, L.W. A Numerical Investigation of the Impact of Low Boundary Walls on Pedestrian Exposure to Air Pollutants in Urban Street Canyons. *Sci. Total Environ.* **2009**, *407*, 760–769. [CrossRef] [PubMed]
18. Gallagher, J.; Gill, L.W.; McNabola, A. Numerical Modelling of the Passive Control of Air Pollution in Asymmetrical Urban Street Canyons Using Refined Mesh Discretization Schemes. *Build. Environ.* **2012**, *56*, 232–240. [CrossRef]
19. Gromke, C.; Jamarkattel, N.; Ruck, B. Influence of Roadside Hedgerows on Air Quality in Urban Street Canyons. *Atmos. Environ.* **2016**, *139*, 75–86. [CrossRef]
20. Tominaga, Y.; Stathopoulos, T. CFD Simulation of Near-Field Pollutant Dispersion in the Urban Environment: A Review of Current Modeling Techniques. *Atmos. Environ.* **2013**, *79*, 716–730. [CrossRef]

21. Menter, F.R. *Best Practice: Scale-Resolving Simulations in ANSYS CFD Version 2.00*; ANSYS Germany GmbH: Nuremberg, Germany, 2015.

22. Hernádi, Z.; Kristóf, G. Prediction of Pressure Drop and Heat Transfer Coefficient in Helically Grooved Heat Exchanger Tubes Using Large Eddy Simulation. *Proc. Inst. Mech. Eng. Part A J. Power Energy* **2013**, *228*, 317–327. [CrossRef]

23. Liu, C.-H.; Barth, M.C. Large-Eddy Simulation of Flow and Scalar Transport in a Modeled Street Canyon. *J. Appl. Meteorol.* **2002**, *41*, 660–673. [CrossRef]

24. Liu, C.-H.; Barth, M.C.; Leung, D.Y.C. Large-Eddy Simulation of Flow and Pollutant Transport in Street Canyons of Different Building-Height-to-Street-Width Ratios. *J. Appl. Meteorol.* **2004**, *43*, 1410–1424. [CrossRef]

25. Liu, C.-H.; Barth, M.C.; Leung, D.Y.C. On the Prediction of Air Pollutant Exchange Rates in Street Canyons of Different Aspect Ratios Using Large Eddy Simulation. *Atmos. Environ.* **2005**, *39*, 1567–1574. [CrossRef]

26. So, E.S.P.; Chan, A.T.Y.; Wong, A.Y.T. Large Eddy Simulation of Wind Flow and Pollutant Dispersion in a Street Canyon. *Atmos. Environ.* **2005**, *39*, 3573–3582. [CrossRef]

27. Michioka, T.; Sato, A.; Takimoto, H.; Kanda, M. Large Eddy Simulation for the Mechanism of Pollutant Removal from a Two-Dimensional Street Canyon. *Bound. Layer Meteorol.* **2011**, *138*, 195–213. [CrossRef]

28. Baker, J.; Walker, H.L.; Cai, X. A Study of the Dispersion and Transport of Reactive Pollutants in and above Street Canyons—A Large Eddy Simulation. *Atmos. Environ.* **2004**, *38*, 6883–6892. [CrossRef]

29. Kikumoto, H.; Ooka, R. A Study on Air Pollutant Dispersion with Bimolecular Reactions in Urban Street Canyons Using Large Eddy Simulations. *J. Wind Eng. Ind. Aerodyn.* **2012**, *104–106*, 516–522. [CrossRef]

30. Schalkwijk, J.; Griffith, E.J.; Post, F.H.; Jonker, H.J.J. High-Performance Simulations of Turbulent Clouds on a Desktop PC: Exploiting the GPU. *Bull. Am. Meteor. Soc.* **2012**, *93*, 307–314. [CrossRef]

31. Onodera, N.; Aoki, T.; Shimokawabe, T.; Kobayashi, H. Large-Scale LES Wind Simulation Using Lattice Boltzmann Method for a 10km × 10km Area in Metropolitan Tokyo. *TSUBAME e-Sci. J. Glob. Sci. Inf. Comput. Cent.* **2013**, *9*, 2–8.

32. Boundary Layer Profile—Wind tunnel No. 4—Laboratory of Building- and Environmental Aerodynamics. Available online: http://www.windforschung.de/bilder_orginale/CODA/approaching%20flow%20wind%20tunnel.htm (accessed on 18 September 2018).

33. Kastner-Klein, P.; Berkowitz, R.; Ritter, R. The Influence of Street Architecture on Flow and Dispersion in Street Canyons. *Meteorol. Atmos. Phys.* **2004**, *87*, 121–131. [CrossRef]

34. DePaul, F.T.; Sheih, C.M. Measurements of Wind Velocities in a Street Canyon. *Atmos. Environ.* **1986**, *20*, 455–459. [CrossRef]

35. Hoydysh, W.G.; Dabberdt, W.F. Kinematics and Dispersion Characteristics in Asymmetric Street Canyons. *Atmos. Environ.* **1988**, *22*, 2677–2689. [CrossRef]

36. Hunter, L.J.; Watson, I.D.; Johnson, G.T. Modelling Air Flow Regimes in Urban Street Canyons. *Energy Build.* **1990**, *15*, 315–324. [CrossRef]

37. Belcher, S.E. Mixing and Transport in Urban Areas. *Philos. Trans. R. Soc. Lond. Math. Phys. Eng. Sci.* **2005**, *363*, 2947–2968. [CrossRef] [PubMed]

38. Kellnerova, R.; Janour, Z. The Flow Instabilities within an Urban Intersection. *Int. J. Environ. Pollut.* **2008**, *47*, 268–277. [CrossRef]

39. Franke, J. The European COST Action 732—Quality Assurance and Improvement of Micro-Scale Meteorological Models. In Proceedings of the Seventh Asia-Pacific Conference on Wind Engineering, Taipei, Taiwan, 8–12 November 2009.

40. Kellnerova, R.; Fuka, V.; Uruba, V.; Jurcakova, K.; Mosek, S.; Chaloupecka, H.; Janour, Z. On Street-Canyon Flow Dynamics: Advanced Validation of LES by Time-Resolved PIV. *Atmosphere* **2018**, *9*, 161. [CrossRef]

atmosphere

MDPI

Article

Natural Ventilation of a Small-Scale Road Tunnel by Wind Catchers: A CFD Simulation Study

Shanhe Liu [1,2], Zhiwen Luo [3,*], Keer Zhang [1,2] and Jian Hang [1,2,*]

[1] School of Atmospheric Sciences, Sun Yat-sen University, Guangzhou 510275, China;
    liushh9@mail2.sysu.edu.cn (S.L.); zhangker@mail2.sysu.edu.cn (K.Z.)
[2] Guangdong Province Key Laboratory for Climate Change and Natural Disaster Studies, Sun
    Yat-sen University, Guangzhou 510275, China
[3] School of the Built Environment, University of Reading, Reading RG6 6UR, UK
[*] Correspondence: z.luo@reading.ac.uk (Z.L.); hangj3@mail.sysu.edu.cn (J.H.);
    Tel.: +44(0)-118-378-5219 (Z.L.); +86-20-8411-2436 (J.H.)

Received: 29 July 2018; Accepted: 15 October 2018; Published: 20 October 2018

Abstract: Providing efficient ventilation in road tunnels is essential to prevent severe air pollution exposure for both drivers and pedestrians in such enclosed spaces with heavy vehicle emissions. Longitudinal ventilation methods like commercial jet fans have been widely applied and confirmed to be effective for introducing external fresh air into road tunnels that are shorter than 3 km. However, operating tunnel jet fans is energy consuming. Therefore, for small-scale (~100 m–1 km) road tunnels, mechanical ventilation methods might be highly energetically expensive and unaffordable. Many studies have found that the use of wind catchers could improve buildings' natural ventilation, but their effect on improving natural ventilation in small-scale road tunnels has, hitherto, rarely been studied. This paper, therefore, aims to quantify the influence of style and arrangement of one-sided flat-roof wind catchers on ventilation performance in a road tunnel. The concept of intake fraction (*IF*) is applied for ventilation and pollutant exposure assessment in the overall tunnel and for pedestrian regions. Computational fluid dynamics (CFD) methodology with a standard k-epsilon turbulence model is used to perform a three-dimensional (3D) turbulent flow simulation, and CFD results have been validated by wind-tunnel experiments for building cross ventilation. Results show that the introduction of wind catchers would significantly enhance wind speed at pedestrian level, but a negative velocity reduction effect and a near-catcher recirculation zone can also be found. A special downstream vortex extending along the downstream tunnel is found, helping remove the accumulated pollutants away from the low-level pedestrian sides. Both wind catcher style and arrangement would significantly influence the ventilation performance in the tunnel. Compared to long-catcher designs, short-catchers would be more effective for providing fresh air to pedestrian sides due to a weaker upstream velocity reduction effect and smaller near-catcher recirculation zone. In long-catcher cases, *IF* increases to 1.13 ppm when the wind catcher is positioned 240 m away from the tunnel entrance, which is almost twice that in short-catcher cases. For the effects of catcher arrangements, single, short-catcher, span-wise, shifting would not help dilute pollutants effectively. Generally, a design involving a double short-catcher in a parallel arrangement is the most recommended, with the smallest *IF*, i.e., 61% of that in the tunnel without wind catchers (0.36 ppm).

Keywords: road tunnel; natural ventilation; wind catcher; intake fraction

---

1. Introduction

With the development of urban transportation networks, urban road tunnels have gained popularity and can be commonly found in many large cities. These tunnels provide an efficient transportation process for vehicles and offer a much safer public traveling experience for pedestrians.

Besides, they are core connections between different transportation systems like metro stations and bus stops, coupled with the provision of shade and commercial opportunities.

A road tunnel, as a semi-enclosed space, accommodating both vehicles and pedestrians, is a place of concern for exposure to air pollution. According to the city air quality report for the Barbican, London, UK [1], air pollution is quite severe in vehicle-pedestrian tunnels, with the concentration of pollutants like carbon monoxide (CO) and nitrogen dioxide (NO_2) dramatically exceeding the EU mean annual target. The highest monthly mean NO_2 concentration in the Beech Street tunnel in the City of London was found in November (30% higher than the EU target). Keyte et al. [2] conducted field measurements of polycyclic aromatic hydrocarbons (PAH) in tunnels in Birmingham and Paris which found that the mean PAH concentration in a tunnel was approximately 4.5 times higher than that in the ambient environment. Similar results of severe air pollution in road tunnels are also revealed by Kim et al. [3] and Wingfors et al. [4]. Thus, improvement on ventilation performance in the vehicle-pedestrian tunnels is necessary. Ventilation performance in tunnels is always poor due to the semi-enclosed space, and therefore supplying clean external air into the inner tunnel has become the priority for road tunnel ventilation design.

Many effective methodologies for enhancing tunnel ventilation have been proposed and investigated by researchers. Betta et al. [5] studied the optimal pitch angle of the tunnel jet fans, which could induce the greatest amount of fresh air from the outdoor environment. The possibility of improving the inner-tunnel ventilation by a set of spaced jet fans fixed on the top of the tunnel has also been explored [6–8]. The results showed that the performance of the commercial jet fans varies dramatically with the number, pitch angle, and position of jet fans, whilst the best pitch angle varied along with traffic conditions: When there is no traffic, the best pitch angle is 6 degrees and, during traffic jams, the pitch angle should be maintained between 2 and 4 degrees. Besides, ventilation performance in curved tunnels with jet fans was also investigated by Wang et al. [9]. This kind of longitudinal ventilation scheme with jet fans is effective and commonly used in tunnels less than 3 km [5,7]. However, the operation of such ventilation engines is energy consuming. According to Peeling et al. [10], most of the energy consumption in tunnels is due to lighting and ventilation, and the surveys from sample tunnels in Australia, Norway, and the Netherlands indicate that tunnel consumption could reach 356 kWh per meter of tunnel per annum. What is more, tunnel jet fans would also generate severe noise in the tunnel, which is not favored by pedestrians. Hence, new types of effective and energy-efficient ventilation schemes should be explored. Natural ventilation could possibly be the most energy-saving method for inner-tunnel ventilation in small-scale tunnels. Harish [11] investigated the performance of ceiling openings in a tunnel for removing smoke and hot gases during fire emergencies. This kind of natural ventilation approach is commonly studied by researchers into buoyancy-driven tunnel ventilation [12–15], but it tends to be ineffective for providing wind-driven ventilation.

Wind catchers—a traditional building design—are commonly used for providing natural ventilation for buildings in hot and arid or humid areas. Dehghani-sanij and Soltani [16,17] gave a specific review of proposed wind catcher designs for buildings. Both traditional and the most advanced design of wind towers have been introduced and compared by reviewers. While traditional styles of design like one-sided and two-sided wind towers would be very efficient for providing indoor ventilation and cooling effects in windy regions, the design still has limitations in facility protection and operation. The effect of wind catchers on building ventilation performance has been widely studied. An experimental study conducted by Afshin et al. [18] showed that the ventilation capacity of wind catchers increases dramatically with increases in the external wind speed. Two-sided or multi-opening wind catchers can provide more effective air exchange capacity between inner rooms and the outdoor environment. However, the flow rate would decrease with the increase in the wind incident angle and the number of catcher openings [18,19]. Optimal arrangements and shapes of wind catcher have also been investigated [19–22]. The wind catchers with a curved or inclined roof are able to supply higher ventilation rates compared with the ones with a flat roof. Catcher-induced flow is sensitive to the wind incident angle and the wind catchers in a staggered arrangement would induce a much stronger wind by preventing the blockage effect from the upstream flow. The thermal effects of wind catchers, like indoor thermal comfort and cooling capacity, have also been explored by Hosseini et al. [23] and Calautit and Hughes [24]. Bhadori introduced wind towers with wet columns or wet surfaces [25]. In this design, water evaporation helps increase the cooling potential of wind towers and helps introduce much cooler air into the building. Soltani et al. investigated the proposed design of wind towers with wetted surfaces by analyzing velocity, total pressure, and the pressure coefficient in different wind speed conditions [17]. Results show that the ventilation performance of wind towers is much greater at much higher wind velocities due to relatively smaller eddies within the wind tower.

Haghighi et al. [26] studied the possibility of applying wind catchers to an integrated indoor ventilation and cooling system in a complex indoor construction. Calautit et al. [22] first simulated and analyzed pollutant distribution in an indoor environment and discussed its dilution with the arrangement of wind catchers. The literature reported that wind catchers could help dilute indoor pollutants but were also likely to recirculate pollutants from upstream regions. Generally, wind catchers have been proved to be an effective natural ventilation and cooling approach for various indoor environments. However, the possibility of applying such types of architectural design in small-scale tunnels has not yet been investigated. This research aims to explore the ventilation capability of wind catchers in vehicle-pedestrian tunnels and propose a new avenue for natural ventilation in small-scale tunnels.

In order to evaluate the impacts of different types of ventilation design on air pollutants within a tunnel environment, this research adopts the concept of the intake fraction (*IF*), which is defined as the fraction of the total emission inhaled by the population [27–30]. In this research, CFD methodology with a validation process is applied for investigation. A total of 35 cases including the base case (without wind catcher) are built to investigate the wind catcher's ventilation performance for different wind catcher arrangements and styles. In every single case series, the effects of wind catcher position on ventilation performance are presented, and the effects of the catcher style are addressed for different case series. The structure of the remaining paper is as follows: Section 2 introduces the numerical method of computational fluid dynamics (CFD) and the models built in this study. Section 3 introduces the concept of the intake fraction (*IF*). The results and discussion are presented in Section 4 and the conclusions given in Section 5.

2. CFD Methodology and Model Description

2.1. CFD Methodology

Computational Fluid Dynamic (CFD) is applied in this study to understand the aerodynamics and pollutant dispersion in the tunnel environment using Fluent 6.3.26 [31]. The road tunnel environment could be regarded as a special kind of indoor environment but on a much larger scale than an indoor room, it approximately fits the scale of urban street canyons. The standard *k*-ε model can well-simulate the airflow and pollutant dispersion with a performance which has been highly validated in urban scale studies [27]. Ghadiri [32] investigated the simulation performance of different numerical models in cases of two-sided rectangular wind catchers. Though SSG RSM models are regarded as being superior to standard *k*-ε models in terms of predicting the lateral volumetric airflow rate, the standard *k*-ε model is still widely used to study wind catcher performance because of its proven accuracy in wind catcher ventilation [20,22,24,32]. Hence, the standard *k*-ε model is employed to solve the steady state isothermal problem in this study.

The governing equations are discretized using the finite volume method (FVM) and are built into the model with a second order upwind scheme. Additionally, a Semi-implicit Method for a Pressure-linked Equation (SIMPLE) algorithm is applied in the steady-state simulation. The governing equations in the standard *k*-ε model are presented as follows:

Mass conservation equation [33–35]:

$$\frac{\partial \overline{u}_i}{\partial x_i} = 0 \tag{1}$$

Momentum equation:

$$\overline{u}_j \frac{\partial \overline{u}_i}{\partial x_j} = -\frac{1}{\rho}\frac{\partial \overline{p}}{\partial x_i} + \frac{\partial}{\partial x_j}\left(v\frac{\partial \overline{u}_i}{\partial x_j} - \overline{u_i'' u_j''}\right) \tag{2}$$

Transport equation for turbulent kinetic energy (k) and dissipation rate (ε):

$$\overline{u}_i \frac{\partial k}{\partial x_i} = \frac{\partial}{\partial x_i}\left[\left(v + \frac{v_t}{\sigma_k}\right)\frac{\partial k}{\partial x_i}\right] + \frac{1}{\rho}P_k - \varepsilon \tag{3}$$

$$\overline{u}_i \frac{\partial \varepsilon}{\partial x_i} = \frac{\partial}{\partial x_i}\left[\left(v + \frac{v_t}{\sigma_\varepsilon}\right)\frac{\partial \varepsilon}{\partial x_i}\right] + \frac{1}{\rho}C_{\varepsilon 1}\frac{\varepsilon}{k}P_k - C_{\varepsilon 2}\frac{\varepsilon^2}{k} \tag{4}$$

In the equation, $\overline{u}_j$ is the time-averaged velocity components and u_i'' is the fluctuation of velocity components. v and $v_t = C_\mu \frac{k^2}{\varepsilon}$ are the kinematic viscosity and kinematic eddy viscosity respectively. $-\overline{u_i'' u_j''} = v_t\left(\frac{\partial \overline{u}_i}{\partial x_j} + \frac{\partial \overline{u}_j}{\partial x_i}\right) - \frac{2}{3}k\delta_{ij}$ is the Reynold stress tensor and. $P_k = v_t \times \frac{\partial \overline{u}_i}{\partial x_j}\left(\frac{\partial \overline{u}_i}{\partial x_j} + \frac{\partial \overline{u}_j}{\partial x_i}\right)$ is the turbulence production term. $C_{\varepsilon 1}$, $C_{\varepsilon 2}$ and C_μ are empirical constants, which are derived from experiment and equal to 1.44, 1.92 and 0.09 respectively. $\sigma_k = 1.0$ and $\sigma_\varepsilon = 1.3$ are Prandtl numbers in the equations of turbulent kinetic energy and its dissipation rate.

The under-relaxation factors for the pressure term, momentum term, k and ε terms are 0.3, 0.7, 0.8 and 0.8 respectively. The solution process does not stop until all residuals stop decreasing. Typical residuals at convergence are less than: 5×10^{-11}, 2×10^{-11}, 3×10^{-11} for stream-wise velocity, span-wise velocity, and vertical velocity respectively, and 2×10^{-10}, 2×10^{-8}, 1×10^{-7}, 3×10^{-9} for k, ε, continuity, and CO respectively. Figure 1 summarizes the numerical methodology employed in this study.

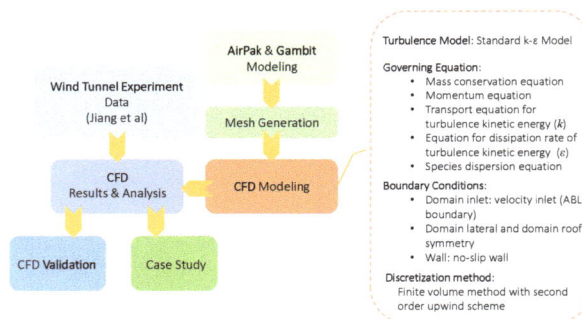

Figure 1. Computational Fluid Dynamic (CFD) modeling flowchart.

2.2. Model Description

Figure 2a depicts the 3D domain of the tunnel-catcher model in the CFD simulation (single long-catcher case with catcher positioned at $x = 150$ m). Figure 2b–d presents the details of the three basic catcher arrangement styles, which correspond to three different case series: Single long-catcher series, single short-catcher series and double short-catcher series, and the specified cases are listed in Table 1. The idea of the basic catcher style comes from the flat-roof catcher design that has been widely adopted for improving ventilation in some enclosed indoor environments. For investigating the effects of the catcher length, models of a long catcher and a short catcher were built respectively, and for exploring the influence of the integrated effect of two short wind catchers, double short-catcher cases are built. A total of 35 cases including the base case (without wind catcher) were built to investigate the catcher's ventilation performance in different arrangements and catcher styles. In every single case series, the effects of catcher position on ventilation performance would be emphasized, and between different case series, the effects of the catcher style could be addressed.

Table 1. List of Cases.

Base Case	Case Name	x (m)	x/L		
-	Case [Base]	-	-		
Series of Single Long-Catchers	**Case Name**	**x (m)**	**x/L**		
	Case_sl [10]	10	0.04		
	Case_sl [15]	15	0.06		
	Case_sl [20]	20	0.07		
	Case_sl [25]	25	0.09		
	Case_sl [30]	30	0.11		
	Case_sl [40]	40	0.15		
	Case_sl [50]	50	0.19		
	Case_sl [60]	60	0.22		
	Case_sl [90]	90	0.33		
	Case_sl [120]	120	0.44		
	Case_sl [150]	150	0.56		
	Case_sl [180]	180	0.67		
	Case_sl [210]	210	0.78		
	Case_sl [240]	240	0.89		
Series of Single Short-Catchers	**Case Name**	**x (m)**	**x/L**	**y (m)**	
	Case_ss [30]	30	0.11	0	
	Case_ss [60]	60	0.22	0	
	Case_ss [90]	90	0.33	0	
	Case_ss [120]	120	0.44	0	
	Case_ss [150]	150	0.56	0	
	Case_ss [150 (2.5)]	150	0.56	2.5	
	Case_ss [150 (5.0)]	150	0.56	5.0	
	Case_ss [180]	180	0.67	0	
	Case_ss [210]	210	0.78	0	
	Case_ss [240]	240	0.89	0	
Series of Double Short-Catchers	**Case Name**	**x_1 (m)**	**x_1/L**	**x_2 (m)**	**x_2/L**
	Case_ds [30, 30]	30	0.11	30	0.11
	Case_ds [60, 60]	60	0.22	60	0.22
	Case_ds [90, 90]	90	0.33	90	0.33
	Case_ds [120, 120]	120	0.44	120	0.44
	Case_ds [150, 150]	150	0.56	150	0.56
	Case_ds [120, 180]	120	0.44	180	0.67
	Case_ds [90, 210]	90	0.33	210	0.78
	Case_ds [180, 180]	180	0.67	180	0.67
	Case_ds [210, 210]	210	0.78	210	0.78
	Case_ds [240, 240]	240	0.89	240	0.89

Rules for case names: **Case_Series [CP-l (dl), CP-r (dl)]**; **Case_sl**: Series of single-long catcher; **Case_ss**: Series of single-short catcher; **Case_ds**: Series of double-short catcher; **CP-l**: Left catcher position (m); **CP-r**: Right catcher position (m) (default: No catcher on the right); **dl**: Distance (m) from left tunnel wall (default value: 0 m).

The tunnel model was built on a 1:1 real scale in accordance with the real-world vehicle-pedestrian tunnel in the Barbican district, City of London, UK. The tunnel height is $H = 5$ m and the length of the tunnel is $L = 270$ m. The long catcher was designed to be 3 m tall ($0.6H$) and 5 m wide (H) with a length of 15 m ($3H$) (Figure 2b). Meanwhile, the short catcher was designed to be 2.5 m in length (Figure 2c). The thickness of the building was neglected to reduce the mesh number and improve the computational efficiency due to its relatively small influence on flow at this scale, which was confirmed in some pre-modeling tests in this research (Figures A1 and A2). The preliminary test is specifically discussed in Appendix A.

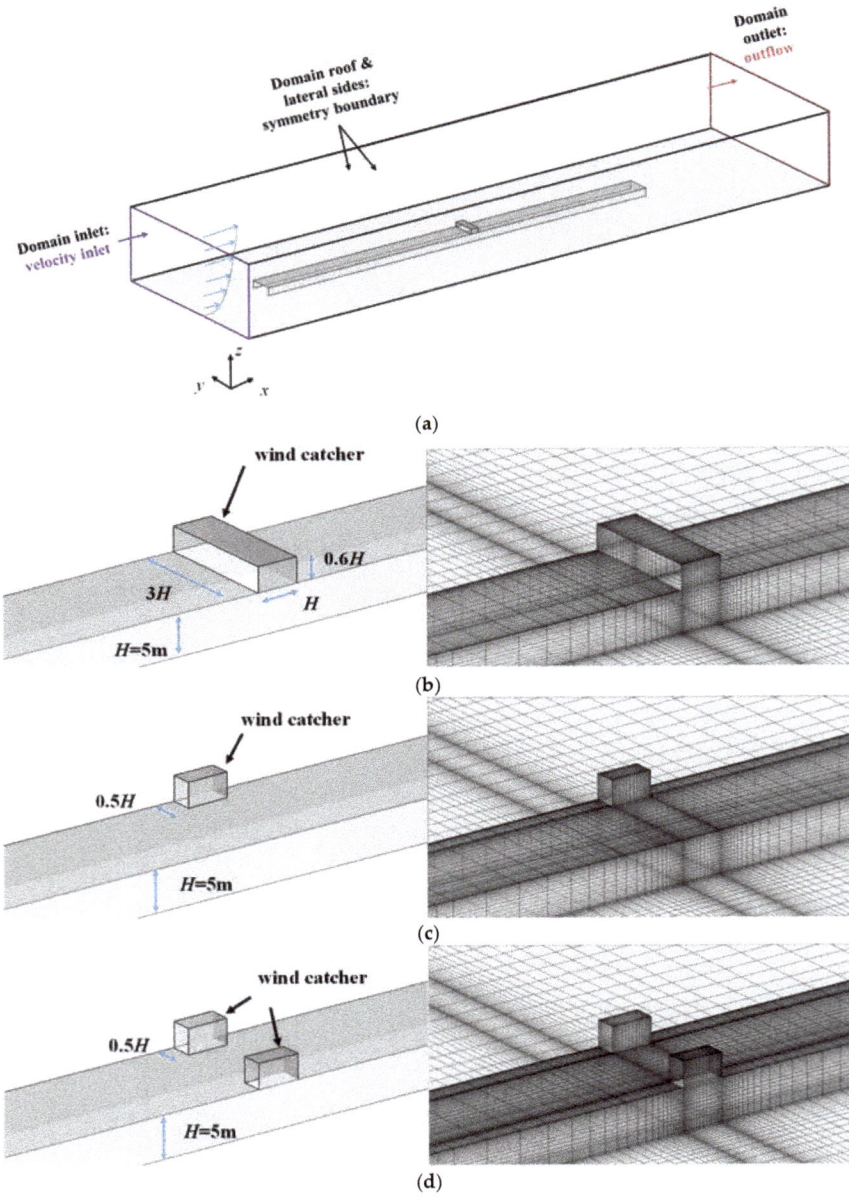

(a)

(b)

(c)

(d)

Figure 2. *Cont.*

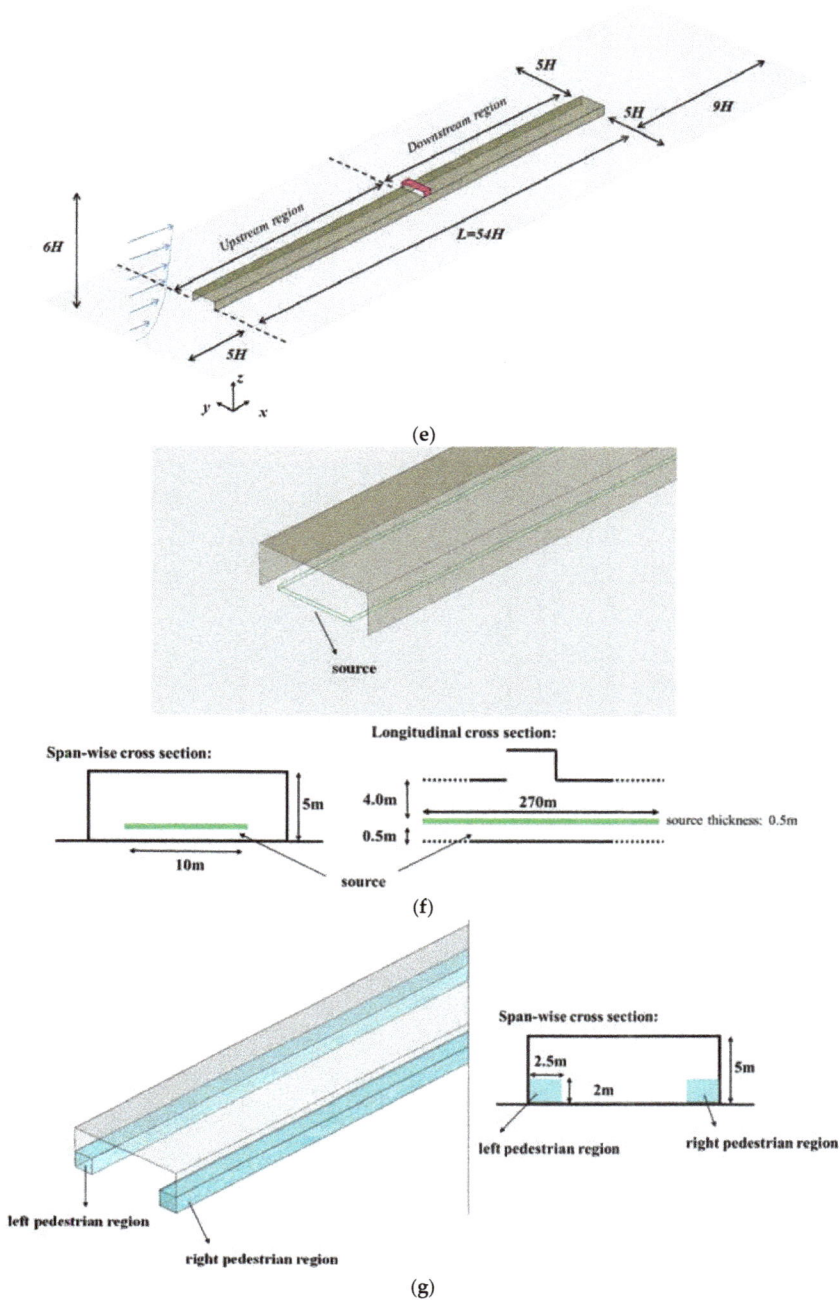

Figure 2. (**a**) Physical model and boundary condition setups in tunnel-catcher simulation cases. Model and grid distribution of (**b**) single long-catcher cases; (**c**) single short-catcher cases and (**d**) double short-catcher cases; (**e**) setup in computational domain; (**f**) source setup in cases; (**g**) setup of pedestrian regions.

2.3. CFD Domain, Boundary Conditions and Grid Arrangements

As displayed in Figure 1, the general blockage ratio of the computational domain in this research was smaller than 3.0%, which satisfies the requirement in the urban wind environment simulation CFD guideline [36]. Furthermore, according to this CFD guidance, $5H$ was reserved between the target building (i.e., tunnel) and the laterals, front and top boundaries to ensure full development of the inlet wind profile (Figure 2e).

The medium grid arrangements with the minimum grid size of 0.1 m at wall surfaces were well-generated with the grid expansion ratio of 1.2 toward regions away from the tunnel model. The total hexahedral cell number ranged from 1,463,365 to 2,225,069 in all test cases. The standard k-ε model was employed as a no-slip wall boundary condition was applied in the near wall treatment. The normalized distance from wall surfaces (y+) ranged from 30 to 500 at most regions of the wall surfaces, i.e., the first grid point near the wall surface located in the fully-developed turbulent region [33]. Zero normal gradient conditions were employed at the two lateral domain boundaries, the domain roof (symmetry boundary), and the domain outlet boundary (i.e., outflow) (Figure 2a). The power-law velocity profile was employed at the domain inlet [36]:

$$U_0(z) \; = \; U_{ref}\left(\frac{z}{z_{ref}}\right)^{\alpha} \tag{5}$$

$$k(z) \; = \; \frac{U_*{}^2}{\sqrt{C_\mu}} \tag{6}$$

$$\varepsilon(z) \; = \; \frac{U_*^3}{\kappa(z+z_0)} \tag{7}$$

where U_{ref} is the reference velocity and equals 3 m/s at the reference height z_{ref} = 16 m. $U_* = \kappa U_{ref}/ln\left(\frac{z_{ref}+z_0}{z_0}\right)$ is the friction velocity, and aerodynamic roughness length z_0 = 1.0 m, which fits the land roughness of the urban surface. The von Karman constant κ = 0.41, and α is set to 0.27 according to the study by Irwin [37]. Turbulence has been fully developed at the entrance of the computational domain at reference level according to the setup of the power law.

All CFD setups including computational domain size, boundary conditions, and grid arrangements etc. satisfy the CFD guideline requirements (Tominaga et al. [36]). It is also worth mentioning that some researchers use the uniform inlet velocity profile instead of the atmospheric boundary layer (ABL) in catcher cross-ventilation simulations [23,38]. Though the latter requires much larger computational areas and more meshes than the former, the uniform inlet velocity profile would possibly underestimate the general flow rate inside the room [23]. Thus, to generate more accurate flow features in tunnels, the ABL boundary was applied in this research.

2.4. CFD Setup in Pollutant Dispersion Modelling

Carbon monoxide (CO) is one of the most common inert pollutants among vehicle emissions in London, UK [1]. Thus, CO was chosen as the pollutant source in this research. The emission rate was determined according to the research by Ng and Chau [39], with the realistic emission rate of 36.1 g/h/m per unit street length. The CO source was above road, with a thickness of 0.5 m and span-wise width of 10 m (Figure 2f). The emission was homogeneous and was controlled by the constant emission rate that is defined in the model. The governing equation of the CO dispersion model is [33]:

$$\bar{u}_j\frac{\partial C}{\partial x_j} - \frac{\partial}{\partial x_j}\left[(D_m + D_t)\frac{\partial C}{\partial x_j}\right] \; = \; S \tag{8}$$

C is the time-averaged CO concentration. D_m is the molecular diffusivity and D_t is the turbulent diffusivity. S is the pollutant emission rate. Here, $D_t = \nu_t/Sc_t$ and Sc_t is the turbulent Schmidt number whose value is set as $Sc_t = 0.7$ following the literature [27,40,41].

In particular, pedestrian regions were defined at the two lateral sides of the tunnel, with 2 m in height and 2.5 m in width along the whole tunnel (Figure 2g). The CO mean concentration in pedestrian regions is used when the intake fraction of the walking pedestrians in the tunnel is calculated, and the results will be covered in Section 3.

2.5. Grid Independence Study and Model Validation by Wind Tunnel Experiments

The natural ventilation feature of the wind catcher was validated by a wind tunnel experiment by Jiang et al. [42]. A 10:1 3D cubic building model with two openings was built, with the scale similar to the catcher-tunnel models. The corresponding parameters of the building model are presented in Figure 2a, i.e., building height of 2.5 m and opening size of 0.84 m wide and 1.25 m tall. At the domain inlet, the vertical profiles of the stream-wise velocity and turbulent quantities are defined as follows [43]:

$$U(y) = (u_*/\kappa)ln(y/y_0) \tag{9}$$

$$k(y) = u_*{}^2/\sqrt{C_\mu} \tag{10}$$

$$\varepsilon(y) = C_\mu{}^{3/4}k^{3/2}/(\kappa y) \tag{11}$$

where u_* is the friction velocity and equals 1.068 m/s according to Jin et al. [43], y_0 is the roughness length = 0.05 m in the wind tunnel experiment, velocity components in the y and z directions are zero. The parameter roughness height Ks and roughness constant Cs and their relationship with y_0 is $K_s = 9.793y_0/C_s$, which has been built into models in Fluent [44]. The Reynolds number can be calculated by Equation (12):

$$Re = \frac{U_{ref}H}{\nu} \tag{12}$$

where the reference velocity U_{ref} = 10 m/s, the characteristic length H is 0.25 m and ν is the kinetic viscosity which is 1.54×10^{-5} m²/s. Re reached approximately 162,000 in wind tunnel experiments. In the validation model, U_{ref} was defined as 3 m/s and the length was 2.5 m. Thus, Re reached about 487,000, verifying that the Reynolds numbers in both cases satisfy the Reynolds independence requirement (Re $\gg$ 11,000), ensuring that the turbulence was fully developed. It is worth mentioning that the friction velocity u_* was 0.314 m/s, which could be calculated by Equation (13):

$$u_* = \frac{\kappa U_{ref}}{ln\left(\frac{H}{y_0}\right)} \tag{13}$$

The cross-ventilation building model for CFD validation was built with fine grid (0.05 m), medium grid (0.1 m) and coarse grid (0.2 m) near wall surfaces and inside building models respectively. The data of the RNG k-ε model were derived from a validation study conducted by Jin et al. [43]. A similar validation process has also been done by Ai, Mak and Niu [45,46]. The standard k-ε model with enhanced wall function was employed in this validation study, and the boundary conditions were exactly the same as in the research cases (illustrated in Section 2.3). The validation model is presented in Figure 3a.

Normalized wind speed profile (U/U_{ref}) of three feature lines x = $-25/H$, $H/2$ and $3H/2$ at the center plane of the model are presented in Figure 3b–d respectively. Results show that both the RNG k-ε model and the Standard k-ε model predict the flow well in the cross ventilation cases, with the profiles generally fitting the experiment data. For better evaluation of the model simulation accuracy, a series of statistical values were calculated, including normalized mean error (*NMSE*), the fraction bias (*FB*) and the correlation coefficient (*R*). The results of the statistical metrics are listed in Table 2. According to Santiago et al. [47], a credible simulation model should satisfy the criterion of the statistical metrics: $NMSE \leq 1.5$ and $-0.3 \leq FB \leq 0.3$. All of the correlation coefficients were significant with 95% confidence. Data shows that all of the statistical values satisfy the criterion, while both the RNG k-ε model and the Standard k-ε model could predict the flow with an acceptable accuracy. The medium grid attains some different CFD results from the coarse grid, but CFD results change little if the fine grid is used (Figure 3b–d, Figure 4). Thus, the medium grid arrangements with a minimum grid of 0.1 were adopted for the following CFD simulations. In addition, the standard k-ε model was employed in our study owing to it being the most widely adopted and accurate predictor validated by the literature [21,23,25,33].

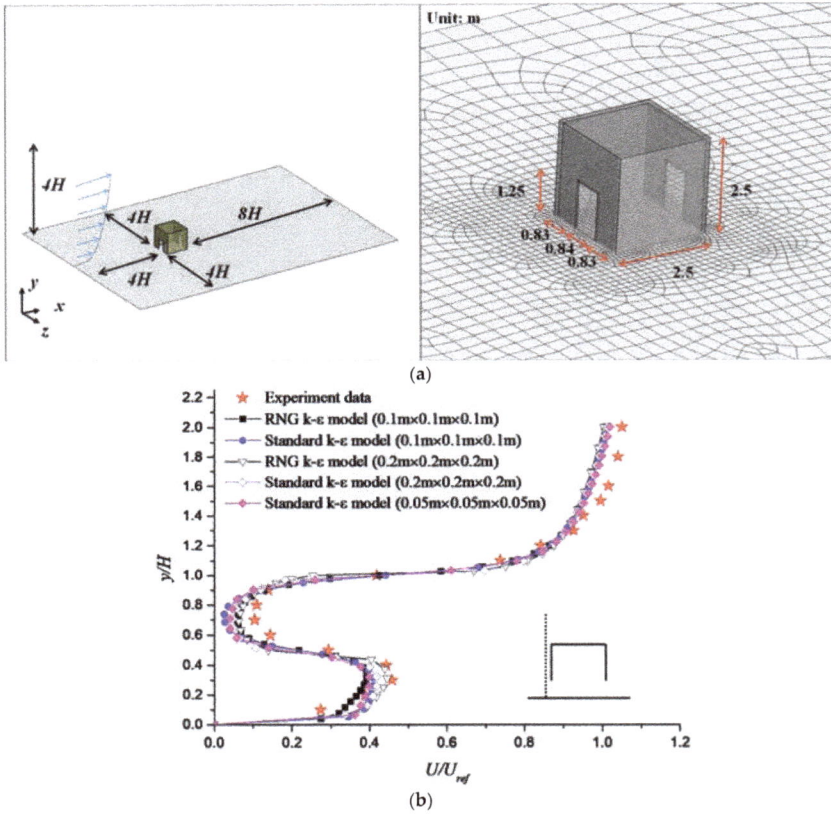

(a)

(b)

Figure 3. *Cont.*

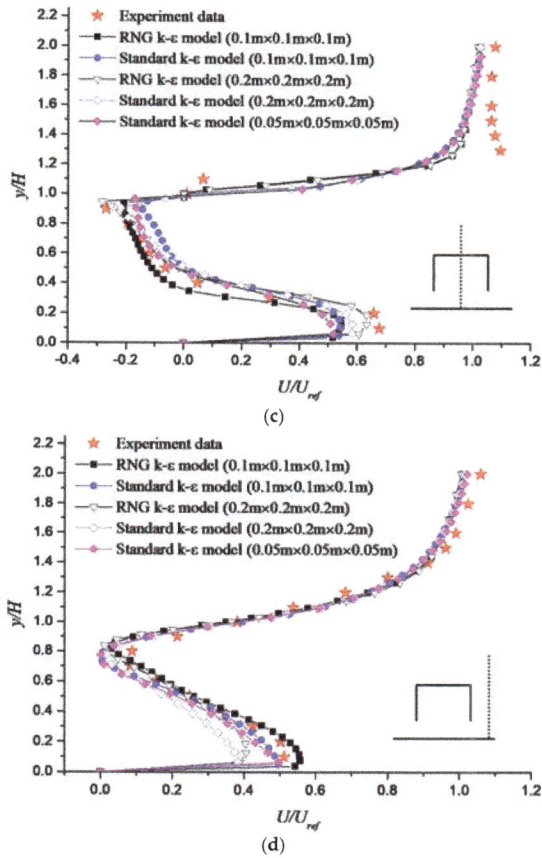

Figure 3. (**a**) Domain setup and grid arrangement in validation models; (**b–d**) vertical profile of U/U_{ref} along lines at the center cross section of the building models. (**b**) $x = -25/H$; (**c**) $x = H/2$; (**d**) $x = 3H/2$.

Table 2. Statistical metrics for the validation cases.

Values	Mesh Size (m)	$x = -25/H$		$x = H/2$		$x = 3H/2$	
		RNG	Standard	RNG	Standard	RNG	Standard
NMSE	$0.1 \times 0.1 \times 0.1$	0.005	0.007	0.034	0.059	0.005	0.005
	$0.2 \times 0.2 \times 0.2$	0.011	0.013	0.033	0.054	0.010	0.014
	$0.05 \times 0.05 \times 0.05$	-	0.007	-	0.054	-	0.007
FB	$0.1 \times 0.1 \times 0.1$	0.050	0.044	0.088	−0.028	0.002	0.036
	$0.2 \times 0.2 \times 0.2$	0.053	0.066	−0.023	−0.017	0.046	0.079
	$0.05 \times 0.05 \times 0.05$	-	0.050	-	0.014	-	0.049
R	$0.1 \times 0.1 \times 0.1$	0.994	0.991	0.976	0.961	0.991	0.993
	$0.2 \times 0.2 \times 0.2$	0.984	0.985	0.977	0.961	0.986	0.985
	$0.05 \times 0.05 \times 0.05$	-	0.991	-	0.961	-	0.993

3. Intake Fraction (*IF*)

The outdoor intake fraction (*IF*) has been widely used for the assessment of pollutant exposure for nearby populations in many urban climate studies [27–30], and it has also been used for indoor environment evaluation [48]. The intake fraction is defined as the fraction of total intake by a certain group of population among the total emission. For example, *IF* is 1 ppm (part per million) when 1 g is inhaled by a group of population in 1 ton pollutant emission. *IF* for urban climate studies is related to factors including the local population (*P*), breathing rate (*Br*), time spent in the environment (Δt), environmental pollutant concentration (*Ce*), and the pollutant emission ($\dot{m}$). It is defined as below [27]:

$$IF_{urban} = \sum_i^N \sum_j^M P_i \times Br_{i,j} \times \Delta t_{i,j} \times Ce_j / \dot{m} \qquad (14)$$

where *M* is the number of the investigated microenvironments e.g., indoor environment, near vehicle environment etc., and *N* is the number of the studied population group like the elderly, adults, children etc. Breathing rate, staying time and the environment pollutant concentration would vary with the studied population groups (*i*) and the studied microenvironments (*j*). The intake fraction for different population groups and microenvironments is listed in Table 3.

Table 3. Breathing rate for various age groups in different microenvironments [49].

Breathing Rate *Br* (m³/day)	Indoor at Home	Near Vehicle
Children	12.5	14
Adults	13.8	15.5
Elderly	13.1	14.8

The intake fraction was introduced in this research to assess the influences of pollutant distribution on a certain group of people regardless of the complex pattern of pollutant distribution in real cases. A new version of the intake fraction for tunnel research is defined as the fraction of the total inhaled by a person and the total emission during the time the person passes through the tunnel. This is presented as follows:

$$IF_{tunnel} = \frac{Vol_{br} \times \sum C_i}{\dot{m}} = \frac{Br \times t \times \overline{C}}{\dot{m}} \qquad (15)$$

where Vol_{br} is the intake volume of each breath (m³) and C_i is the spatial CO concentration (mg/m³) at the time he breathes, and $\dot{m}$ is the total emission during the time he passes through the tunnel (mg). The equation is transferred into the form that could be calculated by the breathing rate (*Br*, m³/day) and the spatial averaged CO concentration at the pedestrian region (*C*) (Equation (14)), and *t* is the time the person passes through the tunnel. Only adults are considered in this study, and it is assumed that the walking speed is 1.4 m/s, so the time taken is about 193 s in this case. Since the pedestrian region could be regarded as the near vehicle region, according to Allan et al. [49] (Table 3), the breathing rate was 15.5 m³/day, approximately 1.79×10^{-4} m³/s. The intake fraction would finally convert into ppm units, which means the value of the inhaled parts per million emission parts.

4. Result and Discussion

4.1. Wind Aerodynamics in a Tunnel

As shown in Table 1, 35 cases were studied to explore the effects of wind catcher design on pollutant concentration and exposure in tunnels. The simulation results show that the flow pattern and CO dispersion mechanism would not change significantly for stream-wise variations in wind catcher position, while significant differences in flow and pollutant dispersion pattern could be found for span-wise changes using a single short-catcher (Case_ss [150 (2.5)] and Case_ss [150 (5.0)]) and two short catchers changed from parallel to staggered arrangements (Case_ds [120, 180] and Case_ds [90, 210]). All of these wind aerodynamics and significant shifts in flow patterns among cases are specifically discussed in the following subsections.

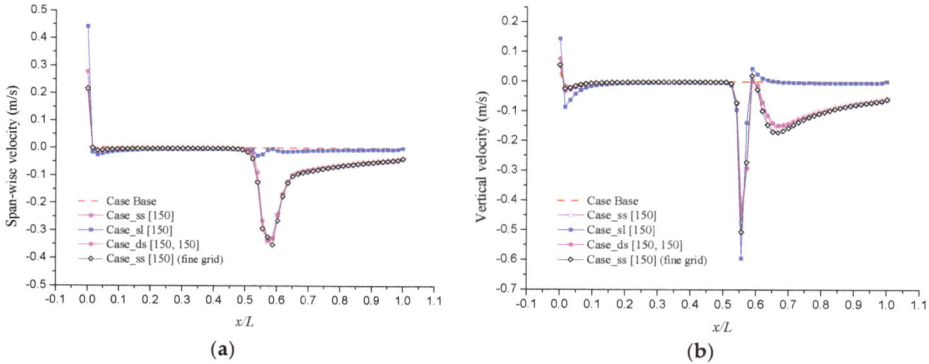

Figure 4. Velocity along the left pedestrian side (**a**) span-wise velocity under three basic catcher arrangements; (**b**) vertical velocity under three basic catcher arrangements.

4.1.1. Effects of Catcher Style on Wind Aerodynamics in the Tunnel

Three basic styles of wind catcher arrangement were considered, i.e., one single long catcher, one single short catcher above the left pedestrian side and two short catchers in a parallel arrangement above both pedestrian sides (Table 1). According to the simulation results, the wind in tunnels is dominated by longitudinal velocities, whose magnitude is 1.8 to 2.3 m/s higher than that of span-wise and vertical velocities, and in every single catcher arrangement series, flow patterns inside the tunnel tend to be similar when the catcher positions only vary stream-wise. Hence, in this section, discussion of the flow pattern will be based on the cases with the catcher located at $x = 150$ m ($x/L = 0.55$) with $y = 0$ in three basic catcher arrangement cases.

Airflow Structures in Horizontal Views

Figure 5 presents the horizontal velocity magnitude with streamline on slices at pedestrian level ($z = 1.6$ m). In the single long-catcher case (Case_sl [150], Figure 5b), the velocity magnitude is significantly intensified at the back of the wind catcher, with the highest value reaching 2.8 m/s. However, considerable reductions in velocity magnitude could also be found at the upstream regions of the tunnel when compared with the results in the base case (Figure 5a). This reduction is mainly due to the negative influence of the strong catcher-driven flow, which is observed as a strong downward vertical flow in the near-catcher regions, with the strong vertical velocity magnitude reaching -0.63 m/s at the catcher position in Case_sl [150] (Figure 4b). This intense catcher-driven flow works as a barrier, dissipating the kinetic energy of the domain-driven flows and preventing the flow from passing through the tunnel to the downstream region. According to the simulation results, this kind of velocity reduction effect exists for all of the catcher cases, and would be much stronger when the wind catcher is positioned closer to the tunnel exit. The design of the short wind catchers better diminishes this kind of reduction effect (Figure 5c–e). Compared to Case_sl [150], much higher velocity magnitudes could be found at the upstream region of the tunnel in Case_ss [150], Case_ss [150 (2.5)] and Case_ds [150, 150], with the value reaching around 1.6 m/s. However, the velocity strengthening effects at the downstream region of the tunnel decline in short-catcher cases: A single short-catcher on the left pedestrian side tends to intensify the airflow just behind where the wind catcher is positioned, while a double short-catcher above the two pedestrian sides would also strengthen the flows in the middle of the tunnel, where the vehicle roadway is located.

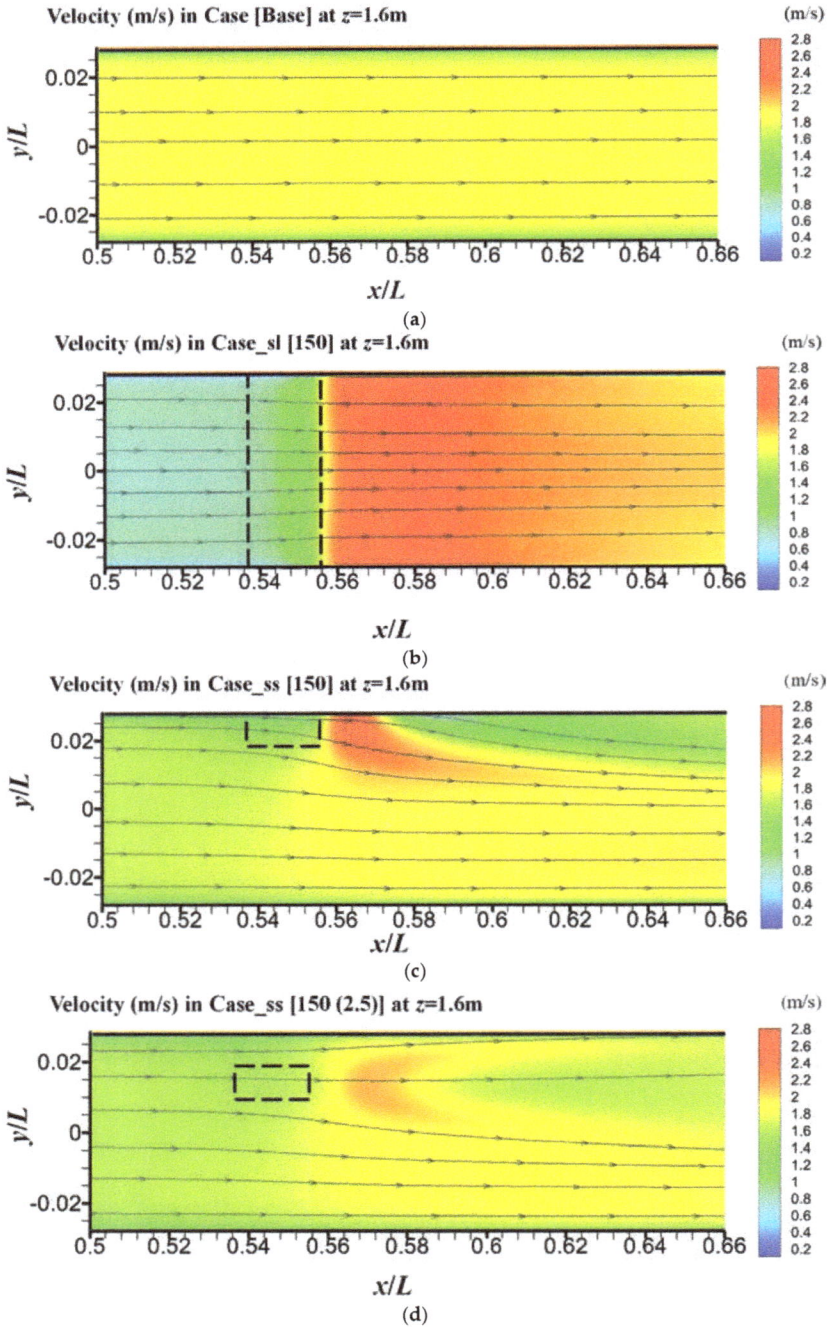

Figure 5. *Cont.*

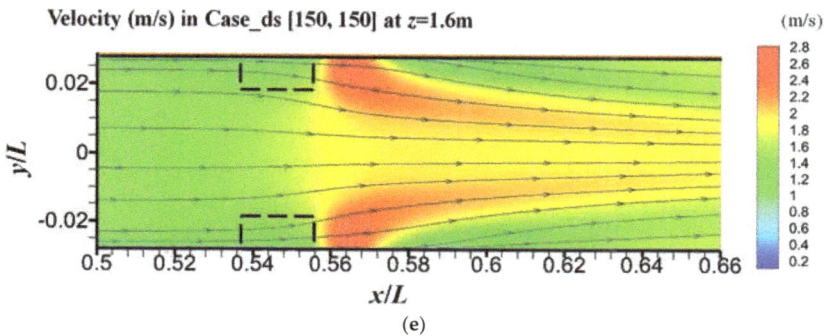

Figure 5. Velocity magnitude with streamtrace on horizontal cross section from 135 m (*x/L* = 0.5) to 178 m (*x/L* = 0.66) at pedestrian level (*z* = 1.6 m). (**a**) Base Case; (**b**) Case_sl [150]; (**c**) Case_ss [150]; (**d**) Case_ss [150, 2.5]; (**e**) Case_ds [150, 150].

Figure 4a presents the span-wise velocity magnitude along the left pedestrian side at level $z = 1.6$ m. The velocity shows similar patterns in Case_ss [150] and Case_ds [150]. Distinguishable negative values of span-wise wind could be observed at $x/L = 0.50$ (about 13.5 m ahead of where the catcher is located) for both of the cases. The value peaks at -0.35 m/s at catcher position ($x/L = 0.55$), and the intensity of the flow decreases sharply the farther it goes away from the catcher position between $x/L = 0.55$ and 0.70, and then it becomes stable, maintaining at around -0.15 m/s from $x/L = 0.70$ to the end of the tunnel. However, significant strong span-wise velocity is absent in single long-catcher cases where its magnitude only peaks at -0.05 m/s and lasts for a very short distance due to lower span-wise pressure differences. This branch of the span-wise airflow plays a vital role in hindering vehicle pollutants dispersing from the vehicle side to the pedestrian side, which is specifically analyzed in Section 4.2.

Airflow Structures Viewed Vertically

Figure 6 shows the velocity magnitude with streamlines on a vertical slice located at the middle of the left pavement (1.5 m from the left tunnel wall). As revealed by many other researchers, a wind catcher would cause two large recirculation zones in an enclosed room, i.e., one occurs behind the catcher and the other in front of the catcher [22]. Velocity would sharply reduce to about 0.2 m/s on the edge of the zone in this case. However, unlike the results in enclosed room models in many other studies, a strong longitudinal wind dominates the flow in a tunnel, thus only one intensified recirculation zone could be found in a tunnel at the back of the catcher just beneath the roof (Figure 6a). This recirculation zone is smaller in size but much stronger than that in enclosed rooms. Airflow is introduced to the lower level of the tunnel, and due the impinging effects, the velocity magnitude increases rapidly to 2.7 m/s at pedestrian level just behind the catcher position. Compared to the long-catcher cases, the recirculation zone in short-catcher cases is relatively smaller, with much weaker impinging effects at pedestrian level. Besides, the upstream velocity of the tunnel region is generally larger in the short-catcher case, where the reduction effect is much weaker compared with that of the long-catcher cases (Figure 6a,b). To sum up, stronger impinging effects would effectively intensify the velocity in the downstream region but would also cause much stronger upstream velocity reduction.

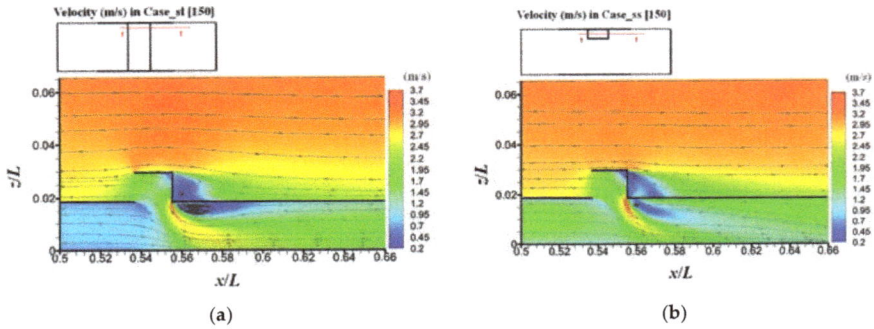

Figure 6. Velocity magnitude with streamlines in (**a**) Case_sl [150] and (**b**) Case_ss [150] on a vertical cross section at the central left-pedestrian region.

A special downstream vortex could be found in short-catcher cases in a vertical plane (Figure 7). The vortex in not fully developed in the near catcher region, and it shows a pattern of a high span-wise velocity center just beneath the catcher entrance, which peaks at 0.7 m/s in Case_ss [150] and Case_ds [150, 150] (Figure 7b,c). Due to the blockage of the lateral tunnel walls, catcher-induced flows move to the middle of the tunnel. The vortex is more obvious in the downstream region: One main anticlockwise vortex is found in single short-catcher cases (Case_ss [150] in Figure 7b), and two counter-rotating vortexes could be found in double short-catcher cases (Case_ds [150, 150] in Figure 7c).

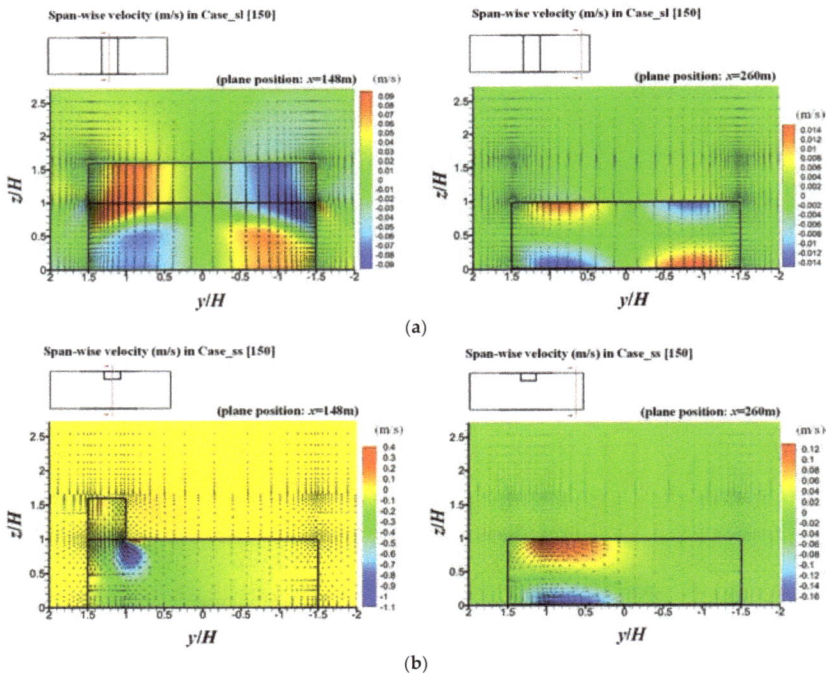

Figure 7. *Cont.*

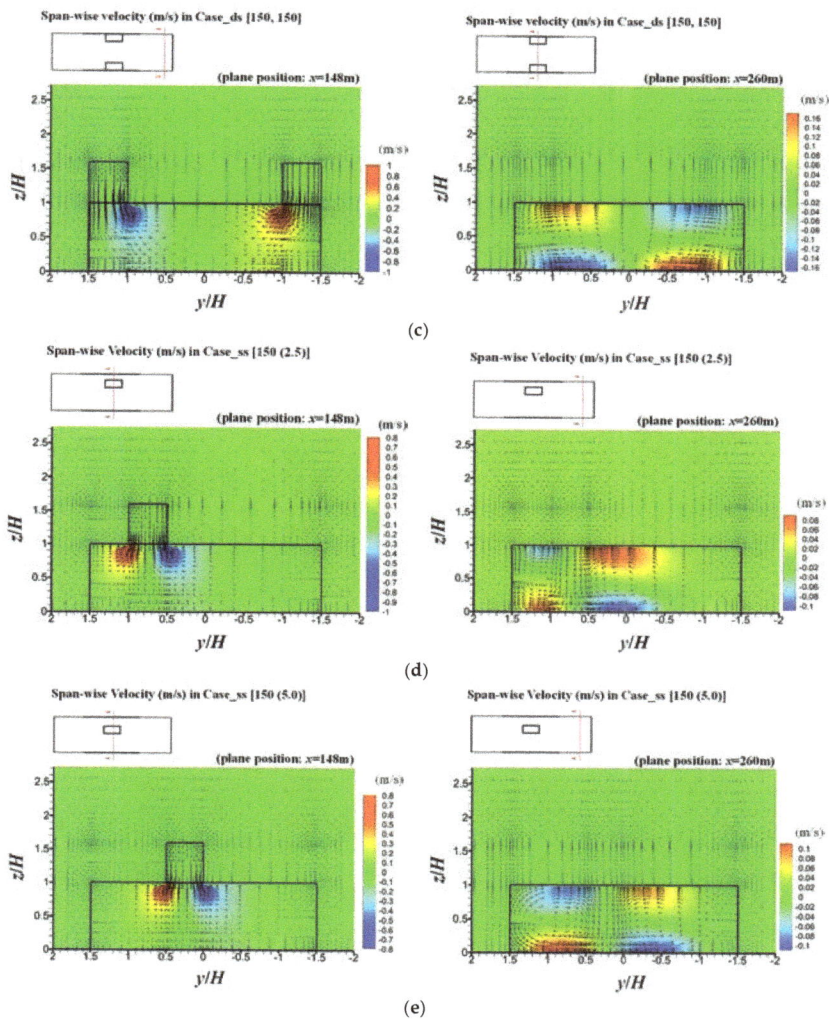

Figure 7. Span-wise velocity on a vertical cross section at $x = 148$ m ($x/L = 0.54$) and $x = 260$ m ($x/L = 0.96$) respectively. (**a**) Case_sl [150]; (**b**) Case_ss [150]; (**c**) Case_ds [150, 150]; (**d**) Case_ss [150 (2.5)]; (**e**) Case_ss [150 (5.0)].

Figure 4 also reveals the feature of the vortex development: The constant negative value of span-wise velocity along the pedestrian sides corresponds to the span-wise branch of the anticlockwise vortex at pedestrian level, while the vertical velocity corresponds to the downward branch of the vortex. The magnitude of the span-wise velocity is in a relatively steady state with the value reducing from -0.35 m/s to -0.15 m/s at $x/L = 0.65$. Distinguishable negative values of vertical velocity also become stable at the same position ($x/L = 0.65$) in both single short-catcher and double short-catcher cases. This reveals that the left anticlockwise vortex just occurred at $x/L = 0.6$, which is 13.5 m away from the catcher position. Additionally, the vortex develops into the strongest state at pedestrian level at $x/L = 0.65$, where the vertical velocity peaks. The vortex exists along the tunnel and extends driven by the strong longitudinal flow.

4.1.2. Effects of Wind Catcher Arrangement on Wind Aerodynamics in a Tunnel

As has been shown in Table 1, Case_ss [150 (2.5)] and Case_ss [150 (5.0)] were built to investigate the influences of the single short-catcher when positioned farther away from the left tunnel wall. Additionally, Case_ds [120, 180] and Case_ds [90, 210] were built to investigate the impacts of parallel and staggered arrangements for double short-catchers. From the results of the cases (Figure 7d–e), short wind catchers would have little influence on the airflow in the upstream region of the tunnel, but dramatically change the vortex in the downstream region, hence influencing the longitudinal ventilation significantly. When two short wind catchers are in a staggered arrangement, the situation becomes more complicated. The intensity of the vortexes is not equally distributed at any x position, and this would also be covered in this section.

Span-wise variations in single short-catcher positions

From the vertical cross section at $x = 148$ m (Figure 7d), different from the results in cases with the catcher positioned immediately above the left pedestrian side ($y = 0$), two distinguishable branches of span-wise flow are observed just beneath the catcher entrance, heading to the two sides of the tunnel. The areas of strengthened velocity in Case_ss [150 (2.5)] and Case_ss [150 (5.0)] are significantly smaller than that in Case_ss [150] (Figure 7b,d,e). A downstream-vortex also exists (Figure 7d,e), but one more clockwise vortex occurs on the left side of the tunnel when compared to the result in Case_ss [150]. Due to the blockage of the left tunnel wall, the clockwise vortex on the left is much weaker than the anticlockwise vortex on the right in Case_ss [150 (2.5)]. However, when the catcher is set in a more central position (Case_ss [150 (5.0)]), the two vortexes would gradually share the same intensity. The span-wise velocity is generally weaker than that in cases with the catcher positioned immediately above the left pedestrian side, and this also indicates relatively weaker downstream vortexes.

Double Short-Catchers in a Staggered Arrangement

Figure 8 depicts the span-wise and vertical velocity along the tunnel pedestrian sides in Case_ds [120, 180]. On the left pedestrian side, comparison is made with Case_ds [120, 120], where both of the short catchers are positioned at $x = 120$ m. According to the results, the curve of the span-wise velocity along the left pedestrian side fits quite well with the curve in Case_ds [120, 120], with the velocity magnitude peaking at around -0.35 m/s at $x = 120$ m. This trend maintains at the right pedestrian side (but with positive values), where the maximum span-wise velocity could be found at $x = 180$ m in both parallel and staggered cases. However, a staggered arrangement tends to have little influence on the vertical velocity when compared to the parallel arrangement cases. For all of the short-catcher cases: The downstream vortex does not reach the ground when it forms right behind the catcher entrance. It will gradually develop in shapes and reach the pedestrian level at around 13.5 m away from the catcher, at which position the second peak of the vertical velocity could be found in Figure 8. In staggered arrangement cases, due to the asymmetrical arrangement of short-catchers, two opposite vortexes unequally develop along the tunnel, with the anticlockwise vortex on the left being much weaker than the clockwise one on the right because the left catcher is much closer to the tunnel entrance.

4.2. CO Concentration Distribution and the Pollutant Dispersion Mechanism

Airflow directly influences the pollutant distribution inside the tunnel. Figure 9 presents the CO concentration for slices at $z = 1.6$ m (pedestrian level) and $z = 4.0$ m (upper level) for each case. Due to the strong stream-wise driven flows, CO accumulates in the downstream region of the tunnels near the exit. Compared with the results of the base case, it is apparent that among all of the cases with wind catchers, high CO concentration could be found in the upper level of the tunnel. In addition, this phenomenon demonstrates that CO might not be dispersed directly out of the tunnel with the intensified winds induced by the catchers, and that the ventilation efficiency of the tunnel is possibly reduced by the catchers. Designers should be particularly aware of this when arranging the catchers within the tunnels.

(a)

(b)

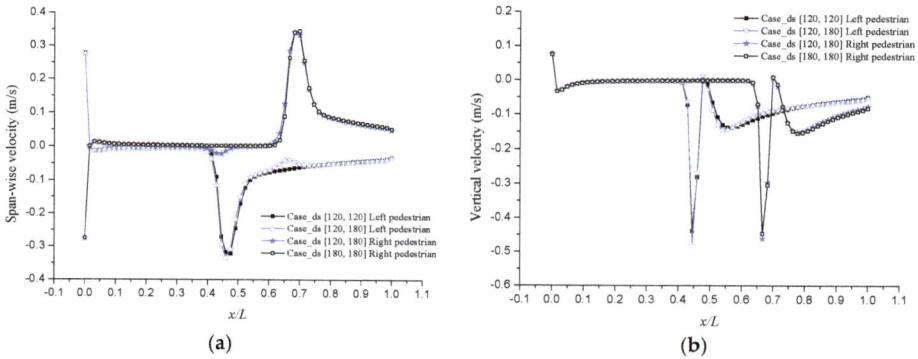

Figure 8. (**a**) Comparison of span-wise velocity between Case_ds [120, 120], Case_ds [120, 180] and Case_ds [180, 180]; (**b**) comparison of vertical velocity between Case_ds [120, 120], Case_ds [120, 180] and Case_ds [180, 180].

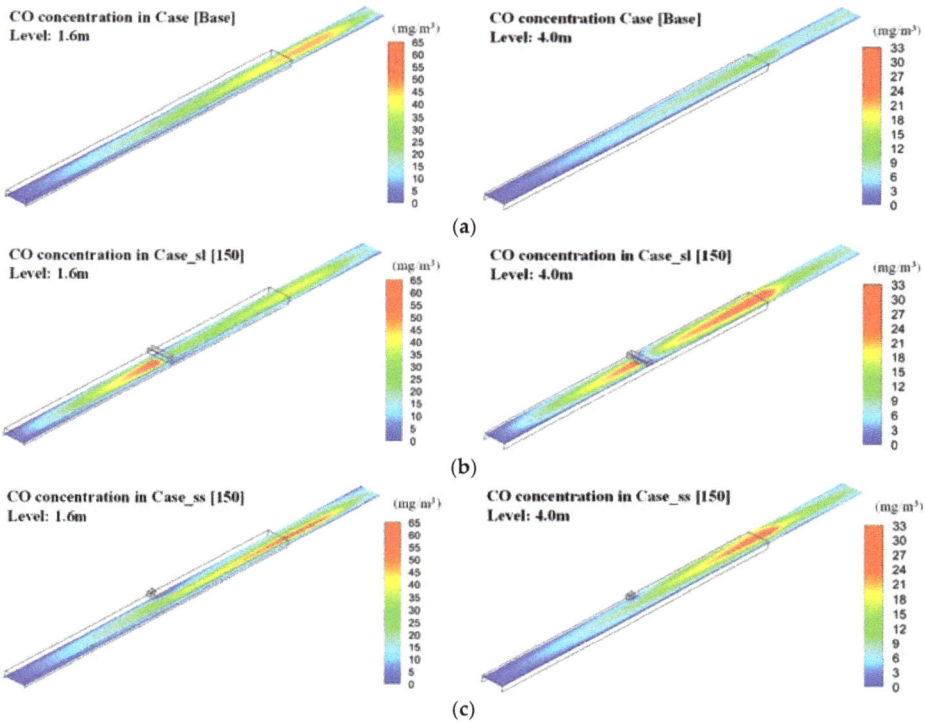

(a)

(b)

(c)

Figure 9. *Cont.*

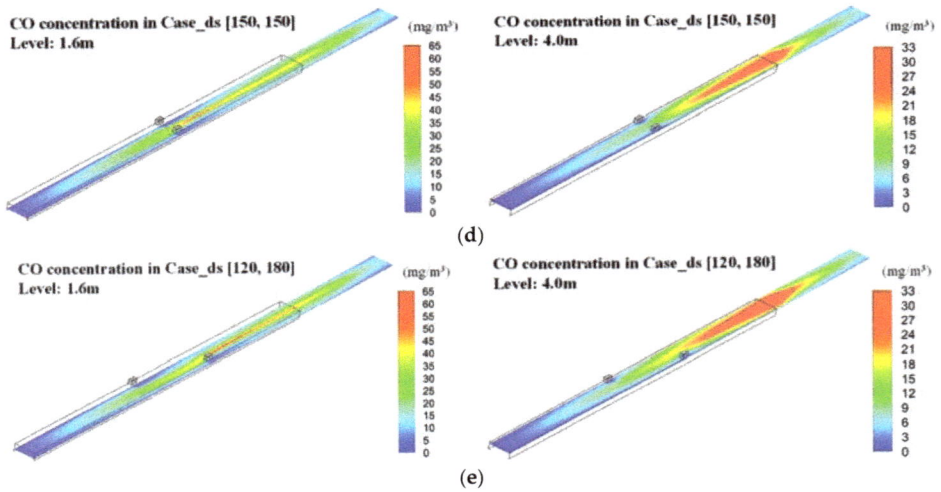

Figure 9. CO concentration on slices at z = 1.6 m and 4.0 m. (**a**) Case [Base]; (**b**) Case_sl [150]; (**c**) Case_ss [150]; (**d**) Case_ds[150, 150]; (**e**) Case_ds [120, 180].

In single long-catcher cases (Figure 9b), due to strong velocity reduction effects in the upstream region of the tunnel, CO concentration is relatively higher at the region right in front of the long catcher at both the pedestrian (65 mg/m³) and upper levels (35 mg/m³). Significant pollutant removal shows at the downstream region of the tunnel at pedestrian level, with the CO concentration reaching around 30 mg/m³ on average. In the downstream region, CO concentration shows a much higher value at the upper level compared with the result in the base case. This indicates that the ventilation efficiency for the downstream region is distinguishably reduced by the long catcher, and the long-catcher style might not be efficient for pollutant removal in tunnels.

In short-catcher cases, at pedestrian level, CO removal along the pavement is quite effective compared with the result in the base case (Figure 9c–e). A clear low-concentration band could be found behind where the wind catcher is positioned. As has been discussed previously, a special downstream-vortex exists in short-catcher cases, and the vortex would help transport the dispersed pollutants above the pavement from lower to higher levels in the tunnel. Figure 10 reveals how the downstream vortex influences the CO in Case_ss [150], ss [150 (2.5)] and ds [150, 150]. The vortex is not fully developed very close to the catcher position (x = 150 m), where the vortex just exists beneath the tunnel roof, with little influence on flows and CO at lower levels. As has been discussed before, the vortex would fully develop and reach the ground at about 13.5 m away from the catcher position, and the pollutants would be transported to higher levels and recirculate with the vortex. When a single short-catcher is positioned above the left pedestrian side, the high-value center of the CO concentration moves from the middle to the right and squeezes in shape, with the maximum concentration reaching 80 mg/m³ at x = 260 m (Figure 10a). Some of the CO recirculates with the anticlockwise vortex and mostly accumulates at the left-top of the tunnel. With two short-catchers arranged in parallel at x = 150 m above both pedestrian sides, the CO would recirculate and gradually fill the whole cross section at the end of the tunnel, with most of the pollutants accumulating in the middle and the top (Figure 10c). This kind of catcher arrangement would be quite efficient at reducing the CO concentration at the low-level pedestrian sides, where low values of CO concentration are always observed (Figure 10c).

When a single short-catcher is put away from the left tunnel wall (Case_ss [150 (2.5)] & ss [150 (5.0)]), another clockwise vortex occurs at the left pedestrian side. CO will not be transported upwards in the very first place but splits up with two branches of span-wise flows and accumulates at the pedestrian level (Figure 10b). Hence, this kind of catcher arrangement is not efficient for pollutant dispersion at both left and right pedestrian sides. Specified data is presented in Figure 11, in single short-catcher cases, relatively higher CO concentration could be found at the right pedestrian side, where no catcher is positioned. According to the results, in Case_ss [150], the catcher is quite effective for pollutant removal in the left pedestrian side (Figure 11a) by transporting the pollutant to the upper level of the tunnel (Figure 11b), while CO accumulates more on the lower-left pedestrian side in Case_ss [150 (2.5)], and much more severe negative effects on the right pedestrian side could also be found in this case (Figure 11a,b).

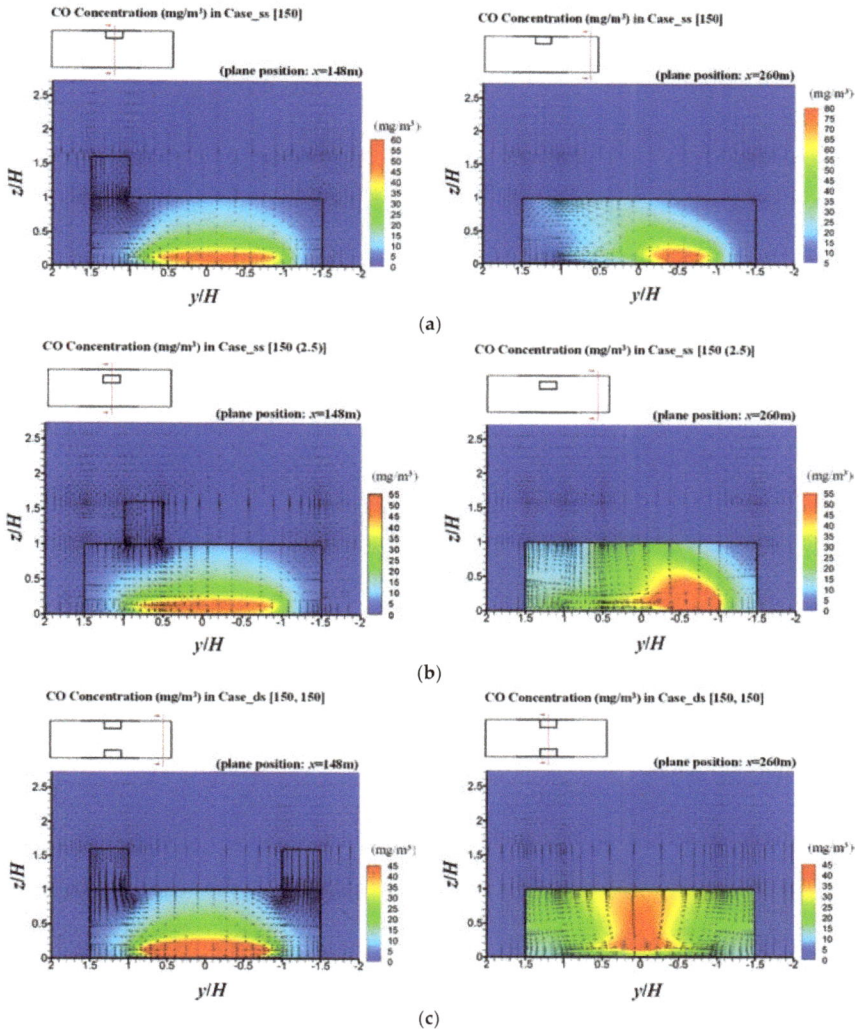

Figure 10. CO concentration on vertical cross section at $x = 148$ m ($x/L = 0.54$) and $x = 260$ m ($x/L = 0.96$) respectively. (**a**) Case_ss [150]; (**b**) Case_ss [150 (2.5)]; (**c**) Case_ds [150, 150].

General comparison is made among the cases with these three basic catcher styles, and the results are presented in Figure 12. The features can be summarized as follows:

- Higher pollutant concentration in upstream regions could be found for all of the wind catcher cases, and such an effect is more obvious in single long-catcher cases.
- Pollutant removal for the left pedestrian side is quite efficient in single short-catcher cases with the wind catcher being positioned immediately above the left pedestrian side.
- Much more severe recirculation is found in double short-catcher cases, with much higher CO concentration occupying the end of the tunnel.

Figure 9e presents the CO concentration in the staggered arrangement case (Case_ds [120, 180]) at two different layers, where significant asymmetrical distribution of CO could be found. A much clearer low-concentration band exists on the right pedestrian side at $z = 1.6$ m, but more pollutants accumulate at the upper level of the tunnel compared with the results in Case_ds [150, 150] (Figure 9d).

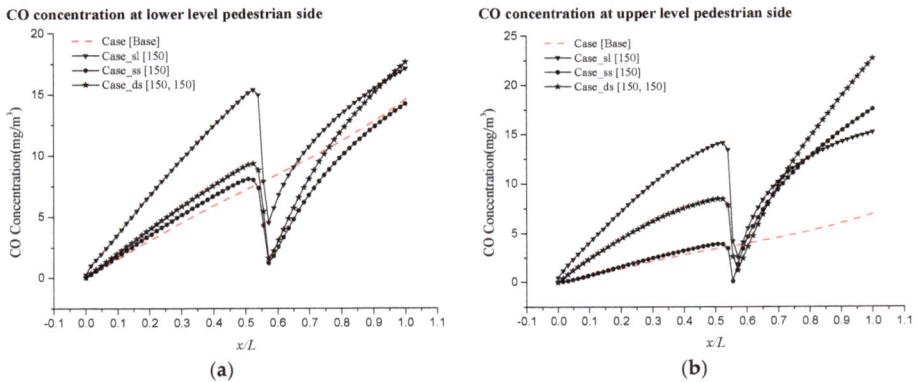

Figure 11. CO concentration along the left-pedestrian side in cases with three basic catcher arrangements CO concentration at (**a**) lower pedestrian side and (**b**) upper level pedestrian side.

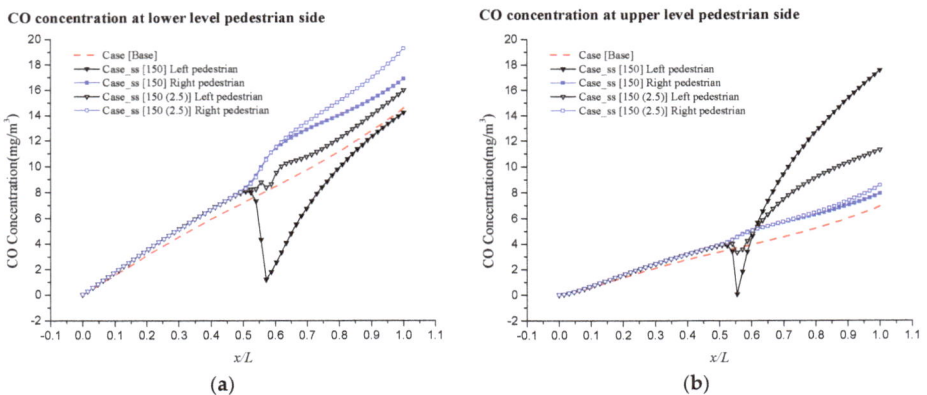

Figure 12. CO concentration in Case_ss [150] and Case_ss [150 (2.5)] at: (**a**) Lower level pedestrian side and (**b**) upper level pedestrian side.

4.3. Intake Fraction (IF)

Table 4 presents the intake fraction (*IF*) in single long-catcher cases and compares them with *IF* in the base case. Due to the symmetrical features of the models, *IF* values along both of the pedestrian sides tend to be the same in each case, so the average *IF* along both pedestrian sides is presented. *IF* will increase when the wind catcher is positioned farther away from the tunnel entrance, i.e., from 0.46 ppm (catcher position: $x = 10$ m) to 1.13 ppm (catcher position: $x = 240$ m).

This trend can also be found among all of the short-catcher cases, and this is mainly because the upstream velocity reduction effect would be intensified when the wind catcher is positioned much farther from the tunnel entrance. Comparing with the *IF* in the base case, a single long-catcher should better be arranged before $x = 30$ m ($x/L = 0.11$), otherwise negative effects on pollutant removal at the pedestrian sides would occur. Additionally, it also indicates that such a long catcher might not be efficient for pollutant dispersion along the pavements though it may induce a quite intense longitudinal wind in the near catcher region.

Table 4. Intake Fraction (*IF*) in cases with a single long catcher.

	Case Name	Catcher Position x (m)	IF (ppm)
-	Case [Base]	-	0.49
	Case_sl [10]	10	0.46
	Case_sl [15]	15	0.47
	Case_sl [20]	20	0.47
	Case_sl [25]	25	0.48
	Case_sl [30]	30	0.49
	Case_sl [40]	40	0.50
	Case_sl [50]	50	0.51
	Case_sl [60]	60	0.66
	Case_sl [90]	90	0.64
	Case_sl [120]	120	0.66
	Case_sl [150]	150	0.65
	Case_sl [180]	180	0.69
	Case_sl [210]	210	0.74
	Case_sl [240]	240	1.13

Table 5 shows the *IF* in cases with a single short catcher being placed immediately above the left pedestrian side. Better air quality could be found at the left pedestrian region. In Case_ss [30], *IF* is reduced to 0.30 ppm, which is only 61% of the *IF* in the base case (0.49 ppm). However, the percentage increases to 94% when the catcher is positioned at 240 m from the tunnel entrance. Poorer ventilation conditions are found at the right pedestrian region, where the *IF* reaches 0.47 ppm in Case_ss [30] and increases to 0.58 ppm in Case_ss [240]. However, it is still relatively lower than that in long-catcher cases. When the single short catcher is moved away from the left tunnel wall, more CO would accumulate with the recirculation at both pedestrian sides, with the *IF* sharply increasing to 0.57 and 0.89 ppm in Case_ss [150 (2.5)] and Case_ss [150 (5.0)] respectively (Table 6).

Table 5. Intake Fraction (*IF*) in cases with a single short catcher above the left-pedestrian side.

Case Name	x (m)	IF (ppm)		
		Left-Ped	Right-Ped	Average
Case [Base]	-	0.49	0.49	0.49
Case_ss [30]	30	0.30	0.47	0.39
Case_ss [60]	60	0.35	0.51	0.43
Case_ss [90]	90	0.38	0.54	0.46
Case_ss [120]	120	0.40	0.57	0.49
Case_ss [150]	150	0.40	0.60	0.50
Case_ss [180]	180	0.41	0.60	0.51
Case_ss [210]	210	0.43	0.60	0.51
Case_ss [240]	240	0.46	0.58	0.52

For double parallel short-catcher cases, air quality at both pedestrian sides would be improved (Table 7). When wind catchers are positioned before 210 m (*x/L* = 0.78), better air quality occurs at both pedestrian regions compared with that in the base case. The smallest CO exposure is found in Case_ds [30, 30], i.e., 0.32 ppm, only 63% of the *IF* in the base case, where both of the wind catchers are placed 30 m from the tunnel entrance. When it comes to staggered cases, for example, in Case_ds [120, 180], *IF* at the left pedestrian side is 0.47 ppm, which is slightly lower compared with that of Case_ds [120, 120] (0.48 ppm) but relatively larger than that in Case_ss [120] (0.40 ppm). At the right pedestrian side, *IF* increases to 0.50 ppm in staggered Case_ds [120, 180], and it is a little larger than *IF* in Case_ds [180, 180] (0.48 ppm) but much smaller than that in Case_ss [120] (0.57 ppm). So two catchers would effectively avoid the negative effects that exist on the right pedestrian side in single short-catcher cases, but ventilation capacity at the left pedestrian side would be relatively poorer. Besides, even though CO concentration asymmetrically distributes along the tunnel in staggered cases, the *IF*s in Case_ds [120, 180] turn out to be approximately similar to those in corresponding parallel cases (i.e., Case_ds [120, 120] and Case_ds [180,180]), which might indicate that air condition at one pedestrian side in double short-catcher cases is not sensitive to the catcher position on the other side, as long as they are positioned immediately above the pavements.

Table 6. Intake Fraction (*IF*) in cases with a single small catcher at 150 m.

Case Name	y (m)	IF (ppm)		
		Left-Ped	Right-Ped	Average
Case [Base]	-	0.49	0.49	0.49
Case_ss [150]	0	0.40	0.60	0.50
Case_ss [150 (2.5)]	2.5	0.57	0.63	0.60
Case_ss [150 (5.0)]	5.0	0.89	0.80	0.84

Table 7. Intake Fraction (*IF*) in cases with two short catchers.

	Case Name	Catcher Position			*IF* (ppm)
-	Case [Base]	-			0.4900
Parallel arrangement		*x* (m)			Averaged
	Case_ds [30, 30]	30			0.32
	Case_ds [60, 60]	60			0.39
	Case_ds [90, 90]	90			0.45
	Case_ds [120, 120]	120			0.48
	Case_ds [150, 150]	150			0.48
	Case_ds [180, 180]	180			0.48
	Case_ds [210, 210]	210			0.49
	Case_ds [240, 240]	240			0.52
Staggered arrangement		x_1 (m)	x_2 (m)	**Left-Ped**	**Right-Ped**
	Case_ds [120, 180]	120	180	0.47	0.50
	Case_ds [90, 210]	90	210	0.45	0.49

5. Conclusions

This research investigates the possibility of applying wind catchers for providing natural ventilation and reducing pedestrian exposure to air pollution in short road tunnels. Computational fluid dynamics (CFD) methodology with the standard *k-ε* turbulence model was used to perform three-dimensional (3D) turbulent flow simulations which were validated with wind tunnel experiment data. Thirty-five cases with long and short catchers with different wind catcher arrangements were built to investigate the effects of the catcher style and arrangement on tunnel ventilation performance. The intake fraction was applied as an assessment of the in-tunnel ventilation conditions and pollutant exposure at pedestrian regions. The simulation results and outcomes can be summarized as follows:

1. Effects of catcher style:

    *   Ventilation performance is much worse in cases with a long-catcher design due to extremely strong velocity reduction effects in the upstream region and large recirculation zones behind the catcher entrance.
    *   A downstream vortex could be found in short-catcher cases, which will fully develop and reach the ground 13.5 m behind where the wind catcher is positioned. This vortex would help transport CO from one pedestrian side to the other and from lower to upper levels. Consequently, pollutants will accumulate at the left-top (or right-top) of the tunnel.

2. Effects of catcher arrangement:

    *   Among all cases, the closer wind catchers are positioned to the tunnel entrance, the better ventilation for the inner-tunnel environment they can provide.
    *   A single short-catcher above one pedestrian side will lead to a negative impact on the opposite side; when the single short-catcher is positioned away from the lateral wall, it will result in much worse ventilation performance at both pedestrian sides.
    *   Double short-catchers in a parallel arrangement have generally good performance by transporting pollutants to the middle and the top regions of the tunnel. In cases with wind catchers of staggered arrangement, relatively lower *IF* could be found at the left pedestrian side, because the catcher is positioned much closer to the tunnel entrance on this side. CO distribution in staggered cases is asymmetrical, but the intake fraction at one pedestrian side is nearly

the same as the value in the corresponding parallel case with the same catcher position. This indicates that *IF* is not sensitive to the catcher position at the opposite side.

From the results above, it can be concluded that design of double short-catcher in parallel arrangement is recommended for providing natural ventilation in short road tunnels, such as the Barbican in the City of London, with the smallest *IF* being only 61% of the base case.

Author Contributions: Conceptualization, S.L, Z.L. and J.H.; Methodology, S.L, Z.L. and K.Z.; Software, S.L.; Formal Analysis, S.L.; Investigation, S.L.; Resources, Z.L, and J.H.; Writing-Original Draft Preparation, S.L.; Writing-Review & Editing, Z.L., S.L. K.Z. and J.H.; Visualization, S.L.

Funding: This study was funded by National Key R&D Program of China [grant No. 2016YFC0202206 and 2016YFC0202205]; the National Natural Science Foundation of China (grant No. 51478486 and 11471343) and National Natural Science Foundation—Outstanding Youth Foundation (grant No. 41622502) as well as Science and Technology Program of Guangzhou, China (grant No. 201607010066 and 2014B020216003).

Conflicts of Interest: The authors declare no conflict of interest.

Appendix A

In the preliminary study, we have conducted tests on a real shape model, which is presented in Figure A1. Columns at the opening of the wind catcher and the thickness of the walls are considered in the test case. The results have been compared with that in Case_ss [150] (Figure A2). Results show that span-wise velocity and vertical velocity along the left pedestrian side show similar patterns and fit quite well with each other. Small differences only appear in near catcher regions due to situations where fluctuation and blockage of the ground strongly impinge. Since the focus of this research is on the relative ventilation performance of tunnel-catcher models, the small differences in the near catcher region can be neglected. Besides, in real shape models, over 4,000,000 meshes are built, and this figure is nearly double that in simplified models. Hence, we employed the simplified models in our case study without considering the effect of catcher columns and wall thickness. Similar study methods are found in the literature [20].

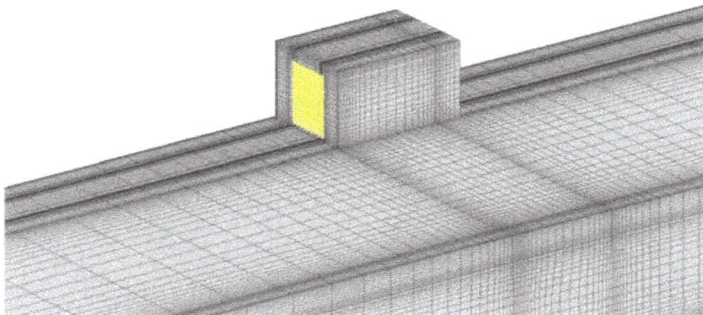

Figure A1. Model and grid distribution in real shape case in preliminary test.

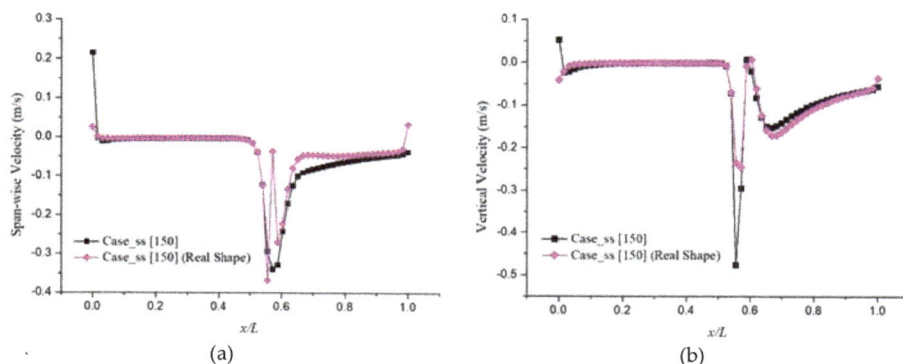

Figure A2. Span-wise velocity and vertical velocity along left pedestrian side in Case_ss [150] and Case_ss [150] (real shape). (**a**) Span-wise velocity; (**b**) vertical velocity.

References

1. Francis, L.; Stockwell, H. Science in the City: Barbican Report. Available online: https://www.cityoflondon.gov.uk/business/environmental-health/environmental-protection/air-quality/Pages/Air-Quality-Champions.aspx (accessed on 12 October 2013).
2. Keyte, I.J.; Albinet, A.; Harrison, R.M. On-road traffic emissions of polycyclic aromatic hydrocarbons and their oxy- and nitro-derivative compounds measured in road tunnel environments. *Sci. Total. Environ.* **2016**, *566–567*, 1131–1142. [CrossRef] [PubMed]
3. Kim, J.Y.; Lee, J.Y.; Kim, Y.P.; Lee, S.-B.; Jin, H.C.; Bae, G.-N. Seasonal characteristic of the gaseous and particulate PAHs at a roadside station in Seoul, Korea. *Atmos. Res.* **2012**, *116*, 142–150. [CrossRef]
4. Wingfors, H.; Sjödin, A.; Haglund, P.; Brorström-Lundén, E. Characterisation and determination of profiles of polycyclic aromatic hydrocarbons in a traffic tunnel in Gothenburg Sweden. *Atmos. Environ.* **2001**, *35*, 6361–6369. [CrossRef]
5. Betta, V.; Cascetta, F.; Musto, M.; Rotondo, G. Numerical study of the optimization of the pitch angle of an alternative jet fan in longitudinal tunnel ventilation system. *Tunn. Undergr. Space Technol.* **2009**, *24*, 164–172. [CrossRef]
6. Bogdan, S.; Birgmajer, B.; Kovačić, Z. Model predictive and fuzzy control of a road tunnel ventilation system. *Transp. Res. C-Emerg. Technol.* **2008**, *16*, 574–592. [CrossRef]
7. Betta, V.; Cascetta, F.; Musto, M.; Rotondo, G. Fluid dynamic performances of traditional and alternative jet fans in tunnel longitudinal ventilation systems. *Tunn. Undergr. Space Technol.* **2010**, *25*, 415–422. [CrossRef]
8. Eftekharian, E.; Dastan, A.; Abouali, O.; Meigolinedjad, J.; Ahmandi, G. A numerical investigation into the performance of two types of jet fans in ventilation of an urban tunnel under traffic jam condition. *Tunn. Undergr. Space Technol.* **2014**, *44*, 56–67. [CrossRef]
9. Wang, F.; Wang, M.N.; He, S.; Zhang, J.S.; Deng, Y.Y. Computational study of effects of jet fans on the ventilation of a highway curved tunnel. *Tunn. Undergr. Space Technol.* **2010**, *25*, 382–390. [CrossRef]
10. Peeling, J.; Waymann, M.; Mocanu, I.; Nitsche, P.; Rands, J.; Potter, J. Energy efficient tunnel solutions. In Proceedings of the 6th Transport Research Arena, Warsaw, Poland, 18–21 April 2016.
11. Harish, R.; Venkatasubbaiah, K. Effects of buoyancy induced roof ventilation systems for smoke removal in tunnel fires. *Tunn. Undergr. Space Technol.* **2014**, *42*, 195–205. [CrossRef]
12. Tang, F.; Mei, F.Z.; Li, L.J.; Chen, L.; Ding, J.X.; Wang, Q.; Xu, X.Y. Ceiling smoke front velocity in a tunnel with central mechanical exhaust system: Comparison of model predictions with measurements. *Appl. Therm. Eng.* **2017**, *127*, 689–695. [CrossRef]
13. Tang, F.; Mei, F.Z.; Wang, Q.; He, Z.; Fan, C.G.; Tao, C.F. Maximum temperature beneath the ceiling in tunnel fires with combination of ceiling mechanical smoke extraction and longitudinal ventilation. *Tunn. Undergr. Space Technol.* **2017**, *68*, 231–237. [CrossRef]

14. Tanaka, F.; Kawabata, N.; Ura, F. Effects of a transverse external wind on natural ventilation during fires in shallow urban road tunnels with roof openings. *Fire Saf. J.* **2016**, *79*, 20–36. [CrossRef]

15. Chen, L.F.; Hu, L.H.; Zhang, X.L.; Zhang, X.Z.; Zhang, X.C.; Yang, L.Z. Thermal buoyant smoke back-layering flow length in a longitudinal ventilated tunnel with ceiling extraction at difference distance from heat source. *Appl. Therm. Eng.* **2015**, *78*, 129–135. [CrossRef]

16. Dehghani-sanij, A.R.; Soltani, M.; Raahemifar, K. A new design of wind tower for passive ventilation in buildings to reduce energy consumption in windy regions. *Renew. Sustain. Energy Rev.* **2015**, *42*, 182–195. [CrossRef]

17. Soltani, M.; Dehghani-Sanij, A.; Sayadnia, A.; Kashkooli, F.M.; Gharali, K.; Mahbaz, S.B.; Dusseault, M. Investigation of airflow patterns in a new design of wind tower with a wetted surface. *Energy* **2018**, *11*, 1100. [CrossRef]

18. Afshin, M.; Sohankar, A.; Dehghan-Manshadi, M.; Kazemi-Esfeh, M. An experimental study on the evaluation of natural ventilation performance of a two-sided wind-catcher for various wind angles. *Renew. Energy* **2016**, *85*, 1068–1078. [CrossRef]

19. Montazeri, H. Experimental and numerical study on natural ventilation performance of various multi-opening wind catchers. *Build. Environ.* **2011**, *46*, 370–378. [CrossRef]

20. Montazeri, H.; Montazeri, F. CFD simulation of cross-ventilation in buildings using rooftop wind-catchers: Impact of outlet openings. *Renew. Energy* **2018**, *118*, 502–520. [CrossRef]

21. Dehghan, A.A.; Kazemi Esfen, M.; Dehghan Manshadi, M. Natural ventilation characteristics of one-sided wind catchers: Experimental and analytical evaluation. *Energy Build.* **2013**, *61*, 366–377. [CrossRef]

22. Calautit, J.K.; O'Connor, D.; Hughes, B.R. Determining the optimum spacing and arrangement for commercial wind towers for ventilation performance. *Build. Environ.* **2014**, *82*, 274–287. [CrossRef]

23. Hosseini, S.H.; Shokry, E.; Ahmadian Hosseini, A.J.; Ahmadi, G.; Calautit, J.K. Evaluation of airflow and thermal comfort in buildings ventilated with wind catchers: Simulation of conditions in Yazd City, Iran. *Energy Sustain. Dev.* **2016**, *35*, 7–24. [CrossRef]

24. Calautit, J.K.; Hughes, B.R. A passive cooling wind catcher with heat pipe technology: CFD, wind tunnel and field-test analysis. *Appl. Energy* **2016**, *162*, 460–471. [CrossRef]

25. Bahadori, M.N.; Mazidi, M.; Dehghani, A.R. Experimental investigation of new designs of wind towers. *Renew. Energy* **2008**, *33*, 2273–2281. [CrossRef]

26. Haghighi, A.P.; Pakdel, S.H.; Jafari, A. A study of a wind catcher assisted adsorption cooling channel for natural cooling of a 2-storey building. *Energy* **2016**, *102*, 118–138. [CrossRef]

27. He, L.J.; Hang, J.; Wang, X.M.; Lin, B.R.; Li, X.H.; Lan, G.D. Numerical investigations of flow and passive pollutant exposure in high-rise deep street canyons with various street aspect ratios and viaduct settings. *Sci. Total Environ.* **2017**, *584–585*, 189–206. [CrossRef] [PubMed]

28. Luo, Z.W.; Li, Y.G.; Nazaroff, W.W. Intake fraction of nonreactive motor vehicle exhaust in Hong Kong. *Atmos. Environ.* **2010**, *44*, 1913–1918. [CrossRef]

29. Marshall, J.D.; Teoh, S.K.; Nazaroff, W.W. Intake fraction of nonreactive vehicle emissions in US urban areas. *Atmos. Environ.* **2005**, *39*, 1363–1371. [CrossRef]

30. Zhou, Y.; Levy, J.I. The impact of urban street canyons on population exposure to traffic-related primary pollutants. *Atmos. Environ.* **2008**, *42*, 3087–3098. [CrossRef]

31. FLUENT V6.3. 2006. Available online: http://fluent.com (accessed on 1 December 2006).

32. Ghadiri, M.H.; Lukman, N.; Ibrahim, N.; Mohamed, M.F. Computational analysis of wind-driven natural ventilation in a two sided rectangular wind catcher. *Int. J. Vent.* **2013**, *12*, 51–62. [CrossRef]

33. ANSYS, Inc. *ANSYS Fluent Theory Guide*; ANSYS, Inc.: Canonsburg PA, USA, 2013; pp. 47–48. Available online: http://fluent.com (accessed on 1 November 2013).

34. Mehryan, S.A.M.; Kashkooli, F.M.; Soltani, M. Comprehensive study of the impacts of surrounding structures on the aero-dynamic performance and flow characteristics of an outdoor unit of split-type air conditioner. *Build. Simul.* **2018**, *11*, 325–337. [CrossRef]

35. Zargar, B.; Kashkooli, F.M.; Soltani, M.; Wright, K.; Ijaz, M.K.; Sattar, S. Mathematical modeling and simulation of bacteria distribution in an aerobiology chamber using computational fluid dynamics. *Am. J. Infect. Control* **2016**, *44*, S127–S137. [CrossRef] [PubMed]

36. Tominaga, Y.; Mochida, A.; Yoshie, R.; Kataoka, H.; Nozu, T.; Yoshikawa, M.; Shirasawa, T. AIJ guidelines for practical applications of CFD to pedestrian wind environment around buildings. *J. Wind Eng. Ind. Aerodyn.* **2008**, *96*, 1749–1761. [CrossRef]

37. Irwin, J.S. A theoretical variation of the wind profile power-law exponent as a function of surface roughness and stability. *Atmos. Environ.* **1979**, *13*, 191–194. [CrossRef]

38. Lu, H.F.; Huang, K.; Fu, L.D.; Zhang, Z.H.; Wu, S.J.; Lyu, Y.; Zhang, X.L. Study on leakage and ventilation scheme of gas pipeline in tunnel. *J. Nat. Gas Sci. Eng.* **2018**, *53*, 347–358. [CrossRef]

39. Ng, W.Y.; Chau, C.K. A modelling investigation of the impact of street and building configurations on personal air pollutant exposure in isolated deep urban canyons. *Sci. Total Environ.* **2014**, *468–469*, 429–448. [CrossRef] [PubMed]

40. Habilomatis, G.; Chaloulakou, A. A CFD modelling study in an urban street canyon for ultrafine particles and population exposure: The intake fraction approach. *Sci. Total Environ.* **2015**, *530–531*, 227–232. [CrossRef] [PubMed]

41. Hang, J.; Luo, Z.W.; Wang, X.M.; He, L.J.; Wang, B.M.; Zhu, W. The influence of street layouts and viaduct settings on daily CO exposure and intake fraction in idealized urban canyons. *Environ. Pollut.* **2017**, *220*, 72–86. [CrossRef] [PubMed]

42. Jiang, Y.; Alexander, D.; Jenkins, H.; Arthur, R.; Chen, Q.Y. Natural ventilation in buildings: Measurement in a wind tunnel and numerical simulation with large-eddy simulation. *J. Wind Eng. Ind. Aerodyn.* **2003**, *91*, 331–353. [CrossRef]

43. Jin, R.Q.; Hang, J.; Liu, S.S.; Wei, J.J.; Liu, Y.; Xie, J.L.; Sandberg, M. Numerical investigation of wind-driven natural ventilation performance in a multi-story hospital by coupling indoor and outdoor airflow. *Indoor Built Environ.* **2015**, *25*, 1–22. [CrossRef]

44. Blocken, B.; Stathopoulos, T.; Carmeliet, J. CFD simulation of the atmospheric boundary layer: Wall function problems. *Atmos. Environ.* **2007**, *41*, 238–252. [CrossRef]

45. Ai, Z.T.; Mak, C.M. A study of interunit dispersion around multistory buildings with single-sided ventilation under different wind directions. *Atmos. Environ.* **2014**, *88*, 1–13. [CrossRef]

46. Ai, Z.T.; Mak, C.M.; Niu, J.L. Numerical investigation of wind-induced airflow and interunit dispersion characteristics in multistory residential buildings. *Indoor Air* **2013**, *23*, 417–429. [CrossRef] [PubMed]

47. Santiago, J.L.; Martilli, A.; Martin, F. CFD simulation of airflow over a regular array of cubes. Part I: Three dimensional simulation of the flow and validation with wind-tunnel measurements. *Bound.-Lay. Meteorol.* **2007**, *122*, 609–634. [CrossRef]

48. Nazaroff, W.W. Inhalation intake fraction of pollutants from episodic indoor emissions. *Build. Environ.* **2008**, *43*, 269–277. [CrossRef]

49. Allan, M.; Richardson, G.M.; Jones-Otazo, H. Probability density functions describing 24-hour inhalation rates for use in human health risk assessments: An update and comparison. *Hum. Ecol. Risk Assess.* **2008**, *14*, 372–391. [CrossRef]

atmosphere

MDPI

Article

Effects of Unstable Stratification on Ventilation in Hong Kong

Tobias Gronemeier [1,*], Siegfried Raasch [1] and Edward Ng [2,3,4]

[1]   Insitute of Meteorology and Climatology, Leibniz Universität Hannover, 30419 Hannover, Germany;
     raasch@muk.uni-hannover.de
[2]   School of Architecture, The Chinese University of Hong Kong, Hong Kong, China; edwardng@cuhk.edu.hk
[3]   Institute of Future City (IOFC), The Chinese University of Hong Kong, Hong Kong, China
[4]   Institute of Energy, Environment and Sustainability (IEES), The Chinese University of Hong Kong,
     Hong Kong, China
*   Correspondence: gronemeier@muk.uni-hannover.de; Tel.: +49-511-762-3232

Received: 3 August 2017; Accepted: 6 September 2017; Published: 8 September 2017

Abstract: Ventilation in cities is crucial for the well being of their inhabitants. Therefore, local governments require air ventilation assessments (AVAs) prior to the construction of new buildings. In a standard AVA, however, only neutral stratification is considered, although diabatic and particularly unstable conditions may be observed more frequently in nature. The results presented here indicate significant changes in ventilation within most of the area of Kowloon City, Hong Kong, included in the study. A new definition for calculating ventilation was introduced, and used to compare the influence of buildings on ventilation under conditions of neutral and unstable stratification. The overall ventilation increased due to enhanced vertical mixing. In the vicinity of exposed buildings, however, ventilation was weaker for unstable stratification than for neutral stratification. The influence on ventilation by building parameters, such as the plan area index, was altered when unstable stratification was considered. Consequently, differences in stratification were shown to have marked effects on ventilation estimates, which should be taken into consideration in future AVAs.

Keywords: convective boundary layer; LES (large-eddy simulations); street-level ventilation

1. Introduction

Air ventilation is a crucial factor of city climate and has a major impact on the well being of the urban population. The wind field within a city, and hence ventilation, is markedly influenced by the actual building setup (e.g., [1–3]). Accordingly, local governments, particularly those of larger cities, have started to regulate the construction of new buildings to maintain or improve ventilation. As a consequence, an air ventilation assessment (AVA) is usually required to obtain approval for large building projects [2]. These AVAs typically only require wind tunnel experiments. However, wind tunnel experiments have the disadvantage of usually only being capable of reproducing neutrally stratified atmospheric conditions, and the effects of diabatic stratification on ventilation are neglected. This is often justified when focusing on high wind speeds, where mechanically induced turbulence has a greater influence on ventilation than turbulence generated by buoyancy, which is only present if diabatic stratification is considered.

However, the most crucial situations for city ventilation are those where only a low wind speed is present, which drastically reduces ventilation. For such situations, the building setup must be well organized to ensure sufficient ventilation of the whole city area. However, the assumption that buoyancy-driven turbulence only has a minor influence on ventilation is not true in such low-wind situations. Particularly for regions with regularly occurring low wind speeds, an AVA that solely focuses on neutral conditions may not include the actual ventilation effects of planned buildings within the study area. Therefore, the validity of such an AVA would be limited.

This study was performed to identify the limitations if only neutral conditions are considered when analyzing air ventilation under weak-wind conditions. We computed and compared the ventilation under neutral and unstable atmospheric conditions. Ventilation analyses were performed for Kowloon City, Hong Kong, by large-eddy simulations (LES).

The LES technique predicts the wind field within a building array more accurately than a Reynolds-averaged Navier-Stokes (RANS) simulation [4]. LES resolves the large energy-containing turbulence elements (eddies) and only parameterizes the small (sub-grid)-scale turbulence, while the RANS technique parameterizes the whole turbulence spectrum. This also has the advantage that the convective up- and downdrafts are directly simulated within the LES instead of being parameterized, which tends to give more realistic results.

Although there have been many studies regarding ventilation within large cities, particularly Hong Kong (e.g., [5–9]), there have been few high-resolution LES studies dealing with large real urban areas. Letzel et al. [10] investigated the ventilation in two areas within Kowloon City for neutral conditions using LES. They concluded that the urban morphology has a marked impact on ventilation. The ventilation of a single city quarter can be affected by its surroundings, which implies that neglecting the surrounding city area may lead to inaccuracies in ventilation analysis. Ref. Park et al. [11] utilized an LES model to study the ventilation in a region of roughly 7 km^2 within the densely built-up metropolitan area of Seoul, South Korea. Their results showed good ventilation of wide streets and at intersections, while poor ventilation was observed in densely built-up areas.

However, the above studies only analyzed neutral conditions excluding thermal buoyancy effects. Park and Baik [12] included thermal effects by surface heating, and found that the spanwise flow is stronger within an idealized building array compared to the non-heated case. Yang and Li [13] focused on the influence of stratification on ventilation considering a very simplified building array with a maximum of 21 blocks to simulate Hong Kong city. Turbulence was fully parameterized by the RANS model used in their study. Generally, higher ventilation was reported for unstable stratification in the case of weak background wind compared to neutral stratification. However, the simplifications (building setup and fully parameterized turbulence) allowed only a general evaluation of ventilation.

In this study, LES was used to analyze and compare ventilation in a large real metropolitan area (Kowloon City) for neutral and unstable stratification, whereas previous studies only focused on idealized building setups (e.g., [13]). There are a number of additional challenges for unstable stratification, including that a large model domain is required to catch all relevant turbulent structures, which are considerably larger than for neutral stratification, while the grid size must be kept small to sufficiently resolve the street-canyon flows (e.g., [10]). This substantially increases the computational expense of the simulations. Special attention must also be given to the choice of lateral boundary conditions such that they do not alter the ventilation results within the analyzed area. The comparison of ventilation for neutral and unstable stratification purely focused on the differences in how buildings influence ventilation under different stratification conditions. Ventilation effects due to differential heating between the city and the surroundings (e.g., sea-breeze effects) have not been studied and model setups have been chosen in a way that they are explicitly excluded. To the best of our knowledge, this is the first LES study to encompass a large realistic city domain with high (2 m) resolution and to compare ventilation results for different stratification.

The local government of Hong Kong initiated its AVA program focusing on summer weak-wind conditions [2]. These are the most hazardous conditions, as the weak wind in summer leads to a rapid increase in heat stress on the population and pollutants accumulate quickly in the streets because of reduced ventilation. This study therefore focused on these conditions.

The following text is divided into three parts. First, Section 2 presents a description of the simulation setup and the methods used; and Section 3 discusses the simulation results and compares two cases with different stratification. Finally, the conclusions are given in Section 4.

2. Model and Case Description

2.1. LES Model

The LES model PALM [14,15], version 4.0 revision 1746 (developed at the Insitute of Meteorology and Climatology of Leibniz Universität Hannover, Hannover, Germany), was used to perform the simulations in this study. It has been previously used to simulate the atmospheric boundary layer in densely built-up areas (e.g., [10,11,16]), and was successfully validated against wind-tunnel measurements within urban-like building arrays using neutrally and unstable stratified conditions [12,16,17]. PALM solves the non-hydrostatic incompressible Boussinesq equations.

2.2. Simulation Setup

Two cases were simulated with different types of atmospheric stratification. The first case featured neutral stratification, while the second had unstable stratification.

The simulation domains for the neutral and unstable cases are shown in Figure 1a,b, respectively. The simulated city area is detailed in Figure 1c and included an area of about 3.9 km × 4.7 km of Kowloon City. The same city area was used for both cases. Orographic features within the city area and the surroundings were neglected in order to limit ventilation effects purely to buildings and atmospheric stratification. The city was oriented south-north along the x (streamwise) direction and east-west along the y (spanwise) direction due to model constraints.

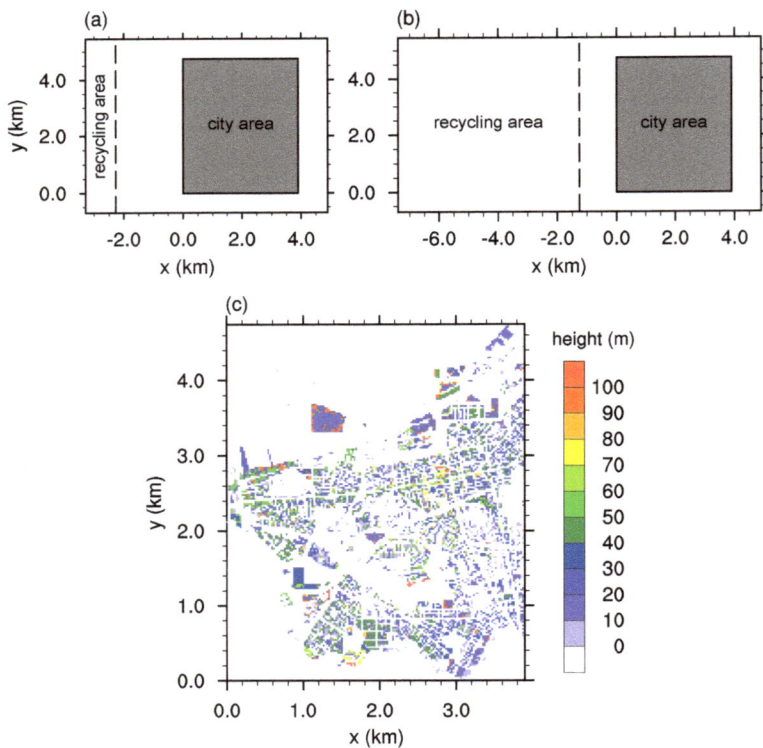

Figure 1. Domain setup for the (**a**) neutral case and (**b**) unstable case; building height information is depicted in (**c**). The dashed line marks the recycling area. The gray rectangle marks the city area shown in detail in (**c**).

A buffer region around the city ensured that the city area was not influenced by the domain boundaries. The buffer region had a width of 500 m at both sides (spanwise direction) and a length of 1000 m at the windward and leeward sides of the city. In the neutral case, the buffer region in front of the city had to be enlarged to 2000 m. The reduced mixing in the neutral case caused the blocking effects of the buildings to reach further upstream than in the unstable case, which required a larger windward buffer region.

To ensure a realistic turbulent inflow, a turbulence recycling method was used at the upstream boundary, which is described in more detail in Section 2.3. This method required an additional recycling domain at the inflow boundary with its length depending on the sizes of the turbulent structures. In the neutral case, a length of 1 km was sufficient for the recycling domain, while a length of 6 km was required in the unstable case. Upstream topographic features like Hong Kong Island were not considered. The outflow boundary condition used at the downstream boundary is described in Section 2.4. Cyclic boundary conditions were used in the spanwise direction.

The total domain sizes summed to 8 km (neutral case) and 12 km (unstable case) in the streamwise direction, 6 km in the spanwise direction, and 2.6 km in the vertical direction. Due to restrictions of the model grid, the exact domain size was 8192 m × 6144 m × 2653 m in the neutral case and 12.288 m × 6144 m × 2653 m in the unstable case.

The grid size was set to 2 m in each direction according to the results of a grid sensitivity study (see Appendix A). Starting at a height of 1100 m, the vertical grid size increased by 4% at each height level up to a maximum grid size of 40 m to reduce the computational time.

A Dirichlet boundary condition was applied at the top boundary and the Monin–Obukhov similarity theory was used at the bottom boundary as well as at the building walls. The roughness length was set to 0.1 m at each surface to account for roughness elements in the streets such as billboards, cars, and so forth. At the top boundary, Rayleigh damping was used to prevent the reflection of gravity waves.

In the unstable case, a constant near-surface heat flux of $0.165\,\mathrm{K\,m\,s^{-1}}$ (approximately $200\,\mathrm{W\,m^{-2}}$) was applied at every horizontal surface while a heat flux of 0 was set at all vertical building walls. This setup was comparable to the situation at noon in the summer with the sun situated in the zenith heating only horizontally oriented surfaces. Any additional heat release from buildings, for example, by air conditioning systems, was neglected. In addition, no distinction was made for different land use or land covers like water bodies or roads (i.e., no difference in surface heat flux). This simplification prevented the development of secondary circulations like sea-breeze, which would otherwise affect city ventilation [18].

The geostrophic wind was set to $1.5\,\mathrm{m\,s^{-1}}$ with a southerly wind direction (along a positive x direction), and the potential temperature θ was set to a constant value of 308 K within the boundary layer. This corresponded to daytime summer weak-wind conditions. To initialize the unstable case, a capping inversion layer was set above a height of 700 m where θ increased with a vertical gradient of $0.01\,\mathrm{K\,m^{-1}}$. A large-scale subsidence velocity was used to limit the growth of the boundary layer during the simulation. This prevented drifts in average wind speed and turbulence characteristics within the boundary layer. The large-scale subsidence velocity was set to zero at the surface and decreased linearly until it reached $-0.025\,\mathrm{m\,s^{-1}}$ at a height of 700 m from which it remained constant. After a spin-up time of 2 h, the boundary layer reached a height of 900 m and increased by only 60 m during the analysis period. The Coriolis force was taken into consideration.

Each simulation was initialized with turbulent three-dimensional velocity and temperature fields received from a precursor simulation with cyclic boundaries and a flat surface, and otherwise used the same setup as the main simulations.

Both cases were integrated over 6 h. Within the first 2 h, the turbulent fields adjusted to the urban surface and the simulations reached a quasi-stationary state. After this spin-up time, both cases were simulated for an additional 4 h to gain stable average values for the analysis. Simulations were performed using the Cray-XC40 supercomputer of the North-German Supercomputing Alliance

(HLRN) in Hannover, Germany. Due to the large domains and high spatial resolution, more than 7×10^9 and 12×10^9 grid points had to be used for the neutral and unstable case, respectively. The simulations required between 20 h and 60 h of computing time using more than 12,000 cores.

2.3. Inflow Boundary Conditions

To impose realistic inflow conditions on the LES, a turbulence recycling method was used at the inflow boundary following the method proposed by Lund et al. [19] and Kataoka and Mizuno [20]. A subdomain, named "recycling area" in Figure 1, was included in the simulation domain at the inflow boundary. Within this recycling area, the turbulence information Ψ' was recycled. Ψ' is defined as

$$\Psi' = \Psi - \langle \Psi \rangle_y , \tag{1}$$

where $\langle \dots \rangle_y$ denotes the spatial average along the spanwise or y direction, and is calculated at the downwind boundary of the recycling area. At the inflow boundary, Ψ' was added to a fixed mean inflow profile. Ψ was one of u, v, w or e, which were the wind velocity components in streamwise (x), spanwise (y), and vertical (z) directions, and subgrid-scale turbulent kinetic energy, respectively.

In the case of potential temperature θ, the method was altered such that, instead of the turbulent signal, the instantaneous value θ was copied from the downwind boundary of the recycling area and pasted to the inflow boundary. This ensured that the temperature level at the inflow boundary was equal to that in the simulation domain. Using the standard recycling method instead would cause a horizontal temperature gradient because the vertical temperature profile at the inlet would be fixed, while θ increased due to surface heating in the model domain. Then, this gradient would trigger a secondary circulation and hence alter the ventilation within the whole simulation domain. Although such a secondary circulation does occur in reality due to different surface heat-fluxes between Kowloon City and the surrounding bay, this effect on ventilation was omitted as it was not within the aim of the study.

The size of the recycling area was bound to the size of the turbulent structures present in the atmosphere. To ensure that the turbulent structures are not restricted by the size of the recycling area, its size must be large enough to enable the development of several turbulent structures of the largest occurring size but at least double the boundary layer height. In the neutral case, the boundary layer reached a height of 500 m. Hence, the size of the recycling area was set to 1 km in the streamwise direction, which proved to be sufficient. In the unstable case, the diameter of the convective cells, which was about 2 km, defined the size of the recycling area. To ensure that the convective cells could develop freely without being restricted by the boundaries of the recycling area, its size was set to three times the convective cell diameter, which was 6 km. Due to technical restrictions of the model grid, the actual size of the recycling domain was set to 1024 m and 6144 m in the neutral and unstable cases, respectively.

2.4. Outflow Boundary Condition

In PALM, a radiation boundary condition [21,22] was set as the standard outflow condition in the non-cyclic boundary case. However, this could not be used in the current study. The radiation boundary condition required a positive outflow (i.e., $u > 0$, at all times). This is not a problem if a sufficient background wind is considered like Park et al. [11] or Gryschka et al. [23] did. In this study, however, the weak background wind did not ensure that no negative u values occurred at the outflow boundary, particularly in the unstable case where strong turbulent motions were present. Once negative velocities occurred at the outflow, they were artificially strengthened by the radiation condition. This led to strong inward-directed artificial winds at the outflow boundary, which persisted in time.

To prevent these strong artificial winds, a new technique was introduced to handle the outflow. The instantaneous values of u, v, w, θ, and e were copied from a vertical plane (source plane), positioned

500 m in front of the outflow boundary, and then pasted to the outflow boundary. As the values were taken from within the simulation domain, they changed according to the flow field around the source plane and negative values were not artificially strengthened. This way, negative values were possible at the outflow boundary. It should be noted that this method was a technical workaround and did not represent the actual physics. However, the modification to the flow field was limited to the area close to the outflow boundary. A buffer zone of 1 km width ensured that the analysis area was unaffected by the boundary condition.

3. Results

The following analysis showed the differences in ventilation in varying atmospheric conditions. The analysis focused on the pedestrian height level 2 m above ground. All of the data are presented as averages during the last 4 h of each simulation unless otherwise stated.

First, to verify that the above-mentioned boundary conditions produced reasonable results, particularly from a meteorological viewpoint, Figure 2 shows the non-averaged vertical velocity component w and potential temperature θ for the unstable case at a height of 100 m at the last time step of the simulation. At this time point, the boundary layer reached a height of 960 m. The hexagonal structures visible in these figures had a size of about 2 km, which is within the typical range of 2–3 times the boundary layer height for a convective boundary layer [24]. Consequently, the size of the recycling area chosen was large enough to enable the development of several convective structures. Furthermore, no general horizontal temperature gradient was visible within the streamwise direction. This is because the temperature profile at the inflow boundary was constantly updated to the temperature level within the model domain as described in Section 2.3. Finally, none of the fields depicted in Figure 2 showed any visual effects due to the newly introduced outflow boundary condition, but retained their characteristics throughout the whole simulated domain.

Figure 3a,b depicts the magnitude of the time-averaged three-dimensional wind vector at 2 m height

$$V_{2m} = \left.\overline{\sqrt{u^2 + v^2 + w^2}}\right|_{z=2m} \tag{2}$$

for the neutral and unstable cases, respectively. It is obvious that V_{2m} was significantly higher throughout the whole city area and its surroundings in the unstable case compared to the neutral case. The average of V_{2m} in the unstable case was about 0.68 m s^{-1} or 1.9 times higher than that in the neutral case. The higher wind velocity in the unstable case was related to the greater downward vertical transport of momentum from above the city due to buoyancy-induced turbulence. It reflected the typical increase in near-surface winds during daytime (unstable stratification) compared to nighttime (neutral or stable stratification).

Figure 4 shows the vertical profile of the time and horizontally averaged horizontal wind vector $|\overline{v}_h| = \sqrt{\overline{u}^2 + \overline{v}^2}$ for both cases within the recycling area (i.e., without building effects). The strong vertical mixing led to a higher velocity near the surface in the unstable case than in the neutral case. As a result, the unstable case gave better ventilation than the neutral case with regard to ventilation solely at local near-surface wind speed. This was not related to any building effect and only reflected the change due to differences in stratification.

Usually, ventilation within the city is quantified using the velocity ratio

$$V_r = \frac{V_{2m}}{V_{ref}}, \tag{3}$$

where V_{ref} denotes a reference velocity often defined as $|\overline{v}_h|$ at a height well above the city area [2,10]. However, this definition of V_{ref} resulted in a higher V_r in the unstable case than in the neutral case according to general differences in the vertical distribution of horizontal wind speed (higher wind speed near the surface, lower wind speed within the boundary layer in the unstable case than in the neutral case, cf. Figure 4). This effect made it extremely difficult to detect and analyze the separate effects of the buildings on V_r for the different stratifications. To eliminate this trivial, purely stratification-related difference between the two cases, V_{ref} was redefined as V_{2m} calculated over the flat surface in front of the city area.

This adapted definition of V_r excluded the differences in vertical profiles between both cases, as now V_{2m} and V_{ref} are both calculated at the same height level within and outside the city region. Consequently, only differences in ventilation caused by the buildings under different stratification were emphasized. A $V_r < 1$ indicated reduced wind speed (low ventilation), while $V_r > 1$ was related to a higher wind speed (high ventilation) compared to that outside the city.

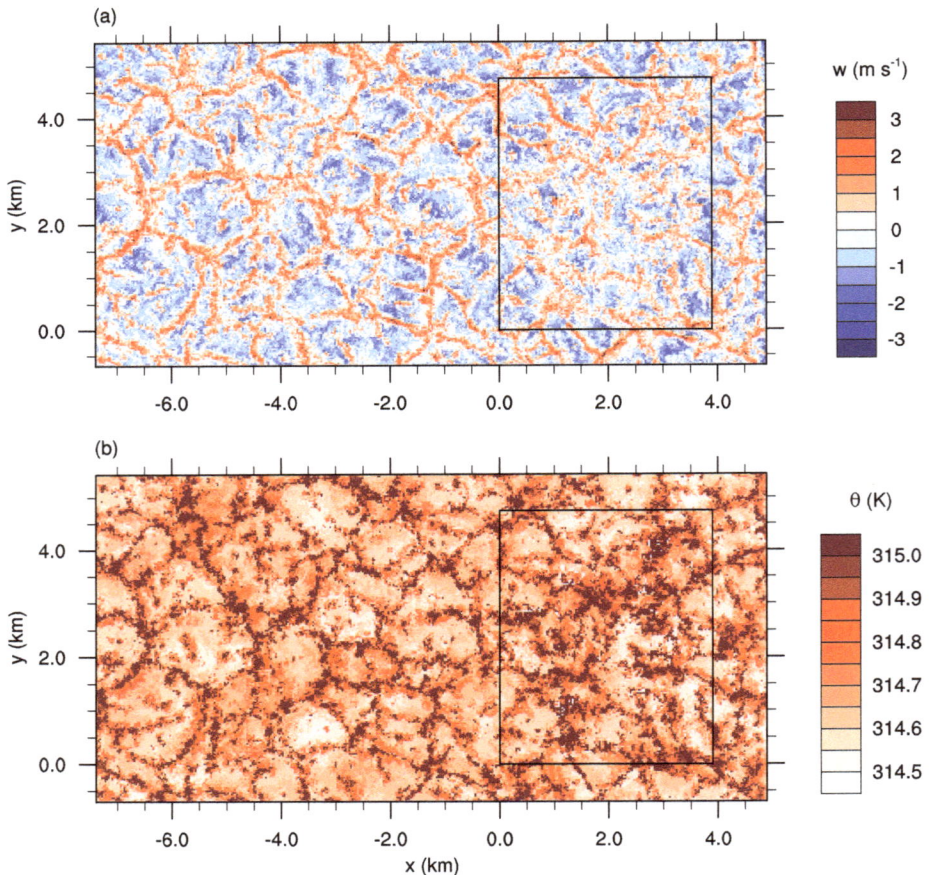

Figure 2. Instantaneous vertical wind velocity (**a**) and potential temperature (**b**) for the unstable case after a simulation time of 6 h at a height of 100 m. The solid inner rectangle marks the city area.

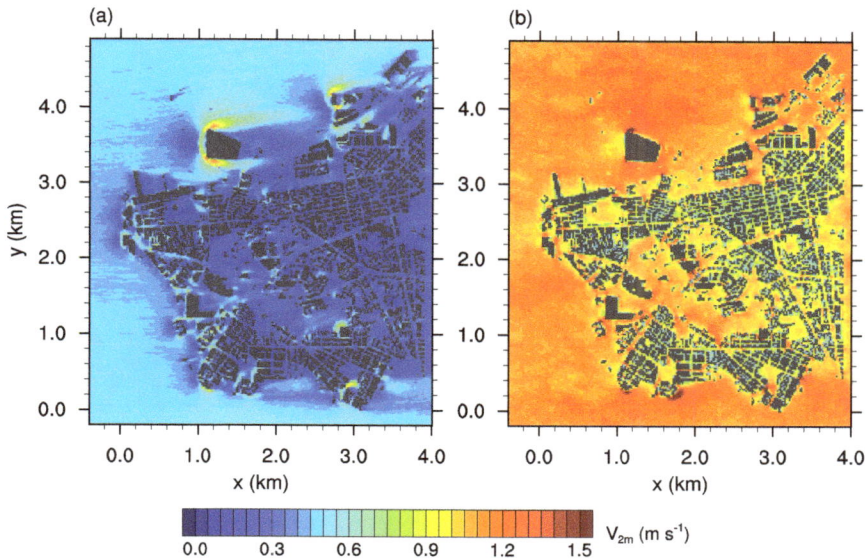

Figure 3. Averaged three-dimensional wind velocity at 2 m height for the (**a**) neutral case and (**b**) unstable case. Buildings are shown in gray.

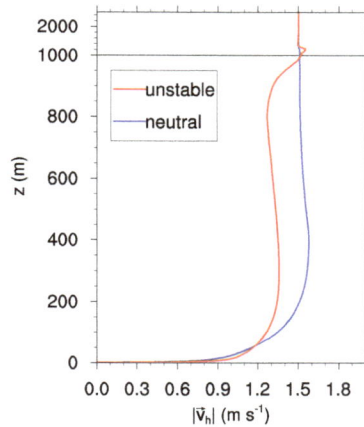

Figure 4. Vertical profile of the wind speed of the mean horizontal wind vector $|\vec{v}_h|$ within the recycling area.

Figure 5a,b shows the newly defined V_r for the neutral and unstable cases, respectively. In the neutral case, V_r was significantly less than 1 within the city area. The low V_r, which was related to a decrease in V_{2m} between the surroundings and the city, resulted from the buildings that were blocking the airflow within the city area. However, at the corners of exposed buildings, V_r increased as the air was forced to move around the buildings, which was consistent with well-known flow patterns around bluff bodies (e.g., [25]). In the unstable case (Figure 5b), such local influences of the buildings on V_r were significantly reduced in general, resulting in a more uniform distribution of V_r throughout the domain analyzed. Areas of high V_r in the neutral case still showed $V_r > 1$ in the unstable case, but the magnitude of the increase in V_r was significantly lower (e.g., at the edges of the large building complex at position $(x, y) = (1.2 \text{ km}^2, 3.6 \text{ km}^2)$.

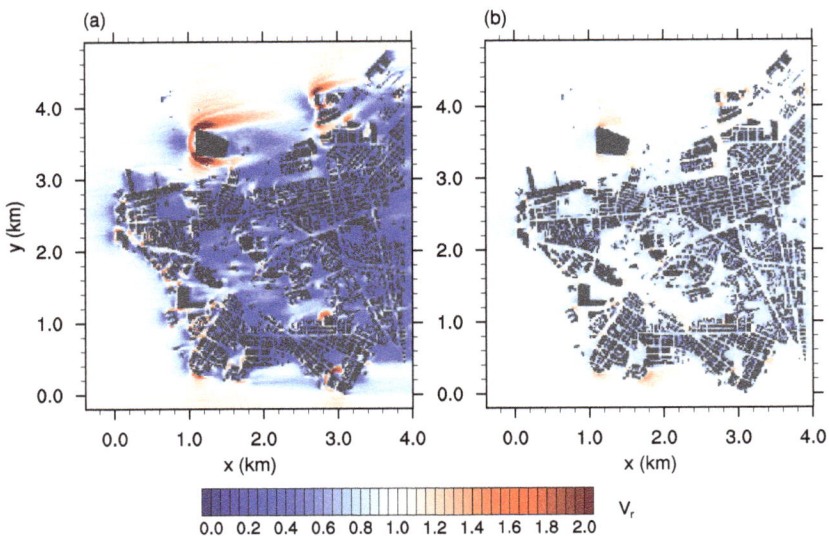

Figure 5. Velocity ratio V_r for the (**a**) neutral case and (**b**) unstable case.

A better view of the differences in V_r is shown in Figure 6, which shows the normalized velocity ratio

$$V_{r,\mathrm{norm}} = \frac{V_r(\mathrm{unstable})}{V_r(\mathrm{neutral})}. \tag{4}$$

Within the city, V_r was up to four times higher in the unstable case than in the neutral case. However, at the edges of the exposed buildings and at the windward edge of the city, V_r was only about 0.6 times its value in the neutral case. In the neutral case, the buildings blocked the airflow and forced the air to circulate around them horizontally. For reasons of continuity, the air was accelerated at the edges and decelerated at the leeward side of the buildings. In the unstable case, however, stratification made it much easier for the blocked air to flow over the buildings. This significantly reduced the air volume that was forced around the buildings, thereby preventing a strong increase in V_r around the exposed buildings. Furthermore, the enhanced vertical exchange in momentum due to convection led to higher V_r at the leeward side of these buildings and was also responsible for the strong increase in V_r within the city. The average V_r was about twice as high in the unstable case than in the neutral case.

The increase in V_r within the city area for the unstable case compared to the neutral case was also found by Yang and Li [13], who reported that flow rates through street canyons were higher in an unstable stratified atmosphere compared to those under conditions of neutral stratification. However, the reduction of V_r in the vicinity of exposed buildings was not reported, which was because Yang and Li [13] only used a very simplified building setup that did not include exposed buildings.

Further effects of a change in stratification on the ventilation could be derived from correlations of V_r with different building parameters. Figure 7a,b shows the scatter plot for V_r and the average building height H_{avg} and for V_r and the plan area index λ_p (building area divided by total area), respectively. For this, the city was divided into non-overlapping $100\,\mathrm{m} \times 100\,\mathrm{m}$ patches. The data points represent the average values within each of the patches. Data for the neutral and unstable cases are depicted by blue dots and red crosses, respectively.

Figure 7 shows that $V_r(\mathrm{unstable}) > V_r(\mathrm{neutral})$ as most data points for the neutral case lay within 0.1 and 0.7, while, for the unstable case, the majority of data points were within 0.5 and 1.1. However, no significant correlation was found between H_{avg} and V_r (Figure 7a). For both stratifications, the coefficient of determination is $R^2 < 0.1$. This means that H_{avg} had almost no influence on V_r in the neutral or unstable case. By contrast, Hang et al. [26] observed higher wind speed in a tall idealized building array compared to a shallower building array considering neutral stratification. The dependency found by Hang et al. [26] was related to perfectly aligned buildings channeling the flow within the idealized building arrays. In this case, higher buildings improved the channeling effect as a larger air volume was blocked at the front of the building array and forced into the streets. This effect was not observed in the simulations in this study, in which streets were randomly aligned to the wind direction.

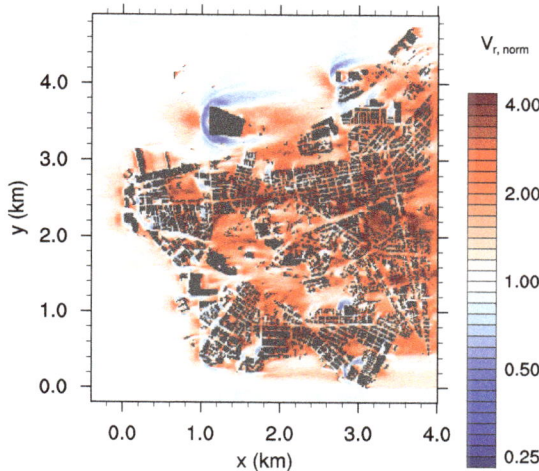

Figure 6. Normalized velocity ratio $V_{r,norm}$. Values above 1 indicate higher V_r in the unstable case, while values below 1 indicate higher V_r in the neutral case.

Figure 7b shows the scatter plot for V_r and λ_p. In the neutral and unstable cases, V_r and λ_p show a negative correlation. This was also reported by Hang et al. [26] for the neutral case. An area with high λ_p described a dense built-up area with narrow streets where the wind velocity was significantly reduced resulting in a low V_r. By contrast, a mostly open-space area, where λ_p was low, had less influence on the wind field and therefore V_r was high in such areas.

However, determination of the correlation varied between the neutral case and the unstable case. In the neutral case, the correlation was weak as V_r showed a high level of variation for a specific λ_p. This resulted in a low R^2 of 0.104. In the unstable case, the impact of λ_p on ventilation was significantly higher than in the neutral case as R^2 was 0.511, which resulted from less variation in V_r for a given λ_p.

This difference in correlation between the two cases can be explained when considering the influence of the wind direction. In the neutral case, the ventilation was highly dependent on the orientation of the wind direction with regard to the orientation of the streets [27]. The wind direction changed only slightly at a given point in the neutral case. Figure 8 shows representative measurements of wind direction for both cases at position $(x, y) = (2713 \, m, \, 2671 \, m)$, which is the center of a street crossing within the city center. The wind direction varied between 54° and 188° for the neutral case, while the wind came from all directions in the unstable case. At the same time, the orientation of the streets within Kowloon was heterogeneous, with many patches existing with the same λ_p but different street orientations. Therefore, in the neutral case, some patches experienced good ventilation as their streets were oriented favorably for the given wind direction, while other patches constantly experienced unfavorable winds and therefore were poorly ventilated. Therefore, V_r varied markedly in the neutral case for a given λ_p. As stated above, in the unstable case, ventilation was dominated by vertical mixing. Due to the strong convective motions and low background wind speed, the wind direction changed frequently (Figure 8). Therefore, the actual orientation of streets according to the wind direction was less important in the unstable case than in the neutral case. The main parameter determining the ventilation for the unstable case was therefore the amount of void space where convective motions could develop. This was related to λ_p, as it gave the ratio of occupied area to the total area. As $R^2 > 0.5$, λ_p was the key parameter determining the ventilation for the unstable case.

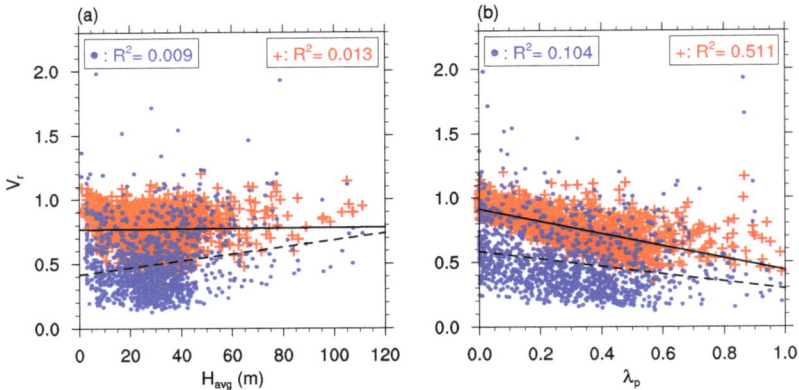

Figure 7. Scatter plot for V_r and (**a**) the average building height H_{avg} and (**b**) the plan area index λ_p. Each point represents an average value inside a $100\,\text{m} \times 100\,\text{m}$ area within the city. Blue dots represent the neutral case and red crosses represent the unstable case.

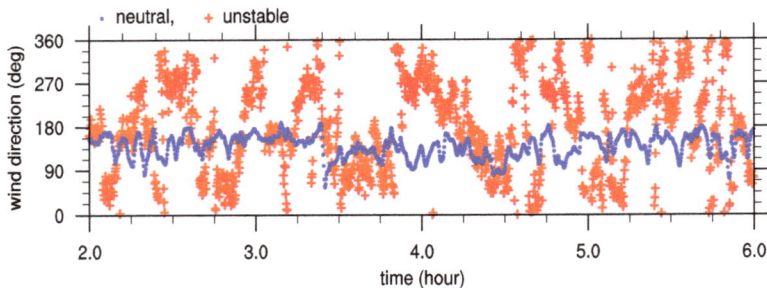

Figure 8. Time series of wind direction at position $(x, y) = (2713\,\text{m}, 2671\,\text{m})$ in the city center. Blue dots represent the neutral case and red crosses represent the unstable case.

4. Conclusions

This study compared ventilation in Kowloon City, Hong Kong, under conditions of neutral and unstable atmospheric stratification. For the comparison, a summer weak-wind situation with a geostrophic wind of $1.5\,\text{m}\,\text{s}^{-1}$ was chosen, as this condition was also the primary target of AVAs for Kowloon City.

An alternative definition of ventilation via the velocity ratio V_r was presented. The standard definition considers the vertical distribution of wind velocity, and therefore depends on the stratification. The new definition neglected this vertical distribution and purely focused on the impact of obstacles under conditions of varying stratification, as it was directly calculated from the reduction of the wind velocity due to blockage of airflow by buildings. This enabled a better comparison of the influence of building on ventilation under different stratification conditions.

Atmosphere **2017**, *8*, 168

The averaged ventilation in the unstable case was about double that in the neutral case. This was due to the large convective eddies in the unstable case, which created a high degree of variation in wind direction and strong up- and downdrafts. The strong vertical motions reduced the decelerating impact of buildings on the flow field, as it was easier for the air to flow over them. However, this also reduced acceleration effects at the side edges of exposed buildings, which appeared in the neutral case. In these areas, V_r was reduced to 0.6 times its value in the neutral case. Consequently, considering only neutral stratification when analyzing the ventilation of a city area was insufficient, as the ventilation appeared to be significantly changed, positively and negatively, under conditions of unstable stratification.

A linkage between the plan area index λ_p and V_r was found similar to other studies for neutral stratification where low V_r corresponds to high λ_p. However, the correlation between both variables was stronger in the unstable case than in the neutral case. In the neutral case, the correlation was reduced because, apart from λ_p, the orientation of the buildings in relation to the wind direction also influenced ventilation. In the unstable case, no distinct wind direction was present as it changed frequently due to the convective motions. This led to a smaller influence of the building orientation and a greater influence of λ_p on ventilation. Therefore, for cities where convective low-wind conditions are often present, such as Hong Kong, city planning should focus more on reducing λ_p to improve city ventilation than on the orientation of buildings and streets.

In contrast to other studies, no correlation was found between the average building height H_{avg} and V_r. As these other studies only focused on idealized homogeneous block arrays, it is possible that the idealized cases overestimated the channeling induced by these idealized building arrays.

The results of this study indicated that AVAs should not focus purely on neutral stratification but should also consider unstable stratification, particularly when these conditions in combination with low wind speed are observed frequently, as in Hong Kong. When focusing on summer weak-wind conditions, a complete view on the ventilation of a city area can only be obtained if both neutral and unstable stratification are included in the analysis. For strong-wind conditions, the influence of mechanically induced turbulence may become stronger than that of thermally induced turbulence, which was already found by Yang and Li [13] for a very simplified city case.

The impact of stable stratification was not covered in this study but should be examined in future analyses. The first inspection of the impact of stable stratification on ventilation was made by Yang and Li [13] for a simplified city case. However, these results should be tested with a more realistic setup. The results of this study revealed differences between a simple building case and a more realistic setup for unstable stratification. Therefore, it is possible that results would also differ for a stable case with a more sophisticated building setup.

Surface elevation and land cover were neglected in this study to extract the pure influence of buildings on ventilation under different stratification conditions. Recent studies by Wolf-Grosse et al. [28] and Ronda et al. [18], however, showed the importance of topographic effects and consideration of water bodies at city scale. Therefore, these effects will be considered in a follow-up study, focusing on their combined influence on ventilation. In particular, the difference in land cover with resulting heterogeneous surface heat flux may have a large impact on ventilation as this leads to small-scale secondary circulations, such as sea breeze. A further step towards reality will also be to use an urban surface model available in PALM [29,30], which allows to accurately calculate surface temperatures based on a building-energy-balance model.

Supplementary Materials: The modified code parts used in addition to the standard code base of PALM are available online at www.mdpi.com/2073-4433/8/9/168/s1.

Acknowledgments: The authors thank Weiwen Wang, School of Architecture, Chinese University of Hong Kong, for providing the building data. The study was supported by a research grant (14408214) from General Research Fund of Hong Kong Research Grants Council (HK RGC-GRF). Tobias Gronemeier was supported by MOSAIK, which is funded by the German Federal Ministry of Education and Research (BMBF) under grant 01LP1601A within the framework of Research for Sustainable Development (FONA; http://www.fona.de). The simulations were performed with resources provided by the North-German Supercomputing Alliance (HLRN). NCL (The NCAR Command Language, Version 6.1.2, 2013 (Software), Boulder, Colorado: UCAR/NCAR/CISL/VETS, http://dx.doi.org/10.5065/D6WD3XH5) was used for data analysis and visualization. The PALM code can be accessed under https://palm.muk.uni-hannover.de. The publication of this article was funded by the Open Access fund of Leibniz Universität Hannover.

Author Contributions: T.G., S.R. and E.N. conceived and designed the simulations; T.G. performed the simulations; T.G. and S.R. analyzed the data; T.G. wrote the paper.

Conflicts of Interest: The authors declare no conflict of interest.

Appendix A

A grid sensitivity study was conducted to determine an appropriate grid width for the main simulations. Four simulations with grid sizes Δ of 1 m, 2 m, 4 m, and 8 m were compared. The domain used in the sensitivity study included 1 km^2 of Kowloon City. Cyclic boundary conditions and neutral stratification were used. As turbulent structures are generally larger in the unstable case than in the neutral case, the latter defined the minimum grid size to be used.

Figure A1 depicts the cumulative distribution function of the 1 h averaged 3-dimensional wind velocity $V = \sqrt{u^2 + v^2 + w^2}$ at a height of 4 m for each simulation. Due to differences in grid level height between the simulations, data were linearly interpolated to $z = 4$ m. Significant differences in the distribution of V could be observed between 2 m, 4 m and 8 m grid sizes. The distribution of low V decreased if a smaller grid size was used. The test statistic $A(a, b)$ of Kuiper's test [31], where $A(a, b)$ compares the distribution function of case a to that of case b, yields $A(8, 4) = 0.11$, $A(4, 2) = 0.10$, and $A(2, 1) = 0.06$. Thus, the distribution of V changed significantly less if the grid size was reduced from 2 m to 1 m compared to a reduction from 8 m to 4 m or from 4 m to 2 m. Consequently, a reduction of grid size from 2 m to 1 m only slightly improved the quality of the representation of the wind field within the city. As a reduction of the grid size by a factor of 2 increased the computational load by a factor of 16 (double the grid points in each dimension multiplied by 2 for double the amount of time steps needed), a grid size of 2 m was selected for the main simulations as a compromise between accuracy and computational cost.

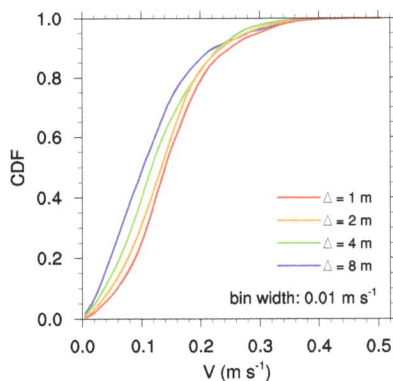

Figure A1. Cumulative distribution function of three-dimensional wind velocity V at 4 m height. Data are averaged over 1 h.

References

1. Xie, X.; Huang, Z.; Wang, J.S. Impact of building configuration on air quality in street canyon. *Atmos. Environ.* **2005**, *39*, 4519–4530.
2. Ng, E. Policies and technical guidelines for urban planning of high-density cities—Air ventilation assessment (AVA) of Hong Kong. *Build. Environ.* **2009**, *44*, 1478–1488.

3. Ng, E.; Yuan, C.; Chen, L.; Ren, C.; Fung, J.C.H. Improving the wind environment in high-density cities by understanding urban morphology and surface roughness: A study in Hong Kong. *Landsc. Urban Plan.* **2011**, *101*, 59–74.

4. Cheng, Y.; Lien, F.; Yee, E.; Sinclair, R. A comparison of large Eddy simulations with a standard k–ϵ Reynolds-averaged Navier–Stokes model for the prediction of a fully developed turbulent flow over a matrix of cubes. *J. Wind Eng. Ind. Aerodyn.* **2003**, *91*, 1301–1328.

5. Ashie, Y.; Kono, T. Urban-scale CFD analysis in support of a climate-sensitive design for the Tokyo Bay area. *Int. J. Climatol.* **2011**, *31*, 174–188.

6. Tominaga, Y. Visualization of city breathability based on CFD technique: Case study for urban blocks in Niigata City. *J. Vis.* **2012**, *15*, 269–276.

7. Yang, L.; Li, Y. City ventilation of Hong Kong at no-wind conditions. *Atmos. Environ.* **2009**, *43*, 3111–3121.

8. Yim, S.H.L.; Fung, J.C.H.; Lau, A.K.H.; Kot, S.C. Air ventilation impacts of the "wall effect" resulting from the alignment of high-rise buildings. *Atmos. Environ.* **2009**, *43*, 4982–4994.

9. Yuan, C.; Ng, E.; Norford, L.K. Improving air quality in high-density cities by understanding the relationship between air pollutant dispersion and urban morphologies. *Build. Environ.* **2014**, *71*, 245–258.

10. Letzel, M.O.; Helmke, C.; Ng, E.; An, X.; Lai, A.; Raasch, S. LES case study on pedestrian level ventilation in two neighbourhoods in Hong Kong. *Meteorol. Z.* **2012**, *21*, 575–589.

11. Park, S.B.; Baik, J.J.; Lee, S.H. Impacts of mesoscale wind on turbulent flow and ventilation in a densely built-up urban area. *J. Appl. Meteorol. Climatol.* **2015**, *54*, 811–824.

12. Park, S.B.; Baik, J.J. A Large-Eddy Simulation Study of Thermal Effects on Turbulence Coherent Structures in and above a Building Array. *J. Appl. Meteorol. Climatol.* **2013**, *52*, 1348–1365.

13. Yang, L.; Li, Y. Thermal conditions and ventilation in an ideal city model of Hong Kong. *Energy Build.* **2011**, *43*, 1139–1148.

14. Raasch, S.; Schröter, M. PALM-A large-eddy simulation model performing on massively parallel computers. *Meteorol. Z.* **2001**, *10*, 363–372.

15. Maronga, B.; Gryschka, M.; Heinze, R.; Hoffmann, F.; Kanani-Sühring, F.; Keck, M.; Ketelsen, K.; Letzel, M.O.; Sühring, M.; Raasch, S. The Parallelized Large-Eddy Simulation Model (PALM) version 4.0 for atmospheric and oceanic flows: Model formulation, recent developments, and future perspectives. *Geosci. Model Dev.* **2015**, *8*, 2515–2551.

16. Lo, K.W.; Ngan, K. Characterising the pollutant ventilation characteristics of street canyons using the tracer age and age spectrum. *Atmos. Environ.* **2015**, *122*, 611–621.

17. Park, S.B.; Baik, J.J.; Ryu, Y.H. A Large-Eddy Simulation Study of Bottom-Heating Effects on Scalar Dispersion in and above a Cubical Building Array. *J. Appl. Meteorol. Climatol.* **2013**, *52*, 1738–1752.

18. Ronda, R.; Steeneveld, G.; Heusinkveld, B.; Attema, J.; Holtslag, B. Urban fine-scale forecasting reveals weather conditions with unprecedented detail. *Bull. Am. Meteorol. Soc.* **2017**, doi:10.1175/BAMS-D-16-0297.1.

19. Lund, T.S.; Wu, X.; Squires, K.D. Generation of Turbulent Inflow Data for Spatially-Developing Boundary Layer Simulations. *J. Comput. Phys.* **1998**, *140*, 233–258.

20. Kataoka, H.; Mizuno, M. Numerical flow computation around aeroelastic 3D square cylinder using inflow turbulence. *Wind Struct.* **2002**, *5*, 379–392.

21. Orlanski, I. A simple boundary condition for unbounded hyperbolic flows. *J. Comput. Phys.* **1976**, *21*, 251–269.

22. Miller, M.J.; Thorpe, A.J. Radiation conditions for the lateral boundaries of limited-area numerical models. *Q. J. R. Meteorol. Soc.* **1981**, *107*, 615–628.

23. Gryschka, M.; Fricke, J.; Raasch, S. On the impact of forced roll convection on vertical turbulent transport in cold air outbreaks. *J. Geophys. Res. Atmos.* **2014**, *119*, 12513–12532.

24. Atkinson, B.W.; Wu Zhang, J. Mesoscale shallow convection in the atmosphere. *Rev. Geophys.* **1996**, *34*, 403–431.

25. Larousse, A.; Martinuzzi, R.; Tropea, C. Flow Around Surface-Mounted, Three-Dimensional Obstacles. In Proceedings of the Turbulent Shear Flows 8: Selected Papers from the Eighth International Symposium on Turbulent Shear Flows, Munich, Germany, 9–11 September 1991; Durst, F., Friedrich, R., Launder, B.E., Schmidt, F.W., Schumann, U., Whitelaw, J.H., Eds.; Springer: Berlin/Heidelberg, Germany, 1993; pp. 127–139.

26. Hang, J.; Li, Y.; Sandberg, M. Experimental and numerical studies of flows through and within high-rise building arrays and their link to ventilation strategy. *J. Wind Eng. Ind. Aerodyn.* **2011**, *99*, 1036–1055.

27. Kim, J.J.; Baik, J.J. A numerical study of the effects of ambient wind direction on flow and dispersion in urban street canyons using the RNG k-e turbulence model. *Atmos. Environ.* **2004**, *38*, 3039–3048.

28. Wolf-Grosse, T.; Esau, I.; Reuder, J. Sensitivity of local air quality to the interplay between small- and large-scale circulations: A large-eddy simulation study. *Atmos. Chem. Phys.* **2017**, *17*, 7261–7276.

29. Resler, J.; Krč, P.; Belda, M.; Juruš, P.; Benešová, N.; Lopata, J.; Vlček, O.; Damašková, D.; Eben, K.; Derbek, P.; et al. A new urban surface model integrated in the large-eddy simulation model PALM. *Geosci. Model Dev. Discuss.* **2017**, *2017*, 1–26.

30. Yaghoobian, N.; Kleissl, J.; Paw, U.K.T. An Improved Three-Dimensional Simulation of the Diurnally Varying Street-Canyon Flow. *Bound.-Layer Meteorol.* **2014**, *153*, 251–276.

31. Kuiper, N.H. Tests concerning random points on a circle. *Indag. Math.* **1960**, *63*, 38–47.

atmosphere

MDPI

Article

Numerical Study on the Urban Ventilation in Regulating Microclimate and Pollutant Dispersion in Urban Street Canyon: A Case Study of Nanjing New Region, China

Fan Liu, Hua Qian *, Xiaohong Zheng, Lun Zhang and Wenqing Liang

School of Energy and Environment, Southeast University, Nanjing 210096, Jiangsu, China;
18252712528@163.com (F.L.); xhzheng@seu.edu.cn (X.Z.); zhanglun@seu.edu.cn (L.Z.);
lwq21cn@hotmail.com (W.L.)
* Correspondence: qianh@seu.edu.cn

Received: 30 July 2017; Accepted: 26 August 2017; Published: 29 August 2017

Abstract: Urban ventilation plays an important role in regulating city climate and air quality. A numerical study was conducted to explore the ventilation effectiveness on the microclimate and pollutant removal in the urban street canyon based on the rebuilt Southern New Town region in Nanjing, China. The RNG $k - \varepsilon$ turbulence model in the computational fluid dynamics (CFD) was employed to study the street canyon under parallel and perpendicular wind directions, respectively. Velocity inside of the street canyon and temperature on the building envelopes were obtained. A novel pressure coefficient was defined, and three methods were applied to evaluate the urban ventilation effectiveness. Results revealed that there was little comfort difference for the human body under two ventilation patterns in the street canyon. Air stagnation occurred easily in dense building clusters, especially under the perpendicular wind direction. In addition, large pressure coefficients ($C_P > 1$) appeared at the windward region, contributing to promising ventilation. The air age was introduced to evaluate the "freshness" of the air in the street canyon and illustrated the ventilation effectiveness on the pollutant removal. It was found that the young air distributed where the corresponding ventilation was favorable and the wind speed was large. The results from this study can be useful in further city renovation for the street canyon construction and municipal planning.

Keywords: street canyon; computational fluid dynamics (CFD); ventilation effectiveness; the age of air

---

1. Introduction

The outdoor thermal and wind environment are of great significance to the living conditions and people's daily activities [1]. With the rapid urbanization, street canyons appear in cities as more and more high-rise buildings spring up, resulting in poor city ventilation [2]. Stagnant air caused by poor ventilation in street canyons leads to the massive accumulation of pollutants and anthropogenic heat, exacerbating the outdoor air quality [3–5] and the Urban Heat Island effect (UHI) [6–9]. Outdoor environment is essential not only for people directly exposed to the circumstance [10,11], but it also has an important influence on the indoor environment [12]. From the last decade, study on microclimate in a street canyon has become a hot issue focussing on ventilation, temperature, and pollutant dispersion. Numerical and experimental validation studies on street canyons indicate that ventilation is a vital element in these related indices on urban microclimate [12]. Air in the city can be well mixed and exchanged under the windy condition that introduces the fresh air from the suburbs into the city. The heat and pollutant generated from the city is likely to be removed by the airflow [1]. It is noted that the effectiveness of the heat and pollutant removal is strongly affected by the airflow patterns inside street canyons.

Many studies have been carried out to investigate the airflow pattern around buildings by wind tunnel experiments [13,14] and field measurements [15]. With the development of computer technology and numerical computation, the computational fluid dynamics (CFD) method has been gradually employed attributed to its low expense and high flexibility. Combining the simulation method and the architectural design strategy, it enables architects to better design and evaluate the city wind environment. Studies on outdoor environment by CFD method mainly focus on airflow pattern analysis [16,17], human comfort evaluation at the pedestrian level around buildings [18], and the prediction of pollution dispersion [19]. Over the last few decades, many attempts have been made to link urban ventilation research with the urban design in a given condition. The ventilation has been found to have a close relationship with the building layout, which subsequently affects the air quality in street canyons [20]. Moreover, the canyon aspect ratio and the orientation of street canyons are two crucial parameters informing the urban microclimates and influencing the thermal comfort of pedestrians, which is more obvious in full-scale high-rise deep street canyons [19,21]. The vertical microclimate and mass-exchange have also been investigated in different street canyons with different geometries [22]. It has been noted that the street canyon is a key factor in maintaining urban microclimates by previous studies. The mechanisms of the airflow patterns and the ventilation effectiveness on heat reduction and pollutant dispersion in street canyons have been fully understood. However, only simple construction models [23,24] and a typical wind direction in street canyons [25] are considered in most existing numerical studies. Limited investigations are conducted to study the urban ventilation and its effects on the microclimate. Moreover, qualified indexes for the evaluation of the urban ventilation effectiveness have not been appropriately considered up until now.

To illustrate ventilation effectiveness in cities with street canyons, a numerical study has been carried out in this paper based on the project of Southern New Town region, Nanjing. A long street altered from a military airport runway is centered in this region, with a large number of building clusters including residential areas, innovative, and military districts on both sides. The airflow inside the street canyon and the temperature distribution at building facades were discussed under different local metrological parameters. The ventilation effectiveness was evaluated by three commonly used methods on the heat and pollutant removal. Results from this study can be helpful for further the city construction and urban environment maintenance with urban street valleys.

2. Study Case

The Southern New Town region, located at the southeast Nanjing, is designed to be built on the location of a military airport. The original airport runway will be altered into a long street at the center of this region. Surrounding areas are designed to three particular sub-regions (residential areas, innovative, and military districts) as well as some ecological areas and open space. The long street canyon goes through the region from southeast to northwest, as shown in Figure 1.

Figure 1. Planning diagram of the Southern New City region.

As planned, the total area is 12.88 km^2, with construction land of 5.93 km^2. The local wind rose diagrams monitored are shown in Figure 2. Table 1 presents the local meteorological parameters, obtained at the height of 10 m above the ground for numerical simulations, such as air temperature, wind velocity, and prevailing direction. The prevailing wind directions are south-southeast with the average speed of 2.4 m/s in summer, and east-northeast with the average speed of 2.7 m/s in winter. Table 2 shows the calorific intensity on the building surface and in the interspaces in summer. These parameters are all used as boundary conditions for simulation scenarios.

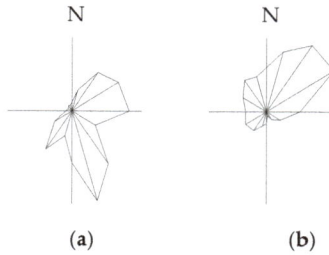

(a) (b)

Figure 2. Wind rose diagrams of Nanjing: (**a**) Summer; (**b**) Winter.

Table 1. Meteorological data for simulations.

Scenario	Season	Temperature (°C)	Prevailing Wind Direction	Wind Speed (m/s)
1	Summer	20~37	SSE	2.4
2	Winter	−2~10	ENE	2.7

SSE: South-Southeast; ENE: East-Northeast.

Table 2. The calorific intensity in summer.

Items	Military Districts	Innovative Streets	Residential Areas
Building area (m^2)	167.0	410.3	488.4
Body heat (W·m^{-3})	1.6	4.3	4.5
Surface heat (W·m^{-2})	91.3	155.6	147.0

Three commonly used methods are adopted to evaluate the ventilation effectiveness on the heat and pollutant removal, i.e., wind scales and human body comfort, wind pressure, and the age of air. To evaluate the pollutant dispersion in the street canyons, the metric, "local mean age of air" is used in this paper. The local mean age of air was firstly proposed as an indicator of the air quality in indoor spaces [26,27], which has been confirmed to be applicable in outdoors [28,29]. It serves as an indicator of outdoor air quality in this paper to help recognize the stagnant areas or polluted region where contaminants easily accumulate.

3. Computational Settings and Parameters

The Fluent calculation module in the Ansys 17.0 software (ANSYS, Inc., Canonburg, PA, USA) is chosen to conduct the numerical simulations. According to the three-dimensional architectural model of the Southern New Town region, an appropriate computational domain is selected and meshed, and the entire airflow field is solved with mathematical models under corresponding boundary and initial conditions. The wind field, temperature distribution, and pollutant dispersion are calculated and analyzed.

3.1. Geometric Model, Computational Domain and Meshing Generation

It is essential to simplify the geometric model for meshing, reducing computational nodes, and accelerating the convergence speed before calculation. Small convex and concave structures of buildings are neglected, and all curved shapes are regularized in the process. The simplified model is shown in Figure 3. In terms of the size of the computational domain, the top boundary is set as 4 H_{max} (H_{max} is the height of the tallest building), lateral boundaries is set as 5 W (W is the width of the target region), and the inlet and outflow boundaries are set as 6 L (L is the length of the target region).

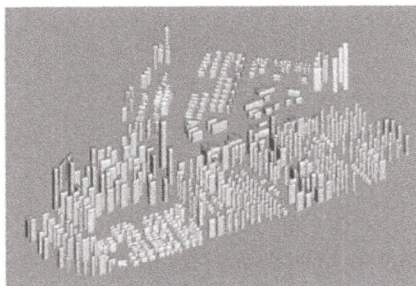

Figure 3. Physical model in computational fluid dynamics (CFD).

In order to predict the airflow around buildings in acceptable accuracy, the most important thing is to correctly reproduce the separating airflows near the corners and the walls. Therefore, a fine grid arrangement is required for calculation. Before the meshing, the whole domain is divided into several connected areas according to construction shapes and sizes. Unstructured tetrahedral cells are used to construct the whole computational domain. For the present simulation, the minimum grid resolution is set to be 0.5–5.0 m ($\frac{1}{10}$ of the building scale [30]) within the target region, and increases gradually with distance above 100 m in the computational domain. The resulting grid contains 7,890,431 control volumes (see Figure 4), and is based on a grid-sensitivity analysis focused on the mean wind speed in the entrance.

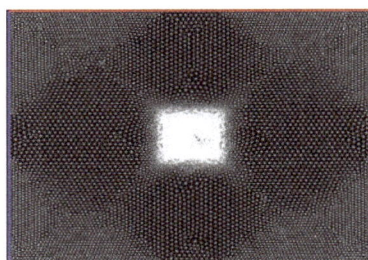

Figure 4. Computational grid on the bottom surface the domain. Total number of cells is 7,890,431.

3.2. Methodology

The outdoor airflow is governed by conservations of mass and momentum in each flow direction. An incompressible tairflow with constant properties is assumed in the numerical calculation. The Reynolds-Averaged Navier-Stokes (RANS) equations are used and written as follows [31].

The mass conservation equation,

$$\text{div } \mathbf{U} = 0 \tag{1}$$

The equations for conservation of momentum,

$$\text{div}(UU) = -\frac{1}{\rho}\frac{\partial P}{\partial x} + \gamma\text{div}(\text{grad}(U)) + \frac{1}{\rho}\left[\frac{\partial\left(-\overline{\rho u'^2}\right)}{\partial x} + \frac{\partial\left(-\overline{\rho u'v'}\right)}{\partial y} + \frac{\partial\left(-\overline{\rho u'w'}\right)}{\partial z}\right] + S_{Mx} \quad (2a)$$

$$\text{div}(VU) = -\frac{1}{\rho}\frac{\partial P}{\partial y} + \gamma\text{div}(\text{grad}(V)) + \frac{1}{\rho}\left[\frac{\partial\left(-\overline{\rho u'V'}\right)}{\partial x} + \frac{\partial\left(-\overline{\rho v'^2}\right)}{\partial y} + \frac{\partial\left(-\overline{\rho v'w'}\right)}{\partial z}\right] + S_{My} \quad (2b)$$

$$\text{div}(WU) = -\frac{1}{\rho}\frac{\partial P}{\partial z} + \gamma\text{div}(\text{grad}(W)) + \frac{1}{\rho}\left[\frac{\partial\left(-\overline{\rho u'w'}\right)}{\partial x} + \frac{\partial\left(-\overline{\rho v'w'}\right)}{\partial y} + \frac{\partial\left(-\overline{\rho w'^2}\right)}{\partial z}\right] + S_{Mz} \quad (2c)$$

with three normal stresses,

$$\tau_{xx} = -\overline{\rho u'^2},\ \tau_{yy} = -\overline{\rho v'^2},\ \tau_{zz} = -\overline{\rho w'^2} \quad (3a)$$

and three shear stresses,

$$\tau_{xy} = \tau_{yx} = -\overline{\rho u'v'},\ \tau_{xz} = \tau_{zx} = -\overline{\rho u'w'},\ \tau_{yz} = \tau_{zy} = -\overline{\rho v'w'} \quad (3b)$$

These extra turbulent stresses in Equation (3) are the Reynolds stresses. $S_{Mx} = 0$, $S_{My} = 0$, and $S_{Mz} = -\rho g$ denote projections of the body force in x, y, and z directions, respectively. The density variation is calculated by the Boussinesq approximation, i.e., $\rho = \rho_0[1 - \beta(T - T_0)]$. The energy equation,

$$\text{div}(TU) = \frac{1}{\rho}\text{div}(\alpha\text{grad}T) - \left(\frac{\partial\overline{u't'}}{\partial x} + \frac{\partial\overline{u't'}}{\partial y} + \frac{\partial\overline{u't'}}{\partial z}\right) + S_T \quad (4)$$

where the flow variables $\mathbf{u}$ (hence also u, v, and w), p and t are described by the sum of a mean and fluctuating component, i.e., $\mathbf{u} = \mathbf{U} + \mathbf{u}'$ ($u = U + u'$, $v = V + v'$ and $w = W + w'$), $p = P + p'$, $t = T + t'$. γ is the kinematic viscosity, α the thermal diffusion coefficient, ρ_0 the reference density, T_0 the reference temperature.

Closure is obtained by the Renormalization Group (RNG) $k - \varepsilon$ turbulence model with enhanced wall treatment [32]. The transport equations for turbulent kinetic energy (TKE) k and dissipation rate ε are given as:

$$\text{div}(\rho k U) = \text{div}(\alpha_k \mu_{\text{eff}}\text{grad}k) + \tau_{ij}{\cdot}S_{ij} - \rho\varepsilon \quad (5)$$

$$\frac{\partial}{\partial x_j}(\rho\varepsilon U) = \text{div}(\alpha_\varepsilon \mu_{\text{eff}}\text{grad}\varepsilon) + C_{1\varepsilon}^*\frac{\varepsilon}{k}\tau_{ij}{\cdot}S_{ij} - C_{2\varepsilon}\rho\frac{\varepsilon^2}{k} \quad (6)$$

with

$$\tau_{ij} = -\overline{\rho u_i' u_j'} = 2\mu_t S_{ij} - \frac{2}{3}\rho k\delta_{ij},\ S_{ij} = \frac{1}{2}\left(\frac{\partial u_i}{\partial x_j} + \frac{\partial u_j}{\partial x_i}\right) \quad (7)$$

and

$$\mu_{\text{eff}} = \mu + \mu_t,\ \mu_t = \rho C_\mu\frac{k^2}{\varepsilon} \quad (8)$$

$$C_{1\varepsilon}^* = C_{1\varepsilon} - \frac{\eta(1 - \eta/4.377)}{1 + 0.012},\ \eta = \frac{k}{\varepsilon}\sqrt{2S_{ij}{\cdot}S_{ij}} \quad (9)$$

The convention of the suffix notation is that i or j = 1 corresponds to the x-direction, i or j = 2 the y-direction, i or j = 3 the z-direction. μ is the molecular viscosity, μ_t the turbulent viscosity, δ_{ij} the Kronecker delta ($\delta_{ij} = 1$ if i = j and $\delta_{ij} = 0$ if i $\neq$ j). $C_\mu = 0.0845$, $\alpha_k = \alpha_\varepsilon = 1.39$, $C_{1\varepsilon} = 1.42$, $C_{2\varepsilon} = 1.68$ [33].

The age of air is utilized to demonstrate the effects of the ventilation corridor on the "freshness" of outdoor air in the Southern New Town region. The definition of the age of air φ_n at a point within a zone is described in Figure 5.

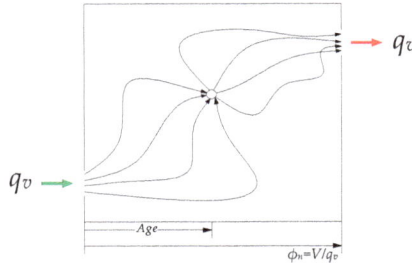

Figure 5. Definition of the age of air [27].

As shown in Figure 5, Age is a nominal time constant which is relevant with the volume V and flux q_v of airflow. The local mean age of air, Φ, is obtained from additional transport equations in the CFD method, which can be formulated as [28]:

$$\text{div}(\Phi U) = \frac{1}{\rho}\text{div}(Dgrad\Phi) - \left(\frac{\partial \overline{u'\varphi'}}{\partial x} + \frac{\partial \overline{u'\varphi'}}{\partial y} + \frac{\partial \overline{u'\varphi'}}{\partial z}\right) + 1 \tag{10}$$

where D is the diffusivity of pollutant.

3.3. Boundary Conditions and Solver Settings

The wind speed and wind direction given in Table 1 are regarded as the input conditions of the simulation. According to Prandtl mixing-length theory, the inlet wind speed has an exponential distribution with height [15]:

$$u = u_0 \left(\frac{z}{z_0}\right)^a \tag{11}$$

where u is the wind speed at the height z, u_0, the wind speed at the reference height z_0, and a = 0.3 is the constant reflecting the surface roughness in this paper. Therefore, velocity-inlet boundary conditions are input by creating user-defined functions based on Formula (11).

Supposing the outflow has been fully developed and the flow has been restored to normal flow without building obstruction, the relative pressure of the outflow boundary is 0.

SIMPLEC algorithm is employed for coupling pressure and momentum equations. The second-order schemes are used to discretize the convection and diffusion terms. Convergence is achieved when all scaled residuals are less than 10^{-6}.

4. Results and Discussion

4.1. Velocity and Temperature Distribution in the Street Canyon

4.1.1. Velocity Distribution in the Perpendicular Prevailing Wind Direction

In summer, the south-southeast wind prevails and the coming airflow is perpendicular to the street canyon. The contour and vector plot of air velocities are presented in Figure 6. The air velocity is found to distribute unevenly in the whole region. Higher air velocity appears in residential areas and innovative districts located at the edge of the city. The velocity is obviously weakened due to the resistance of high-rise building clusters in the front, as shown in dark blue in Figure 6a. The air velocity distributing in the north direction of the street is less than 1 m/s, while in innovative districts the air velocity is up to 3 m/s. Branches of the main street play an important role in ventilation when the coming wind direction is perpendicular to the street canyon. The city ventilation is enhanced by airflows inside the new urban region, as shown in Figure 6b. The central street then brings the airflow to residential areas along downstream with the air velocity of 1 m/s.

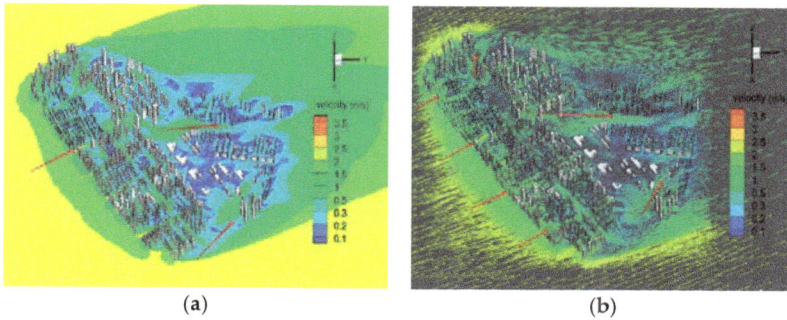

(a)   (b)

Figure 6. Urban wind velocity distribution in summer: (**a**) Contour plot; (**b**) Vector plot.

4.1.2. Velocity Distribution in the Parallel Prevailing Wind Direction

In winter, the east-northeast wind prevails in this city, parallel to the street canyon. The contour and vector plot of the air velocity are illustrated in Figure 7. Results demonstrate that wind stream from ecological areas and open space is mostly induced by the central long street into building clusters and then disperses into dense areas downstream through branches.

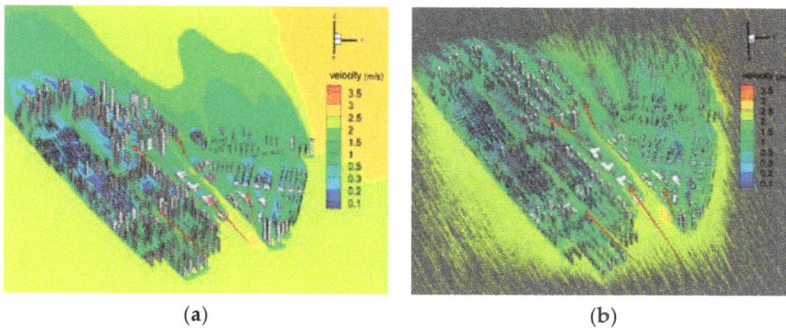

(a)   (b)

Figure 7. Urban wind velocity distribution in winter: (**a**) Contour plot; (**b**) Vector plot.

Figure 7 indicates that the air velocity in the central street ranges from 2 m/s to 2.5 m/s, higher than that in other regions. Different from that in summer, the central street plays the major role in the urban ventilation in winter. What is more, both military and innovative districts are located upwind with a high air velocity so that pollutants produced here will disperse quickly. However, the air velocity decreases in residential areas due to the drag force around high-rise buildings in the front, which means that the high-rise buildings potentially act as windbreaks in winter.

4.1.3. Temperature Distribution on Building Facades under the Ventilation in the Street Canyon

The formation of heat accumulation is strongly closely related to the poor urban ventilation. A well-designed street canyon is not only an energy-saving way to optimize the wind environment, but an effective method to reduce heat accumulations. Among the heat generated from the whole region, the emission from air conditioning systems in densely-populated residential areas, innovative districts, and public service areas on both sides contribute chiefly to the heat accumulation. The main emission is simplified as the body heat source inside each sub-regions. In addition, the significant heat radiation from sunshine in summer is simplified as surface heat source (see Table 2). The inflow is set with the same temperature with near suburbs.

Figure 8 demonstrates air temperature on building facades in the street canyon in summer. The maximum temperature appears near the center of the underlying surface in the region, especially in dense building clusters. A wider range of high temperature appears in the leeward side. It can be seen that the temperature in the southeast (33~35 °C at windward) is relatively low where the wind flows into the city from open spaces. Because of the low height of buildings, effective ventilation through branches occurs in residential districts and innovative areas, and heat elimination can be easily realized and consequently downgrades the temperature. When the air flows through the central street, the originally emitted heat is reduced greatly due to favorable ventilation. As a result, the temperature in downstream districts decreases, especially in some residential areas. In comparison, airflows diffuse freely without any obstruction, such as buildings around the central street, expelling the heat inside the city quickly and maintaining a low temperature.

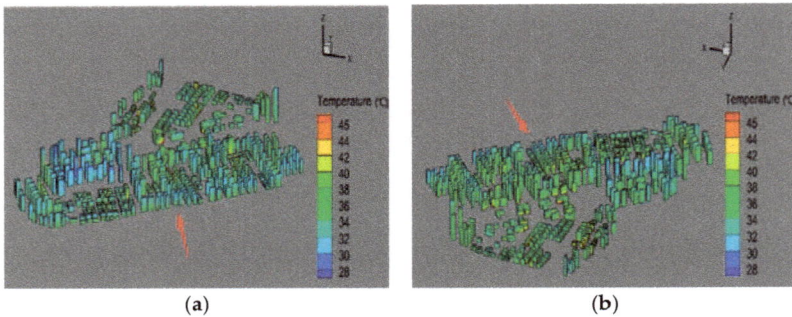

(a) (b)

Figure 8. Temperature distributions on building façades in summer: (**a**) Windward; (**b**) Leeward.

The vertical variations of temperatures inside the street canyon are calculated and compared under given prevailing wind speeds, as shown in Figure 9.

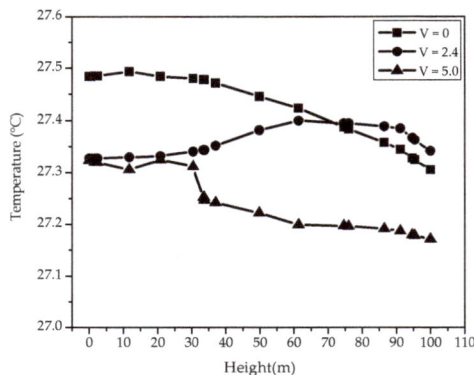

Figure 9. Vertical temperature inside the street canyon (V is the velocity of prevailing wind, $m \cdot s^{-1}$).

It is obviously seen that the temperature in the street canyon varies from wind speeds and height. When the height is lower than 30 m, the temperature varies similarly from height under two coming wind speeds, and keeps smaller values as compared to circumstance without ventilation. Subsequently, a larger the coming wind speed is found to be associated with a lower air temperature, when the height is lower than 80 m, approximately the height of most buildings around the monitored site. When the height is above 80 m, the temperature naturally decreases with height and without ventilation.

Larger airflow (V > 0) in the street canyon greatly benefits the removal of the heat inside of the city, which then reduces the local air temperature. Driven by temperature difference, airflows spread from high-temperature areas to low-temperature areas, and the heat can be partially carried away by the convection. The temperature difference near the ground surface (< 30 m) of different airflow is ignorable and larger airflow plays a vital role in higher spaces (30 m < V < 80 m). Besides, the central long street enables cooler and fresher air to be introduced from the suburb area to the central building clusters area, improving the ventilation and weakening the heat accumulation effect in this region.

4.2. Outdoor Wind Speeds and Human Body Comfort

It is reported that the physical environment about 1.5 m above ground affects how a person feels most directly [16]. Therefore, airflow velocity at 1.5 m is employed to assess the ventilation effectiveness in this region. Table 3 illustrates the different comfort levels of the human body at different wind speed levels. Generally, the outdoor wind speed from 1.0 m/s to 5.0 m/s is appreciated to be the most comfortable environment for human outdoors.

Table 3. Wind scales and human body comfort [34].

Wind Scale	Human Body Comfor
<1.0 m/s	Breezeless
1.0~5.0 m/s	Comfortable
5.0~10.0 m/s	Uncomfortable with movement affected
10.0~15.0 m/s	Very uncomfortable with movement greatly affected
15.0~20.0 m/s	Intolerable
>20.0 m/s	Dangerous

To evaluate the urban ventilation and to compare its effects on human bodies under two prevailing wind directions, wind speeds at the height of 1.5 m in two seasons are calculated, respectively (see Table 4), and analyzed below.

Table 4. The percentage of the wind speed, with the observed minimum and maximum speed values in parentheses.

Velocity (m/s)	Percentage (%)	
	SSE (Summer)	ENE (Winter)
0 (air stagnation)	10.44	5.78
0~0.5	45.25 (0.04–0.50)	20.17 (0.01–0.50)
0.5~1.0	18.22 (0.51–0.98)	46.13 (0.51–0.10)
1.0~3.0	26.09 (1.01–2.89)	27.92 (1.01–2.78)

Results show that the air stagnation in the whole region occupies 10.44% in summer, larger than 5.78% in winter, which means that worse air circulation occurs more easily when the prevailing wind is perpendicular to the street canyon. In winter, compared with street branches, the central long street brings better airflows. Moreover, the velocity varying within 0~0.5 m/s in summer accounts for 45.25%, considerably larger than the percentage (20.17%) in winter while the proportion of the velocity within 0.5~1.0 m/s in summer (18.2%) is greatly lower than that in winter (46.13%). This further emphasizes relatively worse ventilations induced by when the coming wind is perpendicular to the street canyon. On the contrary, the prevailing wind direction parallel to the street canyon easily accelerate airflows, especially inside building clusters. It is interesting to note from the data in Table 2 that the velocity in summer ranging from 1.0 m/s to 3.0 m/s constitutes 26.09%, which is similar to the percentage 27.92% in winter. In respect to wind velocity, little comfort difference is found for human body between the prevailing wind direction of perpendicular or parallel to the street canyon.

From the perspective of human body comfort, the ventilation conditions are shown to be favorable in both patterns in the street canyon where the central long street dominates the whole wind field, especially when the wind comes from its parallel direction. When the wind direction is perpendicular to the street canyon, the combination of branches of the street also effectively promotes the ventilation. However, air stagnation occurs more easily in the street canyon under the perpendicular wind, leading to adverse contaminant accumulation.

4.3. Wind Pressure on Building Envelopes

Natural ventilation is driven by the wind pressure difference and the temperature difference in air [35]. When the temperature difference between the indoor and outdoor environment is too small to provide drive force, the natural ventilation needs to rely on the wind pressure on building envelopes [36]. The natural ventilation is formed when the wind pressures differences exist on both sides of certain openings (i.e., all kinds of holes, doors, windows, and gaps). The outdoors pressure difference also determines the indoor ventilation effectiveness, as greater pressure difference providing more favorable ventilation.

The flux of airflow through openings can be calculated by Bernoulli equation [37]:

$$Q = \alpha A \sqrt{\frac{2\Delta p}{\rho}} \tag{12}$$

The flux of airflow has an increase tendency with the increase of pressure on building surfaces, while the atmospheric pressure maintains a constant P_0. Here, convert the surface pressure to dimensionless ones, the coefficient of wind-pressure C_p can be derived as:

$$C_p = \frac{P_0}{\rho v^2 / 2} \tag{13}$$

where ρ is air density and v is airflow velocity.

The distribution of coefficients of wind pressure is shown in Figure 10. Large coefficients $(C_p > 1)$ generally appear on the windward side of high-rise buildings and the lower reaches of the street canyon. In these areas, the pressure on building envelopes is higher than that of atmospheric pressure, which contributes to the airflow circulation from the street canyon to the suburbs. Besides, large pressure on building facades can achieve a favorable ventilation for indoor individuals. Conversely, in these areas with low pressure coefficients $(C_p < 1)$, little pressure difference between forward and backward of buildings has a critical influence on the formation of shade of wind, followed by the contaminant accumulation in the street canyon. Meanwhile, for indoor dwellers, the reasonable interior arrangement, apartment layout, and window installation are suggested to make a good ventilation.

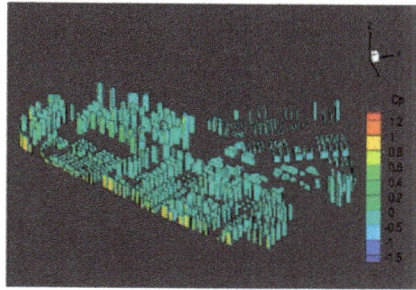

Figure 10. Coefficients of wind pressure.

4.4. Air-Age-Based Discussions on the "Freshness" of Outdoor Air

The air age in the street canyon is used to evaluate the "freshness" of outdoor air and lower values suggest fresher air. The boundary conditions for the air age governing equation are 0 at the air inlet, and the gradient is 0 in both the air exit and the wall surfaces of the computational domain.

Figure 11 shows the calculative air age and its corresponding air velocity. It is obviously seen that the maximum ages (4000~6000 s) appear in military districts where it is difficult for fresh air to flow into due to its poor ventilation. Similarly, in those areas perpendicular to the coming wind and spaces among high-rise buildings in innovative districts and residential areas, larger air age and lower velocity is found (see Figure 11b). In lower reaches of innovative districts and leeside of military districts, where air velocities are mostly lower than 0.2 m/s, the local maximum of air age is up to 7000 s. Low air ages (< 3000 s) distribute in most part of this region, mainly in residential areas and innovative streets and gets older along downstream gradually. Relatively high air velocities appear in the streets that are parallel to the direction of the wind, indicating promising ventilation and quick air exchange in those. Derived data illustrates that the ages of air in the street canyon are mostly twice the minimum values in upstream of residential areas and innovative streets. In military districts and some downstream building clusters, the age of air becomes larger with a local maximum of up to 3 times larger than the minimum. Correspondingly, fresher air exists in most parts of this region and older air only distributes in local districts with poor ventilation, especially in downstream regions behind dense buildings.

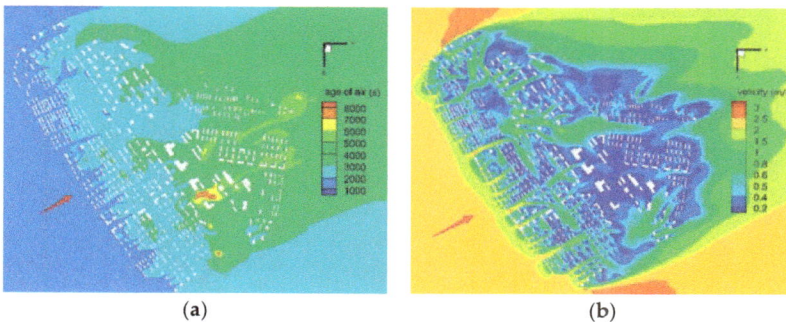

(a)   (b)

Figure 11. Stagnation of air in the Southern New City region in summer: (**a**) Age of air; (**b**) Airflow velocity.

5. Conclusions

Based on the planned Southern New Town region in Nanjing, China, this study focused on the microclimate and pollutant dispersion in urban street canyons under two typical seasons (i.e., summer and winter). The flow field, temperature field, and the age of air were calculated with the CFD method and analyzed under two typical prevailing winds. Results revealed that there was little comfort difference for the human body between the prevailing wind direction of perpendicular or parallel to the street canyon. Air stagnation occurred easily in dense building clusters, especially under the perpendicular wind direction. In addition, large pressure coefficients ($C_P > 1$) appeared at the windward region, contributing to the promising ventilation. The air age was introduced to evaluate the "freshness" of the air in the street canyon and illustrate the ventilation effectiveness on the pollutant removal. It was found that the air was updated easily where the ventilation was favorable and wind speed was large.

Following conclusions were to be drawn:

(1) The planned region was basically well-ventilated whether the coming wind direction was parallel or perpendicular to the street canyon. However, the air stagnation easily occurred in summer when the prevailing wind was perpendicular to the street canyon.

(2) The favorable ventilation comfort appeared both patterns in the street canyon where the central long street dominated the whole wind field, especially when the wind came from its parallel direction.

(3) Pressure coefficients indicated that the outdoor wind environment for building clusters located at the upper reaches were more favorable than those in lower reaches of the coming wind.

(4) The age of air showed that the poor ventilation will cause the old air detained and make it difficult to exchange the fresh air, resulting in bad air quality in the street canyon, especially in downstream regions behind dense buildings.

However, there are still limitations in field measurements to improve on the study of street canyons in this region, and the wind environment in the area outside the street canyon is worth discussing and studying.

Acknowledgments: We sincerely appreciate the financial supported by the National Key Research and Development Program of China 2016YFC0700500.

Author Contributions: Fan Liu and Hua Qian conceived and designed the research content; Xiaohong Zheng, Lun Zhang and Wenqing Liang collected the data; Fan Liu analyzed the data; Hua Qian contributed to the discussion of data analysis; Fan Liu wrote the paper.

Conflicts of Interest: The authors declare no conflict of interest.

References

1. Fernando, H.J.S.; Lee, S.M.; Anderson, J.; Princevac, M.; Pardyjak, E.; Grossman-Clarke, S. Urban Fluid Mechanics: Air Circulation and Contaminant Dispersion in Cities. *Environ. Fluid Mech.* **2001**, *1*, 107–164. [CrossRef]
2. Huang, Y.D.; He, W.R.; Kim, C.N. Impacts of shape and height of upstream roof on airflow and pollutant dispersion inside an urban street canyon. *Environ. Sci. Pollut.* **2014**, *22*, 2117–2137. [CrossRef] [PubMed]
3. Kwak, K.H.; Baik, J.J.; Lee, K.Y. Dispersion and photochemical evolution of reactive pollutants in street canyons. *Atmos. Environ.* **2013**, *70*, 98–107. [CrossRef]
4. Baklanov, A.; Molina, L.T.; Gauss, M. Megacities, air quality and climate. *Atmos. Environ.* **2016**, *126*, 235–249. [CrossRef]
5. Jin, X.; Yang, L.; Du, X.; Yang, Y. Particle transport characteristics in the micro-environment near the roadway. *Build. Environ.* **2016**, *102*, 138–158. [CrossRef]
6. Hsieh, C.M.; Huang, H.C. Mitigating urban heat islands: A method to identify potential wind corridor for cooling and ventilation. *Comput. Environ. Urban* **2016**, *57*, 130–143. [CrossRef]

7. Takebayashi, H.; Moriyama, M. Relationships between the properties of an urban street canyon and its radiant environment: Introduction of appropriate urban heat island mitigation technologies. *Sol. Energy* **2012**, *86*, 2255–2262. [CrossRef]

8. Giridharan, R.; Ganesan, S.; Lau, S.S.Y. Daytime urban heat island effect in high-rise and high-density residential developments in Hong Kong. *Energy Build.* **2004**, *36*, 525–534. [CrossRef]

9. Georgakis, C.; Zoras, S.; Santamouris, M. Studying the effect of "cool" coatings in street urban canyons and its potential as a heat island mitigation technique. *Sustain. Cities Soc.* **2014**, *13*, 20–31. [CrossRef]

10. Mei, S.J.; Liu, C.W.; Liu, D.; Zhao, F.Y.; Wang, H.Q.; Li, X.H. Fluid mechanical dispersion of airborne pollutants inside urban street canyons subjecting to multi-component ventilation and unstable thermal stratifications. *Sci. Total Environ.* **2016**, *565*, 1102–1115. [CrossRef] [PubMed]

11. Li, S.S.; Williams, G.; Guo, Y.M. Health benefits from improved outdoor air quality and intervention in China. *Environ. Pollut.* **2016**, *214*, 17–25. [CrossRef] [PubMed]

12. Liu, J.L.; Niu, J.L. CFD simulation of the wind environment around an isolated high-rise building: An evaluation of SRANS, LES and DES models. *Build. Environ.* **2016**, *96*, 91–106. [CrossRef]

13. Yoshie, R.; Mochida, A.; Tominaga, Y.; Kataoka, H.; Harimoto, K.; Nozu, T.; Shirasawa, T. Cooperative project for CFD prediction of pedestrian wind environment in the architectural institute of Japan. *J. Wind Eng. Ind. Aerodyn.* **2007**, *95*, 1551–1578. [CrossRef]

14. Ricci, A.; Burlando, M.; Freda, A.; Repetto, M.P. Wind tunnel measurements of the urban boundary layer development over a historical district in Italy. *Build. Environ.* **2017**, *111*, 192–206. [CrossRef]

15. Tominaga, Y.; Mochida, A.; Yoshie, R.; Kataoka, H.; Nozu, T.; Yoshikawa, M.; Shirasawa, T. AIJ guidelines for practical applications of CFD to pedestrian wind environment around buildings. *J. Wind Eng. Ind. Aerodyn.* **2008**, *96*, 1749–1761. [CrossRef]

16. Ramponi, R.; Blocken, B.; de Cooa, L.B.; Janssen, W.D. CFD simulation of outdoor ventilation of generic urban configurations with different urban densities and equal and unequal street widths. *Build. Environ.* **2015**, *92*, 152–166. [CrossRef]

17. Tong, Z.M.; Chen, Y.J.; Malkawi, A. Defining the Influence Region in neighborhood-scale CFD simulations for natural ventilation design. *Appl. Energy* **2016**, *182*, 625–633. [CrossRef]

18. Blocken, B.; Janssen, W.D.; van Hooff, T. CFD simulation for pedestrian wind comfort and wind safety in urban areas: General decision framework and case study for the Eindhoven University campus. *Environ. Model. Softw.* **2012**, *30*, 15–34. [CrossRef]

19. He, L.J.; Hang, J.; Wang, X.M.; Lin, B.R.; Li, X.H.; Lan, G.D. Numerical investigations of flow and passive pollutant exposure inhigh-rise deep street canyons with various street aspect ratios and viaduct settings. *Sci. Total Environ.* **2017**, *584–585*, 189–206. [CrossRef] [PubMed]

20. Yang, F.; Gao, Y.W.; Zhong, K.; Kang, Y.M. Impacts of cross-ventilation on the air quality in street canyons with different building arrangements. *Build. Environ.* **2016**, *104*, 1–12. [CrossRef]

21. Chatzidimitriou, A.; Yannas, S. Street canyon design and improvement potential for urban open spaces; the influence of canyon aspect ratio and orientation on microclimate and outdoor comfort. *Sustain. Cities Soc.* **2017**, *33*, 85–101. [CrossRef]

22. Perret, L.; Blackman, K.; Fernandes, R.; Savory, E. Relating street canyon vertical mass-exchange to upstream flow regime and canyon geometry. *Sustain. Cities Soc.* **2017**, *30*, 49–57. [CrossRef]

23. Tominaga, Y.; Stathopoulos, T. CFD modeling of pollution dispersion in a street canyon: Comparison between LES and RANS. *J. Wind Eng. Ind. Aerodyn.* **2011**, *99*, 340–348. [CrossRef]

24. Park, S.J.; Choi, W.; Kim, J.J.; Kim, M.J.; Park, R.J.; Han, K.S.; Kang, G. Effects of building-roof cooling on the flow and dispersion of reactive pollutants in an idealized urban street canyon. *Build. Environ.* **2016**, *109*, 175–189. [CrossRef]

25. Ai, Z.T.; Mak, C.M. CFD simulation of flow in a long street canyon under a perpendicular wind direction: Evaluation of three computational settings. *Build. Environ.* **2016**, *114*, 293–306. [CrossRef]

26. Sandberg, M. What is ventilation efficiency? *Build. Environ.* **1981**, *16*, 123–135.

27. Etherldge, D.; Sandberg, M. *Building Ventilation: Theory and Measurement*; John Wiley & Sons: Chichester, UK, 1996.

28. Hang, J.; Sandberg, M.; Li, Y.G. Age of air and air exchange efficiency in idealized city models. *Build. Environ.* **2009**, *44*, 1714–1723. [CrossRef]

29. Hang, J.; Li, Y.G. Age of air and air exchange efficiency in high-rise urban areas and its link to pollutant dilution. *Atmos. Environ.* **2011**, *45*, 5572–5585. [CrossRef]

30. Mochida, A.; Tominaga, Y.; Murakami, S.; Yoshie, R.; Ishihara, T.; Ooka, R. Comparison of various k–emodels and DSM applied to flow around a high-rise building—Report on AIJ cooperative project for CFD prediction of wind environment. *Wind Struct.* **2002**, *5*, 227–244. [CrossRef]

31. Versteeg, H.K.; Malalasekera, W. *An Introduction to Computational Fluid Dynamics: The Finite Volume Method*, 2nd ed.; Beijing World Publishing Corporation: Beijing, China, 2007; pp. 87–88.

32. Yakhot, V.; Orszag, S.A.; Thangam, S.; Gatski, T.B.; Speziale, C.G. Development of turbulence models for shear flows by a doble expansion technique. *Phys. Fluids A* **1992**, *4*, 1510–1520. [CrossRef]

33. Yakhot, V.; Orszag, S.A. Renormalization-group analysis of turbulence. *Phys. Rev. Lett.* **1986**, *57*, 6396–6409. [CrossRef] [PubMed]

34. Sini, J.F.; Anquetin, S.; Mestayer, P.G. Pollutant dispersion and thermal effects in urban street canyons. *Atmos. Environ.* **1996**, *30*, 2659–2677. [CrossRef]

35. Burnett, J.; Bojic, M.; Yik, F. Wind-induced pressure at external surfaces of a high-rise residential building in Hong Kong. *Build. Environ.* **2005**, *40*, 765–777. [CrossRef]

36. Yuan, C.S. The effect of building shape modification on wind pressure differences for cross-ventilation of a low-rise building. *Int. J. Vent.* **2007**, *6*, 167–176. [CrossRef]

37. Kelley, J.B. The extended Bernoulli equation. *Am. J. Phys.* **1950**, *18*, 202–204. [CrossRef]

atmosphere

MDPI

Article

The Impact of Planting Trees on NO$_x$ Concentrations: The Case of the Plaza de la Cruz Neighborhood in Pamplona (Spain)

Jose-Luis Santiago [1,*], Esther Rivas [1], Beatriz Sanchez [1], Riccardo Buccolieri [2] and Fernando Martin [1]

[1]   Environment Department, Research Center for Energy, Environment and Technology (CIEMAT), Madrid 28040, Spain; esther.rivas@ciemat.es (E.R.); beatriz.sanchez@ciemat.es (B.S.); fernando.martin@ciemat.es (F.M.)
[2]   Dipartimento di Scienze e Tecnologie Biologiche ed Ambientali, University of Salento, S.P. 6 Lecce-Monteroni, Lecce 73100, Italy; riccardo.buccolieri@unisalento.it
*   Correspondence: jl.santiago@ciemat.es; Tel.: +34-91-346-6206

Received: 4 May 2017; Accepted: 20 July 2017; Published: 22 July 2017

Abstract: In this paper, the role of trees on airborne pollutant dispersion in a real neighborhood in Pamplona (Spain) is discussed. A Computational Fluid Dynamics (CFD) model is employed and evaluated against concentrations measured during the last part of winter season at a monitoring station located in the study area. Aerodynamic and deposition effects of trees are jointly considered, which has only been done in few recent studies. Specifically, the impact on NO$_x$ concentration of: (a) tree-foliage; and (b) introducing new vegetation in a tree-free street is analyzed considering several deposition velocities and Leaf Area Densities (LAD) to model deciduous and evergreen vegetation. Results show that the higher the LAD, the higher the deposition (concentration reduction) and the blocking aerodynamic effect (concentration increase). Regardless of foliage or deposition rates, results suggest the predominance of aerodynamic effects which induce concentration increases up to a maximum of 7.2%, while deposition induces concentration decreases up to a maximum of 6.9%. The inclusion of new trees in one street modifies the distribution of pollutant, not only in that street, but also in nearby locations with concentration increase or decrease. This finding suggests that planting trees in street with traffic as an air pollution reduction strategy seems to be not appropriate in general, highlighting the necessity of ad hoc studies for each particular case to select the suitable location of new vegetation.

Keywords: street vegetation; CFD; aerodynamic and deposition; tree scenarios; urban planning

1. Introduction

In urban areas, air quality problems usually occur due to reduced ventilation and high pollutant concentrations. Traffic emissions generally constitute the major source of air pollution and roadside barriers can be employed to influence flow patterns and, thus, the resulting levels of concentrations. Advantages and disadvantages of using several barriers, such as trees and vegetation, noise barriers, low boundary walls, and parked cars, have been recently reviewed by Gallagher et al. [1].

In urban areas, vegetation has been shown to exert several ecosystem services, such as carbon sequestration, micro-climate regulation, noise reduction, rainwater drainage, improvement of mental health, and recreational values, as well as changes in air pollution. A recent research overview on the impacts of urban trees on water, heat, and pollution cycles has been given by Livesley et al. [2]. They summarized 14 studies attempting to provide a global perspective on the ecological services of trees in towns and cities from five continents. The complexity of the ecosystem service valuation has still prevented comprehensive investigations for specific areas and thus, further studies in urban areas are still needed [3].

Among the ecosystem services, the impact on urban air pollution, which is the focus of the present paper, has been documented in several studies, but not yet completely established. Flow and pollutant dispersion in the presence of vegetation (mainly trees) is an up-to-date research using field (e.g., [4–13]) and wind tunnel (e.g., [14–17]) investigations. Several studies, recently reviewed by Janhäll [18], Salmond et al. [19], Grote et al. [3], and Abhijith et al. [20], have shown the potential of vegetation in mitigating air pollution, but also has left open questions on the the impact that street trees have on air quality in urban areas and street canyons, since they may lead to increased or decreased concentrations.

Specifically, as summarized by Grote et al. [3], positive impacts of trees on air quality occur due to the deposition of pollutants on plant surfaces and stomatal uptake. If the stomata are closed, gaseous, and particle deposition mostly occurs at leaf surfaces. Together with pollutants that are bind to or destroyed at the outer surface, uptake into leaves occurs through the stomata and such a mechanism is enhanced if compounds are removed from intercellular spaces. In general, deposition rates depend on pollutant concentrations, meteorological conditions, air movement through the crown, transfer through the boundary layer adjacent to surfaces, and absorption capacity of surfaces, which also depend on stomatal conductance. In turn, these depend on species, arrangement, crown, and foliage characteristics. Pollutant removal from the atmosphere also occurs through the influence on microclimates as temperature reductions by shade and evapotranspiration may change the rate of chemical reactions, leading to reduced concentrations of ozone.

On the other hand, the negative impacts of trees on air quality are due to the release of allergenic particulates and harmful volatile organic compounds that can act as a precursor to smog or ozone formation, particularly when NO_x is present and climatic conditions are favorable. Further, vegetation, and in particular trees, may obstruct the air exchange and dispersion of traffic-related pollutants, and increase concentrations in the lower region below the crowns of trees, especially when they are characterized by high leaf area density (LAD). One of the pioneering experiments was that performed in the wind tunnel of the University of Karlsruhe. Aerodynamic effects of trees were found to increase wall-averaged concentrations of isolated symmetric street canyons up to about 100%. Results also showed that street-level concentrations depend on wind direction and aspect ratio [14]. On the other hand, Gromke et al. [17] showed a reduction up to 60% at pedestrian level in the presence of continuous hedgerows using the same wind tunnel set-up.

Based on field and wind tunnel investigations, which were also used for validation purposes, several modelling techniques, especially Computational Fluid Dynamics (CFD), were also applied. Both aerodynamic [21–25] and deposition effects of trees [26–30] were considered within idealized and real scenarios. Modelling such effects of trees in microscale models is always a challenge since several mechanisms have to be taken into account simultaneously—for particles, emitted gases, and ozone, the challenges are different. As for aerodynamic effects, trees are usually considered as porous media and additional terms are added to the momentum and turbulence equations, while the deposition is modelled as a volumetric sink term in the transport equation of pollutants. This term is proportional to LAD, deposition velocity, and air pollutant concentration. As mentioned above, the values of deposition velocity depend on the type of vegetation and pollutant. Many discrepancies between published values are found [18]. Deposition velocities for vegetated surfaces are usually less than 1 cm s^{-1} for some gases to several cm s^{-1} for particles.

Many modelling studies found that aerodynamic effects of trees are more significant than deposition [26,27,31], even though Santiago et al. [28] reported decreased concentrations close to the ground up to 60% in several idealized arrays of different packing density depending on tree location, LAD, and deposition velocity. They showed that the deposition effects are also crucial in determining the final concentration levels. Positive effects were also reported by Jeanjean et al. [26,27], who found that trees are beneficial from a purely dynamic point of view, with a concentration decrease of 7% on average at pedestrian height in the neighbourhoods of Leicester and London (UK).

Even though challenges and strategies for urban green-space planning in compact cities have also been proposed [32], it can be argued that the effects of urban vegetation strictly depend on their interaction with the city morphology and meteorological conditions. Currently, studies which account for the main effects of trees (aerodynamic, deposition and thermal) are still poor in the literature and thus, comprehensive strategies on the use of urban vegetation for air quality purposes are still missing [1].

It is worth noting that in the modelling studies aerodynamic and deposition effects of trees have been separately investigated using simple geometries, and only few recent studies bring them together. However, these studies have not quantified the relative influences of both effects on pollutant concentration in real scenarios. In this perspective, the purpose of this work is: (a) to determine the influence of aerodynamic and deposition effects on NO_x concentrations in a real neighborhood; and (b) to evaluate the impact of introducing new trees as mitigation strategy of air pollution. A CFD model with a Reynolds-Averaged Navier-Stokes (RANS) closure evaluated against data monitored from an air quality station of the Regional Government of Navarra (Spain) network is used to achieve this purpose. The focus is on rush hour conditions in winter, because higher levels of NO_x are usually found for these cases. Specifically, starting from the real scenario of deciduous trees (low LAD), the impact on NO_x concentration of: (a) tree-foliage of different LAD; and (b) new trees planted in the neighborhood, as well as the importance of deposition and aerodynamic effects in each case, is analyzed.

2. Description of Study Area, Modelling Set-Up and Investigated Scenarios

2.1. The Study Area and Modeling Set-Up

The study area is located at the *II Ensanche* neighbourhood of Pamplona (Spain), whose diameter is about 1.3 km (Figure 1a). The height of buildings ranges from 11 m to 51 m, with a mean height of 20 m. An air quality (AQ) monitoring station is located in a square in the centre of the neighbourhood. The extent of vegetation in the zone, in terms of plan area (i.e., the extent of vegetation projected in a horizontal plane respect to the total plan area of the streets and squares), is 13.8%. There are trees in most of the streets and small parks (Figure 1a). Due to the lack of specific tree data, the mean height of trees was estimated through satellite images from Google Earth® and ranges from 5 m to 12 m.

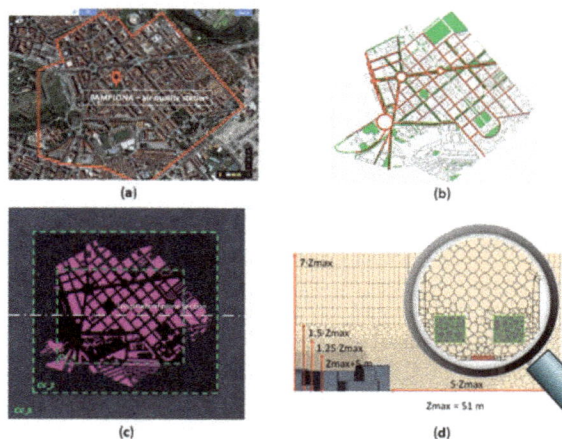

Figure 1. (a) *II Ensanche* neighbourhood of Pamplona (Google Maps® satellite image [33]), with indication of the modelled domain in red; (b) Computational Fluid Dynamic (CFD) 3D model of buildings, trees (green), and traffic emissions (red); (c) CFD mesh model; (d) Zoom at the longitudinal plane section of CFD mesh: typical sizes as function of the highest building in the domain, Zmax.

The CFD model used was the code Star-CCM+ from CD-Adapco (London, UK) [24,28,34] solves Reynolds-averaged Navier-Stokes (RANS) equations with the Realizable k-ε turbulence model, where k is the turbulent kinetic energy and ε is the dissipation of turbulent kinetic energy. A transport equation is used to simulate the dispersion of nitrogen oxides (NO_x), where diffusivity is related to turbulent viscosity divided by Schmidt number (Sc). Dispersion of NO_x (regarded as $NO + NO_2$) is modelled in order to avoid the inclusion of chemical reactions in the CFD simulations since NO_x can be considered as a non-reactive gas [24,35,36]. The aerodynamic effects of vegetation are modelled by means of a sink in the momentum equation (Su_i, Equation (1)) and sinks/sources in turbulence equations (S_k and S_ε, Equations (2) and (3)). In addition, the fraction of pollutant removed from air by means of the deposition to the leaves is represented as a mass sink in the transport equation (S_d, Equation 4). This approach to model vegetation has been evaluated by Santiago et al. [24], Krayenhoff et al. [37], and Santiago et al. [28], and it is also similar to those employed in other CFD studies [23,25,31]. The mathematical expressions are the following,

$$Su_i = -\rho LAD c_d U u_i \tag{1}$$

$$S_k = -\rho LAD c_d \left(\beta_p U^3 - \beta_d U k\right) \tag{2}$$

$$S_\varepsilon = -\rho LAD c_d (C_{\varepsilon 4}\beta_p \frac{\varepsilon}{k} U^3 - C_{\varepsilon 5}\beta_d U\varepsilon) \tag{3}$$

$$mS_d = -LAD\, v_{dep}C(x,y,z) \tag{4}$$

where ρ is the air density, c_d is the sectional drag coefficient for vegetation (=0.2), U is the wind speed, u_i is the velocity component in direction i, β_p is the fraction of mean kinetic energy converted into turbulent kinetic energy, β_d is the dimensionless coefficient for the short-circuiting of turbulent cascade, $C_{\varepsilon 4}$ and $C_{\varepsilon 5}$ are model constant, v_{dep} is deposition velocity and $C(x,y,z)$ is the concentration at position (x, y, z). Values of β_d, $C_{\varepsilon 4}$ and $C_{\varepsilon 5}$ are based on analytical expressions [38] with $\beta_p = 1$ as in Santiago et al. [28].

$$\beta_d = C_\mu^{0.5}\left(\frac{2}{\alpha}\right)^{\frac{2}{3}}\beta_p + \frac{3}{\sigma_k} \tag{5}$$

$$C_{\varepsilon 4}(= C_{\varepsilon 5}) = \sigma_k\left(\frac{2}{\sigma_\varepsilon} - \frac{C_\mu^{0.5}}{6}\left(\frac{2}{\alpha}\right)^{\frac{2}{3}}(C_{\varepsilon 2} - C_{\varepsilon 1})\right) \tag{6}$$

We assume $C_{\varepsilon 4} = C_{\varepsilon 5}$ and use $\alpha = 0.05$ and $(C_\mu, \sigma_k, \sigma_\varepsilon, C_{\varepsilon 1}, C_{\varepsilon 2}) = (0.09, 1, 1.3, 1.44, 1.92)$.

The geometry of each building has been obtained from a 2D map of the city in CAD format where each building is extruded considering its height. This real neighbourhood configuration has been imported to the CFD model. Specifically, geometry models for trees and traffic sources have been set up from satellite images from Google Earth® [39]. For simplicity, only rows of trees have been considered instead of individual trees. Trees are placed through the streets, and the base and the top of the crown are located depending on the type of tree within each street by using Google Earth® information. The bases and the tops of tree crowns range from 2 m to 4 m, and 5 m to 12 m, respectively. In the virtual case (i.e., where new vegetation is introduced in one tree-free street), the base and the top of the crown are located at 4 m and 10 m, respectively, which are consistent with those of trees located within the parallel street. Traffic emissions are distributed along the streets and the width is determined by the number of lanes (e.g., 3.5 m wide for one-way street, 7 m wide for two-way street and so on). Also, it is assumed that traffic emissions height is 1 m above the ground (Figure 1b) in order to take the initial dispersion into account.

The domain size has been built according to the best practice guidelines [40]. The height of the domain is 7 Zmax, where Zmax is the height of the tallest building (50 m). The distance between buildings and inlet and outlet boundaries are larger than 8 times the building heights. Note that, except the tallest building, the average height of most of the buildings is around 20 m.

The choice of the mesh has been made based on grid sensitivity tests. The domain has been discretized using polyhedral cells. It is made up of 3 control volumes (CV_1, CV_2, and CV_3) of characteristic sizes: 2.7 m, 6.7 m, and 10 m, respectively (Figure 1c). Further, a prism layer of hexahedral cells around buildings (of about 1 m) and ad hoc refinements in the narrowest streets have been added. Polyhedral and structure grids are combined in order to optimize the number of grid cells and save computational cost. Figure 1d shows the gradual shifting between control volumes at a longitudinal plane. The total number of cells is 7.4×10^6 cells. In order to check the independence of numerical results on the grid size, two finer meshes have been evaluated: mesh 2 with control volumes characteristic sizes of 2 m, 5 m, and 10 m, respectively. Mesh 3 which characteristic sizes are 1.5 m, 3.8 m, and 10 m, respectively. Vertical profiles of wind speed and turbulent kinetic energy in three different locations have been analysed in this test. The differences found against the results from the three grid resolutions are insignificant, and the first mesh is considered appropriate.

Concerning boundary conditions, building, and ground are modelled as walls. At the top of domain symmetry, boundary conditions are considered to establish zero normal velocity and zero normal gradients of all variables at this plane. Neutral inlet profiles of velocity, turbulent kinetic energy (k), and dissipation of turbulent kinetic energy (ε) are computed by the following equations [21,24,41,42]:

$$u(z) = \frac{u_*}{\kappa} \ln\left(\frac{z + z_0}{z_0}\right) \tag{7}$$

$$k = \frac{u_*^2}{\sqrt{C_\mu}} \tag{8}$$

$$\varepsilon = \frac{u_*^3}{\kappa(z + z_0)} \tag{9}$$

where u_* is the friction velocity, z_0 is the roughness length, C_μ is a model constant (=0.09) and κ is von Karman's constant ($\kappa = 0.4$). This approach is acceptable for winter season [24,42].

2.2. Description of Investigated Scenarios

Several scenarios, both real and virtual, have been investigated. First, the real scenario is evaluated against data monitored during two weeks from the AQ monitoring station (located at 3 m a.g.l.). For this, time average NO_x concentrations are computed.

Winter has been selected because levels of NO_x measured at the AQ station in the study area are usually higher than in other seasons. As for LAD, over the year, LAD of deciduous vegetation changes, and for instance in winter, is almost 0. With this in mind, the month of March 2015 has been chosen, when LAD (=0.1 m^2m^{-3}) is low, but not 0, and NO_x levels are still high. This LAD has been selected according to a preliminary study performed by Rivas et al. [43] in the same area. They evaluated NO_x concentration during different time periods considering LAD = 0 (no trees), 0.1 and, 0.5 m^2m^{-3}, concluding that the fit with experimental values was better for cases with LAD = 0.1 m^2m^{-3} in winter and with LAD = 0.5 m^2m^{-3} in summer. Note that the values of LAD used in the study are slightly low in comparison with the literature [44]. This is because trees have been modelled by means of rows of trees and not as individual trees. Therefore, this value includes LAD for trees and the gap between them, which is different to the real LAD of an individual tree.

Second, the study focuses on the worst case in terms of air quality. The maximum values of NO_x concentrations have been found at 8 a.m., which correspond to the maximum of traffic emissions in these streets during the day. Taking this into account, the focus is on NO_x dispersion under these adverse conditions. Meteorological data obtained from a station located close to the neighborhood have been used to simulate the typical meteorological conditions at this hour. Under these adverse conditions, the impact of deposition and aerodynamic effects of trees on pollutant concentrations has been analyzed, as well as the effect of increasing LAD or planting new trees. To analyze a wide range of deposition velocities, four deposition values were considered: 0, 0.005, 0.01, and 0.03 m s^{-1}.

The objective is not to employ a specific and accurate value of deposition velocity, but to analyze the impact of vegetation on pollutant concentration for several scenarios with different realistic deposition velocities. This allows us to generalize the results for different deposition conditions as performed in Santiago et al. [25] for idealized scenarios. Specifically, the analysis focuses on:

a) The effects of tree-foliage on concentration. LAD = 0.1 and 0.5 m^2m^{-3} have been used to model deciduous vegetation in real cases and evergreen vegetation in virtual cases, respectively;

b) The effects on concentration of introducing new vegetation in a tree-free street.

Table 1 summarizes the studied scenarios which we expect to provide a decision support to urban planners for the selection of appropriate tree species and planting new trees.

Table 1. Description of investigated scenarios.

Scenario	Location of Vegetation	Type of Vegetation	Deposition Velocity (m s^{-1})
Current-1.a			0
Current-1.b	Current location	Deciduous	0.005
Current-1.c		(LAD = 0.1 m^2m^{-3})	0.01
Current-1.d			0.03
Current-2.a			0
Current-2.b	Current location	Evergreen	0.005
Current-2.c		(LAD = 0.5 m^2m^{-3})	0.01
Current-2.d			0.03
New-1.a			0
New-1.b	New trees in one	Deciduous	0.005
New-1.c	tree-free street	(LAD = 0.1 m^2m^{-3})	0.01
New-1.d			0.03
New-2.a			0
New-2.b	New trees in one	Evergreen	0.005
New-2.c	tree-free street	(LAD = 0.5 m^2m^{-3})	0.01
New-2.d			0.03

3. CFD Modelling Evaluation

3.1. Previous Validation Studies

A detailed validation of the CFD-RANS simulations using real local data has not been possible, since only one AQ monitoring station from Regional Government of Navarra is located in the study area, specifically in a square in the centre of the neighbourhood. Therefore, the modelling approach employed here has been evaluated with data available from wind tunnel experiments by Brunet et al. [45] and Raupach et al. [46] for a "continuous" forest and a "forest" edge. These experiments have been extensively used to validate simulations with RANS by Foudhil et al. [47] and with Large Eddy Simulations by Dupont and Brunet [48]. Our validation results have been presented in Krayenhoff et al. [37] and Santiago et al. [28]. In addition, the current modelling of urban vegetation was evaluated by using CODASC wind-tunnel dataset (COncentration DAta of Street Canyons - www.windforschung.de/CODASC.htm) [49,50] by simulating a street canyon with and without vegetation. Two different tree porosities were used with a pressure loss coefficient (λ) of 80 and 200 m^{-1} (0.53 and 1.33 m^{-1} at full scale). Considering a drag coefficient of 0.2, these values correspond to LAD = 2.6 and 6.6 m^2m^{-3}, respectively. Overall, a slight overestimation of concentration was obtained [28]. Similar behavior was found by other studies using RANS [51–53] and LES [54]. Better results were obtained for a small Schmidt number (Sc = 0.3).

Based on the validation studies mentioned above, we are confident that the CFD model employed is able to reproduce the NO_x dispersion in the real scenario with vegetation and thus, the impact of several tree foliage densities or planting new trees has been quantified.

3.2. Current Validation Study

To further get confidence in the use of CFD-RANS, the real scenario of this neighborhood has been here evaluated against data during two weeks in March 2015 (from 1st to 14th) using data from the AQ station. A LAD = 0.1 $m^2 m^{-3}$ was considered because all trees are deciduous as already mentioned in Section 2.2 [43]. In principle, we are aware that it is necessary to evaluate air quality during large periods of time, but it is usually not affordable to perform unsteady CFD simulations of several days due to large computational costs. For this reason, here the methodology WA CFD-RANS (weighted average CFD-RANS simulations) [42] has been employed: it uses CFD simulations for several wind directions (16 following the wind rose) to compute a time-averaged concentration map, taking into account that the concentration is inversely proportional to wind speed [42]. However, being pollutant deposition considered in this study, this fact is not fulfilled and thus, WA CFD-RANS methodology has been modified as follows:

-   A deposition velocity of 0.01 m s^{-1} has been considered, which is a high value for NO_x, but still within the range of realistic values [25]. This selection has been done in order to analyze the case where the reduction of concentration by means of vegetation is maximum;
-   Three different ranges of inlet wind speeds were considered to simulate the corresponding scenarios for each wind direction, so 48 simulations have been carried out (Table 2);
-   Then, depending on wind speed measured by the meteorological station located close to the neighborhood, at each hour the corresponding simulation was selected and the concentrations were computed.

Table 2. Ranges of inlet wind speed at 10 m used in the WA CFD-RANS methodology.

Ranges of Inlet Wind Speed at 10 m	v_{ref}/v_{dep}
$v_{ref} > 4.5$ m·s^{-1}	640
2 m·s$^{-1} < v_{ref} < 4.5$ m·s^{-1}	320
$v_{ref} < 2$ m·s^{-1}	107

Results have thus been evaluated against the AQ monitored data (Figure 2). From the figure, small differences between results with and without deposition are observed. The time average difference is 2 µg m^{-3} with a maximum of 11 µg m^{-3} during this time period. Both time series of modelling results have an acceptable correlation with monitored values (R = 0.71). Normalized Mean Square Error (NMSE) and the fraction of predictions within a factor 2 of observation (FAC2) are computed. NMSE is 0.27 and 0.28, and 0.05, and FAC2 is 0.73 and 0.72 for the cases with and without deposition. These values indicate a good agreement between monitored and modelled results with a slightly better fit when deposition is considered and confirm the accuracy of modelling approach for the evaluation of flow and NO_x dispersion within the investigated area in winter period. We are aware that the current methodology has been evaluated for a winter period. During summer when LAD is higher, thermal effects of urban surfaces and trees could be important, and neglecting such effects, as is typically done in many previous studies mainly due to a lack of a common methodology, could introduce more uncertainties in the model results. Furthermore, the evaluation of model simulations with only one measurement point in summer, since there are no other appropriate available data, is not reliable. However, we have confidence at least in the application of the model in cases with higher LAD trees, because it has been validated against CODASC measurements (see Section 3.1). For these reasons, we have focused this paper in winter and we think it can provide useful insights on the use of trees as mitigation strategies.

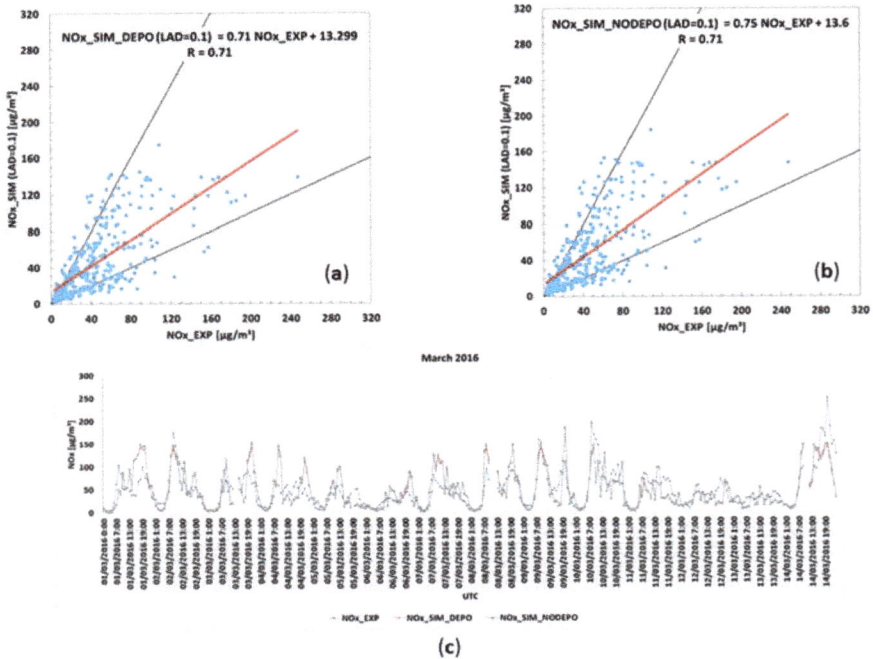

Figure 2. (a) Scatter plots of modelling results considering deposition; (b) Scatter plots of modelling results when deposition is neglected; (c) Time series of concentrations at the air quality (AQ) monitoring station position from 1st March to 14th March 2015. NO_x_EXP and NO_x_SIM are experimental and modelled NO_x concentrations, respectively. $NO_x\_SIM\_DEPO$ and $NO_x\_SIM\_NODEPO$ are modelled NO_x concentration with and without considering deposition.

Considering the time-averaged concentrations in the domain (Figure 3) at 3 m (height of the AQ monitoring station), the effect of deposition in the whole neighborhood is limited and the differences found considering and neglecting deposition are <7 $\mu g \ m^{-3}$. In addition, these differences are located only close to the zones characterized by a large amount of vegetation or high concentration of NO_x combined with some trees. For example, there is a difference of 3 $\mu g \ m^{-3}$ in the square (dash line (**A**) in Figure 3b), or there are differences slightly higher in some parts of the avenue North of the domain (dot line (**B**) in Figure 3b). Then, only in the zones with high amount of vegetation, the differences in time-averaged concentration by including deposition reaches 10%. Therefore, the error in modelling NO_x concentration of considering or neglecting the deposition effect of trees for winter vegetation (LAD = 0.1 $m^2 m^{-3}$) is low.

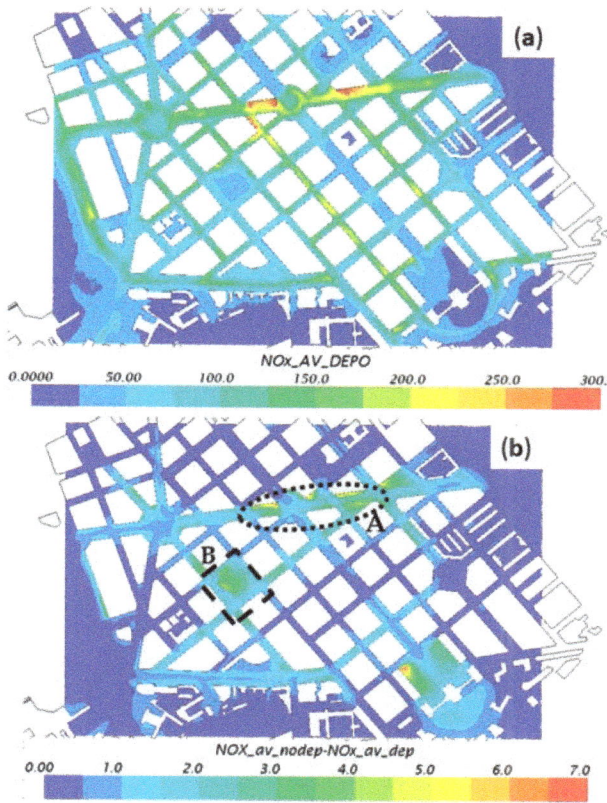

Figure 3. (**a**) Time-averaged NO_x concentration taking into account pollutant deposition; (**b**) Differences of time-averaged NO_x concentration due to considered deposition effect.

From these results, it can be concluded that deposition seems to play a minor role on time-averaged concentration with respect to the aerodynamic effects of trees, even using a high deposition velocity value for NO_x (1 cm s^{-1}). Then, the modelling of this process could be neglected under these conditions.

4. Impact of Tree-Foliage on NO_x Concentration: Influence of Deposition and Aerodynamic Effects

In the next sections, the study focuses on conditions of maximum NO_x concentrations. These adverse conditions correspond to the maximum of traffic emissions in the streets during the day (8 a.m.). The inlet meteorological conditions have been taken from the meteorological station located close to the neighborhood. Predominant wind direction (North-West) and average wind speed were computed from these data and used to simulate the typical meteorological conditions at this hour. The impact of tree foliage on urban air quality is thus analyzed, which could provide useful information to urban planners for the selection of appropriate tree species. Then, the objective is to quantify the relative contribution of deposition and aerodynamic effects of vegetation on NO_x concentration in cases with different tree-foliage. LAD = 0.1 and 0.5 m^{2}m^{-3} have in particular been used to model deciduous vegetation from 1st to 14th March 2015 (real cases corresponding to Current-1 scenarios in Table 1) and evergreen vegetation (virtual cases corresponding to Current-2 scenarios in Table 1), respectively.

4.1. The Effects of Deposition

Firstly, Current-1 cases are compared to quantify tree deposition for LAD = 0.1 m^2m^{-3}. Figure 4 shows maps of concentration at 3 m considering vegetation with no deposition (Current-1a) and the absolute and relative differences in a percentage when a deposition velocity of 0.01 m s^{-1} is used (Current-1a–c). Figure 4b shows decreases of concentration of about 10 μg m^{-3} in some areas close to vegetation and the maximum of reduction is just located within vegetation. Relative percentage differences can reach 10% (Figure 4c). In order to quantify the size of area at 3 m height affected by this concentration reduction, two different criteria have been defined: (1) the zones where the concentration is reduced more than 5 μg m^{-3} (Reduction zone 1); and (2) the zones where the concentration is higher than 50 μg m^{-3} and is reduced more than 5% (Reduction zone 2). Following these criteria, it is found that the Reduction zone 1 and 2 are only 0.9% and 0.7% of the total neighborhood area simulated, respectively. In addition, the overall decrease of spatial-averaged concentrations of the domain with respect to the no-deposition scenario is 0.54%. Deposition thus has no effect on spatial-averaged concentration of the zone. We found that the plan area of vegetation is 13.8% of the domain.

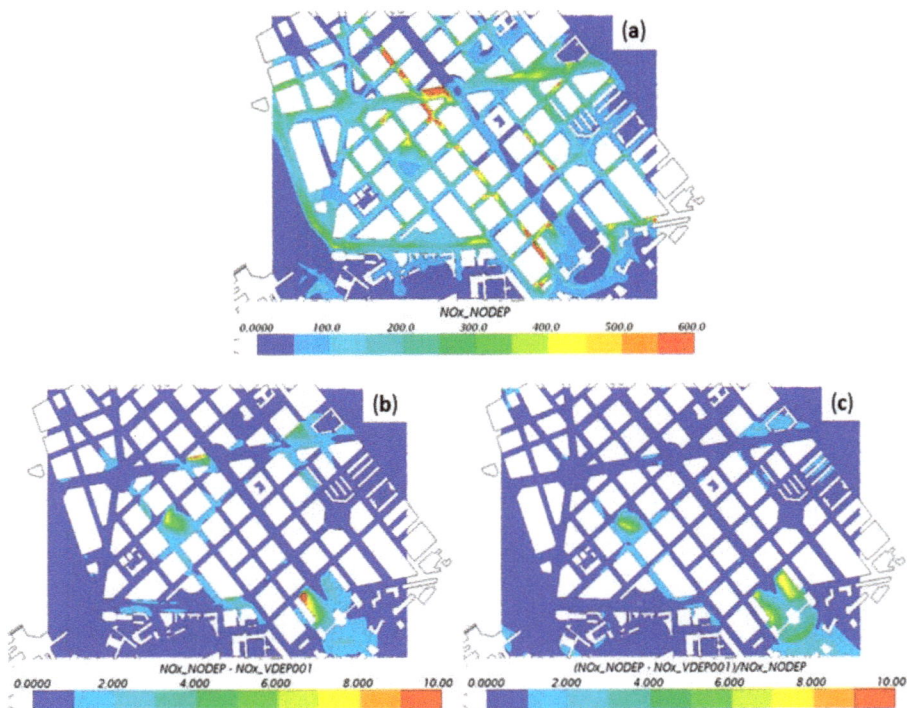

Figure 4. Current-1 scenarios: (**a**) NO_x concentration map obtained by neglecting deposition; (**b**) Absolute differences of NO_x concentration between considering and neglecting deposition for v_{dep} = 0.01 m s^{-1}; (**c**) Same as (**b**), but in terms of relative percentage differences.

Similar maps are obtained using other deposition velocities. As expected, the reduction of concentration is almost proportional to the deposition velocity. Table 3 shows the reduction parameters for each case. The table shows that, for this type of vegetation, only the effect of pollutant deposition on air quality is not negligible in few zones close to trees.

Table 3. Concentration reduction for vegetation with LAD = 0.1 m^2m^{-3} and 3 different deposition velocities (Current-1 scenarios).

Deposition Velocity	Maximum of Reduction ($\mu g\ m^{-3}$)	Maximum of Relative Reduction	Reduction Zone 1 (%)	Reduction Zone 2 (%)	Spatial-Averaged Concentration ($\mu g\ m^{-3}$)	Reduction of Spatial-Averaged Concentration (%)
0.005	6.9	4.5	0.07	0	105.0	0.27
0.01	13.4	8.7	0.9	0.7	104.7	0.54
0.03	35.6	25	7	4	103.7	1.54

The increase of tree-foliage induces an increase of pollutant deposition since more surface (leaves) is available for pollutant deposition. In these scenarios (Current-2), evergreen vegetation is modelled with an increase of LAD from 0.1 to 0.5 m^2m^{-3}. The concentration maps at 3 m for evergreen vegetation with no deposition (Current-2a) and the absolute and relative differences in percentage considering a deposition velocity of 0.01 m s^{-1} (Current-2a–c) are shown in Figure 5. In these cases, the deposition increases due to vegetation is denser (deposition is proportional to LAD). Then, comparing the results considering and not considering deposition, it can be observed that the decrease of concentration (Current-2a–c) is higher for this LAD (Figure 5b,c). For example, in some zones, the relative reduction can reach 49% and the spatial-averaged concentration of the domain decreases of 2.8% for a deposition velocity of 0.01 m s^{-1} (Current-2c). Note that the maximum reduction is located within vegetation. In addition, the Reduction zones 1 and 2 increase until 17% and 9.2%, respectively. Results for the other deposition velocities are shown in the Table 4.

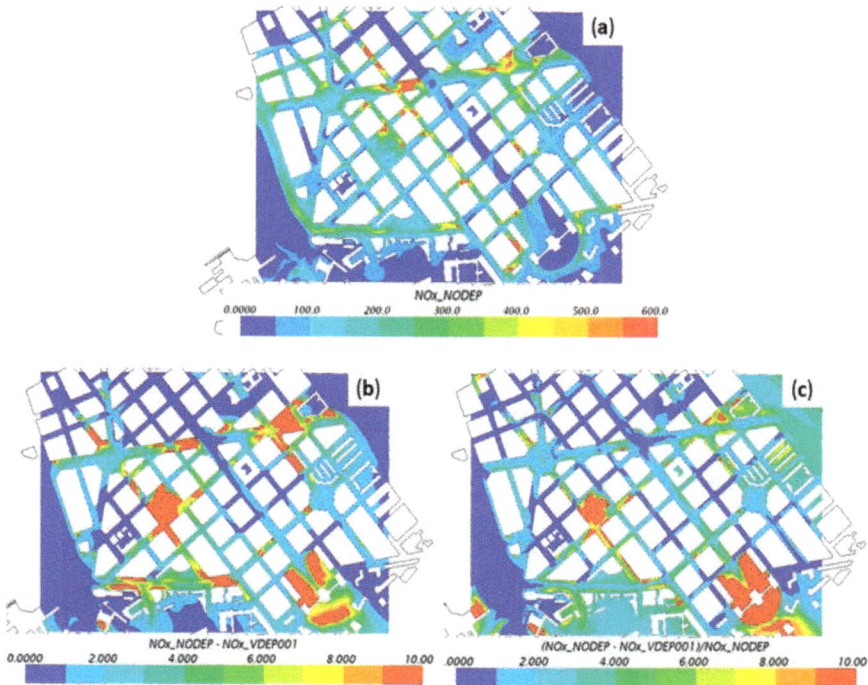

Figure 5. Same as Figure 4, but for Current-2 scenarios (LAD = 0.5 m^2m^{-3}). Note that red color indicates zones characterized by values higher than 600 $\mu g\ m^{-3}$, 10 $\mu g\ m^{-3}$, and 10%, respectively. The color scale is the same of that used in Figure 4 for sake of comparison.

Table 4. Concentration reduction for vegetation with LAD = 0.5 m^2m^{-3} and 3 different deposition velocities (Current-2 scenarios).

Deposition Velocity	Maximum of Reduction ($\mu g\ m^{-3}$)	Maximum of Relative Reduction	Reduction Zone 1 (%)	Reduction Zone 2 (%)	Spatial-Averaged Concentration ($\mu g\ m^{-3}$)	Reduction of Spatial-Averaged Concentration (%)
0.005	38	31	8	3.6	111.2	1.5
0.01	66	49	17	9.2	109.7	2.8
0.03	147	74	40	30.5	105.1	6.9

4.2. The Relative Contribution of Aerodynamic and Deposition Effects

The increase of tree-foliage induces not only a greater deposition, but also a greater modification of street ventilation (higher aerodynamic effects). Comparing concentrations obtained for the Current-2 scenarios (LAD = 0.5 m^2m^{-3}) with those of the Current-1 scenarios (LAD = 0.1 m^2m^{-3}), it can be noted that the increase of LAD strongly affects the ventilation of the streets inducing different concentration patterns (see Figures 4 and 5). Focusing on aerodynamic effects, differences between Current-2a and Current-1a scenarios are compared (Figure 6a). The figure shows that the concentration increases in some zones or decreases in others and a pedestrian street (street without traffic emissions, see emissions in Figure 1b) is not affected by the increase of LAD. In some other streets, the maximum and minimum of differences are close due to slight displacements of recirculation and stagnation zones. However, the zones where the concentration is higher for LAD = 0.5 m^2m^{-3} are wider than the zones where it is reduced. This is also observed considering a deposition velocity 0.01 m s^{-1} (Figure 6b). For example, the area where the concentration increases of 20 $\mu g\ m^{-3}$ or more is 1.81 greater than the area where it decreases of 20 $\mu g\ m^{-3}$ or more (2.01 when no deposition is considered), and the average of differences is 5 $\mu g\ m^{-3}$ (6.2 $\mu g\ m^{-3}$ when no deposition is considered). Moreover, the spatial-averaged concentration increases as LAD increases and the effect of deposition is not enough to cancel out the reduction of ventilation in these cases. Figure 7 shows the spatial-averaged concentrations in the neighborhood for Current-1 and Current-2 scenarios. Here, only for v_{dep} = 0.03 m s^{-1} (note that this value is very high and does not seem to be realistic for NO_x) the deposition almost cancels out the effect of ventilation reduction in terms of spatial-averaged concentration. In all cases, the NO_x concentration is clearly higher (1–8%) for Current-2 cases (LAD = 0.5 m^2m^{-3}) than for Current-1 (LAD = 0.1 m^2m^{-3}). Thus, it can be concluded that in these cases the aerodynamic effects of vegetation on air pollutant concentration are more important than the deposition.

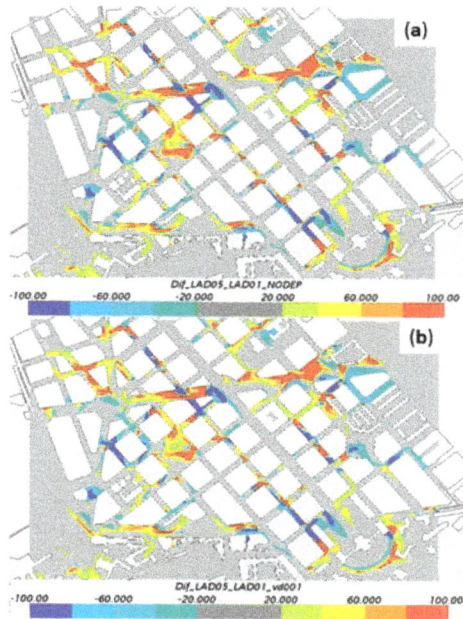

Figure 6. Differences of NO_x concentrations between results from: (**a**) Current-2a and Current-1a (**b**) Current-2c and Current-1c. Red indicates that the concentration is higher for evergreen vegetation (LAD = 0.5 $m^2 m^{-3}$, Current-2 cases). Grey indicates similar concentration (± 20 µg m^{-3}).

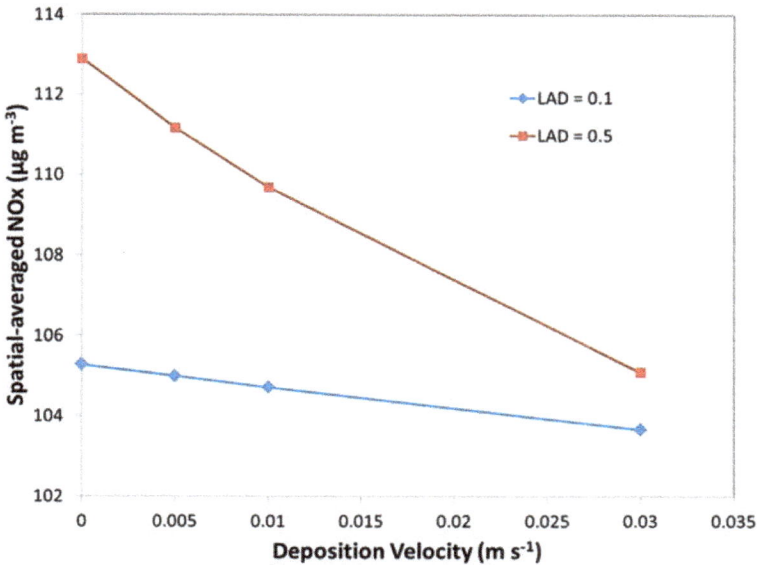

Figure 7. Variation of spatial-averaged concentrations in the neighborhood with deposition velocity for real-tree and evergreen-tree cases.

5. Impact of New Vegetation on NO$_x$ Concentration: Influence of Deposition and Aerodynamic Effects

As in the previous section, the focus is on the adverse conditions characterized by maximum values of NO$_x$. The effects on concentration due to the introduction of new vegetation in one tree-free street have been investigated. Since the introduction of trees modifies wind flow and changes pollutant distribution within the neighborhood, the main objective of this section is to assess whether the decision of planting new trees could be considered as a mitigation measure of pollutant concentration in this specific study case.

In the investigated neighborhood, there is, in particular, a tree-free street (Tafalla Street, see Figure 8) and virtual scenarios including trees with different foliage (New-1 with LAD = 0.1 m^{2}m^{-3} and New-2 with LAD = 0.5 m^{2}m^{-3}) scenarios have been simulated and compared with Current-1 and Current-2 cases (see Table 1). These new trees have modelled with the same features as trees of the parallel street. By introducing such trees, the surface covered by vegetation increases from 13.8 to 14.8% of the domain.

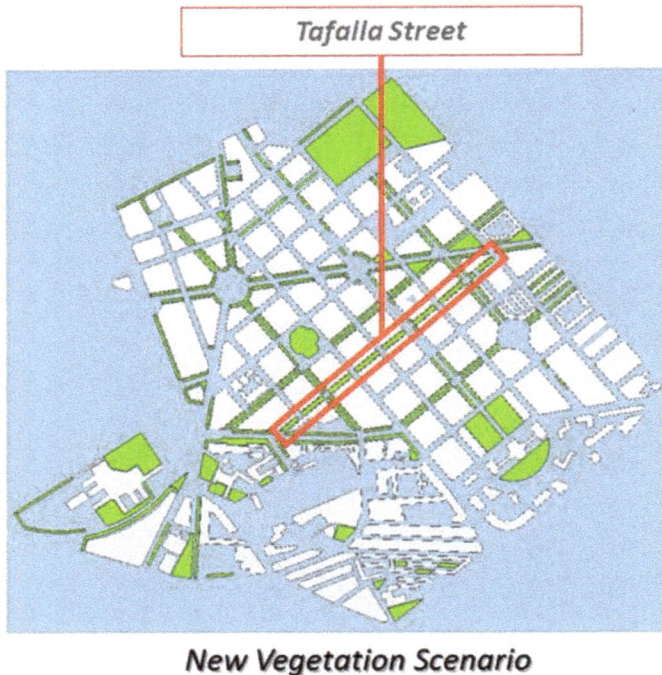

Figure 8. Location of modelled vegetation of New-1 and New-2 cases.

The concentration map of the New-1c scenario is shown in Figure 9. Comparing with results obtained for the Current-1 scenarios (Figure 4), it can be noted that the distribution of pollutant is slightly different. To better illustrate the effects of vegetation, these differences are plotted in Figure 10, which shows that the modification of concentration is a local effect, being the differences of average concentration in the whole neighborhood less than 0.01%. In general, there is an increase of concentration within this street due to the reduction of ventilation. Also, the deposition over these new trees is negligible compared to aerodynamic effects. Only in one area of Tafalla Street, the new trees induce a reduction of pollutant concentration. However, this fact is more due to aerodynamic effects, rather than the deposition. This zone is close to two junctions of streets and one of them is very close to the main avenue at the North. Here, the trees modify the distribution of pollutant increasing in one area and decreasing in other nearby area–it is displacement of the maximum concentration in the same street with negligible average effects. In addition, close to this street (see dash line area in Figure 10), the concentration increases because the presence of trees there induces a recirculation and stagnation zone, as shown in Figure 11. Further, the aerodynamic effects of vegetation seem to be more important than deposition due to the general increases of concentration around Tafalla Street, even though the effects are local. In addition, these effects could affect four streets away from Tafalla Street.

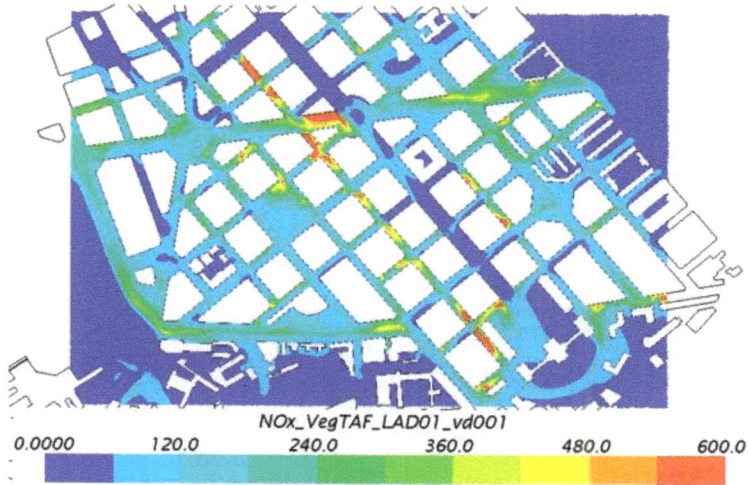

Figure 9. NO_x concentration map for New-1c case.

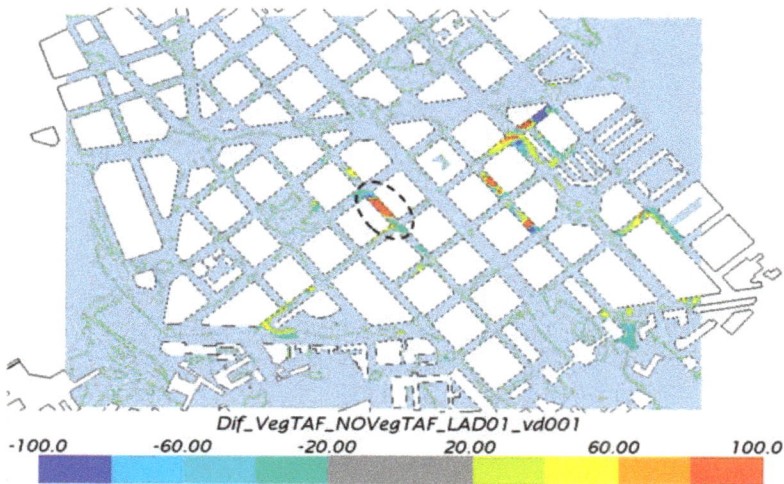

Figure 10. Differences of concentration comparing Current-1c case with New-1c case. Red indicates that the concentration increases in the new vegetation scenario and grey indicates the zone where the differences are lower than 20 $\mu g\ m^{-3}$.

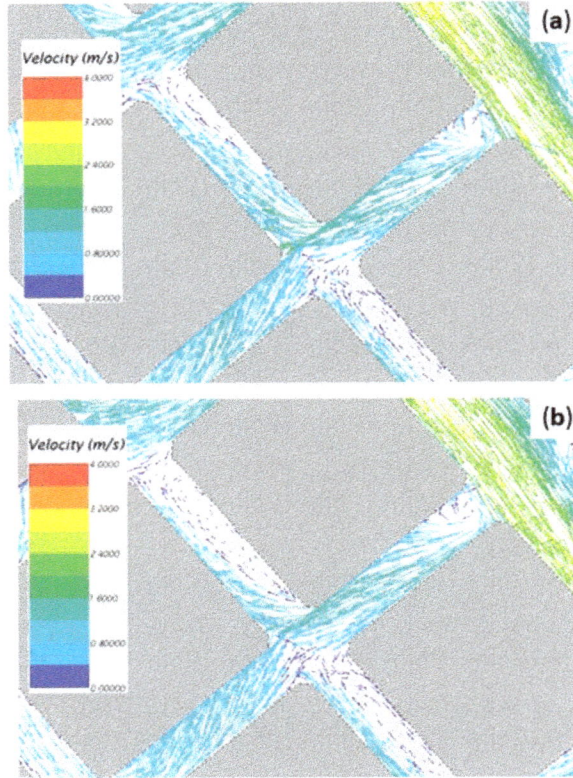

Figure 11. (**a**) Flow pattern in Current-1 case; (**b**) Flow pattern in New-1 case within the area limited by dashed line in Figure 10.

This vegetation configuration is also analyzed for LAD = 0.5 m^{2}m^{-3} (New-2 cases). The concentration map for New-2c case is shown in Figure 12. As shown in the previous section, the effects of vegetation are more intense for higher LAD. However, the differences of average concentration in the whole neighborhood are negligible as for low LAD. In this case, the introduction of new vegetation in Tafalla Street induces more significant modifications in the concentration patterns due to greater drag exerted by new trees (Figure 13). Zones further from Tafalla Street are also affected by these trees. In addition, in the area delimited by dashed line (Figure 13), the concentration also increases due to the air recirculation created by the inclusion of trees (Figure 14).

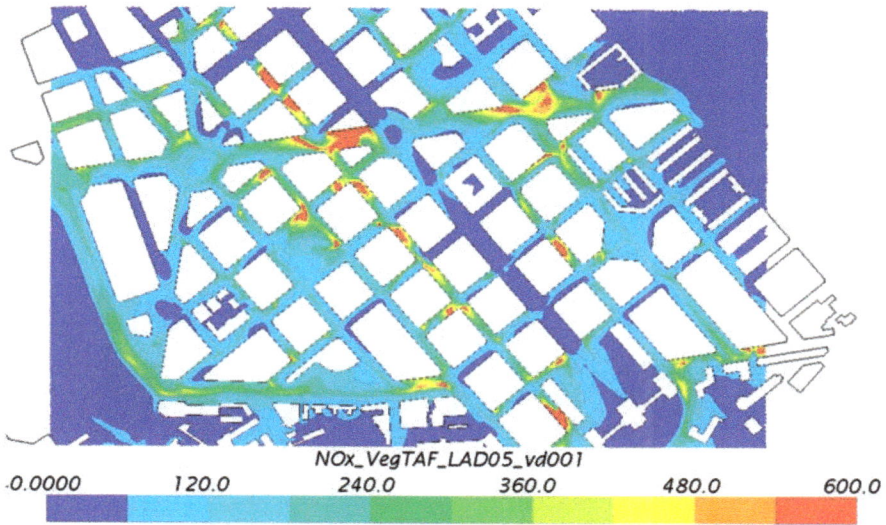

Figure 12. Concentration map for New-2c case.

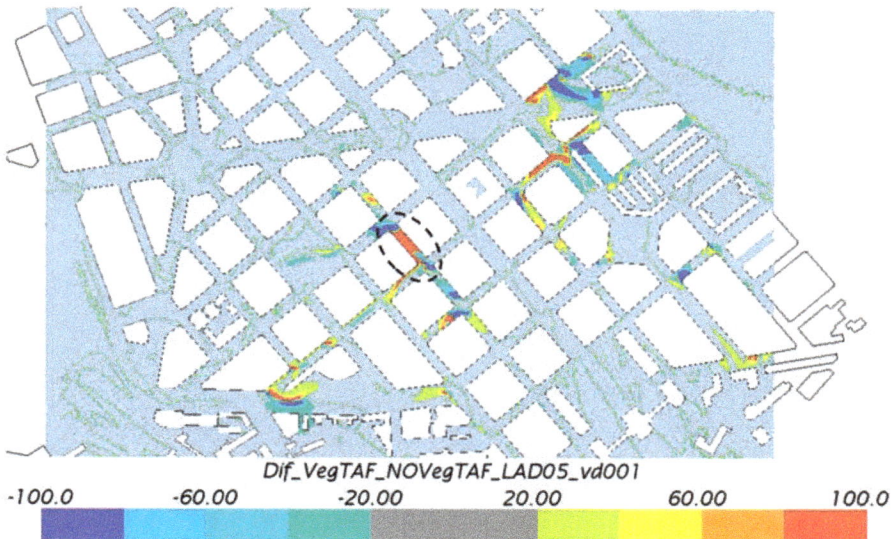

Figure 13. Differences of concentration comparing Current-2c case with New-2c case. Red indicates that the concentration increases in the new vegetation scenario and grey indicates the zone where the differences are lower than 20 µg m^{-3}.

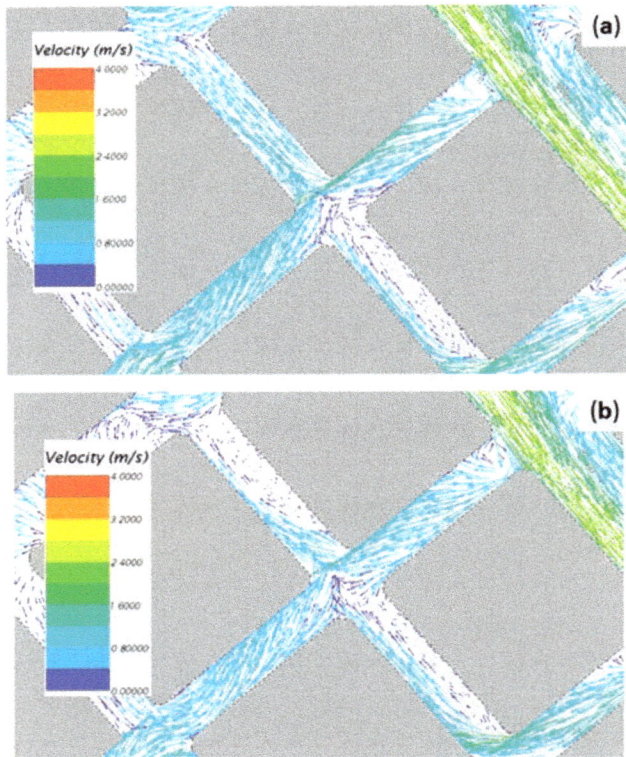

Figure 14. (a) Flow pattern in Current-2 case; (b) Flow pattern in New-2 case in the area limited by dashed line in Figure 13.

Results above show that any variation in the distribution of vegetation should be done with caution because the levels of resulting concentrations might change by a modification of the flow also in surrounding areas, and this modification increases with increasing LAD.

6. Summary and Conclusions

In this study, the impact of urban trees has been evaluated in a real neighborhood, which has only been done in few recent studies. A methodology to compute modelled concentration during a long period of time (e.g., several weeks, months, or a year) considering deposition (WA CFD-RANS) was applied and enhanced to account for both the aerodynamic and deposition effects of trees. Modelled results were evaluated with data recorded at an air quality monitoring station. Concentration maps related to adverse conditions (high NO_x concentrations) in winter have been studied in a real case, considering different deposition velocities and for other virtual vegetation scenarios with different tree-foliage and including new trees in one tree-free street. This study can be interpreted as a decision support for urban planners. It is observed that measures which are supposed to indicate the service value of trees could not be used as a general mitigation strategy of pollutant concentration in streets with traffic, since the deposition plays a minor role with respect to aerodynamic effects of trees.

Specifically, the main conclusions achieved from this study are:

- The global decrease of concentration at 3 m in the neighborhood due to deposition is small for cases with low LAD (deciduous). For example, in the real-tree case comparing spatial-averaged concentrations from no deposition simulation with simulation considering a deposition velocity of 0.03 m s^{-1} (very high deposition velocity), differences of less than 2% are observed. A slightly higher effect (6.9%) is obtained for LAD = 0.5 m^{2}m^{-3}; however, deposition effects could be locally higher in certain zones, especially for higher LADs;

- The aerodynamic effects of vegetation induce a general increase of concentration which dominates versus the decreasing of concentration due to deposition. Comparing cases with different LADs, deposition increases with increasing LAD—however spatial-averaged concentrations are always higher for high LADs (dense foliage);

- The inclusion of new trees in one street modifies the distribution of pollutant, not only in that street, but also in nearby locations. Global effects in pollutant concentration are small, however, locally differences much greater of 20 μg m^{-3} are found when comparing Current cases with New cases. In some zones, the concentration increases with the new trees, but decreases in others. Also, the use of vegetation as an air pollution reduction strategy within the streets seems to not be appropriate in general, and local studies would be necessary for each particular case to select the suitable location of new vegetation planted.

This work confirms previous findings about the predominance of the aerodynamic effects of vegetation on deposition. In addition, this study has been applied to a complex geometry (real scenario) which is different from idealized cases that are commonly investigated in the literature, since extra turbulence mixing caused by surrounding buildings complicates flow and dispersion within the investigated streets. These conclusions are restricted to the configurations investigated here, but can be extrapolated to other cities with similar street layout and similar tree species, for example, typical Mediterranean cities.

An important assumption made here is to treat vegetation in rows within the streets. This was done due to the absence of individual tree data (geo data or inventory of trees) in this Pamplona neighborhood. Vegetation was thus modelled as rows of trees. In order to take into account the gap between trees, we used a value of LAD for the whole row lower than the LAD corresponding to an individual tree. This approach could locally affect the results, however the general impact of trees is captured by the model. As an example, Buccolieri et al. [55] studied the impact of stand density of trees on pollutant levels and distribution by using wind-tunnel experiments and CFD simulations and no significant impact was found, suggesting that stand density has a minor impact with respect to the street geometry and meteorological conditions. As a future line of research, it would be interesting to carry out more studies about the effect of tree geometry (e.g., considering gap between trees, LAD changing with height, etc.) on pollutant dispersion. Other urban properties such as layout of buildings (e.g., ratio between height of building and width of the street, packing density, distribution of buildings) and the relative location between emissions and trees should be analyzed in each studied district in order to provide information about the appropriate green infrastructure.

In future, more field experimental data are needed to support these modelled results. These new experiments should help the modelling community to improve vegetation modelling in order to gain confidence in CFD modelling as a useful planning tool. In addition, it would be important to investigate the effect of shape and size of tree crowns and improve their representation (and modelling) in CFD models.

Acknowledgments: This study has been supported by the project LIFE+ RESPIRA (LIFE13 ENV/ES/000417) funded by EU. Authors thank Extremadura Research Centre for Advances Technologies (CETA-CIEMAT) by helping in using its computing facilities for simulations. CETA-CIEMAT belongs to CIEMAT and the Government of Spain and is funded by European Regional Development Fund (ERDF). Authors also wish to thank the University of Navarra for providing data about building geometry.

Author Contributions: Jose-Luis Santiago, Esther Rivas, Beatriz Sanchez and Fernando Martin conceived and designed the simulated scenarios. Esther Rivas and Jose-Luis Santiago did the simulations and analyzed the data. The main part of the paper was written by Jose-Luis Santiago. but Riccardo Buccolieri and Esther Rivas wrote some parts of the article. All authors contributed to the discussion of the results and have read and approved the final manuscript.

Conflicts of Interest: The authors declare no conflict of interest.

References

1. Gallagher, J.; Baldauf, R.; Fuller, C.H.; Kumar, P.; Gill, L.W.; McNabola, A. Passive methods for improving air quality in the built environment: A review of porous and solid barriers. *Atmos. Environ.* **2015**, *120*, 61–70. [CrossRef]
2. Livesley, S.J.; McPherson, E.G.; Calfapietra, C. The urban forest and ecosystem services: Impacts on urban water, heat, and pollution cycles at the tree, street, and city scale. *J. Environ. Qual.* **2016**, *45*, 119–124. [CrossRef] [PubMed]
3. Grote, R.; Samson, R.; Alonso, R.; Amorim, J.H.; Cariñanos, P.; Churkina, G.; Fares, S.; Le Thiec, D.; Niinemets, Ü.; Mikkelsen, T.N. Functional traits of urban trees in relation to their air pollution mitigation potential: A holistic discussion. *Front. Ecol. Environ.* **2016**, *14*, 543–550. [CrossRef]
4. Mao, Y.; Wilson, J.D.; Kort, J. Effects of a shelterbelt on road dust dispersion. *Atmos. Environ.* **2013**, *79*, 590–598. [CrossRef]
5. Salmond, J.A.; Williams, D.E.; Laing, G.; Kingham, S.; Dirks, K.; Longley, I.; Henshaw, G.S. The influence of vegetation on the horizontal and vertical distribution of pollutants in a street canyon. *Sci. Total Environ.* **2013**, *443*, 287–298. [CrossRef] [PubMed]
6. Jin, S.; Guo, J.; Wheeler, S.; Kan, L.; Che, S. Evaluation of impacts of trees on $PM_{2.5}$ dispersion in urban streets. *Atmos. Environ.* **2014**, *99*, 277–287. [CrossRef]
7. Chen, X.; Pei, T.; Zhou, Z.; Teng, M.; He, L.; Luo, M.; Liu, X. Efficiency differences of roadside greenbelts with three configurations in removing coarse particles (PM_{10}): A street scale investigation in Wuhan, China. *Urban For. Urban Green.* **2015**, *14*, 354–360. [CrossRef]
8. Mori, J.; Sæbø, A.; Hanslin, H.M.; Teani, A.; Ferrini, F.; Fini, A.; Burchi, G. Deposition of traffic-related air pollutants on leaves of six evergreen shrub species during a Mediterranean summer season. *Urban For. Urban Green.* **2015**, *14*, 264–273. [CrossRef]
9. Tong, Z.; Whitlow, T.H.; MacRae, P.F.; Landers, A.J.; Harada, Y. Quantifying the effect of vegetation on near-road air quality using brief campaigns. *Environ. Pollut.* **2015**, *201*, 141–149. [CrossRef] [PubMed]
10. Di Sabatino, S.; Buccolieri, R.; Pappaccogli, G.; Leo, L.S. The effects of trees on micrometeorology in a real street canyon: Consequences for local air quality. *Int. J. Environ. Pollut.* **2015**, *58*, 100–111. [CrossRef]
11. Chen, L.; Liu, C.; Zou, R.; Yang, M.; Zhan, Z. Experimental examination of effectiveness of vegetation as bio-filter of particulate matters in the urban environment. *Environ. Pollut.* **2016**, *208*, 198–208. [CrossRef] [PubMed]
12. Hofman, J.; Bartholomeus, H.; Janssen, S.; Calders, K.; Wuyts, K.; Van Wittenberghe, S.; Samson, R. Influence of tree crown characteristics on the local PM_{10} distribution inside an urban street canyon in Antwerp (Belgium): A model and experimental approach. *Urban For. Urban Green.* **2016**, *20*, 265–276. [CrossRef]
13. Tong, Z.; Whitlow, T.H.; Landers, A.; Flanner, B. A case study of air quality above an urban roof top vegetable farm. *Environ. Pollut.* **2016**, *208*, 256–260. [CrossRef] [PubMed]
14. Gromke, C.; Ruck, B. Pollutant Concentrations in Street Canyons of Different Aspect Ratio with Avenues of Trees for Various Wind Directions. *Bound.-Layer Meteorol.* **2012**, *144*, 41–64. [CrossRef]
15. Huang, C.W.; Lin, M.Y.; Khlystov, A.; Katul, G. The effects of leaf area density variation on the particle collection efficiency in the size range of ultrafine particles (UFP). *Environ. Sci. Technol.* **2013**, *47*, 11607–11615. [CrossRef] [PubMed]
16. Räsänen, J.V.; Holopainen, T.; Joutsensaari, J.; Ndam, C.; Pasanen, P.; Rinnan, A.; Kivimäenpää, M. Effects of species-specific leaf characteristics and reduced water availability on fine particle capture efficiency of trees. *Environ. Pollut.* **2013**, *183*, 64–70. [CrossRef] [PubMed]
17. Gromke, C.; Jamarkattel, N.; Ruck, B. Influence of roadside hedgerows on air quality in urban street canyons. *Atmos. Environ.* **2016**, *139*, 75–86. [CrossRef]
18. Janhäll, S. Review on urban vegetation and particle air pollution—Deposition and dispersion. *Atmos. Environ.* **2015**, *105*, 130–137. [CrossRef]
19. Salmond, J.A.; Tadaki, M.; Vardoulakis, S.; Arbuthnott, K.; Coutts, A.; Demuzere, M.; Dirks, K.N.; Heaviside, C.; Lim, S.; Macintyre, H.; et al. Health and climate related ecosystem services provided by street trees in the urban environment. *Environ. Health.* **2016**, *15* (Suppl. 1), 36. [CrossRef] [PubMed]

20. Abhijith, K.V.; Kumar, P.; Gallagher, J.; McNabola, A.; Baldauf, R.; Pilla, F.; Broderick, B.; Di Sabatino, S.; Pulvirenti, B. Air pollution abatement performances of green infrastructure in open road and built-up street canyon environments—A review. *Atmos. Environ.* **2017**, *162*, 71–86. [CrossRef]

21. Buccolieri, R.; Salim, S.M.; Leo, L.S.; Di Sabatino, S.; Chan, A.; Ielpo, P.; Gromke, C. Analysis of local scale tree–atmosphere interaction on pollutant concentration in idealized street canyons and application to a real urban junction. *Atmos. Environ.* **2011**, *45*, 1702–1713. [CrossRef]

22. Wania, A.; Bruse, M.; Blond, N.; Weber, C. Analysing the influence of different street vegetation on traffic-induced particle dispersion using microscale simulations. *J. Environ. Manag.* **2012**, *94*, 91–101. [CrossRef] [PubMed]

23. Amorim, J.H.; Rodrigues, V.; Tavares, R.; Valente, J.; Borrego, C. CFD modelling of the aerodynamic effect of trees on urban air pollution dispersion. *Sci. Total Environ.* **2013**, *461*, 541–551. [CrossRef] [PubMed]

24. Santiago, J.L.; Martín, F.; Martilli, A. A computational fluid dynamic modelling approach to assess the representativeness of urban monitoring stations. *Sci. Total Environ.* **2013**, *454–455*, 61–72. [CrossRef] [PubMed]

25. Gromke, C.; Blocken, B. Influence of avenue-trees on air quality at the urban neighborhood scale. Part II: Traffic pollutant concentrations at pedestrian level. *Environ. Pollut.* **2015**, *196*, 176–184. [CrossRef] [PubMed]

26. Jeanjean, A.P.R.; Monks, P.S.; Leigh, R.J. Modelling the effectiveness of urban trees and grass on $PM_{2.5}$ reduction via dispersion and deposition at a city scale. *Atmos. Environ.* **2016**, *147*, 1–10. [CrossRef]

27. Jeanjean, A.; Buccolieri, R.; Eddy, J.; Monks, P.; Leigh, R. Air quality affected by trees in real street canyons: The case of Marylebone neighbourhood in central London. *Urban For. Urban Green.* **2017**, *22*, 41–53. [CrossRef]

28. Santiago, J.-L.; Martilli, A.; Martin, F. On Dry Deposition Modelling of Atmospheric Pollutants on Vegetation at the Microscale: Application to the Impact of Street Vegetation on Air Quality. *Bound.-Layer Meteorol.* **2017**, *162*, 451–474. [CrossRef]

29. Selmi, W.; Weber, C.; Rivière, E.; Blond, N.; Mehdi, L.; Nowak, D. Air pollution removal by trees in public green spaces in Strasbourg city, France. *Urban For. Urban Green.* **2016**, *17*, 192–201. [CrossRef]

30. Tong, Z.; Baldauf, R.W.; Isakov, V.; Deshmukh, P.; Max Zhang, K. Roadside vegetation barrier designs to mitigate near-road air pollution impacts. *Sci. Total Environ.* **2016**, *541*, 920–927. [CrossRef] [PubMed]

31. Vos, P.E.; Maiheu, B.; Vankerkom, J.; Janssen, S. Improving local air quality in cities: To tree or not to tree? *Environ. Pollut.* **2013**, *183*, 113–122. [CrossRef] [PubMed]

32. Haaland, C.; van den Bosch, C.K. Challenges and strategies for urban green-space planning in cities undergoing densification: A review. *Urban For. Urban Green.* **2015**, *14*, 760–771. [CrossRef]

33. Google Maps®. Available online: https://www.google.es/maps (accessed on 4 May 2017).

34. CD-adapco. User Guide STAR-CCM+ Version 7.04. Cd-adapco, 2012. Available online: http://www.cd-adapco.com/products/star-ccm%C2%AE (accessed on 21 July 2017).

35. Sanchez, B.; Santiago, J.L.; Martilli, A.; Palacios, M.; Kirchner, F. CFD modeling of reactive pollutant dispersion in simplified urban configurations with different chemical mechanisms. *Atmos. Chem. Phys.* **2016**, *16*, 12143–12157. [CrossRef]

36. Sanchez, B.; Santiago, J.L.; Martilli, A.; Martin, F.; Borge, R.; Quaassdorff, C.; de la Paz, D. Modelling NO_x concentrations through CFD-RANS in an urban hot-spot using high resolution traffic emissions and meteorology from a mesoscale model. *Atmos. Environ.* **2017**, *163*, 155–165. [CrossRef]

37. Krayenhoff, E.S.; Santiago, J.L.; Martilli, A.; Christen, A.; Oke, T.R. Parametrization of drag and turbulence for urban neighbourhoods with trees. *Bound.-Layer Meteor.* **2015**, *156*, 157–189. [CrossRef]

38. Sanz, C. A note on $k - \varepsilon$ modeling of vegetation canopy air-flows. *Bound.-Layer Meteor.* **2003**, *108*, 191–197. [CrossRef]

39. Google Earth®. Available online: https://www.google.com/intl/es/earth/ (accessed on 4 May 2017).

40. Franke, J.; Schlünzen, H.; Carissimo, B. *Best Practice Guideline for the CFD Simulation of Flows in the Urban Environment. COST Action 732—Quality assurance and improvement of microscale meteorological models;* Distributed by University of Hamburg (Germany), Meteorological Institute: Hamburg, Germany, 2007; ISBN 3-00-018312-4.

41. Richards, P.; Hoxey, R. Appropriate boundary conditions for computational wind engineering models using the k-turbulence model. *J. Wind Eng. Ind. Aerodyn.* **1993**, *46*, 145–153. [CrossRef]

42. Santiago, J.L.; Borge, R.; Martin, F.; de la Paz, D.; Martilli, A.; Lumbreras, J.; Sanchez, B. Evaluation of a CFD-based approach to estimate pollutant distribution within a real urban canopy by means of passive samplers. *Sci. Total Environ.* **2017**, *576*, 46–58. [CrossRef] [PubMed]

43. Rivas, E.; Santiago, J.L.; Martin, F.; Sánchez, B.; Martilli, A. CFD study of induced effects of trees on air quality in a neighborhood of Pamplona. 10th International Conference on Air Quality—Science and Application, Milan, Italy, 14–18 March 2016; Available online: https://drive.google.com/drive/folders/0B2iFZ3L-H5pRazh6QjlBWGVWcHc (accessed on 3 March 2017).

44. Lalic, B.; Mihailovic, D.T. An empirical relation describing leaf-area density inside the forest for environmental modeling. *J. Appl. Meteor.* **2004**, *43*, 641–645. [CrossRef]

45. Brunet, Y.; Finnigan, J.J.; Raupach, M.R. A wind tunnel study of air flow in waving wheat: Single-point velocity statistics. *Bound.-Layer Meteor.* **1994**, *70*, 95–132. [CrossRef]

46. Raupach, M.R.; Bradley, E.F.; Ghadiri, H. *Wind Tunnel Investigation Into the Aerodynamic Effect of Forest Clearing on the Nesting of Abbott's Booby on Christmas Island*; Internal report; CSIRO Centre for Environmental Mechanics: Canberra, Australia, 1987.

47. Foudhil, H.; Brunet, Y.; Caltagirone, J.P. A Fine-Scale $k-\varepsilon$ Model for Atmospheric Flow over Heterogeneous Landscapes. *Environ. Fluid Mech.* **2005**, *5*, 247–265. [CrossRef]

48. Dupont, S.; Brunet, Y. Edge flow and canopy structure: A large-eddy simulation study. *Bound.-Layer Meteor.* **2008**, *126*, 51–71. [CrossRef]

49. Gromke, C.; Ruck, B. Influence of trees on the dispersion of pollutants in an urban street canyon-experimental investigation of the flow and concentration field. *Atmos. Environ.* **2007**, *41*, 3287–3302. [CrossRef]

50. Gromke, C.; Ruck, B. On the impact of trees on dispersion processes of traffic emissions in street canyons. *Bound.-Layer Meteor.* **2009**, *131*, 19–34. [CrossRef]

51. Gromke, C.; Buccolieri, R.; Di Sabatino, S.; Ruck, B. Dispersion study in a street canyon with tree planting by means of wind tunnel and numerical investigations—Evaluation of CFD data with experimental data. *Atmos. Environ.* **2008**, *42*, 8640–8650. [CrossRef]

52. Balczó, M.; Gromke, C.; Ruck, B. Numerical modeling of flow and pollutant dispersion in street canyons with tree planting. *Meteorologische. Zeitschrift.* **2009**, *18*, 197–206. [CrossRef]

53. Vranckx, S.; Vos, P.; Maiheu, B.; Janssen, S. Impact of trees on pollutant dispersion in street canyons: A numerical study of the annual average effects in Antwerp, Belgium. *Sci. Total Environ.* **2015**, *532*, 474–483. [CrossRef] [PubMed]

54. Moonen, P.; Gromke, C.; Dorer, V. Performance assessment of large eddy simulation (LES) for modelling dispersion in an urban street canyon with tree planting. *Atmos. Environ.* **2013**, *75*, 66–76. [CrossRef]

55. Buccolieri, R.; Gromke, C.; Di Sabatino, S.; Ruck, B. Aerodynamic effects of trees on pollutant concentration in street canyon. *Sci. Total Environ.* **2009**, *407*, 5247–5256. [CrossRef] [PubMed]

atmosphere

MDPI

Article

Assessment of Natural Ventilation Potential for Residential Buildings across Different Climate Zones in Australia

Zijing Tan [1] and Xiang Deng [2],*

[1] School of Civil Engineering, Chang'an University, Xi'an 710064, China; tanzijing01@163.com
[2] Sustainable Buildings Research Centre, University of Wollongong, Wollongong, NSW 2500, Australia
* Correspondence: xd902@uowmail.edu.au; Tel.: +61-4-1548-4106

Received: 3 September 2017; Accepted: 16 September 2017; Published: 20 September 2017

Abstract: In this study, the natural ventilation potential of residential buildings was numerically investigated based on a typical single-story house in the three most populous climate zones in Australia. Simulations using the commercial simulation software TRNSYS (Transient System Simulation Tool) were performed for all seasons in three representative cities, i.e., Darwin for the hot humid summer and warm winter zone, Sydney for the mild temperate zone, and Melbourne for the cool temperate zone. A natural ventilation control strategy was generated by the rule-based decision-tree method based on the local climates. Natural ventilation hour (NVH) and satisfied natural ventilation hour (SNVH) were employed to evaluate the potential of natural ventilation in each city considering local climate and local indoor thermal comfort requirements, respectively. The numerical results revealed that natural ventilation potential was related to the local climate. The greatest natural ventilation potential for the case study building was observed in Darwin with an annual 4141 SNVH out of 4728 NVH, while the least natural ventilation potential was found in the Melbourne case. Moreover, summer and transition seasons (spring and autumn) were found to be the optimal periods to sustain indoor thermal comfort by utilising natural ventilation in Sydney and Melbourne. By contrast, natural ventilation was found applicable over the whole year in Darwin. In addition, the indoor operative temperature results demonstrated that indoor thermal comfort can be maintained only by utilising natural ventilation for all cases during the whole year, except for the non-natural ventilation periods in summer in Darwin and winter in Melbourne. These findings could improve the understanding of natural ventilation potential in different climates, and are beneficial for the climate-conscious design of residential buildings in Australia.

Keywords: natural ventilation; residential building; climate zone; thermal comfort; natural ventilation hour

1. Introduction

Rapid urbanisation has led to a significant increase in building energy usage, which accounts for nearly one third of the total primary energy consumption worldwide [1]. As a key solution to the efficient operation of buildings, natural ventilation plays a significant role in maintaining an acceptable indoor environment [2,3]. The benefits of natural ventilation include, but are not limited to, improved indoor thermal comfort, reductions in occupant illness associated with indoor environmental quality (IEQ), and increased work productivity with low energy consumption and greenhouse gas (GHG) emissions [4–6].

Natural ventilation potential was defined to evaluate the possibility of ensuring an acceptable indoor air quality and thermal comfort naturally [7]. Determined by both the indoor and outdoor environment, natural ventilation potential can be influenced by local climate, urban form and building characteristics (geometrical and thermal) [8,9]. An early preliminary study conducted by Teitel and Tanny [10] investigated the impact of building structure and window height on the natural ventilation of greenhouses, and revealed that the effect of natural ventilation (e.g., air change rate) increased with the height of the window opening and the wind speed. Fordham [11] indicated that the heat effect due to building thermal capacity should not be ignored in the design of natural ventilation. Meanwhile, the influence of internal heat gain on natural ventilation performance has also been intensively studied, and heat source geometries [12–14] and transient characters [15,16] have been proven to be the most significant factors influencing ventilation rate and airflow pattern. In the last few decades, the effect of the urban environment on natural ventilation potential has become of increasing concern. Ghiaus et al. [17] conducted a series of field measurements and quantified the effect of urban phenomena on natural ventilation. Han et al. [18] compared the thermal comfort performance in urban and rural residential buildings in a naturally ventilated environment, and claimed that high-density urban settings could reduce the cooling effect of natural ventilation. Hang et al. [19] adopted a CFD model to investigate the effect of semi-open street roofs on the natural ventilation of urban canopy layers.

Since building detail and urban configuration are known factors, natural ventilation potential depends greatly on the suitability of local climate [20]. In recent years, the natural ventilation potential in different climate zones has been investigated around the world. Wang and Greenberg [21] evaluated the thermal comfort and energy performance of three major cities with different climates in the US to identify the available natural ventilation time through the EnergyPlus simulation. Su et al. [22] analysed thermal comfort conditions in Shanghai, located in the hot summer and cold winter district of China, in a naturally ventilated residential building and found natural ventilation was not inappropriate in winter and transition seasons. Calautit et al. [23] investigated using CFD simulation the indoor air change rate of a traditional row house using a wind tower in the hot and arid Middle East. Artmann et al. [24] studied the potential of night-time ventilation in commercial buildings all over Europe, with the results showing that northern Europe had the most potential for passive cooling during night time, while in southern Europe night-time ventilation could only be used as an auxiliary cooling method.

Although numerous studies have been conducted to detail thermal-comfort-related natural ventilation over the world, there has been little research related to natural ventilation feasibility evaluation across different climate zones in Australia. It is, therefore, necessary to evaluate the suitability of natural ventilation across the diverse climates of this country. This study presents a computational methodology for the analysis of the climate suitability of natural ventilation for residential buildings in three Australian climates. A typical multi-zone residential building model with a rule-based window control strategy was adopted as the research platform. This research can contribute to a more sophisticated approach to understanding natural ventilation potential across different climate zones, and hence to achieving the free cooling predesign purpose.

2. Research Methodology

2.1. Climate Zones in Australia

Because Australia is not subject to the movements of frigid polar air from the South Pole due to its separation by the Antarctic Ocean, the climate is generally temperate: most of the country receives more than 3000 h of sunshine a year. The temperature difference between summer and winter can be relatively small compared to the northern continents. Nevertheless, having a vast interior, many areas are characterised by particular climate conditions. Based on a set of definitions relating to summer and winter temperature and humidity conditions, eight key zones across Australia from north to south are categorised as hot humid summer and warm winter, warm humid summer and mild winter, hot dry summer and warm winter, hot dry summer and cool winter, warm temperate, mild temperate, cool temperate and alpine (Figure 1) [25].

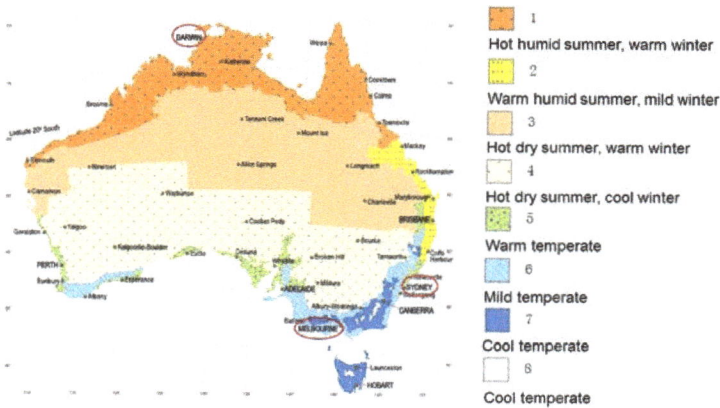

Figure 1 legend:
1 — Hot humid summer, warm winter
2 — Warm humid summer, mild winter
3 — Hot dry summer, warm winter
4 — Hot dry summer, cool winter
5 — Warm temperate
6 — Mild temperate
7 — Cool temperate
8 — Cool temperate

Figure 1. Australian climate zones [25]. Reproduced with permission from Australian Bureau of Statistics, Year Book Australia, 2012; published by ABS Canberra: Canberra, Australia, 2012.

Since the majority of Australia's population lives in coastal instead of central areas (dry and desert regions), three coastal cities, Darwin, Sydney and Melbourne, were selected to represent the three most populous climate zones. Darwin is the capital of the Northern Territory located in the hot humid summer, warm winter zone with a tropical climate. The coldest month's average air temperature is 20 °C. The hottest monthly average air temperature is 32 °C. Sydney is the capital of the State of New South Wales, located in the mild temperate zone with a humid subtropical monsoon climate. The coldest month's average air temperature is 17.2 °C. The hottest monthly average air temperature is 36.4 °C. Melbourne is the capital of the State of Victoria, located in the cool temperate zone with a temperate marine climate. The climate details of the three cities are illustrated in Table 1.

Table 1. Climatic characteristics of the representative cities [25].

City	Climate Zone	Annual Temperature (°C)		Average Annual Relative Humidity (%)	Average Annual Wind Speed (m/s)
		Average High	Average Low		
Darwin	Tropical	32.0	23.2	53.4	4.3
Sydney	Warm temperate	22.5	14.5	56.2	3.1
Melbourne	Mild temperature	20.4	11.4	51.8	2.9

2.2. Residential Building Model

Description of the Case Study Building

The AS/NZS building code published by Standards Australia [26] specifies universal material requirements for thermal insulation of residential buildings in different climate zones in Australia. Given this, building models with the same envelope conditions could be employed for potential natural ventilation analysis in different climate zones. A multi-zone single-story house model based on a typical residential building was adopted.

It consists of five main parts: a master bedroom at the northeast corner, a study/guest bedroom at the southeast corner, an open-plan living room connected with the dining room and kitchen, a laundry space, and a shower space (Figure 2). Both the external envelope and internal enclosure are insulated by glasswool. Two types of windows are used in this house, including three top-hung clerestories on the south wall of the living room, and six side-hung windows in the rest of the walls. All windows are equipped with low-e double argon-filled glazing. This building is located in a flat yard with no significant wind obstructions.

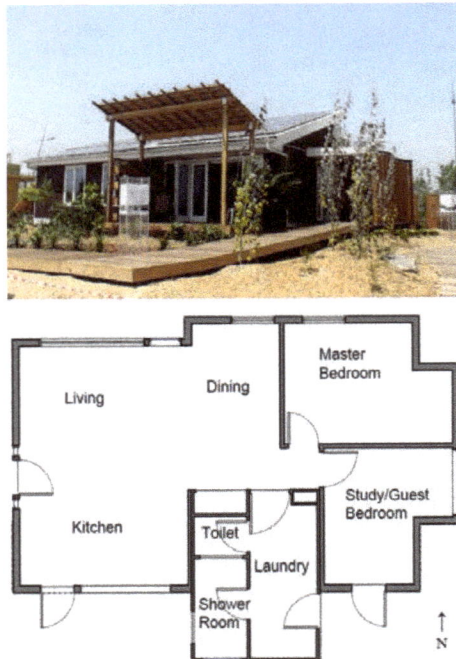

Figure 2. Case study building.

2.3. Simulation Approach

The commercial simulation software TRNSYS (Transient System Simulation Tool) combined with COMIS (Conjunction of Multi-zone Infiltration Specialists) was used for the indoor airflow and thermal modelling of the case study building, as well as for the natural ventilation control strategy proposed in the following section of this study. The control strategy was programmed in MATLAB and integrated within the simulation platform through the TRNSYS component Type 22. The schematic of the airflow-thermal modelling in TRNSYS is presented in Figure 3.

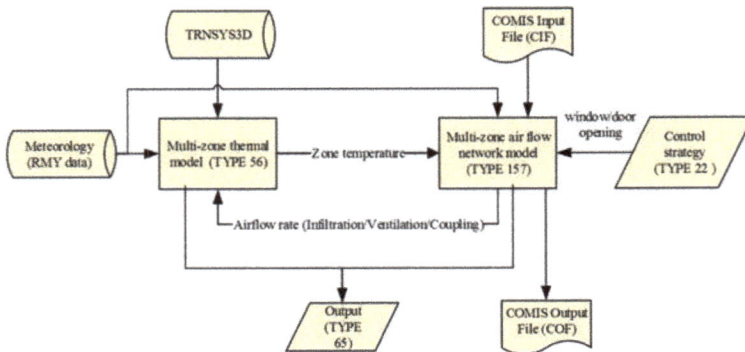

Figure 3. Integration of COMIS into TRNSYS simulation.

To translate the physical layout of the house into a building model with a node-based airflow network (Component Type 56), several assumptions were made as follows:

- The cracks for closed windows and doors were set as 1 mm based on the estimated crack dimensions of the house;
- Zones were defined by rooms of the house and each zone was assumed to have a uniform air temperature distribution and pressure distribution;
- The door between the laundry and the living room was assumed to be closed. The laundry and shower spaces were not considered in the airflow network for the multi-zone ventilation calculation but were modelled in thermal simulations;
- Internal doors that connect the bedroom and the guest room to the living room were assumed to be normally open.

Although actual local temperature and velocity distributions might be non-uniform and even contribute to thermal comfortability divergence in a room, this study was mostly concerned with the flow rate and ventilation efficiency over a whole residential building. As a highly efficient and widely used ventilation model, the multi-zone airflow network was adopted in the current simulation.

The airflow network of the case study building (Component Type 157) is shown in Figure 4. The components of external openings were highlighted in red. As there were three clerestory windows in the living room, the buoyancy effect was taken into account by introducing a virtual horizontal opening component, highlighted in green.

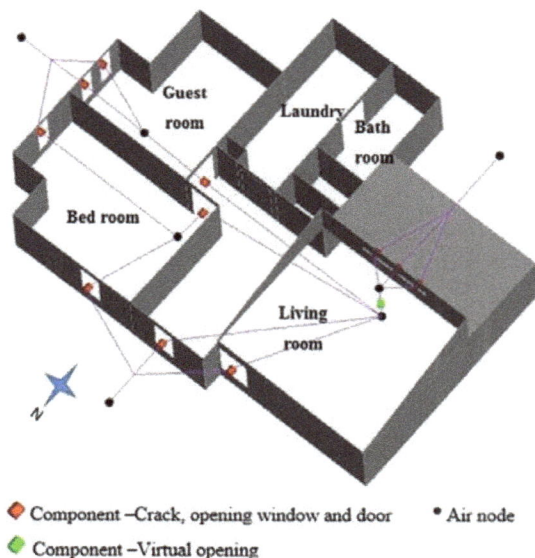

◆ Component –Crack, opening window and door   • Air node

◆ Component –Virtual opening

Figure 4. Schematic of the airflow nodal network.

Since the wind pressure coefficients of the openings in the case study building were required for wind-induced ventilation calculation, CFD simulation was conducted as one of the prime sources [27] (simulation results were shown in Appendix A Table A1). The reference meteorological year (RMY) data of the three selected cities was selected for the performance evaluation. The internal heat gains from occupants and equipment activities were set as the recommended value referred to in the ASHRAE handbook [28] (Appendix A Table A2).

2.4. Model Validation

The current numerical approach was validated against a field measurement in the case study building located in the Sydney area. Indoor air temperature, as a joint result of the airflow and thermal transfer process, was measured and collected hourly in each room over eight consecutive days (27 July 2015–3 August 2015) during the daytime. During the measurements, three window conditions, i.e., fully open, half open (50% open) and fully closed, were performed. The temperature data was continuously tested for three days under each scenario. As a main functional area of the residential building, the master bedroom was selected for the validation. Indoor temperatures predicted by numerical and measurement methods are compared in Figure 5.

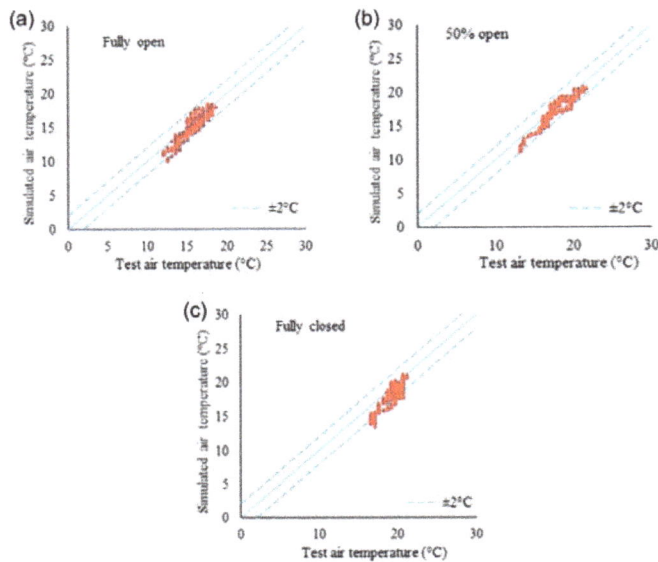

Figure 5. Comparison of indoor air temperatures of the master bedroom between numerical simulated methods and measured data under three window conditions: (**a**) fully open; (**b**) half open; (**c**) fully closed.

In general, the numerical prediction of the indoor temperature under all window conditions agreed well with that of the field measurement data. The deviations of the simulated data points from the bisector (e.g., the test points) mostly fell within ± 2 °C. Only slightly divergence were observed for the fully closed condition. These could be attributed to the diversity of numerous factors influencing indoor thermal conditions (e.g., instantaneous meteorological conditions, surrounding obstacles, ground roughness, etc.). In the numerical model, the ambient wind condition was assumed to be constant in half an hour, and obstacles around the building were neglected. The value divergence was of the same level as previous studies [29–31], where the error band for the predicted indoor temperature based on the TRNSYS was shown to be about ± 2 °C.

Overall, the reasonable level of agreement found in these comparisons demonstrate that the proposed numerical approach is capable of predicting the indoor thermal environment and airflow with fair accuracy.

2.5. Natural Ventilation Control Strategy

2.5.1. Decision-Tree Model

Actual thermal comfort is dependent on environmental factors, such as air temperature, air velocity, relative humidity and the uniformity of conditions and personal factors such as clothing and metabolic heat. However, it is very complex to assess indoor thermal comfort by considering all these variables (see predicted mean vote), and a simpler measure can be more useful in practice. In practice, operative temperature derived from air temperature, mean radiant temperature and air speed is widely used as a reasonable indicator of thermal comfort. Operative temperature is defined as:

$$T_o = \left(T_r + \left(T_a \times \sqrt{10v}\right)\right) / \left(1 + \sqrt{10v}\right) \tag{1}$$

where T_a is the air temperature; T_r is the mean radiant temperature; and v is the air speed. In this study, the indoor operative temperature was selected as the indoor thermal comfort index for natural ventilation control.

In this paper, the decision-tree induction method was used to generate a rule-based window control algorithm in order to determine whether natural ventilation could be used under local climate conditions. For a decision-tree model, a reversed tree-like structure is built with several nodes and branches. Each internal node and leaf node represents a test condition with an attribute and a classification prediction, respectively. Meanwhile, the outcomes of the test are presented by branches [32]. The process to generate the decision-tree model employed in this study is presented in Figure 6.

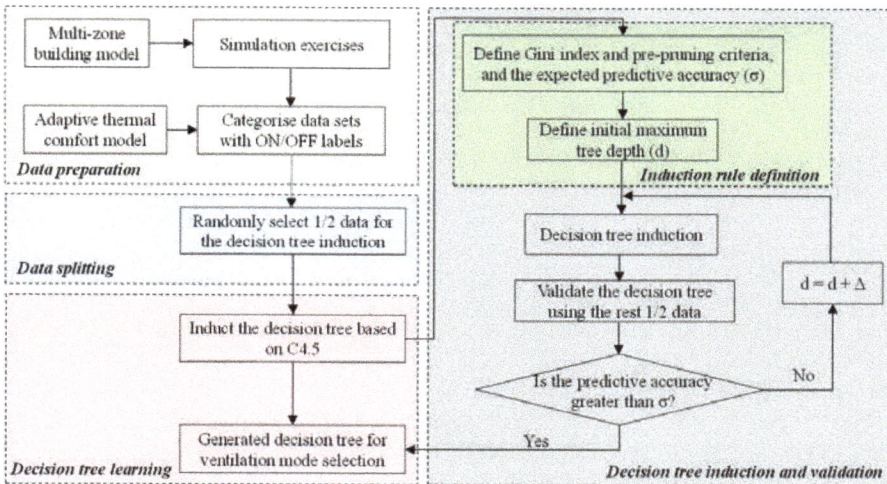

Figure 6. Illustration of the decision-tree generation process.

The data used for the decision-tree induction and validation was first generated based on the hourly simulation with the multi-zone building model under different window opening percentages (i.e., 25%, 50%, 75% and 100%). Then, the simulated indoor operative temperatures under different window and weather conditions (including ambient air temperature, relative humidity, solar radiation, wind speed and wind direction) were prepared as the data sets for decision-tree induction and validation. Based on these data sets, the applicability of natural ventilation under different window and weather conditions was assessed by an adaptive thermal comfort model (the 80% thermal comfort band) developed by De Dear [33]. The equation is given below [34]:

$$T_{up} = 0.31 T_{out} + 17.8 + \frac{1}{2} \Delta T_{80\%} \tag{2}$$

$$T_{low} = 0.31 T_{out} + 17.8 - \frac{1}{2} \Delta T_{80\%} \tag{3}$$

where T_{up} and T_{low} represent the upper and lower thresholds of temperature varided by month, T_{out} is the monthly average outdoor temperature, and $\Delta T_{80\%}$ is the mean comfort zone temperature band for 80% acceptability.

If indoor operative temperature was within the 80% acceptable indoor operative temperature thresholds [35], it was considered that natural ventilation could be used for this particular condition and the specific data set was then labelled as "ON". Otherwise, the data set would be labelled as "OFF". Half of the labelled data was randomly selected as the training data for the decision-tree induction using the C4.5 algorithm [36]. C4.5 inducts the decision tree based on the concept of Shannon entropy in order to measure the unpredictability or the impurity of the information content [37]. The impurity of the attribute partition decreases with the decrease of Shannon entropy. If a set of training data was allocated to a node S, and the probability distribution of the target attributes was $D_i = (D_1, D_2 \cdots D_n)$, Shannon entropy for the training data carried by this distribution is defined as Equation (4).

$$Entropy(D, S) = - \sum_{i=1}^{n} (D_i \times \log_2(D_i)) \tag{4}$$

In the decision-tree induction, rules including the Gini index, pre-pruning criteria, and the minimal expected predictive accuracy were defined first. In order to balance the decision-tree scale and splitting accuracy (i.e., the ratio between the correctly labelled training datasets and all the training datasets), the Gini index was used to measure the impurity of a node [38]. Meanwhile, pre-pruning criteria including the minimal gain, minimal leaf size and minimal size were adopted to avoid overfitting of the decision tree. As it represented the expected ratio between the correctly labelled testing data sets and all the testing data sets, the expected predictive accuracy was related to the data quality. Thus, a reasonable expected predictive accuracy was selected through trial and error, set as 0.93. The details of induction rule settings were given in the Appendix A (Table A3).

On account of the defined Gini index, pre-pruning criteria, and the expected predictive accuracy, an initial maximum tree depth can be assigned for decision-tree induction and validation. To control the size of the decision tree, the depth should start with a relatively small value. The values for the tree induction were used according to [39].

If the predictive accuracy of the decision tree validated by the testing data was larger than the minimum expected value, the decision-tree learning process would be terminated. Then, the generated decision tree would be used for ventilation control. Otherwise, a new tree would be generated by increasing the maximum tree depth so as to improve the predictive accuracy.

2.5.2. Natural Ventilation Strategy Based on the Decision-Tree Model

The decision tree for the case study building was introduced by the open source data mining software RapidMiner. A total of 3000 hourly data sets for each of the four window opening conditions (i.e., 25%, 50%, 75% and 100%) were obtained during the whole RMY for decision-tree generation and validation.

A final decision-tree model for the case study building in Sydney is depicted in Figure 7. The decision tree consisted of 55 nodes, among which 27 yellow rectangular nodes presented the categorical parameters, and 12 blue and 16 red ovals at the bottom denoted the classification results. The outdoor air temperature nodes accounted for 1/3 of the total internal (i.e., a node between input and output) nodes, indicating that the outdoor air temperature was one of the most critical parameters for natural ventilation. By using this decision tree, each data record was assigned to a leaf node that was associated with a specific window condition, and a window opening prediction could be made. In addition, no internal node related to the window opening percentage was found in this decision tree, implying that window opening percentage had less influence on ventilation mode selection when compared to the outdoor climate.

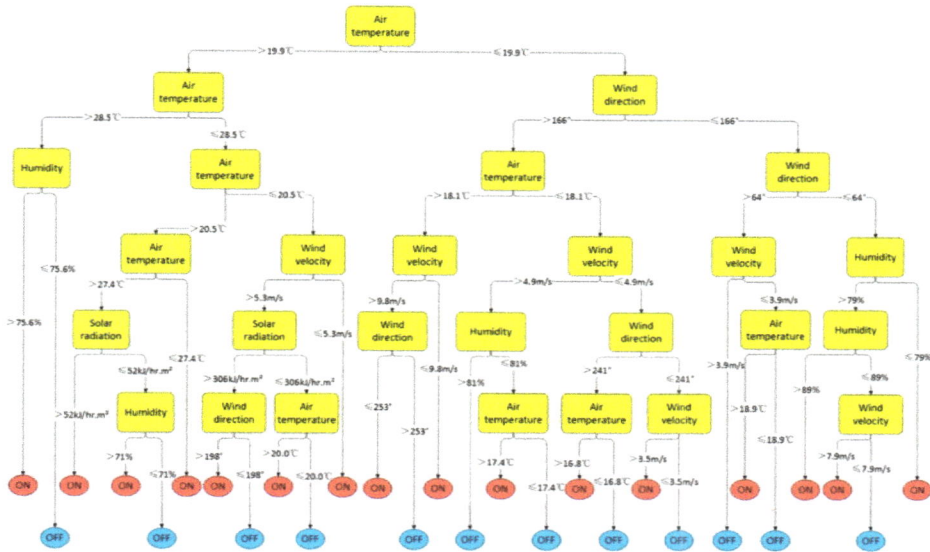

Figure 7. The generated decision tree for ventilation mode selection in Sydney.

3. Results and Discussion

3.1. Natural Ventilation Hour

The natural ventilation hour (NVH) [1] was employed as an indicator to measure the natural ventilation potential for the case study building at each location. It is defined as the number of hours in a typical year (8760 h) or a typical season (2190 h) during which outdoor weather conditions are suitable for utilising natural ventilation. The NVH of the studied residential building in the three representative cities were simulated by the proposed TRNSYS simulation approach. The results for spring, summer, autumn and winter at the three locations were presented in Table 2, respectively.

Table 2. Natural ventilation potential under different seasons.

City	Natural Ventilation Hour				
	Spring	Summer	Autumn	Winter	Total
Darwin	1010.5	996	1109.5	1612	4728
Sydney	915	1898.5	1240.5	125.5	4179.5
Melbourne	352.5	1150	571	41.5	2115

As expected, the building located in Darwin in the hot humid summer and warm winter zone showed the greatest potential for utilising natural ventilation, with the NVH number of 4728. While, the least natural ventilation potential with 2115 NVH was observed at Melbourne in the cool temperate zone. Additionally, the seasonal NVH results showed that summer and transition seasons (spring and autumn) were the optimal periods for utilising natural ventilation, while not being applicable in winter in Sydney and Melbourne. By contrast, roughly similar NVHs (about 1000 of 2190 h) were observed in Darwin for summer and transition seasons, while a higher NVH was found in winter. This can be attributed to Darwin's tropical climate, with remarkably similar outdoor weather conditions during summer and transition seasons, yet more temperate conditions in winter [40].

The seasonal and annual NVH results of the three cities shown above demonstrated that the potential of utilising natural ventilation was significantly altered by local climate. Natural ventilation was applicable for more than half the year in the hot humid summer and warm winter zone and mild temperate zone, while only for a quarter of the year in the cool temperate zone.

3.2. Satisfied Natural Ventilation Hour

Since maintaining indoor thermal comfort without consuming industrial energy is the most significant advantage of natural ventilation, the potential for utilising natural ventilation meeting indoor thermal comfort requirements should be a significant concern. To quantify this potential, an index named the satisfied natural ventilation hour (SNVH) is proposed in this study. It is defined as the number of hours in a typical year (8760 h) or a typical season (2190 h) when indoor operative temperature under natural ventilation conditions could meet the thermal comfort requirements (i.e., within the 80% acceptable thermal comfort band). As both outdoor weather and building conditions (geometry, thermal performance) are considered in the definition, SNVH measures the maximum number of hours when the outdoor weather is favourable for the natural ventilation of a specific building based on indoor thermal comfort requirements. The SNVH numbers of the studied residential building in the three representative cities were calculated based on the proposed TRNSYS simulation.

Statistical data of SNVH during four seasons is presented in Table 3. Similar seasonal results were obtained for the SNVH and the formentioned NVH. The maximum SNVH value for the whole year occurred in Darwin, while the minimum value was observed in Melbourne. This demonstrates that the residential building located in Darwin, the hot humid summer and warm winter zone, has the greatest natural ventilation potential for maintaining thermal comfort. Due to the contribution of the hot humid summer and warm winter climate, the average air temperature of Darwin is about 20 °C in winter. Such pleasantly cool outdoor conditions improve the applicability of natural ventilation for maintaining indoor thermal comfort in Darwin. A relatively higher SNVH number was found in Darwin in winter. Significant seasonal difference for SNVH was found in Sydney and Melbourne. The greatest value of SNVH occurred in summer, while it was quite small and negligible in winter. However, the SNVH value for Melbourne during all seasons was dramatically lower than that of the other two cities. In addition, the annual SNVH to NVH ratio was calculated and is shown in the last column in Table 3. As a quantitative index representing the correlation between SNVH and NVH, the ratio was found to be 88% for Darwin, 84% for Sydney and 85% for Melbourne. In this study, the natural ventilation control strategy was constructed via the rule-based decision-tree prediction model based on outdoor climate data. These results implied that, in spite of the fact that natural ventilation potential for maintaining thermal comfort was significantly influenced by outdoor climate (embodied by the ANVH to NVH ratio), the effect of building characteristics such as layout, internal heat gain and the thermal performance of the envelope on indoor thermal comfort could not be underestimated when using natural ventilation. The factor of building characteristics should be considered in the rule-based decision-tree prediction model in future studies to further improve predictive accuracy.

Table 3. Thermal comfort potential during different seasons.

City	Satisfied Natural Ventilation Hours					
	Spring	**Summer**	**Autumn**	**Winter**	**Total**	**SNVH$_{total}$/NVH$_{total}$ (%)**
Darwin	862	875	977.5	1426.5	4141	88
Sydney	787.5	1561.5	1036	125.5	3510.5	84
Melbourne	322.5	913.5	522.5	41.5	1800	85

3.3. Thermal Comfort Level or Indoor Operative Temperature

As mentioned, summer is the best season for utilising natural ventilation in all the three cities. To quantify the performance of indoor operative temperature during summer, hourly averaged data for the hottest month (January) was plotted for the three cities (Figure 8).

As revealed in the analysis of the decision-tree generation, outdoor temperature was the essential factor influencing indoor operative temperature when utilising natural ventilation. This result was confirmed by the temperature profiles shown in Figure 8. For all three cases, the fluctuations of outdoor air temperature and average indoor operative temperature were nearly consistent all the time, and the peak and trough values of average indoor operative temperature appeared simultaneously during natural ventilation

Although consistent variation was found in average indoor operative temperature and outdoor air temperature, there were distinctive natural ventilation performances during January for the three cities. For Darwin and Melbourne cases (Figure 8a,c), natural ventilation was utilised with relatively regular time intervals during the whole month, resulting in virtually equal hours of natural ventilation and non-ventilation. By contrast, the natural ventilation schedule of Sydney was irregular in January, with a long and continuous ventilation period (Figure 8b). Thus, the natural ventilation time during January for the Sydney case was observed to be much longer than that for the Darwin and Melbourne cases.

The indoor operative temperature for the Darwin case was higher than the upper thermal comfort threshold during most non-natural ventilation periods (Figure 8a). However, this was maintained in the thermal comfort band for the Sydney and Melbourne cases (Figure 8b,c). Although outdoor temperature during most non-natural ventilation periods was dramatically lower than the thermal comfort threshold for the Melbourne case, the indoor operative temperature remained in the comfort band (Figure 8c). This can be attributed to the coupling effect of outdoor conditions and internal heat gain during natural ventilation and non-ventilation periods. For the Darwin case, the hot summer climate with discomforting outdoor conditions and internal heat gain were all disadvantages when it came to indoor thermal comfort. However, due to the heating effect of the internal heat source, the indoor operative temperature could be kept in the thermal comfort band with much lower outdoor temperatures during the non-ventilation periods in Melbourne.

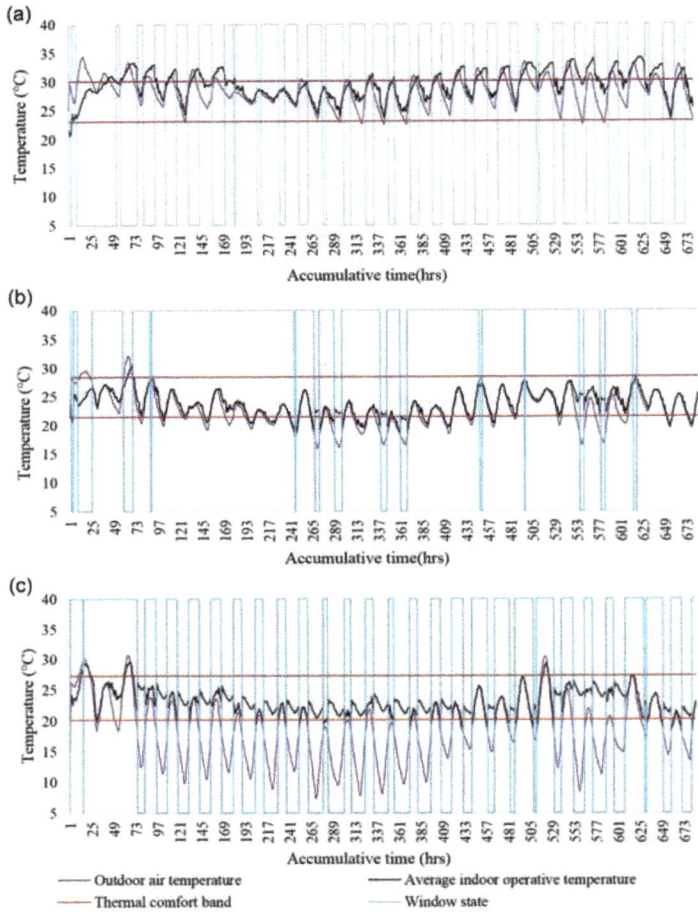

Figure 8. Indoor operative temperature variation during January in: (**a**) Darwin; (**b**) Sydney; and (**c**) Melbourne.

The indoor operative temperature results in January shown above implied that an air-conditioning system should be used in summer during non-ventilation periods in Darwin, while indoor thermal comfort requirements can be met only by utilising natural ventilation in Sydney and Melbourne.

Further qualitative analysis was also conducted to evaluate the average indoor operative temperature during winter for the three cases. The hourly outdoor temperature, indoor operative temperature, and thermal comfort band, were plotted against window opening conditions in Figure 9. In general, the fluctuation tendency of the indoor operative temperature was in keeping with that of the outdoor temperature for all the three cases. However, the thermal comfort level during July was found to be significantly different across Darwin, Sydney and Melbourne. Due to the warm winter climate, natural ventilation was applicable during the whole month in Darwin. Consequently, a regular ventilation schedule was adopted during July with longer natural ventilation hours than non-natural ventilation hours (Figure 9a). By contrast, during July the cold winter climate in Sydney and Melbourne reduced the suitability of natural ventilation. Thus, the natural ventilation hours were nearly zero for both the Sydney and Melbourne cases (Figure 9b,c). In summary, the average indoor operative temperature that was observed remained in the thermal comfort band during most periods in Darwin and Sydney, although the natural ventilation periods for Sydney could almost be ignored; while the indoor operative temperature was found to be frequently lower than the lower thermal comfort threshold during the whole month for the Melbourne case. These results indicate that indoor thermal comfort could be achieved during winter with reasonable ventilation control in Darwin and Sydney, but the heating system would be needed for improving thermal comfort in Melbourne.

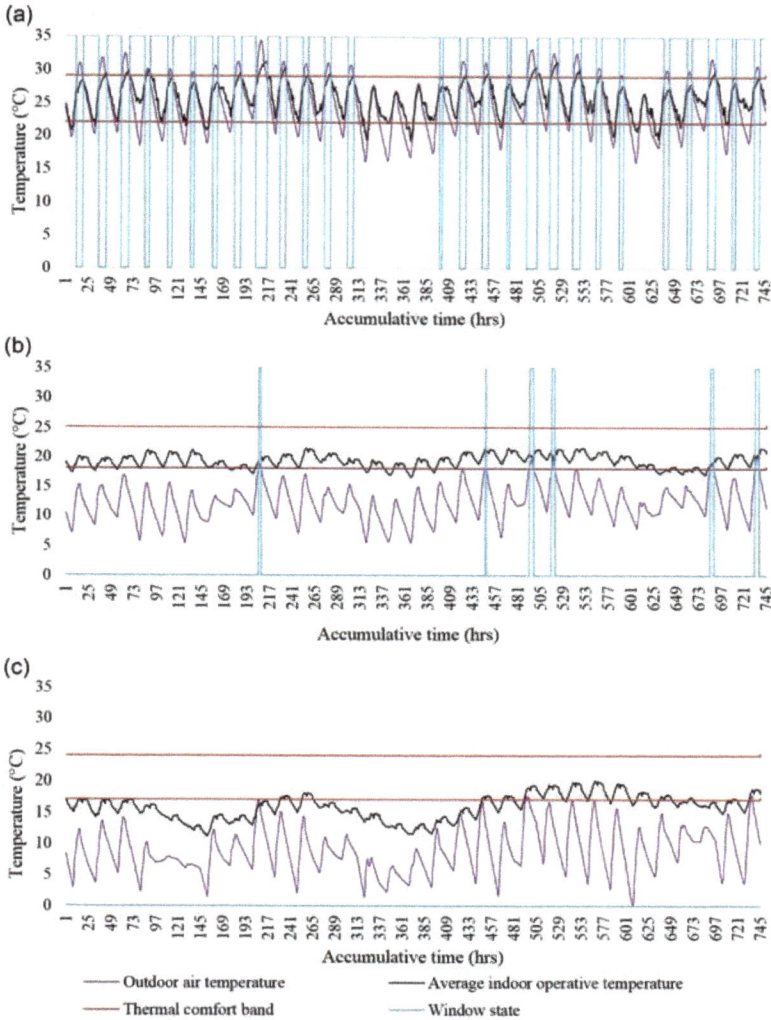

Figure 9. Indoor operative temperature variation during July in: (a) Darwin; (b) Sydney; and (c) Melbourne.

4. Conclusions

The natural ventilation potential of residential buildings located in three different Australian climate zones (hot humid summer and warm winter zone, mild temperate zone, and cool temperate zone) were investigated for a typical single-story house model. Three events were modelled in the representative cities of Darwin, Sydney and Melbourne. Numerical simulations considering a natural ventilation strategy based on an approach combined with TRNSYS and COMIS were conducted for all the seasons. A rule-based decision-tree model was generated to modulate the window state for natural ventilation. Through a comparison of the numerical results with onsite experimental data, the accuracy of the proposed simulation approach was validated. The conclusions yielded from this study are summarised as follows:

The greatest natural ventilation potential was observed in Darwin (hot humid summer, warm winter zone) with the largest annual NVH and SNVH numbers 4728 and 4141, respectively; while, the least natural ventilation potential was found for the Melbourne case in the cool temperate zone. Natural ventilation was applicable during the whole year in Darwin, while summer and transition seasons (spring and autumn) were found to be the optimal periods in Sydney and Melbourne for utilising this. The potential for utilising natural ventilation to maintain indoor thermal comfort was altered by both local climate and building conditions. The indoor operative temperature was higher than the upper thermal comfort threshold during most non-natural ventilation periods in January in the case of Darwin, but out of the lower thermal comfort band for most of the time during July for Melbourne. This indicates that, except for non-natural ventilation periods in summer in Darwin and winter in Melbourne, indoor thermal comfort requirements can be met only by utilising natural ventilation in all the three cities over the whole year.

The potential for utilising natural ventilation for residential buildings in Australia is significantly altered by climate. As the current study considered only the three most populous climate zones with three representative cities, further research is needed to investigate the variation of natural ventilation potential across the whole of Australia. Despite existing limitations, the findings and simulation approach presented in this study can assist architects and policy makers in quantifying the potential for utilising natural ventilation under the three most representative climates; and can be used as a reference guideline for the natural ventilation design of residential buildings in Australia.

Acknowledgments: The financial support of the Fundamental Research Funds for the Central Universities (Project ID: 310828171006) is gratefully acknowledged. The contribution of Dr. Zhenjun Ma on simulation design is gratefully acknowledged.

Author Contributions: X.D. and Z.T. conceived and designed the experiments; X.D. performed the experiments; Z.T. and X.D. analyzed the data; Z.T. and X.D. wrote the paper.

Conflicts of Interest: The authors declare no conflict of interest.

Appendix A

Table A1. Pressure coefficient in the openings of the case study building.

Angle	Facade-E1	Facade-E2	Facade-E3	Facade-E4	Facade-N1	Facade-N2	Facade-N3	Facade-N4	Facade-S1	Facade-S2	Facade-S3	Facade-S4	Facade-S5	Facade-S6	Facade-S7	Facade-W1
0	−0.333	−0.437	−0.47	−0.797	0.817	0.864	0.835	0.827	−0.34	−0.339	−0.312	−0.243	−0.24	−0.258	−0.329	−0.353
45	−0.629	0.085	0.127	0.404	0.527	0.449	0.08	0.432	−0.434	−0.696	−1.371	−0.278	−0.476	−0.636	−0.642	−0.263
90	−0.15	0.942	0.956	0.946	−0.608	−0.403	−0.197	−0.136	−0.115	−0.137	−0.178	−0.22	−0.282	−0.251	−0.614	−0.102
135	0.792	0.327	0.328	0.151	−0.45	−0.388	−0.329	−0.333	−0.046	0.161	0.538	0.183	0.399	−0.009	0.739	−0.267
180	0.196	−0.664	−0.718	−0.402	−0.314	−0.209	−0.301	−0.307	0.289	0.345	0.17	0.769	0.933	0.555	0.661	−0.435
225	−0.429	−0.402	−0.395	−0.404	−0.338	−0.349	−0.354	−0.306	0.224	0.365	0.053	0.85	0.858	0.958	−0.278	0.161
270	−0.181	−0.092	−0.098	−0.08	−0.122	−0.176	−0.493	−0.674	−0.577	−0.523	−0.487	−0.51	−0.482	0.18	−0.190	0.357
315	−0.216	−0.291	−0.303	−0.274	0.25	0.287	0.797	0.712	−0.47	−0.47	−0.453	−0.38	−0.395	−0.432	−0.195	0.151

Table A2. Occupant activity and equipment usage.

	Heat Gain	Schedule
Occupant 1	60 w	23:00–08:00 in the master room; 08:00–17:00 in the study room; and 17:00–23:00 in the living room
Occupant 2	60 w	23:00–08:00 in the master room and 17:00–23:00 in the living room
Light	4 w/m^2	18:00–23:00 in all rooms
Computer	90 w	09:00–17:00 in the study room
TV	120 w	19:00–23:00 in the living room
Fridge	60 w	24 h in the living room
Oven	1500 w	08:00–08:30, 12:30–13:00 and 18:00–18:30 in the living room
Washing machine	300 w	09:00–10:00 in the laundry

Table A3. Induction rules setting.

Minimal Gain	Minimal Leaf Size	Minimal Size for Splitting	Initial Maximum Tree Depth
10%	4	2	3

References

1. Chen, Y.; Tong, Z.; Malkawi, A. Investigating natural ventilation potentials across the globe: Regional and climatic variations. *Build. Environ.* **2017**, *122*, 386–396. [CrossRef]
2. Haase, M.; Amato, A. An investigation of the potential for natural ventilation and building orientation to achieve thermal comfort in warm and humid climates. *Sol. Energy* **2009**, *83*, 389–399.
3. Chenari, B.; Carrilho, J.D.; Silva, M.G.D. Towards sustainable, energy-efficient and healthy ventilation strategies in buildings: A review. *Renew. Sustain. Energy Rev.* **2016**, *59*, 1426–1447.
4. Jomehzadeh, F.; Nejat, P.; Calautit, J.K.; Yusof, M.B.M.; Zaki, S.A. A review on windcatcher for passive cooling and natural ventilation in buildings, Part 1: Indoor air quality and thermal comfort assessment. *Renew. Sustain. Energy Rev.* **2017**, *70*, 736–756. [CrossRef]
5. Stabile, L.; Dell'Isola, M.; Russi, A.; Massimo, A.; Buonanno, G. The effect of natural ventilation strategy on indoor air quality in schools. *Sci. Total Environ.* **2017**, *595*, 894. [CrossRef] [PubMed]
6. Dimitroulopoulou, C. Ventilation in European dwellings: A review. *Build. Environ.* **2012**, *47*, 109–125. [CrossRef]
7. Ghiaus, C.; Allard, F.; Wilson, M. *Natural Ventilation in the Urban Environment Assessment and Design*; Earthscan: London, UK, 2005.
8. Luo, Z.; Zhao, J.; Gao, J.; He, L. Estimating natural-ventilation potential considering both thermal comfort and IAQ issues. *Build. Environ.* **2007**, *42*, 2289–2298. [CrossRef]
9. Yang, L.; Zhang, G.; Li, Y.; Chen, Y. Investigating potential of natural driving forces for ventilation in four major cities in China. *Build. Environ.* **2005**, *40*, 738–746. [CrossRef]
10. Teitel, M.; Tanny, J. Natural ventilation of greenhouses: Experiments and model. *Agric. For. Meteorol.* **1999**, *96*, 59–70. [CrossRef]
11. Fordham, M. Natural ventilation. *Renew. Energy* **2000**, *19*, 17–37.
12. Kaye, N.B.; Hunt, G.R. Heat source modelling and natural ventilation efficiency. *Build. Environ.* **2007**, *42*, 1624–1631. [CrossRef]
13. Chen, Z.D.; Li, Y.; Mahoney, J. Natural ventilation in an enclosure induced by a heat source distributed uniformly over a vertical wall. *Build. Environ.* **2001**, *36*, 493–501. [CrossRef]
14. El-Agouz, S.A. The effect of internal heat source and opening locations on environmental natural ventilation. *Energy Build.* **2008**, *40*, 409–418. [CrossRef]
15. Bolster, D.; Maillard, A.; Linden, P. The response of natural displacement ventilation to time-varying heat sources. *Energy Build.* **2008**, *40*, 2099–2110. [CrossRef]
16. Coomaraswamy, I.A. *Natural Ventilation of Buildings: Time-Dependent Phenomena*; University of Cambridge: Cambridge, UK, 2011.
17. Ghiaus, C.; Allard, F.; Santamouris, M.; Georgakis, C.; Nicol, F. Urban environment influence on natural ventilation potential. *Build. Environ.* **2006**, *41*, 395–406. [CrossRef]
18. Han, J.; Yang, W.; Zhou, J.; Zhang, G.; Zhang, Q.; Moschandreas, D.J. A comparative analysis of urban and rural residential thermal comfort under natural ventilation environment. *Energy Build.* **2009**, *41*, 139–145. [CrossRef]
19. Hang, J.; Luo, Z.; Sanberg, M.; Gong, J. Natural ventilation assessment in typical open and semi-open urban environments under various wind directions. *Build. Environ.* **2013**, *70*, 318–333. [CrossRef]
20. Kolokotroni, M.; Giannitsaris, I.; Watkins, R. The effect of the London urban heat island on building summer cooling demand and night ventilation strategies. *Sol. Energy* **2006**, *80*, 383–392. [CrossRef]
21. Wang, L.; Greenberg, S. Window operation and impacts on building energy consumption. *Energy Build.* **2015**, *92*, 313–321. [CrossRef]
22. Su, X.; Zhang, X.; Gao, J. Evaluation method of natural ventilation system based on thermal comfort in China. *Energy Build.* **2009**, *41*, 67–70. [CrossRef]
23. Calautit, J.K.; Hughes, B.R.; Ghani, S.A. A numerical investigation into the feasibility of integrating green building technologies into row houses in the Middle East. *Archit. Sci. Rev.* **2013**, *56*, 279–296. [CrossRef]
24. Artmann, N.; Manz, H.; Heiselberg, P. Climatic potential for passive cooling of buildings by night-time ventilation in Europe. *Appl. Energy* **2007**, *84*, 187–201. [CrossRef]
25. Australian Bureau of Statistics. *Year Book Australia, 2012*; ABS Canberra: Canberra, Australia, 2012.

26. *BD-058, AS/NZS 4859.1:2009 Materials for the Thermal Insulation of Buildings—General Criteria and Technical Provisions*; SAI Global: Sydney, Australia, 2009.

27. Montazeri, H.; Blocken, B. CFD simulation of wind-induced pressure coefficients on buildings with and without balconies: Validation and sensitivity analysis. *Build. Environ.* **2013**, *60*, 137–149. [CrossRef]

28. ASHRAE. *Applications (SI)*; ASHRAE Inc.: Atlanta, GA, USA, 2011.

29. Mei, L.; Infield, D.; Eicker, U.; Fux, V. Thermal modelling of a building with an integrated ventilated PV façade. *Energy and Build.* **2003**, *35*, 605–617. [CrossRef]

30. Lu, S.; Zhao, Y.; Fang, K.; Li, Y.; Sun, P. Establishment and experimental verification of TRNSYS model for PCM floor coupled with solar water heating system. *Energy Build.* **2017**, *140*, 245–260. [CrossRef]

31. Aparicio-Fernández, C.; Vivancos, J.L.; Ferrer-Gisbert, P.; Royo-Pastor, R. Energy performance of a ventilated façade by simulation with experimental validation. *Appl. Therm. Eng.* **2014**, *66*, 563–570. [CrossRef]

32. Han, J.; Kamber, M.; Pei, J. *Data Mining: Concepts and Techniques: Concepts and Techniques*; Elsevier: Amsterdam, The Netherlands, 2011.

33. Deuble, M.P.; de Dear, R.J. Mixed-mode buildings: A double standard in occupants' comfort expectations. *Build. Environ.* **2012**, *54*, 53–60. [CrossRef]

34. Tong, Z.; Chen, Y.; Malkawi, A. Estimating natural ventilation potential for high-rise buildings considering boundary layer meteorology. *Appl. Energy* **2017**, *193*, 276–286. [CrossRef]

35. ASHRAE. *Thermal Environmental Conditions for Human Occupancy*; Standard 55-2013; ASHRAE: Atlanta, GA, USA, 2013.

36. Pang-Ning, T.; Steinbach, M.; Kumar, V. *Introduction to Data Mining*; Library of Congress: Washington, DC, USA, 2006.

37. Yu, Z.; Haghighat, F.; Fung, B.C.M.; Yoshino, H. A decision tree method for building energy demand modeling. *Energy Build.* **2010**, *42*, 1637–1646. [CrossRef]

38. Rokach, L.; Maimon, O. *Data Mining with Decision Trees: Theory and Applications*; World Scientific: Singapore, 2014.

39. Shri, A.; Shri, A.; Sandhu, P.S.; Gupta, V.; Anand, S. Prediction of reusability of object oriented software systems using clustering approach. *World Acad. Sci. Eng. Technol.* **2010**, *43*, 853–856.

40. Sturman, A.P.; Tapper, N.J. *Weather and Climate of Australia and New Zealand*; Oxford University Press: Oxford, UK, 2006.

![atmosphere logo] *atmosphere*

MDPI

Article

Inter-Building Effect and Its Relation with Highly Reflective Envelopes on Building Energy Use: Case Study for Cities of Japan

Jihui Yuan *, Craig Farnham and Kazuo Emura

Department of Housing and Environmental Design, Graduate School of Human Life Science,
Osaka City University, 3-3-138, Sugimoto, Sumiyoshi-ku, Osaka 558-8585, Japan;
farnham@life.osaka-cu.ac.jp (C.F.); emura@life.osaka-cu.ac.jp (K.E.)
* Correspondence: yuanjihui@hotmail.co.jp; Tel.: +81-6-6605-2820

Received: 19 September 2017; Accepted: 27 October 2017; Published: 28 October 2017

Abstract: The built environment with respect to building envelope designs and the surrounding micro-environment significantly affects building energy use. The influence of the inter-building effect (IBE) on building energy use cannot be ignored and thermal properties of building envelopes also largely affect building energy use. In order to evaluate the influence of IBE and its relation with highly-reflective (HR) building envelopes on building energy use, the building energy use under three simulated scenarios was quantitatively analyzed using the building energy optimization software "BEopt" for five cities of Japan. Analysis indicated that when the simulated building is neighbored by other buildings, an envelope coated with HR material is more effective than lowly-reflective (LR) material to reduce building energy use. A simulated single building without surrounding buildings and a LR envelope has the highest building energy use among the three simulated scenarios. This study also showed the influence of IBE on building energy savings is stronger in cities with lower latitudes.

Keywords: Japan cities; building energy use; inter-building effect; highly-reflective building envelope; BEopt analysis

1. Introduction

The rapidly growing world energy use has already raised concerns over supply difficulties, exhaustion of energy resources, and heavy environmental impacts, such as ozone layer depletion, global warming, climate change, etc. Previous research has indicated that the global contribution from buildings towards energy consumption, both residential and commercial, has steadily increased to between 20% and 40% in developed countries, and has exceeded the other major sectors: industrial and transportation [1]. The nation's 114 million households and more than 4.7 million commercial buildings consume more energy than the transportation or industry sectors, accounting for nearly 40% of total U.S. energy use [2]. Cities represent the highest concentration of energy use; they occupy 2% of the Earth's surface. However, their inhabitants consume about 75% of the world's resources [3]. A report showed that buildings account for the largest proportion of energy consumption, with as much as 32% of total final energy consumption and nearly 40% of primary energy consumption [4]. The numbers are especially prominent in developed countries. According to the latest report from Ministry of Economy, Trade and Industry (METI), the carbon dioxide emissions from energy consumption in the consumer sector accounts for approximately two-third of the carbon dioxide emissions of Japan [5].

Many measures related to buildings have been implemented for energy conservation and sustainability. Existing UHI mitigation and energy conservation strategies are outlined by Akbari et al. [6]; such as development of highly reflective (HR) materials, development of cool and green roof technologies, development of cool pavement technologies and urban trees that can decrease ambient and surface temperatures in cities. Furthermore, UHI mitigation leads to energy savings, improves urban air quality and ambient conditions, and helps to counter global warming (GW).

1.1. State of Research on HR and Retro-Reflective Envelopes

Among these technologies for UHI mitigation, HR envelopes are being researched globally [7,8]. A review showed that the HR roof can reduce the roof daily heat gain between 11% and 60%, reduce the indoor air temperatures around 1–7 °C and decrease daily cooling energy consumption between 1% and 80% [9]. A research of quantitative estimation on the impact of white reflective and colored reflective materials indicated that the reflective white roofs without insulation lead to net savings over a 10-year life-cycle cost analysis between 3.0 $/m^2 and 67.4 $/m^2 compared to gray roofs in cities in Mexico. When thermally insulated, reflective white roofs lead to positive net savings, just in warmer locations; such savings ranged between 0.1 $/m^2 and 13.6 $/m^2 [10]. In addition, a field measurement on the change in solar reflectivity of HR roofing sheet was implemented [11]. It showed that the solar reflectivity of the HR roofing sheet installed on the rooftop decreased by about 0.04 after approximately one-year exposure due to the dirt accumulated on the roof surface, and its solar reflectivity was restored after cleaning the roof surface with distilled water. In recent years, possible application of retro-reflective (RR) materials to building facades has been studied by many scholars [12–14]. It revealed that the RR envelope is more effective to mitigate UHI and increase urban albedo compared to the diffuse reflection of HR envelopes.

1.2. Urbanization and Inter-Building Effect

As the same time, substantial global trend has emerged to impede us from a more sustainable built environment. A report by the United Nations indicated that a significant shift in population from rural areas to urban areas will continue through 2050 [15]. The population in urban areas is predicted to increase by approximately 40% and the population in rural areas will decrease by about the same percentage. If the population in urban areas is increasing, it will be reasonable to expect the morphology of urban areas to involve tighter spatial interrelationships among buildings. When buildings evolved to be in closer proximity due to the increase of the urban population, the current modeling approaches that treat buildings as stand-alone entities may not accurately represent building energy performance because they often do not consider the close proximity of other buildings in an urban environment and the energy implications that this phenomenon could cause [16]. With consideration of the inter-building effect (IBE) on building energy use, a numerical simulation tool for predicting the influence of the outdoor thermal environment on building energy performance was proposed by He et al. [17]. This indicated that when assessing energy use of buildings, not only thermophysical properties of the external wall surface are considered, but also the impact of surrounding environment is taken into account. Other research on the IBE was implemented via analyzing the mutual shading among buildings by Shaviv and Yezioro [18]. It showed how to calculate the impact of shading on solar gains of building envelopes.

1.3. Simulation

Simulation tools offer powerful functionalities to predict and improve building energy consumption for both research and design purposes. EnergyPlus is an energy analysis and thermal load simulation engine distributed by the U.S. Department of Energy (DOE) and it has become a popular building energy performance simulation owing to its sophisticated and validated functions [19]. In order to understand such complex mutual impact of the IBE within spatially proximal buildings, the influence of mutual shading and mutual reflection within a network of buildings was disaggregated and quantified using EnergyPlus [20]. This showed that the shading effect caused by surrounding buildings plays a more significant role in terms of impact on energy consumption. In addition, a methodology for evaluating a building energy performance was proposed by enlarging the assessment perspective from a single building to a network of buildings [16]. It showed that buildings can mutually impact the energy dynamics of other buildings and this effect varies by climatological context and by season. The IBE analysis and the specific proposed methodology revealed energy requirement modeling inaccuracies of up to 42% in summer in Miami and up to 22% in winter in Minneapolis.

1.4. Aims of This Research

For the aim of evaluating the influence of the IBE and its relation with HR building envelopes on the building energy use, the analysis for five representative cities (Sapporo, Tokyo, Nagoya, Osaka, and Kagoshima) of Japan was implemented using "Building Energy Optimization (BEopt)" in this study, considering the annualized energy use, annualized energy related costs, and annualized utility bills of a simulated building over a setting building envelope lifetime of 15-year.

2. Experiments

2.1. Simulated Building and Geographical Locations

According to the building standard law of Japan, a three-story building with a gable roof (roof pitch of 6:12) was simulated as shown in Figure 1 in this study. Its construction area is approximately 92 m^2, and its total floor area is approximately 273 m^2. The orientation of the simulated building is south–north facing. In order to evaluate the influence of IBE and its relation with the HR building envelope on the energy use of buildings, the simulated building was defined for three different scenarios. Structures of the three simulated scenarios are detailed in Table 1. Scenario A is a simulated single building which is not surrounded by buildings and its envelope coated with a lowly-reflective (LR) brick material (solar reflectivity: 0.12) [21]; Scenario B is a simulated building which is surrounded by buildings on four sides and its envelope coated with a LR brick material the same as Scenario A; Scenario C is a simulated building which is surrounded by buildings and its envelope coated with a brick material painted with HR paint (solar reflectivity: 0.7) [22]. In addition, a white color HR roofing sheet (solar reflectivity: 0.7) [23] is installed on the roofs for all three simulated scenarios. It is worth mentioning that all costs of materials applied to external walls and rooftops are more uniformly moderately priced than other building materials in the Japanese market. The left and right distance between two buildings must be more than 0.5 m according to the building standard law of Japan, thus the distance between two buildings is set to 2.5 m for Scenarios B and C in this study.

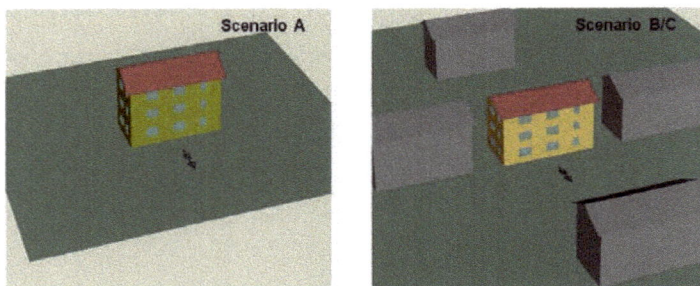

Figure 1. The simulated building (Scenario A: a single building without surrounding buildings and its envelope coated with LR materials; Scenario B: a building with surrounding buildings and its envelope coated with LR materials; and Scenario C: a building with surrounding buildings and its envelope covered with HR materials).

For cooling and heating equipment used in the simulated building, central air-conditioning (AC) was used for cooling in summer, and a gas furnace with an annual fuel utilization efficiency of 78% was used for heating in winter [24]. In addition, gas water heaters with an energy factor of 0.59 [25], and light-emitting diode (LED) lighting fixtures with an efficacy of 55 lm/W were used [26]. Detailed conditions of equipment usage in the simulated building are shown in Table 2. This study focuses on the effect of the building coating structure and surrounding buildings on building energy use, thus, the effects of indoor human and equipment usage factors are not investigated. The same indoor human and equipment usage conditions are set for the three scenarios in this study.

The average consumer electricity price in Japan is 0.24 \$/kWh for the year 2016 [27]. The average consumer gas price in Japan is 0.89 \$/m^3 for the year 2016 [28]. The homeowner costs calculated in this study assume a 15-year mortgage at a 4.0% interest rate with a 2.4% general inflation rate and a 3.0% real discount rate, which is chosen as one of the common forms of housing loan in Japan.

Five representative cities of Japan were selected from high-latitude to low-latitude to analyze energy use of the simulated building in this study. The five representative cities are detailed in Table 3.

Table 1. Orientations and structures of three simulated scenarios.

	Scenario A: Base (Single Building with LR Envelope)	Scenario B: Building Group (IBE with LR Envelope)	Scenario C: Improved Building (IBE with HR Envelope)
Orientation	South-north facing	Same as Scenario A	Same as Scenario A
Neighbors	None	Left/Right at 2.5 m Front/Back at 8.0 m	Same as Scenario B
Exterior wall	■ Standard steel framed walls with cavity insulation + Brick ■ Solar reflectivity of envelope: 0.12 ■ R-value of exterior wall structure: 2.9 m^{2}K/W ■ Cost per unit external wall: 9.2 \$/m^2	Same as Scenario A	■ Standard steel framed walls with cavity insulation + Brick painted with HR paint ■ Solar reflectivity of envelope: 0.70 ■ R-value of exterior wall structure: 2.9 m^{2}K/W ■ Cost per unit external wall: 18.9 \$/m^2
Interior wall	Mortar + Ordinary concrete	Same as Scenario A	Same as Scenario A
Roof	■ Gypsum board + Hollow layer + Ordinary concrete + Fiberglass batt + Mortar + Asphalt + White color HR roofing sheet ■ Solar reflectivity: 0.70 ■ R-value of roof: 3.0 m^2 K/W; ■ Cost per unit: 12.5 \$/m^2	Same as Scenario A	Same as Scenario A
Window	Window area: (Front/Back: 15 m^2; Left/Right: 8.5 m^2); Total window-wall ratio: 0.15; Low-E double with U-value of 2.1 W/m^2 K	Same as Scenario A	Same as Scenario A

Table 2. Detailed conditions of equipment usage in simulated building.

Cooling Equipment	Central Air-Conditioning
Heating equipment	Furnace (gas, 78% annual fuel utilization efficiency)
Space conditioning schedules	Cooling set point: 24.4 °C; Heating point: 21.7 °C; Humidity set point: 45%; 24-h conditioning
Water heater	Gas (energy factor: 0.59)
Lighting	Light-emitting diodes (LED efficacy: 55 lm/W)

Table 3. Details of five representative cities in Japan.

City	Location (Latitude, Longitude)	Climate	Average High Temperature (°C) (August)	Average Low Temperature (°C) (January)
Sapporo	43.1° N, 141.4° E	Humid continental climate with a wide range of temperature between the summer and winter	26.4	−7.0
Tokyo	35.7° N, 139.9° E	Humid subtropical climate zone with hot humid summers and generally mild winters with cool spells	30.8	0.9
Nagoya	35.2° N, 136.9° E	Humid subtropical climate with hot summers and cool winters	32.8	0.8
Osaka	34.7° N, 135.5° E	Humid subtropical climate zone with four distinct seasons	33.4	2.8
Kagoshima	31.6° N, 130.5° E	Humid subtropical climate with hot, wet summers and cool, relatively dry winters	32.5	4.6

2.2. BEopt

An analysis tool, BEopt, was used to analyze the energy consumption of the simulated building with three different scenarios. It provides detailed simulation-based analysis based on specific house characteristics, such as size, architecture, occupancy, vintage, location, and utility rates. Discrete envelope and equipment options, reflecting realistic construction materials and practices, are evaluated. It can be used to analyze both new construction and existing home retrofits, as well as single-family detached and multi-family buildings, through evaluation of single building designs, parametric sweeps, and cost-based optimizations [29]. The BEopt analysis method was developed to determine the least-cost path to zero net energy (ZNE) homes, based on evaluating the marginal costs of different combinations of energy efficiency and renewable-energy options [30]. It is known that the sequential search technique has several advantages. First, it finds intermediate optimal points all along the path, i.e., minimum-cost building designs at different target energy saving levels, not just the global optimum or the ZNE optimum. Second, discrete rather than continuous, building options are to be evaluated to reflect realistic construction options. Third, some near-optimal alternative designs are identified (that can serve as a starting point for generating a more complete set by permutations).

This method has recently been applied to determine the most cost-effective approaches to achieve the near-term and long-term performance targets for the DOE Building America Program [31]. The BEopt calls the DOE2 and TRNSYS simulation programs and automates the optimization process (see Figure 2). The optimization method involves sequentially searching for the most cost-effective option across a range of categories (wall type, ceiling type, window glass type, HVAC type, etc.) to identify optimal building designs along the path to ZNE.

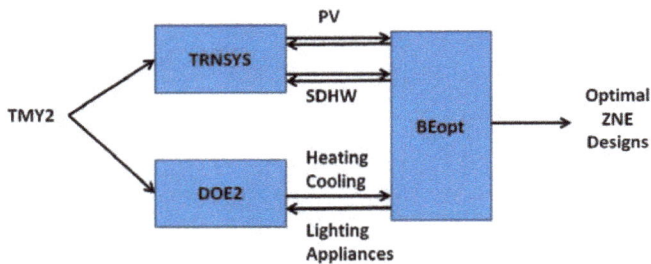

Figure 2. Optimization with multiple simulation programs.

2.3. Input and Output Definition

According to the definition of BEopt 2.6.0.0 [32], the input items in this study include: building (geographical location, climate, geometry, orientation, neighbors, floor area, etc.), wall structure (insulation, etc.), roof structure, thermal mass, windows, space conditioning, water heating, lighting, appliances & fixtures, local mortgage, utility rates, etc. (see Figure 3).

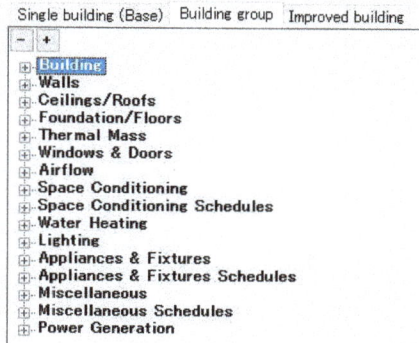

Single building (Base) Building group Improved building

- +

⊞ Building
⊞ Walls
⊞ Ceilings/Roofs
⊞ Foundation/Floors
⊞ Thermal Mass
⊞ Windows & Doors
⊞ Airflow
⊞ Space Conditioning
⊞ Space Conditioning Schedules
⊞ Water Heating
⊞ Lighting
⊞ Appliances & Fixtures
⊞ Appliances & Fixtures Schedules
⊞ Miscellaneous
⊞ Miscellaneous Schedules
⊞ Power Generation

Figure 3. Input items interface of BEopt software.

The output items used in this study are defined as following:

- Annualized source energy: is the total energy consumed to meet the energy needs of the building, including that used in transmission, delivery and production, in MMBtu/yr or kWh/yr.
- Annualized site energy: is electricity or fuel consumed within a property boundary (e.g. a home), which is reflected in the utility bills at the site, in MMBtu/yr or kWh/yr.
- Annualized utility bills: is divided into two parts, the gas charge and the electricity charge, in $/yr.
- Annualized energy related cost: is calculated by annualizing the energy related cash flows over the analysis period, in $/yr.

The annualized cost is then subtracted from the reference for every cash flow but utility bills (i.e., the values displayed in the cost/energy graph are full annualized utility bills plus incremental annualized values for every other cash flow). The annualized energy related cost is calculated with Equation (1),

$$C_e = U + C_m + C_r + C_{etc} \tag{1}$$

where C_e is the annualized energy related cost, $/yr; U is the annualized utility bills, $/yr; C_m is the incremental mortgage cost, $/yr; C_r is the incremental replacement costs, $/yr; and C_{etc} is the other incremental costs, $/yr.

Thus, in general, the C_e will be greater than U. However, this will not be true if any of the incremental cash flows are negative. If the incremental costs (C_m, C_r and C_{etc}) are ignored, the C_e will be equal to U.

3. Results

3.1. Annualized Energy Use and Energy Savings

The annualized source energy use and annualized site energy use of the simulated building with three different scenarios in five representative cities of Japan are calculated and detailed in Table 4. As an example, the annualized source energy use and annualized site energy use of the simulated building with three different scenarios for Tokyo are shown in Figures 4 and 5, respectively.

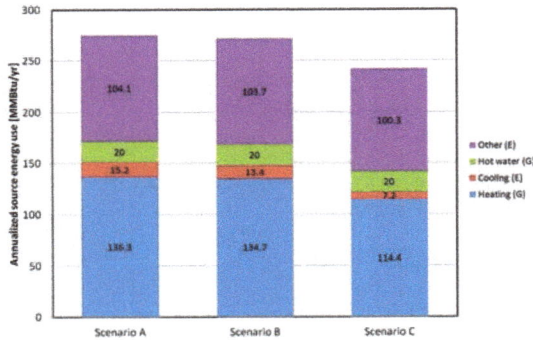

Figure 4. Annualized source energy (E: electricity and G: gas) use of simulated building with three different scenarios in Tokyo (1 MMBtu = 293 kWh).

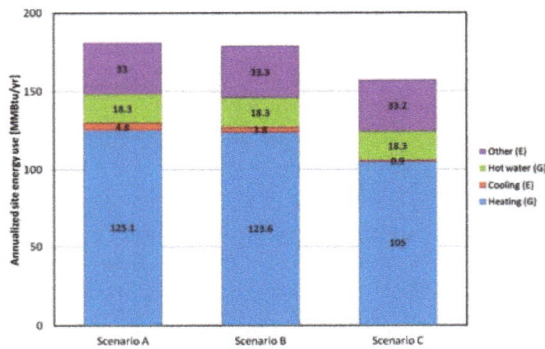

Figure 5. Annualized site energy (E: electricity and G: gas) use of simulated building with three different scenarios in Tokyo (1 MMBtu = 293 kWh).

Compared to the total annualized source energy use of Scenario A (detailed in Table 4) in five representative cities of Japan, that of Scenario B in Kagoshima decreased the largest, about 2.5% (cooling: 9.4%; heating: 1.3%), followed by Nagoya, Osaka, Tokyo, and Sapporo. Compared to the total annualized site energy use of Scenario A (in Table 4), that of Scenario B in Nagoya decreased the most, by about 2.1% (cooling: 10.2%; heating: 2.3%), followed by Kagoshima, Osaka, Tokyo, and Sapporo.

Compared to the total annualized source energy use of Scenario A (detailed in Table 4) in five representative cities of Japan, that of Scenario C in Kagoshima decreased the most, by about 14.7% (cooling: 42.9%; heating: 17.7%), followed by Nagoya, Osaka, Tokyo, and Sapporo. Compared to the total annualized site energy use of Scenario A (in Table 4), that of Scenario C in Nagoya decreased the most, by about 14.5% (cooling: 43.2%; heating: 17.7%), followed by Kagoshima, Osaka, Tokyo and, Sapporo.

Table 4. Annualized source energy use and site energy use (for total, heating, and cooling) of a simulated building with three different scenarios in five representative cities of Japan.

			Scenario A	Scenario B	Scenario C	Δ(B−A)%	Δ(C−A)%
Sapporo	Source energy use [MMBtu/yr]	Total	362.0	360.3	317.8	0.5%	12.2%
		Heating	224.9	224.3	189.0	0.3%	16.0%
		Cooling	4.0	3.0	0.6	25.0%	85.0%
	Site energy use [MMBtu/yr]	Total	263.2	262.4	227.7	0.3%	13.5%
		Heating	206.3	205.8	173.4	0.2%	15.9%
		Cooling	0.6	0.1	0.05	83.3%	91.7%
Tokyo	Source energy use [MMBtu/yr]	Total	275.6	271.8	241.9	1.4%	12.2%
		Heating	136.3	134.7	114.4	1.2%	16.1%
		Cooling	15.2	13.4	7.2	11.8%	52.6%
	Site energy use [MMBtu/yr]	Total	181.2	179.0	157.4	1.2%	13.1%
		Heating	125.1	123.6	105.0	1.2%	16.1%
		Cooling	4.8	3.8	0.9	20.8%	81.3%
Nagoya	Source energy use [MMBtu/yr]	Total	259.6	254.0	223.9	2.2%	13.8%
		Heating	108.5	106.1	89.4	2.2%	17.6%
		Cooling	27.8	24.7	15.7	11.2%	43.5%
	Site energy use [MMBtu/yr]	Total	158.7	155.4	135.8	2.1%	14.5%
		Heating	99.6	97.3	82.0	2.3%	17.7%
		Cooling	8.8	7.9	5.0	10.2%	43.2%
Osaka	Source energy use [MMBtu/yr]	Total	265.2	260.1	229.4	1.9%	13.5%
		Heating	102.7	101.1	84.9	1.6%	17.3%
		Cooling	31.5	28.4	18.1	9.8%	42.5%
	Site energy use [MMBtu/yr]	Total	158.2	155.6	136.2	1.6%	13.9%
		Heating	94.2	92.8	77.9	1.5%	17.3%
		Cooling	10.0	9.0	5.8	10.0%	42.0%
Kagoshima	Source energy use [MMBtu/yr]	Total	221.4	215.9	188.8	2.5%	14.7%
		Heating	60.9	60.1	50.1	1.3%	17.7%
		Cooling	39.4	35.7	22.5	9.4%	42.9%
	Site energy use [MMBtu/yr]	Total	117.0	114.7	100.2	2.0%	14.4%
		Heating	55.9	55.1	46	1.4%	17.7%
		Cooling	12.5	11.3	7.2	9.6%	42.4%

(1 MMBtu = 293 kWh).

3.2. Annualized Energy Related Costs and Annualized Utility Bills

According to Section 2.3, the annualized energy related costs can be calculated using Equation (1). In this study, the incremental costs (C_m, C_r and C_{etc}) were ignored; thus, the annualized energy-related cost is equal to the annualized utility bill. The annualized energy-related costs, or annualized utility bills (electricity charge and gas charge), of the simulated building in three different scenarios in five representative cities of Japan is shown in Table 5.

Table 5. Annualized energy-related costs or annualized utility bills (electricity charge and gas charge) of a simulated building with three different scenarios in five representative cities of Japan.

	Annualized Energy Related Costs or Annualized Utility Bills [$/yr]	Scenario A	Scenario B	Scenario C	Δ(B−A)%	Δ(C−A)%
Sapporo	Total	3561	3545	3162	0.4%	11.2%
	Gas	2127	2122	1818	0.2%	14.5%
	Electricity	1242	1231	1163	0.9%	6.4%
Tokyo	Total	2827	2788	2512	1.4%	11.1%
	Gas	1345	1331	1156	1.0%	14.1%
	Electricity	1291	1265	1163	2.0%	9.9%
Nagoya	Total	2720	2664	2376	2.1%	12.6%
	Gas	1093	1072	928	1.9%	15.1%
	Electricity	1435	1400	1255	2.4%	12.5%
Osaka	Total	2789	2737	2440	1.9%	12.5%
	Gas	1061	1047	908	1.3%	14.4%
	Electricity	1536	1497	1340	2.5%	12.8%
Kagoshima	Total	2415	2358	2086	2.4%	13.6%
	Gas	669	662	577	1.0%	13.8%
	Electricity	1554	1503	1317	3.3%	15.3%

The results (Table 5) show that the annualized energy related costs or annualized utility bills in Sapporo is the largest, approximately 3561 $/yr (gas charge: 2127 $/yr; electricity charge: 1242 $/yr) for Scenario A, approximately 3545 $/yr (gas charge: 2122 $/yr; electricity charge: 1231 $/yr) for Scenario B, and approximately 3162 $/yr (gas charge: 1818 $/yr; electricity charge: 1163 $/yr) for Scenario C, followed by Tokyo, Osaka, Nagoya and Kagoshima.

As an example of Tokyo, the annualized utility bills of the simulated building with three different scenarios are shown in Figure 6.

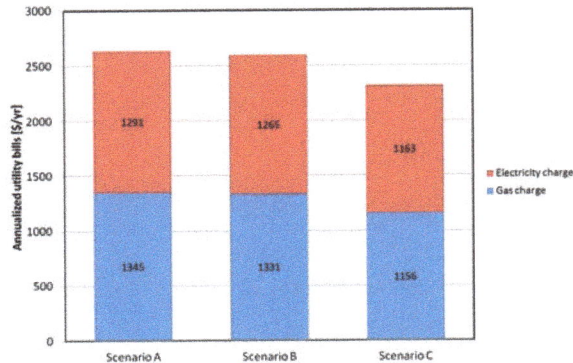

Figure 6. Annualized utility bills of the simulated building with three different scenarios in Tokyo.

Compared to the annualized utility bills (or annualized energy related costs) of Scenario A (detailed in Table 5) in five representative cities of Japan, that of Scenario B in Kagoshima decreased the largest, about 2.4% (electricity charge: 3.3%; gas charge: 1.0%), followed by Nagoya, Osaka, Tokyo, and Sapporo.

Compared to the annualized utility bills of Scenario A (in Table 5) in five representative cities of Japan, that of Scenario C in Kagoshima decreased the most, by about 13.6% (electricity charge: 15.3%; gas charge: 13.8%), followed by Nagoya, Osaka, Tokyo. and Sapporo.

4. Discussion

4.1. Influence of IBE with LR Building Envelopes on Energy Use

Through the comparison between Scenario A (a single building with LR building envelope) and Scenario B (IBE with LR building envelope), we can see that Scenario B with neighboring buildings can contribute to annualized energy savings and annualized utility bill savings, compared to a single building (Scenario A). It is considered that the IBE including mutual shading and mutual reflection within a network of buildings is effective to reduce energy use [20]. In addition, we can see that the lower the latitude of Japanese cities, the larger the annualized energy savings, annualized energy related cost savings, and annualized utility bill savings, due to the impact of IBE. However, this changing trend is not absolute: it also depends on regional climate conditions and other factors.

4.2. Influence of IBE with HR Building Envelopes on Energy Use

Through the comparison between Scenario B (IBE with the LR building envelope) and Scenario C (IBE with HR building envelope), we can see that the IBE with the HR building envelope (Scenario C) can better contribute to annualized energy savings and annualized utility bill savings, compared to IBE with LR building envelope (Scenario B). The reason for this is considered to be that the HR building envelope can reflect more sunlight to the surrounding environment than the LR building envelope, thus, the solar radiation absorbed by the HR building envelope will be less and the heat transfer from the exterior wall to the interior wall will also be less, resulting in a cooling load reduction during summer. As the reflected sunlight from the HR envelope and the mutual reflection with the surrounding buildings will heat the ambient air, the outdoor radiant temperature in Scenario C will be higher than that in Scenario B, thus, the heating load during winter also decreased in this study.

In addition, it also indicated that the lower the latitude of Japanese cities, the larger the annualized energy savings and the annualized utility bill savings, due to the building envelopes with HR performance. The reason is considered that the HR envelopes are more effective in terms of reducing annualized thermal loads and energy use in locations of Japan with lower latitudes and hotter summers, according to previous research [33].

4.3. Comparison of Annualized Energy Related Cost

Comparing the annualized energy related cost of the simulated building with three different scenarios in five representative cities of Japan, it showed that the annualized energy related cost of Scenario A is the largest, followed by Scenario B, and Scenario C has the lowest annualized energy related cost for each city of Japan. The annualized energy related cost of the simulated building decreased when varying the city from high-latitude to low-latitude for each scenario. The influence of IBE within a network of buildings on annualized energy related cost became a bit greater when varying the Japan cities from high-latitude to low-latitude. The IBE with HR building envelopes is the most effective to reduce the annualized energy-related costs among three simulated scenarios.

4.4. Model Verification

In order to verify the analyzed results obtained by BEopt, we compared the annualized site energy use between actual office buildings [34] and the simulated building in five representative cities. The result (Figure 7) showed that the average annualized site energy use of actual office buildings is about 90 kWh/m^2-yr higher than that of the simulated building.

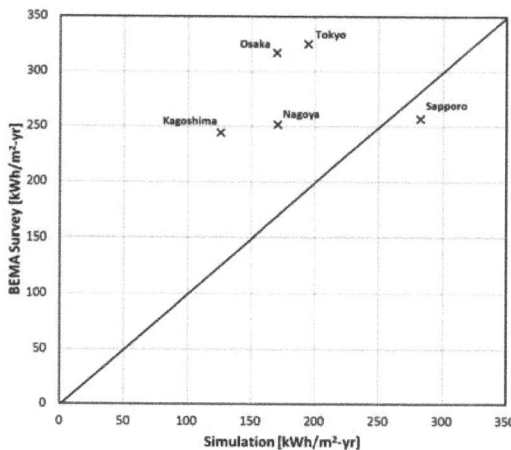

Figure 7. Comparison of annualized site energy use between BEMA survey and BEopt's simulation results for five representative cities of Japan.

The deviation between the actual survey and simulation is considered to be due to the following reasons: (i) the simulated building construction is not completely consistent with the actual building construction, the average floor space of the actual surveyed office buildings was 19340 m^2 over 444 cases, much larger than the simulated case of 273 m^2. The much smaller simulated three-floor building gains far more heating from solar radiation per square meter of floor space than the tall buildings with more floors in the survey; and (ii) the usage conditions of the actual buildings are likely different from the simulated building, i.e., the number of people in the building, use of equipment in the building, not using 24-h air conditioning as in the simulation, and other human factors, etc.

According to the above problems, a much deeper verification should be carried out by importing the measured energy use data of the actual buildings respectively in five representative cities of Japan, with consideration of the coincidence of actual building and simulated building, and the influence of human behavior factors, etc.

5. Conclusion and Future Work

In order to evaluate the influence of the IBE with HR building envelope on building energy use, the BEopt analysis method has been used to evaluate the annualized energy use, annualized energy related costs and annualized utility bills of simulated building with three different scenarios located in five representative cites of Japan in this study.

The comparison between Scenario A (base: a single building with LR building envelope) and Scenario B (IBE with LR building envelope) by BEopt analysis indicated that the building with surrounding buildings is more effective than that without surrounding buildings in terms of reducing annualized energy use, annualized utility bills and annualized energy related costs for five representative cities of Japan. Thus, it is necessary to consider the influence of IBE (Scenario B), including mutual shading and mutual reflection within a network of buildings when evaluating the energy consumption of buildings. In addition, it showed that the annualized energy savings, annualized energy related cost savings, and annualized utility bill savings increased when varying the Japanese cities from high-latitude to low-latitude, due to the impact of IBE. However, this changing trend is not absolute: it also depends on regional climate conditions and other factors.

The comparison between Scenario B (IBE with LR building envelope) and Scenario C (IBE with HR building envelope) demonstrated that the IBE with HR building envelope can better contribute to annualized energy savings and annualized utility bill savings in five representative cities of Japan.

The future work will focus on evaluating the influence of IBE with HR envelope design on the outdoor micro-environment within a network of buildings while, at the same time, monitoring the energy use of actual buildings. The increasing use of solar panels on rooftops could influence the results. Here, the rooftop parameters were kept uniform in all three scenarios, but net heat flux through the roof could change. Furthermore, the influence of wind fields around buildings, humidity, etc., on the building energy use will also evaluated.

Acknowledgments: The authors are sincerely grateful to the BEopt Principal Investigator: Craig Christensen (National Renewable Energy Laboratory) for developing the advanced building energy optimization software.

Author Contributions: All authors contributed equally in the preparation of this manuscript.

Conflicts of Interest: The authors declared that there is no conflict of interest with respect to the research, authorship, and/or publication of this article.

References

1. Pérez-Lombard, L.; Ortiz, J.; Pout, C. A review on buildings energy consumption information. *Energy Build.* **2008**, *40*, 394–398.
2. U.S. Department of Energy, Energy Efficiency and Renewable Energy. Energy Efficiency Trends in Residential and Commercial Buildings. 2008. Available online: https://www1.eere.energy.gov/buildings/publications/pdfs/corporate/bt_stateindustry.pdf (accessed on 28 October 2017).
3. Hui, S.C.M. Low energy building design in high density urban cities. *Renew. Energy* **2001**, *24*, 627–640. [CrossRef]
4. International Energy Agency. Energy Efficiency. Available online: http://www.iea.org/aboutus/faqs/energyefficiency (accessed on 14 March 2017).
5. Ministry of Economy, Trade and Industry. Carbon dioxide emissions information. Available online: http://www.meti.go.jp (accessed on 14 March 2017).

6. Akbari, H.; Cartalis, C.; Kolokotsa, D.; Muscio, A.; Pisello, A.L.; Rossi, F.; Santamouris, M.; Synnefa, A.; Wong, N.H.; Zinzi, M. Local climate change and urban heat island mitigation techniques—The state of the art. *J. Civ. Eng. Manag.* **2016**, *22*, 1–16. [CrossRef]

7. Gobakis, K.; Kolokotsa, D.; Maravelaki-Kalaitzaki, N.; Perdikatsis, V.; Santamouris, M. Development and analysis of advanced inorganic coatings for buildings and urban structures. *Energy Build.* **2016**, *89*, 196–205. [CrossRef]

8. Cozza, E.S.; Alloisio, M.; Comite, A.; di Tanna, G.; Vicini, S. NIR-reflecting properties of new paints for energy-efficient buildings. *Sol. Energy* **2015**, *116*, 108–116. [CrossRef]

9. Hernández-Pérez, I.; Álvarez, G.; Xamán, J.; Zavala-Guillén, I.; Arce, J.; Simá, E. Thermal performance of reflective materials applied to exterior building components—A review. *Energy Build.* **2014**, *80*, 81–105. [CrossRef]

10. Hernández-Pérez, I.; Xamán, J.; Macías-Melo, E.V.; Aguilar-Castro, K.M. Chapter 4: Reflective Materials for Cost-Effective Energy-Efficient Retrofitting of Roofs. In *Cost-Effective Energy Efficient Building Retrofitting*; Woodhead Publishing Limited: Cambridge, UK, 2017; pp. 119–139.

11. Yuan, J.; Emura, K.; Sakai, H. Evaluation of the Solar Reflectance of Highly Reflective Roofing Sheets Installed on Roofs. *J. Build. Phys.* **2012**, *37*, 170–184. [CrossRef]

12. Nishioka, M.; Inoue, S.; Sakai, K. Retroreflective properties calculating method based on geometrical-optics analysis. *J. Environ. Eng.* **2008**, *73*, 1249–1254. (In Japanese) [CrossRef]

13. Rossi, F.; Castellani, B.; Presciutti, A.; Morini, E.; Filipponi, M.; Nicolini, A.; Santamouris, M. Retroreflective façades for urban heat island mitigation: Experimental investigation and energy evaluations. *Appl. Energy* **2015**, *145*, 8–20. [CrossRef]

14. Yuan, J.; Farnham, C.; Emura, K. Development of a retro-reflective material as building coating and evaluation on albedo of urban canyons and building heat loads. *Energy Build.* **2015**, *103*, 107–117. [CrossRef]

15. United Nations. World Urbanization Prospects: The 2007 Revision Population Database, Panel 1: Urban and Rural Areas. Available online: http://esa.un.org/unup/ (accessed on 15 March 2017).

16. Pisello, A.L.; Taylor, J.E.; Xu, X.; Cotana, F. Inter-building effect: Simulating the impact of a network of buildings on the accuracy of building energy performance predictions. *Build. Environ.* **2012**, *58*, 37–45. [CrossRef]

17. He, J.; Hoyano, A.; Asawa, T. A numerical simulation tool for predicting the impact of outdoor thermal environment on building energy performance. *Appl. Energy* **2009**, *86*, 1596–1605. [CrossRef]

18. Shaviv, E.; Yezioro, A. Analyzing mutual shading among buildings. *Sol. Energy* **1997**, *59*, 83–88. [CrossRef]

19. U.S. Department of Energy. EnergyPlus Engineering Reference: The Reference to EnergyPlus Calculations. Available online: https://energyplus.net/ (accessed on 15 March 2017).

20. Han, Y.; Taylor, J.E. Disaggregate Analysis of the Inter-Building Effect in a Dense Urban Environment. *Energy Procedia* **2015**, *75*, 1348–1353. [CrossRef]

21. Brick Exterior Walls. Available online: http://ansin-tosou.com/gaiheki/renga.html (accessed on 15 March 2017).

22. Highly Reflective Exterior Wall Materials. Available online: http://www.radiant88.com/500/50040/ (accessed on 16 March 2017).

23. Highly Reflective Roofing Sheet. Available online: http://ansin-tosou.com/gaiheki/renga.html (accessed on 15 March 2017).

24. Gas Furnace. Available online: http://rinnai.jp/products/living/fan_heater/fh (accessed on 16 March 2017).

25. Gas Water Heater. Available online: http://rinnai.jp/products/waterheater/gas/ (accessed on 16 March 2017).

26. Light-Emitting Diode Lighting. Available online: http://www.mitsubishielectric.co.jp/ldg/ja/products/lighting/lineup/fixture/index.html (accessed on 16 March 2017).

27. Calculation of Electricity Rate in Japan. Available online: http://www.denki-keisan.info (accessed on 16 March 2017).

28. Calculation of Gas Rate in Japan. Available online: https://www.e5.osakagas.co.jp/custserv/ryokinhyo1001.html (accessed on 16 March 2017).

29. National Renewable Energy Laboratory, BEopt. Available online: https://beopt.nrel.gov (accessed on 6 October 2017).

30. Christensen, C.; Barker, G.; Horowitz, S. A Software Tool for Identifying Optimal Building Designs on the Path to Zero Net Energy. In Proceedings of the Solar 2004 Conference: Including Proceedings of 33rd ASES Annual Conference, Proceedings of 29th National Passive Solar Conference, Portland, OR, USA, 11–14 July 2004.

31. Anderson, R.; Christensen, C.; Horiwitz, S. Analysis of Residential Systems Targeting Least-Cost Solutions Leading to Net Zero Energy Homes. In Proceedings of the 2006 ASHRAE Annual Meeting, Quebec City, QC, Canada, 24–28 June 2006.

32. National Renewable Energy Laboratory. BEopt 2.6.0.0 Help. 2006. Available online: https://beopt.nrel.gov/sites/beopt.nrel.gov/files/help/prntdoc/BEopt.pdf (accessed on 3 April 2017).

33. Yuan, J.; Farnham, C.; Emura, K.; Alam, A. Proposal for optimum combination of reflectivity and insulation thickness of building exterior walls for annual thermal load in Japan. *Build. Environ.* **2016**, *103*, 228–237. [CrossRef]

34. *Report on Building Energy Consumption Survey*; 36th Report; The Building-Energy Manager's Association of Japan (BEMA): Tokyo, Japan, 2014.

atmosphere

MDPI

Article

Ventilation and Air Quality in City Blocks Using Large-Eddy Simulation—Urban Planning Perspective

Mona Kurppa [1,*], Antti Hellsten [2], Mikko Auvinen [1,2], Siegfried Raasch [3], Timo Vesala [1,4] and Leena Järvi [1]

[1] Institute for Atmospheric and Earth System Research/Physics, Faculty of Science, 00014 University of Helsinki, Finland; mikko.auvinen@helsinki.fi (M.A.); timo.vesala@helsinki.fi (T.V.); leena.jarvi@helsinki.fi (L.J.)

[2] Finnish Meteorological Institute, 00101 Helsinki, Finland; antti.hellsten@fmi.fi

[3] Institute of Meteorology and Climatology, Leibniz Universität Hannover, 30419 Hannover, Germany; raasch@muk.uni-hannover.de

[4] Institute for Atmospheric and Earth System Research/Forest Sciences, Faculty of Agriculture and Forestry, 00014 University of Helsinki, Finland

* Correspondence: mona.kurppa@helsinki.fi; Tel.: +358-400-991-944

Received: 2 January 2018; Accepted: 11 February 2018; Published: 13 February 2018

Abstract: Buildings and vegetation alter the wind and pollutant transport in urban environments. This comparative study investigates the role of orientation and shape of perimeter blocks on the dispersion and ventilation of traffic-related air pollutants, and the street-level concentrations along a planned city boulevard. A large-eddy simulation (LES) model PALM is employed over a highly detailed representation of the urban domain including street trees and forested areas. Air pollutants are represented by massless and passive particles (non-reactive gases), which are released with traffic-related emission rates. High-resolution simulations for four different city-block-structures are conducted over a 8.2 km² domain under two contrasting inflow conditions with neutral and stable atmospheric stratification corresponding the general and wintry meteorological conditions. Variation in building height together with multiple cross streets along the boulevard improves ventilation, resulting in 7–9% lower mean concentrations at pedestrian level. The impact of smaller scale variability in building shape was negligible. Street trees further complicate the flow and dispersion. Notwithstanding the surface roughness, atmospheric stability controls the concentration levels with higher values under stably stratified inflow. Little traffic emissions are transported to courtyards. The results provide urban planners direct information to reduce air pollution by proper structural layout of perimeter blocks.

Keywords: LES; ventilation; urban planning; dispersion; air quality

1. Introduction

Decreased air quality is one of the major environmental challenges urban areas are facing today. An increasing number of people are exposed to high air pollution levels due to the ongoing intense urbanisation [1]. Exposure to air pollution has several acute and chronic health effects including respiratory and cardiovascular diseases that further increase mortality (e.g., [2,3]). To illustrate, in 2016 around 4.1 million premature deaths worldwide [4], of which almost 10% in Europe [5], were linked to exposure to elevated concentrations of ambient air pollution.

In urban areas, road traffic accounts for a significant share of the local air pollutant emissions (e.g., [6]). If the prevailing wind conditions within the street canyons lead to inefficient transport and mixing of air, the traffic emissions can accumulate and have longer residence times near the ground level. Consequently, the highest pollution levels within the street canyons are commonly observed at pedestrian level [7]. In order to reduce human exposure to air pollutants within urban areas, the mechanisms affecting the dispersion conditions at street level must be understood and thereby examined in greater detail. To this end, the study of ventilation becomes of critical importance as it addresses the capacity with which a densely built urban structure is capable of replacing the contaminated air with ambient fresh air. Here, ventilation is recognised as a transport process that improves local air quality and closely relates to the term breathability [8,9]. The efficiency at which street canyon ventilation occurs depends on the complex interaction between the atmospheric boundary layer (ABL) flow and the local urban structures. Recognising that the structural layout of urban landscape plays a critical role in determining the local urban air quality, opens up the possibility for the urban planners to incorporate air quality considerations into structural urban design. However, this necessitates sufficiently detailed modelling solutions to address the problem of street canyon ventilation at the scale of individual structures, such as perimeter blocks which are ubiquitous in European cities.

The current capacity to examine street canyon ventilation relies mainly upon parametric models, wind tunnel simulations and field measurements [7]. Parametric models are computationally inexpensive but have decreased accuracy and their applicability is limited to very simplified model set-ups. Reduced-scale wind tunnel simulations, in turn, can suffer from similarity constraints, and field measurements from uncontrollable and unrepeatable boundary conditions and inability to account for spatial variability [10]. Recently, along with rapidly advancing computer power, the application of computational fluid dynamics (CFD) using either models based on the Reynolds-averaged Navier-Stokes equations (RANS) with parametrised turbulence or the large-eddy simulation (LES) method has increased. CFD models provide complete and full-scale three-dimensional flow and concentration fields, which is a major advantage in real and complex urban areas. Majority of the previous urban CFD studies have employed the RANS method, owing to its lower computational costs. Nonetheless, LES outperforms above a complex urban surface due to its ability to resolve instantaneous turbulence structures (e.g., [11–13]) also when taking street trees into account [14].

Most of the urban LES studies have considered an idealistic two-dimensional street canyon or a simplified urban topography without including the aerodynamic impact of street trees (e.g., [15–17]). This simplification can be dangerous as porous trees decelerate the flow and generate turbulence (e.g., [18]) as well as influence the canyon vortex [19] that is commonly observed in idealised street canyons [20]. Thereby, trees can modify the vertical transport of air pollutants within street canyons [21–24]. Thanks to the availability of detailed urban topography and land use datasets [25,26], the application of LES to real urban environments (e.g., [27–31]) and also to directly support urban planning [32] has become possible over the last decade. In these cases the computational domain has several requirements to meet. Firstly, the computational domain has to be large enough to capture all relevant turbulence scales (e.g., [33]) and to minimize uncertainties related to the domain boundary conditions [11]. Furthermore, it is assumed that the whole ABL should be included vertically. Secondly, the grid spacing has to be small enough to explicitly resolve the turbulence scales containing most of the energy [31], which requires large computational resources.

The few ventilation studies conducted over a real urban surface [30,32,34] have applied a simple, indirect ventilation indicator, velocity ratio $v_r = v_p/v_\infty$, i.e., the ratio between the wind velocities at pedestrian level v_p (height $z = 2$ m above ground level (a.g.l.)) and at the top of the model domain v_∞, where the flow is no longer influenced by the urban surface, rather than having examined the pollutant concentration, and processes of ventilation and dispersion together in more detail. This study employs an LES model, coupled with a Lagrangian stochastic particle model, to perform high-resolution urban flow simulations in order to answer: How does the structural layout of densely arranged building blocks along a city boulevard influence the ventilation of the local traffic emissions? The accumulation and ventilation of traffic-related air pollutants are investigated in four virtual city-block-design alternatives that are immersed within a real complex urban environment. The question is approached by quantifying and comparing the ventilation efficiency within perimeter blocks using three different measures of air pollutant ventilation and dispersion that are more sophisticated than v_r: concentration, vertical turbulent transport and dilution rate of Lagrangian particles that represent non-reactive gaseous air pollutants. The ventilation measures are further examined in parallel. Simulations are conducted applying two contrasting meteorological conditions for the inflow.

2. Methods

2.1. Model Description

The LES model employed is the Parallelized Large-Eddy Simulation Model (PALM) version 4.0 (revision 1904) for atmospheric and oceanic flows [35,36], which solves the three-dimensional fields of wind and scalar variables (e.g., potential temperature and scalar concentrations). PALM has been applied to various types of ABL studies, for example to study cloudy boundary layers (e.g., [37]), the aerodynamic impact of a plant canopy (e.g., [38]) and stable boundary layer [39]. The performance of PALM over an urban-like surface has been validated against wind tunnel simulations, previous LES studies and field measurements [28–30,40]. The aerodynamic impact of vegetation is taken into account in PALM by means of an embedded canopy model. Vegetation decelerates the flow due to the form and viscous drag forces, and the decelerating force depends on the wind velocity and plant area density (PAD, m^2 m^{-3}). Despite the model has previously been developed for continuous vegetation, it has also been used for individual trees (e.g., [19,41]). For this study, the canopy model was revised to allow a heterogeneous distribution of tree canopy.

Technical specifications of PALM are represented in Appendix A. As a new feature in PALM, a full three-dimensional two-way self-nesting is utilised for the first time [42]. In the nesting approach, a "child" computational domain with chosen grid spacing and dimensions is defined inside the "parent" computational domain. PALM is run in parallel in both domains with respective computational set-ups (Table 1) and the domains communicate with each other. Nesting enables to have both a large computational domain and high enough resolution in the main area of interest without making the simulation computationally too expensive.

Table 1. The boundary conditions of the model runs. Details of the conditions can be found in Maronga et al. [35].

Boundary	Domain	
	Parent	**Child**
Bottom and solid walls	No-slip condition for the horizontal wind components u and v (i.e., $u = v = w = 0$ m s^{-1}). For potential temperature θ, the vertical gradient $\partial\theta/\partial z = 0$ K m^{-1}. Monin-Obukhov similarity theory (MOST) is applied between any solid-wall boundary and the first grid level normal to the respective boundary surface.	Same as for the parent.
Top	Dirichlet condition, i.e., $u = U_g$ (geostrophic wind speed) and $v = w = 0$ m s^{-1}. θ is extrapolated using the initial gradient of θ from a precursor run.	Two-way nesting. Boundary conditions obtained from the parent domain.
Horizontal: Lateral	Cyclic boundary conditions	Two-way nesting. Boundary conditions obtained from the parent domain.
Horizontal: streamwise	Non-cyclic. A time-dependent turbulent inflow is produced by a turbulence recycling method [43]. Requires a precursor run that is carried out over a domain of the same vertical extent as the parent domain and 1/16 in area.	Two-way nesting. Boundary conditions obtained from the parent domain.

Ventilation of traffic-related air pollutants is studied by applying a Lagrangian stochastic particle model (LPM) [35]. LPM allows studying pollutant accumulation inside urban structures as concentration values can become higher than at the source unlike in the commonly used Eulerian method (e.g., [31]) due to the maximum principle for scalar conservation laws (e.g., [44]). In LPM, Lagrangian particles are released inside assigned source volumes at selected moments in time, after which they are transported inside the computational domain by the flow field solved by PALM. In this study, Lagrangian particles are defined passive and massless, and hence they represent non-reactive gaseous air pollutants. As for reactive gases and aerosol particles, chemical reactions and dynamic processes manipulating the concentrations should be included. Furthermore, deposition of air parcels on surfaces is not considered here. LPM in PALM was revised for this study to allow for the horizontal heterogeneity and relative strengths of air pollutant sources, i.e., streets with traffic. Particles are released at a selected height with a constant release rate Q ($\mathrm{s^{-1}\,m^{-2}}$), whereas the horizontal locations and group numbers i ($1 \leq i \leq 10$) of different sources are read from an input file. In order to consider the difference in emission strengths, each source group i can be given a weighting factor w_i and the weighted particle concentration pc_{weight} is then calculated as

$$pc_{\mathrm{weight}} = \sum_{}^{N_i} w_i pc_i \tag{1}$$

where pc_i is the particle concentration inside a grid box. Weighting method is applied in order to utilise the same, sufficiently high, particle release rate throughout the source area. This approach allows LPM to represent a nearly continuous emission source everywhere without the need to utilise an excessive number of particles there where emissions are high.

2.2. Model Construction

2.2.1. Modelling Area

The simulations are conducted in western Helsinki on a city boulevard that is planned to be built in the new City Plan of Helsinki for 2050. The boulevard will be framed by a densely built neighbourhood, and in this study, four different city-block-design versions are being investigated. The design versions V_{type}, where type is either par (parallel), per (perpendicular), perHV (perpendicular with height variation) or parJJ (parallel with Jin-Jang shape variation) are visualised in Figure 1 and their specific characteristics are listed in Table 2. In all design versions, the width of the boulevard is 54 m and the total length is around 3.3 km (Figure 2). The average floor area is set to a constant value and the average number of floors is eight. The different city-block-designs are described at the level of detail which does not take into account balconies and bay windows, for example. Furthermore, all planned buildings along the boulevard have flat roofs.

Table 2. The specific characteristics of different city-block-design alternatives.

V_{type}	Characteristics
V_{par}	Building blocks by the boulevard are oriented so that the longest side is parallel to the boulevard. Building heights are fixed to 30 m.
V_{per}	Building block by the boulevard are oriented so that the longest side is perpendicular to the boulevard. Building heights are fixed to 30 m.
V_{perHV}	The orientation of the building blocks is similar to V_{per} but the building height varies. The highest buildings are situated at the nodal points of the public transport, whereas the lowest buildings as well as open urban spaces are situated between the nodal points. Buildings on the eastern side of the boulevard are generally higher.
V_{parJJ}	A so-called "Jin-Jang" block model, in which the buildings are similar to those in V_{par} but the base height is lower and tower-like structures set above the base. Thus the building shape and height are very irregular.

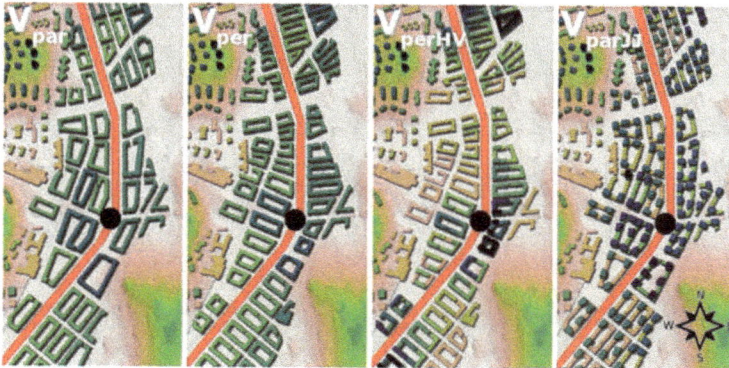

Figure 1. The city-block-design alternatives. From left to right: V_{par} (the longer sides of buildings parallel to the boulevard), V_{per} (the longer sides perpendicular to the boulevard), V_{perHV} (similar to V_{per} but the building height varies) and V_{parJJ} (similar to V_{par} but with Jing-Jang building structure variation). The boulevard is marked in red and the major junction with a black dot.

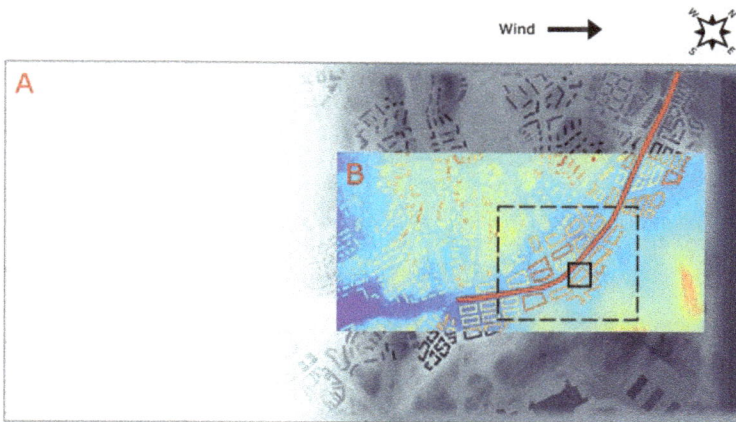

Figure 2. The computational domain for southwesterly ($WD = 225°$) inflow conditions and V_{par} where the longer sides of building blocks are parallel to the boulevard. The domain is separated into child (**B**) and parent (**A**) domains, where the white colour stands for the topography elevation $Z = 0$ m. The data output domains are marked with rectangles: the small domain with a solid black and the large domain with a black dashed line. The boulevard is marked in red.

The dimensions of the parent domain are 4096 m × 2048 m × 384 m in the *x*-, *y*- and *z*-directions. Inside, a child domain of 2048 m × 1024 m × 96 m (Figure 2) or 1536 m × 1536 m × 96 m (not shown) for southwesterly or easterly inflow conditions, respectively, is defined. The grid spacing in all directions (x, y, z) is 1.0 m in the child and 2.0 m in the parent domain.

2.2.2. Urban Surface Data

The information on the surface elevation and cover for each *x,y*-pixel is fed to the model as two-dimensional ASCII-formatted raster files, with separate files for the topography (Figure 2) and tree canopy (Figure 3). The topography data contains topography elevation Z (m) above sea level (a.s.l.) and solid flow obstacles that have a volume of at least one grid box, i.e., buildings but not vehicles.

Correspondingly, the tree canopy data contains the tree canopy height a.s.l. No overhanging structures are accepted in the current model version.

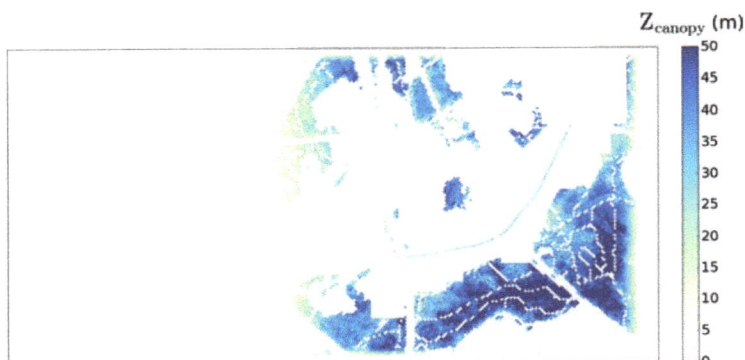

Figure 3. A map of the tree canopy height Z_{canopy} (m) for the whole computational domain. Orientation as in Figure 2.

The landform information is obtained from the archive of the National Land Survey of Finland with a grid resolution of 2 m and the mean forest height from the archive of the Natural Resources Institute of Finland with a grid resolution of 20 m. All the rest, including the existing buildings, planned city-blocks and their corresponding street network alternatives, modifications in the landform and height of the street trees to be planted, is provided by the City Planning Department of Helsinki. Topography and building information, and mean forest and street tree heights, are superimposed on a single raster map file, respectively, which is then pivoted according to the prevailing geostrophic wind direction in each simulation. Over the whole computational domain, Z is given in a vertical resolution of 2 m. The non-cyclic inflow condition applied in the simulation (see Table 1) requires a topography-free zone of around 1.4 km in width, starting from the inflow boundary, for turbulence recycling. In addition, buffer regions of 20 m and 80 m in width, where Z is smoothed towards 20 m and 35 m (see boundaries in Figure 2), are added at the outflow and remaining lateral boundaries, respectively. This is done in order to avoid computational instabilities and to satisfy the periodic boundary conditions at the lateral boundaries [45].

2.2.3. Tree Canopy Model

The tree canopy includes two rows of planted street trees in the middle of the boulevard and the surrounding forested areas (Figure 3). At each surface x,y-grid box with vegetation, a vertical profile of the plant area density (PAD) is defined (Figure 4). Street trees are modelled as 20-m-tall Tilia × vulgaris trees, for which the vertical PAD profile is constructed based on an experimental 5-year mean summertime value of the leaf area density $LAD = 5.3$ m^2 m^{-3} for Tilia × vulgaris trees in Viikki, Helsinki [46]. In addition to leaf area, PAD considers also the trunk and stems. The trees are assumed to have a circular cone shape with the maximum PAD at the height of the lowest branches at 6.5 m and a surface value of 0.3 m^2 m^{-3}. Furthermore, the winter-time leaf-off period (see Section 2.3) is considered by decreasing PAD by 80% following previous studies (e.g., [47,48]). Here, the main interest is on the impact of the street trees along the boulevard. Thus, the PAD profiles of the trees outside the boulevard are constructed based on their height relative to the street trees. In all simulations, a drag coefficient $C_D = 0.2$ is assigned following previous LES studies considering ABL flows individual trees (e.g., [41]) and forest canopies (e.g., [38,49]).

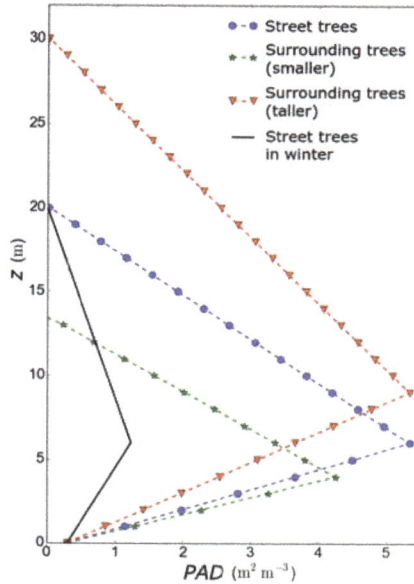

Figure 4. The vertical profile of PAD ($m^2\,m^{-3}$) for the street along the boulevard (blue dashed line with dots). Example profiles for smaller and taller trees outside of the boulevard plotted in green (dashed line with stars) and red (dashed line with triangles), respectively. Additionally, the winter time PAD profile for the street trees is given in black (solid line).

2.2.4. Lagrangian Stochastic Particle Model (LPM)

Equal to the surface cover information, the horizontal location and group of particle sources are given to LPM as a *x,y*-pixel raster file. The planned street network in the vicinity of the boulevard is defined as the particle source area (Figure 5). Hence, only the local traffic emissions are taken into account similar to previous numerical ventilation studies (e.g., [50]). This assumption is adequate enough since ventilation depends on the short-term temporal variation of the concentration of a substance, which is mainly governed by local sources.

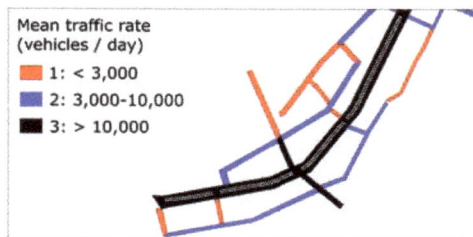

Figure 5. The particle source area (in colours) in the block-design alternatives V_{par} and V_{parJJ}. Source areas are divided into groups 1, 2 and 3 based on the estimated mean traffic rates in year 2025 (see legends). Street surfaces below trees (white dots) are omitted as source areas. Orientation as in Figure 2.

Particles are released at a constant rate of $0.25\,m^{-2}\,s^{-1}$ within each $1\,m \times 1\,m \times 1\,m$ source grid box. No particles are released below the street trees. The street network is divided into three particle groups ($i = 1$–3) according to the estimated traffic rates in year 2025 provided by the City Planning Department of Helsinki (Figure 5). Group 1 covers minor side streets with less than 3000 vehicles per day, group 2 medium streets with 3000 to 10,000 vehicles per day and group 3 the boulevard itself and a major crossroads with over 10,000 vehicles per day. To take into account the proportional difference of traffic rates, pc_{weight} (see Equation (1)) is calculated by defining $w_i = 2i - 1$ for each particle group. In order to remove the particles stuck in the computational domain and unnecessarily increasing the computational load, their maximum age is set to 800 s, after which particles disappear from the domain. This is a loosely calculated expected time for a particle to travel across the child domain with a speed of 2 m s^{-1}.

LPM is applied only inside the child domain (Figure 2). Particles are restrained from entering the parent domain by setting an absorption condition at all boundaries, except at the bottom where particles are reflected. Detailed information about LPM can be found in [51].

2.3. Computational Set-Up

The different boundary conditions used for the parent and child domains are listed in Table 1. The roughness length z_0 applied in the wall treatment using MOST is set to 0.05 m at all solid surfaces. A sufficiently developed turbulent inflow, with which the main runs are initialised, is created with a precursor runs. The characteristics of the inflow, including u, θ and Reynolds stress $\overline{u'w'}$, are illustrated in Figure 6. The ABL depth is set in the precursor run by imposing a large vertical gradient to the potential temperature θ, i.e., a strong temperature inversion, from the ABL height up to the total height of the computational domain. This inversion prevents the ABL from increasing with time.

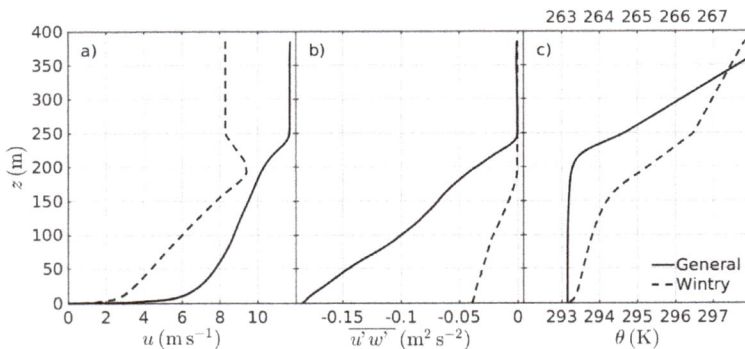

Figure 6. The vertical profile of (**a**) streamwise velocity u (m s^{-1}), (**b**) Reynolds stress $\overline{u'w'}$ (m^2 s^{-2}), and (**c**) potential temperature θ (K) at the inflow boundary for the general (solid line) and wintry (dashed line) inflow conditions. The top x-axis for θ is for the wintry conditions.

In order to examine the influence of the meteorological conditions of the inflow on pollutant dispersion and ventilation, two contrasting conditions are employed in the simulations. The first runs are conducted applying "general" meteorological conditions in Helsinki with a neutrally stratified ABL of 200 m in height and a geostrophic wind of $U_g = 10$ m s^{-1} from the southwest ($WD = 225°$) [52]. The surface heat flux is set to zero to maintain the neutral atmospheric stability. Densely built urban areas tend to be unstably stratified in Helsinki (e.g., [53,54]), and thus applying a neutral stability is a conservative choice as the removal of air pollutants by turbulence is found to improve under unstable conditions [55,56]. The second runs are performed applying "wintry" meteorological conditions with a moderately stable ABL of around 160 m in height and a geostrophic wind of $U_g = 8$ m s^{-1} from the east ($WD = 90°$). This simulates conditions that usually lead to worst air quality events in Helsinki in winter if the Siberian high is prevailing. The atmospheric stability and vertical wind profile applied are follow those in Basu and Porté-Agel [57], and they are attained by initializing the precursor run with these profiles and applying a surface heat exchange rate of $-0.006\ \mathrm{K\,m\,s^{-1}}$ to maintain the atmospheric stability. Simulating a stably stratified ABL is a challenging task as turbulent eddies are small, and a relatively high grid resolution is needed in order to resolve most of the energy containing turbulence [39]. Furthermore, two supportive simulations were conducted for all design versions: under a neutral stratification but with an easterly wind, and under stable stratification but with a southwesterly wind. Results of the simulations are shown in Appendix B, D and E.

2.4. Simulations and Data Output

Complete simulations are done in several parts. First, one precursor run per each inflow condition is carried out over one hour (3600 s) and the final state of the run is applied to initialize the main runs. The main run is carried out in batches of 55 min (batch 1) and 6 min (batch 2). The second batch starts from the final state of the first batch. The release of particles starts after 5 min from the start of batch 1 and is stopped at 56 min, i.e., one minute after the start of batch 2.

Data are output in three sequences over two different domains of 0.5 km² and 0.04 km² (Figure 2) after the transients formed in the initialisation of the main run have subdued. Data output 1 with a time interval of 5 s is collected over the large data output domain over a time span of 40 min starting at 15 min. The high frequency data output 2 starts after 50 min with an interval equal to the integration time step of the simulation of ca. 0.07 s and 0.15 s under the neutrally and stably stratified inflow, respectively. Due to the high logging frequency, the data output 2 covers only the small data output domain. The data output 3 is collected over the large data output domain at a time interval of 5 s during the last 5 min of the simulation when the release of particles has been stopped. The vertical resolution is 1.0 m for all data output whereas the horizontal resolution is 1.0 m for the data output 2 and 2.0 m for the outputs 1 and 3.

2.5. Ventilation and Dispersion Measures

Ventilation and dispersion of pollutants in street canyons and courtyards is assessed by three different measures: number concentration pc (m^{-3}), which from hereafter equals pc_{weight} for the sake of simplicity, the vertical turbulent flux density F_p (m^{-2} s^{-1}) and dilution rate D (m^{-3} s^{-1}) of particles. Various other measures have also been proposed in the literature, for instance the exchange velocity [58], the local purging flow rate and visitation frequency [59], the mean tracer age and age distribution [60], the net escape velocity [61], and the particle exchange rate which assumes horizontal homogeneity of p_c [50]. These measures are however applicable only for simplified urban areas. The velocity ratio [34], on the other hand, does not directly measure pollutant ventilation, and therefore is not used here.

F_p is calculated as the covariance between the vertical wind velocity and the particle number concentration as follows:

$$F_p(x,y,z) = \overline{w'(t,x,y,z)\,pc'(t,x,y,z)} \qquad (2)$$

where $w'(t,x,y,z)$ and $pc'(t,x,y,z)$ are the instantaneous fluctuating vertical velocity and particle number concentration at point (x,y,z) at time t, and the overbar denotes the time average over a selected averaging period. Before calculating the covariance, linear de-trending is applied on both time series. Positive F_p indicates upward transport, i.e., ventilation, and negative downward, i.e., re-entrainment of particles from air above. Hence, the higher F_p the higher ventilation. F_p has previously been observed to determine pollutant removal from street canyons in CFD studies over idealized street canyons (e.g., [15,17,62–64]) as well as in wind tunnel simulations (e.g., [65]). By contrast, advective flux densities are shown to govern the redistribution of pollutants below the roof level [17]. Therefore, to study pollutant ventilation, only the turbulent transport is examined here. F_p is calculated at $z = 20$ m, which is the minimum roof height of all city-block-design versions and well above the surface. Furthermore, F_p is calculated both from the high frequency data output 2 ($F_{p,HF}$) and the low frequency data output 1 ($F_{p,LF}$). As regards the low temporal frequency of 0.2 Hz of the data output 1, a proportion of the total F_p may be missed and thus the exact values of $F_{p,LF}$ should be treated with caution. On the other hand, representativeness of $F_{p,HF}$ calculated only over the small data output domain is limited. Therefore, $F_{p,HF}$ is compared to $F_{p,LF}$ over the same domain in order to show the applicability of $F_{p,LF}$ in this comparative study.

Particle dilution rate D is calculated after the particle source has been switched of at 56 min in batch 2 as

$$D(t, x, y, z) = -\frac{pc(t, x, y, z) - pc(t - \Delta t, x, y, z)}{\Delta t} \tag{3}$$

where Δt is the output time step. The higher D, the higher ventilation. According to the scalar conservation equation (e.g., Equation (5) in [35]), D equals to the sum of the advective and turbulent transport terms if no sources are present. D is calculated from the data output 3 inside an air volume below 20 m a.g.l. First, to minimize the dependence of D on the initial particle concentration, the initial total particle concentration values $pc_{tot}(t = 0)$ inside the analysis domain in each block-design version are normalized relative to that of design version V_{par}. The analysis of D is based on two measures: volume average $\langle D(t) \rangle_V$ over a volume of 0.015 km^3 (around 2.4 million grid cells) and time-column average $\langle D(x, y) \rangle_{t,z}$ within each 2 m × 2 m × 20 m vertical column. The dilution of particles is observed to occur very rapidly as the total particle concentrations drop to half within 60–120 s. Hence, only the first 50 s of the data output 3 are selected for the analysis, which is also the averaging period for $\langle D(x, y) \rangle_{t,z}$. Sensitivity of $\langle D(x, y) \rangle_{t,z}$ to the averaging period was tested by altering the averaging period by ±20 s (see Appendix F).

3. Results

From hereafter, all heights are given in heights above ground level (a.g.l.) unless otherwise specified. The major junction refers to the junction marked in Figure 1. Along with spatial visualisation, the different measures (pc, F_p and D) are analysed separately over the boulevard (marked in Figure 1), other street canyons excluding the boulevard, courtyards, and surroundings, which is the surface area classified neither as a street canyon nor a courtyard, covering around 50% of the large data output domain.

3.1. Particle Concentration pc

The 40-min temporal mean and 90th percentile values of pc are calculated for two layers, $z = 3$–5 m and $z = 9$–11 m, in both inflow conditions. The mean values for both heights separated to the areas defined above are shown in Figure 7, and the horizontal distributions at $z = 4$ m are displayed in Figure 8. The mean distributions at $z = 10$ m and 90th percentiles at both heights follow closely the mean values at $z = 4$ m, which is hence considered to represent the horizontal variation of pc.

In general, the spatial variability of pc within the study domain is notable with higher values along the boulevard than in other areas. Under the general inflow conditions (Figure 8a), the concentration patterns in the southern and northern parts of the boulevard are distinctly different, as the mean wind direction relative to the boulevard changes from nearly parallel to oblique, respectively. The perpendicular wind component results in accumulation to the upwind side of the boulevard due to particle transport by canyon vortex circulation (e.g., [15]). This accumulation is however weak in V_{perHV}. Even though the largest concentrations are commonly seen at the boulevard, some strong accumulation is formed along the side street west of the major junction. Accumulation here is particularly notable in V_{perHV} with building height decreasing downwind, which is attributable to a typically weaker canyon vortex inside this type of step-down canyons (e.g., [65]). In the northern part, the hotspots of pc on the leeward side of the boulevard are nearly continuous along the boulevard in V_{par} and V_{parJJ}, whereas in V_{per} and V_{perHV} ventilation from the cross streets breaks the accumulation patterns.

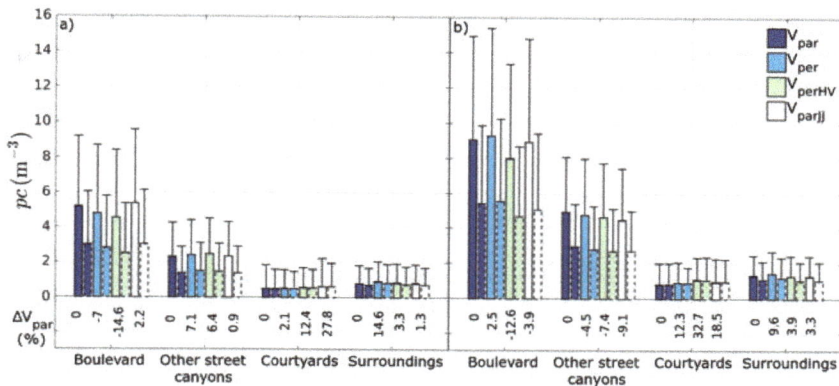

Figure 7. 40-min horizontal mean particle concentrations pc (m^{-3}) under the (**a**) general and (**b**) wintry inflow conditions separately above the boulevard, other street canyons, courtyards and surroundings at height $z = 4$ m (bars with solid lines) and $z = 10$ m (bar with dashed lines) for all runs. 90th percentile values are given with errorbars. The mean difference to the values in V$_{par}$ is given in percentages (ΔV$_{par}$).

Under the wintry inflow conditions (Figure 8b), the mean wind direction is close to perpendicular to the boulevard, especially north of the major junction. A large forested area lies to the east of the boulevard, and hence its cross streets provide fresh, unpolluted air and decrease the mean concentrations along those streets and the boulevard. This is particularly emphasised in V$_{per}$ and V$_{perHV}$ which have more cross streets. In the southern part of the boulevard, the smallest particle accumulation on the leeward side of the boulevard is seen in V$_{perHV}$. Yet, pc north of the major junction are not distinctly higher in V$_{perHV}$ than in V$_{per}$ despite building height decreasing downwind in V$_{perHV}$ in this wind direction, which is generally expected to weaken the canyon vortex and ventilation (e.g., [65]).

Notwithstanding the lower *PAD* values of street trees in the simulations under the wintry inflow conditions, pc at $z = 4$ m are approximately two-fold compared to the general inflow conditions, which demonstrates how strongly meteorology influences pollutant dispersion. Under both inflow conditions, pc values are of the same magnitude in V$_{par}$ and V$_{parJJ}$. The courtyards are notably clean and pc remains low throughout all simulations.

The vertical dispersion of pc and mean flow structure inside the boulevard street canyon for two vertical cross Sections 1 and 2 marked in Figure 8 are shown in Figure 9a,b under the general, and Figure 9c,d under the wintry inflow conditions. In Figure 9a (cross Section 1), the wind is close to parallel to the canyon, i.e., towards the picture, and due to flow channelling and advective particle transport the concentrations remain low. In Figure 9b (cross Section 2), instead, the mean wind has a strong perpendicular component creating optimal conditions for a canyon vortex. However, the street trees block the flow and break the vortex in two. This explains the more consistent dispersion of pc inside the canyon and accumulation on both sides of the boulevard. Under the wintry inflow conditions, on the other hand, a uniform canyon vortex is formed, and it is emphasised in V$_{par}$ and V$_{parJJ}$ with longer street canyons along the boulevard. Yet the exact location varies between the city-block versions. Vertical maxima of pc under both inflow conditions are seen right below the maximum of the *PAD* profile at $z = 6.5$ m.

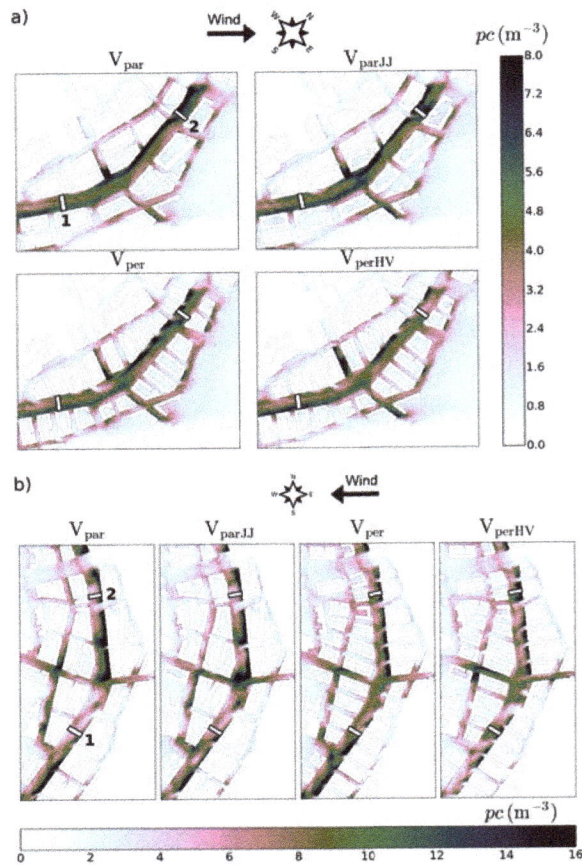

Figure 8. 40-min temporal mean of particle concentration pc (m^{-3}) at height $z = 4$ m under the (**a**) general and (**b**) wintry inflow conditions. Notice the orientation of the mean wind and different scales of pc. Cross sections in Figure 8 are marked with white lines.

In spite of the great spatial variability in pc between the block-design versions, the concentration characteristics of different block-design versions are similar at both 4 and 10 m levels in almost all cases, as shown in Figure 7. Above the boulevard, both the mean values and 90th percentiles are lowest in V_{perHV} under both general and wintry inflow conditions, whereas the highest values are observed in V_{parJJ} and V_{per} under the general and wintry inflow conditions, respectively. In other street canyons, no systematic pattern is seen while in the courtyards and surroundings, V_{par} has on average the lowest mean concentrations.

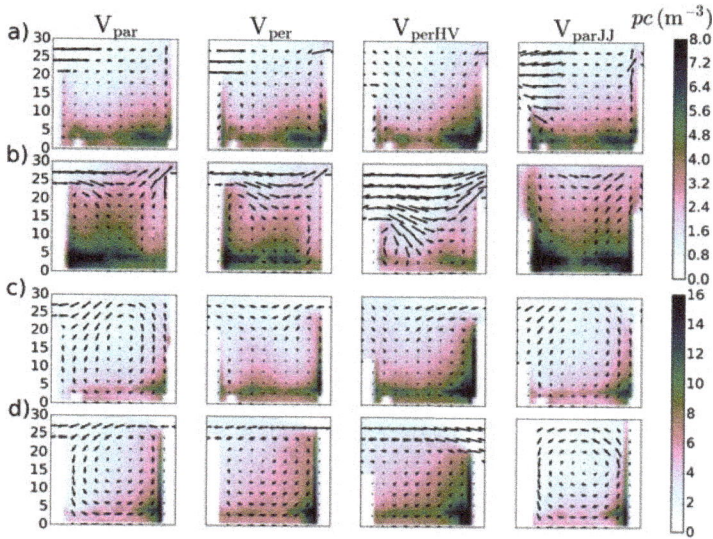

Figure 9. 40-min mean particle concentration pc (m^{-3}) across the boulevard. (Cross Sections 1 and 2, which are marked in Figure 8, are shown in (**a,b**) under the general inflow conditions, and in (**c,d**) under the wintry inflow conditions, respectively. Cross sections are viewed from the south. Lengths of the wind arrows are relative to the wind speeds normal to the cross section. Height in m a.g.l. are given in the left. Notice the different scales of pc.)

3.2. Turbulent Particle Flux F_p

Due to data constrains, $F_{p,HF}$ was computed only for the small data output domain, which is visualised Appendix C. Furthermore, the lower resolution $F_{p,LF}$ was also calculated in order to compare the turbulent particle flux with the other metrics used in the study. To estimate possible uncertainties originating from the lower temporal resolution, Table 3 summarizes the mean values of both $F_{p,HF}$ and $F_{p,LF}$ in the same small data output domain separately for the boulevard, other street canyons, courtyards and surroundings. Overall, $F_{p,HF}$ and $F_{p,LF}$ show rather similar orders of magnitude between the design-versions except above the boulevard under the general inflow conditions and in the surroundings under the wintry inflow conditions. $F_{p,LF}$ generally overestimates flux density values, apart from above the courtyards and above the other street canyons under the general inflow conditions. However, a Student's one-sample t-test for the difference of the values of $F_{p,HF}$ and $F_{p,LF}$ shows that the difference is insignificant at 95% confidence level ($p = 0.069$). $F_{p,HF}$ and $F_{p,LF}$ values agree better under the wintry than the general inflow conditions, owing most likely to the nearly three times longer time step of the wintry simulations. The difference between $F_{p,HF}$ and $F_{p,LF}$ becomes even less significant ($p = 0.18$), if $F_{p,HF}$ under the general inflow conditions is calculated using only every other time step of the data output 2 when the data logging frequency is around 7 Hz. As a conclusion, $F_{p,LF}$ can be applied when comparing ventilation in different city-block-design versions.

Table 3. Horizontal mean of the high-frequency vertical turbulent particle flux density $F_{p,HF}$ ($m^{-2}\,s^{-1}$) separately for the boulevard, other street canyons, courtyards and surroundings at $z = 20$ m under both general and wintry inflow conditions. Horizontal mean of the low-frequency vertical turbulent particle flux density $F_{p,LF}$ over the same domain is given in brackets.

Inflow Conditions	V_{par}	V_{per}	V_{perHV}	V_{parJJ}
Boulevard				
General	0.193 (0.233)	0.162 (0.238)	0.169 (0.292)	0.213 (0.247)
Wintry	0.221 (0.294)	0.168 (0.209)	0.159 (0.211)	0.178 (0.191)
Other street canyons				
General	0.085 (0.054)	0.078 (0.056)	0.247 (0.138)	0.117 (0.102)
Wintry	0.115 (0.120)	0.141 (0.120)	0.180 (0.229)	0.102 (0.111)
Courtyards				
General	2.2×10^{-3} (-4.5×10^{-3})	-3.5×10^{-3} (0.8×10^{-3})	1.1×10^{-3} (1.2×10^{-3})	17.1×10^{-3} (16.7×10^{-3})
Wintry	-0.0×10^{-3} (0.5×10^{-3})	-0.4×10^{-3} (-2.0×10^{-3})	2.0×10^{-3} (-4.3×10^{-3})	-5.6×10^{-3} (-7.7×10^{-3})
Surroundings				
General	0.038 (0.026)	0.039 (0.040)	0.058 (0.071)	0.045 (0.047)
Wintry	0.043 (0.085)	0.041 (0.059)	0.045 (0.057)	0.037 (0.074)

The horizontal distribution and mean values of $F_{p,LF}$ above the separate areas for all block-design versions under both inflow conditions are visualised in Figures 10 and 11. For all model runs, $F_{p,LF}$ appears to be on average positive indicating upward transport of particles and ventilation. This can be expected since the particle source is constantly maintaining pc at street level. In general, $F_{p,LF}$ inside the courtyards is close to zero or even negative, as the only sources are re-entrainment from above or advection from the building openings. $F_{p,LF}$ is slightly increased when the building height increases and correspondingly decreased when the building height decreases downwind in V_{perHV}. Furthermore, $F_{p,LF}$ values are smaller under the wintry than the general inflow conditions. Under the general inflow conditions, $F_{p,LF}$ shows most positive values along the boulevard in V_{perHV}, whereas in the other design versions areas of negative flux density (re-entrainment) also appear. Under the wintry inflow conditions, $F_{p,LF}$ is most positive in V_{per} and V_{perHV}. In V_{par}, a vast area of negative flux density is observed south of the major junction. The same is observed in V_{parJJ} as well but less pronounced. $F_{p,LF}$ is decreased on the leeward side of buildings and especially behind the corners of buildings and higher towers in V_{parJJ}, which is reflected to particle concentrations.

Horizontal mean values in Figure 10 correspond well to the visual analysis based on Figure 11. As noted, values are higher under the general inflow conditions. $F_{p,LF}$ is mainly largest in V_{perHV} and smallest in V_{par} under both inflow conditions. Both V_{per} and V_{parJJ} show higher values than V_{par} regardless of weather conditions. Furthermore, F_p over the courtyards in V_{parJJ} depends strongly on the inflow.

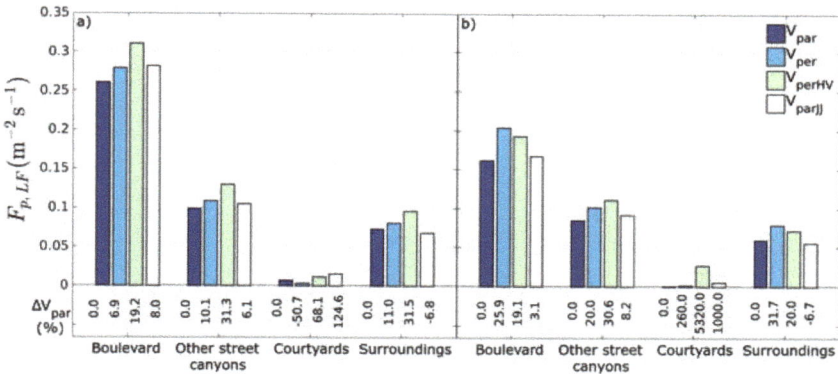

Figure 10. Horizontal mean of the low-frequency vertical turbulent particle flux density $F_{p,LF}$ ($\mathrm{m^{-2}\,s^{-1}}$) under the (**a**) general and (**b**) wintry inflow conditions separately for the boulevard, other street canyons, courtyards and surroundings at $z = 20$ m for all runs. The difference to the value in V_{par} is given in percentages (ΔV_{par}).

3.3. Particle Dilution Rate D

The time series of $\langle D(t) \rangle_V$ under the general (a) and wintry (b) inflow conditions are illustrated in Figure 12. Values are given relative to $\langle D(t) \rangle_V$ in V_{par}. Overall, $\langle D(t) \rangle_V$ shows high variability with time notwithstanding the spatial averaging of D over around 2.4 million grid points. Figure 12a for the general inflow conditions reveals no significant differences between V_{per}, V_{perHV} and V_{parJJ} whereas they all show mostly smaller values than V_{par}. Under the wintry inflow conditions (Figure 12b), $\langle D(t) \rangle_V$ in V_{per} and particularly in V_{perHV} remain mostly above that in V_{par} whereas $\langle D(t) \rangle_V$ in V_{parJJ} varies between above and below.

The horizontal distribution and mean values of $\langle D(x,y) \rangle_{t,z}$ above the boulevard, other street canyons, courtyards and surroundings under both inflow conditions are displayed in Figures 13 and 14. The analysis area is the same as for pc and $F_{p,LF}$. In general, $\langle D(x,y) \rangle_{t,z}$ increases downwind under both general and wintry inflow conditions as particles in the upwind columns are transported quickly by the horizontal advection. Furthermore, values are twofold under the wintry compared to the general inflow conditions. This also shown in the supportive simulations in Appendix E. Especially V_{perHV} shows efficient dilution to the surroundings. Sensitivity tests show how the averaging period over which $\langle D(x,y) \rangle_{t,z}$ is calculated has minor impact on the results (Appendix F).

Horizontal mean values in Figure 14 quantify the differences in $\langle D(x,y) \rangle_{t,z}$ between the design versions. Along the boulevard, dilution is most efficient in V_{par} and V_{parJJ} under the general inflow conditions, and in V_{perHV} under the wintry inflow conditions. The efficient dilution to the surroundings in V_{perHV} is seen also in the mean values, especially in the wintry inflow conditions. Moreover, V_{perHV} shows mostly higher values than V_{per}. Dilution inside the courtyards is weakest in V_{per}.

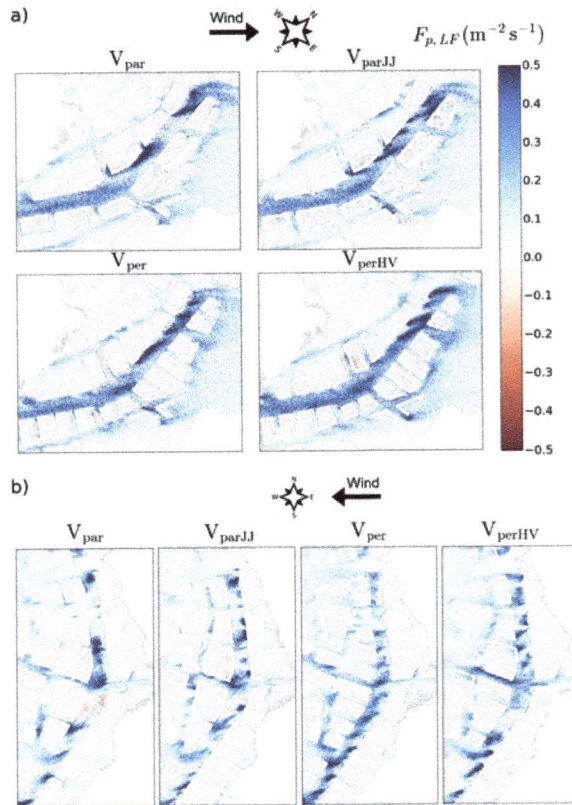

Figure 11. 40-min mean low-frequency vertical turbulent particle flux density $F_{p,LF}$ ($m^{-2}\,s^{-1}$) at $z = 20$ m under the (**a**) general and (**b**) wintry inflow conditions. Positive values indicate upward flux.

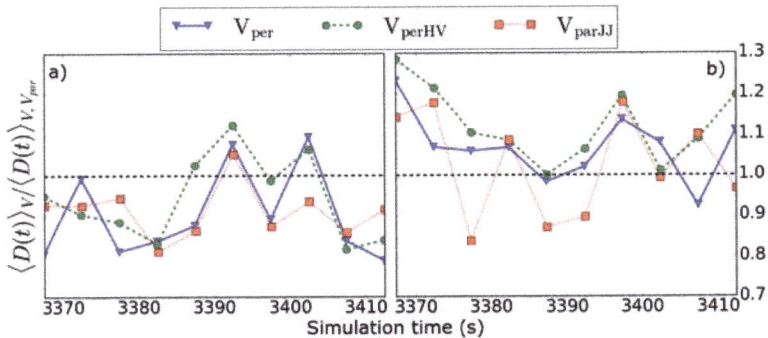

Figure 12. The volume averaged particle dilution rate $\langle D(t) \rangle_V$ ($m^{-3}\,s^{-1}$) between $z = 1$–20 m under the (**a**) general and (**b**) wintry inflow conditions. Results are represented relative to V_{par} ($\langle D(t) \rangle_{V,V_{par}}$).

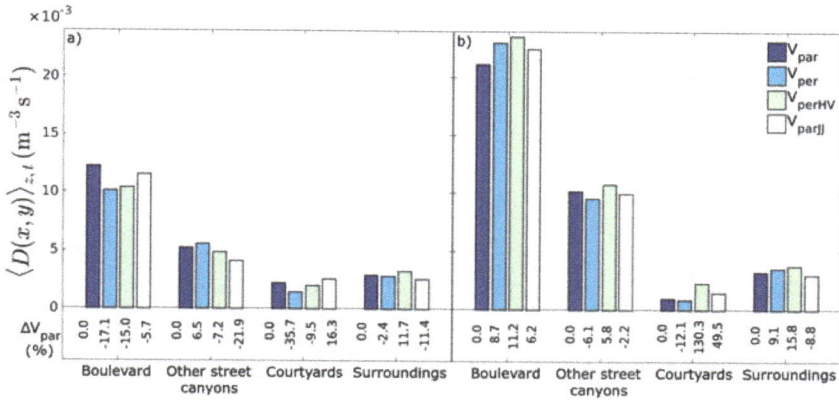

Figure 13. Mean $\langle D(x,y)\rangle_{t,z}$ ($\times 10^{-3}$ m^{-3} s^{-1}) under the (**a**) general and (**b**) wintry inflow conditions separately for the boulevard, other street canyons, courtyards and surroundings between $z = 1$–20 m for all runs. D is calculated using data from the first 50 s after the particle source has been switched off. The difference to the value in V_{par} is given in percentages (ΔV_{par}).

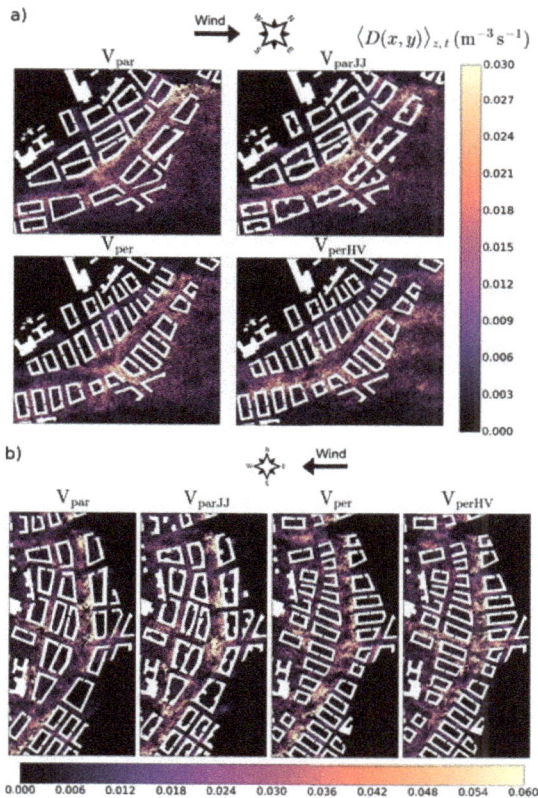

Figure 14. The temporal mean particle dilution rate $\langle D(x,y)\rangle_{t,z}$ (m^{-3} s^{-1}) between $z = 1$–20 m for the first 60 s after the particle has been switched off: (**a**) general and (**b**) wintry inflow conditions.

4. Discussion

Pollutant dispersion is strongly governed by inflow conditions: wind parallel to a street canyon sweeps the particles away due to particle transport by horizontal advection, as also shown by Moon et al. [66], whereas perpendicular or oblique winds result in accumulation to the leeward side. Mean concentrations are two-fold under the wintry compared to the general inflow conditions, which stems from less efficient vertical mixing and pollutant transport in stable ABL (see also Appendix B). On the other hand, Figure 10 and Appendix D show that vertical transport is notably higher under a southwesterly wind despite the atmospheric stability. This is likely due to more efficient mechanical production of turbulence under a southwesterly wind as the inflow travels over a longer stretch of urban area before entering the area of interest, unlikely when the wind is from the east with mostly homogeneous vegetated areas.

Urban morphology is shown to have a significant role in the dispersion of particles, which is illustrated in Figures 7 and 8. Greatest concentrations along the boulevard and weakest vertical transport are observed when the longer sides of building blocks are parallel to the boulevard, which supports previous studies performed over idealised street canyons [63,67,68]. Non-uniform building height in V_{perHV} is shown to decrease pc accumulation on the leeward side of the boulevard street canyon, especially under a southwesterly wind when the building height mainly increases downwind [65,69,70]. Apart from buildings, street trees can alter the flow field and canyon vortex, and thereby decrease pollutant ventilation. Therefore, a detailed representation of trees in urban flow and air quality simulations is of high importance.

Of all the design options, V_{perHV} with building height variation results on average in the greatest ventilation rates and lowest street level concentrations above the street canyons, with around 7–9% lower concentrations at $z = 4$ m than the city-block-design alternative with highest concentrations inside street canyons. The result supports previous studies conducted over simplified urban topographies (e.g., [60,65,71–74]). The building shape variability in V_{parJJ}, instead, is mainly shown to increase $F_{p,LF}$ and improve ventilation in vertical, whereas the mean values of pc and D display equally efficient or even weaker ventilation than in V_{par}. The result indicates that the aerodynamically rougher city-blocks in V_{parJJ} do not notably improve the air quality near the pedestrian level. The irregularities in the building shape in V_{parJJ} probably destroy the canyon vortex [75], which reduces advective particle transport. Further in–depth studies are needed to explain the last-mentioned result.

Applying a high-resolution LES model over a real urban topography of a vast extent provides a large amount of detailed information about the flow and concentration fields. However, the amount of output data is high and the complexity of the topography further complicates the analysis. The ventilation measures chosen to be applied in this study cover the pollutant dispersion mechanisms broadly and complement each other. Based on the results, the following generalisation can be deduced. The pedestrian level concentrations are lowest when F_p are highest. Furthermore, under both inflow conditions the highest mean values of $F_{p,LF}$ are observed in the V_{perHV} and lowest in V_{par}, suggesting that F_p is strongly determined by the urban morphology. Generally, D is correlated to pc, and the differences between their horizontal distributions stem from the flow changing the location of particles but not necessarily removing them from the study domain. A comparison of the horizontal patterns of $\langle D(x,y)\rangle_{t,z}$ and $F_{p,LF}$ points out areas where the dilution in vertical is notable and vice versa. On the whole, however, the patterns are unlike, indicating that D measures mainly horizontal advection. This advective nature of D can also be concluded from the higher values when the flow is parallel to the longer sides of buildings as the flow is horizontally less blocked and disturbed by flow structures created by building corners. Last of all, pc as well as F_p and D are smaller in courtyards and surroundings than inside street canyons indicating little particle transport from the streets. The vertical exchange above courtyards is shown to be low when the angle between the wind and principal courtyard axis is around 90°, supporting results by Moonen et al. [76].

Owing to the large extent of the study domain, the analysis was confined to the temporal mean values of the analysis measures. Moreover, the number of different inflow conditions applied was limited to two due to the high computational expenses of the simulations. They were, however, carefully chosen to represent typical and worst-case conditions at the modelling site, but also for the results to apply as guidelines in all urban areas. Last of all, the study focused on the aerodynamic impacts and omitted, for instance, air pollutant chemistry and dynamics and anthropogenic heat sources.

Explicit rules can be difficult to give to optimize of flow and pollutant dispersion over complex urban terrains [71]. To yield detailed information for a specific case, similar studies have to be conducted separately for each urban planning solution. However, the outcome of this study, related firstly to the most efficient dispersion in V_{perHV} with variable building height and short wall facing the main road and secondly to the insignificant improvement of ventilation by building shape variation in V_{parJJ}, can be used as a general guideline also for other areas. Such information is vital for urban planners in order to design dense urban areas where the level of pollution exposure is minimized. For decision making, a separate model could be used in which each of the ventilation measures and areas of interest would get a weighting factor. Based on the factors that are considered important, for instance good ventilation inside courtyards, one could define the most suitable city-block-design alternative.

5. Conclusions

This study examines how the structural arrangement and orientation of perimeter blocks affect the ventilation and dispersion of traffic-related emissions within their street canyons under two contrasting meteorological conditions of the inflow. The principal objective of this numerical study is to demonstrate means to generate information that enables future urban planners to improve the pedestrian level air quality within real urban areas. The study is comprised of simulations that feature four different virtual building-block arrangements that are proposed for a real city boulevard site. In addition to buildings, the representation of the urban area is highly realistic including surrounding landform as well as forested areas and street trees. The simulations are conducted using a LES model PALM with an embedded Lagrangian stochastic particle model. The Lagrangian particles representing massless and inert air pollutant tracers are released over streets with a source strength (i.e., weighted particle release rate) that is relative to the local traffic volumes. The study employs a full three-dimensional two-way self-nesting functionality, which is newly implemented to PALM.

The analysis is founded on three different measures: particle concentration pc, vertical turbulent particle transport F_p and particle dilution rate D. Variation in building height is shown to enhance both F_p and D, and thus decrease accumulation of pc at pedestrian level. Furthermore, short canyons by the boulevard are shown preferential for F_p. However, building shape variability of smaller scale does not result in improved ventilation. D is governed by horizontal advection and is thus strongly determined by the horizontal wind direction. It is shown how the design version with the shortest wall parallel to the boulevard with variable building height creates on average the most optimal air quality at pedestrian level along the boulevard, but the spatial variability within the street is highly variable depending on the pollutant transport and dispersion. Courtyards remain clean throughout all simulations, implying that in general traffic-related pollutants are not easily transported to there. Despite the high roughness and complexity of the urban surface that leads to efficient mechanical production of turbulent, pollutant concentrations are clearly higher under stably stratified inflow conditions.

This is the first LES study over a vast, urban area applying sophisticated measures to assess pollutant dispersion and ventilation. The numerical methods are novel and highly developed providing realistic estimations for the removal of nonreactive gaseous air pollutants from the pedestrian level. The results of this study provide unique information about the transport of traffic-related pollutants in this specific urban environment and the results can directly be applied by local urban planners, but naturally the findings support urban planning also in other cities.

Supplementary Materials: The modified code parts of PALM revision 1904 used in this study are available online at www.mdpi.com/xxx/s1.

Acknowledgments: This study was commissioned and funded by the City Planning Department (current Urban Environment Division) of the City of Helsinki. We also acknowledge the Doctoral Programme in Atmospheric Sciences (ATM-DP, University of Helsinki), Helsinki Metropolitan Region Urban Research Program and the Academy of Finland (181255, 277664) for financial support.

Author Contributions: M.K. constructed the model set-up, carried out the simulations, analysed the data and wrote the manuscript; A.H. and M.A. planned the experiments together with M.K., supervised her in performing the simulations and analysing the data, and commented on the manuscript; S.R. and T.V. provided critical feedback on the manuscript and helped shape the research; L.J. supervised the project and took actively part in writing the manuscript.

Conflicts of Interest: The authors declare no conflict of interest. The founding sponsors had no role in the design of the study; in the collection, analyses, or interpretation of data; in the writing of the manuscript, and in the decision to publish the results.

Abbreviations

The following abbreviations are used in this manuscript:

ABL	Atmospheric boundary layer
CFD	Computational fluid dynamics
LAD	Leaf area density
LES	Large-eddy simulaion
LPM	Lagrangian stochastic particle model
MOST	Monin-Obukhov similarity theory
PAD	Plant area density
PALM	Parallelized Large-Eddy Simulation Model
par	parallel
parJJ	paraller with Jin-Jang shape variation
per	perpendicular
perHV	perpendicular with height variation
RANS	Reynolds-averaged Navier-Stokes equations

Appendix A. Technical Specifications

Table A1. Technical specifications for the simulations conducted with the LES model PALM.

V_{type}	Characteristics
Programming language	Fortran 95/2003
Discretization	Arakawa staggered C-grid [77,78]
Parallelization	Two-dimensional domain decomposition (e.g., [36]). Communication between processors realized using Message Passing Interface (MPI).
Sub-grid closure	1.5-order scheme based on Deardorff [79]
Advection scheme	5th-order advection scheme by Wicker and Skamarock [80]
Pressure solver	Iterative multigrid scheme (e.g., [81])
Time step closure	3rd-order Runge-Kutta approximation [82]
Boundary condition between the surface and the first grid level	Monin-Obukhov similarity theory [83]

Appendix B. Particle Concentration *pc* for the Supportive Simulations

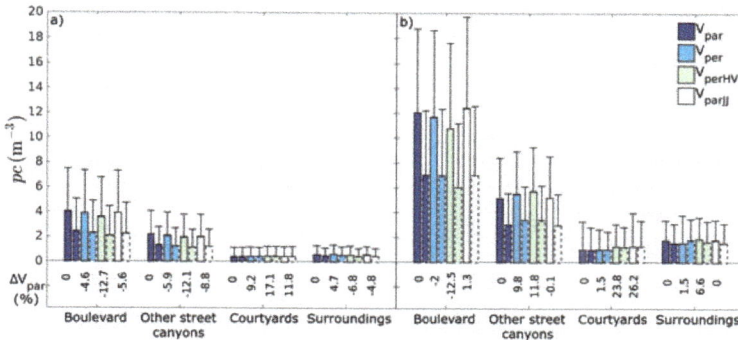

Figure A1. 40-min horizontal mean particle concentrations *pc* (m^{-3}) for the (**a**) neutral run with an easterly wind and (**b**) stable run with a southwesterly wind separately above the boulevard, other street canyons, courtyards and surroundings at height z = 4 m (bars with solid lines) and z = 10 m (bar with dashed lines) for all runs. 90th percentile values are given with errorbars. The mean difference to the values in V$_{par}$ is given in percentages (ΔV$_{par}$).

Appendix C. 5-Minute Mean High-Frequency Vertical Turbulent Particle Flux Density $F_{p,HF}$

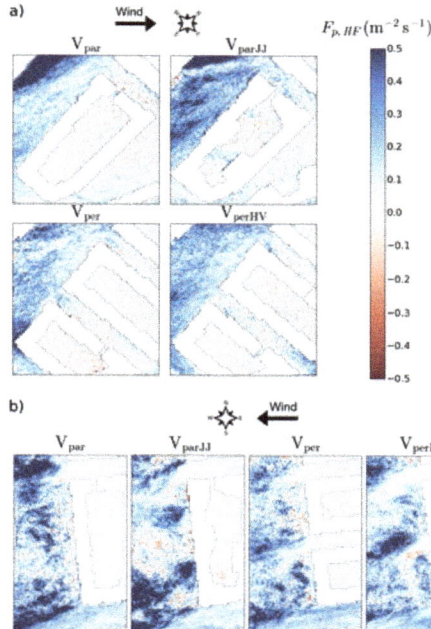

Figure A2. 5-min mean high-frequency vertical turbulent particle flux density $F_{p,HF}$ (m^{-2} s^{-1}) at z = 20 m under the (**a**) general and (**b**) wintry inflow conditions. The analysis area is marked in Figure 2 with a black solid line.

Appendix D. Low-Frequency Vertical Turbulent Particle Flux Density $F_{p,LF}$ for the Supportive Simulations

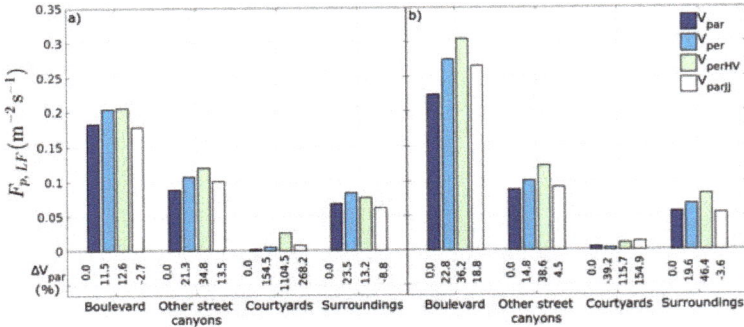

Figure A3. Horizontal mean of the low-frequency vertical turbulent particle flux density $F_{p,LF}$ $(m^{-2}\,s^{-1})$ for the (**a**) neutral and (**b**) stable runs separately for the boulevard, other street canyons, courtyards and surroundings at $z = 20$ m for all runs. The difference to the value in V_{par} is given in percentages (ΔV_{par}).

Appendix E. Column-Averaged Dilution Rate $\langle D(x,y)\rangle_{t,z}$ for the Supportive Simulations

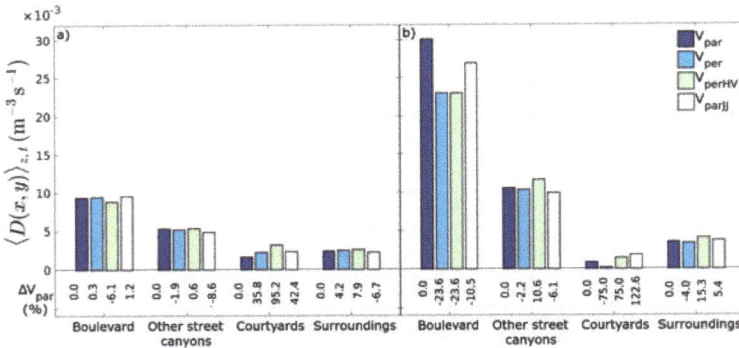

Figure A4. Mean $\langle D(x,y)\rangle_{t,z}$ $(\times 10^{-3}\ m^{-3}\,s^{-1})$ for the (**a**) neutral run with an easterly wind and (**b**) stable run with a southwesterly wind separately for the boulevard, other street canyons, courtyards and surroundings between $z = 1$–20 m for all runs. D is calculated using data from the first 50 s after the particle source has been switched off. The difference to the value in V_{par} is given in percentages (ΔV_{par}).

Appendix F. Column-Averaged Dilution Rate $\langle D(x,y)\rangle_{t,z}$ with Different Averaging Periods

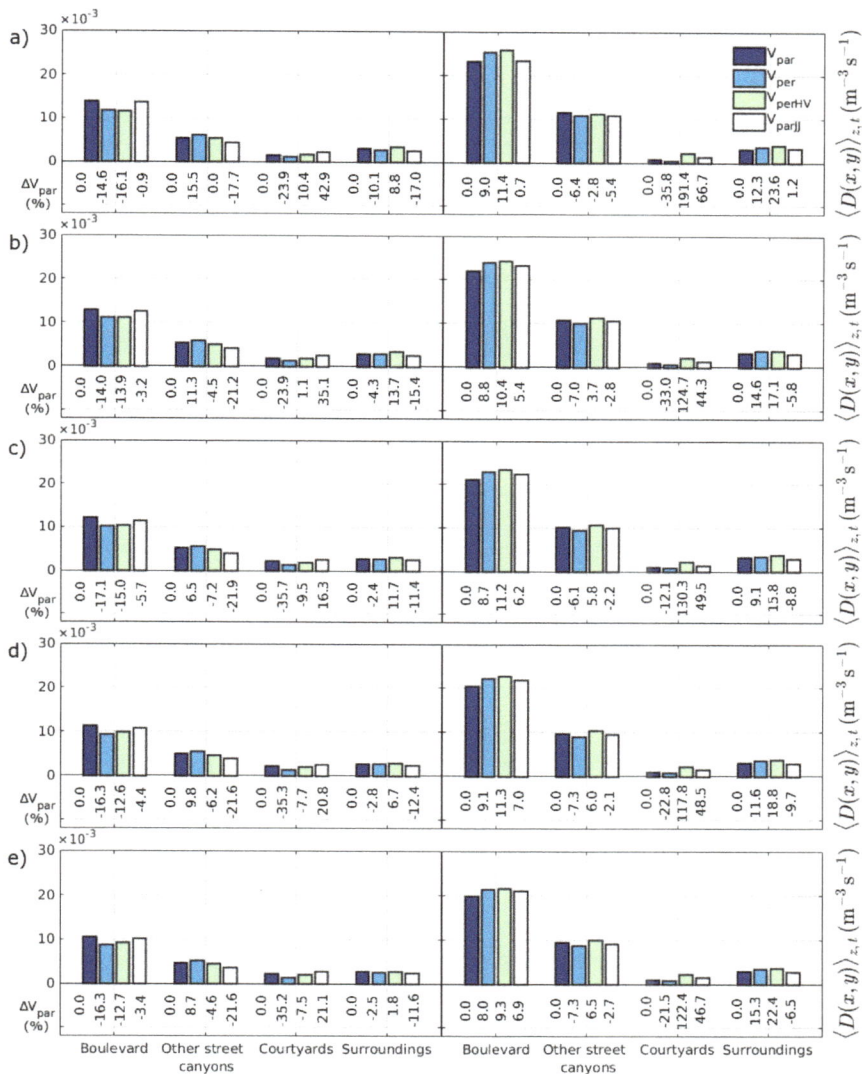

Figure A5. Mean $\langle D(x,y)\rangle_{t,z}$ ($\times 10^{-3}$ m^{-3} s^{-1}) under the general (**left**) and wintry (**right**) inflow conditions separately for the boulevard, other street canyons, courtyards and surroundings between $z = 1$–20 m for all runs. D is calculated for different averaging periods: (**a**) 30 s; (**b**) 40 s; (**c**) 50 s; (**d**) 60 s and (**e**) 70 s. The difference to the value in V$_{par}$ is given in percentages (ΔV$_{par}$).

References

1.   United Nations. *World Urbanization Prospects: The 2014 Revision, Highlights*; Department of Economic and Social Affairs, Population Division: New York, NY, USA, 2014.
2.   Brunekreef, B.; Holgate, S.T. Air pollution and health. *Lancet* **2002**, *360*, 1233–1242.

3. Pope, C.A.I.; Dockery, D.W. Health effects of fine particulate air pollution: Lines that connect. *J. Air Waste Manag. Assoc.* **2006**, *56*, 709–742.

4. Gakibou, E.; Afshin, A.; Abajobir, A.A.; Abate, K.H.; Abbafati, C.; Abbas, K.M.; Abd-Allah, F.; Abdulle, A.M.; Abera, S.F.; Aboyans, V.; et al. Global, regional, and national comparative risk assessment of 84 behavioural, environmental and occupational, and metabolic risks or clusters of risks, 1990–2016: A systematic analysis for the Global Burden of Disease Study 2016. *Lancet* **2017**, *390*, 1345–1422.

5. European Environmental Agency. *Air Quality in Europe—2017 Report*; EEA Report No 13/2017; Publications Office of the European Union: Luxembourg, 2017.

6. Fenger, J. Air pollution in the last 50 years–From local to global. *Atmos. Environ.* **2009**, *43*, 13–22.

7. Vardoulakis, S.; Fisher, B.E.; Pericleous, K.; Gonzalez-Flesca, N. Modelling air quality in street canyons: A review. *Atmos. Environ.* **2003**, *37*, 155–182.

8. Britter, R.E.; Hanna, S.R. Flow and dispersion in urban areas. *Ann. Rev. Fluid Mech.* **2003**, *35*, 469–496.

9. Buccolieri, R.; Sandberg, M.; Sabatino, S.D. City breathability and its link to pollutant concentration distribution within urban-like geometries. *Atmos. Environ.* **2010**, *44*, 1894–1903.

10. Tominaga, Y.; Stathopoulos, T. Ten questions concerning modeling of near-field pollutant dispersion in the built environment. *Build. Environ.* **2016**, *105*, 390–402.

11. Giometto, M.G.; Christen, A.; Meneveau, C.; Fang, J.; Krafczyk, M.; Parlange, M.B. Spatial Characteristics of Roughness Sublayer Mean Flow and Turbulence over a Realistic Urban Surface. *Bound. Layer Meteorol.* **2016**, *160*, 425–452.

12. Gousseau, P.; Blocken, B.; Stathopoulos, T.; van Heijst, G. CFD simulation of near-field pollutant dispersion on a high-resolution grid: A case study by LES and RANS for a building group in downtown Montreal. *Atmos. Environ.* **2011**, *45*, 428–438.

13. Tominaga, Y.; Stathopoulos, T. CFD modeling of pollution dispersion in a street canyon: Comparison between LES and RANS. *J. Wind Eng. Ind. Aerodyn.* **2011**, *99*, 340–348.

14. Salim, S.M.; Cheah, S.C.; Chan, A. Numerical simulation of dispersion in urban street canyons with avenue-like tree plantings: Comparison between RANS and LES. *Build. Environ.* **2011**, *46*, 1735–1746.

15. Baik, J.J.; Kim, J.J. On the escape of pollutants from urban street canyons. *Atmos. Environ.* **2002**, *36*, 527–536.

16. Cai, X.M.; Barlow, J.; Belcher, S. Dispersion and transfer of passive scalars in and above street canyons—Large-eddy simulations. *Atmos. Environ.* **2008**, *42*, 5885–5895.

17. Li, X.X.; Liu, C.H.; Leung, D.Y. Numerical investigation of pollutant transport characteristics inside deep urban street canyons. *Atmos. Environ.* **2009**, *43*, 2410–2418.

18. Raupach, M.R.; Finnigan, J.J.; Brunet, Y. *Boundary-Layer Meteorology 25th Anniversary Volume, 1970–1995: Invited Reviews and Selected Contributions to Recognise Ted Munn's Contribution as Editor over the Past 25 Years*; Chapter Coherent Eddies and Turbulence in Vegetation Canopies: The Mixing-Layer Analogy; Springer: Dordrecht, The Netherlands, 1996; pp. 351–382.

19. Gromke, C.; Blocken, B. Influence of avenue-trees on air quality at the urban neighborhood scale. Part II: Traffic pollutant concentrations at pedestrian level. *Environ. Pollut.* **2015**, *196*, 176–184.

20. Oke, T.R. *Boundary Layer Climates*, 2nd ed.; Routledge: London, UK, 1987.

21. Buccolieri, R.; Gromke, C.; Sabatino, S.D.; Ruck, B. Aerodynamic effects of trees on pollutant concentration in street canyons. *Sci. Total Environ.* **2009**, *407*, 5247–5256.

22. Gromke, C.; Ruck, B. On the Impact of Trees on Dispersion Processes of Traffic Emissions in Street Canyons. *Bound. Layer Meteorol.* **2009**, *131*, 19–34.

23. Mochida, A.; Lun, I.Y. Prediction of wind environment and thermal comfort at pedestrian level in urban area. *J. Wind Eng. Ind. Aerodyn.* **2008**, *96*, 1498–1527.

24. Vos, P.E.; Maiheu, B.; Vankerkom, J.; Janssen, S. Improving local air quality in cities: To tree or not to tree? *Environ. Pollut.* **2013**, *183*, 113–122.

25. Lindberg, F.; Grimmond, C.S.B. Continuous sky view factor maps from high resolution urban digital elevation models. *Clim. Res.* **2010**, *42*, 177–183.

26. Nordbo, A.; Karsisto, P.; Matikainen, L.; Wood, C.R.; Järvi, L. Urban surface cover determined with airborne lidar at 2 m resolution—Implications for surface energy balance modelling. *Urban Clim.* **2015**, *13*, 52–72.

27. Gousseau, P.; Blocken, B.; Stathopoulos, T.; van Heijst, G. Near-field pollutant dispersion in an actual urban area: Analysis of the mass transport mechanism by high-resolution Large Eddy Simulations. *Comput. Fluids* **2015**, *114*, 151–162.

28. Kanda, M.; Inagaki, A.; Miyamoto, T.; Gryschka, M.; Raasch, S. A New Aerodynamic Parametrization for Real Urban Surfaces. *Bound. Layer Meteorol.* **2013**, *148*, 357–377.

29. Letzel, M.O.; Krane, M.; Raasch, S. High resolution urban large-eddy simulation studies from street canyon to neighbourhood scale. *Atmos. Environ.* **2008**, *42*, 8770–8784.

30. Park, S.B.; Baik, J.J.; Lee, S.H. Impacts of Mesoscale Wind on Turbulent Flow and Ventilation in a Densely Built-up Urban Area. *J. Appl. Meteorol. Climatol.* **2015**, *54*, 811–824.

31. Xie, Z.T.; Castro, I.P. Large-eddy simulation for flow and dispersion in urban streets. *Atmos. Environ.* **2009**, *43*, 2174–2185.

32. Keck, M.; Raasch, S.; Letzel, M.O.; Ng, E. First Results of High Resoltuion Large-Eddy Simulations of the Atmospheric Boundary Layer. *J. Heat Island Inst. Int.* **2014**, *9*, 39–43.

33. Cui, Z.; Cai, X.; Baker, C.J. Large-eddy simulation of turbulent flow in a street canyon. *Q. J. R. Meteorol. Soc.* **2004**, *130*, 1373–1394.

34. Letzel, M.O.; Helmke, C.; Ng, E.; An, X.; Lai, A.; Raasch, S. LES case study on pedestrian level ventilation in two neighbourhoods in Hong Kong. *Meteorol. Z.* **2012**, *21*, 575–589.

35. Maronga, B.; Gryschka, M.; Heinze, R.; Hoffmann, F.; Kanani-Sühring, F.; Keck, M.; Ketelsen, K.; Letzel, M.O.; Sühring, M.; Raasch, S. The Parallelized Large-Eddy Simulation Model (PALM) version 4.0 for atmospheric and oceanic flows: Model formulation, recent developments, and future perspectives. *Geosci. Model Dev.* **2015**, *8*, 2515–2551.

36. Raasch, S.; Schröter, M. PALM—A large-eddy simulation model performing on massively parallel computers. *Meteorol. Z.* **2001**, *10*, 363–372.

37. Hoffmann, F.; Raasch, S.; Noh, Y. Entrainment of aerosols and their activation in a shallow cumulus cloud studied with a coupled LCM–LES approach. *Atmos. Res.* **2015**, *156*, 43–57.

38. Kanani-Sühring, F.; Raasch, S. Spatial Variability of Scalar Concentrations and Fluxes Downstream of a Clearing-to-Forest Transition: A Large-Eddy Simulation Study. *Bound. Layer Meteorol.* **2015**, *155*, 1–27.

39. Beare, R.J.; Macvean, M.K.; Holtslag, A.A.M.; Cuxart, J.; Esau, I.; Golaz, J.C.; Jimenez, M.A.; Khairoutdinov, M.; Kosovic, B.; Lewellen, D.; et al. An Intercomparison of Large-Eddy Simulations of the Stable Boundary Layer. *Bound. Layer Meteorol.* **2006**, *118*, 247–272.

40. Razak, A.A.; Hagishima, A.; Ikegaya, N.; Tanimoto, J. Analysis of airflow over building arrays for assessment of urban wind environment. *Build. Environ.* **2013**, *59*, 56–65.

41. Giometto, M.G.; Christen, A.; Egli, P.E.; Schmid, M.F.; Tooke, R.T.; Coops, N.C.; Parlange, M.B. Effects of trees on mean wind, turbulence and momentum exchange within and above a real urban environment. *Adv. Water Resour.* **2017**, *106*, 154–168.

42. Hellsten, A.; Ketelsen, K.; Barmpas, F.; Tsegas, G.; Moussiopoulos, N.; Raasch, S. Nested multi-scale system in the PALM large-eddy simulation model. In Proceedings of the 35th International Technical Meeting on Air Pollution Modelling and its Application, Chania, Crete, Greece, 3–7 October 2016; Springer: Cham, Switzerland, 2017; pp. 287–292.

43. Kataoka, H.; Mizuno, M. Numerical flow computation around aeroelastic 3D square cylinder using inflow turbulence. *Wind Struct.* **2002**, *5*, 379–392.

44. Zhang, X.; Shu, C.-W. On Maximum-principle-satisfying High Order Schemes for Scalar Conservation Laws. *J. Comput. Phys.* **2010**, *229*, 3091–3120.

45. Auvinen, M.; Järvi, L.; Hellsten, A.; Rannik, U.; Vesala, T. Numerical framework for the computation of urban flux footprints employing large-eddy simulation and Lagrangian stochastic modeling. *Geosci. Model Dev.* **2017**, *10*, 4187–4205.

46. Riikonen, A.; Järvi, L.; Nikinmaa, E. Environmental and crown related factors affecting street tree transpiration in Helsinki, Finland. *Urban Ecosyst.* **2016**, *19*, 1693–1715.

47. Groenendijk, M.; Dolman, A.J.; Ammann, C.; Arneth, A.; Cescatti, A.; Dragoni, D.; Gash, J.H.C.; Gianelle, D.; Gioli, B.; Kiely, G.; et al. Seasonal variation of photosynthetic model parameters and leaf area index from global Fluxnet eddy covariance data. *J. Geophys. Res. Biogeosci.* **2011**, *116*, G04027.

48. Muraoka, H.; Saigusa, N.; Nasahara, K.N.; Noda, H.; Yoshino, J.; Saitoh, T.M.; Nagai, S.; Murayama, S.; Koizumi, H. Effects of seasonal and interannual variations in leaf photosynthesis and canopy leaf area index on gross primary production of a cool-temperate deciduous broadleaf forest in Takayama, Japan. *J. Plant Res.* **2010**, *123*, 563–576.

49. Dupont, S.; Brunet, Y. Edge Flow and Canopy Structure: A Large-Eddy Simulation Study. *Bound. Layer Meteorol.* **2008**, *126*, 51–71.

50. Liu, C.; Leung, D.; Barth, M. On the prediction of air and pollutant exchange rates in street canyons of different aspect ratios using large-eddy simulation. *Atmos. Environ.* **2005**, *39*, 1567–1574.

51. Hellsten, A.; Luukkonen, S.M.; Steinfeld, G.; Kanani-Suehring, F.; Markkanen, T.; Jarvi, L.; Lento, J.; Vesala, T.; Raasch, S. Footprint Evaluation for Flux and Concentration Measurements for an Urban-Like Canopy with Coupled Lagrangian Stochastic and Large-Eddy Simulation Models. *Bound. Layer Meteorol.* **2015**, *157*, 191–217.

52. Pirinen, P.; Simola, H.; Allto, J.; Kaukoranta, J.P.; Karlsson, P.; Ruuhela, R. *Climatological Statistics of Finland 1981–2010*; Finnish Meteorological Institute: Helsinki, Finland, 2012.

53. Karsisto, P.; Fortelius, C.; Demuzere, M.; Grimmond, C.S.B.; Oleson, K.W.; Kouznetsov, R.; Masson, V.; Järvi, L. Seasonal surface urban energy balance and wintertime stability simulated using three land-surface models in the high-latitude city Helsinki. *Q. J. R. Meteorol. Soc.* **2016**, *142*, 401–417.

54. Kurppa, M.; Nordbo, A.; Haapanala, S.; Järvi, L. Effect of seasonal variability and land use on particle number and CO_2 exchange in Helsinki, Finland. *Urban Clim.* **2015**, *13*, 94–109.

55. Cheng, W.; Liu, C.H. Large-eddy simulation of turbulent transports in urban street canyons in different thermal stabilities. *J. Wind Eng. Ind. Aerodyn.* **2011**, *99*, 434–442.

56. Li, X.X.; Britter, R.; Norford, L.K. Effect of stable stratification on dispersion within urban street canyons: A large-eddy simulation. *Atmos. Environ.* **2016**, *144*, 47–59.

57. Basu, S.; Porté-Agel, F. Large-Eddy Simulation of Stably Stratified Atmospheric Boundary Layer Turbulence: A Scale-Dependent Dynamic Modeling Approach. *J. Atmos. Sci.* **2006**, *63*, 2074–2091.

58. Bentham, T.; Britter, R. Spatially averaged flow within obstacle arrays. *Atmos. Environ.* **2003**, *37*, 2037–2043.

59. Kato, S.; Ito, K.; Murakami, S. Analysis of visitation frequency through particle tracking method based on LES and model experiment. *Indoor Air* **2003**, *13*, 182–193.

60. Lo, K.; Ngan, K. Characterising the pollutant ventilation characteristics of street canyons using the tracer age and age spectrum. *Atmos. Environ.* **2015**, *122*, 611–621.

61. Lim, E.; Ito, K.; Sandberg, M. New ventilation index for evaluating imperfect mixing conditions—Analysis of Net Escape Velocity based on RANS approach. *Build. Environ.* **2013**, *61*, 45–56.

62. Liu, C.H.; Barth, M.C.; Leung, D.Y.C. Large-Eddy Simulation of Flow and Pollutant Transport in Street Canyons of Different Building-Height-to-Street-Width Ratios. *J. Appl. Meteorol.* **2004**, *43*, 1410–1424.

63. Michioka, T.; Takimoto, H.; Sato, A. Large-Eddy Simulation of Pollutant Removal from a Three-Dimensional Street Canyon. *Bound. Layer Meteorol.* **2014**, *150*, 259–275.

64. Walton, A.; Cheng, A. Large-eddy simulation of pollution dispersion in an urban street canyon—Part II: Idealised canyon simulation. *Atmos. Environ.* **2002**, *36*, 3615–3627.

65. Nosek, Š.; Kukačka, L.; Kellnerová, R.; Jurčáková, K.; Jaňour, Z. Ventilation Processes in a Three-Dimensional Street Canyon. *Bound. Layer Meteorol.* **2016**, *159*, 259–284.

66. Moon, K.; Hwang, J.M.; Kim, B.G.; Lee, C.; Choi, J.I. Large-eddy simulation of turbulent flow and dispersion over a complex urban street canyon. *Environ. Fluid Mech.* **2014**, *14*, 1381–1403.

67. Dabberdt, W.F.; Hoydysh, W.G. Street canyon dispersion: Sensitivity to block shape and entrainment. *Atmos. Environ. Part A Gen. Top.* **1991**, *25*, 1143–1153.

68. Ossanlis, I.; Barmpas, P.; Moussiopoulos, N. The Effect of the Street Canyon Length on the Street Scale Flow Field and Air Quality: A Numerical Study. In *Air Pollution Modeling and Its Application XVII*; Borrego, C., Norman, A.L., Eds.; Springer: Boston, MA, USA, 2007; pp. 632–640.

69. Hoydysh, W.G.; Dabberdt, W.F. Kinematics and dispersion characteristics of flows in asymmetric street canyons. *Atmos. Environ.* **1988**, *22*, 2677–2689.

70. Xiaomin, X.; Zhen, H.; Jiasong, W. The impact of urban street layout on local atmospheric environment. *Build. Environ.* **2006**, *41*, 1352–1363.

71. Chan, A.T.; Au, W.T.; So, E.S. Strategic guidelines for street canyon geometry to achieve sustainable street air quality—Part II: Multiple canopies and canyons. *Atmos. Environ.* **2003**, *37*, 2761–2772.

72. Gu, Z.L.; Zhang, Y.W.; Cheng, Y.; Lee, S.C. Effect of uneven building layout on air flow and pollutant dispersion in non-uniform street canyons. *Build. Environ.* **2011**, *46*, 2657–2665.

73. Hang, J.; Wang, Q.; Chen, X.; Sandberg, M.; Zhu, W.; Buccolieri, R.; Sabatino, S.D. City breathability in medium density urban-like geometries evaluated through the pollutant transport rate and the net escape velocity. *Build. Environ.* **2015**, *94 Pt 1*, 166–182.

74. Hu, T.; Yoshie, R. Indices to evaluate ventilation efficiency in newly-built urban area at pedestrian level. *J. Wind Eng. Ind. Aerodyn.* **2013**, *112*, 39–51.

75. Yang, F.; Gao, Y.; Zhong, K.; Kang, Y. Impacts of cross-ventilation on the air quality in street canyons with different building arrangements. *Build. Environ.* **2016**, *104*, 1–12.

76. Moonen, P.; Dorer, V.; Carmeliet, J. Evaluation of the ventilation potential of courtyards and urban street canyons using RANS and LES. *J. Wind Eng. Ind. Aerodyn.* **2011**, *99*, 414–423.

77. Harlow, F.H.; Welch, J.E. Numerical calculation of Time-Dependent viscous incompressible flow of fluid with free surface. *Phys. Fluids* **1965**, *8*, 2182–2189.

78. Arakawa, A.; Lamb, V.R. Computational Design of the Basic Dynamical Processes of the UCLA General Circulation Model. In *General Circulation Models of the Atmosphere*; Chang, J., Ed.; Methods in Computational Physics: Advances in Research and Applications; Elsevier: Amsterdam, The Netherlands, 1977; Volume 17, pp. 173–265.

79. Deardorff, J.W. Stratocumulus-capped mixed layers derived from a three-dimensional model. *Bound. Layer Meteorol.* **1980**, *18*, 495–527.

80. Wicker, L.; Skamarock, W. Time-splitting methods for elastic models using forward time schemes. *Mon. Weather Rev.* **2002**, *130*, 2088–2097.

81. Hackbusch, W. *Multi-Grid Methods and Applications*, 1st ed.; Springer: Berlin/Heidelberg, Germany, 1985.

82. Williamson, J.H. Low-storage Runge-Kutta schemes. *J. Comput. Phys.* **1980**, *35*, 48–56.

83. Monin, A.S.; Obukhov, A. Basic laws of turbulent mixing in the surface layer of the atmosphere (in Russian). *Contrib. Geophys. Inst. Acad. Sci. USSR* **1954**, *24*, 163–187.

atmosphere

MDPI

Article

On-Road Air Quality Associated with Traffic Composition and Street-Canyon Ventilation: Mobile Monitoring and CFD Modeling

Kyung-Hwan Kwak [1], Sung Ho Woo [2], Kyung Hwan Kim [2,3], Seung-Bok Lee [2,*], Gwi-Nam Bae [2,3], Young-Il Ma [4], Young Sunwoo [4] and Jong-Jin Baik [5]

[1] School of Natural Resources and Environmental Science, Kangwon National University, Chuncheon 24341, Korea; khkwak@kangwon.ac.kr

[2] Center for Environment, Health and Welfare Research, Korea Institute of Science and Technology, Seoul 02792, Korea; sungho2236@nate.com (S.H.W.); khkim@kist.re.kr (K.H.K.); gnbae@kist.re.kr (G.-N.B.)

[3] Center for Particulate Air Pollution and Health, Korea Institute of Science and Technology, Seoul 02792, Korea

[4] Department of Environmental Engineering, Konkuk University, Seoul 05029, Korea; bluesky@udi.re.kr (Y.-I.M.); ysunwoo@konkuk.ac.kr (Y.S.)

[5] School of Earth and Environmental Sciences, Seoul National University, Seoul 08826, Korea; jjbaik@snu.ac.kr

* Correspondence: sblee2@kist.re.kr; Tel.: +82-29-585-821

Received: 1 February 2018; Accepted: 24 February 2018; Published: 2 March 2018

Abstract: Mobile monitoring and computational fluid dynamics (CFD) modeling are complementary methods to examine spatio-temporal variations of air pollutant concentrations at high resolutions in urban areas. We measured nitrogen oxides (NO_x), black carbon (BC), particle-bound polycyclic aromatic hydrocarbons (pPAH), and particle number (PN) concentrations in a central business district using a mobile laboratory. The analysis of correlations between the measured concentrations and traffic volumes demonstrate that high emitting vehicles (HEVs) are deterministically responsible for poor air quality in the street canyon. The determination coefficient (R^2) with the HEV traffic volume is the largest for the pPAH concentration (0.79). The measured NO_x and pPAH concentrations at a signalized intersection are higher than those on a road between two intersections by 24% and 25%, respectively. The CFD modeling results reveal that the signalized intersection plays a role in increasing on-road concentrations due to accelerating and idling vehicles (i.e., emission process), but also plays a countervailing role in decreasing on-road concentrations due to lateral ventilation of emitted pollutants (i.e., dispersion process). It is suggested that the number of HEVs and street-canyon ventilation, especially near a signalized intersection, need to be controlled to mitigate poor air quality in a central business district of a megacity.

Keywords: on-road air quality; traffic composition; high emitting vehicles; street canyon; mobile laboratory; CFD model

1. Introduction

Vehicle emission is one of the major sources of air pollution in urban areas. Exposure to air pollutants that are emitted from vehicles can cause human health problems, such as cardiovascular, respiratory, and allergic diseases [1–3]. Diesel exhaust was recently classified as Group 1 carcinogen by the International Agency for Research on Cancer [4].

On-road emissions of primary pollutants (e.g., nitrogen oxides ($NO_x = NO + NO_2$), carbon monoxide (CO), and particulate matter) have been estimated based on nationwide annual databases and are allocated with specified temporal and spatial profiles. In estimating on-road emissions, an emission factor or emission rate of air pollutant is calculated using various factors. Across a nation, the number of registered vehicles, traffic volume and composition, fuel-use fraction, and vehicle age are well reported. For example, on-road emission studies have stated that heavy-duty diesel vehicles (HDDVs), or heavy emitters, dominantly contribute up to 50% of NO_2, 41% of NO_x, 51% of CO, 77% of black carbon (BC), 70% of polycyclic aromatic hydrocarbons (PAHs), 50% of particle number (PN), and 60% of particulate matter emissions [5–10]. It is challenging to represent the effects of actual driving situations on on-road emissions because driving mode (i.e., idling, accelerating, cruising, and decelerating) and road configuration (i.e., intersection, upslope, and downslope) spatio-temporally vary at small scale, particularly in urban areas [11,12]. Among various driving situations, congestion with frequent accelerations at low vehicle speeds [13–15], signalized traffic intersection [16,17], and upslope [7,18] are found to aggravate on-road air quality.

Intra-urban variability of on-road air quality is largely associated with not only vehicle emission distribution [19,20], but also urban built environment [21–23]. A street canyon is a space that is enclosed by road and roadside buildings on both sides and is often characterized by heavy traffic, poor ventilation, and a large floating population. The street canyon has attracted many researchers for pollutant emission and dispersion studies because of the direct impacts on human health and the excellence as an experimental space similar to a road tunnel. To investigate the distribution of near-or on-road air quality, mobile measurements have been utilized to overcome certain limitations to the representation of spatial variability when using stationary measurements [24–29]. A few studies using mobile measurements have focused on the street-canyon environment in the central business districts of New York City [30], Hong Kong [31], Seoul [32], and Thessaloniki [33], where situations are highly complex. The first aim of this study is to investigate on-road air quality in a street canyon in a central business district using a mobile laboratory (ML) and the relationship between the on-road air quality and individual traffic compositions that are heavily responsible for pollutant emissions.

Air quality in the street canyon is significantly influenced by dispersion characteristics due to in-canyon ventilation along with ambient wind as well as vehicle emission [34,35]. To take the dispersion characteristics into account suitably, numerical models have been utilized to examine on-road emission and dispersion in the urban built environment [36–40]. Computational fluid dynamics (CFD) models can explicitly resolve roads and buildings at high resolutions [41–43]. Mobile monitoring and CFD modeling are complementary methods to examine the temporal and spatial patterns of air quality in street canyons. CFD modeling requires a real-world pollutant emission rate on roads as a boundary condition for any pollutant concentration, while mobile monitoring results need to be interpreted by considering pollutant dispersion with the presence of building and ambient wind effects. The second aim of this study is to investigate the horizontal distribution of on-road air quality in a street canyon in a central business district by applying CFD modeling results for analyzing mobile monitoring results.

2. Methods

2.1. Site Description

The study area of interest is a street canyon of eight-lane Teheran road with a southwest-northeast (i.e., 249° and 69° from the due north) orientation in a central business district of Seoul, which is a megacity in Republic of Korea (Figure 1a). The areas surrounding Teheran road are highly-developed commercial and residential areas, and a vegetated area is located in the north. Many high-rise buildings are densely located on both sides in the 1.8 km-long street canyon crossing with four-lane and six-lane roads at signalized intersections A and B, respectively (Figure 1b,c). The elevation of road gradually decreases from the southwest to the northeast by a few meters.

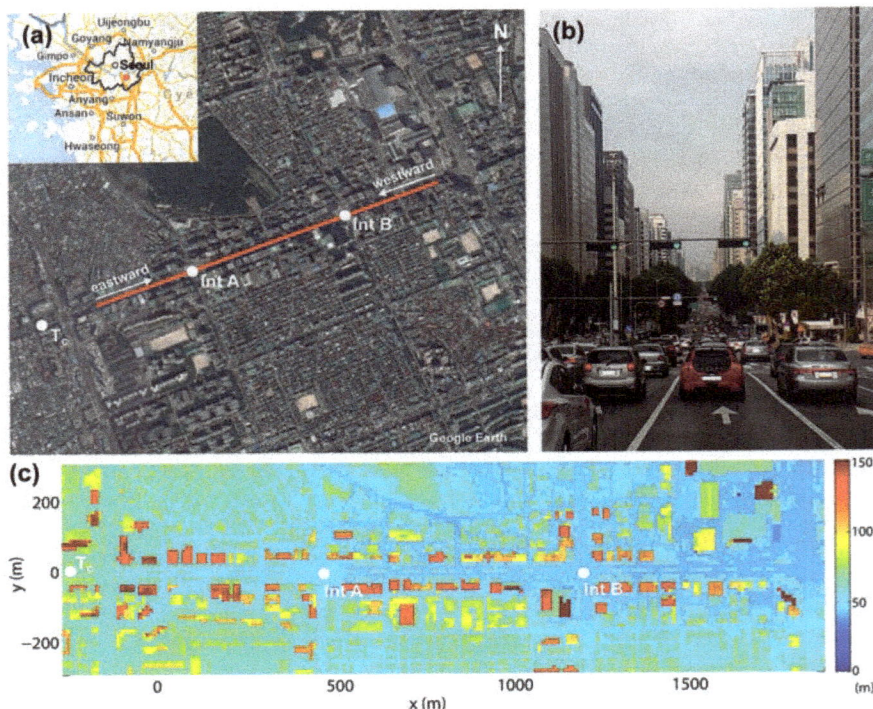

Figure 1. (a) Satellite image (Google Earth) covering a mobile monitoring route and a traffic count location (T_c). Inset on the left-top corner is a map of Seoul metropolitan area with the monitoring area indicated by a red dot. **(b)** Photo of Teheran road at a location near T_c. **(c)** Building-top and topographical heights in and around Teheran road in the CFD model domain.

The traffic volumes of private cars, taxis, RV/SUVs, trucks, vans, buses, and motorcycles were manually counted at T_c. At first, traffic flows on the road were continuously recorded using a video camera in a roadside building for a week. Then, the number of vehicles in each traffic composition was counted for 15 min consecutively. Note that diesel and gasoline vehicles were not distinguished during this process. The diurnal variations of hourly traffic volumes on 5–8 November 2013 are shown in Figure 2. The traffic pattern on Teheran road is different from the typical bimodal pattern with maxima during the morning and evening rush-hours in urban areas. The daily traffic volume is over 90,000 vehicles on weekdays, and the maximum hourly traffic volume appears in the middle of working hours (i.e., 11:00–12:00 LT). The hourly traffic volume larger than 4000 vehicles consistently appears from 08:00 to 23:00 LT. Among the seven traffic compositions, private car (50%) accounts for the largest portion of traffic volume followed by taxi (28%), RV/SUV (9%), and truck (4%). Here, RV/SUVs, trucks, vans, and buses are grouped into high emitting vehicles (HEVs) based on their conventional vehicle emissions. The daily HEV traffic volume is 19% of the daily total traffic volume. The HEV portion of hourly traffic volume reaches a maximum during the morning rush hour (07:00–08:00 LT) (29%) and remains larger than 20% (06:00–17:00 LT) till the working hours.

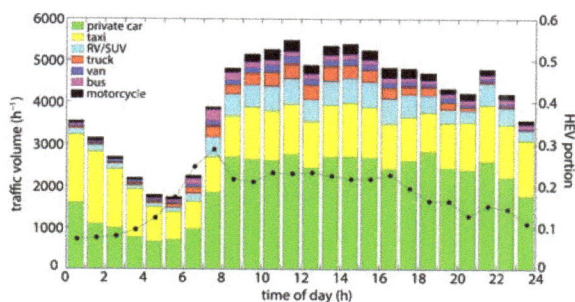

Figure 2. Hourly traffic volumes of private car, taxi, RV/SUV, truck, van, bus, and motorcycle and hourly HEV portion (black dot) counted at T_c for 72 h from 04:00 LT on 5 November to 04:00 LT on 8 November 2013.

2.2. Mobile Monitoring

The mobile monitoring was repeatedly conducted in the 1.8 km-long street canyon of Teheran road on 5–8 November 2013 (Table 1). During the three monitoring periods, the predominant wind directions were the along-canyon direction (period 1), opposite along-canyon direction (period 2), and direction oblique to along-canyon direction (period 3). An ML has been developed and operated for the last few years to monitor on-road concentrations of gaseous and particulate air pollutants in urban areas [16,32]. The sampling inlet in front of the ML at a 2-m height consists of a Teflon tube and a stainless steel pipe for gaseous and particulate pollutants, respectively. Table 2 provides the brief information on the instruments equipped in the ML. NO_x, BC, particle-bound PAH (pPAH), and PN (>5 nm) concentrations were measured using a chemiluminescence NO_x analyzer (AC32M, Environmental S.A., Poissy, France), an aethalometer (AE42, Magee Scientific, Berkeley, CA, USA) with an impactor of 2.5 μm cut-off diameter, a photoelectric aerosol sensor (PAS2000, EcoChem Analytics, League City, TX, USA), and a condensation particle counter (CPC model 5.403, GRIMM, Ainring, Germany), respectively. In the pre-processing step, the NO_x analyzer was calibrated using pure air and standard NO gas, and the instruments for particulate pollutants were initialized and/or zero checked. The time in the data acquisition system was synchronized with the global positioning system (GPS) data that was recorded every 1 s by a GPS data logger (GPS742, Ascen, Seoul, Republic of Korea). In the post-processing step, delay times of instruments in the ML were adjusted.

Table 1. Overview of mobile monitoring using the ML. The dates of monitoring period 1, 2, and 3 were 5, 5, and 7–8 November, respectively.

Monitoring Period	Time of Day	Number of Trips	Wind Speed [a] (m s^{-1})	Predominant Wind Direction [a]
1	04:21 to 09:00 LT	13	1.4	ENE (parallel)
2	17:13 to 23:00 LT	13	1.7	WSW (parallel)
3	23:02 to 04:00 LT	13	2.0	WNW (diagonal)

[a] Wind speed and direction were observed at the nearest automatic weather station with a 950-m distance from Teheran road.

Table 2. Overview of instruments equipped in the ML.

Pollutant	Instrument	Flow Rate (L min^{-1})	Time Resolution (s)	Delay Time (s)
NO_x	AC32M, Environmental S.A.	1	5	19
BC	AE42, Magee Scientific	5	30	28
pPAH	PAS2000, EcoChem Analytics	2	6	13
PN	CPC model 5.403, GRIMM	1.5	1	18

The on-road measurements of NO_x, BC, pPAH, and PN concentrations during the three periods are divided into every two-way trip starting from the eastern end of the street canyon. Thirteen trips for each period are valid after excluding incomplete and interrupted measurements. In every trip, the median concentration is calculated in every 100-m section in the street canyon for each driving direction. A 100-m spacing is acceptable for studying the spatial variation of air pollutant near a major road [44]. For calculating the median concentration in a 100-m section, the measured concentrations with a vehicle speed higher than 20 km h^{-1} are used to avoid the effects of exhaust plumes of a vehicle directly ahead [45].

2.3. CFD Modeling

The CFD model used in this study is a Reynolds-averaged Navier–Stokes equations (RANS) model with the renormalization group (RNG) *k-ε* turbulence closure scheme [46]. The RANS model has been validated and used for simulating urban flow and pollutant dispersion, particularly in street canyons [43,47,48]. Numerical simulations are performed for the three dispersion scenarios corresponding to mobile monitoring periods 1, 2, and 3. The isothermal condition is considered because of sufficiently low air temperature (7–17 °C) and reduced or absent daylight during the mobile monitoring periods. The size of the domain covering the 1.8-km long street canyon oriented in the *x*-direction and surrounding commercial and residential areas is 4000 m in the *x*-direction, 1800 m in the *y*-direction, and 997 m in the *z*-direction. The grid size is 10 m in the *x*- and *y*-direction and 4 m in the *z*-direction up to *z* = 200 m, with an increasing ratio (1.05) of vertical grid size above the height. The surface geometries in the domain were obtained from the airborne light detection and ranging (LIDAR) measurement (Figure 1c). The logarithmic vertical wind profile is applied at the inflow boundaries. Inflow wind speed at the top boundary (*z* = 997 m) and direction are 1.4 m s^{-1} and 69° from the due north for the period-1 scenario, 1.7 m s^{-1} and 249° from the due north for the period-2 scenario, and 2.0 m s^{-1} and 292° from the due north for the period-3 scenario. A zero-gradient boundary condition is applied at the outflow boundaries. At all of the boundaries, the vertical velocity is specified to be zero. The inflow pollutant concentration is set to zero, and the inert pollutant emission is identically considered at a same emission rate for all the lowest grids on major roads (the number of lanes $\geq$ 2) in the domain. As the vertical size of lowest grids on roads is 4 m, it is accordingly assumed that the emitted pollutants at the tailpipe are instantaneously mixed by the vehicle-induced turbulence within the extents of grids. On Teheran road, four rows of grids corresponding to the 40-m width in the *y*-direction are considered to be the pollutant emission points. It is noteworthy that the emission setting is deviated from the real-world emission distribution in order to solely examine the dispersion of emitted pollutants on roads in different dispersion scenarios. The model integration time is 120 min, with a time step of 1 s. Pollutant emission is activated after 10-min integration.

3. Results and Discussion

3.1. Association with Traffic Composition

The measured NO_x, BC, pPAH, and PN concentration for a trip is obtained by averaging their median concentrations of all 100-m sections. Similar to the concentrations, the traffic volumes collected for 15 min consecutively during the mobile monitoring periods are assigned to the corresponding trips. Then, the traffic volumes of each composition are averaged for each trip. The associations of the measured concentrations with the traffic volumes of individual compositions are analyzed by calculating the determination coefficients (R^2) based on the linear regressions for 39 trips in the street canyon (Table 3). Although private car has the largest portion of traffic volumes, the correlations with the NO_x, BC, pPAH, and PN concentrations are intermediate (0.36, 0.26, 0.41, and 0.42, respectively). Taxi has the second largest portion of traffic volumes and is more weakly correlated with the concentrations than private car. This is because the majority of private car and taxi are gasoline-and liquefied petroleum gas (LPG)-powered vehicles, respectively. In contrast to private car and taxi,

RV/SUV for which the portion of traffic volumes is only 9% is strongly correlated with the NO_x, BC, pPAH, and PN concentrations (R^2 = 0.52, 0.44, 0.62, and 0.44, respectively). The emission amount of a RV/SUV is undoubtedly smaller than that of a truck, van, or bus. However, the 2–4 times larger traffic volume of RV/SUV than those of truck, van, and bus results in stronger or similar correlations with the concentrations. The first and second largest determination coefficients among the seven traffic compositions for any measured pollutant correspond to one of the HEV compositions (i.e., RV/SUV, truck, van, or bus). It is thought that the on-road emission capacities of individual compositions grouped into HEVs are comparable to each other and obviously larger than those of private car, taxi, and motorcycle in the street canyon.

Table 3. Determination coefficients (R^2) between NO_x, BC, pPAH, and PN concentrations and traffic volumes of individual compositions based on the linear regression.

Pollutant	Private Car	Taxi	RV/SUV	Truck	Van	Bus	Motorcycle
NO_x	0.36	0.27	**0.52**	0.34	0.21	**0.53**	0.08
BC	0.26	0.21	**0.44**	0.33	0.28	0.38	0.05
pPAH	**0.41**	0.14	**0.62**	0.38	**0.44**	0.32	0.07
PN	**0.42**	0.21	**0.44**	**0.43**	0.31	0.30	0.16

Bold numbers indicate R^2 larger than 0.4.

The scatter diagrams (Figure 3) exhibit that the measured NO_x, BC, pPAH, and PN concentrations are well correlated with the traffic volumes. Among the measured pollutants, the highest correlations with the traffic volumes are shown for the pPAH concentration (R^2 = 0.41 and 0.79 for the total and HEV traffic volumes, respectively), whereas the lowest ones are shown for the BC concentration (R^2 = 0.23 and 0.61 for the total and HEV traffic volumes, respectively). The highest correlations for the pPAH concentration are attributed to the dominant contribution of local mobile sources to the concentration in a roadside environment [49]. Regardless of pollutants, the correlations with the HEV traffic volume are significantly higher than those with the total traffic volume. This implies that only 19% of the total traffic volume is capable of describing the on-road pollutant concentrations in the street canyon. The relative importance of HEVs to the on-road pollutant concentrations is assessed with the linear regression lines in the scatter diagrams. When the pollutant concentration is ideally correlated with the traffic volume, the slope and y-intercept of a regression line indicate an increase in concentration per an increase in traffic volume and a base concentration with no vehicle emission on a road, respectively. The slopes of the regression lines for the HEV traffic volume are distinctly 4–5 times larger than those for the total traffic volume. The increase in concentration responding to the increase in HEV traffic volume is 266 ppb for NO_x, 5.6 $\mu g\ m^{-3}$ for BC, 207 ng m^{-3} for pPAH, and 3.28 $\times 10^4\ cm^{-3}$ for PN per 1000 HEVs. Although the increase in concentration is not entirely attributed to the HEV emission, the concentration gradient with respect to the HEV traffic volume can be a useful indicator to the on-road NO_x, BC, pPAH, and PN concentrations in the street canyon.

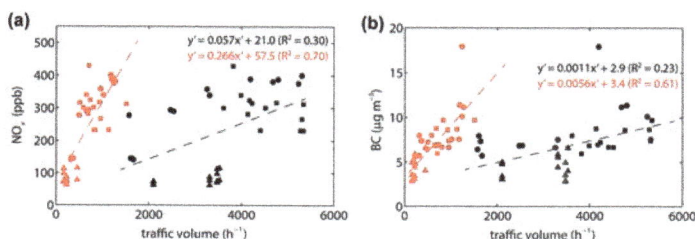

Figure 3. *Cont.*

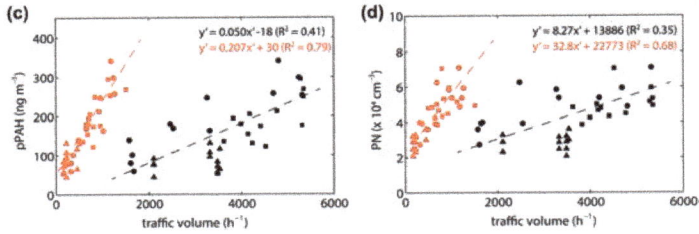

Figure 3. Scatter diagrams between measured (**a**) NO$_x$, (**b**) BC, (**c**) pPAH, and (**d**) PN concentrations and total (black) and HEV (red) traffic volumes during mobile monitoring periods 1 (circle), 2 (square), and 3 (triangle) with linear regression lines for total (black dashed line) and HEV (red dashed line) traffic volumes, separately.

3.2. Association with Street-Canyon Ventilation

The measured NO$_x$, BC, pPAH, and PN concentrations are unevenly distributed in the 1.8-km street canyon. The unevenly distributed concentrations in the street canyon are attributed to the combination of uneven driving modes due to traffic signals and congestion and irregular street-canyon ventilation. In the following analysis, the distributions of measured concentrations in the street canyon are presented first and later converted by taking account of the influence of pollutant dispersion through CFD model simulations for the three dispersion scenarios.

Figure 4 shows the horizontal distributions of measured NO$_x$, BC, pPAH, and PN concentrations and ML speed in the street canyon. The median NO$_x$, BC, pPAH, and PN concentrations and ML speed in 100-m sections for 13 trips are averaged over each mobile monitoring period. Among the three periods, all of the measured concentrations in 100-m sections are the lowest for period 3 (i.e., nighttime) when the traffic volume is the smallest. The average NO$_x$, BC, and pPAH concentrations that were measured during the westward driving are generally higher than those that were measured during the eastward driving. This is because of slight upsloping and also congested traffic flow in the westward driving direction. The average PN concentrations tend to be invariable, regardless of driving direction. The NO$_x$ and pPAH concentrations at the two signalized intersections are higher than those between the intersections by up to 24% and 25%, respectively (Table 4). The higher NO$_x$ and pPAH concentrations are more significant at signalized intersection B than at signalized intersection A. This is because vehicle emissions are larger at signalized intersection B due to accelerating and idling vehicles, not only on Teheran road, but also on the six-lane road that perpendicularly cross with the street canyon. The differences between concentrations at the intersections and those between the intersections for BC and PN are smaller (<±10%) than those for NO$_x$ and pPAH (Table 4). The *y*-intercepts of linear regression lines for HEV traffic volumes in Figure 3 may explain the base NO$_x$, BC, pPAH, and PN concentrations (57.5 ppb, 3.4 µg m^{-3}, 30 ng m^{-3}, and 22,773 cm^{-3}, respectively). The NO$_x$ and pPAH concentrations at the intersections are 4–6 times higher than the base concentrations, while BC and PN concentrations at the intersections are only about two times higher than the base concentrations. The BC and PN concentrations are less localized than the NO$_x$ and pPAH concentrations at the signalized intersections partly because the time resolution for the BC measurement is relatively large, and also because some influencing sources outside other than HEVs are considerably involved in determining on-road BC and PN concentrations.

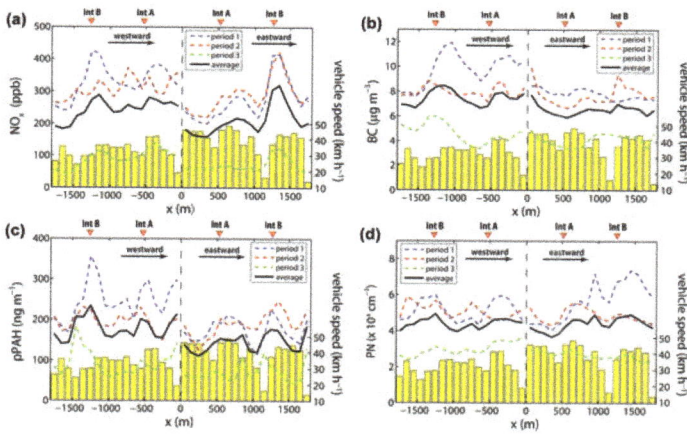

Figure 4. Horizontal distributions of measured (**a**) NO_x, (**b**) BC, (**c**) pPAH, and (**d**) PN concentrations during mobile monitoring periods 1 (blue dashed line), 2 (red dashed line), and 3 (green dashed line) and averaged over the three monitoring periods (black solid line). The horizontal distribution of vehicle speed averaged over the three monitoring periods is given by yellow bars. The concentrations with negative (positive) distance in x on the left (right) side were measured by the ML driven westward (eastward).

Table 4. Comparison of NO_x, BC, pPAH, and PN concentrations at the two signalized intersections with those between the intersections (Btw Int). The 300-m and 400-m sections corresponding to signalized intersections A and B are centered at ± 550 and ± 1300 m, respectively. The numbers in parentheses indicate the relative percentages of concentrations at the intersections to those between the intersections.

Pollutant	Measured Concentration			Converted Concentration		
	Btw Int	Int A	Int B	Btw Int	Int A	Int B
NO_x (ppb)	214	226 (6)	265 (24)	212	227 (7)	333 (57)
BC ($\mu g\ m^{-3}$)	6.98	6.57 (–6)	7.34 (5)	–	–	–
pPAH (ng m^{-3})	151	167 (11)	189 (25)	157	198 (26)	225 (43)
PN ($\times 10^4\ cm^{-3}$)	4.32	4.30 (–1)	4.68 (8)	–	–	–

Scalar dispersion for the three different scenarios corresponding to mobile monitoring periods 1, 2, and 3 is numerically examined through the CFD model simulations with an identical distribution of on-road pollutant emission. Scalar dispersion has been commonly interpreted as dispersion of chemically inactive gases or ultrafine particles at small scale such as road environment [50]. Figure 5 shows the horizontal distributions of normalized on-road wind speed and pollutant concentration (hereafter, model coefficient) and of building height on the driving side in the street canyon for the three dispersion scenarios. The on-road wind speeds are normalized by the ambient (inflow) wind speed for each simulation. On the other hand, the on-road concentrations are normalized by the average on-road concentration over the three simulations. By doing this, the influence of ambient wind direction is included in both horizontal distributions of normalized on-road wind speeds and concentrations, whereas the influence of ambient wind speed is solely included in the horizontal distribution of normalized on-road concentrations. Therefore, the average of all the model coefficients for the three simulations is unity. The normalized on-road wind speed is generally higher for the period-1 and period-2 dispersion scenarios than for the period-3 dispersion scenario, since the ambient wind directions are fairly parallel to the canyon orientation for the period-1 and period-2 dispersion scenarios. The higher on-road wind speed for the parallel ambient wind directions is attributed to channeling flows that formed in the street canyon [51,52]. However, the normalized on-road pollutant concentration is higher for the period-1 and period-2 dispersion scenarios than for the period-3 dispersion scenario. This is because the emitted pollutants hardly escape from the street canyon for the parallel ambient wind directions. The ambient wind speed is slightly lower during periods 1 and 2 than during period 3, which also contributes to the inverse-proportional increase in on-road pollutant concentration. As a result, the ambient wind direction as well as the ambient wind speed is one of the determining factors in estimating the on-road pollutant concentrations in the street canyon. The normalized on-road wind speed is obviously high at the intersections where roadside buildings rarely obstruct lateral ventilation. The normalized on-road pollutant concentrations are consequently low at the intersections, particularly at intersection B wider than intersection A.

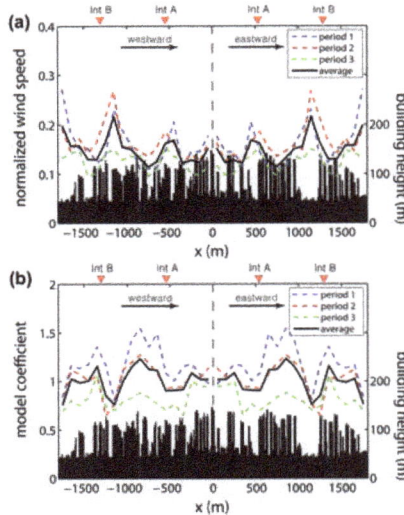

Figure 5. Horizontal distributions of normalized (**a**) wind speed and (**b**) scalar concentration (i.e., model coefficient) for the period-1 (blue dashed line), period-2 (red dashed line), and period-3 (green dashed line) dispersion scenarios and averaged over the three dispersion scenarios (black solid line). The horizontal distributions of normalized wind speed and scalar concentration are symmetric with $x = 0$ m as the centerline of symmetry. The horizontal distribution of building height on the driving side is given by dark gray bars.

It is interesting that emission and dispersion processes play counterbalancing roles in determining any pollutant concentration at a signalized intersection in the street canyon. At a signalized intersection, pollutants are intensively emitted from idling and accelerating vehicles and are ventilated out of the street canyon. Figure 6 exhibits the horizontal distributions of converted NO_x and pPAH concentrations in the street canyon. The converted concentration is the measured concentration (shown in Figure 4) multiplied by a reciprocal of the model coefficient (shown in Figure 5b). In the process, the influence of pollutant dispersion is excluded from the measured concentration. In the calculation of converted concentrations, the BC and PN concentrations are not taken into account because their base concentrations seem to be comparable to their on-road concentrations. The horizontal variations of converted NO_x and pPAH concentrations become more pronounced since the converted NO_x and pPAH concentrations are elevated at intersections A and B. The converted NO_x and pPAH concentrations are higher than those between the intersections by 7% and 26% at signalized intersection A, respectively, and by 57% and 43% at signalized intersection B, respectively (Table 4). The higher converted concentrations than the measured concentrations are more evident at signalized intersection

B than at signalized intersection A. This reveals that signalized intersection B is a distinct hotspot where the dispersion process acts to alleviate severe air pollution due to the emission process. In conclusion, the emission process aggravates on-road air quality in the street canyon, and is efficiently compensated by the dispersion process at signalized intersection B. Later, the horizontal distributions of converted NO_x and pPAH concentrations can be utilized as the real-world on-road emission for realistic CFD modeling.

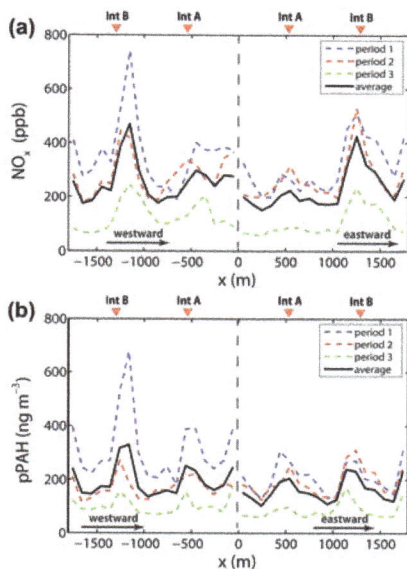

Figure 6. Horizontal distributions of converted (**a**) NO_x and (**b**) pPAH concentrations during mobile monitoring periods 1 (blue dashed line), 2 (red dashed line), and 3 (green dashed line) and averaged over the three monitoring periods (black solid line).

4. Conclusions

Spatio-temporal variations of air pollutant concentrations in a street canyon in a central business district of Seoul, Republic of Korea were investigated on multiple days based on complementary approaches using an ML and a CFD model. The ML monitored primary pollutants that were emitted from on-road vehicles, and traffic volume and composition were recorded at the same time. The CFD model simulates pollutant dispersion in the presence of proper inflow boundary conditions and real building morphology. In the emission and dispersion processes of on-road air pollutants, the HEV portion and the street-canyon ventilation are the determining factors of the spatio-temporal variations in their on-road concentrations. Among the seven traffic compositions, RV/SUV appears to be the most responsible for poor air quality in the street canyon. A signalized intersection that is commonly characterized as a traffic hotspot exhibits countervailing roles between the emission and dispersion processes. In the street canyon, air quality at a signalized intersection is aggravated by up to 25% over that between signalized intersections due to the emission increase that is partially compensated by efficient lateral ventilation. Consequently, controlling the number of HEVs and the in-canyon ventilation near signalized intersections can effectively manage on-road air quality in street canyons.

Acknowledgments: This work was supported by the Korea Auto-Oil Program (13-04-10) and the Institutional Program (2E28160) of the Korea Institute of Science and Technology.

Author Contributions: Kyung-Hwan Kwak conducted CFD modeling and wrote the paper; Sung Ho Woo, Kyung Hwan Kim, and Seung-Bok Lee conducted mobile monitoring; Seung-Bok Lee and Gwi-Nam Bae managed the research project and contributed to the discussion of mobile monitoring results; Young-Il Ma and Young Sunwoo analyzed the traffic data; Jong-Jin Baik contributed to CFD modeling.

Conflicts of Interest: The authors declare no conflict of interest.

References

1. Bernard, S.M.; Samet, J.M.; Grambsch, A.; Ebi, K.L.; Romieu, I. The potential impacts of climate variability and change on air pollution-related health effects in the United States. *Environ. Health Perspect.* **2001**, *109*, 199–209. [CrossRef] [PubMed]

2. Kim, B.-J.; Lee, S.-Y.; Kwon, J.-W.; Jung, Y.-H.; Lee, E.; Yang, S.I.; Kim, H.-Y.; Seo, J.-H.; Kim, H.-B.; Kim, H.-C.; et al. Traffic-related air pollution is associated with airway hyperresponsiveness. *J. Allergy Clin. Immunol.* **2014**, *133*, 1763–1765. [CrossRef] [PubMed]

3. Halonen, J.I.; Blangiardo, M.; Toledano, M.B.; Fecht, D.; Gulliver, J.; Anderson, H.R.; Beevers, S.D.; Dajnak, D.; Kelly, F.J.; Tonne, C. Long-term exposure to traffic pollution and hospital admissions in London. *Environ. Pollut.* **2016**, *208*, 48–57. [CrossRef] [PubMed]

4. IARC. *Diesel Engine Exhaust Carcinogenic*; International Agency for Research on Cancer, World Health Organization: Lyon, France, 2012; Available online: http://www.iarc.fr/en/media-centre/pr/2012/pdfs/pr213_E.pdf (accessed on 1 March 2018).

5. Riddle, S.G.; Robert, M.A.; Jakober, C.A.; Hannigan, M.P.; Kleeman, M.J. Size-resolved source apportionment of airborne particle mass in a roadside environment. *Environ. Sci. Technol.* **2008**, *42*, 6580–6586. [CrossRef] [PubMed]

6. Wang, X.; Westerdahl, D.; Wu, Y.; Pan, X.; Zhang, K.M. On-road emission factor distributions of individual diesel vehicles in and around Beijing, China. *Atmos. Environ.* **2011**, *45*, 503–513. [CrossRef]

7. Wang, X.; Westerdahl, D.; Hu, J.; Wu, Y.; Yin, H.; Pan, X.; Zhang, K.M. On-road diesel vehicle emission factors for nitrogen oxides and black carbon in two Chinese cities. *Atmos. Environ.* **2012**, *46*, 45–55. [CrossRef]

8. Dallmann, T.R.; DeMartini, S.J.; Kirchstetter, T.W.; Herndon, S.C.; Onasch, T.B.; Wood, E.C.; Harley, R.A. On-road measurement of gas and particle phase pollutant emission factors for individual heavy-duty diesel trucks. *Environ. Sci. Technol.* **2012**, *46*, 8511–8518. [CrossRef] [PubMed]

9. Tan, Y.; Lipsky, E.M.; Saleh, R.; Robinson, A.L.; Presto, A.A. Characterizing the spatial variation of air pollutants and the contributions of high emitting vehicles in Pittsburgh, PA. *Environ. Sci. Technol.* **2014**, *48*, 14186–14194. [CrossRef] [PubMed]

10. Lau, C.F.; Rakowska, A.; Townsend, T.; Brimblecombe, P.; Chan, T.L.; Yam, Y.S.; Močnik, G.; Ning, Z. Evaluation of diesel fleet emissions and control policies from plume chasing measurements of on-road vehicles. *Atmos. Environ.* **2015**, *122*, 171–182. [CrossRef]

11. Durbin, T.D.; Johnson, K.; Miller, J.W.; Maldonado, H.; Chernich, D. Emissions from heavy-duty vehicles under actual on-road driving conditions. *Atmos. Environ.* **2008**, *42*, 4812–4821. [CrossRef]

12. Maness, H.L.; Thurlow, M.E.; McDonald, B.C.; Harley, R.A. Estimates of CO_2 traffic emissions from mobile concentration measurements. *J. Geophys. Res. Atmos.* **2015**, *120*, 2087–2102. [CrossRef]

13. Shah, S.D.; Johnson, K.C.; Miller, J.W.; Cocker, D.R., III. Emission rates of regulated pollutants from on-road heavy-duty diesel vehicles. *Atmos. Environ.* **2006**, *40*, 147–153. [CrossRef]

14. Chen, C.; Huang, C.; Jing, Q.; Wang, H.; Pan, H.; Li, L.; Zhao, J.; Dai, Y.; Huang, H.; Schipper, L.; et al. On-road emission characteristics of heavy-duty diesel vehicles in Shanghai. *Atmos. Environ.* **2007**, *41*, 5334–5344. [CrossRef]

15. Zhang, K.; Batterman, S. Air pollution and health risks due to vehicle traffic. *Sci. Total Environ.* **2013**, *450–451*, 307–316. [CrossRef] [PubMed]

16. Kim, K.H.; Lee, S.-B.; Woo, S.H.; Bae, G.-N. NO_x profile around a signalized intersection of busy roadway. *Atmos. Environ.* **2014**, *97*, 144–154. [CrossRef]

17. Goel, A.; Kumar, P. Characterisation of nanoparticle emissions and exposure at traffic intersections through fast–response mobile and sequential measurements. *Atmos. Environ.* **2015**, *107*, 374–390. [CrossRef]

18. Sun, K.; Tao, L.; Miller, D.J.; Khan, M.A.; Zondlo, M.A. On-road ammonia emissions characterized by mobile, open-path measurements. *Environ. Sci. Technol.* **2014**, *48*, 3943–3950. [CrossRef] [PubMed]

19. Beevers, S.D.; Kitwiroon, N.; Williams, M.L.; Carslaw, D.C. One way coupling of CMAQ and a road source dispersion model for fine scale air pollution predictions. *Atmos. Environ.* **2012**, *59*, 47–58. [CrossRef] [PubMed]

20. Klompmaker, J.O.; Montagne, D.R.; Meliefste, K.; Hoek, G.; Brunekreef, B. Spatial variation of ultrafine particles and black carbon in two cities: Results from a short-term measurement campaign. *Sci. Total Environ.* **2015**, *508*, 266–275. [CrossRef] [PubMed]

21. Wu, H.; Reis, S.; Lin, C.; Beverland, I.J.; Heal, M.R. Identifying drivers for the intra-urban spatial variability of airborne particulate matter components and their interrelationships. *Atmos. Environ.* **2015**, *112*, 306–316. [CrossRef]

22. Ghassoun, Y.; Ruths, M.; Löwner, M.-O.; Weber, S. Intra-urban variation of ultrafine particles as evaluated by process related land use and pollutant driven regression modelling. *Sci. Total Environ.* **2015**, *536*, 150–160. [CrossRef] [PubMed]

23. Choi, W.; Ranasinghe, D.; Bunavage, K.; DeShazo, J.R.; Wu, L.; Seguel, R.; Winer, A.M.; Paulson, S.E. The effects of the built environment, traffic patterns, and micrometeorology on street level ultrafine particle concentrations at a block scale: Results from multiple urban sites. *Sci. Total Environ.* **2016**, *553*, 474–485. [CrossRef] [PubMed]

24. Kittelson, D.B.; Watts, W.F.; Johnson, J.P. Nanoparticle emissions on Minnesota highways. *Atmos. Environ.* **2004**, *38*, 9–19. [CrossRef]

25. Weijers, E.P.; Khlystov, A.Y.; Kos, G.P.A.; Erisman, J.W. Variability of particulate matter concentrations along roads and motorways determined by a moving measurement unit. *Atmos. Environ.* **2004**, *38*, 2993–3002. [CrossRef]

26. Westerdahl, D.; Fruin, S.; Sax, T.; Fine, P.M.; Sioutas, C. Mobile platform measurements of ultrafine particles and associated pollutant concentrations on freeways and residential streets in Los Angeles. *Atmos. Environ.* **2005**, *39*, 3597–3610. [CrossRef]

27. Hagler, G.S.W.; Thoma, E.D.; Baldauf, R.W. High-resolution mobile monitoring of carbon monoxide and ultrafine particle concentrations in a near-road environment. *J. Air Waste Manag. Assoc.* **2010**, *60*, 328–336. [CrossRef] [PubMed]

28. Brantley, H.L.; Hagler, G.S.W.; Kimbrough, E.S.; Williams, R.W.; Mukerjee, S.; Neas, L.M. Mobile air monitoring data-processing strategies and effects on spatial air pollution trends. *Atmos. Meas. Tech.* **2014**, *7*, 2169–2183. [CrossRef]

29. Baldwin, N.; Gilani, O.; Raja, S.; Batterman, S.; Ganguly, R.; Hopke, P.; Berrocal, V.; Robins, T.; Hoogterp, S. Factors affecting pollutant concentrations in the near-road environment. *Atmos. Environ.* **2015**, *115*, 223–235. [CrossRef]

30. Zwack, L.M.; Paciorek, C.J.; Spengler, J.D.; Levy, J.I. Characterizing local traffic contributions to particulate air pollution in street canyons using mobile monitoring techniques. *Atmos. Environ.* **2011**, *45*, 2507–2514. [CrossRef]

31. Rakowska, A.; Wong, K.C.; Townsend, T.; Chan, K.L.; Westerdahl, D.; Ng, S.; Močnik, G.; Drinovec, L.; Ning, Z. Impact of traffic volume and composition on the air quality and pedestrian exposure in urban street canyon. *Atmos. Environ.* **2014**, *98*, 260–270. [CrossRef]

32. Kim, K.H.; Woo, D.; Lee, S.-B.; Bae, G.-N. On-road measurements of ultrafine particles and associated air pollutants in a densely populated area of Seoul, Korea. *Aerosol Air Qual. Res.* **2015**, *15*, 142–153. [CrossRef]

33. Argyropoulos, G.; Samara, C.; Voutsa, D.; Kouras, A.; Manoli, E.; Voliotis, A.; Tsakis, A.; Chasapidis, L.; Konstandopoulos, A.; Eleftheriadis, K. Concentration levels and source apportionment of ultrafine particles in road microenvironments. *Atmos. Environ.* **2016**, *129*, 68–78. [CrossRef]

34. Kumar, P.; Fennell, P.; Britter, R. Measurements of particles in the 5–1000 nm range close to road level in an urban street canyon. *Sci. Total Environ.* **2008**, *390*, 437–447. [CrossRef] [PubMed]

35. Kwak, K.-H.; Lee, S.-H.; Seo, J.M.; Park, S.-B.; Baik, J.-J. Relationship between rooftop and on-road concentrations of traffic-related pollutants in a busy street canyon: Ambient wind effects. *Environ. Pollut.* **2016**, *208*, 185–197. [CrossRef] [PubMed]

36. Oanh, N.T.K.; Martel, M.; Pongkiatkul, P.; Berkowicz, R. Determination of fleet hourly emission and on-road vehicle emission factor using integrated monitoring and modeling approach. *Atmos. Res.* **2008**, *89*, 223–232. [CrossRef]

37. Solazzo, E.; Vardoulakis, S.; Cai, X. A novel methodology for interpreting air quality measurements from urban streets using CFD modeling. *Atmos. Environ.* **2011**, *45*, 5230–5239. [CrossRef]

38. Pu, Y.; Yang, C. Estimating urban roadside emissions with an atmospheric dispersion model based on in-field measurements. *Environ. Pollut.* **2014**, *192*, 300–307. [CrossRef] [PubMed]

39. Hang, J.; Wang, Q.; Chen, X.; Sandberg, M.; Zhu, W.; Buccolieri, R.; Di Sabatino, S. City breathability in medium density urban-like geometries evaluated through the pollutant transport rate and the net escape velocity. *Build. Environ.* **2015**, *94*, 166–182. [CrossRef]

40. Zhai, W.; Wen, D.; Xiang, S.; Hu, Z.; Noll, K.E. Ultrafine-particle emission factors as a function of vehicle mode of operation for LDVs based on near-roadway monitoring. *Environ. Sci. Technol.* **2016**, *50*, 782–789. [CrossRef] [PubMed]

41. Liu, Y.S.; Cui, G.X.; Wang, Z.S.; Zhang, Z.S. Large eddy simulation of wind field and pollutant dispersion in downtown Macao. *Atmos. Environ.* **2011**, *45*, 2849–2859. [CrossRef]

42. Wang, Y.J.; Nguyen, M.T.; Steffens, J.T.; Tong, Z.; Wang, Y.; Hopke, P.K.; Zhang, K.M. Modeling multi-scale aerosol dynamics and micro-environmental air quality near a large highway intersection using the CTAG model. *Sci. Total Environ.* **2013**, *443*, 375–386. [CrossRef] [PubMed]

43. Kwak, K.-H.; Baik, J.-J.; Ryu, Y.-H.; Lee, S.-H. Urban air quality simulation in a high-rise building area using a CFD model coupled with mesoscale meteorological and chemistry-transport models. *Atmos. Environ.* **2015**, *100*, 167–177. [CrossRef]

44. Batterman, S.; Chambliss, S.; Isakov, V. Spatial resolution requirements for traffic-related air pollutant exposure evaluations. *Atmos. Environ.* **2014**, *94*, 518–528. [CrossRef] [PubMed]

45. Woo, S.-H.; Kwak, K.-H.; Bae, G.-N.; Kim, K.H.; Kim, C.H.; Yook, S.-J.; Jeon, S.; Kwon, S.; Kim, J.; Lee, S.-B. Overestimation of on-road air quality surveying data measured with a mobile laboratory caused by exhaust plumes of a vehicle ahead in dense traffic areas. *Environ. Pollut.* **2016**, *218*, 1116–1127. [CrossRef] [PubMed]

46. Kim, J.-J.; Baik, J.-J. A numerical study of the effects of ambient wind direction on flow and dispersion in urban street canyons using the RNG k–ε turbulence model. *Atmos. Environ.* **2004**, *38*, 3039–3048. [CrossRef]

47. Baik, J.-J.; Kwak, K.-H.; Park, S.-B.; Ryu, Y.-H. Effects of building roof greening on air quality in street canyons. *Atmos. Environ.* **2012**, *61*, 48–55. [CrossRef]

48. Kwak, K.-H.; Baik, J.-J. Diurnal variation of NO_x and ozone exchange between a street canyon and the overlying air. *Atmos. Environ.* **2014**, *86*, 120–128. [CrossRef]

49. Kim, B.M.; Lee, S.-B.; Kim, J.Y.; Kim, S.; Seo, J.; Bae, G.-N.; Lee, J.Y. A multivariate receptor modeling study of air-borne particulate PAHs: Regional contributions in a roadside environment. *Chemosphere* **2016**, *144*, 1270–1279. [CrossRef] [PubMed]

50. Tong, Z.; Wang, Y.J.; Patel, M.; Kinney, P.; Chrillrud, S.; Zhang, K.M. Modeling spatial variations of black carbon particles in an urban highway-building environment. *Environ. Sci. Technol.* **2012**, *46*, 312–319. [CrossRef] [PubMed]

51. Baik, J.-J.; Park, S.-B.; Kim, J.-J. Urban flow and dispersion simulation using a CFD model coupled to a mesoscale model. *J. Appl. Meteorol. Climatol.* **2009**, *48*, 1667–1681. [CrossRef]

52. Weber, S.; Kordowski, K.; Kuttler, W. Variability of particle number concentration and particle size dynamics in an urban street canyon under different meteorological conditions. *Sci. Total Environ.* **2013**, *449*, 102–114. [CrossRef] [PubMed]

atmosphere

MDPI

Article

Impacts of Traffic Tidal Flow on Pollutant Dispersion in a Non-Uniform Urban Street Canyon

Tingzhen Ming [1], Weijie Fang [1], Chong Peng [2,*], Cunjin Cai [1], Renaud de Richter [3], Mohammad Hossein Ahmadi [4] and Yuangao Wen [1]

[1] School of Civil Engineering and Architecture, Wuhan University of Technology, No. 122 Luoshi Road, Hongshan District, Wuhan 430070, China; tzming@whut.edu.cn (T.M.); fangweijie@whut.edu.cn (W.F.); cunjincai@163.com (C.C.); wenyg2000@126.com (Y.W.)
[2] School of Architecture and Urban Planning, Huazhong University of Science and Technology, No. 1037, Luoyu Road, Hongshan District, Wuhan 430074, China
[3] Tour-Solaire.Fr, 8 Impasse des Papillons, F34090 Montpellier, France; renaud.derichter@gmail.com
[4] Faculty of Mechanical Engineering, Shahrood University of Technology, Shahrood 3619995161, Iran; mohammadhosein.ahmadi@gmail.com
* Correspondence: pengchong@hust.edu.cn; Tel.: +86-159-2752-1021

Received: 5 January 2018; Accepted: 22 February 2018; Published: 25 February 2018

Abstract: A three-dimensional geometrical model was established based on a section of street canyons in the 2nd Ring Road of Wuhan, China, and a mathematical model describing the fluid flow and pollutant dispersion characteristics in the street canyon was developed. The effect of traffic tidal flow was investigated based on the measurement results of the passing vehicles as the pollution source of the CFD method and on the spatial distribution of pollutants under various ambient crosswinds. Numerical investigation results indicated that: (i) in this three-dimensional asymmetrical shallow street canyon, if the pollution source followed a non-uniform distribution due to the traffic tidal flow and the wind flow was perpendicular to the street, a leeward side source intensity stronger than the windward side intensity would cause an expansion of the pollution space even if the total source in the street is equal. When the ambient wind speed is 3 m/s, the pollutant source intensity near the leeward side that is stronger than that near the windward side (R = 2, R = 3, and R = 5) leads to an increased average concentration of CO at pedestrian breathing height by 26%, 37%, and 41%, respectively. (R is the ratio parameter of the left side pollution source and the right side pollution source); (ii) However, this feature will become less significant with increasing wind speeds and changes of wind direction; (iii) the pollution source intensity exerted a decisive influence on the pollutant level in the street canyon. With the decrease of the pollution source intensity, the pollutant concentration decreased proportionally.

Keywords: street canyon; traffic tidal flow; numerical simulation; vehicular pollution; non-uniform distribution of the pollution source

1. Introduction

The recent urbanization process continues to advance all over the world. The rapid growth of vehicle ownership leads to motor vehicle exhaust emissions being one of the main sources of air pollution in cities [1]. Streets become increasingly canyon-style in modern cities due to the increasing frequency of tall buildings. Traffic growth and related congestion results in increased pollution emissions. The construction of high-rise buildings and the increase of building density have caused the deterioration of the urban ventilation environment. Consequently, the pollutants emitted from vehicles are difficult to be diluted and disperse slowly. These factors severely endanger travelers who are directly exposed to the atmosphere and this furthermore has a severe impact on the indoor air quality of street buildings.

Environmental pollution is a severe problem, threatening the health and survival of human beings [2–4]. Primarily due to these health-related issues, extensive research on identifying and understanding the physical processes that both drive and influence the near-field pollutant dispersion in urban environments has experienced a substantial progress over the last three decades. Most of these studies began with the basic unit of any city—the Street Canyon. This has been defined as a relatively narrow street space formed by successive buildings on both sides of a city street [5]. Relevant studies mainly relied on field measurements [6,7], wind tunnel experiments [8–10], and numerical simulations [11–13]. Based on previous studies, important factors that influence the flow patterns and the dispersion mechanism of pollutant can be grouped into the following categories: Inflow conditions (such as wind speed, wind direction [14–16], turbulence intensity [17]); Geometric conditions of building structures (such as building aspect ratio [18,19] and the street canyon aspect ratio [20,21]); Ground surface and building surface conditions (such as building surface roughness and hot or cold conditions [22–25]); The impact of turbulences caused by vehicle movement [11]. Although the research results in this field are substantial and mature, Lateb et al. [26] clearly concluded that "the topic of micro-scale dispersion still requires further investigation to understand the effect of all parameters on wind flow and pollutant dispersion in urban areas".

At present, both in large and medium-sized cities, by the influence of urban planning layout and the increasing price of central area land, a new pattern of work unit has formed that focuses on the city center area, while the residential areas are mainly concentrated in peripheral regions. In the morning, a large number of motorized vehicles enters the city on one side of the roads, while in the afternoon, a similar number of motorized vehicles leaves the city on the other side of the roads. Usually, this phenomenon is called traffic tidal flow as shown in Figure 1. Traffic tidal flow will increase the pollution source on one side of the road than on the other side because the motor vehicles on one side of the road are far more numerous than on the other side. However, most previous studies on the definition of pollution source are too idealistic, often assuming the pollutant emission source as the constant point source or line source in the center of the road. In addition, geometric models often assume an ideal canyon type with a uniform building roof height. Using even and uneven roof height along each courtyard building's wall of a regular urban array, Nosek et al. [27] showed that the pollutant fluxes and pollutant removal capabilities through the street-canyon roof top are strongly affected by the roof-height arrangement. Gu et al. [28] furthermore highlighted that the roof-height non-uniformities along both street-canyon walls are able to either improve or worsen the air quality of the street-canyon with regard to the source position and above-roof wind direction. Recently, Nosek et al. [29] employed two street canyon models with either uniform building roof height or non-uniform building roof height to analyze the dispersion characteristics of pollutants. The results showed that the buildings' roof-height variability at the intersections plays an important role for the resulting dispersion of traffic pollutants within the canyons. These studies provided insight into the pollutant dispersion within street networks formed by blocks of non-uniform height. Therefore, our study focused on the impact of traffic tidal flow on pollutant dispersion in a non-uniform urban street canyon.

The present study is principally an extension of the study of urban traffic pollution and the aims are as follows: (i) to investigate the pollutant exchange processes in a 3D asymmetrical street canyon; (ii) to analyze the impacts of traffic tidal flow on pollutant concentration distribution characteristics; (iii) to find the correlations between source intensity and pollutant concentration levels in the street space. The pollution was simulated via homogeneously emitted passive gas (CO) from a ground-level volume source. Both volume sources were positioned along the two traffic-ways of the investigated street canyon. Field measurements of the traffic flow in two parallel traffic-ways at different time intervals (6:00 a.m. to 8:00 p.m.) were conducted and the results formed the pollution source intensity of the Computational Fluid Dynamics (CFD) method.

Figure 1. Tidal phenomenon of urban road traffic (Photo was taken on Wednesday, 9 November 2016 at 7:23 a.m.).

2. Model Description

2.1. Geometric Model

A three-dimensional geometrical model was established based on a section of the street canyon in the 2nd Ring Road of Wuhan, China, as an apparent traffic tidal phenomenon regularly occurs in this road. The street length is 220 m. There are six standard motorways in the middle of the road (three on each side) and the width of each lane is 4 m. Either side of the motorway has a non-motorized lane with a single width of 5 m. The total width of the street is 34 m. Street buildings are four main middle-rise residential buildings and a large number of low-rise shops. Not only is this particular road very busy, there are also many pedestrians on both sides of the street. Following appropriate simplification principles we built the geometric model using the Gambit 2.4.6 software (FLUENT INC., Lebanon, NH, USA). Considering the non-uniform distribution of the pollution source under the influence of traffic tidal flow, two pollution sources were set up. The three motorways on the left side became source 1 and those on the right became source 2. This was the main difference to previous studies.

The maximum height of the street buildings was 30 m. Because the maximum height of the buildings (30 m) on both sides of the road still remained below the street width (34 m), the canyon type belonged to the shallow street canyon, where the wind blows easily to the bottom of the street. According to the technical guidance of the Japan AIJ building outdoor wind environment CFD simulation [30], the computing area inlet met 5 H from the windward building boundary, and the lateral boundary was 5 H from the edge of the building. The top boundary was set to be 6 H from the ground, and the outlet boundary was located 15 H from the leeward building edge. H was the target height. This study used $H = 30$ m as the maximum height. The area of the building covered less than 3% of the total computing area. Based on the above principles, the total size of the computing domain $X \times Y \times Z$ was 658 m × 520 m × 180 m. The computing domain model and the boundary conditions are shown in Figure 2 below.

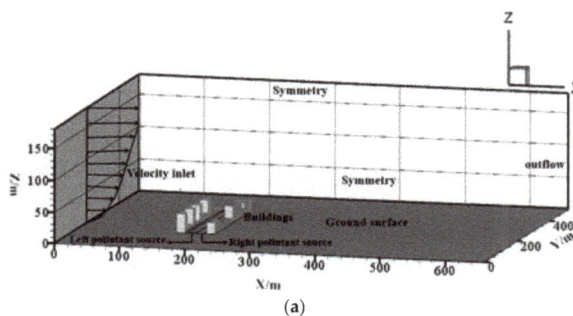

(a)

Figure 2. *Cont.*

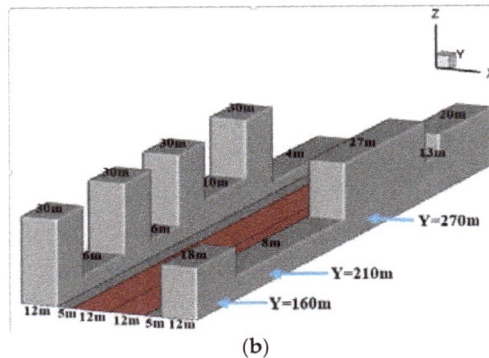

(b)

Figure 2. Street canyon model (**a**) computing domain model and the boundary conditions; (**b**) building geometric structure.

2.2. Mathematical Model

Pollutant dispersion in built environments is a both central and complex issue. Complex flow patterns control the wind flow around the buildings; therefore, the dispersion of pollutants in a street canyon is a typical turbulence dispersion issue [31,32]. Consequently, the choice of an accurate CFD turbulence model is a prerequisite for calculation accuracy.

At present, the Reynolds-averaged Navier–Stokes (RANS) turbulence model is widely used for numerical simulations of street canyons. Many scholars have compared different RANS models [33–35]. The standard k-ε model (SKE) is less able to show separation flow because it overestimates the turbulence kinetic energy near the windward corner of buildings. In addition, when there is a source exit in the recirculation area of roof and wall, the concentration can be predicted to be low. The renormalization group k-ε (RNG) turbulence model is a modification of the standard k-ε model, which performs well in the prediction of pollutant concentration [36]. Compared to the LES model that has been favored by scholars in recent years, the resulting simulation results show a good agreement [37]. The RNG k-ε model has been widely used to simulate the complex flow of air in construction groups or urban areas [38–40]. Therefore, to guarantee calculation accuracy and to ensure that computer resources remain as small as possible, this study employed the RNG k-ε turbulence model proposed by Yakhot and Orszag [41] to simulate the influence of turbulence. The flow of viscous incompressible fluids is commonly described with the Navier-Stokes equations [15]. The solution control equations are as follows:

Continuity Equation:

$$\frac{\partial (U_i)}{\partial x_i} = 0 \tag{1}$$

Momentum conservation Equation (N-S):

$$U_j \frac{\partial U_i}{\partial x_j} = -\frac{1}{\rho} \frac{\partial p}{\partial x_i} + \frac{\partial}{\partial x_j} \left(\nu \left(\frac{\partial U_i}{\partial x_j} + \frac{\partial U_j}{\partial x_i} \right) - \overline{u_i' u_j'} \right)$$

$$\overline{u_i' u_j'} = \nu_t \left(\frac{\partial U_i}{\partial x_j} + \frac{\partial U_j}{\partial x_i} \right) - \frac{2}{3} \delta_{ij} k \tag{2}$$

Turbulence kinetic Equation k:

$$\frac{\partial k}{\partial t} + U_i \frac{\partial k}{\partial x_i} = \frac{\partial}{\partial x_i} \left[\left(\nu + \frac{\nu_t}{\sigma_k} \right) \frac{\partial k}{\partial x_i} \right] + G - \varepsilon$$

$$v_t = C_\mu \frac{k^2}{\varepsilon} \tag{3}$$

Turbulent kinetic energy dissipation rate ε:

$$\frac{\partial \varepsilon}{\partial t} + U_i \frac{\partial \varepsilon}{\partial x_i} = \frac{\partial}{\partial x_i}\left[(v + \frac{v_t}{\sigma_\varepsilon})\frac{\partial \varepsilon}{\partial x_i}\right] + C_{1\varepsilon}\frac{\varepsilon}{k}G - \left[C_{2\varepsilon} + \frac{C_\mu \rho \eta^3 (1 - \eta/\eta_0)}{1 + \beta \eta^3}\right]\frac{\varepsilon^2}{k} \tag{4}$$

In the above Equation, U_i and U_j represent the average velocity components in the i and j direction coordinates, respectively; ρ represents the air density and P represents the air pressure; v represents the kinematic viscosity; $\overline{u'_i u'_j}$ represents the Reynolds stress term; δ_{ij} represents the Kronecker function; k and ε represent the turbulent kinetic energy and turbulent dissipation rate, respectively; G represents the production of turbulent kinetic energy due to the average velocity gradient. v_t represents the turbulence eddy viscosity. C_μ, $C_{\varepsilon 1}$ and $C_{\varepsilon 2}$ represent empirical constants which can be assumed to be 0.09, 1.44, and 1.92, respectively; σ_k and σ_ε represent the turbulent Prandtl numbers corresponding to the turbulent kinetic energy and the turbulent dissipation rate, respectively, which can be assumed to be 1.0 and 1.3, respectively. $\eta = (k/\varepsilon)(G/v_t)0.5$, $\beta = 0.012$, $\eta_0 = 4.38$ [42].

The species (pollutant) transport Equation:

$$u_j \frac{\partial \bar{c}}{\partial x_j} = \frac{\partial}{\partial x_j}(K_c \frac{\partial \bar{c}}{\partial x_j}) + S_c \tag{5}$$

$\bar{c}$ represents the pollutant concentration (kg/m^3), S_c represents the pollutant emission rate (kg/m^3 s). K_c represents the turbulent eddy diffusivity of pollutants. Here $K_c = v_t/Sc_t$, v_t represents the turbulent eddy viscosity, Sc_t represents the turbulent Schmidt number, which represents the ratio of momentum diffusivity and mass (or pollutants) diffusivity. Here we use $Sc_t = 0.7$ according to Hang et al. [43].

2.3. Boundary Conditions

2.3.1. Inlet Boundary

The vertical characteristics of the wind speed are affected by terrain and have a close relationship with the roughness of the terrain. Due to the roughness of the ground, wind flow often occurs in the form of gradient winds (see Figure 2). For the inlet boundary conditions, either the exponential law or the logarithmic law can be used as an expression of the wind velocity profile [44,45]. As a result, this paper employed the exponential law for the wind speed profile expression:

$$U(z) = U_S \left(\frac{z}{z_S}\right)^\alpha \tag{6}$$

where U_S represents the average wind speed at the reference altitude z_S and α represents the ground roughness index. The reference height is usually 10 m above the ground. When z increases beyond a certain height z_H, different ground conditions lead to different ground roughness index value. According to Wang et al. [46], the simulation value of α was chosen to be 0.22.

The inlet boundary turbulence is also an important factor affecting flow characteristics. The following expression describes the turbulent kinetic energy and the dissipation rate at the inlet boundary:

$$k = \frac{U_*^2}{\sqrt{C_\mu}} \tag{7}$$

$$\varepsilon = \frac{U_*^3}{\kappa z} \tag{8}$$

where U_* represents the friction velocity, which is the square root of the ratio of turbulent shear stress to air density. κ refers to the von Karman constant 0.4; here, the value of C_μ was 0.09.

For this study, wind speeds of 1.5 m/s, 3 m/s, 4.5 m/s, and 6 m/s at the reference height of 10 m and wind directions of $0°$, $30°$, $45°$, $60°$ and $90°$ were employed.

2.3.2. Outlet Boundary

After taking the effects of ambient wind into account, two vertical surfaces in the model can be considered as outlet boundary, upper and right borders, respectively. To avoid the influence of the "reflection source" formed by the pressure boundary on the calculation convergence, the right exit uses a free flow boundary; however, the upper boundary had a relatively static pressure of zero; therefore, the boundary of the symmetry plane was used.

2.3.3. Lateral Boundary

The two sides and the top of the domain were far away from the building wall, and the wind flow was parallel to the lateral surface. We assumed the speed gradient along the lateral surface to be zero. Therefore, the symmetry boundary was used for lateral of the domain. The speed of both the ground surface and building wall was zero and consequently, a no-slip wall boundary was adopted.

The k-ε model is generally a high Reynolds number turbulent model, which is only effective for the full development of turbulence. The Reynolds number is low in the near-wall area and the turbulence is not sufficiently developed; therefore, it was necessary to utilize the near-wall treatment. The standard wall function method has a good simulation effect for the actual flow of many projects. The principle is as follows:

$$F^* = \frac{1}{\kappa} \ln(E^* y^*) \tag{9}$$

where $F^* = (u_p c_\mu^{1/4} k_p^{1/2})/(\tau_w/\rho)$, $y^* = (\rho c_\mu^{1/4} \kappa_p^{1/2} y_p)/\mu$, κ represents the constant of Von Karman; E^* represents the experimental constant 9.81; u_p represents the average velocity of the fluid at point p; k_p represents the turbulent kinetic energy of point p; τ_w represents the wall shear stress and ρ represents the air density; y_p represents the distance from point p to the wall; μ represents the viscosity coefficient of the fluid.

2.4. Measurement of the Traffic Pollution Source in the Street

The source intensity of the pollutants is typically expressed in $kg/(m^3 \, s)$, which is generally affected by the type of vehicles in the street, the emission rate of these vehicles, and the traffic flow. Due to the large proportion of CO emissions as part of the vehicle exhaust, which does not easily react with other components in the air, we chose CO as pollution source to calculate and analyze its dispersion characteristics. The calculation formula of the pollution source intensity can be described with Equation (10). Single vehicle exhaust emissions have a strong relationship with their speed. In this study, the method of calculating the average emission rate in a certain driving speed range proposed by Zhang et al. [47] was adopted to calculate the emission intensity of road pollutants. Table 1 shows the pollutant emission rate for a specific speed range.

$$Q = \frac{N \times E}{V_s \times 10^6} \tag{10}$$

here, Q represents the source intensity of the road pollutant ($kg/(m^3 \, s)$); N represents the total number of vehicles on the road per unit of time (vehicle); E represents the average CO emission rate under mixed traffic flow (mg/veh s); and V_s represents the volume of source intensity (m^3).

Table 1. Relationship between driving speed and emission rate [47].

Speed Range km/h	Emission Rate mg/(veh s)		
	NO_x	HC	CO
0–10	0.20013	0.53241	5.90124
10–20	0.67005	1.01235	19.86452
20–30	1.65470	1.05106	22.14546
30–40	2.03404	1.10454	23.14653
40–50	3.10247	1.22414	26.15460
50–60	2.75461	1.45127	24.15641
60–70	4.05120	1.10021	14.48432
70–80	4.16471	1.01145	9.08424
80–90	3.45153	1.02104	4.11461
90–100	2.13451	1.01412	2.21457
>100	1.02465	1.01214	0.94564

The driving statuses of all different vehicle types on the road are complicated and changeable. Although the changes are complex, they have a certain characteristic tendency, which can be revealed through extensive observations and analysis. To obtain the data of traffic flow and vehicle speed, the area of the road section was measured for one month by means of taking photos and videos. Photos were taken every 30 s to obtain the number of vehicles on the road, thus counting 120 times per hour. Then, we calculated the average value during the corresponding time period. Within a limited distance of 220 m, we marked a vehicle to obtain the time it takes to pass this fixed distance to calculate its speed. The average speed value was obtained through a large number of measurements. Measurement results showed that the traffic on the left side of the road was significantly denser than on the right side from 6:00 to 8:30 a.m. The average number of vehicles was 62, with an average speed of 11.2 km/h. The average number of vehicles on the right side was 20 and their average speed was 27.6 km/h. Table 1 shows that the CO emission rates were 19.86 mg/s and 22.15 mg/s in their respective speed ranges. By multiplying the number of vehicles and then dividing the pollution source volume (220 m × 12 m × 0.3 m), we calculated the pollution source intensity for the left side as 9.93×10^{-7} kg/(m^3 s) and for the right side as 5.54×10^{-7} kg/(m^3 s). Based on this method, the source intensity during noon (11:30–13:00), evening (17:30–20:00), and at other times (8:30–11:30 and 13:00–17:30) could be obtained. The results are shown in Table 2.

Table 2. Experimental values of pollution source intensity at different times during one day.

	Time	Morning	Evening	Noon	Other Times
		6:00–8:30	17:30–20:00	11:30–13:00	8:30–11:30 13:00–17:30
	Traffic flow on the left side (vehicle)	62	30	30	16
	Driving speed on the left side (km/h)	11	25	25	40
Experimental Value	CO emission rate (mg/s)	19.86	22.14	22.15	26.15
	Pollution source intensity on the left side (kg/m^3 s)	9.93×10^{-7}	5.5×10^{-7}	5.5×10^{-7}	3.3×10^{-7}
	Traffic flow on the right side (vehicle)	28	72	30	16
	Driving speed on the right side (km/h)	25	11	25	40
	CO emission rate (mg/s)	22.14	19.86	22.15	26.15
	Pollution source intensity on the right side (kg/m^3 s)	5.54×10^{-7}	1.15×10^{-6}	5.5×10^{-7}	3.3×10^{-7}

The experimental results show that a significant traffic tidal phenomenon exists between the morning and evening, due to migration between working places and residential areas. Non-uniform distribution of vehicles results in non-uniform distribution of pollution sources. Via field experiments, we can conclude that the magnitude of the source intensity was about 10^{-7}. To further explore the pollutant dispersion characteristics under the case of non-uniform pollution source distribution, we added several pollution source settings for CFD simulation based on the experimental value. The specific simulation settings are shown in Table 3.

Table 3. Simulation setting value of pollution source on both sides.

Pollution Source at the Left Side (kg/m^3 s)	Pollution Source at the Right Side (kg/m^3 s)
2×10^{-7}	1×10^{-7}
3×10^{-7}	1×10^{-7}
5×10^{-7}	1×10^{-7}
1×10^{-7}	2×10^{-7}
1×10^{-7}	3×10^{-7}
1×10^{-7}	5×10^{-7}

To facilitate the processing and analysis of results, we defined R as the ratio parameter of the left side pollution source (near the leeward side) and the right side pollution source (near the windward side). It can be written as follows:

$$R = \frac{Q_L}{Q_R} \tag{11}$$

where Q_L represents the left side (near leeward side) pollution source and Q_R represents the right side (near winward side) pollution source.

2.5. Meshing Skills and Computational Procedure

Computational area discretization is a critical step in computational fluid dynamics. In general, for the same meshing zones, the hexahedral (HEX) meshing method is more economical and is more efficient at reducing false dispersion than the tetrahedral method. Thanks to the support of Gambit software and grid adaptive technology, the fine grids near the building walls and ground surfaces are particularly concentrated in these locations compared to places that are not proximal to walls or surfaces. This method can save computing resources in case of limited computer hardware, and obtain accurate flow field, concentration field, and other characteristics in the shortest amount of time. Figure 3 shows a schematic of the grid demarcation in a cross-section.

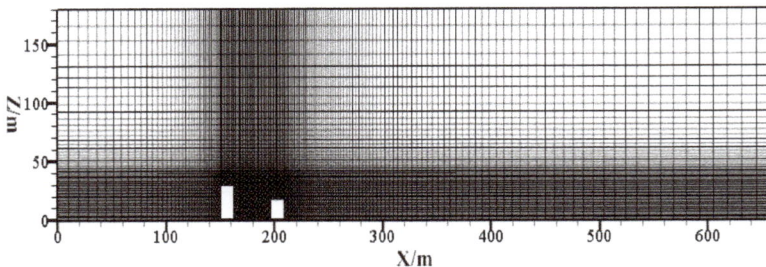

Figure 3. Schematic diagram of grid demarcation in a cross-section.

CFD code ANSYS FLUENT 15.0 (ANSYS INC., Pittsburgh, PA, USA) was used to conduct the numerical simulations. FLUENT is a multipurpose commercial CFD software that has widely been used to model flow and dispersion for urban applications. The inflow wind is treated with a user-defined-function (UDF) method. The discretization methods of both convection and dispersion items were selected in QUICK and second-order windward formats. The air is moving at a relatively low speed and can thus be considered incompressible. The solution control equation has no pressure term; therefore, the pressure field distribution has to be obtained via the method of pressure-velocity coupling. The coupling method used the SIMPLE algorithm proposed by Patankar [48]. The iterations were continued until the relative error in the conservation equation was below 1×10^{-5} and the energy equation was below 1×10^{-8}.

To capture the effects of grid independence to simulate the results, for this study, three types of grids (numbers 2906760, 4596120, and 6054320) were selected and calculated with the same computer under the same working conditions (wind speed $v = 1.5$ m/s, perpendicular to the street, and a source ratio of R = 2/1). We then selected a specific point in the computing domain, and found wind speeds of 1.571 m/s, 1.593 m/s, and 1.610 m/s under the three different grid numbers. The concentration levels of pollutants were 8.06×10^{-9}, 7.84×10^{-9}, and 7.67×10^{-9}, respectively. In addition, the velocity magnitude and turbulent kinetic energy profile at a height of 2 m along the $Y = 160$ m section and the street centerline were compared in Figure 4. As can be seen from the figure, the maximum error appears at the turbulent kinetic energy profile at $Y = 350$ m, about 6%. Apart from this, Other errors are less than 5%. These results showed that further increasing the number of grids does not lead to apparent deviations of velocity and pollutant concentration, thus demonstrating precision of the numerical solution and irrelevance of the number of grids. The grid independence test results provide a good foundation for the remainder of the paper. Thus, mesh cells 4596120 were selected as the basic mesh system of this study.

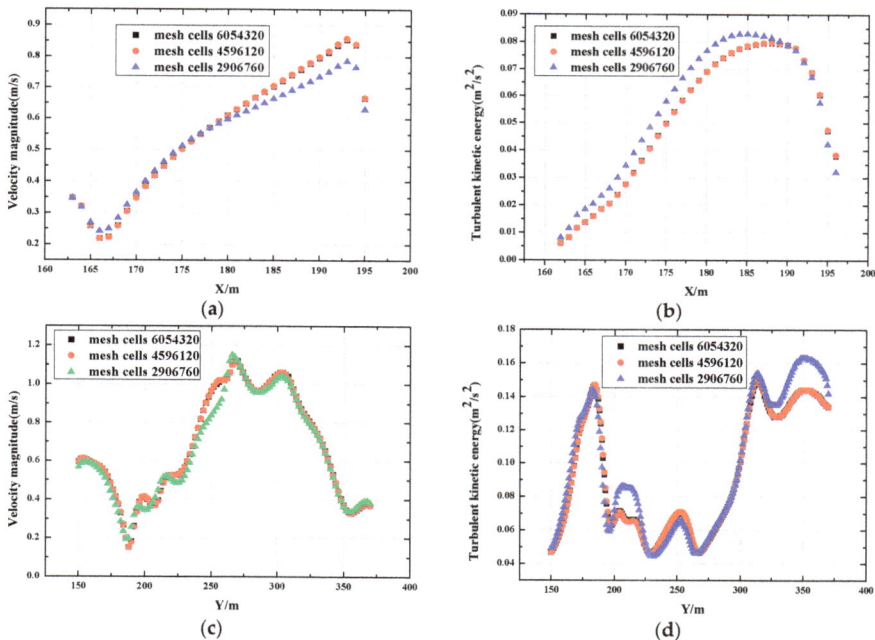

Figure 4. Velocity magnitude and turbulent kinetic energy profile at a height of 2 m along the $Y = 160$ m section and the street centerline. (**a**) velocity magnitude profile along the $Y = 160$ m section; (**b**) turbulent kinetic energy profile along the $Y = 160$ m section; (**c**) velocity magnitude profile along the street centerline; (**d**) turbulent kinetic energy profile along the street centerline.

2.6. Model Validation

To ensure accuracy and reliability of the numerical simulation, model validation is indispensable. Generally, to test the validity of a mathematical model that describes the fluid flow and the heat transfer characteristics of a specific system, the most feasible method is to compare numerical results to experimental results by providing both identical boundary and working conditions. However, many geometric models are too complex to find effective wind tunnel experimental data for comparison. The validation of the accuracy is aimed at testing the applicability of the RNG k-ε numerical model employed in this paper. Therefore, to validate the accuracy of the RNG k-ε numerical model we established another geometric model identical to the wind tunnel experiment model conducted by Takenobu et al. [49]. Experiments were conducted in a wind-tunnel facility (TWINNEL: twinned wind tunnel) at the Central Research Institute of Electric Power Industry (CRIEPI). The utilized wind tunnel is a closed-circuit. The experiment was conducted in the larger test section with dimensions of 17.0 m × 3.0 m × 1.7 m in streamwise (*X*), spanwise (*Y*), and vertical (*Z*) directions, respectively. Seven Irwin-type vortex generators with a height of 0.65 m were placed at the entrance of the test sections and three L-shaped cross-sections were located on the wind-tunnel floor from *X* = 0 to *X* = 5.0 m at equal intervals of 1.5 m to generate turbulent motions near the surface. A series of regularly spaced bars of 1.56 m (13 H) × 0.12 m (1 H) × 0.12 m (1 H) were set on the floor at equal intervals of H (0.12 m) from 10.5 m downwind of the entrance, normal to the wind direction. The origin of the coordinate axis forms the centre of the floor at the leeward wall of the 25th block, which is 16.62 m downwind of the entrance of the test section. The streamwise and vertical velocities at *X/H* = 0.25, 0.5, and 0.75 were measured in the vertical direction at *Y/H* = 0 using a laser Doppler velocimeter (LDV). The reference streamwise velocity at *X/H* = 0 and *Z/H* = 2.0 is 1.15 m/s. The pollutant emission rate Q is represented by a ground-level continuous pollutant line source of length Ly placed parallel to the spanwise axis at *X/H* = 0.5. The tracer gas ethane (it is both used in wind-tunnel experiment and current numerical simulation) is only emitted from a line source within the canyon. The upstream boundaries of the domain C | $_{X=0}$ = 0 is applied and the stream at the outlet boundary emits the tracer gas out of the domain. That is, the free stream at the inlet boundary is free of pollutant. The concentration C is normalized by the freestream mean velocity U_0 (1 m/s), block height *H* (0.12 m), line source length *L* (1.56 m), and total emission Q (10^{-7} kg/m^3 s) as:

$$C^* = \frac{CU_0HL}{Q} \tag{12}$$

Using the same geometry and boundary conditions, we used the RNG k-ε numerical model for numerical calculation and compared the obtained results to experimental data. Vertical distributions of the streamwise velocity and normalized mean concentration at *X/H* = 0.50 are shown in Figure 5. Figure 5 shows that numerical simulation results of normalized mean concentration agree well with the experiment data. However, there is a small deviation in the vertical distribution of the streamwise velocity. The main reason may be due to the flow generated by the vortex generators in the wind tunnel experiment, which differs from the wind flow in the numerical simulation. Despite some slight differences, the RNG k-ε turbulence model is feasible for solving fluid flow and pollutant dispersion.

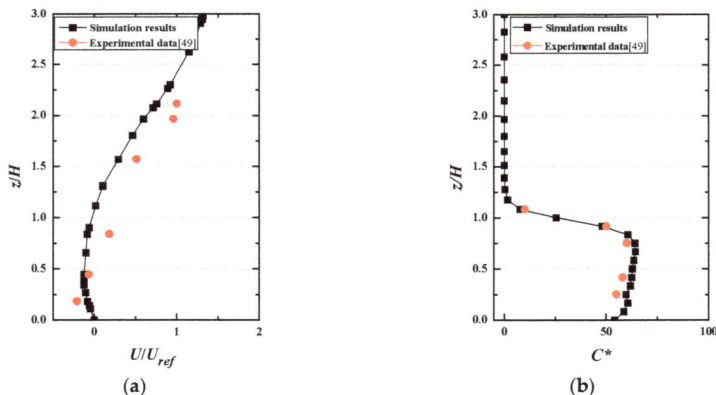

Figure 5. Vertical distributions of streamwise velocity (**a**) and normalized mean concentration (**b**) at *x/H* = 0.50.

3. Results and Discussion

3.1. Impact Analysis of Non-Uniform Pollution Source

During both morning and evening, a two parallel road pollution source non-uniform distribution is caused by the impact of traffic tidal flow. This part focuses on the dispersion characteristics of pollutants under this specific case.

3.1.1. Concentration Distribution of CO at Pedestrian Breathing Height

The wind speed 3 m/s, wind direction 90°, pollutant emission ratios 1/2, 1/3, 1/5, and the corresponding 2, 3, and 5 were chosen to show the distribution of CO concentration at the level $Z = 1.5$ m, which represents the pedestrian breathing height. Due to the asymmetrical arrangement of buildings in this street canyon, the flow field in the canyon becomes increasingly complicated. Figure 6 shows that CO mainly accumulates on the leeward side under the action of the perpendicular inlet wind. In some areas where the upstream building is higher than the downstream, a local high pollution level will appear. For the source distribution of $R = 1/3$ and $R = 3$, it can be found that even though the total source intensity in the street is equal, the stronger source at the left side in comparison to the right side ($R = 3$) will cause the expansion of the high pollution area. The results are found to be consistent when compared to the case of $R = 1/2$, $R = 2$ and $R = 1/5$, $R = 5$. Furthermore, this phenomenon becomes more apparent with increasing difference of source distribution between the two sides. Therefore, in the horizontal area of $Z = 1.5$ m in the street canyon, the increase of the pollution source near the leeward side will increase the range of the high pollution concentration area. Figure 7 shows the average concentration of CO at pedestrian breathing height under different pollution source distributions. A source intensity near the leeward side stronger than near the windward side ($R = 2$, $R = 3$, and $R = 5$) leads to an increase of the average concentration of CO by 26%, 37%, and 41%, respectively. The underlying causes of this phenomenon will be explained in Section 3.1.3.

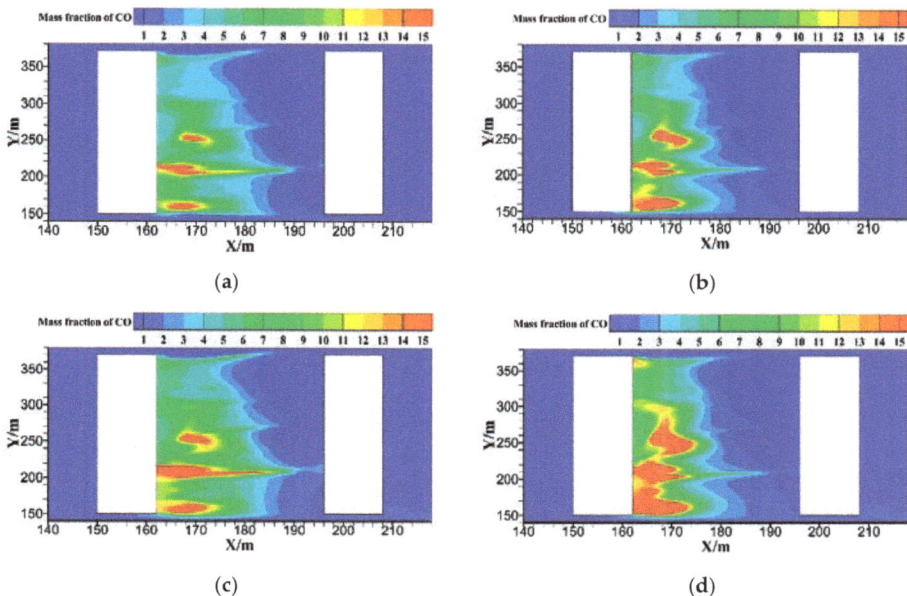

(a)

(b)

(c)

(d)

Figure 6. *Cont.*

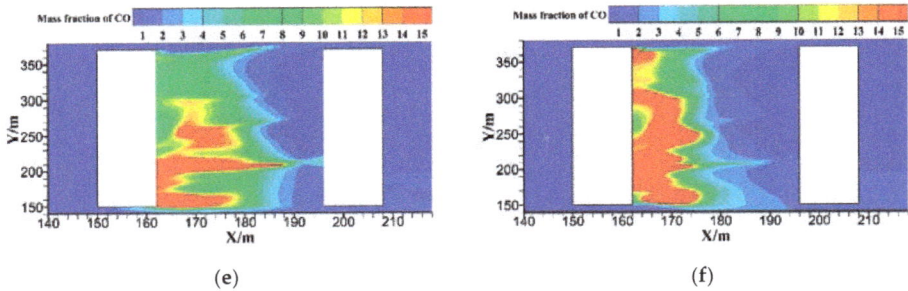

(e) (f)

Figure 6. CO concentration field profiles (10^{-9}) in the street space for different pollution source distributions under a wind speed of 3 m/s. (**a**) R = 1/2; (**b**) R = 2; (**c**) R = 1/3; (**d**) R = 3; (**e**) R = 1/5; (**f**) R = 5.

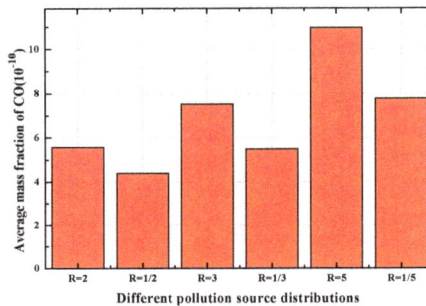

Figure 7. The average concentration of CO at pedestrian breathing height under different pollution source distributions.

3.1.2. Influence of Height Variation of Street Building on Non-Uniform Distribution Characteristics of the Pollution Source

Through the above analysis of different source distributions in the street space at a horizontal height of 1.5 m, we found that high pollution areas appeared locally near the leeward surface. This was mainly related to the height of the buildings on both sides. To further analyze the influence of different building heights on the CO distribution characteristic, the section of Y = 160 m (the difference of building height on both sides is small), Y = 210 m (typical step-up canyon) and Y = 270 m (typical step-down canyon), pollution source intensity R = 1/3 and R = 3 were chosen. Under the lower wind speed condition, the pollutant dispersion was mainly affected by the geometric structure of the canyon; then, the wind speed was chosen to be 1.5 m/s.

In the Y = 160 m section, both sides are residential buildings, belonging to a densely populated area. From Figure 8a,b, it can be seen that the pollutant dispersion characteristics are very different even if the total source intensity would be equal in the street. The specific performance for the pollution source intensity for the left side (near the leeward side) was stronger than for the right side (near the windward side). The high pollution area presents approximately a quarter of a circle at the leeward side. However, when the source intensity at the right side was stronger than at the left side, the high pollution area presents a triangular trend from the leeward side to the center of the street. For a step-down canyon, the increase of source intensity at the right side expands the high pollution concentration area to the windward surface, which greatly increased the whole street space pollution. However, for the step-up canyon, due to the lower upstream building, the wind flow can soon eject pollutants out of the street. Therefore, in the step-up canyon, the impact of the pollution source non-uniform distribution on the pollutant dispersion characteristics is no longer significant.

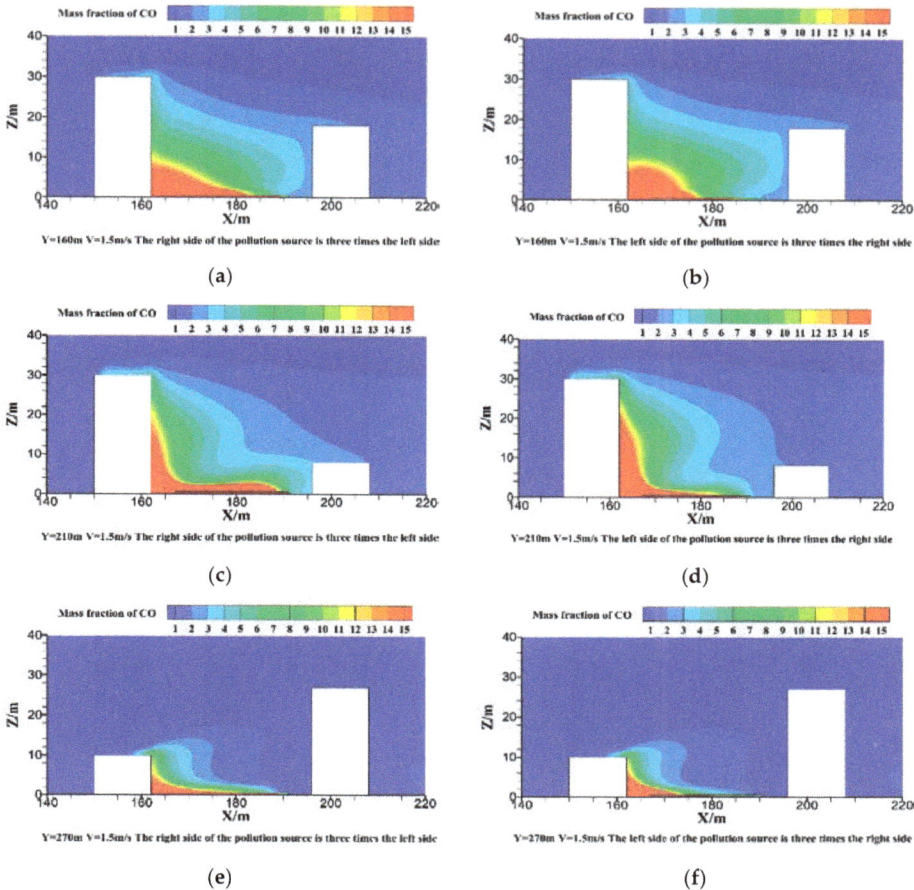

Figure 8. CO concentration field profile (10^{-9}) in three different sections at a wind speed of 1.5 m/s and emission ratios of R = 1/3 and R = 3. (**a**) Y = 160 m, R = 1/3; (**b**) Y = 160 m, R = 3; (**c**) Y = 210 m, R = 1/3; (**d**) Y = 210 m, R = 3; (**e**) Y = 270 m, R = 1/3; (**f**) Y = 270 m, R = 3.

This phenomenon shown above is affected by flow field characteristics and the vortex intensity within the canyon. Figure 9a,b show the distribution of the streamlines in the street cross-section of Y = 210 m (typical step-up canyon) and Y = 270 m (typical step-down canyon) at a wind speed of 3 m/s. For the step-down canyon (Figure 9a), under the action of air transportation, CO accumulates in the leeward side of the street. Due to the increasing height of upstream buildings, the transportation ability of wind flow gradually decreases with increasing height in the leeward building surface. This leads to an accumulation of pollutants in the leeward side. For the step-up canyon (Figure 9b), a clockwise rotation vortex exists in the canyon due to the downstream building block air flow. The vortex center is located at half of the height of the downstream building. Under the action of vortex, the pollutants emitted by motor vehicles are transported from the windward side to the leeward side. Due to the relatively lower height of upstream buildings, CO can be quickly transported to the roof of the leeward surface, and will soon be diluted by the wind flow. The CO moving out of the canyon barely returns to the street. Thus, in the step-up canyon, the impact of the pollution source non-uniform distribution on the pollutant dispersion characteristics is no longer significant.

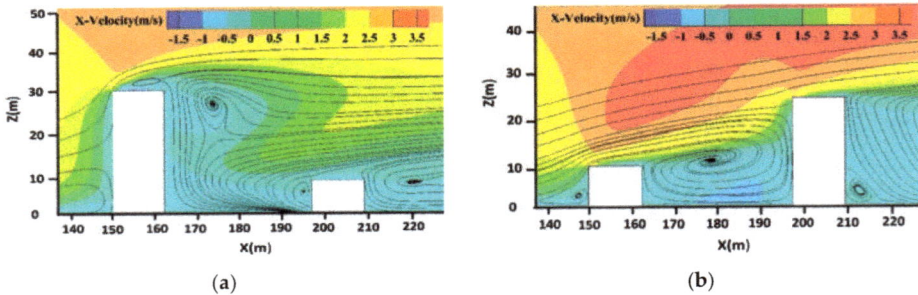

Figure 9. Schematic diagram of the streamline at a wind speed of 3 m/s: (**a**) $Y = 210$ m section; (**b**) $Y = 270$ m section.

3.1.3. Analysis of Spatial Distribution Characteristics of CO Concentration in Specific Section

From the analysis in Sections 3.1.1 and 3.1.2, it can be seen that both the pollution source distribution and the height variation of the buildings on both sides of the street influence the dispersion characteristics of the pollutant. To better understand the distribution of pollutants in the street space when the pollution source is non-uniform distribution, it is necessary to quantitatively analyze the pollutant concentration in the horizontal and vertical directions within the street canyon. However, due to the large computational domain, quantitative analysis is difficult to implement sequentially. Therefore, this paper selected the typical $Y = 160$ m cross-section in the street with the vertical direction $X = 180$ m (street center), $X = 165$ m, and $X = 193$ m (non-motorized road center in left and right side) and the horizontal direction $Z = 1.5$ m, $Z = 5$ m, and $Z = 10$ m as shown in Figure 10, to plot CO concentration curves at a wind speed of 1.5 m/s and wind direction 90°. Then, the distribution characteristics of CO in the horizontal and vertical street space were analyzed.

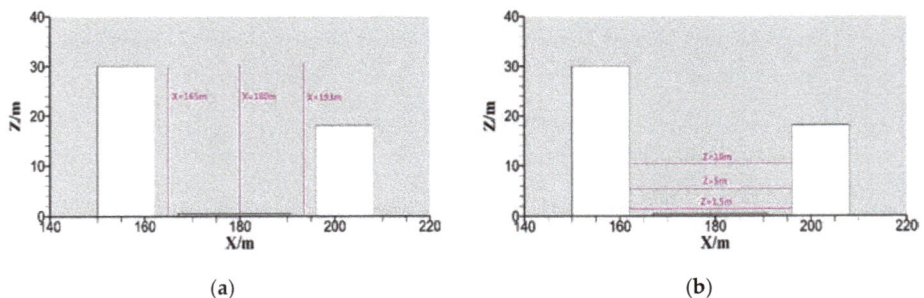

Figure 10. Selected line diagram of vertical and horizontal directions in the $Y = 160$ m section. (**a**) Vertical direction; (**b**) Horizontal direction.

Firstly, the dispersion characteristics of CO in the vertical direction were analyzed when the pollution sources were non-uniformly distributed. Figure 11a shows the distribution of CO along the vertical direction of the non-motorized lane center near the leeward side. This analysis shows that, even though the total pollution source intensity in the street remained the same, the source intensity near the leeward side was stronger than near the windward side ($R = 2$, $R = 3$, and $R = 5$), increasing the CO concentration compared to the corresponding R = 1/2, R = 1/3, and R = 1/5 within the vertical direction of a 10 m height. Furthermore, a larger ratio leads to a more apparent difference in CO concentration. This was mainly because the source intensity near the leeward side had a decisive effect on the concentration of CO. Above a height of 10 m, the influence of non-uniform pollution source distribution basically disappeared. Both CO concentration curves of the same source intensity were basically consistent. With gradually increasing height, the impact of the total source intensity becomes inconspicuous. The CO level obtains a minimum value near the windward side. As the street center $X = 180$ m was close to the windward side, the CO concentration under the source distribution of R = 1/2, 1/3, and 1/5 was higher than the source distribution of R = 2, 3, and 5 within a height of 1 m. Since the wind force near the ground is relatively small, the CO concentration is mainly affected by the source intensity. However, soon the impact of both source intensity and non-uniform distribution have ceased. Since the vortex formed in the section, the pollutants are blown away from the windward side to the leeward side. Part of them spreads out of the canyon, while part of it accumulates at the leeward surface. Therefore, the CO concentration level in the vertical direction of the street center is neither affected by the source intensity nor by the non-uniform distribution. The concentration of CO is much smaller in the vicinity of the windward surface ($X = 193$ m), but the effect of pollution source non-uniform distribution is still present: When the source close to the leeward surface is strong, the CO concentration at the same height increases. The greater the ratio, the more obvious the difference in the concentration level will become (as shown in Figure 11c R = 1/5 and R = 5).

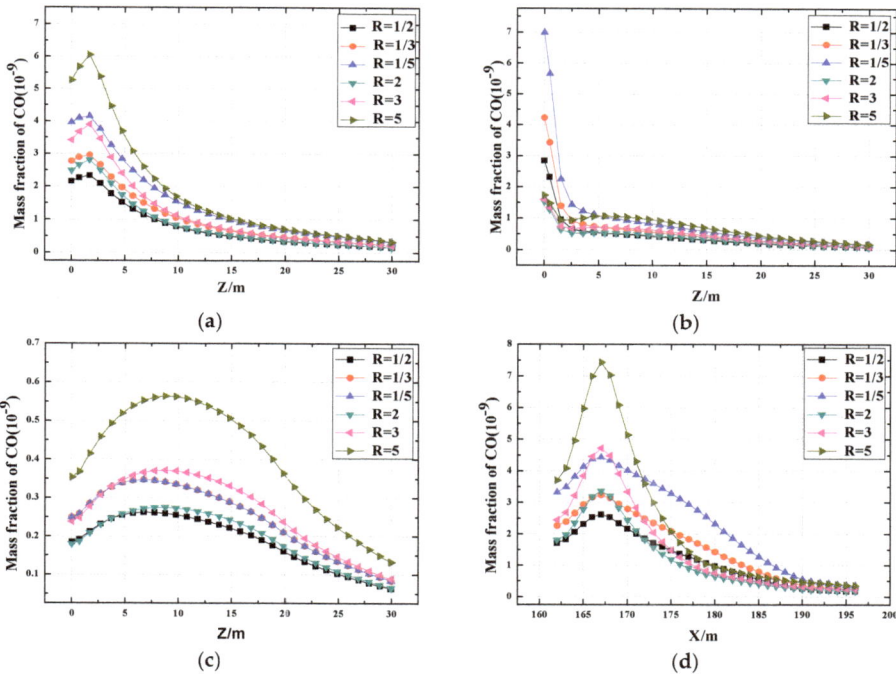

Figure 11. *Cont.*

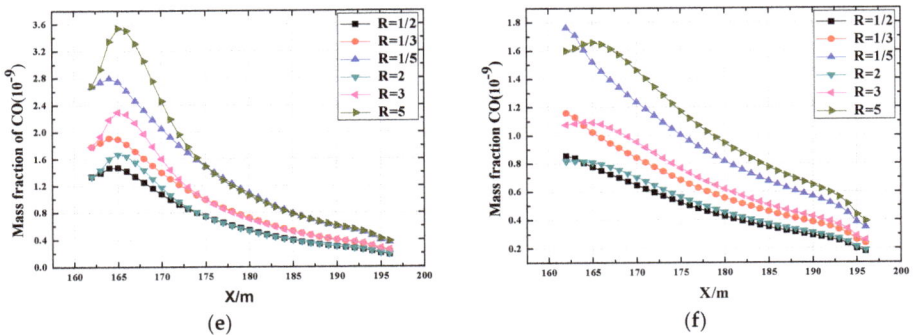

Figure 11. CO concentration profile at different spatial positions at the wind speed 1.5 m/s: (**a**) $X = 165$ m; (**b**) $X = 180$ m; (**c**) $X = 193$ m; (**d**) $Z = 1.5$ m; (**e**) $Z = 5$ m; (**f**) $Z = 10$ m.

Figure 11d–f represent the CO concentration profiles at horizontal heights of $Z = 1.5$ m, $Z = 5$ m, and $Z = 10$ m, respectively. At the lower horizontal height ($Z = 1.5$ m), the concentration of CO is mainly dominated by the source intensity of both sides. In the vicinity of the leeward area (162 m < X < 172 m), the concentration of CO in the horizontal direction is high due to the stronger source intensity on the leeward side. With approaching the windward surface, the concentration of CO begin to change: the concentration of CO in the horizontal direction near the windward surface is high due to the stronger source intensity on the windward side. Finally, when the distance from the windward surface is very short, the impacts of pollution source intensity and non-uniform distribution cease. The concentration of CO in the windward surface reaches a minimum. In general, a stronger source intensity closer to the leeward than to the windward side will cause the expansion of the high pollution area. With increasing horizontal height ($Z = 5$ m), the CO concentration in the area near the leeward side is mainly dominated by the source intensity on the left side. However, as an approach to the windward surface, the impact of source non-uniform distribution basically disappeared. Furthermore, the two CO concentration distribution curves with the same total source intensities are beginning to be consistent. At the horizontal height of $Z = 10$ m (Figure 11f), a source intensity at left side stronger than at the right side (R = 2, 3, and 5) results in higher CO concentrations even though the total sources are equal.

The above analysis shows that when the pollution source is non-uniformly distributed, the concentration level of CO is mainly affected by the source intensity on both sides at a low horizontal height. As the height increases, the influence of source intensity gradually fades. Finally, a source intensity near the leeward side that is stronger than near the windward side will have a higher CO concentration level. This is mainly caused by the air vortex that formed in this section. The stream flows down along the windward building and begins to flow upward after passing the ground block. Pollutants near the windward area are more likely to be blown out of the street, while part of the pollutants near the leeward area enter the relatively small secondary-vortex which lies at the corner of the leeward surface. Pollutants in the small secondary-vortex are difficult to spread out of the canyon and result in a wide range of pollution areas within the street space. A further reason is due to the wind force near the windward side, which is usually stronger than the wind force near the leeward side. Then, pollutants near the windward side are more easily spread out of the canyon. Furthermore, when the source intensity near the leeward side is stronger, it will cause a large amount of pollutants to remain in the street canyon. Therefore, we can conclude that in the case of low wind speeds, a stronger source intensity near the leeward side easily forms a relatively large street pollution space.

3.2. Influence of Wind Speed and Wind Direction on the Non-Uniform Distribution of Pollution Sources

The above analyses about the dispersion characteristics of CO in the street canyon under non-uniform distribution of pollution source are based on a relatively low wind speed and for when the wind direction is perpendicular to the street. Therefore, it is necessary to analyze whether the influence of a non-uniformly distributed pollution source on the dispersion characteristics of pollutant remains significant under different wind speed and wind direction. We employed the pollution source emission ratios $R = 1/5$ and $R = 5$, wind speeds of $V = 1.5$ m/s, $V = 3$ m/s, $V = 4.5$ m/s, and $V = 6$ m/s, and a wind direction of 90° to analyze the impact of wind speed. Also, the pollution source emission ratios $R = 1/5$ and $R = 5$, wind speed $V = 3$ m/s, and wind direction 0°, 30°, 45°, and 60° were employed to analyze the impact of the wind direction. Then, the average concentration of CO at pedestrian breathing height $Z = 1.5$ m under different cases are shown in Figure 12.

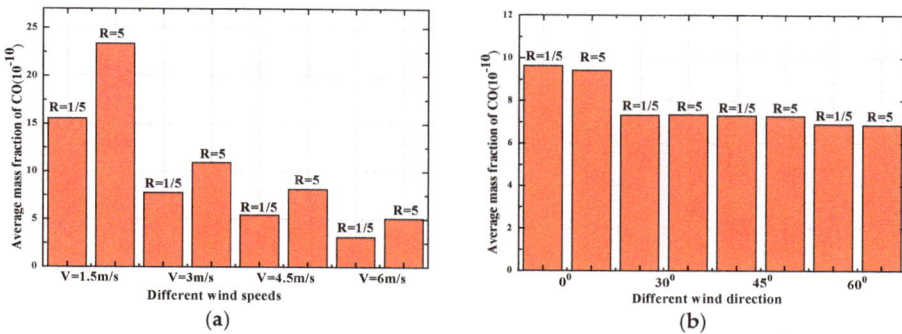

Figure 12. Average concentration of CO at pedestrian breathing height under different wind speeds (**a**) and wind direction (**b**).

According to Figure 12a, we firstly found that with increasing wind speeds, the average concentration of CO at the zone of pedestrian breathing height shows a decreasing trend. Furthermore, the characteristics of larger pollution space caused by stronger source intensity near the leeward side are also no longer significant with increasing wind speeds. I.e., the increasing wind speeds will weaken the effect of a non-uniform distribution of the pollution source on the dispersion characteristics of CO. The increasing wind speeds enhance the effect of wind force transportation and dilution, and accelerate the air exchange rate (ACH) between street canyon and the upper atmosphere. In addition, the wind force near the leeward side also enhanced with increasing wind speeds, and consequently, the pollutants near the leeward side are more likely to spread out of the canyon. This causes the phenomenon that larger pollution space caused by stronger source intensity near the leeward side is less significant with increasing wind speeds. Furthermore, when the wind direction is not perpendicular to the street, the effect of the non-uniform pollution source distribution on the dispersion characteristics of CO is also not significant. The average CO concentration remains the same under the two pollution source emission ratios of $R = 1/5$ and $R = 5$ as shown in Figure 12b. The wind direction determines the pollutant transport direction. For a wind direction of 0° or if it has a certain angle to the street axis, e.g., 30°, 45°, and 60°, then there is a velocity component on the Y-axis compared to the 90° wind direction. Then, the velocity component on the Y-axis will lead to a downward movement of pollutants along the street. Due to the absence of blocks downstream of the street, pollutants are easily dispersed. The parallel and oblique wind direction strengthens the spanwise (Y) direction dispersion of pollutants. Therefore, the effect of the non-uniform distribution of the pollution source on the dispersion characteristics of CO is no longer significant.

3.3. Influence of Source Intensity on Pollution Levels in Street Space

To analyze the influence of different pollution source intensities on the concentration of CO in the street, we employed the source intensity during noon (11:30–13:00) and other time (8:30–11:30 and 13:00–17:30) for the CFD numerical simulation. Then, the CO concentration levels along the center of two non-motorized lanes were plotted at a wind speed of 3 m/s. We mainly focused on the pedestrian breathing height of Z = 1.5 m. Through the field measurement we have obtained the source intensity during noon and at other times. These were 5.5×10^{-6} (kg/m^3 s) and 3.3×10^{-7} (kg/m^3 s).

Figure 13 shows a diagram of CO concentration levels along the center of two non-motorized lanes. The pollution source intensity significantly impacted the level of CO concentration. Through analysis of the CO concentration data, we found that with decreasing pollution source intensity, the average level of CO concentration at pedestrian breathing height decreased proportionally. At the left non-motorized center, the CO concentration decreased by about 30%. Furthermore, it decreased more at the right non-motorized center by about 40%. Thus, for the control of urban street vehicle pollution, the reduction of pollution source intensity is the most direct and effective way.

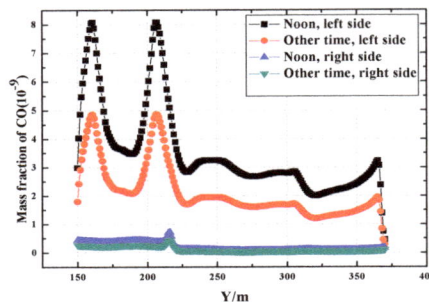

Figure 13. CO concentration profile at a height of Z = 1.5 m at the left and right sides of the non-motorized center.

4. Conclusions

For this study, a three-dimensional geometrical model was established based on a street canyon section with a typical traffic tidal phenomenon in the 2nd Ring Road of Wuhan, China. A mathematical model describing the fluid flow and pollutant dispersion characteristics in the street canyon was developed. The number and driving speeds of vehicles on the road during different periods of one day were measured; then, the emission rate of pollutants in two parallel roads was calculated with the CFD simulation method. The results of the numerical simulation indicate that:

In this three-dimensional asymmetrical shallow street canyon, when pollution sources in the street are non-uniformly distributed and when the wind flow is perpendicular to the street, a stronger source intensity near the leeward side than near the windward side will cause the expansion of pollution space even though the total source intensity remains equal. For example, when the wind speed is 3 m/s, a source intensity near the leeward side stronger than near the windward side (R = 2, R = 3, and R = 5) increases the average concentration of CO at pedestrian breathing height increased by 26%, 37%, and 41%, respectively.

However, with increasing wind speeds and when the wind direction is not perpendicular to the street, the concentration of pollutants in the whole street space shows a decreasing trend. Furthermore, the characteristics of a larger pollution space caused by stronger source intensity near the leeward side were also less significant. I.e., the increase in wind speeds and changes in wind direction weakens the effect of pollution source non-uniform distribution on pollutant dispersion characteristics.

The source intensity significantly impacts the level of pollutant concentration in the street canyon. With decreasing source intensity, the level of pollutant concentration at pedestrian breathing level decreased proportionally. In the left non-motorized center, the CO concentration decreased by about 30%, while it decreased more at the right non-motorized center by about 40%. Thus, for the control of urban street vehicle pollution, a reduction of pollution source intensity is the most direct and effective way.

Acknowledgments: This research was supported by the National Natural Science Foundation of China (Nos. 51778511, 51778253), Key Project of ESI Discipline Development of Wuhan University of Technology (WUT No. 2017001), the Scientific Research Foundation of Wuhan University of Technology (No. 40120237).

Author Contributions: Tingzhen Ming and Chong Peng discussed and aroused the idea; Weijie Fang and Cunjin Cai performed the simulation; Weijie Fang drafted the manuscript; Tingzhen Ming, Renaud de Richter, Mohammad Hossein Ahmadi, and Yuangao Wen finalized the manuscript.

Conflicts of Interest: The authors declare no conflict of interest.

References

1. Gong, T.; Ming, T.; Huang, X.; de Richter, R.K.; Wu, Y.; Liu, W. Numerical analysis on a solar chimney with an inverted U-type cooling tower to mitigate urban air pollution. *Sol. Energy* **2017**, *147*, 68–82. [CrossRef]
2. Song, J.; Guang, W.; Li, L.; Xiang, R. Assessment of Air Quality Status in Wuhan, China. *Atmosphere* **2016**, *7*, 56. [CrossRef]
3. Deng, Q.; Lu, C.; Yu, Y.; Li, Y.; Sundell, J.; Norbäck, D. Early life exposure to traffic-related air pollution and allergic rhinitis in preschool children. *Respir. Med.* **2016**, *121*, 67–73. [CrossRef] [PubMed]
4. Deng, Q.; Chan, L.; Wei, J.; Zhao, J.; Deng, L.; Xiang, Y. Association of outdoor air pollution and indoor renovation with early childhood ear infection in China. *Chemosphere* **2017**, *169*, 288–296. [CrossRef] [PubMed]
5. Nicholson, S.E. A pollution model for street-level air. *Atmos. Environ.* **1975**, *9*, 19. [CrossRef]
6. Andreou, E.; Axarli, K. Investigation of urban canyon microclimate in traditional and contemporary environment. Experimental investigation and parametric analysis. *Renew. Energy* **2012**, *43*, 354–363. [CrossRef]
7. Prajapati, S.K.; Tripathi, B.D.; Pathak, V. Distribution of vehicular pollutants in street canyons of Varanasi, India: A different case. *Environ. Monit. Assess.* **2009**, *148*, 167–172. [CrossRef] [PubMed]
8. Salizzoni, P.; Soulhac, L.; Mejean, P. Street canyon ventilation and atmospheric turbulence. *Atmos. Environ.* **2009**, *43*, 5056–5067. [CrossRef]
9. Allegrini, J.; Dorer, V.; Carmeliet, J. Wind tunnel measurements of buoyant flows in street canyons. *Build. Environ.* **2013**, *59*, 315–326. [CrossRef]
10. Stabile, L.; Arpino, F.; Buonanno, G.; Russi, A.; Frattolillo, A. A simplified benchmark of ultrafine particle dispersion in idealized urban street canyons: A wind tunnel study. *Build. Environ.* **2015**, *93*, 186–198. [CrossRef]
11. Wang, Y.; Huang, Z.; Liu, Y.; Yu, Q.; Ma, W. Back-Calculation of Traffic-Related PM_{10} Emission Factors Based on Roadside Concentration Measurements. *Atmosphere* **2017**, *8*, 99. [CrossRef]
12. Chew, L.; Nazarian, N.; Norford, L. Pedestrian-Level Urban Wind Flow Enhancement with Wind Catchers. *Atmosphere* **2017**, *8*, 159. [CrossRef]
13. Zhou, Y.; Deng, Q. Numerical simulation of inter-floor airflow and impact on pollutant transport in high-rise buildings due to buoyancy-driven natural ventilation. *Indoor Built Environ.* **2014**, *23*, 246–258. [CrossRef]
14. Soulhac, L.; Mejean, P.; Perkins, R.J. Modelling the transport and dispersion of pollutants in street canyons. *Int. J. Environ. Pollut.* **2001**, *16*, 404–416. [CrossRef]
15. Kim, J.J.; Baik, J.J. A numerical study of the effects of ambient wind direction on flow and dispersion in urban street canyons using the RNG—Turbulence model. *Atmos. Environ.* **2004**, *38*, 3039–3048. [CrossRef]
16. Balogun, A.A.; Tomlin, A.S.; Wood, C.R.; Barlow, J.F.; Belcher, S.E.; Smalley, R.J.; Lingard, J.J.N.; Arnold, S.J.; Dobre, A.; Robins, A.G. In-Street Wind Direction Variability in the Vicinity of a Busy Intersection in Central London. *Bound.-Layer Meteorol.* **2010**, *136*, 489–513. [CrossRef]
17. Kim, J.J.; Baik, J.J. Effects of inflow turbulence intensity on flow and pollutant dispersion in an urban street canyon. *J. Wind Eng. Ind. Aerodyn.* **2003**, *91*, 309–329. [CrossRef]

18. Chan, A.T.; So, E.S.; Samad, S.C. Erratum to "Strategic guidelines for street canyon geometry to achieve sustainable street air quality" (Atmospheric Environment 35 (24) 4089–4098). *Atmos. Environ.* **2001**, *35*, 5679. [CrossRef]

19. Chan, A.T.; So, E.S.; Samad, S.C. Strategic guidelines for street canyon geometry to achieve sustainable street air quality. *Atmos. Environ.* **2001**, *35*, 5681–5691. [CrossRef]

20. Chang, C.H.; Meroney, R.N. Concentration and flow distributions in urban street canyons: Wind tunnel and computational data. *J. Wind Eng. Ind. Aerodyn.* **2003**, *91*, 1141–1154. [CrossRef]

21. Liu, C.; Barth, M.C.; Leung, D.Y.C. Large-Eddy Simulation of Flow and Pollutant Transport in Street Canyons of Different Building-Height-to-Street-Width Ratios. *J. Appl. Meteorol.* **2004**, *43*, 1410–1424. [CrossRef]

22. Kang, Y.S.; Baik, J.J.; Kim, J.J. Further studies of flow and reactive pollutant dispersion in a street canyon with bottom heating. *Atmos. Environ.* **2008**, *42*, 4964–4975. [CrossRef]

23. Kim, J.J.; Baik, J.J. Effects of Street-Bottom and Building-Roof Heating on Flow in Three-Dimensional Street Canyons. *Adv. Atmos. Sci.* **2010**, *27*, 513–527. [CrossRef]

24. Kim, J.J.; Pardyjak, E.; Kim, D.Y.; Han, K.S.; Kwon, B.H. Effects of building-roof cooling on flow and air temperature in urban street canyons. *Asia Pac. J. Atmos. Sci.* **2014**, *50*, 365–375. [CrossRef]

25. Gronemeier, T.; Raasch, S.; Ng, E. Effects of Unstable Stratification on Ventilation in Hong Kong. *Atmosphere* **2017**, *8*, 168. [CrossRef]

26. Lateb, M.; Meroney, R.N.; Yataghene, M.; Fellouah, H.; Saleh, F.; Boufadel, M.C. On the use of numerical modelling for near-field pollutant dispersion in urban environments—A review. *Environ. Pollut.* **2016**, *208*, 271–283. [CrossRef] [PubMed]

27. Nosek, Š.; Kukačka, L.; Kellnerová, R.; Jurčáková, K.; Jaňour, Z. Ventilation Processes in a Three-Dimensional Street Canyon. *Bound.-Layer Meteorol.* **2016**, *159*, 1–26. [CrossRef]

28. Gu, Z.; Zhang, Y.; Cheng, Y.; Lee, S.C. Effect of uneven building layout on air flow and pollutant dispersion in non-uniform street canyons. *Build. Environ.* **2011**, *46*, 2657–2665. [CrossRef]

29. Nosek, Š.; Kukačka, L.; Jurčáková, K.; Kellnerová, R.; Jaňour, Z. Impact of roof height non-uniformity on pollutant transport between a street canyon and intersections. *Environ. Pollut.* **2017**, *227*, 125. [CrossRef] [PubMed]

30. Tominaga, Y.; Mochida, A.; Yoshie, R.; Kataoka, H.; Nozu, T.; Yoshikawa, M.; Shirasawa, T. AIJ guidelines for practical applications of CFD to pedestrian wind environment around buildings. *J. Wind Eng. Ind. Aerodyn.* **2008**, *96*, 1749–1761. [CrossRef]

31. Ming, T.; Gong, T.; Peng, C.; Li, Z. *Pollutant Dispersion in Built Environment*; Springer: Singapore, 2017.

32. de_Richter, R.; Ming, T.; Davies, P.; Liu, W.; Caillol, S. Removal of non-CO_2 greenhouse gases by large-scale atmospheric solar photocatalysis. *Prog. Energy Combust. Sci.* **2017**, *60*, 68–96. [CrossRef]

33. Chan, T.L.; Dong, G.; Leung, C.W.; Cheung, C.S.; Hung, W.T. Validation of a two-dimensional pollutant dispersion model in an isolated street canyon. *Atmos. Environ.* **2002**, *36*, 861–872. [CrossRef]

34. Blocken, B.; Stathopoulos, T.; Saathoff, P.; Wang, X. Numerical evaluation of pollutant dispersion in the built environment: Comparisons between models and experiments. *J. Wind Eng. Ind. Aerodyn.* **2008**, *96*, 1817–1831. [CrossRef]

35. Nazridoust, K.; Ahmadi, G. Airflow and pollutant transport in street canyons. *J. Wind Eng. Ind. Aerodyn.* **2006**, *94*, 491–522. [CrossRef]

36. Blocken, B.; Stathopoulos, T.; Carmeliet, J.; Hensen, J.L.M. Application of computational fluid dynamics in building performance simulation for the outdoor environment: An overview. *J. Build. Perform. Simul.* **2011**, *4*, 157–184. [CrossRef]

37. Tominaga, Y.; Stathopoulos, T. CFD modeling of pollution dispersion in a street canyon: Comparison between LES and RANS. *J. Wind Eng. Ind. Aerodyn.* **2011**, *99*, 340–348. [CrossRef]

38. Xie, X.; Liu, C.H.; Leung, D.Y.C.; Leung, M.K.H. Characteristics of air exchange in a street canyon with ground heating. *Atmos. Environ.* **2006**, *40*, 6396–6409. [CrossRef]

39. Xie, X.; Liu, C.H.; Leung, D.Y.C. Impact of building facades and ground heating on wind flow and pollutant transport in street canyons. *Atmos. Environ.* **2007**, *41*, 9030–9049. [CrossRef]

40. Hang, J.; Li, S. Effect Of Urban Morphology On Wind Condition In Idealized City Models. *Atmos. Environ.* **2009**, *43*, 869–878. [CrossRef]

41. Yakhot, V.; Orszag, S.A. Renormalization group and local order in strong turbulence. *Nucl. Phys. B* **1987**, *2*, 417–440. [CrossRef]

42. Li, X.; Liu, C.H.; Leung, D.Y.C. Development of a k–ε model for the determination of air exchange rates for street canyons. *Atmos. Environ.* **2005**, *39*, 7285–7296. [CrossRef]

43. Hang, J.; Li, Y.; Sandberg, M.; Buccolieri, R.; Sabatino, S.D. The influence of building height variability on pollutant dispersion and pedestrian ventilation in idealized high-rise urban areas. *Build. Environ.* **2012**, *56*, 346–360. [CrossRef]

44. Rajapaksha, I.; Nagai, H.; Okumiya, M. A ventilated courtyard as a passive cooling strategy in the warm humid tropics. *Renew. Energy* **2003**, *28*, 1755–1778. [CrossRef]

45. Leung, K.K.; Liu, C.H.; Wong, C.C.C.; Lo, J.C.Y.; Ng, G.C.T. On the study of ventilation and pollutant removal over idealized two-dimensional urban street canyons. *Build. Simul.* **2012**, *5*, 359–369. [CrossRef]

46. Wang, H.; Chen, Q. A new empirical model for predicting single-sided, wind-driven natural ventilation in buildings. *Energy Build.* **2012**, *54*, 386–394. [CrossRef]

47. Zhang, K.; Yao, L. Research on the relationship between driving behavior and exhaust emission. *Highw. Automot. Appl.* **2014**, *160*, 39–43.

48. Patankar, S.V. *Numerical Heat Transfer and Fluid Flow*; CRC press: Boca Raton, FL, USA, 1980; pp. 125–126.

49. Michioka, T.; Takimoto, H.; Kanda, M. Large-Eddy Simulation for the Mechanism of Pollutant Removal from a Two-Dimensional Street Canyon. *Bound.-Layer Meteorol.* **2011**, *138*, 195–213. [CrossRef]

![atmosphere logo] *atmosphere*

MDPI

Article

Source Apportionment and Data Assimilation in Urban Air Quality Modelling for O₂: The Lyon Case Study

Chi Vuong Nguyen, Lionel Soulhac and Pietro Salizzoni *

Laboratoire de Mécanique des Fluides et d'Acoustique, UMR CNRS 5509 University of Lyon,
Ecole Centrale de Lyon, INSA Lyon, Université Claude Bernard Lyon I, 36, avenue Guy de Collongue,
69134 Ecully, France; cv.nguyen@hotmail.fr (C.V.N.); lionel.soulhac@ec-lyon.fr (L.S.)
* Correspondence: pietro.salizzoni@ec-lyon.fr

Received: 7 September 2017; Accepted: 18 December 2017; Published: 1 January 2018

Abstract: Developing effective strategies for reducing the atmospheric pollutant concentrations below regulatory threshold levels requires identifying the main origins/sources of air pollution. This can be achieved by implementing so called *source apportionment* methods in atmospheric dispersion models. This study presents the results of a source apportionment module implemented in the SIRANE urban air-quality model. This module uses the *tagged species approach* and includes two methods, named SA-NO and SA-NOX, in order to evaluate the sources' contributions to the NO_2 concentrations in air. We also present results of a data assimilation method, named SALS, that uses the source apportionment estimates to improve the accuracy of the SIRANE model results. The source apportionment module and the assimilation method have been tested on a real case study (the urban agglomeration of Lyon, France, for the year 2008) focusing on the O_2 emissions and concentrations. Results of the source apportionment with the SA-NO and SA-NOX models are similar. Both models show that traffic is the main cause of O_2 air pollution in the studied area. Results of the SALS data assimilation method highlights its ability in improving the predictions of an urban atmospheric models.

Keywords: source apportionment; data assimilation; urban air quality modelling

1. Introduction

Obtaining information about the intensity of the pollutant sources is essential in order to determine the main causes of air pollution and to define the relevant actions for its reduction. The assessment of the intensity of the pollutant sources can rely on different criteria, namely on the estimate of the contribution of (i) different typology of sources (e.g., traffic, industry, agriculture or residential-tertiary emissions) [1–5], (ii) sources located in different geographical areas (e.g., emissions from different regions of Europe) [3,6–8], and (iii) emissions occurring at different times. The methods adopted to estimate the contribution of different sources are usually referred to as *source apportionment methods*, and can be classified in three main approaches.

The first is based on the analysis of the chemical composition of the pollutant. This method is essentially applied to particulate matter (PM), which is composed from a variety of chemical elements, some of which are specific to some sources [9–11]. For example, the black carbon reveals the emission by combustion processes [11], whereas the dehydroabietic acid is characteristic of natural sources, as coniferae [9,12].

The second is based on the so called *receptor models* [13,14]. These are statistical approaches, based on mass conservation principles [14], requiring as input data the concentrations measured at the monitoring stations. These models are generally classified into two different categories, known as Chemical Mass Balance (CMB) models [15–19] and multivariate models [17,19–24]. The receptor models are mainly used to evaluate sources of PM pollution [14,17,24,25], but can in principle also be used for other species [26,27].

The third method is based on the use of atmospheric dispersion models. These estimate the concentration field of a pollutant by solving, analytically or numerically, the advection-diffusion equation:

$$\frac{\partial c}{\partial t} + \mathbf{u} \cdot \nabla c = \nabla \cdot (D_t \nabla c) + S, \tag{1}$$

where c is the (time) averaged pollutant concentration, $\mathbf{u}$ is the (time) averaged wind velocity, D_t is the turbulent diffusivity and S represents the source terms (emissions, losses, chemical reactions). Without chemical reactions (i.e., $S = 0$), Equation (1) is linear for the concentration. Therefore, sources' contributions can be assessed by separately addressing emissions from each source. This approach is no longer valid when considering reactive pollutants because chemical reactions induce nonlinear effects. As pointed out by Koo et al. [28], this nonlinearity precludes an exact reconstruction of the sources' contributions, which can therefore be evaluated in several different ways.

One of the simplest methods to evaluate the source effect with an atmospheric dispersion model is the so called *brute force method* (BFM) [3,28,29]. This is carried out in two steps. The first step consists of performing a *reference simulation* including all sources. The second step consists of carrying out simulations excluding some of these sources (or a typology of sources). The difference between the results of the two simulations quantifies the contribution of the sources that have been removed. Nevertheless, this method is computationally expensive because the second step has to be carried out for each source typology considered. A suitable approach to reduce the computational costs is to perform a single simulation with *tagged species* [1,4,5,8,28,30–35]. Tagging the emitted species allows them to be tracked and their origin to be identified: emissions of two sources S_R and S_B, emitting the tagged species CO^R and CO^B, provide concentration fields of CO^R and CO^B, which corresponds to the S_R and S_B contributions, respectively.

Atmospheric dispersion models have been frequently used to evaluate the sources' contributions to PM concentrations [1,5,7,8,28,35]. Studies have also been carried out to estimate the sources' contributions to ozone [8,32–34], CO [36–39] and SO_2 [2,40] concentrations. All of these studies were performed with mesoscale atmospheric dispersion models. In this study, we aim instead at implementing a source apportionment module in an urban dispersion model, simulating pollutant transport at the local scale (a few tens of kilometers).

Results provided by source apportionment methods are here subsequently coupled with a data assimilation method. In the literature, any modelling approach coupling a model and field measurements in order to improve the accuracy of the prediction of a generic physical system is usually referred to as *data assimilation* [41–45]. Data assimilation methods are used since several decades in atmospheric sciences [46–50]. More recently, they have also been applied in air quality studies [51–55], mainly adopting chemical transport models [51–67], and more rarely with urban dispersion models [68]. Here, we present the results of a data assimilation method named *Source Apportionment Least Square* (SALS), developed for the SIRANE urban air quality model [69,70].

In what follows, we first present the SIRANE model and the chemical scheme implemented in it (Sections 2.1 and 2.2). Secondly, we introduce the principles of the source apportionment modules and the data assimilation technique adopted (Section 2.3). Finally, we show the results of a real case study (Section 3), the urban Lyon agglomeration, for the year 2008.

2. Methods

2.1. The SIRANE Model

SIRANE is an operational model to simulate the atmospheric pollutants' dispersion at the urban scale. It is based on the street network concept [69,71] and adopts parametric laws to model the main flow and dispersion processes within an urban area: advection along the street axes, turbulent transfer across the street-atmosphere interfaces, and exchanges at the street intersections. The presence of a roughness sub-layer just above the urban canopy (above roof level) is neglected and the flow is modelled as a boundary layer over a rough surface. There, the pollutants dispersion is modelled by a Gaussian plume, whose standard deviations are parametrised according to the Monin–Obukhov similarity theory. As customary for local scale dispersion models, SIRANE adopts a quasi-steady approach to deal with the unsteadiness of meteorological conditions and pollutant emissions, with an hourly time step. The input data are the urban geometry, the meteorological data, the locations and the modulations of the emissions (represented as point, line, and surface sources) and the hourly evolution of the background concentration, i.e., the concentration due to sources placed outside the domain. More details on the SIRANE model can be found in Soulhac et al. [69] and Soulhac et al. [72].

2.2. Modelling Chemical Reactions

The only chemical reactions taken into account in the SIRANE model concern the NO_2-NO-O_3 cycle. In steady state conditions, i.e., at the photo-stationary equilibrium, this is usually represented by the following set of reactions [73]:

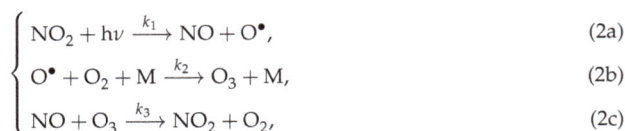

$$\begin{cases} NO_2 + h\nu \xrightarrow{k_1} NO + O^\bullet, & \text{(2a)} \\ O^\bullet + O_2 + M \xrightarrow{k_2} O_3 + M, & \text{(2b)} \\ NO + O_3 \xrightarrow{k_3} NO_2 + O_2, & \text{(2c)} \end{cases}$$

where k_1, k_2, and k_3 are the kinetic constants of reaction and M is a third body species, e.g., O_2 and N_2. Note that the cycle (2) is in reality perturbed by other reactions, as [74]:

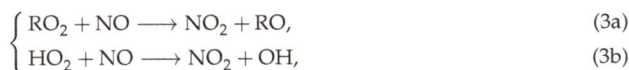

$$\begin{cases} RO_2 + NO \longrightarrow NO_2 + RO, & \text{(3a)} \\ HO_2 + NO \longrightarrow NO_2 + OH, & \text{(3b)} \end{cases}$$

where the RO_2 radical is due to VOC oxidation, occurring over different times scale (larger than those of Equation (2)), depending on VOC chemical lifetime (typically a few hours). Nitroxen oxide concentrations can also be affected by losses due to reactions involving the hydroxyl radical OH, and leading to the production of nitric acid:

$$OH + NO_2 + M \longrightarrow HNO_3 + M. \tag{4}$$

For typical OH concentration in the urban atmosphere ($6 \times 10^6 \, \text{molec cm}^{-3}$), the NO_x chemical lifetime is approximately 4 h, i.e., similar to the time scales of advection across a large urban agglomeration.

As is customary in dispersion models at the local scale [75], reactions (3) and (4) are neglected, and the modelling of chemical transformation of nitrogen oxide relies on Equation (2) only. Since the radical $O^\bullet$ is very reactive, the characteristic time of ozone production (2b) is much smaller than that of the two other reactions, so that the photo-stationary equilibrium (2) can be further simplified as:

$$[NO_2] = \frac{k_3}{k_1}[O_3][NO],\tag{5}$$

where [NO], [NO$_2$] and [O$_3$] represent the NO, NO$_2$, and O$_3$ molar concentrations, respectively. To estimate the NO, NO$_2$ and O$_3$ concentrations, it is then necessary to determine k_1 and k_3 (for each hourly time step), which can be conveniently modelled as [73,76]:

$$\begin{cases} k_1 = \frac{1}{60}\left(0.5699 - [9.056e^{-3}(90-\chi)]^{2.546}\right)\left(1 - 0.75\left[\frac{Cld}{8}\right]^{3.4}\right)[s^{-1}], \\ k_3 = 1.325e^6\exp\left(-\frac{1430}{T}\right)[m^3mole^{-1}s^{-1}], \end{cases}\tag{6}$$

where χ is the solar elevation, Cld is the cloud cover, and T is the temperature.

As already mentioned, Equation (2) represents an over-simplification of the chemical processes occurring in the urban atmosphere. Nevertheless, validation studies [70,72] show good agreement between SIRANE results, obtained adopting this simplified chemical scheme, and on-site measurements.

2.3. Source Apportionment Module

When activating the source apportionment module in the SIRANE model, the sources emit both *classical* species (e.g., NO) and *tagged* species (e.g., NO^g for the source g). In simulating their dispersion in the atmosphere, as a first step, these are both treated as inert species. The role of chemical reactions is then taken into account in a second step of the simulation. The concentration of the species s is denoted as c_s^d at the end of the dispersion phase and as c_s after the chemical reactions phase. Similarly, the concentration of the species s tagged from the source g is denoted $c_{s,g}^d$, after the dispersion phase, and $c_{s,g}$, at the end of the chemical reactions.

2.3.1. Inert Pollutant Species

Tagged species emitted by each source are treated as different species (e.g., NO^{traffic} and NO^{industrial}). In this way, we avoid performing G separate simulations to evaluate the contribution of G sources (or group of sources) for N species, performing instead a single simulation taking into account $G \times N$ species, reducing the computational costs.

2.3.2. Reactive Pollutant Species

The assessment of the sources' contributions for the reactive species is carried out in two steps. The first step consists of determining the sources' contributions after their emission and their atmospheric dispersion, as happens for inert species. The second step consists of evaluating the sources' contributions once the chemical reactions took place. The source module apportionment integrates two models, named SA-NO and SA-NOX. In both, it is assumed that molecules of a same species have the same probability of reacting, independently of their origin [1]. Both take into account the chemical reactions included in Equation (2c) only, but in a different way. In the SA-NO model, we assume that the photo-stationary equilibrium has not been reached, so that the chemical reactions (2c) occurs from the left to the right only. In the SA-NOX model, we assume instead that the photo-stationary equilibrium has already been reached, and that (2c) can be expressed in the form of a dynamical equilibrium:

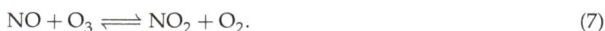

$$NO + O_3 \rightleftharpoons NO_2 + O_2.\tag{7}$$

Model SA-NO

The mass concentration of the specie s (NO, NO$_2$ or O$_3$) after the chemical reactions can be expressed as:

$$c_s = c_s^d + \delta c_s, \tag{8}$$

where δc_s is the variation induced by the chemical reactions. The contribution of a source g for the species s after the chemical reactions is then:

$$c_{s,g} = c_{s,g}^d + \delta c_{s,g}, \tag{9}$$

where $\delta c_{s,g}$ is the contribution of the source g to the concentration variation of the species s due to chemical reactions. The objective of the SA-NO method is to determine $\delta c_{s,g}$. Since NO$_x$ emissions are predominantly NO emissions [77], we assume that the ratio $[c_{NO}^d]/[c_{NO_2}^d]$ after the dispersion modelling phase (before modelling chemical reactions) is higher than the ratio $[c_{NO}]/[c_{NO_2}]$ at the photo-stationary equilibrium (after the chemical reactions), whose achievement requires NO molecules to be consumed and NO$_2$ molecules to be produced. Since the probability that a NO molecule is involved in a chemical reaction is independent of its origin, the relative sources' contributions to the NO concentration variation is equal to their relative contribution to the NO concentration at the end of the dispersion phase:

$$\frac{c_{NO,g}^d}{c_{NO}^d} = \frac{\delta c_{NO,g}}{\delta c_{NO}}. \tag{10}$$

The variation of O$_3$ and NO$_2$ molar concentration induced by (2c) is directly related to the variation of NO moles. The relative sources' contributions to O$_3$ and NO$_2$ concentration variation is then equal to their relative contribution to NO concentration at the end of the dispersion phase:

$$\frac{c_{NO,g}^d}{c_{NO}^d} = \frac{\delta c_{NO_2,g}}{\delta c_{NO_2}} = \frac{\delta c_{O_3,g}}{\delta c_{O_3}} = \frac{\delta c_{s,g}}{\delta c_s}. \tag{11}$$

Thus, the SA-NO method evaluates the contribution of each source g as:

$$c_{s,g} = c_{s,g}^d + \delta c_s \frac{c_{NO,g}^d}{c_{NO}^d}. \tag{12}$$

Note that, with the SA-NO method, the sources' contributions may be negative, namely when the concentration computed after chemical reaction (c_s) is lower than that before (c_s^d). In addition, (12) indicates that a source contributes to NO concentrations only if it emits NO. This constitutes an evident shortcoming of the SA-NO method since, according to the NO$_2$-NO-O$_3$ cycle (2), NO concentrations can be enhanced also by contributing to NO$_2$ and O$_3$ concentration.

Model SA-NOX

Based on the fact that the relative sources' contributions to NO$_x$ concentration (NO + NO$_2$) are the same before and after the chemical reactions, we can write:

$$\frac{[c_{NO,g}^d] + [c_{NO_2,g}^d]}{[c_{NO}^d] + [c_{NO_2}^d]} = \frac{[c_{NO,g}] + [c_{NO_2,g}]}{[c_{NO}] + [c_{NO_2}]}. \tag{13}$$

By further assuming that the relative sources' contributions to NO and NO$_2$ concentrations are the same as their relative contribution to NO$_x$ concentration, we have that:

$$\frac{c_{NO,g}}{c_{NO}} = \frac{c_{NO_2,g}}{c_{NO_2}} = \frac{[c_{NO,g}] + [c_{NO_2,g}]}{[c_{NO}] + [c_{NO_2}]}. \tag{14}$$

The SA-NOX method then evaluates the sources' contributions to NO and NO$_2$ concentration by estimating their contribution to the nitrogen atoms of these molecules:

$$c_{s,g} = c_s \frac{[c_{NO,g}^d] + [c_{NO_2,g}^d]}{[c_{NO}^d] + [c_{NO_2}^d]}. \tag{15}$$

Differently from the SA-NO model, (15) guarantees the contribution of all sources to be positive. Note also that, differently from the SA-NO method, a source can contribute to NO concentrations also by emitting NO$_2$ (or eventually O$_3$).

2.4. Data Assimilation Using Source Apportionment Results

The source apportionment results provide useful information that can be used to improve the performances of the dispersion model by means of the data assimilation techniques. Here, we present a data assimilation method called a *Source Apportionment Least Square method* (SALS). This method consists of modulating, in an *optimal* way, the sources' contributions estimated with a source apportionment method.

We represent the simulated ground level concentration field $c(x, y)$, over n grid points and at a given time t, as a vector $\mathbf{c}_t$ of size n. This vector, called *background*, is then expressed as the sum of different vectors $\mathbf{c}_{g,t}$, each of them representing the modelled contribution to $\mathbf{c}_t$ due to the different g sources:

$$\mathbf{c}_t = \sum_g^G \mathbf{c}_{g,t}, \tag{16}$$

where G is the number of the different sources (or groups of sources). The aim of the SALS method is to obtain estimates of the $\mathbf{c}_t$ as close as possible to their corresponding measured value. More precisely, the objective is estimating a vector, named *analysis* and referred to as $\hat{\mathbf{c}}_t$, defined as a linear combination of the sources' contributions:

$$\hat{\mathbf{c}}_t = \sum_g^G \alpha_{g,t} \mathbf{c}_{g,t}, \tag{17}$$

where $\alpha_{g,t}$ is the (time dependent) modulation coefficient, related to the sources g at the time t. The analysis $\hat{\mathbf{c}}_t$ is evaluated by computing $\alpha_{g,t}$ coefficients minimizing the cost function J (representing the quadratic error):

$$J(\alpha_{1,t}, \alpha_{2,t}, ..., \alpha_{G,t}) = \frac{1}{m_t} \left(\mathbf{y}_t - \sum_g^G \alpha_{g,t} \mathbf{H}_t \mathbf{c}_{g,t} \right)^T \left(\mathbf{y}_t - \sum_g^G \alpha_{g,t} \mathbf{H}_t \mathbf{c}_{g,t} \right), \tag{18}$$

where $\mathbf{y}_t$ is a vector containing the m_t measurement (at the time step t). The matrix $\mathbf{H}_t$, called *observation operator*, is a matrix of size $m_t \times n$ filled of 0 and 1. When applied to the background vector $\mathbf{c}_{g,t}$, it provides a vector of size m_t containing the modelled concentrations at the same location of the measured ones (and at a given time t). The coefficients $\alpha_{g,t}$, considered as uniform over the whole domain (for each time step and each source contribution), are therefore determined by solving the following system:

$$
\begin{pmatrix}
(\mathbf{H}_t \mathbf{c}_{1,t})^{\mathrm{T}}(\mathbf{H}_t \mathbf{c}_{1,t}) & \cdots & (\mathbf{H}_t \mathbf{c}_{1,t})^{\mathrm{T}}(\mathbf{H}_t \mathbf{c}_{G,t}) \\
\vdots & \ddots & \vdots \\
(\mathbf{H}_t \mathbf{c}_{G,t})^{\mathrm{T}}(\mathbf{H}_t \mathbf{c}_{1,t}) & \cdots & (\mathbf{H}_t \mathbf{c}_{G,t})^{\mathrm{T}}(\mathbf{H}_t \mathbf{c}_{G,t})
\end{pmatrix}
\begin{pmatrix}
\alpha_{1,t} \\
\vdots \\
\alpha_{G,t}
\end{pmatrix}
=
\begin{pmatrix}
(\mathbf{y}_t)^{\mathrm{T}}(\mathbf{H}_t \mathbf{c}_{1,t}) \\
\vdots \\
(\mathbf{y}_t)^{\mathrm{T}}(\mathbf{H}_t \mathbf{c}_{G,t})
\end{pmatrix}.
\tag{19}
$$

The resolution of the system (19), that can be typically obtained by solving a least square problem, is here carried out with the method of Lawson and Hanson [78], which guarantees the coefficients $\alpha_{g,t}$ to be positive. The SALS method can be applied only when $m_t \geq G$, i.e., when the measurements number m_t is higher than the number of sources G. The choice of the sources (number and/or type) is a key element in the SALS method. Sources can be grouped based on their corresponding activity sectors (transport, industry, residential-tertiary, agriculture) or on their geographical localisation. Note that, for inert species (or low-reactive species), the application of the SALS method, i.e., the modulation of the sources' contributions, can be interpreted as a method to correct the intensity of the emissions.

3. Case Study—The Lyon Urban Agglomeration

We present an application of the source apportionment module (with the SA-NO and the SA-NOX models) to evaluate the source contribution to NO_2 concentration on the Lyon urban agglomeration (approximately 1.4 million people), for the year 2008. The case study is the same as that used for an extensive analysis of the performances of the SIRANE model, recently presented by Soulhac et al. [72].

We consider the contributions by three typologies of pollutant sources, namely (i) traffic, (ii) industrial sources, and (iii) miscellaneous distributed sources (all other sources not included in the two previous categories, mainly domestic heating), as well as that due to the background concentration (Figure 1a), i.e., related to all sources that are placed outside the studied domain.

Simulations were run over a 36 km × 40 km domain, with a spatial resolution of 10 m. The traffic and industrial emissions are represented by 21833 line sources and 83 point sources, respectively. Miscellaneous distributed sources are represented by surface emissions with a 1 km × 1 km resolution. The annual averaged emissions are represented in Figure 1. Most of the point sources do not exceed $0.05\,\mathrm{g\,s^{-1}}$ (Figure 1b), and the largest emissions are related to chimneys of road tunnels. The traffic emissions are higher in the city centre and along the main roads (Figure 1c). Miscellaneous distributed emissions are higher on the centre of the agglomeration (Figure 1d).

In the simulation, the background concentrations of the different species are assumed to be equal to those measured at the Saint-Exupéry (STE) station, located at about 20 km from the Lyon city center (see Figure 1a).

The meteorological wind field was reconstructed with an hourly time-step and according to the Monin-Obuhkov similarity theory, from data registered at the Météo-France station in Bron (Figure 1a). The dominant wind direction is North–South, with wind speeds that rarely exceed $6\,\mathrm{m\,s^{-1}}$ (Figure 2a). The stability conditions computed by the meteorological pre-processor are presented in (Figure 2b), where we plot the inverse of the Monin–Obukhov length (L_{MO}). The distribution of the ratio $1/L_{MO}$ suggests an equal repartition between stable ($L_{MO} > 0$) and unstable ($L_{MO} < 0$) atmospheric conditions. The high frequency of the condition $1/L_{MO}$ in the range 0.01–0.02 is due to the fact that we have imposed a minimum value for L_{MO}. The purpose of this is to avoid stability conditions (here estimated by means of cloud cover measurements) that rarely occur over an urban area, due to the heat anthropogenic fluxes and the wind shear induced by the presence of the urban canopy. Further details on the input data and the model set-up can be found in Soulhac et al. [72].

Time-series of NO_2 concentration used to evaluate the SALS data assimilation method were collected in different measurement sites over the whole year 2008 by Atmo Auvergne Rhône-Alpes (AURA), the local authority for air quality. These include hourly measurements provided by 16 permanent measurement stations, which have been classified by Atmo AURA into four different categories: suburban stations (Côtière de l'Ain, Genas, Saint-Exupéry and Ternay) placed on high-intensity traffic roads (Berthelot, Grandclément, Lyon périphérique, Mulatière and Vaise), stations close to industrial sites (Feyzin and Saint-Fons) and stations within the urban agglomeration and away from high-intensity traffic roads (Gerland, Lyon centre, Saint-Just and Vaulx-en-Velin). For all these stations, missing hourly data do not exceed 3% over the whole year 2008.

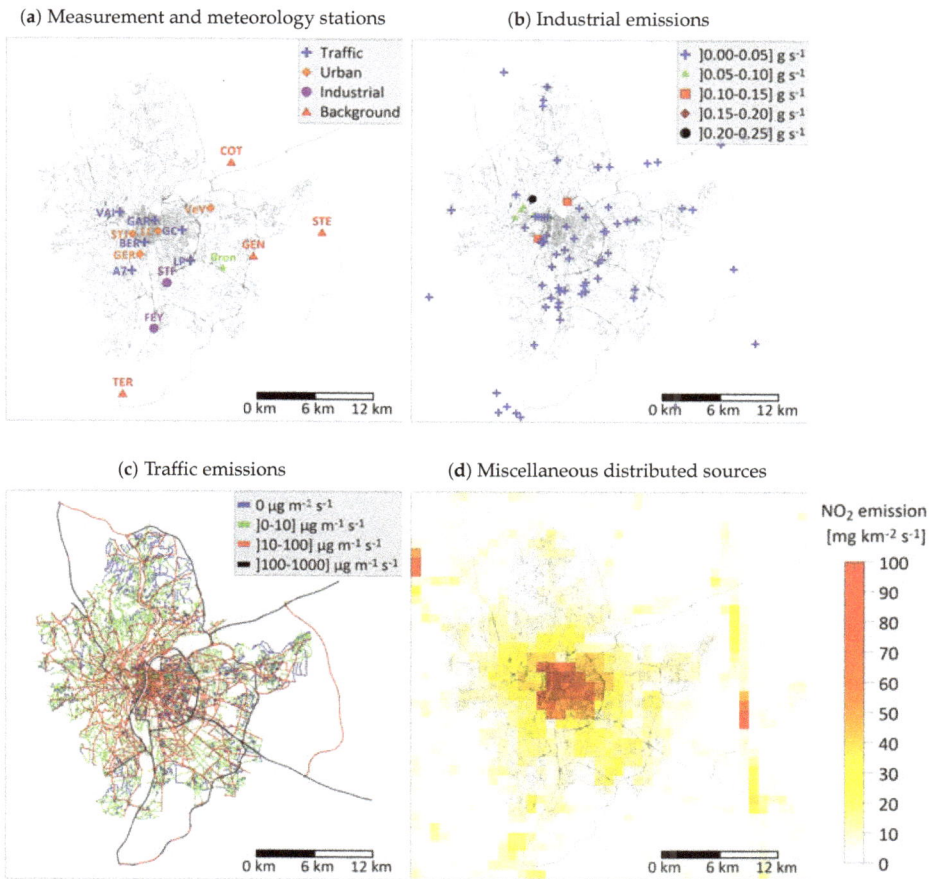

Figure 1. Localisation of the measurements and meteorology (Bron) stations (**a**); annual mean emissions of industries (**b**); traffic (**c**); and distributed miscellaneous sources (mainly domestic heating) (**d**).

(a) Windrose   **(b)** Probability density function of the ratio $1/L_{MO}$

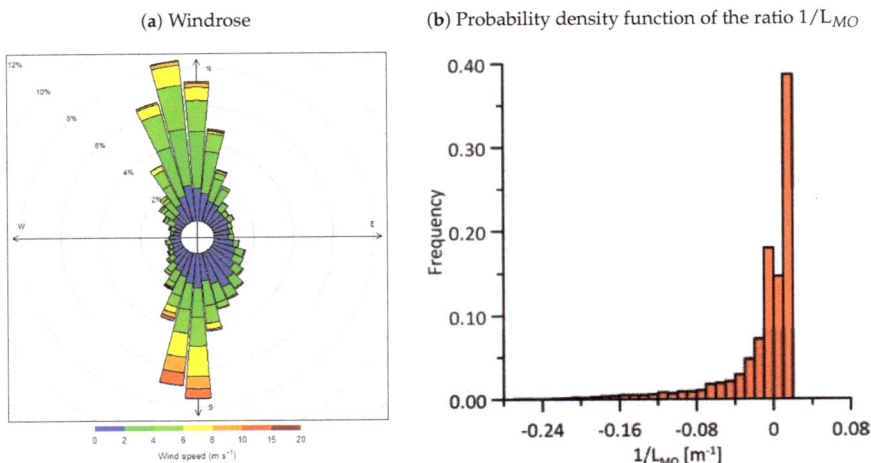

Figure 2. Annual statistics of data collected at the Bron meteorological station in 2008: (**a**) Windrose and (**b**) probability density function of the ratio $1/L_{MO}$.

3.1. Source Apportionment Results

3.1.1. Comparison of the Results Obtained with the SA-NO and SA-NOX Models

The contribution of traffic, miscellaneous distributed sources, industry and background concentration to NO_2 annual mean concentration estimated by the SA-NO and SA-NOX models are shown in Figures 3 and 4. A relevant difference between the two models is related to the negative contributions of the miscellaneous distributed sources estimated by the SA-NO model (Figure 3). According to the SA-NO model formulation (12), these negative contributions indicate that NO_2 concentrations are higher after the first step of the simulation (i.e., before the occurrence of chemical reactions) and that the miscellaneous distributed sources contribute to a consumption of NO_2 molecules larger than its initial (before chemical reactions) contribution. Similar considerations hold for industrial sources (not shown in Figure 3). The contribution of traffic and the miscellaneous distributed sources' contributions are slightly higher when estimated with with the SA-NOX model (on average these are larger of approximately $0.59\,\mu g\,m^{-3}$ and $0.49\,\mu g\,m^{-3}$ for the miscellaneous distributed sources and the traffic, respectively). At the measurement stations' locations, the mean difference is slightly larger than over the whole agglomeration and is equal to $1.90\,\mu g\,m^{-3}$ for the miscellaneous distributed sources, and to $2.01\,\mu g\,m^{-3}$ for traffic. On the other hand, the contribution of the background concentration is slightly higher when estimated with the SA-NO model. On average, these exceed those estimated by means of the SA-NOX model by $1.07\,\mu g\,m^{-3}$, when considering the whole agglomeration, and by $3.96\,\mu g\,m^{-3}$, when considering the measurement stations only.

In summary, we observe a clear tendency of the SA-NO model in overestimating background contribution (with respect to the SA-NOX model), and therefore in underestimating the relative contribution of all other sources. This is due to the fact that, for the background concentration, the relative NO_2 contribution (to the total NO_2 concentration) is inevitably higher than the relative NO_x contribution (to the total NO_x concentration). This enhanced NO_2 contribution of the background concentration is emphasised in the formulation of the SA-NO model, notably by the first term, the r.h.s of Equation (12).

Despite these slight differences, results obtained with the SA-NO and SA-NOX models are very similar. In what follows, we will however exploit only the results of the SA-NOX model, since it guarantees the contribution of all pollutant sources to be positive.

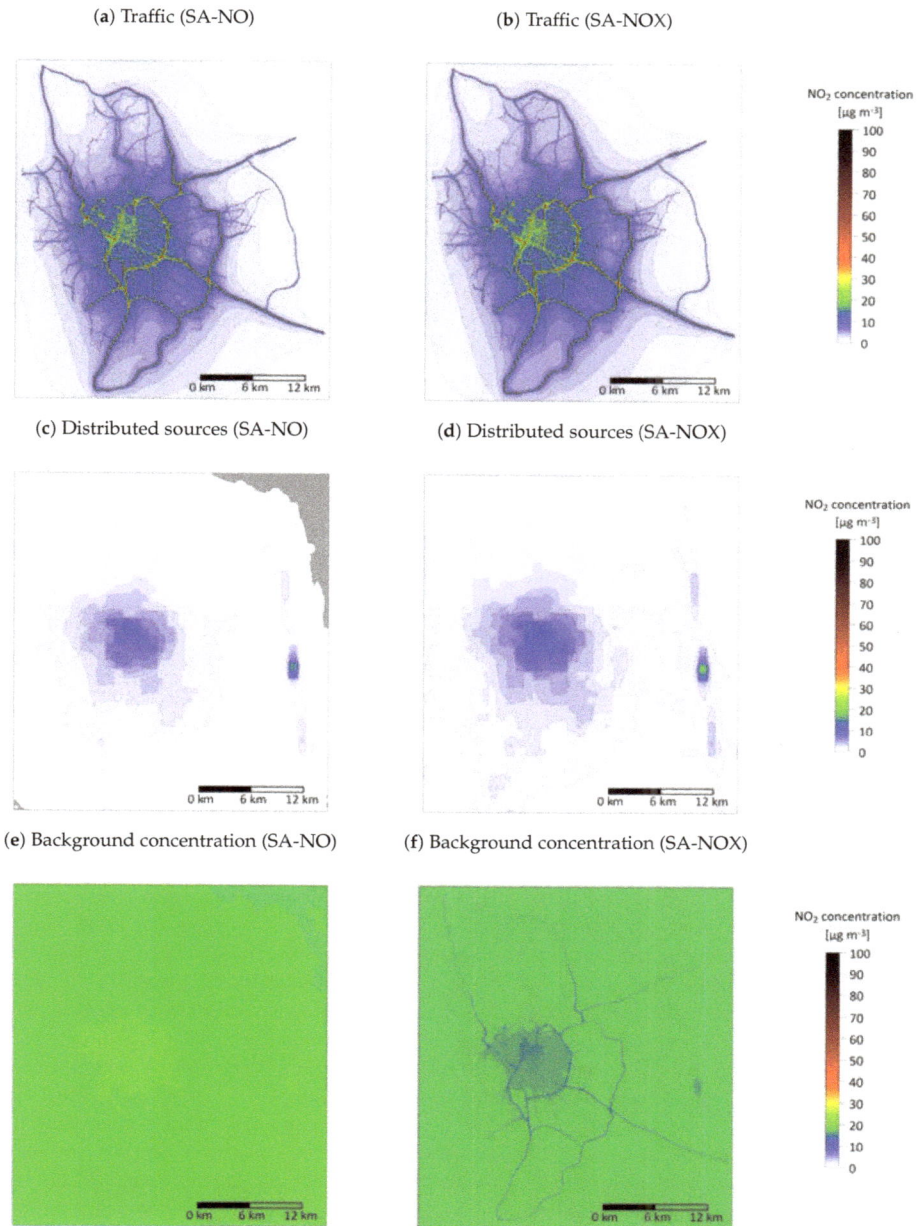

Figure 3. Traffic, miscellaneous distributed sources, background contribution ($\mu g\,m^{-3}$) to the NO_2 annual mean concentration on the Lyon agglomeration in 2008 estimated with the SA-NO model (**a,c,e**) and SA-NOX (**b,d,f**) models (grey areas correspond to negative contributions).

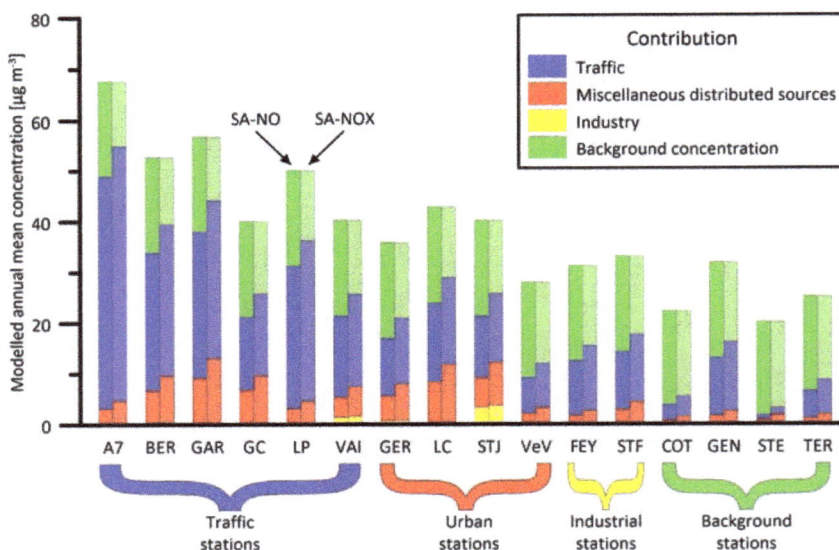

Figure 4. Traffic, miscellaneous distributed sources, industry, and background contribution ($\mu g\, m^{-3}$) to the NO_2 annual mean concentration at the measurement station located in the Lyon agglomeration in 2008 estimated with the SA-NO (left bars) and the SA-NOX (right-handed bars) models.

3.1.2. Estimates of Sources' Contributions

Results show that the industrial sources' contributions to NO_2 mean concentration are very low over the Lyon agglomeration. Those of the miscellaneous distributed sources are generally larger in the city centre (on average 6.42 $\mu g\, m^{-3}$) (Figure 3). Results also show that the traffic contributions are on average higher close to the roads and in the centre of the agglomeration (Figure 3), where NO_2 traffic-induced concentrations can exceed the annual average concentration threshold set by the European Directive 2008/50/EC (40 $\mu g\, m^{-3}$). Conversely, the background concentration contribution is at its lowest in the city centre and close to the main roads (on average 17.43 $\mu g\, m^{-3}$).

The spatial variability of the relative contributions (on percentage) is different (Figure 5). The relative contribution of background concentration (on the total concentration) is globally larger than 50%, except close to some main roads (Figures 5c and 6a). As expected, the traffic contribution is instead higher close to the roads, where its relative contribution exceeds 50% (Figures 5a and 6a). For the miscellaneous distributed sources, the relative contribution is generally higher in the centre of the agglomeration, with some hot spot in the suburbs (Figure 5b). The industrial contribution is spatially homogeneous and relatively low (not shown in Figure 5).

The results of the source apportionment methods allow us to evaluate the contribution to the concentration registered at the monitoring station. We can therefore evaluate ex-post the pertinence of the classification of the monitoring stations adopted by the local air quality authority Atmo AURA. This classification is fully adapted for traffic-type and background-type stations, at which traffic and background contributions, respectively, both exceed 50% (Figure 4). Note also that the industrial, traffic, and miscellaneous distributed sources' contributions are very low for the Saint-Exupéry station, which is therefore representative for the background concentration values.

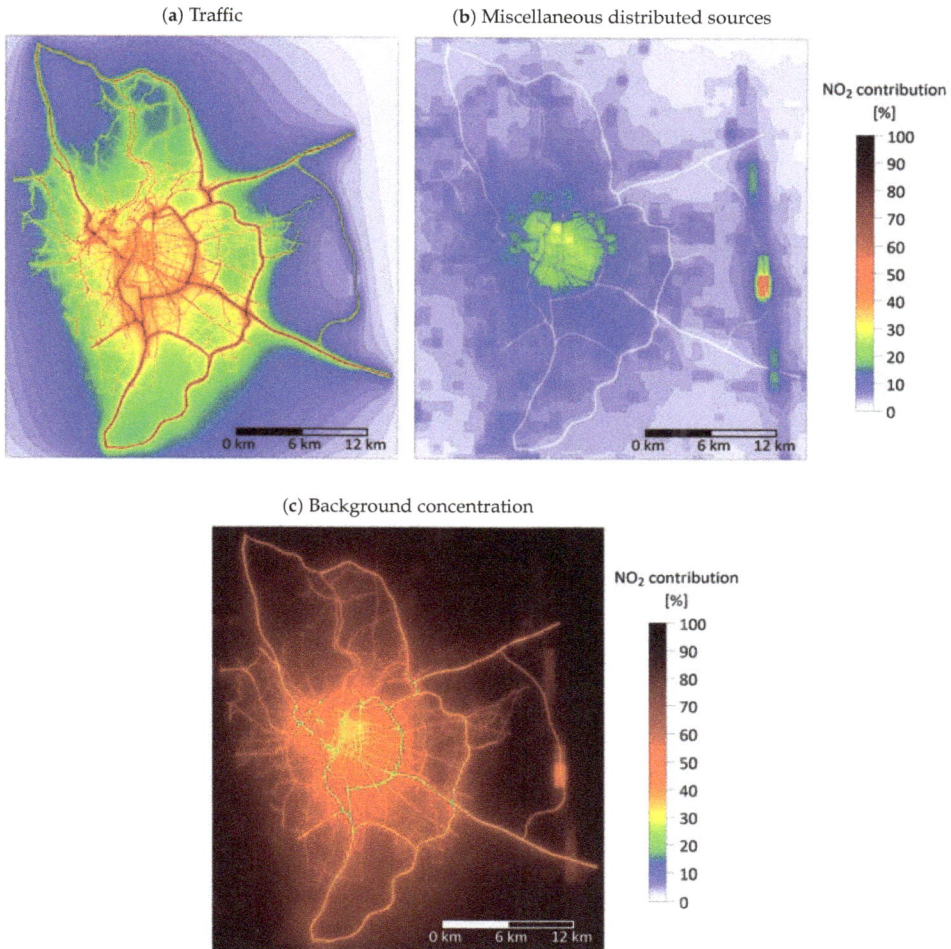

Figure 5. (**a**) Traffic, (**b**) miscellaneous distributed sources, and (**c**) background relative contribution [%] to the NO_2 annual mean concentration on the Lyon agglomeration in 2008 estimated with the SA-NOX model.

A main objective of this kind of analysis is to identify sources having the main impacts on air quality and determine to what extent their emissions have to be reduced in order to attain given concentration threshold. As an example, we show in Figure 6b the contribution(s) to be reduced to lower concentrations below regulatory threshold values. These are determined by successively removing the different contributions, from the largest to the lowest, until reaching a concentration below the threshold value. The analysis suggests that the priority is to reduce the traffic emissions and, to a lesser extent, the emissions from sources placed outside the Lyon urban area (i.e., outside the domain taken into account this simulation), responsible for the background pollution.

(a) Most important contributions   **(b)** Contributions to reduce or eliminate

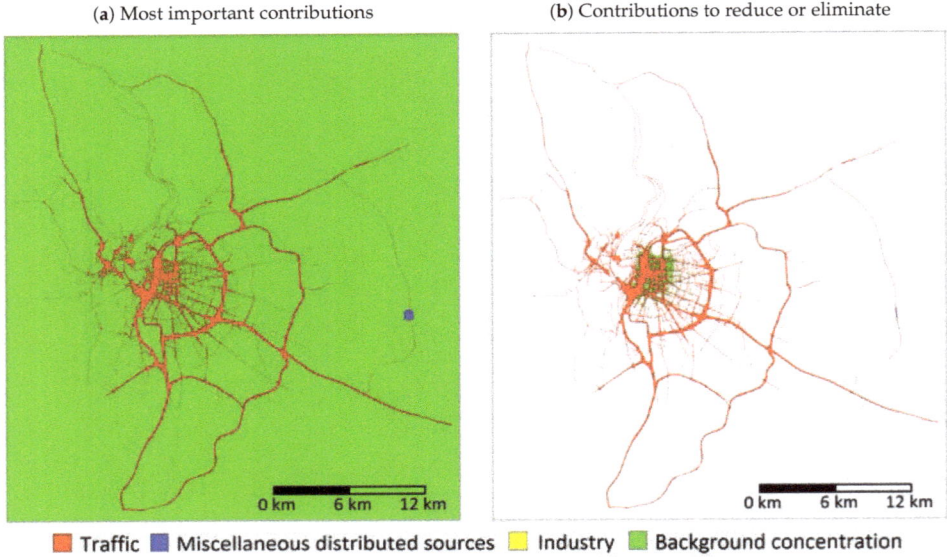

■ Traffic ■ Miscellaneous distributed sources □ Industry ■ Background concentration

Figure 6. Map of (**a**) the most important NO$_2$ contributor on the Lyon urban area in 2008 and (**b**) the NO$_2$ contributions to reduce or eliminate in order to achieve air quality European standard in 2008 (the concentration is (already) below the threshold value in area in white).

3.2. Data Assimilation Results

The SALS method is here applied using three groups of sources: (1) traffic emissions, (2) miscellaneous distributed sources and (3) background concentration and industrial sources.

To evaluate the performances of the method, we compare its results to those provided by the reference simulation (without data assimilation), using the leave-one-out cross-validation approach (LOOCV). This consists of estimating the concentration at one station (at each time step) using all available measured concentrations, except for that associated to that particular station. This procedure is repeated for each of the monitoring stations. The final estimates are compared to the measured concentrations. To evaluate the quality of the model results, we use six statistical indices: the bias, the fractional bias, the root mean square error (RMSE), the normalized mean square error (NMSE), the correlation coefficient (r) and the factor 2 (FAC2) [79]. The definition of these statistical indices are described in Table 1, where c_m is the measured concentration and c_p is the predicted concentration.

The statistical performances associated to the SALS method are given in Table 2. For all stations, statistical performances are good, except for the stations named A7 south Lyon (A7) and Lyon center (LC) (see Figure 1a), at which the correlation coefficient (or the bias) does not satisfy the quality criteria.

Table 1. Statistical indices and quality criteria used to evaluate results quality (c_m is the measured concentration and c_p is the predicted concentration).

	Bias	**RMSE**	**r**
Definition	$\overline{c_m - c_p}$	$\dfrac{\overline{(c_m - c_p)}^2}{\overline{c_m}\,\overline{c_p}}$	$\dfrac{\overline{(c_m - \overline{c_m})\,(c_p - \overline{c_p})}}{\sqrt{\overline{(c_m - \overline{c_m})}^2\,\overline{(c_p - \overline{c_p})}^2}}$
Criteria	$|\text{Bias}| \leq 0.33\,\overline{c_m}$	$\text{RMSE} \leq \overline{c_m}$	$r \geq 0.60$

Table 1. *Cont.*

	FB	NMSE	FAC2
Definition	$\dfrac{2\left(\overline{c_m}-\overline{c_p}\right)}{\overline{c_m}+\overline{c_p}}$	$\dfrac{\overline{\left(c_m-c_p\right)^2}}{\overline{c_m}\,\overline{c_p}}$	Fraction of data that satisfy $0.5 \leq c_m/c_p \leq 2$
Criteria	$|FB| \leq 0.67$	$NMSE \leq 6$	$FAC2 \geq 0.30$

Table 2. Statistical performances of the SALS method ($\overline{c_m}$: mean measured concentration, $\overline{c_p}$: mean modelled concentration). Red values are those that do not respect the quality criteria.

Type	Station	$\overline{c_m}$ (μg m^{-3})	$\overline{c_p}$ (μg m^{-3})	Bias (μg m^{-3})	FB	RMSE (μg m^{-3})	NMSE	r	FAC2
Traffic	A7	79.05	72.39	6.66	0.09	40.33	0.28	0.56	0.79
	BER	52.50	60.14	−7.64	−0.14	18.77	0.11	0.81	0.92
	GAR	74.06	61.78	12.28	0.18	26.80	0.16	0.81	0.95
	GC	47.06	43.41	3.65	0.08	19.39	0.18	0.79	0.90
	LP	50.67	54.26	−3.59	−0.07	23.04	0.19	0.70	0.87
	VAI	59.10	42.63	16.47	0.32	25.60	0.26	0.77	0.82
Urban	GER	38.08	39.47	−1.39	−0.04	9.95	0.07	0.91	0.96
	LC	37.95	50.87	−12.92	−0.29	17.82	0.16	0.90	0.87
	STJ	36.78	45.27	−8.48	−0.21	17.88	0.19	0.83	0.89
	VeV	26.67	31.39	−4.72	−0.16	10.61	0.13	0.89	0.82
Industrial	FEY	33.84	34.23	−0.39	−0.01	12.94	0.14	0.80	0.89
	STF	35.35	35.59	−0.24	−0.01	12.50	0.12	0.88	0.93
Background	COT	23.26	24.59	−1.33	−0.06	11.98	0.25	0.80	0.73
	GEN	33.36	34.73	−1.37	−0.04	13.22	0.15	0.80	0.86
	STE	17.78	22.04	−4.26	−0.21	12.21	0.38	0.79	0.64
	TER	29.41	26.27	3.14	0.11	11.57	0.17	0.82	0.83

The bias, the RMSE and the correlation coefficients (r) of the SALS method are compared with those of the reference simulation in Figure 7. The bias of the SALS method is better than that associated to the reference simulation for half of the stations. Moreover, the absolute value of the least satisfactory bias is similar, with and without data assimilation. The RMSE values of the SALS method are generally better than those related to the reference simulation. However, the worst RMSE is of the same order of magnitude, with and without data assimilation. Similarly, the correlation coefficients associated with the SALS method are better than those related to the reference SIRANE simulation for most of the stations. Note, however, that the worst correlation coefficient does not vary significantly, with and without the application of the SALS method. Note that the bias, evaluated both before and after data assimilation, is generally negative for urban and background stations, therefore revealing a tendency of the model in overpredicting concentrations. This overprediction can be, at least partially, explained by the fact that SIRANE neglects the role of NO$_2$ losses induced by the reactions induced by the hydroxyl radical OH (Equation (4)).

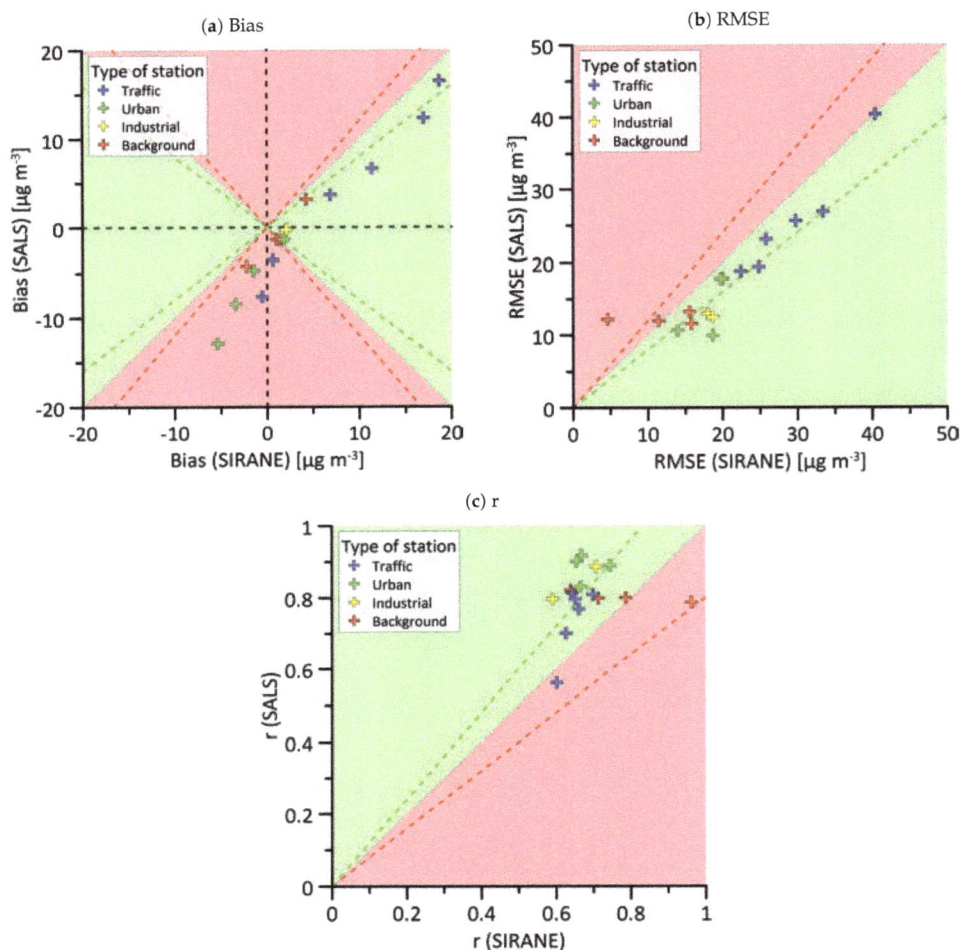

Figure 7. (a) Bias, (b) RMSE, and (c) correlation coefficients before and after data assimilation (the green zone indicates results improved after data assimilation and the red region indicates results worsened after data assimilation. Green (red) dot line indicates improvement (worsening) of 20% after data assimilation).

4. Conclusions

A source apportionment module, using the tagged species approach, has been developed for the SIRANE model in order to estimate the contribution to air pollution of different typologies of pollutant sources. This module includes two methods, named SA-NO and SA-NOX, to evaluate the sources' contributions to NO_2 concentrations. This module has been applied to evaluate the traffic, miscellaneous distributed sources including residential-tertiary sector, industrial sources and background concentration contribution to NO_2 concentration on the Lyon agglomeration in 2008. Overall, the NO_2 contributions evaluated with the SA-NO and SA-NOX models are similar. The contribution of the industrial sources to the NO_2 annual mean concentrations on the Lyon agglomeration is negligible compared to the other emissions' sectors.The traffic emissions are instead the most important contributors. Their reduction is therefore essential to attain the threshold values set by the European regulations on air quality. The evaluation of the sources' contributions allows also for an improvement of the urban air quality simulations' results by means of the data assimilation method called *Source Apportionment Least Square* (SALS). Results highlight usefulness of the source apportionment method as a tool for the assessment of emissions reduction strategies at the local urban scale.

Acknowledgments: The Chi Vuong Nguyen PhD internship was funded by the Région Auvergne-Rhône-Alpes and the Atmo Auvergne-Rhône-Alpes air quality agency.

Author Contributions: Chi Vuong Nguyen, Lionel Soulhac and Pietro Salizzoni conceived and designed the structure of the paper. Chi Vuong Nguyen implemented the algorithms in the SIRANE model, performed the simulations, the literature research, and analyzed the data. All authors contributed to the discussion of the results and the writing of the final manuscript.

Conflicts of Interest: The authors declare no conflict of interest.

Abbreviations

The following abbreviations are used in this manuscript:

BFM	Brut Force Method
CMB	Chemical Mass Balance
PM	Particulate Matter
SALS	Source Apportionment Least Square

References

1. Wagstrom, K.M.; Pandis, S.N.; Yarwood, G.; Wilson, G.M.; Morris, R.E. Development and application of a computationally efficient particulate matter apportionment algorithm in a three-dimensional chemical transport model. *Atmos. Environ.* **2008**, *42*, 5650–5659.
2. Yim, S.H.; Fung, J.C.; Lau, A.K. Use of high-resolution MM5/CALMET/CALPUFF system: SO_2 apportionment to air quality in Hong Kong. *Atmos. Environ.* **2010**, *44*, 4850–4858.
3. Cho, S.; Morris, R.; McEachern, P.; Shah, T.; Johnson, J.; Nopmongcol, U. Emission sources sensitivity study for ground-level ozone and $PM_{2.5}$ due to oil sands development using air quality modelling system: Part II—Source apportionment modelling. *Atmos. Environ.* **2012**, *55*, 542–556.
4. Grewe, V.; Dahlmann, K.; Matthes, S.; Steinbrecht, W. Attributing ozone to NO_x emissions: Implications for climate mitigation measures. *Atmos. Environ.* **2012**, *59*, 102–107.
5. Kwok, R.; Napelenok, S.; Baker, K. Implementation and evaluation of $PM_{2.5}$ source contribution analysis in a photochemical model. *Atmos. Environ.* **2013**, *80*, 398–407.
6. Ying, Q.; Kleeman, M.J. Source contributions to the regional distribution of secondary particulate matter in California. *Atmos. Environ.* **2006**, *40*, 736–752.
7. Yarwood, G.; Morris, R.E.; Wilson, G.M. Particulate matter source apportionment technology (PSAT) in the CAMx photochemical grid model. In *Air Pollution Modeling and Its Application XVII*; Springer: Boston, MA, USA, 2007; pp. 478–492.
8. Wang, Z.S.; Chien, C.J.; Tonnesen, G.S. Development of a tagged species source apportionment algorithm to characterize three-dimensional transport and transformation of precursors and secondary pollutants. *J. Geophys. Res. Atmos.* **2009**, *114*, doi:10.1029/2008JD010846.
9. Pio, C.A.; Alves, C.A.; Duarte, A.C. Identification, abundance and origin of atmospheric organic particulate matter in a Portuguese rural area. *Atmos. Environ.* **2001**, *35*, 1365–1375.
10. Querol, X.; Alastuey, A.; Rodríguez, S.; Plana, F.; Ruiz, C.R.; Cots, N.; Massagué, G.; Puig, O. PM_{10} and $PM_{2.5}$ source apportionment in the Barcelona Metropolitan area, Catalonia, Spain. *Atmos. Environ.* **2001**, *35*, 6407–6419.
11. Putaud, J.P.; Raes, F.; Van Dingenen, R.; Brüggemann, E.; Facchini, M.C.; Decesari, S.; Fuzzi, S.; Gehrig, R.; Hüglin, C.; Laj, P.; et al. A European aerosol phenomenology—2: Chemical characteristics of particulate matter at kerbside, urban, rural and background sites in Europe. *Atmos. Environ.* **2004**, *38*, 2579–2595.

12. Gijzen, M.; Lewinsohn, E.; Savage, T.J.; Croteau, R.B. Conifer monoterpenes: Biochemistry and bark beetle chemical ecology. *ACS Symp. Ser.* **1993**, *525*, 8–22.

13. Watson, J.G. Overview of receptor model principles. *J. Air Pollut. Control Assoc.* **1984**, *34*, 619–623.

14. Viana, M.; Kuhlbusch, T.; Querol, X.; Alastuey, A.; Harrison, R.; Hopke, P.; Winiwarter, W.; Vallius, M.; Szidat, S.; Prévôt, A.; et al. Source apportionment of particulate matter in Europe: A review of methods and results. *J. Aerosol Sci.* **2008**, *39*, 827–849.

15. Held, T.; Ying, Q.; Kleeman, M.J.; Schauer, J.J.; Fraser, M.P. A comparison of the UCD/CIT air quality model and the CMB source–receptor model for primary airborne particulate matter. *Atmos. Environ.* **2005**, *39*, 2281–2297.

16. Subramanian, R.; Donahue, N.M.; Bernardo-Bricker, A.; Rogge, W.F.; Robinson, A.L. Contribution of motor vehicle emissions to organic carbon and fine particle mass in Pittsburgh, Pennsylvania: Effects of varying source profiles and seasonal trends in ambient marker concentrations. *Atmos. Environ.* **2006**, *40*, 8002–8019.

17. Rizzo, M.J.; Scheff, P.A. Fine particulate source apportionment using data from the USEPA speciation trends network in Chicago, Illinois: Comparison of two source apportionment models. *Atmos. Environ.* **2007**, *41*, 6276–6288.

18. Subramanian, R.; Donahue, N.M.; Bernardo-Bricker, A.; Rogge, W.F.; Robinson, A.L. Insights into the primary–secondary and regional–local contributions to organic aerosol and $PM_{2.5}$ mass in Pittsburgh, Pennsylvania. *Atmos. Environ.* **2007**, *41*, 7414–7433.

19. Duvall, R.M.; Norris, G.A.; Burke, J.M.; Olson, D.A.; Vedantham, R.; Williams, R. Determining spatial variability in $PM_{2.5}$ source impacts across Detroit, MI. *Atmos. Environ.* **2012**, *47*, 491–498.

20. Guo, H.; Wang, T.; Louie, P. Source apportionment of ambient non-methane hydrocarbons in Hong Kong: Application of a principal component analysis/absolute principal component scores (PCA/APCS) receptor model. *Environ. Pollut.* **2004**, *129*, 489–498.

21. Almeida, S.M.; Pio, C.A.; Freitas, M.C.; Reis, M.A.; Trancoso, M.A. Approaching $PM_{2.5}$ and $PM_{2.5-10}$ source apportionment by mass balance analysis, principal component analysis and particle size distribution. *Sci. Total Environ.* **2006**, *368*, 663–674.

22. Song, Y.; Xie, S.; Zhang, Y.; Zeng, L.; Salmon, L.G.; Zheng, M. Source apportionment of $PM_{2.5}$ in Beijing using principal component analysis/absolute principal component scores and UNMIX. *Sci. Total Environ.* **2006**, *372*, 278–286.

23. Shi, G.L.; Li, X.; Feng, Y.C.; Wang, Y.Q.; Wu, J.H.; Li, J.; Zhu, T. Combined source apportionment, using positive matrix factorization–chemical mass balance and principal component analysis/multiple linear regression–chemical mass balance models. *Atmos. Environ.* **2009**, *43*, 2929–2937.

24. Escrig, A.; Monfort, E.; Celades, I.; Querol, X.; Amato, F.; Minguillón, M.C.; Hopke, P.K. Application of Optimally Scaled Target Factor Analysis for Assessing Source Contribution of Ambient PM_{10}. *J. Air Waste Manag. Assoc.* **2009**, *59*, 1296–1307.

25. Minguillón, M.C.; Schembari, A.; Triguero-Mas, M.; de Nazelle, A.; Dadvand, P.; Figueras, F.; Salvado, J.A.; Grimalt, J.O.; Nieuwenhuijsen, M.; Querol, X. Source apportionment of indoor, outdoor and personal $PM_{2.5}$ exposure of pregnant women in Barcelona, Spain. *Atmos. Environ.* **2012**, *59*, 426–436.

26. Alier, M.; Felipe-Sotelo, M.; Hernàndez, I.; Tauler, R. Variation patterns of nitric oxide in Catalonia during the period from 2001 to 2006 using multivariate data analysis methods. *Anal. Chim. Acta* **2009**, *642*, 77–88.

27. Alier, M.; Felipe, M.; Hernández, I.; Tauler, R. Trilinearity and component interaction constraints in the multivariate curve resolution investigation of NO and O_3 pollution in Barcelona. *Anal. Bioanal. Chem.* **2011**, *399*, 2015–2029.

28. Koo, B.; Wilson, G.M.; Morris, R.E.; Dunker, A.M.; Yarwood, G. Comparison of source apportionment and sensitivity analysis in a particulate matter air quality model. *Environ. Sci. Technol.* **2009**, *43*, 6669–6675.

29. Hendriks, C.; Kranenburg, R.; Kuenen, J.; van Gijlswijk, R.; Kruit, R.W.; Segers, A.; van der Gon, H.D.; Schaap, M. The origin of ambient particulate matter concentrations in the Netherlands. *Atmos. Environ.* **2013**, *69*, 289–303.

30. Grewe, V. A diagnostic for ozone contributions of various NO_x emissions in multi-decadal chemistry-climate model simulations. *Atmos. Chem. Phys.* **2004**, *4*, 729–736.

31. Held, T.; Ying, Q.; Kaduwela, A.; Kleeman, M. Modeling particulate matter in the San Joaquin Valley with a source-oriented externally mixed three-dimensional photochemical grid model. *Atmos. Environ.* **2004**, *38*, 3689–3711.

32. Grewe, V.; Tsati, E.; Hoor, P. On the attribution of contributions of atmospheric trace gases to emissions in atmospheric model applications. *Geosci. Model Dev.* **2010**, *3*, 487.
33. Butler, T.; Lawrence, M.; Taraborrelli, D.; Lelieveld, J. Multi-day ozone production potential of volatile organic compounds calculated with a tagging approach. *Atmos. Environ.* **2011**, *45*, 4082–4090.
34. Emmons, L.; Hess, P.; Lamarque, J.F.; Pfister, G. Tagged ozone mechanism for MOZART-4, CAM-chem and other chemical transport models. *Geosci. Model Dev.* **2012**, *5*, 1531.
35. Kranenburg, R.; Segers, A.; Hendriks, C.; Schaap, M. Source apportionment using LOTOS-EUROS: Module description and evaluation. *Geosci. Model Dev.* **2013**, *6*, 721–733.
36. Granier, C.; Mueller, J.; Pétron, G.; Brasseur, G. A three-dimensional study of the global CO budget. *Chemosphere-Glob. Chang. Sci.* **1999**, *1*, 255–261.
37. Granier, C.; Pétron, G.; Müller, J.F.; Brasseur, G. The impact of natural and anthropogenic hydrocarbons on the tropospheric budget of carbon monoxide. *Atmos. Environ.* **2000**, *34*, 5255–5270.
38. Lamarque, J.F.; Hess, P. Model analysis of the temporal and geographical origin of the CO distribution during the TOPSE campaign. *J. Geophys. Res. Atmos.* **2003**, *108*, doi:10.1029/2002JD002077.
39. Pfister, G.; Petron, G.; Emmons, L.; Gille, J.; Edwards, D.; Lamarque, J.F.; Attie, J.L.; Granier, C.; Novelli, P. Evaluation of CO simulations and the analysis of the CO budget for Europe. *J. Geophys. Res. Atmos.* **2004**, *109*, doi:10.1029/2004JD004691.
40. Huang, Q.; Cheng, S.; Perozzi, R.E.; Perozzi, E.F. Use of a MM5–CAMx–PSAT modeling system to study SO_2 source apportionment in the Beijing Metropolitan Region. *Environ. Model. Assess.* **2012**, *17*, 527–538.
41. Talagrand, O. Assimilation of observations, an introduction. *J. Meteorol. Soc. Jpn.* **1997**, *75*, 191–209.
42. Rabier, F. Assimilation variationnelle de données météorologiques en présence d'instabilité barocline. *La Météorologie* **1993**, *8*, 57–72.
43. Kalnay, E. *Atmospheric Modeling, Data Assimilation and Predictability*; Cambridge University Press: Cambridge, UK, 2003.
44. Swinbank, R.; Shutyaev, V.; Lahoz, W.A. *Data Assimilation for the Earth System*; Springer Science & Business Media: Berlin, Germany, 2003.
45. Denby, B.; Horálek, J.; Walker, S.E.; Eben, K.; Fiala, J. Interpolation and assimilation methods for European scale air quality assessment and mapping. In *Part I: Review and Recommendations*; European Topic Centre on Air and Climate Change (ETC/ACC): Copenhagen, Denmark, 2005; Volume 7.
46. Morel, P.; Talagrand, O. Dynamic approach to meteorological data assimilation. *Tellus* **1974**, *26*, 334–344.
47. McPherson, R.D. Progress, problems, and prospects in meteorological data assimilation. *Bull. Am. Meteorol. Soc.* **1975**, *56*, 1154–1166.
48. Miyakoda, K.; Umscheid, L.; Lee, D.; Sirutis, J.; Lusen, R.; Pratte, F. The near-real-time, global, four-dimensional analysis experiment during the GATE period, Part I. *J. Atmos. Sci.* **1976**, *33*, 561–591.
49. Miyakoda, K.; Strickler, R.; Chludzinski, J. Initialization with the data assimilation method. *Tellus* **1978**, *30*, 32–54.
50. McPherson, R.; Bergman, K.; Kistler, R.; Rasch, G.; Gordon, D. The NMC operational global data assimilation system. *Mon. Weather Rev.* **1979**, *107*, 1445–1461.
51. Elbern, H.; Schmidt, H.; Ebel, A. Variational data assimilation for tropospheric chemistry modeling. *J. Geophys. Res. Atmos.* **1997**, *102*, 15967–15985.
52. Elbern, H.; Schmidt, H. A four-dimensional variational chemistry data assimilation scheme for Eulerian chemistry transport modeling. *J. Geophys. Res. Atmos.* **1999**, *104*, 18583–18598.
53. Elbern, H.; Schmidt, H.; Talagrand, O.; Ebel, A. 4D-variational data assimilation with an adjoint air quality model for emission analysis. *Environ. Model. Softw.* **2000**, *15*, 539–548.
54. Segers, A.J.; Heemink, A.W.; Verlaan, M.; van Loon, M. A modified rrsqrt-filter for assimilating data in atmospheric chemistry models. *Environ. Model. Softw.* **2000**, *15*, 663–671.
55. Van Loon, M.; Builtjes, P.J.H.; Segers, A.J. Data assimilation of ozone in the atmospheric transport chemistry model LOTOS. *Environ. Model. Softw.* **2000**, *15*, 603–609.
56. Brown, D.; Comrie, A. Spatial modeling of winter temperature and precipitation in Arizona and New Mexico, USA. *Clim. Res.* **2002**, *22*, 115–128.
57. Hooyberghs, J.; Mensink, C.; Dumont, G.; Fierens, F. Spatial interpolation of ambient ozone concentrations from sparse monitoring points in Belgium. *J. Environ. Monit.* **2006**, *8*, 1129–1135.
58. ETC/ACC. *Spatial Mapping of Air Quality for European Scale Assessment*; ETC/ACC: Copenhagen, Denmark, 2007.

59. Lü, C.; Tian, H. Spatial and temporal patterns of nitrogen deposition in China: Synthesis of observational data. *J. Geophys. Res. D Atmos.* **2007**, *112*, doi:10.1029/2006JD007990.

60. Denby, B.; Schaap, M.; Segers, A.; Builtjes, P.; Horálek, J. Comparison of two data assimilation methods for assessing PM_{10} exceedances on the European scale. *Atmos. Environ.* **2008**, *42*, 7122–7134.

61. EEA. *Spatial Assessment of PM_{10} and Ozone Concentrations in Europe (2005)*; European Environment Agency (EEA): Copenhagen, Denmark, 2009.

62. Joseph, J.; Sharif, H.O.; Sunil, T.; Alamgir, H. Application of validation data for assessing spatial interpolation methods for 8-h ozone or other sparsely monitored constituents. *Environ. Pollut.* **2013**, *178*, 411–418.

63. Blanchard, C.L.; Tanenbaum, S.; Hidy, G.M. Spatial and temporal variability of air pollution in Birmingham, Alabama. *Atmos. Environ.* **2014**, *89*, 382–391.

64. Candiani, G.; Carnevale, C.; Pisoni, E.; Volta, M. Assimilation of Chemical Ground Measurements in Air Quality Modeling. In *Large-Scale Scientific Computing*; Lirkov, I., Margenov, S., Waśniewski, J., Eds.; Springer: Berlin/Heidelberg, Germany, 2010; pp. 157–164.

65. Wang, X.; Mallet, V.; Berroir, J.P.; Herlin, I. Assimilation of OMI NO_2 retrievals into a regional chemistry-transport model for improving air quality forecasts over Europe. *Atmos. Environ.* **2011**, *45*, 485–492.

66. Kumar, U.; De Ridder, K.; Lefebvre, W.; Janssen, S. Data assimilation of surface air pollutants (O_3 and NO_2) in the regional-scale air quality model AURORA. *Atmos. Environ.* **2012**, *60*, 99–108.

67. Candiani, G.; Carnevale, C.; Finzi, G.; Pisoni, E.; Volta, M. A comparison of reanalysis techniques: Applying optimal interpolation and Ensemble Kalman Filtering to improve air quality monitoring at mesoscale. *Sci. Total Environ.* **2013**, *458–460*, 7–14.

68. Tilloy, A.; Mallet, V.; Poulet, D.; Pesin, C.; Brocheton, F. BLUE-based NO_2 data assimilation at urban scale. *J. Geophys. Res. Atmos.* **2013**, *118*, 2031–2040.

69. Soulhac, L.; Salizzoni, P.; Cierco, F.X.; Perkins, R. The model SIRANE for atmospheric urban pollutant dispersion; part I, presentation of the model. *Atmos. Environ.* **2011**, *45*, 7379–7395.

70. Soulhac, L.; Salizzoni, P.; Mejean, P.; Didier, D.; Rios, I. The model SIRANE for atmospheric urban pollutant dispersion; PART II, validation of the model on a real case study. *Atmos. Environ.* **2012**, *49*, 320–337.

71. Soulhac, L. ModéLisation de la Dispersion Atmosphérique à L'intérieur de la Canopée Urbaine. Ph.D. Thesis, Ecole centrale de Lyon, Ecully, France, 2000.

72. Soulhac, L.; Nguyen, C.V.; Volta, P.; Salizzoni, P. The model SIRANE for atmospheric urban pollutant dispersion. PART III: Validation against NO_2 yearly concentration measurements in a large urban agglomeration. *Atmos. Environ.* **2017**, *167*, 377–388.

73. Seinfeld, J.H. *Atmospheric Chemistry and Physics of Air Pollution*, 1st ed.; Wiley-Interscience: New York, NY, USA, 1986.

74. Bloss, W. Atmospheric chemical processes of importance in cities. *Issues Environ. Sci. Technol.* **2009**, *28*, 42.

75. Zhong, J.; Cai, X.M.; Bloss, W.J. Modelling the dispersion and transport of reactive pollutants in a deep urban street canyon: Using large-eddy simulation. *Environ. Pollut.* **2015**, *200*, 42–52.

76. Kasten, F.; Czeplak, G. Solar and terrestrial radiation dependent on the amount and type of cloud. *Sol. Energy* **1980**, *24*, 177–189.

77. Vardoulakis, S.; Fisher, B.E.; Pericleous, K.; Gonzalez-Flesca, N. Modelling air quality in street canyons: A review. *Atmos. Environ.* **2003**, *37*, 155–182.

78. Lawson, C.L.; Hanson, R. Linear least squares with linear inequality constraints. *Chap* **1974**, *23*, 158–173.

79. Hanna, S.; Chang, J. Acceptance criteria for urban dispersion model evaluation. *Meteorol. Atmos. Phys.* **2012**, *116*, 133–146.

MDPI

St. Alban-Anlage 66

4052 Basel

Switzerland

Tel. +41 61 683 77 34

Fax +41 61 302 89 18

www.mdpi.com

Atmosphere Editorial Office

E-mail: atmosphere@mdpi.com

www.mdpi.com/journal/atmosphere